Textiles for protection

The Textile Institute and Woodhead Publishing

The Textile Institute is a unique organisation in textiles, clothing and footwear. Incorporated in England by a Royal Charter granted in 1925, the Institute has individual and corporate members in over 90 countries. The aim of the Institute is to facilitate learning, recognise achievement, reward excellence and disseminate information within the global textiles, clothing and footwear industries.

Historically, The Textile Institute has published books of interest to its members and the textile industry. To maintain this policy, the Institute has entered into partnership with Woodhead Publishing Limited to ensure that Institute members and the textile industry continue to have access to high calibre titles on textile science and technology.

Most Woodhead titles on textiles are now published in collaboration with The Textile Institute. Through this arrangement, the Institute provides an Editorial Board which advises Woodhead on appropriate titles for future publication and suggests possible editors and authors for these books. Each book published under this arrangement carries the Institute's logo.

Woodhead books published in collaboration with The Textile Institute are offered to Textile Institute members at a substantial discount. These books, together with those published by The Textile Institute that are still in print, are offered on the Woodhead web site at: www.woodheadpublishing.com. Textile Institute books still in print are also available directly from the Institute's web site at: www.textileinstitutebooks.com

Woodhead Publishing Series in Textiles: Number 44

Textiles for protection

Edited by
Richard A. Scott

The Textile Institute

CRC Press
Boca Raton Boston New York Washington, DC

WOODHEAD PUBLISHING LIMITED
Oxford   Cambridge   New Delhi

Published by Woodhead Publishing Limited in association with The Textile Institute
Woodhead Publishing Limited, Abington Hall, Granta Park, Geat Abington
Cambridge CB21 6AH, UK
www.woodheadpublishing.com

Woodhead Publishing India Private Limited, G-2, Vardaan House, 7/28 Ansari Road,
Daryaganj, New Delhi – 110002, India
www.woodheadpublishingindia.com

Published in North America by CRC Press LLC, 6000 Broken Sound Parkway, NW,
Suite 300, Boca Raton, FL 33487, USA

First published 2005, Woodhead Publishing Limited and CRC Press LLC
Reprinted 2010
© Woodhead Publishing Limited, 2005; Chapter 4 figures © Dr A Shaw; Chapter 21 is
© Crown copyright – Ministry of Defence 2004. Parties wishing to reproduce extracts
from this copyright material work should obtain the prior written consent of the British
Ministry of Defence (operating under delegation of Her Majesty's Stationery Office) at
the following address: IPR-CU, Intellectual Property Rights Group, Poplar 2a #2218,
MoD Abbey Wood, Bristol BS34 8JH; Tel: 0117 9132862; email: ipr-cu@dpa.mod.uk
The authors have asserted their moral rights.

British Library Cataloguing in Publication Data
A catalogue record for this book is available from the British Library.

Library of Congress Cataloging in Publication Data
A catalog record for this book is available from the Library of Congress:

Woodhead Publishing ISBN 978-1-85573-921-5 (book)
Woodhead Publishing ISBN 978-1-84569-097-7 (e-book)
CRC Press ISBN 978-0-8493-3488-7
CRC Press order number: WP3488

The publishers' policy is to use permanent paper from mills that operate a sustainable
forestry policy, and which has been manufactured from pulp which is processed using
acid-free and elementary chlorine-free practices. Furthermore, the publishers ensure that
the text paper and cover board used have met acceptable environmental accreditation
standards.

Printed in the United Kingdom by Lightning Source UK Ltd

Contents

7   Intelligent textiles for protection   176

L VAN LANGENHOVE, R PUERS and D MATTHYS, University of Ghent, Belgium

8   Surface treatments for protective textiles   196

R BUCKLEY, Eastgate Consulting, UK

9   Evaluation of protective clothing systems using manikins   217

E A MCCULLOUGH, Kansas State University, USA

13   Textiles for UV protection   355
A K SARKAR, Colorado State University, USA

14   Textiles for protection against cold   378
I HOLMÉR, Lund Technical University, Sweden

15   Thermal (heat and fire) protection   398
R HORROCKS, University of Bolton, UK

Contributor contact details

(* = main contact)

Introduction
Dr Richard A. Scott
Mirabeau
102 Abbots Road
Colchester
Essex CO2 8BG
UK

Tel: 01206 542766
E-mail: dlo_rascott@hotmail.com

Chapter 1
Professor Dr Wenlong Zhou and Mr Narendra Reddy
Department of Textiles, Clothing and Design
University of Nebraska – Lincoln
234 HE Building
Lincoln
Nebraska 68583-0802
USA

Professor Dr Yiqi Yang*
Department of Textiles, Clothing and Design, and Department of Biological
Systems Engineering
University of Nebraska – Lincoln
234 HE Building
Lincoln
Nebraska 68583-0802
USA

Tel: 001-402-472-5197
Fax: 001-402-472-0640
E-mail: yyang2@unl.edu

Chapter 2
Jürgen Haase
Sächsisches Textilforschungsinstitut e.V.
PO Box 1325
D-09125 Chemnitz
Germany

Tel/Fax: 0049 (0)371 722237
E-mail: j.haase@surf-club.de

Chapter 3
Sandy Black
Reader in Knitwear and Fashion
London College of Fashion
University of the Arts London
20 John Princes Street
London W1G 0BJ
UK

Tel: 0207 514 7440
Fax: 0207 514 7672
E-mail: s.black@fashion.arts.ac.uk

Chapter 4
Dr Anugrah Shaw
Richard Henson Center, Room 2103
University of Maryland Eastern
   Shore
Princess Anne, MD 21853
USA

Tel: 410-651-6064
Fax: 410-651-6285
E-mail ashaw@umes.edu

Chapter 5
Professor John Hearle
The Old Vicarage
Church Road
Mellor
Stockport SK6 5LX
UK

Tel: 0161 427 1149
E-mail: johnhearle@compuserve.com

Chapter 6
Dr Prasad Potluri
Senior Lecturer in Textile
   Composites
Textile Composites Group
School of Materials, Textiles &
   Paper
Sackville Street Building
University of Manchester
Manchester M60 1QD
UK

Tel: (44) 161 200 4128
Fax: (44) 161 955 8128
E-mail:
   Prasad.Potluri@manchester.ac.uk

Chapter 7
Professor Dr ir. Lieva Van
   Langenhove*
Department of Textiles

Ghent University
Technologiepark Zwijnaarde 907
9052 Ghent
Belgium

Tel: +32 9 2645419
E-mail:
   Lieva.vanlangenhove@ugent.be

Professor Dr ir. Robert Puers
Department ESAT-MICAS
Katholieke Universiteit Leuven
Kasteel Park Arenberg 10
3001 Leuven
Belgium

Tel: +32 16 321082
E-mail:
   Robert.Puers@esat.kuleuven.ac.be

Professor Dr Dirk Matthys
Department of Pediatrics and
   Genetics
Ghent University
De Pintelaan 185
9000 Ghent
Belgium

Tel: +32 9 2403585
E-mail: dirk.matthys@ugent.be

Chapter 8
Mr Roy William Buckley
Eastgate Consulting
Eastgate
36 High Street
Halberton
Tiverton
Devon EX16 7AG
UK

Tel: 01884 820 653
Email: roywbuckley@tiscali.co.uk

Chapter 9
Elizabeth McCullough
Kansas State University
Institute for Environmental Research
64 Seaton Hall
Manhattan, KS 66502
USA

Tel: 785-532-2284
E-mail: lizm@ksu.edu

Chapter 10
Dr René Rossi
EMPA Materials Science and
    Technology
Lerchenfeldstrasse 5
CH-9014 St Gallen
Switzerland

Tel: +41 71 274 77 65
Fax: +41 71 274 77 62
E-mail: rene.rossi@empa.ch

Chapter 11
Professor Guowen Song
Department of Human Ecology
331 Human Ecology Building
University of Alberta
Edmonton, AB   T6G 2N1
Canada

Tel: (780)492-0706
Fax: (780)492-4821
E-mail: guowen.song@ualberta.ca

Chapter 12
Jeffrey Stull
International Personnel Protection,
    Inc.
10907 Wareham Court
Austin, TX 78739
USA

Tel/Fax: 512-288-8272
E-mail: Intlperpro@aol.com

Chapter 13
Ajoy K. Sarkar
Colorado State University
Department of Design and
    Merchandising
1574 Campus Delivery
Fort Collins, CO 80523-1574
USA

Tel: 970-491-6740
Fax: 970-491-4376
E-mail: sarkar@cahs.colostate.edu

Chapter 14
Professor Ingvar Holmér
Thermal Environment Laboratory
Department of Design Sciences
Lund Technical University
Box 118
22100 Lund
Sweden

Tel: +46 46 2223932
Fax: +46 46 2224431
E-mail: ingvar.holmer@design.lth.se

Chapter 15
Professor Richard Horrocks
Professor in Textile Science
University of Bolton
Deane Rd
Bolton BL3 5AB
UK

Tel: +44 (0)1204 903831
Fax: +44 (0) 1204 399074
E-mail arh1@bolton.ac.uk

Chapter 16
Dr Karen K. Leonas
Professor of Textile Science
317 Dawson Hall
University of Georgia

Athens, GA 30602
USA

E-mail: kleonas@fcs.uga.edu

Chapter 17
Professor Izabella Krucińska
Technical University of Lodz
90-543 Lodz
ul. Zeromskiego 116
Lodz
Poland

E-mail: ikrucins@mail.p.lodz.pl

Chapter 18
Dr Jose A. Gonzalez
Protective Clothing & Equipment
    Research Facility
Department of Human Ecology
University of Alberta
Edmonton, AB T6G 2N1
Canada

Tel: 780-492-0111
Fax: 780-492-4111
E-mail: jose.gonzalez@ualberta.ca

Chapter 19
Dr Xiaogang Chen
Senior Lecturer
School of Materials, Textiles and
    Paper
The University of Manchester
Sackville Street Building
Manchester M60 1QD
UK

Tel: +44 (0)161 306 4113
Fax: +44 (0)161 955 8164
E-mail:
    xiaogang.chen@manchester.ac.uk

Chapter 20
Quoc Truong
US Army Research, Development
    and Engineering Command
Natick Soldier Center
Individual Protection Directorate
Chemical Technology Team
Kansas Street
Natick, MA 01760-5019
USA

Tel: (508) 233 5484
E-mail: Quoc.Truong@us.army.mil

Eugene Wilusz
US Army Research, Development
    and Engineering Command
Natick Soldier Center
Individual Protection Directorate
Chemical Technology Team
Kansas Street
Natick, MA 01760-5019
USA

Tel: (508) 233 5486
E-mail: Eugene.Wilusz@us.army.mil

Chapter 21
Dr R. A. Scott
RASCOTEX
Mirabeau
102 Abbots Road
Colchester
Essex CO2 8BG
UK

Tel: 01206 542766
E-mail: dlo_rascott@hotmail.com

Chapter 22
Dr Helena Mäkinen
Finnish Institute of Occupational
    Health

Topeliuksenkatu 41 aA
00250 Helsinki
Finland

Tel: +358 304742764
Fax: +358 304742115
E-mail: helena.makinen@ttl.fi

Chapter 23
Mr Paul Fenne
Metropolitan Police
Physical Protection Group
Room 208
40-42 Newlands park
Sydenham
London SE26 5NF
UK

E-mail: Paul.fenne@met.police.uk

Chapter 24
Dr Elizabeth Crown* and
    Professor Linda Capjack
Department of Human Ecology
University of Alberta
Edmonton, AB T6G 2N1
Canada

E-mail: betty.crown@ualberta.ca;
    linda.capjack@ualberta.ca

Chapter 25
Dr Elizabeth Crown*
Department of Human Ecology
University of Alberta
Edmonton, AB T6G 2N1
Canada

E-mail: betty.crown@ualberta.ca

Dr James D. Dale
Department of Mechanical
    Engineering
University of Alberta
Edmonton, AB T6G 2N1
Canada

Chapter 26
Mr Paul Varnsverry
PVA Technical File Services Limited
68 Winchester Road
Delapre
Northampton NN4 8AY
UK

Tel: (0)1604 766077
E-mail: info@pra-ppe.org.uk

Introduction

R A S C O T T, RASCOTEX, UK

The primary aims of this book are to educate, inform, enlighten and to stimulate the reader. T. H. Huxley once wrote in *Science and Education IV*:

> Education is the instruction of the intellect in the laws of nature, under which name I include not merely things and their forces, but men and their ways: and the fashioning of the affections and of the will into an earnest and loving desire to move in harmony with those laws.

Man is a singular creature. He/she has a set of gifts that make them unique among animals. Millions of years of biological evolution has not really fitted man to any specific environment. On the contrary, by comparison with evolved animals he has a rather crude personal survival kit. His imagination, reasoning and toughness make it possible for him to change the environment in which he lives. He is not merely a figure in the landscape, but a shaper of the landscape. So wrote Jacob Bronowski in his book *The Ascent of Man* in 1976.

Man makes plans, inventions and new discoveries by putting different talents together. This book is the fruit of a diverse body of talents, drawing together scientific and technical expertise from all corners of the globe, to produce a valuable source of current knowledge on textile materials and clothing, and their use in the protection of humans in hostile environments.

Historical

About two million years ago man evolved from the hairy ape-like creature into the upright, hairless *homo sapiens* living in a warm Mediterranean or tropical climate. Covering the body to protect against the environment became critical when *homo sapiens* began to move long distances and live in colder climates. Animal skins would constitute the first protective clothing, and wool fibres stuffed into crude footwear probably formed the first nonwoven felt insoles to protect against cold and abrasion.

In Palaeolithic and Mesolithic times (about 10,000 to 5,000 BC) there is evidence of thread and cord made from grasses, reeds, or animal sinews being used to bind or fasten tools or weapons together. The discovery of bone awls and needles indicates that leather was sewn to form protective garments or containers.

The first evidence of spinning and weaving occurred in Neolithic times, when vegetable bast fibres, flax, cotton, silk and wool were available to man, who now lived in agricultural settlements. One of the earliest fragments of woven woollen cloth was dated at about 6,500 BC, and was found in the excavated ruins of Catal Huyuk in the Middle East. The oldest cotton fabric dates from about 2,500 BC and is from the ancient city of Mohenjodaro in what is now the lower Indus river region of Pakistan. Examples of complex brocaded fabrics were found in Neolithic Swiss lake-side villages, along with spindle whorls for hand-spinning, and stone loom weights. (Cole S (1965) *The Neolithic Revolution*, 3rd edn, British Museum Natural History, pp. 45–48.) History indicates that protective textiles were often developed for use by fighters and warriors. Thus, Roman soldiers wore heavy cloth tunics and skirts in conjunction with metal and leather armour. Mediaeval knights wore textile clothing as padding under chain mail or suits of armour to prevent chafing of the skin.

The industrial revolution in the 19th century probably heralded a significant increase in the use of protective clothing against industrial hazards such as heat, fire, blast, impact, cuts, chemical splashes and dirt in the emerging metal, glass, ceramic and chemical industries and crafts. This could be considered to be the first serious interest in personal protective equipment.

The present day

The late 20th century saw an unprecedented increase in emphasis on protection of the human from occupational and recreational hazards. Increasingly complex legislation and regulation in the workplace was the result of the philosophy that it was no longer acceptable for humans to incur injury or death in advanced societies. The range of hazards and the means of combating them continue to grow and become ever more complex. A consequence of this is the development and exploitation of new textile fibres, structures, and clothing systems whose purpose is to provide improved protection, whilst maintaining comfort, efficiency and well-being. The contents of this book cover the major approaches and applications of protective textiles in the present day and for the future.

Technical textiles for protection tend to be complex, high added value speciality products. They tend to be sourced from the scientific and technical communities of North America and Europe, where educated workforces and high labour costs mean that commodity textiles for routine use are not an economically viable part of industry. The production of mass-market utility textiles, such as cotton, nylon, polyester and polyester/cotton blends have graduated towards the Far East.

The protective textiles market

The increasing emphasis on human protection, and the continued introduction of health, safety and environmental legislation means that the technical textile market continues to be buoyant and thrive. The world market for technical textiles in 2005 was about 20 million tons per year. Of this about 280,000 tons was consumed by Protech or technical protective materials, with a value of about $3.3 billion (£1.8 billion). This market was increasing by between 3.3 and 4.0% per annum, so Protech could reach a capacity of 340,000 tons by 2010. (Rigby D (2002) *Non Wovens Report International. http://davidrigbyassociates.co.uk.*)

Hazards to the human

Perhaps the most prevalent hazard encountered in everyday life is the threat from the natural environment. The hazards include heat, cold, rain, snow, wind, ultraviolet light, abrasion, dust, micro-organisms and the effects of static electricity. These can cause problems during occupational, recreational and routine activities. Man-made threats which are caused accidentally include fire, heat, flash, industrial chemicals, bio-hazards, high-velocity and blunt impacts, and cutting/slashing by sharp objects.

Military and emergency personnel face the deliberate use of ballistic projectiles (bullets and bomb fragments), chemical and biological warfare agents, heat, flames and flash, and the threat of detection by wide-spectrum sensors. All these hazards can cause injury or death to the unprotected upper and lower torso, extremities, head, eyes, skin, and respiratory tract.

Overview of book contents

The book is divided into three parts: Part I concerns materials and design issues which are covered in eleven chapters. Part II covers general protective requirements and applications, comprising nine chapters. Part III deals with case studies that detail six specific applications of protective clothing systems. All contributing authors are world-renowned experts in their particular fields of textiles and protective clothing. The majority are scientists from universities and colleges, or with academic backgrounds, whilst others are experts from industry, research organisations, and government departments.

Part I: Materials and design

The starting point for Part I is a complete overview of personal protection by Dr Yiqi Yang, Dr Zhou and Dr Reddy from the University of Nebraska, Lincoln, USA. This includes all the hazards, the types of personal protection available, and includes the design, finishing, manufacture, and testing of materials and protective clothing. This chapter should be read as an introduction to this part of the book.

The second chapter covers the important topic of international legislative and regulative standards for all types of human protection involving textiles. Dr J Haase from the Saxon Technical Research Institute in Germany discusses the markets, requirements, performance and evaluation testing of materials and clothing. The chapter details ISO, European EN, ASTM and other national standards, as they existed in the year 2004. In these respects this information is a valuable reference document.

Chapter 3 on fashion and function is written by a group of experts from the London College of Fashion. Doctors Sandy Black, Frances Geesin, Veronika Kapsali and Jennifer Bougourd present the problem of the balance between functional, fashion and aesthetic requirements for protective clothing, high-lighting the many conflicting factors which need to be reconciled. The classic dilemma which faces protective clothing designers is to provide the required level of protection and comfort, whilst ensuring that the wearer feels that it projects the right image, is fashionable, and meets cultural norms. If these factors are not reconciled there is a tendency for people to avoid wearing the items, or to put forward illogical reasons for rejecting the items during development.

Dr Anugrah Shaw from the University of Maryland, USA, describes a formal process for selecting the appropriate textiles for protective clothing in Chapter 4. Correct selection must be based upon the work environment, economic con-siderations, the performance properties required, and compliance with standards and performance specifications. This chapter includes the demonstration of modern on-line internet knowledge databases, involving drop-down selection menus.

Chapter 5 is a fundamental contribution to this book, dealing as it does with comprehensive details of all technical fibres, and fabrics used in protective clothing. The author is Professor John Hearle, former head of textiles at the University of Manchester Institute of Science and Technology (UMIST) UK, and world renowned for his services to Textile Institute International. He describes the whole range of natural and man-made fibres, from cotton, nylon, and polyester, up to meta and para-aramids, high-modulus polyolefins, carbon fibres, PBO, M5, and other specialist inorganic fibres, chemically resistant polymers, and thermally resistant polymers. Novel electro-spun fibres and carbon nanotubes are also included. This chapter links with all other chapters of this book.

Doctor Prasad Potluri, also from UMIST, Manchester, England, surveys the manufacture of yarns and fabrics using technical textiles. He also deals with the range of protective applications, including environmental, flames and heat, ballistic/blunt impact, and protection from chemicals. State-of-the-art and future possibilities for intelligent textiles are the province of Professors Van Langenhove, Puers, and Matthys from the University of Ghent in Belgium. Students and developers of technical textiles will be fascinated to read of this rapidly developing sector of the industry. It indicates the synergism and integration between textiles, optics, electronics, and advanced engineering techniques. Smart/reactive textiles are described which are able to sense stimuli from the environment, react to them, and adapt to them by integration of functionalities in textile structures. The stimulus as well as the response can have electrical, thermal, chemical, magnetic or other origins.

Chapter 8 is an industrial perspective on the role, range, use and application of surface treatments for protective textiles. Roy Buckley, who worked for many years on textile Research & Development at John Heathcoat and Sons, Tiverton, Devon, UK, discusses flame retardant, water/oil/soil repellent, anti-microbial, UV blocking, reflective, absorptive and reactive finishes. His chapter emphasises and details the industrial application processes and variables used to produce these functional finishes.

Professor Elizabeth McCullough of Kansas State University, USA, has followed a distinguished career in human protection, and writes Chapter 9 on the evaluation of complete protective clothing systems using instrumented manikins that simulate the human thermoregulatory responses. She emphasises the importance of testing clothing systems complete with closures, design features, and auxiliary equipment such as tool belts and respirators. Elizabeth includes thermal insulation and vapour permeability data that is typical of specific protective clothing systems. Heated, sweating, and moving manikins are becoming more widespread throughout the world, emphasising their increasing importance in providing realistic but safe evaluation procedures.

Doctor Rene Rossi from the laboratories of EMPA at St Gallen, Switzerland expounds on the interactions between protection and thermal comfort. The perennial problem of the contradiction between increased protection and thermo-physiological load, and the effects on mobility and efficiency of the wearer are dealt with in Chapter 10. Methods of measuring thermal and water vapour resistance are detailed – from simple plate and cup methods, through skin models, right up to thermal/sweating/moving manikins and human trials. EMPA possess one of the most recent manikins to be completed, called SAM. There are links to Chapter 9 by Elizabeth McCullough, and Chapter 14 by Professor Holmér.

The mathematical modelling of thermal burn injuries to humans is the contribution by Dr Guowen Song from North Carolina State University at Raleigh, USA. Chapter 11 elucidates the mechanisms of heat transfer through protective clothing and materials, and details the analytical and numerical models that have been developed in recent years. The chapter highlights the critical importance of accurate modelling and simulation of the heat threat, given that human trials of flame and heat-protective clothing are obviously not permitted on ethical grounds. Measurement techniques cover the range from simple laboratory tests on materials, up to full-scale system tests on instruments manikins. There is a direct link with Chapter 15 by Professor Horrocks.

Part II: General protection requirements and applications

Doctor Jeffrey Stull of International Protection Inc., USA is well known for his detailed theses and publications on protective clothing. Chapter 12 concentrates on protection from industrial chemicals, with a comprehensive exposition on threats, requirements, and performance of chemical protective clothing. Standards for selection, design, and evaluation of materials and clothing are detailed. There are linkages with Chapter 2 by Haase, Chapter 4 by Shaw, and Chapter 20 by Truong and Wilusz.

Chapter 13 covers textiles for protection of human skin from ultraviolet light, and is the contribution of Dr Ajoy Sarkar of Colorado State University, USA. The last decade of the 20th century saw the incidence of skin cancer increase throughout the world. This is attributed to the depletion of atmospheric ozone by various agents, including halogenated (CFC) compounds. This has been compounded by excessive exposure of skin to sunlight during human leisure activities. Because ozone is a very effective UV absorber, every 1% decrease in ozone concentration gives rise to an increase in skin cancer by 2 to 5%! Doctor Sarkar explains the problem, describes test methods and details the textile solutions to defeat this recent phenomenon.

One of the most prevalent environmental problems encountered is cold weather. Professor Ingvar Holmer from the Thermal Environmental Laboratory at the Lund Technical University, Sweden, is a recognised expert on human protection in cold climates. The environment and its climatic components are defined. The heat exchange between the human body and the environment is presented. Measurements and calculations of thermal insulation and water vapour resistance are described. The effects of fibres, fabrics, and clothing structure and design are discussed. Chapter 14 has links with Chapters 5 by Professor Hearle, 9 by Professor McCullough, and 10 by Dr Rossi.

Professor Richard Horrocks from the Centre for Materials Research and Innovation at the Bolton Institute, UK, is the Chairman of the Textile Institute Council at the time of writing. He is a renowned expert on many technical textile topics, notably protection from flames and heat. In Chapter 15 he defines the threats, presents information on flame retardant fibres and materials, and discusses the performance and evaluation of materials and clothing systems. Professor Horrocks describes the research that his team have carried out on reactive heat protection using intumescent materials. This chapter has links with Chapter 5 by Professor Hearle, Chapter 11 by Dr Song, and Chapter 22 on firefighters' clothing by Dr Mäkinen.

Textiles and their role in protection from micro-organisms such as bacteria and blood-borne pathogens is the specialist contribution from Dr Karen Leonas from the University of Georgia, USA. She describes the need for protective apparel for health care workers, discussing the barrier properties of textile fabrics used in clothing, drapes and patient covers. The increasing use of regulation in health care protective clothing, and the wide variety of test methods to evaluate performance are covered. This chapter has some linkages with Chapter 17 on respiratory protection by Professor Krucińska, and Chapter 20 on chemical/biological protection by Doctors Wilusz and Truong.

Fibrous textiles are utilised in many forms of respiratory protective equipment to filter out or absorb dust, other particulates, and toxic vapours. Professor Krucińska from the Technical University of Lodz, Poland expounds on this special use for textiles, in Chapter 17. She provides a mathematical treatment of air-flow and particle capture theories, to highlight the different mechanisms of filtration by fibres. She goes on to describe the range of non-woven fibrous structures used in respirators, and also the types of treatments and additives which are incorporated to improve the filtration or adsorption of solids and vapours. Test methods which are used to measure the performance of ori-nasal and full-face respirators are detailed in this work.

Textile materials and clothing used in certain hazardous environments need to be chosen or treated to minimise the risk of fire or explosions caused by static electrical charges building up and being released via the wearer. The problem can occur when flammable vapours are present in the air, or where electrically initiated explosives are used. Static electrical charges from the clothed human can also damage sensitive electronic circuits and components during assembly. Doctors Jose Gonzalez and Elizabeth Crown from the University of Alberta, Canada, have studied this phenomenon in recent years. They discuss the factors giving rise to the hazard, and detail solutions to the problem using anti-static fibres and finishes. The alternative approach is to carefully select suitable material combinations, or to earth the wearer through the footwear.

Chapter 19 deals with the threats posed to humans by bullets, bomb fragments and other high-speed projectiles. It is written by Drs Chen and Chaudhry from UMIST, Manchester, UK. They outline the ballistic threats, and give details of the specialist high-modulus fibres used in body armour and helmets. This chapter concentrates on US civilian and police force protection. Test methods for body armour and helmets are presented. It should be read in conjunction with Chapter 21 by Dr Scott, which discusses the military ballistic threats and protective solutions for UK forces.

The threats to military forces and civilian emergency workers posed by chemical and biological warfare agents (CBW) are the subject of Chapter 20 by Dr Eugene Wilusz and Dr Quoc Truong from the US Army Natick Soldier Centre (NSC) in Massachusetts, USA. Both authors are recognised scientific experts in this field. At the time of writing the predicted threat posed by terrorist attacks on civilians using chemical and biological agents is at a high level, so its importance cannot be understated. Their comprehensive dissertation on the subject includes details of the past, present and future chemical and biological threats, and their effects on humans. Different types of textile-based protection systems are described, which include impermeable, air permeable, semipermeable, and selectively permeable materials. Their use in individual protective clothing systems is discussed, highlighting the importance of matching CB material designs to the intended environment and operational scenario. Testing and evaluation of CBW protective materials and clothing systems cover the use of simulants, live agents, toxic industrial chemicals, and biological barrier measurements. Army science and technology programmes past, present, and future are outlined, including the study of smart/reactive materials, nanoscale barriers and self-decontaminating materials. This chapter has links with Chapter 12 on Industrial chemicals by Dr Stull, Chapter 17 on respiratory protection by Professor Krucińska, and Chapter 21 on military protection by the editor, Dr Scott.

Part III: Case studies

Chapter 22 by Dr Helena Mäkinen from the Finnish Institute of Occupational Health, deals with the important subject of fire-fighters' protective clothing. Her comprehensive treatment details the varied work of fire-fighters, and the multiple hazards which they face. These include chemical spills and attacks, vehicle accidents, natural disasters, and explosions leading to destruction of property, in addition to the fighting of urban and wildland fires. Dr Mäkinen goes on to give details of the materials and clothing, including requirements, performance standards, clothing design, and in-service care and maintenance. There are linkages to Chapter 11 by Dr Song, and Chapter 15 by Professor Horrocks.

Protection of military personnel is the specialist area of expertise of the editor, Dr Richard Scott, whose career in the UK Ministry of Defence has concentrated on research and development of combat and special protective materials for clothing systems. The threats facing military forces are wide-ranging, and some of them are unique. Whereas many threats to civilians occur by accident, military personnel face injury and death from the deliberate use of weaponised hazards. However, the natural environment presents the highest priority threat. Battlefield threats include ballistic projectiles, conventional and nuclear explosions, chemical and biological warfare agents, flames, heat and flash, detection by wide-spectrum sensors, and a range of non-lethal weapons which incapacitate rather than kill. The chapter covers technical textiles, clothing systems, and specialist performance requirements. The unique subject of camouflage, concealment and deception is treated here. There are links to Chapter 5 by Professor Hearle, Chapter 14 by Professor Holmér, Chapter 19 by Drs Chen and Chaudhry, and Chapter 20 by Drs Wilusz and Truong.

Protection of police forces is the specialist subject of Chapter 23 by Dr Paul Fenne from the Physical Protection Group of the Metropolitan Police Service in the UK. The work concentrates on protection against knives and other sharp objects, in addition to bullets and fragments. Doctor Fenne details police requirements, the principles, manufacture, and testing of body armour, including the problems of designing for a wide range of male and female body shapes. Future trends analyse improvements in equipment over the short and long term. There are links with Chapter 19 on ballistic protection.

Chapter 24 deals with specialist protective systems for military aviators, including both pilots and other aircrew. Doctors Elizabeth Crown and Linda Capjack from the University of Alberta, Canada detail the hazards associated with military aircraft crashes, highlighting the need to protect aviators from heat and flames. The performance requirements for materials and flight suits are related to tasks and hazards. Specialist aircrew items such as anti-G suits and immersion clothing are briefly discussed. The chapter concludes by providing a glimpse of future integrated clothing systems utilising smart materials. There are obvious links with Chapter 15 by Professor Horrocks, and Chapter 21 by Dr Scott.

Oil and gas industry workers require protection against flames, heat and accidental explosions, whilst the working situation is complicated by the contamination of clothing by oily dirt. Doctor Elizabeth Crown from Alberta University presents the hazards, which also include static electrical charges on clothing – a particular problem in cold, arid climates. High-voltage discharges from the human can increase the risk of large-scale fires and explosions. She then discusses the materials, design, and maintenance requirements relative to the hazards encountered. Current textile and clothing performance requirements are discussed, culminating in a forecast for clothing systems for the future. This has links with Chapter 12 on industrial chemical protection, and Chapter 18 on electrostatic hazards.

The final chapter is presented by Mr Paul Varnsverry of PVA Technical Services Ltd (UK), on the subject of motorcyclists' protection. Mister Varnsverry is an experienced motorcyclist, and has been an active member of the European standardisation organisation dealing with the unique problems which motorcyclists face daily. The main problem is protection from exposure to environmental conditions experienced on the road. The most severe problem concerns traffic accidents or loss of control, resulting in high-speed impact and sliding contact with the road. The history of motorcyclists clothing is followed by the conflicting arguments for 'leather versus textile' clothing over the years. The problems of reconciling different interests to provide a suitable European standard are discussed, including the need for three alternative textile technologies to deliver independently tested and approved clothing systems which are affordable. The chapter concludes with information on current manufacturers of accredited textile clothing for motorcyclists.

Acknowledgements

This book has been made possible because a team of internationally famous authors have contributed a great deal of time, effort, and above all special and significant expertise and knowledge to its preparation. The editor wishes to extend his most sincere thanks to all the authors for their important contributions, cooperation and patience.

Special thanks are also given to the staff at Woodhead Publishing Ltd at Cambridge, especially Emma Starr, Melanie Cotterell, Sarah Whitworth, and Emma Cooper for their persistence, patience, interest, and rapid responses during the preparation of this volume.

Part I

Materials and design

Overview of protective clothing

W ZHOU, N REDDY and Y YANG,
University of Nebraska – Lincoln, USA

1.1 Introduction

Scientific advancements made in various fields have undoubtedly increased the quality and value of human life. It should however be recognized that the technological developments have also exposed us to greater risks and danger of being affected by unknown physical, chemical and biological attacks. One such currently relevant danger is from bioterrorism and weapons of mass destruction. In addition, we continue to be exposed to hazards from fire, chemicals, radiation and biological organisms such as bacteria and viruses. Fortunately, simple and effective means of protection from most of these hazards are available. Textiles are an integral part of most protective equipment. Protective clothing is manufactured using traditional textile manufacturing technologies such as weaving, knitting and non-wovens and also by specialized techniques such as 3D weaving and braiding using natural and man-made fibers.

Protective clothing is now a major part of textiles classified as technical or industrial textiles. Protective clothing refers to garments and other fabric-related items designed to protect the wearer from harsh environmental effects that may result in injuries or death (Adanur, 1995). Today, the hazards that we are exposed to are often so specialized that no single type of clothing will be adequate for protection. Extensive research is being done to develop protective clothing for various regular and specialized civilian and military occupations (Adanur, 1995; Bajaj *et al.*, 1992; Holmes, 2000). Providing protection for the common population has also been taken seriously considering the anticipated disaster due to terrorism or biochemical attacks (Holmes, 2000; Koscheyev and Leon, 1997).

1.2 Market prospects

Protective textiles are a part of technical textiles that are defined as comprising all those textile-based products which are used principally for their performance or functional characteristics rather than their aesthetic or decorative

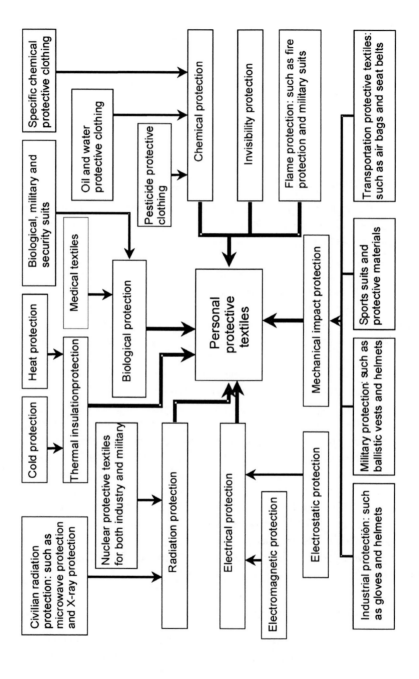

1.1 Schematic classifications of protective textiles.

characteristics (Byrne, 2000). In 2000, technical textiles accounted for about 25% of all textile consumption by weight (David Rigby Associates, 2004). Protective textiles account for 1.4% of the total technical textiles with an estimated value of US$5.2 billion.

Consumption of protective clothing has increased linearly in the last ten years, and in 2010 it is expected that about 340,000 tons of protective clothing will be consumed, an increase of 85% over consumption in 1995. The Americas (mainly USA and Canada) have the highest consumption of protective clothing per annum at about 91 300 tons followed by Europe with 78,200 tons and Asia with 61,300 tons (David Rigby Associates, 2004). All other regions consume only 7,200 tons, 3.0% of total protective textile consumption.

1.3 Classification

Classifying personal protective textiles is complicated because no single classification can clearly summarize all kinds of protection. Overlap of the definitions is common since there are so many occupations and applications that even the same class of protective clothing often has different requirements in technique and protection. Depending on the end use, personal protective textiles can be classified as industrial protective textiles, agricultural protective textiles, military protective textiles, civilian protective textiles, medical protective textiles, sports protective textiles and space protective textiles.

Personal protective textiles can be further classified according to the end-use functions such as thermal (cold) protection, flame protection, chemical protection, mechanical impact protection, radiation protection, biological protection, electrical protection and wearer visibility. Their relationship is illustrated in Fig. 1.1. Unless indicated otherwise, this classification will be used in the following descriptions.

1.3.1 Fire protection

It would have been impossible for humans to survive the primitive age without the use of fire. However, fire could be dangerous. Fire disasters occur frequently resulting in non-fatal and fatal casualties. Of all the accidental fires in dwellings, occupied buildings and outdoor fires, the great majority (79% of the total in 1986) of deaths resulted from fires in dwellings although only 16% of fires happened in dwellings (Bajaj et al., 1992). The most frequently ignited materials were the textiles, especially upholstery and furnishings (Bajaj et al., 1992). It should, however, be noted that the main cause of death in a fire accident is not direct burning but suffocation due to the smoke and toxic gases released during burning. In the UK, 50% of fatalities in fire accidents were directly attributable to this cause (Bajaj et al., 1992). Therefore, the use of non- or low-toxic burning materials is very important for fire protection.

Human tissue (skin) is very sensitive to heat. It is reported that, at 45 °C, the sensation of pain is experienced, and at 72 °C the skin is completely burnt (Bajaj *et al.*, 1992; Panek, 1982). The purpose of fire-protective clothing is to reduce the rate of heating of human skin in order to provide the wearer enough time to react and escape. The time that a wearer stays in flame circumstances and the amount of heat flux produced are important factors for designing the protective stratagem. Under normal conditions, only 3–10 seconds are available for a person to escape from a place of fire with a heat flux of about 130–330 kW/m^2 (Holmes, 2000). Fibers commonly used for textiles are easily burnt. Untreated cotton will either burn (flaming combustion) or smolder (smolder combustion), whenever it is in the presence of oxygen and the temperature is high enough to initiate combustion (360–420 °C) (Wakelyn, 1997).

Protective clothing designed for flame protection must have two functions, i.e., be flame-resistant and form a heat barrier. The latter is a very important factor if the wearer needs to stay near flames for a fairly long time. In fact, the danger of burning lies with the parts of the body not covered by clothing, confirmed by statistics showing that 75% of all firefighter burn injuries in the USA are to the hands and face (Holmes, 2000). Flame-retardant clothing is generally used for occupation uniforms (Holmes, 2000).

Increasing government regulations and safety concerns necessitate that certain classes of garments and home textiles such as children's sleepwear, carpets, upholstery fabrics and bedding be made flame-retardant or resistant (Wakelyn *et al.*, 1998). Using inherently flame-retardant materials such as Kevlar and Nomex, applying a flame-retardant finish or a combination of these methods are commonly used to make clothing and textiles flame retardant.

1.3.2   Heat and cold protection

Basic metabolisms occurring inside our body generate heat that can be life saving or fatal depending on the atmosphere and circumstances that we are in. Normally, human bodies are comfortable to heat in a very narrow temperature range of 28–30 °C (82–86 °F) (Fourt and Hollier, 1970). In summer, we need the heat from our metabolic activity to be transferred outside as soon as possible, while in winter, especially in extremely cold conditions, we must find ways to prevent the loss of heat from our body. Heat stress, defined as the situation where the body cannot dissipate its excess heat to the environment is a serious problem especially during physical working (Bajaj *et al.*, 1992; McLellan, 1996; Muza *et al.*, 1996; Richardson and Capra, 2001; Wasterlund, 1998).

Basically, heat is transferred either as conductive, convective, radiant heat or a combination of these modes depending on the source of heat, the atmosphere the heat-absorbing material is in and the protection available against heat (Bajaj *et al.*, 1992; Fourt and Hollier, 1970). Any heat transfer will have at least one of these modes and heat protection is the method to decrease or increase the rate of heat transfer. For protection from conductive heat, fabric thickness and density are the major considerations, since air trapped between fibers has the lowest thermal conductivity of all materials (Morton and Hearle, 1997). For protection from convective heat (flame hazard in particular), the flame-retardant properties of the fabric are important. As for radiant heat protection, metalized fabrics such as aluminized fabrics are preferred, since metalized fabrics have high surface reflection and also electrical conductivity (Adanur, 1995; Bajaj *et al.*, 1992). Ideal clothing for protection from heat transfer are fabrics with thermo-regulating or temperature-adaptable properties (Bajaj *et al.*, 1992; Pause, 2003). Phase change materials (PCM) are one such example that can absorb heat and change to a high-energy phase in a hot environment, but can reverse the process to release heat in cold situations (Choi *et al.*, 2004b).

Specifically designed protective clothing is necessary to survive and operate in temperatures below −30 °C. Such low-temperature conditions are aggravated in the presence of wind, rain or snow leading to cold stress that may be fatal (Rissanen and Rintamaki, 2000). The most effective method of cold protection is to avoid or decrease conductive heat loss. Clothing designed to protect from cold is usually multi-layered, consisting of a non-absorbent inner layer, a middle insulating layer capable of trapping air but transferring moisture, and an outer layer that is impermeable to wind and water. Temperature-adaptable clothing that can protect from both heat and cold has been developed by fixing poly-ethylene glycol to cotton at different curing temperatures (Bajaj *et al.*, 1992).

1.3.3 Chemical protection

Fortunately, most of us are not involved in handling dangerous and toxic chemicals, since no amount of protection can provide complete isolation from the hazards of chemicals. In recent years, the chemical industry has been facing an ever-increasing degree of regulation to avoid workers being exposed to chemical hazards (Bajaj *et al.*, 1992). Chemical protective clothing (CPC) should be considered the last line of defense in any chemical-handling operation and every effort should be made to use less hazardous chemicals where possible, or to develop and implement engineering controls that minimize or eliminate human contact with chemical hazards (Carroll, 2001, Adanur, 1995).

Protective clothing cannot be made generic for all chemical applications, since chemicals vary in most cases and a particular CPC can protect only against a limited number of specific chemicals (Perepelkin, 2001). Important considerations in designing chemical protective clothing are the amount of chemical permeation, breakthrough time for penetration, liquid repellency, and physical properties of the CPC in specific chemical conditions (Carroll, 2001; Mandel *et al.*, 1996; McQueen *et al.*, 2000; Vo *et al.*, 2001; Park and Zellers, 2000; Singh and Kaur, 1997a,b). Based on the specific requirements and type of clothing, CPC is classified in different ways.

Chemical protective clothing can be categorized as encapsulating or non-encapsulating based on the style of wearing the clothing (Adanur, 1995). The encapsulating system covers the whole body and includes respiratory protection equipment and is generally used where high chemical protection is required. The non-encapsulating clothing is assembled from separate components and the respiratory system is not a part of the CPC. The Environmental Protection Agency (EPA) in the United States classifies protective clothing based on the level of protection from highest to normal protection. CPC is rated for four levels of protection, levels A, B, °C and D from highest protection to normal protection (Adanur, 1995; Carroll, 2001). European standards for CPC are based on the 'type' of clothing based on testing of the whole garment and are classified as types 1 to 7, related to the type of exposure of the CPC such as gas-tight, spray-tight, liquid-tight, etc. (Carroll, 2001). Traditionally, used disposable clothing also offers resistance to a wide range of chemicals and some disposable clothing can be repaired using adhesive patches and reused before being disposed (Adanur, 1995; Carroll, 2001). Chemicals that are in liquid form are more often used than solid chemicals. Therefore, chemical protective clothing should be repellent or impermeable to liquids.

Developing pesticide-resistant clothing has received considerable attention from researchers since exposure of skin to pesticide is a major health hazard to farmers (Zhang and Raheel, 2003). Clothing currently used for pesticide protection does not give adequate protection, especially to the hands and thighs, even if farmers use tractor-mounted boom sprayers with a closed cabin and wear protective clothing with gloves and rubber boots (Fenske *et al.*, 2002, Elmi *et al.*, 1998).

Other important functions of chemical protective clothing are to protect from chemicals present in the air such as toxic and noxious gases or fumes from automobiles, dust and microorganisms present in the air. Safety masks containing activated carbon particles which can absorb the dust present in the atmosphere are commonly used against air pollution.

1.3.4   Mechanical impact protection

Ballistic protection

Ballistic protection is generally required for soldiers, policemen and general security personnel. Ballistic protection involves protection of body and eyes against projectiles of various shapes, sizes, and impact velocities (Adanur, 1995). Historically, ballistic protection devices were made from metals and were too heavy to wear. Textile materials provide the same level of ballistic protection as metals but have relatively low weight and are therefore comfortable to wear. Most of the casualties during military combat or during unintended explosions are from the fragments of matter caused by the explosion hitting the body (Scott, 2000). It is reported that during military combat, only 19% of casualties are caused by bullets, as high as 59% of casualties are caused by fragments, and about 22% are due to other reasons (Scott, 2000). The number of casualties due to ballistic impact can be reduced 19% by wearing helmets, 40% by wearing armor and 65% by wearing armor with helmet (Scott, 2000).

High-performance clothing designed for ballistic protection dissipates the energy of the fragment/shrapnel by stretching and breaking the yarns and transferring the energy from the impact at the crossover points of yarns (Scott, 2000). The ballistic protection of a material depends on its ability to absorb energy locally and on the efficiency and speed of transferring the absorbed energy (Jacobs and Van Dingenen, 2001). One of the earliest materials used for ballistic protection was woven silk that was later replaced by high-modulus fibers based on aliphatic nylon 6,6 having a high degree of crystallinity and low elongation. Since the 1970s, aromatic polyamide fibers, such as Kevlar® (Du Pont) and Twaron® (Enka) and ultra-high-modulus polyethylene (UHMPE) have been used for ballistic protection (Scott, 2000).

Other impact protection

According to the US Labor Department, each year, more than one million workers suffer job-related injuries and 25% of these injuries are to the hands and arms (Adanur, 1995). Gloves, helmets and chain-saw clothing are the main protective accessories used by personnel working in the chemical, construction and other industries (Adanur, 1995). Some examples of non-combat impact protection are the seat belts and air bags used in automobiles. Air bags have reduced the death rate in accidents by 28%, serious injuries by 29% and hospitalization by 24% and seat belts can reduce fatal and serious injuries by 50% (Adanur, 1995, Fung, 2000). A typical seat belt is required to restrain a passenger weighing 90 kg in collision with a fixed object at 50 km/h (about 30 mph). The tensile strength of a seat belt should be at least 30 kN/50mm (Fung, 2000).

Although sports and recreational injuries account for relatively few deaths (0–6% of deaths to those under age 20), these activities are associated with 17% of all hospitalized injuries and 19% of emergency room visits to hospitals (Mackay and Scanlon, 2001). Child and adolescent deaths due to sports and recreational injuries are a major cause of morbidity in Canada (Canadian Institute of Child Health, 1994). In 1995, Canada spent about $4.2 billion in treating unintentional injuries (Mackay and Scanlon, 2001). More than half of the total sports and recreational injuries are attributed to eight activities: ice hockey, baseball, basketball, soccer, jogging, cycling, football and volleyball. Modern sports clothing uses high-performance fabrics that are designed to operate at high speed but are still safe and comfortable to wear (O'Mahony and Braddock, 2002). The most common protective textiles used in sports are in knee braces, wrist braces, ankle braces, helmets and guards.

1.3.5 Biological protection

Most natural textile fibers such as wool, silk and cellulosics are subject to biological degradation by bacteria, dermatophytic fungi, etc. Fortunately, various chemicals and finishing techniques are available that can protect the textile and the wearer from biological attacks. Textiles designed for biological protection have two functions: first, protecting the wearer from being attacked by bacteria, yeast, dermatophytic fungi, and other related microorganisms which cause aesthetic, hygienic, or medical problems; secondly, protecting the textile itself from biodeterioration caused by mold, mildew, and rot-producing fungi and from being digested by insects and other pests (Bajaj *et al.*, 1992; Vigo, 1983).

The antimicrobial properties of silk have been used for many years in medical applications (Choi *et al.*, 2004a). Natural fibers contain lignin and other substances that have inherent antimicrobial properties. Generally, textiles made from natural fibers have better anti-microbial properties than man-made fibers due to the presence of substances such as lignin and pectin. Chemical finishing is most commonly used for imparting anti-microbial properties to natural and man-made textiles by applying functional finishes onto the surface of the fabric or by making fibers inherently resistant to microorganisms.

In high functional fibers that are inherently anti-microbial, the entire surface of the fiber is made from a bioactive material and the bioactivity remains undiminished throughout the useful life of the fiber (Bajaj *et al.*, 1992; Patel *et al.*, 1998, Rajendran and Anand, 2002). In some cases, just providing an anti-microbial finish to the fabrics may not prevent the infection. For example, fungi such as *Aspergillosis* is fatal to about 80% of bone marrow and organ transplant recipients, even with intense hospital and strong antifungal drug treatment (Curtis, 1998). To prevent such trans-infection through fabrics, combined fluid-resistant and anti-microbial finishing have been developed that can avoid fluid penetration through the fabric and decrease the trans-infection (Anonymous, 2003; Belkin, 1999; Kasturiya and Bhargave, 2003; Shekar *et al.*, 2001; Zins, 1998).

Fabrics designed for microbial protection should act as barriers to bacteria and other microorganisms that are believed to be transported from one location to another by carriers such as dust or liquids (Belkin, 1999, 2002; Leonas and Jinkins, 1997). Films generally have high barrier properties against microbes and chemicals. However, films when used with fabrics to provide antimicrobial properties make fabrics impermeable to airflow leading to heat stress and other physiological problems that may be fatal (Wilusz *et al.*, 1997). New membrane structures called 'perm-selective' or 'breathable' membranes have been developed that can prevent airflow through the fabric layer but have high water-vapor permeability. Using these membranes with fabrics provides effective protection from hazardous materials or microbes without causing heat stress (HAZMAT) (Schreuder-Gibson *et al.*, 2003).

Risks and contaminations caused by HIV and other viruses have increased the protective requirements for medical textiles (Rajendran and Anand, 2002; Patel *et al.*, 1998). It is desirable to have anti-microbial finishing even for everyday textiles such as underwear, baby suits, diapers, towels, etc.

1.3.6 Radiation protection

Nuclear radiation protection

Special clothing to prevent exposure to radiation is needed for people working in radioactive environments. Alpha-, beta- and gamma-radiation are the major modes of nuclear radiation. Irradiation injuries by alpha- and some beta-radiation can be prevented by keeping the radioactive dirt off the skin and out of the eyes, nose and mouth. Goggles, respiratory masks, gloves and lightweight protective clothing may be adequate for protection from some alpha- and beta-radiation which have weak penetration (Adanur, 1995). However, gamma- and some beta-radiation have sufficient energy to penetrate through textiles and can affect the human tissue even if the radioactive substance does not contact the human skin. Protection from transmitted radiation depends on the level of contamination control, exposure time, distance from radiation source and the type of radioactive shield available (Adanur, 1995). Shielding is done by placing a dense (heavy) radiation barrier such as lead between the radioactive dirt and the worker.

Woven cotton, polyester/cotton or nylon/polyester fabrics with a twill and sateen weave are the major types of fabric forms used for nuclear protective clothing (Adanur, 1995). Non-woven fabrics used as over- and transit garments in nuclear radiation protection act as a barrier against dangerous particles, shields the main garment against contamination and are disposable when contaminated (Bajaj *et al.*, 1992).

UV radiation protection

The wavelength of solar radiation reaching the Earth's surface spans from 280 to 3,000 nm (Reinert *et al.*, 1997). Ultraviolet (UV) light has the highest energy radiation consisting of UV-A and UV-B, whose radiation is from 320–340 nm and 280–320 nm, respectively. Excessive exposure of the skin to UV-A radiation can be carcinogenic resulting in chronic reactions and injury, accelerated ageing of the skin, promotion of photodermatosis (acne) etc. (Reinert *et al.*, 1997). An overdose of UV-B can lead to acute and chronic reactions, skin reddening (erythema) or sunburn, increasing the risk factor of persons susceptible to melanoma and skin cancer (Gies *et al.*, 1997, 1998; Reinert *et al.*, 1997; Wang *et al.*, 2001). In the last decade, attempts to reduce the incidence of skin cancer were mainly focused on decreasing solar UVR exposure (Gies *et al.*, 1997).

Table 1.1 Main factors affecting UVR protection (Adanur, 1995; Gies *et al.*, 1997, 1998; Reinert *et al.*, 1997; Xin *et al.*, 2004)

Factors	Effectiveness
1. Fiber	Cotton has high permeability to UVR, Wool has high absorption, Polyester has high absorption to UV-B, polyamides are fairly permeable to UVR.
2. Weave	Fabric construction, which determines the porosity and type of weave, is the most important factor affecting UV protection. Tighter the weave, lesser the UVR transmitted.
3. Color	Dark colors absorb UVR more strongly and therefore have high UPFs.
4. Weight	Thicker and heavier fabrics transmit less UVR.
5. Stretch	Greater the stretch, lower the UPF rating.
6. Water	Depends on the moisture absorption capabilities of the fibers/fabrics. Generally, fabrics provide less UVR protection when wet.
7. Finishing	UVR absorbing additives can be used to increase the protection of lightweight summer garments.

Although many terms such as SPF (sun protection factor), and CPF (clothing protection factor) which are generally used in the UK have been used to designate the amount of solar UVR protection of fabrics, UPF (ultraviolet protection factor) is the most commonly used index (Gies *et al.*, 1997, 1998; Hatch, 2002; Wang *et al.*, 2001; Xin *et al.*, 2004). The UPF for clothing with an excellent UV protection should be 40 to 50+ (Gies *et al.*, 1997). But from a clinical viewpoint, a UPF greater than 50 is entirely unnecessary (Gies *et al.*, 1997). Sunscreens, sunglasses, hats and clothing are the main accessories used to protect from UVR. Textiles are excellent materials for UVR protection and most UV can be blocked by common clothing (Reinert *et al.*, 1997). As shown in Table 1.1, the UVR protection of a fabric depends on fiber content, weave, fabric color, finishing processes, the presence of additives, and laundering (Gies *et al.*, 1997, 1998; Wang *et al.*, 2001; Xin *et al.*, 2004).

Electromagnetic-radiation protection

With the development of modern society, people greatly benefit from the electrical and electronic devices used during work and everyday life. However, these devices are capable of emitting radio frequencies that are potential hazards to health. Examples are cell phones with frequencies from 900 to 1,800 MHz, microwave ovens with 2,450 MHz, radar signal communication systems extending from 1 to 10,000 MHz, and so on (Cheng and Lee, 2001; Su and Chern, 2004). Many countries are legislating new regulations so that the manufacturers of electrical and electronic equipment comply with the electromagnetic (EMC) requirement standards (Cheng and Lee, 2001).

When electromagnetic waves enter an organism, they vibrate molecules producing heat that could obstruct a cell's capability for regeneration of DNA and RNA (Su and Chern, 2004). Furthermore, electromagnetic waves can cause abnormal chemical activities that produce cancer cells leading to leukemia and other types of cancer (Su and Chern, 2004).

Traditionally, sheet metals are used for shielding radio frequencies (Cheng and Lee, 2001). In recent years, conductive fabrics have been used for shielding electromagnetic and static charges in defense, the electrical and electronic industries. General textile fibers have sufficient insulating properties with resistivities of the order of 10^{15} Ω/cm^2, much higher than the desirable resistivity for electromagnetic shielding applications (Cheng and Lee, 2001). The desired resistivities for anti-electrostatic, statically dissipated and shielding materials are 10^9 to 10^{13} Ω/cm^2, 10^2 to 10^6 Ω/cm^2 and lower than 10^2 Ω/cm^2 respectively (Cheng and Lee, 2001). Therefore, conductive fabrics are designed according to specific requirements using various techniques such as:

1. Laminating conductive layers onto the surface of the fabric by using conductive coatings, zinc arc sprays, ionic plating, vacuum metallized sputtering, and metal foil binding (Adanur, 1995; Bajaj *et al.*, 1992; Cheng and Lee, 2001; Kirkpatrick, 1973; Last and Thouless, 1971).
2. Adding conductive fillers such as conductive carbon black, carbon fibers, metal fibers (stainless steel, aluminum, copper) or metal powders and flakes (Al, Cu, Ag, Ni) to the insulating material (Bhat *et al.*, 2004; Cheng and Lee, 2001; Miyasaka, 1986).
3. Incorporating conductive fibers and yarns into a fabric. This method provides flexibility in designing the conductive garments (Adanur, 1995; Bajaj *et al.*, 1992; Cheng and Lee, 2001, Su and Chern, 2004).

1.3.7   Electrical protection

Electromagnetic protection

Protection from electromagnetic sources is required because people who work close to power lines and electrical equipment have the possibility of being exposed to electric shocks and acute flammability hazards. Generally, rubber gloves, dielectric hard hats and boots, sleeve protectors, conductive Faraday-cage garments, rubber blankets and non-conductive sticks are used for electromagnetic protection (Adanur, 1995). Conductive protective clothing with flame resistance, known as 'Live line' garments, is necessary for people who work in the vicinity of very high-voltage electrical equipment. A live-line garment which was introduced in the early 1970s is still in use (Adanur, 1995).

Radiation from electro-magnetic fields (EMF) generated by power lines is another potential risk to people working near power lines. There have been reports about the relation between exposure to electromagnetic fields and health hazards like leukemia and brain cancer (Adanur, 1995). A typical electromagnetic protective fabric is woven from conductive material such as spun yarns containing a mixture of fire-retardant textile fibers and stainless steel fibers (8–12 micron diameter). It has been shown that fabrics made of 25% stainless steel fiber/75% wool blend or 25% stainless steel fiber/75% aramid fiber blend can protect the wearer from electromagnetic fields generated by voltages of up to 400 kV (Adanur, 1995). Protection at even higher voltages can be obtained by using a combination of these fabrics in two or more layers (King, 1988).

Electrostatic protection

Electrostatic charges accumulate easily on ordinary textile materials, especially in dry conditions (Holme *et al.*, 1998; Kathirgamanathan *et al.*, 2000; Morton and Hearle, 1997). Charges once accumulated are difficult to dissipate. The dissipation of an electrostatic-charge occurs through shocks and sparks which can be hazardous in a flammable atmosphere. Therefore, the presence of a static charge in textiles can be a major hazard in explosives, paper, printing, electronics, plastics, and the photographic industry (Bajaj *et al.*, 1992). Before the advent of non-flammable anaesthetics and anti-static rubber components in operating theatre equipment there was evidence of static electrically initiated explosions in hospitals (Scott 1981). The charge present in a garment can probably be over 60 kV depending on the balance between the rate of generation and the rate of dissipation of the static charges and the body potential (Holme *et al.*, 1998).

The clinging of garments is a common problem caused due to the presence of electrostatic charges. Electrostatic attraction may impede the opening of parachutes and even lead to catastrophic failure under certain circumstances (Holme *et al.*, 1998). Anti-electrostatic finishes are used for textiles both in civilian and non-civilian applications. The basic principle of making an antistatic garment is to decrease the electrical resistivity or the chance of electrostatic accumulation in a fabric. Examples of the former are spinning yarns containing conductive materials, producing a composite fiber in which at least one element is a conductive material or a fiber containing a conductive material such as metallic or carbon coatings (Holme *et al.*, 1998). Examples of the latter are the addition of a mixture of lubricants and surfactants to the textiles, or anti-static finishing (Holme *et al.*, 1998). It should, however, be noted that electrostatics can be very useful for practical industrial applications. In the textile industry, electrostatics are used as a means of spinning fibers and yarns (Holme *et al.*, 1998; Morton and Hearle, 1997).

1.3.8   Reduced visibility protection

Reduced visibility contributes to fatal pedestrian accidents. It is reported that night-time vehicles hit and kill more than 4000 pedestrians and injure more than 30,000 pedestrians annually in the United States (Adanur, 1995). High-visibility materials (HVM) are believed to be capable of assisting in avoiding worker and pedestrian deaths or serious injuries. HVMs are used by pedestrians, highway workers, cyclists, joggers, hikers, policemen, firemen and other professionals.

Clothing is made highly visible by sewing high-visibility materials or by chemical finishing. There are three major types of high-visibility products:

1. Reflective materials which shine when struck by light; e.g., reflective microprism
2. Photoluminescent materials that can absorb daylight or artifical light, store the energy and emit a green yellow glow in darkness
3. Fluorescent materials (Adanur, 1995).

In some cases, combinations of these methods are used to provide optimum visibility during the night.

1.4   Materials and technologies

As discussed above, there exists a wide variety of personal protective clothing manufactured to suit a particular end use requirement. Protectivity can be imparted to clothing using standard textile manufacturing technologies or by any applicable new technologies. Except for a few items such as safety belts, air bags, safety ropes and parachutes, most personal protective material will be made into apparel. Although the processing technologies for specific protective clothing are different, the main processes as shown in Fig. 1.2 generally include (i) material manufacturing or selection; (ii) producing fabrics and other related items; (iii) finishing, and (iv) clothing engineering.

1.4.1   Fibers/yarn

Chemical structure

Generally, chemical structure determines the properties and performance of any fiber. Natural fibers are one of the main fiber classes used for protective clothing (Adanur, 1995; Bajaj *et al.*, 1992). However, with the emergence of man-made fibers (regenerated and synthetic fibers, especially high-performance fibers), the fiber family has become so wide and resourceful that fibers are available to meet virtually any requirement for protective clothing.

The most important man-made fibers used in personal protective clothing are:

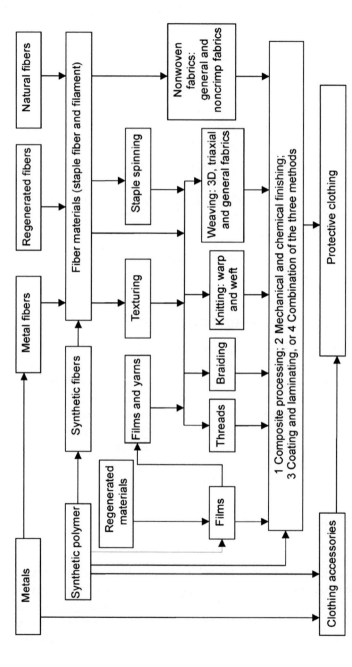

1.2 Schematic of materials and technologies for manufacturing protective textiles.

1. *Synthetic fibers with high mechanical performance.* Fibers in this category have superior strength and high modulus that make protective clothing capable of sustaining high-velocity impacts and retaining their shape during and after impact. High performance polyimide fibers with common commercial names such as Kevlar® (Dupont) and Twaron® (Akzo, now Acordis) are polymerized from a monomer of *para*-aramids using liquid crystalline spinning (Miraftab, 2000; Weinrotter and Seidl, 1993; Doyle 2000). Polyimide fibers have excellent thermal resistance with a high glass transition temperature of about 370 °C and do not melt and burn easily but are prone to photo-degradation (Miraftab, 2000). Another important high mechanical performance fiber is ultra-high molecular weight polyethylene (UHMWPE) fiber that has a modulus in excess of $70\,\mathrm{GNm}^{-2}$ and strength per specific weight is claimed to be 15 times stronger than steel and twice as strong as aromatic polyamides. But, UHMWPE melts at around 150 °C. These fibers are now widely used in strengthening composite materials for mechanical impact protection. PBO (poly-paraphenylene benzobisoxazole) with the trade name Zylon® is another important fiber that possesses superior heat resistance and mechanical properties (Doyle, 2000; Khakhar, 1998).

2. *Combustion-resistant organic fibers.* The limited oxygen index (LOI) is a measure of the resistance of a fiber to combustion. Nomex and Conex, which were produced from *meta*-aramids by Dupont in 1962 and Teijin in 1972 respectively have a LOI of 29. Polybenzimidazole (PBI) produced by Hoechst-Celanese has a LOI of 42. The highest LOI of a fiber realized till now is the PAN-OX, made by RK Textiles, with a LOI of 55. These fibers can be used in flame- and thermal-resistant protective clothing without any chemical finishing (Miraftab, 2000).

3. *High-performance inorganic fibers such as carbon fiber, glass fiber and asbestos.* Carbon fiber has high mechanical properties, is electrically conductive and has high thermal resistance. Carbon fibers can be used as reinforcing fibers in composites and also for electromagnetic and electrostatic protection (Adanur, 1995; Bajaj *et al.*, 1992; Doyle, 2000). Fibers made from aluminosilicate compound mixtures of aluminum oxide (Al_2O) and silicon oxide (SiO_2) can tolerate temperatures from 1,250 to 1,400 °C depending on their composition ratio (Miraftab, 2000). Silicon carbide (SiC) fibers have an outstanding ability to function in an oxidizing condition of up to 1,800 °C (Miraftab, 2000).

4. *Novel fibers.* These fibers were first introduced by Japan in an attempt to reproduce silk-like properties with additional enhanced durability (Miraftab, 2000). The first generation novel fibers were microfibers, fibers with a denier similar to the silk filament. Currently, much thinner fibers have been successfully made and by using these fibers, tight weave fabrics with a density of 30,000 filaments/cm² can be produced (Miraftab, 2000). The

tight weaves make these fabrics impermeable to water droplets, but allow air and moisture vapor circulation. Tightly woven microdenier fabrics are an ideal material for waterproof fabrics and outdoor protective clothing. In addition to microdenier fibers, many functional fibers with superior performance properties can be produced by using multi-component polymer spinning.

Physical structure

Based on their length, fibers can be divided into filaments or staples. Natural fibers generally have an uneven physical structure both in staple and filament (silk) form. The fineness, cross-sectional shape, mechanical properties and even the color are different and vary from fiber to fiber (Morton and Hearle, 1997). The variability among fibers and their non-homogeneity are distinguishing features that provide unique properties to natural fibers. Even man-made fibers are now being produced with properties similar to natural fibers by using techniques such as texturization.

Filament fibers can be directly used for fabric manufacturing, or can be textured prior to being used for weaving. Texturing produces the so-called 'bulked yarns', 'stretch yarns' and 'crimped yarns' that impart synthetic fibers with physical properties similar to those of natural fibers (Hearle *et al.*, 2001). By blending different fibers, yarns can be made to have specific and unique functional properties. For example, blending stainless steel fibers with other fibers produces conductive yarns. Yarns can also be produced from film by first splitting the film and then twisting it into yarns (Tortora, 1978). Twisting can also combine the different yarns to produce novel yarns. Producing bi-component yarns by twisting core yarns with an elastic fiber such as lycra has become one of the main methods to produce elastic yarns and fabrics.

1.4.2 Fabric

Woven and knitted fabrics

Traditional woven fabrics are produced through interlacing of two systems of yarns (warp and weft) at right-angles. A wide variety of different fabric constructions can be made by varying the weave type, density of the yarns and the type of yarns themselves. In knitting, a single yarn or a set of yarns moving in one direction are used instead of two sets of yarns as in weaving (Tortora, 1978). Knitted fabrics are of two types, warp knit and weft knit. Knitted fabrics generally have a soft hand and higher heat-retaining properties compared with that of woven fabrics of a specific thickness or weight. Knitted structures generally have more porosity that can retain more air and therefore provide more warmth. Traditional knits have poor shape retention and are anisotropic in physical performance when compared to woven fabrics. The properties of both woven and knitted fabrics vary in the warp (wale), weft (course) and diagonal directions respectively.

The anisotropic properties of traditional woven and knitted fabrics limit their use in applications where isotropic properties are required. Tri-axial and tetra-axial fabrics have been developed to obtain isotropic properties. Tri-axial fabrics were first developed using a tri-axial weaving machine by Barber Colman Co. under license from Dow Weave and have been further developed by Howa Machinery Ltd., Japan (Road, 2001). Isotropic fabrics have higher tear and burst resistance than traditional woven fabrics because strain is always taken in two directions (Road, 2001).

Non-crimp fabrics

In both woven and knitted fabrics, yarns are crimped due to their interlacing and inter-looping. The crimped structure of yarns makes fabrics change shape relatively easily when external forces are applied to them. To avoid this, non-crimp fabrics have been developed in the last decade using a LIBA system, a modification to multi-axial warp knitting (Adanur, 1995). In the LIBA system, several layers of uncrimped yarns are stacked and stitched together along several axes by knitting needles piercing through the yarn layers (Adanur, 1995). Non-crimp fabrics are a relatively new class of textiles. These fabrics are a form of reinforcement that have the potential to overcome anisotropic deficiencies without affecting other properties (Adanur, 1995).

Braided fabrics

A braid structure is formed by the diagonal intersection of yarns without a definite warp and filling as in woven fabrics (Adanur, 1995). Braiding is one of the major fabrication methods for composite reinforcement structures. Traditional examples of braided structures for industrial applications are electrical wires and cables, hoses, drive belts, etc. (Adanur, 1995). Braiding is also commonly used in manufacturing the accessories used with normal clothing.

Non-woven fabrics

Non-wovens are textile structures produced by bonding and/or interlocking of fibers and other polymeric materials such as films using mechanical, chemical, thermal adhesion or solvents or a combination of these methods (Adanur, 1995; Smith, 2000). For some special applications, fabrics and yarns are also used as parts of a non-woven material. Although there are some exceptions, non-wovens are generally produced in one continuous process directly from the raw material to the finished fabric. This means less material handling than in a traditional textile process and therefore non-wovens are generally cheaper than woven and knitted structures (Smith, 2000). The quality of fibers required for non-wovens is generally not as high as that required for traditional fabrics. Cost advantages have been one of the major reasons for the rapid development of non-wovens in the past few decades (Adanur, 1995).

The use of non-wovens is increasing at a rate of about 11% per annum. Although non-wovens were expected to partially replace woven fabrics in both civilian and non-civilian applications, the poor durability of non-wovens, especially when washed has limited its use for specific applications (Adanur, 1995). However, non-wovens are now widely used in industrial applications such as filtration, geotextiles and medical textiles (Adanur, 1995; Bajaj *et al.*, 1992).

Composite textile materials

Composites can be defined as a combination of dissimilar materials designed to perform a task that neither of the constituent materials can perform individually (Adanur, 1995). In the last few decades, textile composites have made great progress, by imparting novel functions to fabrics or by expanding the scope of textiles, especially in high-tech applications. Textile composites are broadly classified as flexible and rigid materials. Examples of flexible textile composites are coated fabrics, automobile tires and conveyor belts (Adanur, 1995). More often, textile reinforced composites are used as rigid textile materials.

Laminated and coated fabrics

Laminated fabrics can be made by fabric to fabric, fabric to foam, fabric to polymer and fabric to film bonding. Laminating film-like materials to textiles has developed quickly in recent years. Recently, membranes with micropores that are permeable to water-vapor molecules but impermeable to liquids and other organic molecules have been developed. These membranes are called 'perm-selective' membranes, due to the selectivity they exhibit with respect to molecular solubility and diffusion through the polymer structure (Schreuder-Gibson *et al.*, 2003; Wilusz *et al.*, 1997). When used in clothing, membranes are used between the shell fabric and liner fabric providing the clothing with water-vapor permeability but resisting the permeation of organic molecules. Clothing developed using membranes provides protection from hazardous organic chemicals without affecting the comfort properties. Instead of using a membrane, foams are used to make clothing with high warmth retaining properties and also having high vapor and air permeability (Holmes, 2000).

A coated fabric is a composite textile material in which the strength and other properties are improved by applying a suitably formulated polymer composition (Abbott 2001; Adanur, 1995). Coatings used for textiles are largely limited to viscous liquids that can be spread onto the surface of the substrate. The spreading process is followed by a drying or curing process which hardens the coating so that a non-blocking product is produced (Hall, 2000b). Coated fabrics are widely used in chemical or liquid protective clothing, and also in bio-protective clothing (Adanur, 1995, Voronkov *et al.*, 1999).

Textile-reinforced composite materials

Textile-reinforced composite materials are one of the general class of engineering materials called composites (Ogin, 2000). A textile reinforced composite is made from a textile reinforcement structure and a matrix material. Textile reinforcing structures can be made of fiber, yarns and fabrics (which include woven, braided, knitted, non-woven, non-crimp) that can be preformed into various shapes and forms either as molded materials or 3D textiles (Khokar, 2001). Matrix materials can be thermoplastic or thermoset polymers, ceramics or metals.

Textile reinforced composites are most commonly used as technical materials. Main characteristics of a rigid textile composite are high stiffness, high strength and low density. Therefore, textile structural composites have a higher strength-to-weight ratio than metal composites. Another advantage of textile composites is that they can be made anisotropic (Adanur, 1995). With the use of oriented fibers or yarns in bundles or layers, textile composites can be made anisotropic so that they exhibit different properties along different axes.

Textile composites have successfully replaced metals and metal alloys in many applications such as automotives, aerospace, electronics, military and recreation (Adanur, 1995; Ogin, 2000). Whatever these materials are used for, most of them are designed to protect people from being injured against mechanical impact. The most typical textile composites used for protection are made from high-strength and high-modulus fibers for applications such as lightweight armor, ballistic helmets and vests, and add-on car armor (Jacobs and Van Dingenen, 2001). Low-density, high-strength and high-energy absorption capability are the notable characteristics of these products (Jacobs and Van Dingenen, 2001). The US army uses helmets reinforced with Kevlar that are about 15% lighter by weight and have substantially increased protection (ballistic limit (V_{50}) more than 2000 ft/sec) when compared to conventional helmets (Adanur, 1995). Laminating and molding are commonly used techniques to manufacture protective composites, but 3D textiles which are produced via 3D weaving are gaining more importance in reinforcing textile structural materials with improved properties (Khokar, 2001).

1.4.3 Finishing

Textile finishing can be roughly divided into mechanical and chemical finishing. Examples of mechanical finishing are calendering, raising, cropping, compressive shrinkage and heat setting. Chemical processes are those that involve the application of chemicals to the fabrics (Hall, 2000a). Although fibers having inherent functional properties are being commercialized, chemical finishing is still a major technology used for protective clothing due to its cost effectiveness and technological versatility. Chemical finishing can be used to impart fabrics with flame-resistant, liquid-proof, anti-electrostatic, high-visibility, anti-microbial and chemical-protective functions (Adanur, 1995; Bajaj *et al.*, 1992).

1.4.4   Sewing or assembling

Sewing or assembling protective material parts onto clothing is usually the last but a very important process for protective clothing. Most of the protective clothing has specific functions and the requirements for protective clothing may be different even with the same kind of functional protection. Designing protective clothing is a professional job that could determine the level of protection. The most advanced design of protective clothing is probably space suits, which are high-tech integrated systems assembled with many functional parts. The design of a space suit is so perfect that no problems have been related to space suits so far. Protective function, comfort and cost effectiveness are the main criteria in designing a protective clothing system.

In any protective clothing, all accessories used to make the garment should match the protective requirements. For example, in flame-protective clothing, all the accessories such as buttons and threads need flame- or thermal-resistance (Bajaj *et al.*, 1992). Professional designers and equipment are needed to manufacture protective clothing suited for a particular application. For example, in sewing, serged seam is the normal seam for exposure to non-hazardous conditions and bound seam is used as reinforcement with the binding providing strength and tear resistance, and taped seams are reinforced with an adhesive film tape which is capable of resisting water and liquid chemicals. Sealed sleeves and collars are designed to give more protection for operatives during pesticide application (Fenske *et al.*, 2002).

1.5   Future of personal protection

1.5.1   Highly functional clothing with physiological comfort

Protective clothing guards the wearer against the vagaries of nature and against abnormal environments (Fourt and Hollier, 1970). In addition to protection, clothing must also be comfortable so that an energy balance can be maintained within the limits of tolerance for heating or cooling the body (Fourt and Hollier, 1970). When wearing protective clothing while doing hard physical work, metabolic heat is generated by the body that develops heat-stress in the wearer.

Heat-stress or comfort problems have been of great interest to scientists in recent years (Cho *et al.*, 1997; Gibson *et al.*, 2001; McLellan, 1996; Richardson and Capra, 2001; Wasterlund, 1998). Heat-stress increases the rate of heartbeat, body (aural) temperature, blood pressure and fluid loss, that are potential hazards for a wearer's health (McLellan, 1996; Richardson and Capra, 2001).

Newer technologies and materials have made the production of protective clothing with high protective functions and good comfort a reality. The most typical example is the application of breathable membranes in protective clothing (Holmes, 2000; Schreuder-Gibson *et al.*, 2003). Nanotechnology, biotechnology and electronic technology have contributed to developing protective clothing that is more comfortable to wear.

1.5.2 Nanotechnology

Nanotechnology allows inexpensive control of the structure of matter by working with atoms (Wilson *et al.*, 2002). Nanomaterials, sometimes called nanopowders, when not compressed have grain sizes in the order of 1–100 nm in at least one coordinate and normally in three (Wilson *et al.*, 2002). Nanomaterials include nanopowder, nanofiber, nanotube and nanofilms. Nanomaterials are not new. Carbon black is a natural nanomaterial that is used in car tires to increase the life of the tire and provides the black color. Fumed silica, a component of silicon rubber, coatings, sealants and adhesives are also nanomaterials, commercially available since the 1940s (Wilson *et al.*, 2002). However, it was only in the last decade that people began to better understand the basic science of nanotechnology and tried to apply them in engineering (Wilson *et al.*, 2002). Nanomaterials can be made by plasma arcing, chemical vapor deposition, sol-gels, electrodeposition and ball milling (Fan *et al.*, 2003; Wilson *et al.*, 2002).

Nanomaterials are so small in size that most atoms are at the surface. Such structures will exhibit completely different properties from the normal materials in which the atoms are buried in the bulk of the substance (Wilson *et al.*, 2002). Properties of materials change dramatically when made into nanosize. Silicon made into nanotubes will have conductivity similar to metals (Bai *et al.*, 2004). A nanotube fibre made from carbon is tougher than any natural or synthetic organic fiber described so far (Dalton *et al.*, 2003). Nanomaterials such as nanotubes developed either from silicon or carbon would be very useful for producing highly functional protective clothing.

Initial research has proved that nanotechnology will be beneficial to textiles and has tremendous prospects. Nanomaterials can be added to polymers to produce nano-modified polymer fiber or applied during finishing to make nano-finished textiles (Qian, 2004). Polymer-clay nanocomposites have emerged as a new class of materials that have superior properties such as higher tensile strength, heat resistance, and less permeability to gas compared with traditional composites (Krishnamoorti *et al.*, 1996; Tanaka and Goettler, 2002). Polypropylene (PP) fiber is one of the main fibers used for textiles but PP is highly hydrophobic and is inherently undyeable. Fan *et al.* (2003) added nanoclay (montmorillononite, $(OH)_4Si_8Al_4O_{20}$ nH_2O) into polypropylene and succeeded in producing a modified nanoPP which could be dyed with acid and disperse dyes. Nanostructural materials such as nanofiber and films show great prospects for use in textiles (Qian, 2004). A lightweight multifunctional membrane made from electrospun nanofiber exhibits high breathability, elasticity and filtration efficiency (Gibson *et al.*, 2001). Using sol-gel, one of the common methods for manufacturing nanomaterials, a nanolayer of titanium was deposited onto the surface of cotton fibers that gave excellent UV protection. Nanoparticle coatings are also very useful to produce textiles fabrics with special surface effects (Wilson *et al.*, 2002).

Although nanotechnology has provided novel properties to polymers, practical applications in textiles are not yet well established. Nanomaterials have far higher surface-to-bulk ratio than normal materials (Wilson *et al.*, 2002). The high surface energy makes nanomaterials agglomerate, which could greatly reduce the strength of composites. Also, the agglomeration decreases the surface-to-bulk ratio and nanomaterials will have reduced properties.

1.5.3 Biotechnology

Animals have their own effective way of protecting themselves from predators and abnormal climatic conditions. An intriguing example of protection adopted by animals is the changing of color by chameleons to match the color of their surrounding environment. A chameleon has several layers of cells beneath its transparent skin, of which some layers contain pigments while others just reflect light to create new colors (Rohrlich and Rubin, 1975). The most often changed colors of chameleons are between green, brown and gray, which coincidently, often match the background colors of their habitat. Although we are yet to produce a fabric that can change its color with the changing background, camouflage-patterned clothing is an effective way to conceal soldiers in their surrounding environments (Scott, 2000).

Another interesting aspect of color in nature is the vivid and extraordinary fastness of color in the feathers of peacocks. Color production in nature is either due to structural coloration or pigmentation (Zi *et al.*, 2003). The color of peacock feathers is due to the 2D photonic-crystal structure that has the same size as the wavelength of light. This crystal is arranged in lattices in a number of layers called periods that can reflect light to produce colors. The variations in the lattice constants or the number of periods produce the diversified colors (Zi *et al.*, 2003). We are still unable to simulate either the chameleon or peacock color to perfection. Studies on dyes that can change color with changing conditions such as temperature and light have partially succeeded, but the change in the magnitude of color is very narrow.

Natural materials are renowned for their relatively higher strength and toughness. Spider dragline silk has a breaking energy per unit weight two orders of magnitude greater than that of high-tensile steel (Dalton *et al.*, 2003; Smith *et al.*, 1997). Spider silk is stronger than Kevlar and stretches better than nylon, a combination of properties seen in no other fiber (Service, 2002). Spider silk is considered an ideal material for protective ballistic materials (Dalton *et al.*, 2003, Osaki, 1996). Spider silk has been artificially produced by using liquid crystalline spinning (Vollrath and Knight, 2001). By successfully copying the spider's internal processing mechanisms and with precise control over protein folding combined with knowledge of the gene sequences of its spinning dopes, industrial production of silk-based fibers with unique properties can be commercialized (Vollrath and Knight, 2001).

1.5.4   Electronic technology

Wearable electronic systems are a promising area for textiles (Adanur 1995; Barry *et al.*, 2003; Park and Jayaraman, 2003). Wearable electronics are part of the so called 'smart textiles' or 'smart clothing'. A smart material is that which will change its characteristics according to outside conditions or according to a predefined stimulus (Adanur, 1995). Wearable electronics have been success-fully used in some areas such as space suits and in military suits equipped with a GPS (global positioning system) (Adanur, 1995; Barry *et al.*, 2003; Park and Jayaraman, 2003).

Wearable electronic systems are being designed to meet new and innovative applications in military, public safety, healthcare, space exploration, sports and in fitness fields (Park and Jayaraman, 2003). Developments in electronic technology have made it possible to integrate innovation, intelligence and information into a wearable and comfortable infrastructure in a new generation of interactive textiles (Park and Jayaraman, 2003; Barry *et al.*, 2003). An interactive garment called the wearable mother board, or smart shirt has been developed at Georgia Institute of Technology, Georgia, USA. The smart shirt provides an extremely versatile framework for incorporation of sensing, monitoring and information-processing devices (Park and Jayaraman, 2003). Application of electronic technology will surely make protective clothing more reliable, safe and comfortable in future.

1.6   References

Abbott N J (2001), 'Coated fabrics for protective clothing', in Satas D and Tracton A A., *Coatings technology handbook* (2nd edn), Marcel Dekker, 819–823.
Adanur S (1995), *Wellington sears handbook of industrial textiles*, Lancaster, Pennsylvania, Technomic Publishing Company, Inc.
Anonymous (2003), 'Recommended practices for selection and use of surgical gowns and

drapes', *Aorn journal*, 77(1), 206–210.

Bai J, Zeng X, Tanaka H and Zeng J (2004), 'Metallic single-walled silicon nanotube', *Proceedings of the national academy of sciences of the United States of America*, 101(9), 2664–2668.

Bajaj P, Sengupta A K and Tech B (1992), 'Protective clothing', *Textile progress*, 22(2/3/4), 1–117.

Barry J, Hill R, Brasser P, Sobera M, Keleijn C and Gibson P (2003), 'Computational fluid dynamic modeling of fabric systems for intelligent garment design', *MRS bulletin*, 28(8), 568–573.

Belkin N L (1999), 'Effect of laundering on the barrier properties of reusable surgical gown fabrics', *AJIC Am J Infect Control*, 27, 304–308.

Belkin N L (2002), 'A historical review of barrier materials', *Aorn journal*, 76(4), 648–653.

Bhat N V, Seshadri D T and Radhakrishnan S (2004), 'Preparation, characterization, and performance of conductive fabrics: Cotton + PANi', *Textiles Res J*, 74(2), 155–166.

Byrne C (2000), 'Technical textiles market – an overview', in Horrocks A R and Anand S C, *Handbook of technical textiles*, Cambridge, Woodhead, 462–489.

Canadian Institute of Child Health (1994), *The health of Canada's child A CICH profile*, 2nd edn.

Carroll T R (2001), 'Chemical protective clothing', *Occupational health & safety*, 70(8), 36–46.

Cheng K B and Lee M L (2001), 'Electromagnetic shielding effectiveness of stainless steel/polyester woven fabrics', *Textile Res J.*, 71(1), 42–49.

Cho J, Tanabe S and Cho G (1997), 'Thermal comfort properties of cotton and nonwoven surgical gowns with dual functional finish', *Journal of physiological anthropology*, 16(3), 87–95.

Choi H, Bide M, Phaneuf M, Quist and Logerfo F (2004a), 'Antibiotic treatment of silk produces novel infection-resistant biomaterials', *Textile Res. J.*, 74(4), 333–342.

Choi K, Cho G, Kim P and Cho C (2004b), 'Thermal storage/release and mechanical properties of phase change materials on polyester fabrics', *Textiles Res J.*, 74(4), 292–296.

Curtis K (1998), 'Infections in solid organ transplantation', *AJIC Am J Infect Control*, 26, 364–364.

Dalton A B, Collins S, Munoz E, Razal J M, Howard E, Ferraris J P, Coleman J N, Kim B G and Baughman R H (2003), 'Super-tough carbon-nanotube fibres', *Nature*, 423 (12), 703–703.

David Rigby Associates (2004), 'Technical textiles and nonwovens: World market forecasts to 2010', *www.davidrigbyassociates.com*

Doyle B (2000), 'Aramid, carbon, and PBO fibers/yarns in engineered fabrics or membranes', *Journal of industrial textiles*, 30(1), 42–49.

Elmi M, Kalliokoski P, Savolainen K and Kangas J (1998), 'Predictive pesticide exposure model and potato farming', *Occupational hygiene*, 4(3-6), 259–266.

Fan Q, Ugbolue S, Wilson A, Dar Y and Yang Y (2003), 'Nanoclay-modified polypropylene dyeable with acid and disperse dyes', *AATCC review*, 3(6), 25–28.

Fenske R A, Birnbaum S G, Methner M M, Lu C and Nigg H N (2002), 'Fluorescent tracer evaluation of chemical protective clothing during pesticide application in central Florida citrus groves', *Journal of agricultural safety and health*, 8(3), 319–331.

Fourt L and Hollies (1970), *Clothing comfort and function*, New York, Marcel Dekker, Inc.

Fung W (2000), 'Textiles in transportation', in Horrocks A R and Anand S C, *Handbook of technical textiles*, Cambridge, Woodhead, 490–528.

Gibson P, Schreuder-Gibson H and Rivin D (2001), 'Transport properties of porous membranes based on electrospun nanofibers', *Colloids and surfaces A: Physicochemical and engineering aspects*, 187–188, 469–481.

Gies P H, Roy C R, MeLennan A, Diffey B L, Pailthorpe M, Driscoll C, Whillock M, Mckinlay A F, Grainger K, Clark I and Sayre R M (1997), 'UV protection by clothing: An intercomparison of measurements and methods', *Health physics society*, 73 (3), 456–464.

Gies P H, Roy C R, Toomey S and Mclennan A (1998), 'protection against solar ultraviolet radiation', *Mutation research*, 422, 15–22.

Hall M E (2000a), 'Finishing of technical textiles', in Horrocks A R and Anand S C, *Handbook of technical textiles*, Woodhead, Cambridge.

Hall M E (2000b), 'Coating of technical textiles', in Horrocks A R and Anand S C, *Handbook of technical textiles*, Woodhead, Cambridge.

Hatch K L (2002), 'American standards for UV-protective textiles', *Recent results in cancer research*, 160, 42–47.

Hearle J W S, Hollick L and Wilson D K (2001), *Yarn texturing technology*, Woodhead, Cambridge.

Holme I, McIntyre J E and Shen Z J (1998), 'Electrostatic charging of textiles', *Textile Progress*, 28(1), 1–90.

Holmes D A (2000), 'Textiles for survival', in Horrocks A R and Anand S C, *Handbook of technical textiles*, Cambridge, Woodhead, 462–489.

Jacobs M J N and Van Dingenen J L J (2001), 'Ballistic protection mechanisms in personal armour', *Journal of materials science*, 36(13), 3137–3142.

Kasturiya N and Bhargava G S (2003), 'Liquid repellency and durability assessment: a quick technique', *Journal of industrial textiles*, 32(3), 187–222.

Kathirgamanathan P, Toohey M J, Haase J, Holdstock P, Laperre J and SchmeerpLioe G (2000), 'Measurements of incendivity of electrostatic discharges from textiles used in personal protective clothing', *Journal of electrostatics*, 49, 51–70.

Khakhar D V (1998), 'Manufacture of ultra high modulus polymer fibres: The polymerization step', *Journal of scientific & industrial research*, 57(8), 429–440.

Khokar N (2001), '3D-weaving: Theory and practice', *J. Text. Inst. (part 1)*, 92 (2), 193–207.

King M W (1988), 'Thermal protective performance of single-layer and multiple-layer fabrics exposed to electrical flashovers', in ASTM STP 989, Mansdorf S Z, Sager R and Nielsen A P, *Performance of protective clothing: Second symposium,* American Society for Testing and Materials, Philadelphia.

Kirkpatrick S (1973), 'Percolation and conduction', *Rev. Mod. Phys.*, 45(4), 574.

Koscheyev V S and Leon G R (1997), 'Rescue worker and population protection in large-scale contamination disasters', *Clinical & health affairs*, 80(1), 23–27.

Krishnamoorti R, Vaia R A and Giannelis E P (1996), 'Structure and dynamics of polymer-layered silicate nanocomposites', *Chemical materials*, 8(8), 1728–1734.

Last B J and Thouless D J (1971), 'Percolation theory and electrical conductivity', *Phys. Rev. Lett.*, 27(25), 1719.

Leonas K K and Jinkins R (1997), 'The relationship of selected fabric characteristics and the barrier effectiveness of surgical gown fabrics', *AJIC Am J Infect Control*, 25, 16–23.

Mackay M and Scanlan A (2001), *Sports and recreation injury prevention strategies: Systematic review and best practices executive summary*, British Columbia injury research and prevention unit.

Mandel J H, Carr W, Hillmer T, Leonard P R, Halberg J U, Sanderson W T and Mandel J S (1996), 'Factors associated with safe use of agricultural pesticides in Minnesota', *The Journal of Rural Health*, 12(4), 301–310.

McLellan T M (1996), 'Heat strain while wearing the current Canadian or a new hot-weather French NBC protective clothing ensemble', *Aviation, space, and environmental medicine*, 67(11), 1057–1062.

McQueen R H, Laing R M, Niven B E and Webster J (2000), 'Revising the definition of satisfactory performance for chemical protection for agricultural workers', *Performance of protective clothing: Issues and priorities for the 21st century*, 7, 102–116.

Miraftab M (2000), 'Technical fiber', in Horrocks A R and Anand S C, *Handbook of technical textiles*, Woodhead, Cambridge.

Miyasaka K (1986), 'Mechanism of electrical conduction in electrically-conductive filler-polymer composites', *Int. Polym. Sci. Technol.*, 13(6), 41–48.

Morton W E and Hearle J W S (1997), *Physical properties of textile fibres*, The Textile Institute, UK.

Muza S R, Banderet L E and Forte V A (1996), 'Effects of chemical defense clothing and individual equipment on ventilatory function and subjective reaction', *Aviation, space, and environmental medicine*, 67(12), 1190–1197.

Ogin S L (2000), 'Textile-reinforced composite materials', in Horrocks A R and Anand S C, *Handbook of technical textiles*, Woodhead, Cambridge.

O'Mahony M and Braddock S E (2002), *Sporttech, revolutionary fabrics, fashion and design*, Thames & Hudson.

Osaki S (1996), 'Spider silk as mechanical lifeline', *Nature*, 384 (5), 419–419.

Panek C (1982), *Protective clothing*, Shirley Institute, Manchester.

Park J and Zellers E T (2000), 'Determination of solvents permeating through chemical protective clothing with a microsensor array', *J Enviro. Monit*, 2, 300–306.

Park S and Jayaraman S (2003), 'Smart textiles: Wearable electronic systems', *MRS bulletin*, 28(8), 585–590.

Patel S R, Urech D and Werner H P (1998), 'Surgical gowns and drapes into the 21st century', *Journal of theatre nursing*, 8(8), 27, 30–32, 34–27.

Pause B (2003), 'Nonwoven protective garments with thermo-regulating properties', *Journal of industrial textiles*, 33(2), 93–99.

Perepelkin K E (2001), 'Chemical fibers with specific properties for industrial application and personnel protection', *Journal of industrial textiles*, 31(2), 87–102.

Qian L (2004), 'Nanotechnology in textiles: Recent developments and future prospects', *AATCC review*, 4(5), 14–16.

Rajendran S and Anand S C (2002), 'Developments in medical textiles', Textile progress, 32(4), 1–42.

Reinert G, Fuso F, Hilfiker R and Schmidt E (1997), 'UV-protecting properties of textile fabrics and their improvement', *Textile chemist and colorist*, 29(12), 36–43.

Richardson J E and Capra M F (2001), 'Physiological responses of firefighters wearing level 3 chemical protective suits while working in controlled hot environments',

Journal of occupational and environmental medicine, 43(12), 1064–1072.

Rissanen S and Rintamaki H (2000), 'Prediction of duration limited exposure for participants wearing chemical protective clothing in the cold', *International journal of occupational safety and ergonomics*, 6 (4), 451–461.

Road B (2001), 'Technical fabric structures – 1. Woven fabric', in Horrocks A R and Anand S C, *Handbook of technical textiles*, Woodhead, Cambridge.

Rohrlich S T and Rubin R W (1975), 'Biochemical characterization of crystals from the dermal iridophores of a chameleon *Anolis carolinensis*', *Journal of cell biology*, 66(3), 635–645.

Schreuder-Gibson H L, Truong Q, Walker J E, Owens J R, Wander J D and Jones W E Jr. (2003), 'Chemical and biological protection and detection in fabrics for protective clothing', *MRS bulletin*, 28(8), 574–578.

Scott R A (1981) 'Static Electricity in Clothing & Textiles' in Thirteenth Commonwealth Defence Conference on Operational Clothing & Combat Equipment (Malaysia) Colchester, UK, Stores & Clothing Research & Development Estab.

Scott R A (2000), 'Textiles in defence', in Horricks A R and Anand S C, *Handbook of technical textiles*, Cambridge, Woodhead.

Service R F (2002), 'Mammalian cells spin a spidery new yarn', *Science*, 295 (18), 419–421.

Shekar R I, Kasturiya N, Rajand H and Mathur G N (2001), 'Studies on effect of water repellent treatment on flame retardant properties of fabric', *Journal of industrial textiles*, 30(3), 222–253.

Singh O P and Kaur J (1997a), 'Transmission, distribution and analysis of residual pesticides, and barrier and comfort performance of protective clothing: A review (Part I)', *Textile dyer & printer*, (6), 14–16.

Singh O P and Kaur J (1997b), 'Transmission, distribution and analysis of residual pesticides, and barrier and comfort performance of protective clothing: A review (Part II)', *Textile dyer & printer*, (7), 11–13.

Smith B L, Schaffer T E, Viani M, Thompson J B, Frederick N A, Kindt J, Belcher A, Stucky G D, Morse D E and Hansma P K (1997), 'Molecular mechanistic origin of the toughness of natural adhesives, fibers and composites', *Nature*, 399(24), 761–763.

Smith P A (2000), 'Technical fabric structure – 3. Nonwoven fabrics', in Horrocks A R and Anand S C, *Handbook of technical textiles*, Woodhead, Cambridge.

Su C and Chern J (2004), 'Effect of stainless steel-containing fabrics on electromagnetic shielding effectiveness', *Textile Res J*, 74(1), 51–54.

Tanaka G and Goettler L A (2002), 'Predicting the binding energy for nylon 6, 6/clay nanocomposites by molecular modeling', *Polymer*, 43, 541–553.

Tortora P G (1978), *Understanding textiles*, Macmillan Publishing Co. Inc., New York.

Vahdat N and Sullivan V D (2001), 'Estimation of permeation rate of chemicals through elastomeric materials', *Journal of applied polymer science*, 79, 1265–1272.

Vigo T L (1983), 'Protection of textiles from biological attack', in Lewin M and Sello S B, *Chemical processing of fibers and fabrics, Functional finishes Part A*, 368–426.

Vo E, Berardinelli S P and Boeniger M (2001), 'The use of 3M porous polymer extraction discs in assessing protective clothing chemical permeation', *Applied occupational and environmental hygiene*, 16(7), 729–735.

Vollrath F and Knight D P (2001), 'Liquid crystalline spinning of spider silk', *Nature*, 410 (29), 541–548.

Voronkov M G, Chernov N F and Baigozhin A (1999), 'Bioprotective heteroorgano-silicon coatings', *Journal of coated fabrics*, 19 (1), 75–87.

Wakelyn P J (1997), 'Overview of cotton and flammability', in Conference on recent advances in flame retardancy of polymeric materials: Materials, Applications, Industry developments, and market, *Recent advances in flame retardancy of polymeric materials*, Morwalk, CT: Business Communication, Co.

Wakelyn P J, Rearick W and Turner J (1998), 'Cotton and flammability – overview of new developments', *American dyestuff reporter*, 20(2), 13–21.

Wang S Q, Kopf A W, Marx J, Bogdan A, Polsky D and Bart R S (2001), 'Reduction of ultraviolet transmission through cotton T-shirt fabrics with low ultraviolet protection by various laundering methods and dyeing: Clinical implications', *J Am Acad Dermatol*, 44(5), 767–774.

Wasterlund D S (1998), 'A review of heat stress research with application to forestry', *Applied ergonomics*, 29 (3), 179–183.

Weinrotter K and Seidl S (1993), 'High performance polyimide fibers', in Lewin M, *High technology fibers Part C*, Brooklyn, New York.

Wilson M, Kannangara K, Smith G, Simmons M and Raguse B (2002), *Nanotechnology, Basic science and emerging technologies*, Chapman & Hall/CRC.

Wilusz E, Truong Q T, Rivin D and Kendrick C E (1997), 'Development of selectively permeable membranes for chemical protective clothing', *Polymeric materials science and engineering*, 77, 365–365.

Xin J H, Daoud W A and Kong Y Y (2004), 'A new approach to UV-blocking treatment for cotton fabrics', *Textile Res J*, 74(2), 97–100.

Zhang X and Raheel M (2003), 'Statistical model for predicting pesticide penetration in woven fabrics used for chemical protective clothing', *Bull Environ Contam Toxicol*, 70, 652–659.

Zi J, Yu X, Li Y, Hu X, Xu C, Wang C, Wang X, Liu X and Fu R (2003), 'Coloration strategies in peacock feathers', *Proceedings of the national academy of sciences of the United States of America*, 100 (22), 12576–12578.

Zins H M (1998), 'Testing and evaluation of surgical gown fabrics: comments and concerns', *AJIC Am J Infect Control*, 26, 364–365.

Standards for protective textiles

J H A A S E , STFI, Germany

2.1 Introduction

Protective and safety textiles are technical textiles with high-tech character in many cases. They are used for personal protection and protection of objects. In all developed countries they fall under the framework of legal regulations. The standardisation of relevant test methods, safety requirements, quality assurance measures, certification procedures and others plays an extraordinary role in this context.

2.1.1 Market potential of protective textiles

The Techtextil Messe, Frankfurt, Germany, supported by David Rigby Associates, UK (www.davidrigbyassociates.com) defines 12 main end-use markets for technical textiles. These are listed in Table 2.1. These terms and definitions are not universally used and they are not without their problems. However, the Techtextil typology provides the most comprehensive attempt to classify the structure of end-use markets for technical textiles (Chang, 2002). The class 'Protech' presents protection textiles and protective clothing as a product group of technical textiles with high growth rates in the modern textile and garment industries (Byrne, 1997; Davies, 1997; Rigby, 1997, 2002). At least until the year 2010, these annual growth rates will be estimated world-wide between four and five per cent in volume and value. Table 2.2 summarises the forecasts for world technical textiles consumption 1995–2010 in volume terms, split by application area (Rigby, 2002, 2003; Nonwovens Report International, 2002). Table 2.3 shows the forecasts by geographical world region.

The West European consumption of fabrics in protective clothing (flame retardant, dust/particulate, chemical, nuclear/biological, extreme cold, high visibility) in public utilities, military, medical and industry has been estimated for 1996 to more as 200 million m^2 (Davies, 1997). The European market potential of selected protection texile products in 1999 has been estimated to about 2,700 million Euro a year (protective clothing for professional use,

Table 2.1 Full list of application areas and end-use segments included in Techtextils technical textiles classification (Rigby, 1997, 2002)

	Application areas	End-use segments
Agrotech	agriculture, horticulture, forestry and fishing	cover, protection, collection, fishing, tying
Buildtech	building and construction	protection, display, textile construction, building components, reinforcements
Clothtech	functional components of shoes and clothing	shoe components, insulation, structure, sewing products
Geotech	geotextiles and civil engineering	stabilisation, separation, drainage, soil reinforcement, erosion control, linings
Hometech	products used in the home; components of furniture and floor coverings	carpet components, furniture components, cleaning, filtration, tickings, composites
Indutech	filtration, conveying and other products used in industry	filtration, cleaning, lifting, pulling, electrical components, other (e.g. seals)
Medtech	hygiene and medical	cleaning, coverstock, wound care, protection
Mobiltech	transportation, construction, equipment and furnishing	safety, trim, insulation, floor covering, protection, composites, other
Packtech	packaging and storage	bulk packaging, disposable, tying, other
Protech	**personal and property protection**	**particulate protection, chemical protection, flame retardant, cut resistant, outdoor use, other (e.g. safety straps)**
Sporttech	sports and leisure technical components	luggage components, sports equipment, camping equipment, other (e.g. boat covers)
Oekotech	environmental protection	ecological (included above)

protective gloves, lifejackets, body protection for sports (EC, 1999). Based on studies of the Frost & Sullivan Research (Frost and Sullivan, 2001, 2004) (www.it.frost.com) it will be stated that the European market growth for protective gloves is estimated at 2.7 % in the period 2003 (value 1,050 million Euro) until 2010 (value 1,260 million Euro) (Balmer, 2004).

Based on a comprehensive study (LePree, 2002) in the US protective clothing and body armor industry (fire, chemical and bullets) it will be reported (Technical Textiles International, 2003), that corresponding sales in 2001 were 1,600 million US-Dollar, which will climb to about 2,200 million US-dollar in 2006 representing an average annual growth rate of 6.7% in the period 2001-2006. Based on Rigby's studies Kothari (2004) gives an actual report of the high growth potential and prospects of technical textiles and protection textiles in the Indian subcontinent.

Table 2.2 Forecast world technical textiles consumption by application area, 1995–2010, volumes (000 tons) (Rigby, 2002, *Nonwovens Report International*, 2002)

Application area	Years				CAGR%		
	1995	2000	2005	2010	95–00	00–05	05–10
Agrotech	1173	1381	1615	1958	3.3%	3.2%	3.9%
Buildtech	1261	1648	2033	2591	5.5%	4.3%	5.0%
Clothtech	1072	1238	1413	1656	2.9%	2.7%	3.2%
Geotech	196	255	319	413	5.4%	4.6%	5.3%
Hometech	1864	2186	2499	2853	3.2%	2.7%	2.7%
Indutech	1846	2205	2624	3257	3.6%	3.5%	4.4%
Medtech	1228	1543	1928	2380	4.7%	4.6%	4.3%
Mobiltech	2117	2479	2828	3338	3.2%	2.7%	3.4%
Packtech	2189	2552	2990	3606	3.1%	3.2%	3.8%
Protech	**184**	**238**	**279**	**340**	**5.3%**	**3.3%**	**4.0%**
Sporttech	841	989	1153	1382	3.3%	3.1%	3.7%
Totals	13971	16714	19683	23774	3.7%	3.3%	3.8%
Of which Oekotech	161	214	287	400	5.9%	6.0%	6.9%

Note: Oekotech volumes are already part of other application areas and are relatively very small at present; CAGR average annual growth rate

Table 2.3 Forecast world technical textiles consumption by world region , 1995--2010, volumes (000 tons) (Rigby, 2002, *Nonwovens Report International*, 2002). CAGR average annual growth rate

Region	Years				CAGR%		
	1995	2000	2005	2010	95–00	00–05	05–10
Americas	4288	5031	5777	6821	3.2%	2.8%	3.4%
Europe	3494	4162	4773	5577	3.6%	2.8%	3.2%
Asia	5716	6963	8504	10645	4.0%	4.1%	4.6%
Rest of the world	473	558	628	730	3.3%	2.4%	3.1%
Totals	13971	16714	19683	23774	3.7%	3.3%	3.8%

2.1.2   Basis/objectives for standardisation work in the field

Personal protective equipment (PPE), is designed to protect employees from serious workplace injuries or illnesses resulting from contact with chemical, radiological, physical, electrical, mechanical, or other workplace hazards. Besides face shields, safety glasses, hard hats, earplugs, respirators, safety shoes and others, PPE includes a variety of protection textiles and protective clothing, such as garments, coveralls, gloves, safety vests, lifejackets. The foremost aim of standardisation is to facilitate the exchange of goods and services through the elimination of technical barriers to trade. This would include the harmonisation and standardisation of national regulations, harmonised product safety requirements, harmonised conformation procedures and harmonised specifications for testing and certification bodies.

Industries at international and European level, public authorities, institutes, laboratories, user representatives and other non-governmental organisations such as trade unions are interested parties in the standardisation process, since standardised specifications and test methods facilitate trade and help to reduce costs. Small companies, which produce a lot of products pertaining to the specified standards in the field of PPE, have the opportunity to draw on the knowledge of the standards specifications for their manufacturing processes,

Table 2.4 Non-exhaustive guide list of European standards for types of textile PPE, protective clothing

Protective function	Standard code
Protective clothing against heat and flame	EN 531
Protective clothing for use in welding and allied processes	EN 470-1
Protective clothing against mechanical impacts	EN 510
Protective clothing for users of hand-held chainsaws	EN 381-series
Firemen's protective clothing	EN 469
Protective clothing against cold	EN 342
Protective clothing against foul weather (moisture, wind, cold)	EN 343
Protective clothing against radioactive contamination	EN 1073
Protective clothing against electric hazards/electrostatic charges	EN 1149
Protection against thermal hazards of an electric arc (technical specification)	CLC/TS 50354
High-visibility warning clothing	EN 471
Protective clothing for working in an environment of machines	EN 510
Protective clothing against chemical hazards	EN 465, EN 466, EN 467
Lifejackets	EN 395
Protective gloves against mechanical risks	EN 388
Protective gloves against thermal risks (heat and/or fire)	EN 407
Protective gloves against cold	EN 511
Protective clothing for firefighters	EN 659
Protective gloves against chemicals and micro-organisms	EN 374-1
Motorcycle rider's protective clothing	EN 621-1
Buoyancy aids for swimming instruction	EN 393
Body protection for sports	EN 13277-series

although they may not be directly involved in the standardisation process itself. All have a vested interest in a set of standards that creates a terminology for the industry sector, defines generally valid product requirements for protective clothing, hand and arm protection and lifejackets and describes the relevant test methods (EC, 1999).

The statutory basis for product standardisation of PPE in Europe is the Council directive 89/686/EEC (EC, 1989a; EEC – European Economic Community). Protective clothing, hand and arm protection and lifejackets have to comply with the requirements of this directive, which presents the basis for standardisation work in the field. In all member states of the European Community (EC) this directive is transferred at national law level. For example, in Germany this directive has been converted into national law through the 8th Ordinance Regulating the Equipment Safety Act.

2.1.3 Main groups of standards

Corresponding to the aims of industrial safety, protective equipment is mainly used as personal safety equipment. Safety textiles for the protection of objects are used, e.g., for protection against fire, protection against vandalism (cut protection), protection against moisture, equipment protection/facility protection, such as clean-room textiles. Related to contents standards this can be divided into standards for test methods, product requirements, requirements for quality management in testing, production and certification, standards for the evaluation of conformity of products, and competence of test houses. Typical groups of European standardised protective clothing and protection textiles are listed in Table 2.4.

2.2   Requirements
2.2.1 Social and technical factors

Health and safety precautions necessitate the use of protective clothing, hand and arm protection and lifejackets in a large number of workplaces and leisure-time activities. European accident statistics show that more than 50% of all work accidents are to the hand, arm, leg and body area and necessitate absence from work (of at least one day). More than 40% of work disabilities are caused by injuries to the hand, arm, leg and body area (EC, 1999). In the United States studies show that the majority of workplace injuries could be avoided if employees used the proper PPE (Neil, 2004).

In all developed industries the legal authorities for occupational health and safety regard a high level of safety of PPE as a fundamental social factor. This level of safety can be ensured by specifications contained in modern standards based on technical progress in the world. The definition of improved ergonomic design and comfortable physiological parameters in the standards can increase the acceptance of protective clothing, hand and arm protection and lifejackets. The number of fall accidents and related injuries can be reduced considerably.

The technical factor influencing the elaboration of standards for PPE is based in particular on the technological developments, considering advances in the materials used, the further development of legislation in the field of occupational safety and health, and the awareness of the user, that the use of PPE can help to ensure his quality of life. Standardisation in line with these views can be seen as a key component of the market and the workers' safety.

2.2.2   Basic health and safety requirements

European standardisation has gained increasing importance since 1985 with the resolution on a new approach (www.europa.eu.int/comm/enterprise/newapproach) in the field of technical harmonisation and standardisation. This new approach includes four general principles. Only the basic health and safety requirements are specified in directives. Bodies responsible for industrial standardisation prepare European standards to supplement the basic health and safety requirements, taking account of the state of the art. These European standards remain voluntary. Once the creation of a harmonised European standard has been announced in the *Official Journal of the European Communities*, products manufactured in accordance with this standard can be assumed to conform to the basic health and safety requirements. New approach directives are special in that they do not contain technical detail, they contain broad safety requirements. Manufacturers therefore need to translate these broad 'essential' requirements into technical solutions. One of the best ways that manufacturers can do this is to use specially developed European standards. These standards are called harmonised standards and they are said to give a 'presumption of conformity' with the directive for which they have been written.

The Council directive 89/686/EEC (called 'PPE directive' or 'manufacturers directive') (EC, 1989a) lays down the basic safety requirements which PPE must satisfy in order to ensure the health protection and safety of users. For the purposes of this directive, PPE shall mean any device or appliance designed to be worn or held by an individual for protection against one or more health and safety hazards. Table 2.5 shows the basic health and safety requirements and additional requirements in overview.

General requirements applicable to all typs of PPE concern design principles, innocousness of the PPE, comfort and efficiency, and the information supplied by the manufacturer. A lot (more as 300) of European standards for PPE have been developed as the preferred means of demonstrating equipment conformity with the basic health and safety requirements of directive 89/686/EEC (EC 2004, example see Table 2.4). Only equipment which meets these requirements is entitled to carry the CE mark (CE: Communauté Européenne = European Community; see Fig. 2.1), and can be sold for use in the EC; CE marking is the

Table 2.5 Basic health and safety requirements of PPE according to European directive 89/686/EEC (EC, 1989a)

Type of requirements	Sub-groups	Single safety requirements
General requirements applicable to all PPE	Design principles	Ergonomics Levels and classes of protection Absence of risks and other 'inherent' nuisance factors
	Innocuousness of PPE	
	Comfort and efficiency	Adaptation of PPE to user morphology Lightness and design strength Compatibility of different classes or types of PPE designed for simultaneous use
	Information supplied by the manufacturer	
Additional requirements common to several classes or types of PPE	PPE incorporating adjustment systems PPE 'enclosing' the parts of the body to be protected PPE for the face, eyes and respiratory tracts PPE subject to ageing PPE which may be caught up during use PPE for use in explosive atmospheres PPE intended for emergency use or rapid installation and/or removal PPE for use in very dangerous situations PPE incorporating components which can be adjusted or removed by the user PPE for connection to another, external complementary device PPE incorporating a fluid circulation system PPE bearing one or more identification or recognition marks directly or indirectly PPE in the form of clothing capable of signalling the user's presence visually 'Multi-risk' PPE	

Table 2.5 Continued

Type of requirements	Sub-groups	Single safety requirements
Additional requirements specific to particular risks	Protection against mechanical impact	Impact caused by falling or projecting objects and collision of parts of the body with an obstacle
		Falls
		Mechanical vibration
	Protection against (static) compression of part of the body	
	Protection against physical injury (abrasion, perforation, cuts, bites)	
	Prevention of drowning (lifejackets, armbands and lifesaving suits)	
	Protection against the harmful effects of noise	
	Protection against heat and/or fire	PPE constituent materials and other components
		Complete PPE ready for use
	Protection against cold	PPE constituent materials and other components
		Complete PPE ready for use
	Protection against electric shock	
	Radiation protection	Non-ionizing radiation
		Ionizing radiation
	Protection against dangerous substances and infective agents	Respiratory protection. Protection against cutaneous and ocular contact
	Safety devices for diving equipment	

2.1 CE mark-European conformity label, e.g., according to PPE-directive 89/686/EEC.

'Passport for free exchange of goods'. The CE mark is not primarily a sign of quality, it is a conformity label to guarantee correspondence to the harmonised requirements.

PPE must provide adequate protection against all risks encountered. PPE must be so designed and manufactured that in the foreseeable conditions of use for which it is intended the user can perform the risk-related activity normally whilst enjoying appropriate protection of the highest possible level. The optimum level of protection to be taken into account in the design is that beyond which the constraints imposed by the wearing of the PPE would prevent its effective use during the period of exposure to the risk or normal performance of the activity. PPE must be so designed and manufactured as to preclude risks and other nuisance factors under foreseeable conditions of use. PPE must be as light as possible without prejudicing design strength and efficiency.

2.2.3 Additional requirements

Besides the basic requirements a manufacturer of PPE has to consider additional requirements common to several classes or types of PPE and requirements specific to particular risks.

For example, all PPE designed to protect the user against several potentially simultaneous risks must be so designed and manufactured as to satisfy, in particular, the basic requirements specific to each of those risks. PPE designed to protect all or part of the body against the effects of heat and/or fire must possess thermal insulation capacity and mechanical strength appropriate to foreseeable conditions of use. Constituent materials and other components suitable for protection against radiant and convective heat must possess an appropriate coefficient of transmission of incident heat flux and be sufficiently incombustible to preclude any risk of spontaneous ignition under the foreseeable conditions of use.

Legal requirements for use of protective clothing

In Europe PPE are products that must be certified. Since 1st January 1993, the date for the realisation of the European home market, harmonised regulations for admission and use of PPE were validated in the whole European economic area. Personal protective equipment shall be used when the risks cannot be avoided or sufficiently limited by technical means of collective protection or by measures, methods or procedures of work organisation. In Europe a clear distinction is drawn between the manufacture of PPE and the use of PPE products. The use of PPE is governed by EC directive 89/656/EEC (called 'users directive') concerning the minimum safety and health requirements for the use by workers of PPE at the workplace (EC, 1989b).

Employers are obliged to guarantee a series of requirements. Personal protective equipment must comply with the relevant Community provisions. Where the presence of more than one risk makes it necessary for a worker to wear simultaneously more than one item of personal protective equipment, such equipment must be compatible and continue to be effective against the risk or risks in question. The conditions of use of personal protective equipment, in particular the period for which it is worn, shall be determined on the basis of the seriousness of the risk, the frequency of exposure to the risk, the characteristics of the work station of each worker and the performance of the personal protective equipment. Personal protective equipment is, in principle, intended for individual use. Adequate information on each item of PPE shall be provided. Personal protective equipment shall be provided free of charge by the employer, who shall ensure its good working order and satisfactory hygienic condition by means of the necessary maintenance, repair and replacement. The employer shall first inform the worker of the risks against which the wearing of the personal protective equipment protects him. The employer shall arrange for training.

In North America the standardisation in the field of Occupational Safety and Health Standards is mainly carried out by US Department of Labor, Occupational Safety and Health Administration (*www.osha.gov*). Under a regulation of the Occupational Safety and Health Administration (OSHA), an employer must meet specific requirements concerning PPE, analogically to the European regulations. The regulation also gives employees specific rights concerning PPE. OSHA requires employers to survey the workplace to identify hazards, determine whether any hazard requires PPE, pay special attention to working conditions or processes that can produce the hazards, like falling objects, objects that could puncture the skin, objects that could roll over workers' feet, toxic chemicals, heat, harmful dust, radiation. Hazards shall be reassessed whenever necessary, especially when new equipment is installed or following accidents. The employer must select appropriate equipment and ensure that all PPE used is the right kind of equipment for the job, and that it is maintained properly even when workers are using their own equipment. Every employer must ensure that PPE provides a level of protection above the minimum required to protect the worker, all PPE fits properly, no defective or damaged PPE is used, and all PPE is properly cleaned and maintained on a regular basis. The employer must train workers who use PPE.

Table 2.6 General OSHA regulations for US Industry, OSHA – US Department of Labor, Occupational Safety and Health Administration

Standard code	Safety area
29 CFR 1910 Subpart I	Occupational safety and health standards
29 CFR 1910 Subpart I App A	References for further information (non-mandatory)
29 CFR 1910 Subpart I App B	Non-mandatory compliance guidelines for hazard assessment and personal protective equipment selection.
29 CFR 1910.132	General requirements PPE, included protective clothing
29 CFR 1910.133	Eye and face protection
29 CFR 1910.134	Respiratory protection
29 CFR 1910.135	Head protection
29 CFR 1910.136	Occupational foot protection
29 CFR 1910.137	Electrical protective devices
29 CFR 1910.138	Hand Protection

OSHA's primary PPE standards (Table 2.6) are in Title 29 of the Code of Federal Regulations (CFR), Part 1910 Subpart I, and equivalent regulations in States with OSHA approved State plans. The legal citation for the general PPE standard is 29 CFR 1910.132. There are additional standards that cover different specialised types of PPE. Similar and separate PPE standards cover construction (29 CFR 1926.95-106), shipyard, maritime and longshore workers. Public-sector workers for example in New York State are also covered by the PPE standards under PESH (Public Employee Safety & Health).

The basic standard 29 CFR 1910.132 requires that protective equipment, including personal protective equipment for eyes, face, head, and extremities, protective clothing, respiratory devices, and protective shields and barriers, shall be provided, used, and maintained in a sanitary and reliable condition. The standard applies wherever it is necessary by reason of hazards of processes or environment, chemical hazards, radiological hazards, or mechanical irritants encountered in a manner capable of causing injury or impairment to the function of any part of the body through absorption, inhalation or physical contact.

One can find PPE requirements also elsewhere in the General Industry Standards (Personal Protective Equipment (1994, April), divided into six sections:

Section 0 – Intro to 29 CFR Part 1910, PPE for General Industry
Section I – Background
Section II – Workplace hazards involved
Section III – Summary and explanation of the final rule
Section IV – Regulatory impact, regulatory flexibility and environmental assessment of revisions to subpart I, personal protective equipment introduction
Section V – Statutory considerations

Further for example, standard 29 CFR 1910.156 (OSHA's Fire Brigades Standard) has requirements for firefighting gear.

OSHA's general PPE requirements mandate that employers conduct a hazard assessment of their workplaces to determine what hazards are present that require the use of PPE, provide workers with appropriate PPE, and require them to use and maintain it in sanitary and reliable condition. A manual has been designed to help a user to comply with OSHA's PPE Standards 29 CFR 1910-series (Neil, 2004). In the US there are further organisations involved into the development of standards for protective clothing:

- American Society for Testing and Materials (ASTM, West Conshohocken, PA, www.astm.org), Committee F-23 – responsible for the development of test methods to measure the performance of the materials used in these products.
- National Fire Protection Association (NFPA, Quincy, Massachusetts) – responsible for the implementation of the test methods and establishing the acceptance criteria for the clothing ensemble. NFPA, a private non-profit organisation, is the leading authoritative source of technical background, data, and consumer advice on fire protection, problems and prevention. Their web site is (http://www.nfpa.org/).
- National Institute for Occupational Safety and Health (NIOSH, Washington, DC, www.cdc.gov/niosh)
- American National Standard Institute (ANSI, New York, NY, www.ansi.org).
- American Association of Textile Chemists and Colorists (AATCC, www.aatcc.org) Industrial safety equipment association (ISEA, Arlington, Virginia, www.safety equip-ment.org).

2.3   International standards

The standardisation of personal protective clothing takes place on international, European and national levels by different technical committees. Standardisation is a voluntary process based on consensus amongst different economic actors (industry, SMEs, consumers, workers, public authorities, etc.). It is carried out by independent standards bodies, acting at national, European and international level. Originally conceived as an instrument by and for economic operators, standardisation has been used increasingly by authorities.

2.3.1   Standardisation international, ISO

International standardisation of PPE is mainly carried out by ISO (International Standardisation Organisation, Geneva), an affiliation of standardisation organisations from over 90 countries worldwide. Only the principal national standardisation organisation in each country can be a member, i.e., Germany is

Table 2.7 International Standardisation Organisation ISO, Technical Committees in the field of PPE, TC Technical Committee, SC Subcommittee, WG working group

ISO/TC/SC	PPE field
TC 38	Respiratory protective devices
TC 94/SC 1	Head protection
TC 94/SC 3	Foot protection
TC 94/SC 4	Protection against falls from a height
TC 94/SC 12	Hearing protectors
TC 94/SC 13	Protective clothing
TC 188/WG 6	Lifejackets

represented by DIN (Deutsches Institut für Normung). Standardisation of PPE at ISO takes place primarily in Technical Committee (TC) ISO/TC 94 with a series of subcommittees (SC) for different PPE types. Certain other committees, as ISO/TC 83 for sport and leisure equipment and ISO/TC 42/SC 1 for noise, are also of significance for special PPE types. Table 2.7 shows the most important ISO groups related to standardisation of protective textiles and clothing. The standardisation of protective clothing especially against electric risks (shock, arcing heat, electromagnetic radiation) is carried out in general by IEC (International Electrotechnic Commission, Geneva, www.iec.ch).

2.3.2   Standardisation in Europe

European standardisation in the field of PPE mainly will be carried out by CEN (European Committee for Standardisation, Brussels, www.cenorm.be) – which deals with all sectors except the electrotechnology and telecommunication sectors; partially by Cenelec (European Committee for Electrotechnical Standardisation, Brussels, www.cenelec.org) – deals with standards in the electrotechnical field; or ETSI (European Telecommunications Standards Institute, Sophia-Antipolis, France, www.etsi.fr) covers the telecommunications field and some aspects of broadcasting.

European standardisation of PPE take place in the seven technical committees (TC) from CEN (see Table 2.8). These committees are structured into sub committees (SC) or working groups (WG). The TC 162 with 12 working groups is the responsible committee for protective clothing, hand and arm protection and lifejackets. These working groups give an overview of the most important kinds of personal protective clothing, where protective textiles are used. Since 1989, the year of foundation of the technical committees, an enormous standardisation program has been carried out. At present, more as 300 harmonised standards exist in the field of personal protective equipment.

Since the aim is for close interlinking of European, international standardisation work and uniform implementation of international standards,

Table 2.8 European standardisation by CEN (European Committee for Standardisation in the field of PPE, TC Technical Committee, WG Working group)

CEN Committee	PPE field	Secretariat
CEN/TC 79	Respiratory protective devices	DIN (D)
CEN/TC 85	Eye protection	AFNOR (F)
CEN/TC 158	Head protection	BSI (UK)
CEN/TC 159	Hearing protection	SIS (S)
CEN/TC 160	Protection against falls from a height including working belts	DIN (D)
CEN/TC 161	Foot and leg protectors	BSI (UK)
CEN/TC 162	**Protective clothing including hand and arm protection and lifejackets**	DIN (D)
CEN/TC 162/WG 1	General requirements and electrostatics	
CEN/TC 162/WG2	Heat and flame	
CEN/TC 162/WG3	Chemical	
CEN/TC 162/WG4	Cold and foul weather	
CEN/TC 162/WG5	Mechanical	
CEN/TC 162/WG6	Lifejacket	
CEN/TC 162/WG7	High visibility and radioactive contamination	
CEN/TC 162/WG8	Gloves	
CEN/TC 162/WG9	Motorcycle equipment	
CEN/TC 162/WG10	Accessories of flotation for children	
CEN/TC 162/WG11	Sports equipment	
CEN/TC 162/WG12	Diving suits	

an agreement on technical cooperation between ISO and CEN (Vienna Agreement) was concluded in 1991. Advantages are, e.g., the implementation of existing ISO standards by CEN and the cooperation through transfer of work and parallel voting. Since increasing use is being made of the possibility for parallel voting in the field of PPE, international standardisation of PPE is gaining in importance.

2.3.3   Basic standards for all types of protective clothing

In Europe the most important standards for protective clothing which always have to be taken into consideration, are:

- EN 340: 12-2003   Personal protective clothing – general requirements
- EN 420: 09-2003   General requirements for gloves

At international level the standards correspond (but not identically) e.g., to

- ISO 13688: 1998   Personal protective clothing – general requirements
- AS/NZS 2161.2:1998   General requirements (gloves)
- AS/NZS 4501.2:1999   General requirements (protective clothing)

The European standard EN 340 specifies general performance requirements for ergonomics, innocuousness, size, designation, ageing, compatibility and marking of protective clothing, and the information to be supplied by the manufacturer of the protective clothing. Basic health and ergonomic requirements are stated that are relevant for many types of protective clothing. Protective clothing shall not adversely affect the health or hygiene of the user. Protective clothing shall be made of materials such as textiles, leather, rubbers, plastics that have been shown to be chemically suitable. The materials shall not in the foreseeable conditions of normal use release, or degrade to release, substances generally known to be toxic, carcinogenic, mutagenic, allergenic, toxic to reproduction or otherwise harmful. Information claiming that the product is innocuous shall be checked.

Protective clothing should be as light as possible taking into account comfort, water vapour resistance, design, and protection level. Protective clothing should provide users with a level of comfort consistent with the level of protection which is provided against the hazard, the ambient conditions, the level of the user's activity, and the anticipated duration of use of the protective clothing. Protective clothing that imposes significant ergonomic burdens such as heat stress, or is inherently uncomfortable because of the need to provide adequate protection, should be accompanied in the information supplied by the manufacturer by specific advice or warnings. Specific advice on the appropriate duration for continuous use of the clothing in the intended application(s) should be given.

2.3.4   Overview – standards for protective textiles

The actual valid ISO standards fundamentally important in the field of textiles and protective clothing are listed in Tables 2.9 and 2.10. The actual updated list of European standards related to Council directive 89/686/EEC can be found through www.cenorm.be. A comprehensive overview on standardisation activities in the broad field of chemical protective clothing is given by Carroll (2000).

2.4   Certification

2.4.1   EC type examination

In accordance with the European regulation, before placing a PPE model on the market, the manufacturer or his authorized representative established in the Community shall carry out a 'EC declaration of production conformity'. The EC declaration of conformity is the procedure whereby the manufacturer draws up a declaration certifying that the PPE placed on the market is in conformity with the provisions of directive 89/686/EEC and affixes the EC mark of conformity to

Table 2.9 List of international ISO Standards, relevant test methods for protective textiles, ISO/TC 38 textiles, update 12 July 2004

Standard code	Standard title
ISO 105-series A, B, C, D, E, F, G, J, N, P, S, X, Z -	Textiles – Tests for colour fastness (daylight, Xenonlight, etc.)
ISO 139:1973	Textiles – standard atmospheres for conditioning and testing
ISO 675:1979 ISO 675:1979/ Cor 1:1990	Textiles – woven fabrics – determination of dimensional change on commercial laundering near boiling point
ISO 811:1981	Textile fabrics – determination of resistance to water penetration – hydrostatic pressure test
ISO 3175-1:1998 Cor 1:2002 .	Textiles – professional care, dry cleaning and wet cleaning of fabrics and garments – Part 1: assessment of performance after cleaning and finishing
ISO 3175-2:1998 Cor 1: 2002 .	Textiles – professional care, dry cleaning and wet cleaning of fabrics and garments – Part 2: procedure for testing performance when cleaning and finishing using tetrachloroethene
ISO 3175-3:2003	Textiles – professional care, dry cleaning and wet cleaning of fabrics and garments – Part 3: procedure for testing performance when cleaning and finishing using hydrocarbon solvents
ISO 3175-4:2003	Textiles – professional care, dry cleaning and wet cleaning of fabrics and garments – Part 4: procedure for testing performance when cleaning and finishing using simulated wet cleaning
ISO 3758:1991 Suppl:1993	Textiles – care labelling code using symbols
ISO 3759:1994 ISO Cor 1:1999	Textiles – preparation, marking and measuring of fabric specimens and garments in tests for determination of dimensional change
ISO 4880:1997	Burning behaviour of textiles and textile products – vocabulary
ISO 4920:1981	Textiles – determination of resistance to surface wetting (spray test) of fabrics
ISO 5077:1984	Textiles – determination of dimensional change in washing and drying
ISO 5079:1995	Textile fibres – determination of breaking force and elongation at break of individual fibres
ISO 5084:1996	Textiles – determination of thickness of textiles and textile products
ISO 5085-1:1989	Textiles – determination of thermal resistance – Part 1: low thermal resistance
ISO 5085-2:1990	Textiles – determination of thermal resistance – Part 2: high thermal resistance
ISO 6330:2000	Textiles – domestic washing and drying procedures for textile testing
ISO 6940:2004	Textile fabrics – burning behaviour – determination of ease of ignition of vertically oriented specimens
ISO 6941:2003	Textile fabrics – burning behaviour – measurement of flame spread properties of vertically oriented specimens

Table 2.9 Continued

Standard code	Standard title
ISO 7772-1:1998	Assessment of industrial laundry machinery by its effect on textiles – Part1: washing machines
ISO 9073-1:1989	Textiles – test methods for nonwovens – Part 1: determination of mass per unit area
ISO 9073-2:1995 Cor 1:1998	Textiles – test methods for nonwovens – Part 2: determination of thickness
ISO 9073-3:1989	Textiles – test methods for nonwovens – Part 3: determination of tensile strength and elongation
ISO 9073-4:1997	Textiles – test methods for nonwovens – Part 4: determination of tear resistance
ISO 9237:1995	Textiles – determination of the permeability of fabrics to air
ISO 9865:1991	Textiles – determination of water repellency of fabrics by the Bundesmann rain-shower test
ISO 10047:1993	Textiles – determination of surface burning time of fabrics
ISO 10528:1995	Textiles – commercial laundering procedure for textile fabrics prior to flammability testing
ISO 11092:1993	Textiles – physiological effects – measurement of thermal and water vapour resistance under steady-state conditions (sweating guarded-hotplate test)
ISO 12138:1996	Textiles – domestic laundering procedures for textile fabrics prior to flammability testing
ISO 12945-1:2000	Textiles – determination of fabric propensity to surface fuzzing and to pilling – Part 1: pilling box method
ISO 12945-2:2000	Textiles – determination of fabric propensity to surface fuzzing and to pilling – Part 2: modified Martindale method
ISO 12947-1:1998 Cor 1:2002	Textiles – determination of the abrasion resistance of fabrics by the Martindale method – Part 1: Martindale abrasion testing apparatus
ISO 12947-2:1998 Cor 1:2002	Textiles – determination of the abrasion resistance of fabrics by the Martindale method – Part 2: determination of specimen breakdown
ISO 12947-3:1998 I Cor 1:2002	Textiles – determination of the abrasion resistance of fabrics by the Martindale method – Part 3: determination of mass loss
ISO 12947-4:1998 Cor 1:2002	Textiles – determination of the abrasion resistance of fabrics by the Martindale method – Part 4: assessment of appearance change
ISO 13934-1:1999	Textiles – tensile properties of fabrics – Part 1: determination of maximum force and elongation at maximum force using the strip method
ISO 13934-2:1999	Textiles – tensile properties of fabrics – Part 2: determination of maximum force using the grab method
ISO 13935-1:1999	Textiles – seam tensile properties of fabrics and made-up textile articles Part 1: determination of maximum force to seam rupture using the strip method
ISO 13935-2:1999	Textiles – seam tensile properties of fabrics and made-up textile articles Part 2: determination of maximum force to seam rupture using the grab method
ISO 13936-1:2004	Textiles – determination of the slippage resistance of yarns at a seam in woven fabrics – Part 1: fixed seam opening method

Table 2.9 Continued

Standard code	Standard title
ISO 13936-2:2004	Textiles – determination of the slippage resistance of yarns at a seam in woven fabrics – Part 2: fixed load method
ISO 13937-1:2000	Textiles – tear properties of fabrics – Part 1: determination of tear force using ballistic pendulum method (Elmendorf)
ISO 13937-4:2000	Textiles – tear properties of fabrics – Part 4: determination of tear force of tongue-shaped test specimens (double tear test)
ISO 13938-1:1999	Textiles – bursting properties of fabrics – Part 1: hydraulic method for determination of bursting strength and bursting distension
ISO 13938-2:1999	Textiles – bursting properties of fabrics – Part 2: pneumatic method for determination of bursting strength and bursting distension
ISO 14184-1:1998	Textiles – determination of formaldehyde – Part 1: free and hydrolized formaldehyde (water extraction method)
ISO 14184-2:1998	Textiles – determination of formaldehyde – Part 2: released formaldehyde (vapour absorption method)
ISO 14419:1998	Textiles – oil repellency – hydrocarbon resistance test
ISO 15496:2004	Textiles – measurement of water vapour permeability of textiles for the purpose of quality control
ISO 15797:2002 Cor 1:2004	Textiles – industrial washing and finishing procedures for testing of workwear
ISO 15831:2004	Clothing – physiological effects – measurement of thermal insulation by means of a thermal manikin

each PPE. Affixing the CE mark to a piece of personal protective equipment is equivalent to a graphical declaration of conformity with the relevant health and safety requirements of the PPE directive and other relevant directives. PPE products are classified in categories depending on the hazard potential they protect against. Four categories exist, as shown in Table 2.11. This classification is of extraordinary importance for the required set of tests and certifications (conformity assessment) of a product (see overview in Table 2.12.). The conformity assessment proceeding from the EC is modular, and consists of eight basic modules. The modules range from the manufacturer's production control up to a comprehensive quality assurance system. They are identical to a great extent to a quality management system in accordance with ISO 9001.

For category I (simple protective clothing) a self-conformity statement by the manufacturer itself is sufficient. For products of category II (middle hazard potential) the manufacturer has to apply for an EC type examination to a notified body. EC type examination is the procedure whereby the approved inspection body (notified body) establishes and certifies that the PPE model in question satisfies the relevant provisions of this directive. A notified body is designated by the Member States, when a series of conditions to be fulfilled, like independence in carrying out the tests, technical competence, and professional

Table 2.10 List of international ISO Standards, relevant test methods for protective clothing, ISO/TC 94 SC 13 protective clothing, update 12 July 2004

Standard code	Standard title
ISO 2801:1998	Clothing for protection against heat and flame – general recommendations for selection, care and use of protective clothing
ISO 6529:2001	Protective clothing – protection against chemicals – determination of resistance of protective clothing materials to permeation by liquids and gases
ISO 6530:1990	Protective clothing – protection against liquid chemicals – determination of resistance of materials to penetration by liquids
ISO 6942:	2002 Protective clothing – protection against heat and fire – method of test: Evaluation of materials and material assemblies when exposed to a source of radiant heat
ISO 9150:1988	Protective clothing – determination of behaviour of materials on impact of small splashes of molten metal
ISO 9151:1995	Protective clothing against heat and flame – determination of heat transmission on exposure to flame
ISO 9185:1990	Protective clothing – assessment of resistance of materials to molten metal splash
ISO 11393-1:1998	Protective clothing for users of hand-held chain-saws – Part 1: test rig driven by a flywheel for testing resistance to cutting by a chain-saw
ISO 11393-2:1999	Protective clothing for users of hand-held chain-saws – Part 2: test methods and performance requirements for leg protectors
ISO 11393-3:1999	Protective clothing for users of hand-held chain-saws – Part 3: test methods for footwear
ISO 11393-4:2003	Protective clothing for users of hand-held chain-saws – Part 4: test methods and performance requirements for protective gloves
ISO 11393-5:2001	Protective clothing for users of hand-held chain-saws – Part 5: test methods and performance requirements for protective gaiters
ISO/TR 11610:2004	Protective clothing – vocabulary
ISO 11612:1998	Clothing for protection against heat and flame – test methods and performance requirements for heat-protective clothing
ISO 12127:1996	Clothing for protection against heat and flame – determination of contact heat transmission through protective clothing or constituent materials
ISO 13688:1998	Protective clothing – general requirements
ISO 13994:1998	Clothing for protection against liquid chemicals – determination of the resistance of protective clothing materials to penetration by liquids under pressure
ISO 13995:2000	Protective clothing – Mechanical properties – test method for the determination of the resistance to puncture and dynamic tearing of materials
ISO 13996:1999	Protective clothing – Mechanical properties – determination of resistance to puncture

Table 2.10 Continued

Standard code	Standard title
ISO 13997:1999	Protective clothing – Mechanical properties – determination of resistance to cutting by sharp objects
ISO 13998:2003	Protective clothing – aprons, trousers and vests protecting against cuts and stabs by hand knives
ISO 13999-1:1999	Protective clothing – gloves and arm guards protecting against cuts and stabs by hand knives – Part 1: chain-mail gloves and arm guards
ISO 13999-2:2003	Protective clothing – gloves and arm guards protecting against cuts and stabs by hand knives – Part 2: gloves and arm guards made of material other than chain mail
ISO 13999-3:2002	Protective clothing – gloves and arm guards protecting against cuts and stabs by hand knives – Part 3: impact cut test for fabric, leather and other materials
ISO 14460:1999	Protective clothing for automobile racing drivers – protection against heat and flame – performance requirements and test methods
Amd 1:2002	Modified flexion test
ISO 14877:2002	Protective clothing for abrasive blasting operations using granular abrasives
ISO 15025:2000	Protective clothing – protection against heat and flame – method of test for limited flame spread
ISO 16603:2004	Clothing for protection against contact with blood and body fluids – determination of the resistance of protective clothing materials to penetration by blood and body fluids – test method using synthetic blood (available in English only)
ISO 16604:2004	Clothing for protection against contact with blood and body fluids – determination of resistance of protective clothing materials to penetration by blood-borne pathogens – test method using Phi-X 174 bacteriophage (available in English only)
ISO 17491:2002	Protective clothing – protection against gaseous and liquid chemicals – determination of resistance of protective clothing to penetration by liquids and gases
ISO 17492:2003 Cor 1:2004	Clothing for protection against heat and flame – determination of heat transmission on exposure to both flame and radiant heat
ISO 17493:2000	Clothing and equipment for protection against heat – test method for convective heat resistance using a hot-air circulating oven
ISO 22608:2004	Protective clothing – protection against liquid chemicals – measurement of repellency, retention, and penetration of liquid pesticide formulations through protective clothing materials

Table 2.11 Categorisation of personal protective equipment (PPE) according to European directive 89/686/EEC

PPE category	Type of PPE
III	PPE of complex design intended to protect against mortal danger or serious harm to health. Example: fire-fighters' clothing; chemical protective clothing
II	All PPE not mentioned under I and III. Example: welders clothing; high-visibility warning clothing
I	PPE models of simple design, user can himself assess the level of protection. Example: simple gardening gloves
0	PPE excluded from the scope of the PPE directive 89/686/EEC. Example: PPE for use by the armed forces and police

integrity of personnel and others. Each member state shall inform the Commission and the other member states of the approved bodies responsible for the execution of the certification procedures (EC type examination). For information purposes, the Commission shall publish in the Official Journal of the European Communities and keep up to date a list giving the names of these bodies and the distinguishing numbers it has assigned to them. The EC type examination alone is not sufficient for PPE of category III (complex protective clothing with protection against life-threatening or health-threatening hazards). For these applications additional measures for quality assurance during the serial manufacturing process of the product are required. These measures are obligatory and must be controlled by a responsible authority.

Table 2.12 Categorisation of PPE and requirements for the manufacturer according to directive 89/686/EEC

	PPE category		
	I	II	III
Safety requirements	simple	medium	complex
Technical documentation supplied by the manufacturer	yes	yes	yes
EC type examination by notified body	no	yes	yes
Quality-control system at manufacturer for product checked by notified body	no	no	yes
EC declaration of conformity supplied by the manufacturer	yes	yes	yes

2.4.2 Responsibility of the manufacturer

Figure 2.2 shows schematically the procedure and the activities between different groups in the chain of certification. The manufacturer of the PPE stands in the centre and must meet all requirements for the CE marking and the conformity declaration. Beside the CE marking the following information must be presented on the PPE product: business name and full address of the manufacturer, type, name or any other identification of the product, size, the harmonised European standard fulfilled, pictogram for the risk which the product should protect against, including performance classes, and care labels.

2.4.3 Product quality management, risk assessment

EC quality control system for the final product

A manufacturer shall take all steps necessary to ensure that the manufacturing process, including the final inspection of PPE and tests, ensures the homogeneity of production and the conformity of PPE with the type described in the EC type approval certificate and with the relevant basic requirements of the PPE directive. A body to which notification has been given, chosen by a manufacturer, shall carry out the necessary checks. Those checks shall be carried out at random, normally at intervals of at least one year.

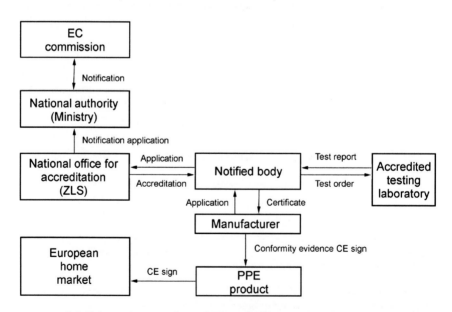

2.2 Schematic procedure of CE-marking according to European Council directive 89/686/EEC, actors and how to work together (ZLS Zentralstelle für Sicherheitstechnik, Munich, German office for national accreditation of notified bodies).

Similar regulations are given by the standard OHSAS 18001 (www.ohsas-18001-occupational-health-and-safety.com, Enclar Compliance Services, Inc., Largo, FL, USA). OHSAS 18001 is a consensus standard developed in 1999 by an independent group of national standards bodies and certification bodies. OHSAS stands for Occupational Health and Safety Assessment Series. OHSAS 18001 is structured the same way as ISO 14001, the environmental management system standard, and has essentially the same elements. It was specifically developed to be compatible with ISO 9001, the quality management system standard, and ISO 14001 to allow companies to develop and register integrated quality, environmental and occupational safety and health management systems.

Risk assessment

The risk assessment procedure should include: identification of the activities to be undertaken by the person(s) who will require to wear PPE, a list of the hazards present, a quantification of the risks that would result from an exposure to the hazards at the foreseeable level, and duration; whether PPE is needed or whether the problem can be solved by other measures; considerations of the protection provided by other control measures; determination of the level and extent of protection required from the PPE (in absolute or relative terms) the environment where the protection has to be worn; additional risks inherent to the use of PPE (ergonomic considerations, heat stress, etc.). Risk assessment should be done by trained personnel. The knowledge and experience of the users of the PPE should be taken into account. A number of risk assessment models may be used to determine the level of risk associated with the activities (see e.g., EN 340: 2003).

2.5   Future trends

The demand for further technical developments and elaboration of new standards is continuing to rise. Stricter legal requirements and the increased threat of insurance liability for employers will further the development and application of improved protective clothing. In some cases new technologies require new types of protective clothing. For example, the development trends for heat protection clothing can be characterised as increased use of high quality flame-retarding/inherently temperature-resistant materials and development of new flame-protection treatments. The development of multifunctional garment assemblies with combined protective functions will be increased. A special goal for further development is the improvement in comfort without loss of protection through lighter, higher-performance heat insulation materials.

Apart from the 'traditional' types of protective clothing there are also a great number of very relevant developments for new types of protective clothing. For example, these concern protective clothing which protects especially against electrical risks, such as electric insulating protective clothing, thermal protective clothing against thermal dangers of arcing during live working activities, shielding clothing for working on live equipment for nominal voltage up to AC 800 kV, protection against high-frequency electro-magnetic fields, complex flame-retardant and antistatic protective clothing. A range of protective clothing types require, apart from a guarantee of non-combustibility, suitable electrical conductivity of the materials, in order to avoid the danger of electrostatic ignition (e.g., explosives, chemicals industry, mining, oil industry, tanker transport, oil extraction, military).

According to the demands of the market many types of protective clothing have to fulfil more then one protective function simultaneously. These special kinds of clothing are called 'functional textiles'. An interesting example is the protection against weather and cold climates, where high protection functionality and good thermo-physiological functionality should be combined. Such textiles are often used in the sports and leisure areas. In everyday use these kinds of textile sometimes called WWWW-textiles (waterproof, water-repellent, windproof, water vapour permeable). These textiles offer good protection against outer moisture, wind and cold, together with good perspiration transport from the inside to the outside simultaneously. The main aim is to prevent condensation of moisture inside the clothing. The technical requirements seem to be mutually exclusive: 'tightness' or proof against harmful effects from external hazards, whilst allowing 'permeability' for vapour and perspiration transport from the body to the outside. From the technical point of view these requirements need compromises, but fibre and textile researchers developed some intelligent and astonishing solutions.

The functional requirements for protective textiles form the basis for future research, which include the development of intelligent, so-called smart textiles. Such developments have already started and will have a decisive impact in the period until the year 2010. Smart textiles constitute an entirely new direction of development with a broad, yet not fully foreseeable performance. They operate on completely new principles and will open up new possibilities in the sector of protective textiles by essentially expanding the desired protective functions to provide ever better protection for man's greatest good – his health. Feasible properties of 'intelligent clothing' are healing, reactive protection, communicating, informing, warning, supervising, supporting, sensitising, caring.

By combining high-performance textile fibres with electronic components and micro-mechanical elements to form one material system, functional features will be developed which have not been exploitable so far. These protective textiles will no longer behave according to a fixed, built-in characteristic but they will actively respond to their environment by communicating with it via integrated sensors. Thus, the textile material becomes an active element which can work with its own control and response mechanism.

In order to maintain the textile properties of protective clothing, the non-textile components such as electronic, mechanical, biological, physical or chemical elements have to behave mechanically in a similar way to textile fibres. In special cases, fibres or fibre modifications shall be used as sensors. This requires new approaches in the development sector as well as in the necessary processing technology and equipment. Already today visible trends can be envisaged where smartly modified clothing will replace the conventional design and where in the near future modified and desirable specification profiles and standards can be put into practice.

2.6   Sources of futher information and advice

2.6.1   Regular conferences related to protective textiles and standardisation

Leading, regularly organised international conferences on the topic of protective textiles and development of standards are: in the USA, the Symposium 'Performance of protective clothing' sponsored by ASTM-committee F23 (www.astm.org) on protective clothing (see e.g., Henry, 2000 – compendium of the seventh symposium 'Performance of protective clothing' Issues and priorities for the 21st century).

In Europe, the European Conference on Protective Clothing (ECPC). Increasing work in the field of protective clothing all over Europe led to the demand for a forum of scientific discussions and also applied solutions. Therefore the NOKOBETEF (Nordic Coordination Group on Protective Clothing), in association with the National Institute for Working Life, Solna (S), organised the first ECPC – held in Stockholm in May 2000. At this conference, the European Society of Protective Clothing (ESPC) was founded. The objective of the new organisation is to widen the former NOKOBETEF body and to create a European network of researchers and experts in the field of protective clothing. The NOKOBETEF as a technical preventive measure is an independent society of professionals. Since its foundation in 1984 six symposia have been organised in different Scandinavian countries. The second ECPC was held at Montreux, Switzerland, in May 2003 by EMPA (Eidgenössische Materialprüfungs- und Forschungsanstalt, St. Gallen, Switzerland, www.empa.ch/ecpc).

Mention is also made of the 'World Congress on Safety and Health at Work' the first of which was organised in Rome in 1955 and the last in Vienna in 2002, organised by the International Social Security Association (ISSA), Geneva, (www.issa.int), the XVIIth World Congress on Safety and Health at Work will be 18–22 September 2005 – Orlando, Florida (safety2005@nsc.org).

Further to direct attention to regular textile conferences with sections on protective textiles, e.g., International Man-Made Fibres Congress Dornbirn/ Austria (yearly, www.dornbirn-fibcon.com/), Textile Conference, Dresden/ Germany (every two years, http://www.tu-dresden.de/mw/itb/itb.html),

Table 2.13 Selected, non exhaustive addresses, information to standardisation of PPE, health and safety

Address	Organisation/partners	Services
http://europa.eu.int/comm/enterprise/	EC European Commission	Personal protective equipment (PPE) legislation, how to apply the directive, proposal for amending, standardisation notified bodies, working structure, international development
www.cenorm.be	CEN, the European Committee for Standardisation	Web portal of CEN, standards and drafts, news, conformity assessment
www.cenorm.be/catweb	CEN, the European Committee for Standardisation, all national bodies	On-line catalogue of European standards
www.idec.gr	IDEC consulting, Greece, Euratex, Centexbel, ITF and other	Web portal, information to EC directives 89/686/EEC and 89/656/EEC EN standards protective clothing and gloves, Pictograms, selection of PPE (EC Leonardo program)
www.hvbg.de/e/bia/	BG-Institute for Occupational Safety and Health – BIA Berufsgenossenschaftliches Institut für Arbeitsschutz – BIA, St. Augustin, Germany	Research and testing of the German Berufsgenossenschaften (BG), the institutions for statutory accident insurance and prevention in Germany, standardisation
www.astm.org	ASTM International, originally known as the American Society for Testing and Materials (ASTM)	ASTM International is one of the largest voluntary standards development organisations in the world, technical standards for materials, products, systems, and services, protective clothing, standards, meetings, symposia, news
www.hse.gov.uk	UK Health and Safety Executive Britain's Health and Safety Commission (HSC) and the Health and Safety Executive (HSE).	Public services health and safety information, health and safety in the textiles, footwear, leather, laundries and dry-cleaning industries
www.ccohs.ca/	Canadian Centre for Occupational Health and Safety (CCOHS) is a Canadian federal government agency based in Hamilton	Work-related illnesses and injuries, cold weather workers/welders health and safety guide: use proper PPE; duties and rights as given in the occupational health and safety legislation

Website	Organization	Description
www.nohsc.gov.au	National Occupational Health and Safety Commission (NOHSC), Canberra and Sydney, Australia	Australia's national body that leads and coordinates national efforts to prevent workplace death, injury and disease in Australia, producing key national standards, worldwide links to international sites
www.hecol.ualberta.ca	University of Alberta, Canada Protective Clothing and Equipment Research Facility (PCERF)	Research and development of devices and test protocols used in the evaluation of materials, clothing and protective equipment
www.kan.de	Commission for Occupational Health and Safety and Standardisation (KAN/Germany) KAN (17 members) brings together the institutions concerned with occupational health and safety in Germany	The KAN website provides information and links relating to occupational health and safety and standardisation
www.occuphealth.fi/ Internet/English/default.htm	Finnish Institute of Occupational Health (FIOH), Helsinki/ Finland	Research and specialist organisation in the sector of occupational health and safety, standardisation
www.stfi.de	Sächsisches Textilforschungsinstitut e.V. (STFI), Chemnitz, Germany	Textile research institute, research and development of devices and test methods in the field of protective textiles, standardisation
www.hohenstein.de	Hohensteiner Institute, Bönnigheim, Germany	Textile research institute, research and development of devices and test methods in the field of protective textiles, standardisation
www.bttg.co.uk	British Textile Technology Group (BTTG), Manchester, UK	Textile research institute, research and development of devices and test methods in the field of protective textiles, standardisation
www.ifth.org	Institut Francais Textile- Habilliment (IFTH) ECULLY CEDEX, France	Textile research institute, research and development of devices and test methods in the field of protective textiles, standardisation

conferences of the American Association of Textile Chemists and Colorists (AATCC, www.aatcc.org).

2.6.2 Organisations worldwide

Table 2.13 shows some important web portals related to protection textiles, standardisation and regulations.

2.7   References

Balmer B (2004) 'The European market for protective gloves – competition from China' *Personal Protection & Fashion* No. 2, May 2004, 58–59 (based on Frost and Sullivan report B284).

Byrne Ch (1997) 'Technical textiles: a model of world market prospects to 2005' Techtextil Symposium '97, Messe Frankfurt, Germany, 12 May 1997.

Carroll T R (2000) 'ANSI/ISEA Draft 103 – a modern success story for ASTM', *Performance of protective clothing: Issues and priorities for the 21st century*: seventh volume, ASTM STP 1386, CN Nelson and NW Henry, eds, ASTM, West Conshohocken, PA, 2000.

Chang W, Kilduff P (2002) Report 'The US market for technical textiles' NC State University, Raleigh, NC, USA (www.sbtdc.org), May 2002.

Davies B (1997) *Trends in the west European protective clothing market: prospects by segment*, 36th International Man-made fibres congress, Dornbirn, Austria 17–19 September 1997.

EC 1989a Council Directive 89/686/EEC of 21 December 1989 on the approximation of the laws of the Member States relating to personal protective equipment, *Official Journal* L 399, 30/12/1989 P. 0018–0038, doc. 31989L0686.

EC 1989b Council Directive 89/656/EEC of 30 November 1989 on the minimum health and safety requirements for the use by workers of personal protective equipment at the workplace *Official Journal* L 393, 30/12/1989 P. 0018–0028, doc. 31989L0656.

EC 1999 Resolution BTC 46/1999, Document *Market, environment and objectives of CEN/TC 162 – Protective clothing including hand and arm protection and lifejackets*.

EC 2004 Summary list of titles and references of harmonised standards related to personal protective equipment, published in *Official Journal of the European Union* OJ C 46 of 2004-02-21, replaces all the previous lists published.

EN 340 (2003) *Protective clothing – General requirements*.

Frost and Sullivan (2001) Report *European Personal Protective Equipment Markets*, 25 May 2001.

Frost and Sullivan (2004) Report B284 *The European Market for Industrial Protective Gloves*, 16 Jan 2004.

Henry N W, Nelson C N (2000) *Performance of protective clothing: Issues and priorities for the 21st century*: seventh volume, ASTM STP 1386.

Kothari V K (2004) 'Technical Textiles – Growth Potential and Prospects in India', Paper on 3rd Indo-Czech Textile Research Conference, June 2004; http://www.ft.vslib.cz/ indoczech-conference/conference_proceedings/abstract/India_02.pdf.

LePree J (2002) Report No.GB-142U *Protective Clothing and Body Armor Industry*:

Fire, Chemicals and Bullets, Business Communications Company, Inc. (BCC), 25 Van Zant St., Norwalk, CT 06855, USA, November 2002.

Neil G (2004) 'PPE Compliance Manual comply with OSHA's PPE Standards' (www.gneil.com) Sunrise, Florida, USA.

Nonwovens report international (2002) 'To 2010 – the technical textiles market is 50% bigger than previously estimated', June 2002, 52–53.

Rigby D (1997) 'The world technical textile industry and its markets: prospects to 2005', Report prepared for Techtextil, Messe Frankfurt, April 1997 http://davidrigbyassociates.co.uk.

Rigby D (2002) 'Technical textiles and nonwovens: world market forecasts to 2010', http://davidrigbyassociates.co.uk.

Rigby D (2003) '150 End-use products in technical textiles and nonwovens, world market forecast to 2010', http://davidrigbyassociates.co.uk.

Technical Textiles International (2003) 'US protective clothing industry to reach nearly $2.3 billion by 2006', 1 January 2003, www.technical-textiles.net/technical-textiles-index/htm/tti_20030101.124897.htm.

Fashion and function – factors affecting the design and use of protective clothing

S BLACK, V KAPSALI, J BOUGOURD and
F GEESIN, London College of Fashion

3.1 Introduction

The design of protective clothing is a subject covering an extremely wide range of circumstances, as indicated by the breadth of activities and diverse contexts included in the present volume. The development of effective protective clothing products and systems is essential for the safe conduct of specifically identified activities spanning, for example: conventional and extreme sports, police and military operations, chemical, agricultural and industrial working, medical and surgical procedures, fire fighting and space exploration. Clothing fulfils a basic human need (Maslow, 1954, 1970), with normal everyday wear providing an acceptable level of protection from environmental and climatic conditions and importantly creating social acceptability. For more extreme conditions, further specific protection is required.

Textiles and clothing systems designed to protect from physiological discomfort have a long history, for example, from clothing designed to protect early motorists and pilots exposed in open vehicles and planes. The last century witnessed enormous developments in textile technology that were driven by the need to improve protective clothing, particularly for military and space applications. Rapid progress in textile development now offers enhanced functionality and responsiveness to changes in environmental conditions, and the recent development of 'smart' and 'intelligent' textiles is accelerating. The transition from natural materials to man-made and engineered high-performance textiles can be illustrated by a comparison between equipment taken on mountaineering expeditions during the 20th century: in the 1930s 'Grenfell' cotton cloth parkas with woollen fleece lining, wolverine fur hoods and three hand-knitted Shetland wool sweaters were worn to scale Nanda Devi; similarly in 1953 the Everest expedition team wore heavy layers of tweeds and string vest, wool shirt, two Shetland pullovers and two pairs of wool socks; in 1970 Chris Bonington took to Annapurna parkas made of 4 oz nylon with open cell foam liner, with bonnet of nylon lined with artificial fur and down-filled suits; by 1982 innovations in modern fibres and finishes, including brands such as Gore-tex®, Thinsulate® and synthetic insulations such as Hollofil®, enabled small teams to survive conditions which would have halted previous expeditions (McCann, 1999: 53, 61, 270).

This chapter examines the product development design process appropriate for protective clothing, the increasing synergy between fashion considerations and functional design and the reconciliation of factors which must be taken into account. It also illustrates some current examples of developments in protective clothing in relation to fashion, and future directions in textiles which will influence both protective clothing and fashion.

3.2 Factors influencing the design development process

The range of hazards to be dealt with by the vulnerable human occupants of protective clothing can be broadly classified into chemical, thermal, mechanical, biological and nuclear or other forms of radiation (Raheel, 1994:1), and include both civilian and military situations. The level of protection required varies enormously within and between each of these categories and is one of the first factors to be determined when responding to a design request. For example, protection may be required from extremes of hot and cold weather, dirt, chemical spillages, fire, bullets, cuts, impact and abrasion, together with safety considerations including visibility and personal protection. Other major factors include the consideration of the entire clothing system in use, the specific functional requirements and the optimising of all components when working together to enable the activity to be successfully carried out whilst maximising protection and eliminating or minimising risk. The notion of fashion and aesthetics in such a context may be deemed by some to be superficial, however, in order to be acceptable to the user, the clothing must balance functional factors with both structural design and aesthetic considerations, including the important subjective element of comfort. Indeed, it is possible for a perfectly functional protective garment to be rejected on the grounds of appearance or perceived discomfort, putting the potential user at risk.

3.2.1 Fashion

In the latter half of the 20th century, fashion has played an increasingly important role in the lifestyle of a wide range of social groups and spread far beyond elitism and the early 'trickle-down' theories of fashion dissemination (Simmel's theory of the fashion process summarised as a cycle of 'adoption, imitation and abandonment' by Sweetman (2001: 61); see also Entwistle (2000)) to a number of other industries, such as the automotive and electronics industries. Since the 1990s, designer-led high fashion has become democratised and some formerly exclusive and aspirational luxury brands (e.g. Gucci, Armani, Burberry) have become widely available. Promotional campaigns and global marketing have played their part in the accessibility and spread of awareness of design and aesthetic considerations in both men's and women's clothing from high street to couture level. At the same time, casual clothing has seen a growing influence from active sportswear and youth culture permeating its design, including the increased use of technical fabrics which were first developed for specific protective needs. Conversely, there is a higher expectation of performance, comfort and durability of textiles used in casual clothing.

Fashion operates on many levels from subtle to extreme. Even uniforms can be subtly customised, showing the desire for individual identity and expression. In her essay *The Dressed Body* cultural theorist Entwistle (2001: 48) argues 'dress is part of the micro-order of social interaction and intimately connected to our (rather fragile) sense of self which is, in turn, threatened if we fail to conform to the standards governing a particular social situation. Dress is therefore a crucial dimension in the articulation of personal identity'. As identified by Joseph (1986: 144) 'While the utilitarian aspect of occupational clothing is undeniably of great importance, nevertheless, it is only one of many functions served by such clothing.' Clothing can represent emotional aspects of behaviour through cultural and peer identification, and use of colour and style. Appropriate clothing engenders and enhances social acceptance and self esteem, which, when coupled with confidence in a product's functionality, will enhance wellbeing, personal confidence and performance productivity in specific working circumstances.

Military and other public service personnel such as fire-fighters and police wear modern branded sportswear for off-duty activities, which they find to be comfortable, practical and fashionable, and are now demanding similar styling together with functionality for their work-based clothing, to improve flexibility and manoeuvrability whilst performing tasks. Military uniforms are not updated very often, and government issue is sometimes substituted by branded sportswear purchases, which although better in fabric quality and comfort, do not necessarily meet all the functionality standards. Referring to motorcyclist's clothing choices in the Metropolitan Police, the technical support manager states:

> Reputation, brand names, features, etc., can form the basis for their selection. The Met issues motorcycle jackets, which have been tested to a very high level. To comply, the jackets are lined with a denser interlining, making the jacket less comfortable (heat retention) than jackets available on the commercial market. The staff does not always appreciate this fact and criticism of 'the Met always provides a cheap option' or even 'an inferior product' is regularly voiced.

Soldiers have also criticised their current clothing from the point of view of comfort, and many remove the protective linings of their jackets because they are too stiff, or buy their own in the retail market, reducing effective protection (authors' interviews, M Campbell, Woolmark Company; R Allen, Shapeanalysis; R von Szaley Metropolitan Police, 2005).

The extent to which fashion considerations impact on the use of protective clothing is clearly dependent on the context of the actual function being carried out. In the case of military or civilian personnel in life-threatening situations, there is no real place for aesthetic considerations, however, the same personnel may be prepared to relax their approach to the complete clothing system when in a non-threatening or routine circumstance, with potentially dangerous consequences. According to the technical support officer of the Metropolitan Police uniform services, 'image is important as it can influence the wearer's perceptions of comfort. Users consider themselves 'specialists in the field'. However, they are strongly influenced by brands and labels. They make choices defying technical advice based on perceived quality advantages. Some manufacturers are well aware of this fact and design accordingly' (author interview R von Szaley). Balancing the range of design requirements across function, performance, protection, comfort and fashion will be discussed in the next section.

3.2.2   Functional design and fashion design

A successful design comprises many elements, and in the case of protective clothing must meet stringent functional requirements and conform to regulatory standards for the specific industry. 'Good' design must have the following properties in this context:

- meets functional requirements
- is appropriate for task and aesthetically pleasing
- is fit for purpose, durable and performs to or exceeds required standards
- is acceptable to both the user and the client (or other interested party) with respect to culture, traditions, specification, manufacturing and costs.

It may, in addition, have the following characteristics, which mark out the best design solutions:

- provides an innovative solution which simplifies existing products, extends norms or breaks new ground in materials, manufacturing or design concept
- adds value by exceeding specifications in functionality, ergonomics, ease of use or other aspect

whereas 'bad' design would clearly fail to meet some of the first set of criteria, and may also possess or produce unwanted characteristics, such as moisture retention, or annoying details which deter use. It would therefore never deliver the additional benefits mentioned above.

Fashion considerations within the design process, including tangible and more intuitive aspects, can be summarised as follows:

- self-perception and identity
- cultural identification with recognisable social groups
- fashion currency; awareness of relevant fashion and lifestyle trends
- feelings and emotions including comfort and wellbeing
- tradition and innovation – impact of emerging technological and fabric trends
- appropriate form, style, materials and colour (overall concept and silhouette)
- market level and costings
- cut, style and proportion (i.e. not 'old-fashioned')
- manufacturing processes and detailing
- functional and fit for purpose
- choice and range availability.

Various models of the design process have been developed in different contexts, particularly in generic product development (see Section 3.2.4). Few, however, relate to fashion or clothing design specifically (although Pitimaneeyakul et al. (2004: 119) propose a model specifically for knitwear development; Lamb and Kallal (1992: 42) presented a conceptual framework for apparel design), and still fewer are offered within protective clothing design (see Watkins (1995: 334) and Raheel (1994: 25)). Much of the focus is on the initial stages of ideation and the models do not develop detailed design criteria. However, McCann (1999) focused specifically on the design requirements for performance sportswear which developed a more detailed and balanced approach, integrating performance and fashion.

Fashion and apparel designers operate in a variety of markets, from mass market to high fashion and designer brand level, or may work with individual clients in the couture or bespoke areas of the industry. This spectrum is reflected in both the volume of garments produced and their related cost bracket, and supply-chain logistics mean the smaller the number produced, the higher the cost. The textile, fashion and clothing industries are a significant global economic force. The fashion industry is well known for being fast-moving in terms of both design and intensive production cycles. This has been accentuated by recent trends towards accelerated cycles and the further compression of time to market, resulting in a new mass market category 'fast fashion', which operates on a turn around of between six to eight weeks, rather than the traditional norm of three to six months.

The fashion design process is initiated by an external brief or by the regular imperative for a new collection stimulated by the fashion seasonal cycles, and is researched within and supported by an international network of trade shows for trends, yarns, fabrics, clothing and high-profile designer fashion catwalk presentations, which traditionally provide 'inspiration' for the high street levels of the market, although the flow of ideas now also works in the opposite direction, from 'streetwear' to designer level. There has in recent years been a polarisation in the marketplace to designer branded goods at one extreme and low priced commodity clothing such as t-shirts, jeans and basic wear at the other, which has seen, particularly in the UK, the retraction of the middle level of the market. At the same time, sportswear has grown in influence and particularly sports footwear brands, such as Nike, have developed as a fashion offering, and certain items of clothing, like the baseball cap, the football shirt, and the trainer, have become iconic and significant symbols of consumer culture. To come full circle, several popular sportswear brands and high street retailers have recently collaborated with high-profile designer names, e.g., Adidas and Yohji Yamamoto; Marks & Spencer and Betty Jackson; Debenhams and John Rocha; Vexed Generation and Puma, Hennes & Mauritz and Karl Lagerfeld.

Watkins (1995) differentiates between the normal processes of clothing and fashion design and the 'functional design' process necessary to achieve the rigorous solutions often required for protective clothing. This is distinguished by an intensive and in-depth period of evaluation of the users needs, involving methodologies including direct observation, interviews and technical testing.

3.2.3   Design process model for the development of protective clothing

Conceptual representations of the product development process have been proposed by several authors, including Carr and Pomeroy (1992) and Lamb and Kallal (1992); a range of features can be identified which are represented in Fig. 3.1 as a cycle of stages.

Problem exploration and analysis

This is the starting point where the characteristics required of the new protective product and its specialised design brief is determined. These break down into four categories: the type of user requiring protection; the activity with which the user is to engage; legislation and standards governing protective clothing, with special reference to the circumstances under consideration; the protective product.

The user

The gender, age, size, shape, lifestyle and socio-economic group of the proposed user group are required, plus an analysis of their clothing preferences. This data may be gathered through interviews, direct and indirect observation and questionnaires.

The activity

Analysis of the activity with which the user will engage is essential to be determined prior to the analysis of the product. Ergonomic and physiological

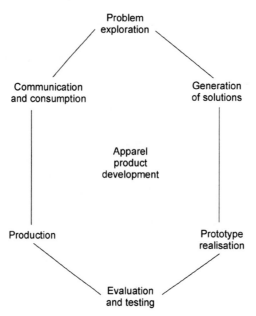

3.1 The process of apparel product development (Source: J Bougourd).

issues, such as range of movement, temperature variations or specific medical condition may be among the intrinsic requirements, whilst extrinsic factors will comprise the environment within which the activity is to be undertaken, and the hazards encountered. The design must meet the challenges encountered in particular occupational activities, and the impact of environmental factors such as airflow, extreme cold or a range of chemical hazards.

Legislation and standards

A review of British, European and international standards for body and product sizing, material testing, garment manufacture and product labelling would need to be undertaken to establish a set of criteria forming the design envelope. For example, the fire-fighters' tunic (section 3.4.2) must conform to standard EN 469, and the fluorocarbon chemical protection fabric coating to EN368. New materials may require the development or adaptation of existing textile testing or new tests and garment manufacturing procedures, prior to their use. For example, textiles incorporating new applications of nanotechnology in coatings may have little previous history and need to be evaluated independently and in context.

The protective apparel product

The investigation may include three aspects: review of earlier products and clothing systems; an analysis of and comparison between existing products and systems; and identification of new and emerging product components. A review of previous products provides an opportunity to identify and evaluate the process followed in reaching previous design solutions. (It would, however, be expected that technological advances – particularly in materials and product components – would now need to meet a wider range of environmental conditions and standards.) Existing products may either be those of competitors and/or drawn from an in-house range. The criteria assembled from the user group and activity analysis determine the level of protection required and the context the apparel is worn in, plus any negative characteristics to be eliminated. The outcome of this stage enables successful product and system characteristics to be identified.

The compilation and comparison of initial information comprises: user profiles; product price, size and colour ranges; new and existing materials, thread and fastening components; manufacturing processes; garment labelling/ aftercare; fashion and design content. Based on research at trade fairs and sampling, a comprehensive study of contemporary fashion, colour, texture, line and silhouette trends, and relevant cultural contexts, informs the final selection of product characteristics, both functional and aesthetic. For example, outdoor clothing company Rohan believe that form still follows function, but that protective clothing must have a design element. Working two years ahead of the retail season, they research and monitor fashion predictions and trends, particularly in materials, and have recently introduced a 10% natural fibres component into their synthetic fabrics for improved drape and handle. They developed their 'baggy' trousers silhouette in response to fashion research, and more 'technology', i.e., functionality is being built into the outer fabric layer (author's interview J. Donaldson).

Outcomes – the design brief

The final outcome of problem exploration and analysis is the design brief. The brief profiles the user group – age, size, shape, culture, lifestyle – from which a body size chart, shape profile and appropriate body form or block pattern can be selected or created. The outcomes of the activity analysis define ranges of movement required whilst wearing the clothing, anthropometric measurements indicate any increases or decreases in body size and shape from which an amended size chart, shape profile and block patterns can be produced. The physiological changes experienced when carrying out the activity will be identified and the range of environmental conditions determined.

The type of product must be appropriate to the activity; for example, whether separate single items such as jackets or trousers, or whole body coverings and consideration of the entire system of clothing and its possible interactions. The ease of putting on and taking off the garments is a crucial consideration for certain activities such as emergency services. A cost price range will be specified for each of the products, which in turn determines the choice of potential components and manufacturing processes. The product components – materials (fibres, fabric constructions and finishes), fastenings and closures – will be selected to be appropriate to the activity, test results, price range, and fashion trends. Manufacturing processes, i.e., whether sewn, bonded or welded, or used in combination, must be determined and will be considered in relation to the activity (the range of movement and environmental conditions), materials (fibres, woven, knitted or non-woven fabrics, and finishes) and aftercare requirements (e.g. whether wash or clean or disposable). The outcomes of the fashion trend analysis will inform material and component selection – colour, texture, line, and silhouette, and influence the overall design concept and the development of individual design elements (e.g. pockets and details). As previously indicated, the culture of the activity and its traditions will have a bearing on design selection and acceptability by the user group.

Generation of solutions

Potential solutions will next be developed using two and three-dimensional visualisations, including both manual sketching and computer-aided design programmes, supported by experimental design developments in which new elements are introduced and variations applied. This is a crucial stage in which innovations can be introduced, and where an open design brief which is more contextually and ergonomically oriented can generate new solutions, as one too narrowly specified will result in only incremental solutions. An evaluation and selection of proposed designs will be made after this stage, which may have several iterations, to ensure the fit to the design brief and suitability of the manufacturing processes specified and that the solution falls within the proposed cost range

Prototype realisation

Preparation for manufacturing the prototype includes the assembly of a modified size chart, shape profile and block pattern incorporating the age and size range of the user group, the range of movement identified in the activity analysis and the product type. These items, together with an appropriate body form and an analytical drawing, will be used to produce the prototype pattern, manufacturing specifications, three-dimensional sample products and full material and manufacturing costing.

Evaluation and testing

The range of products will be evaluated and tested against the set of criteria identified in the design brief, preferably by trials with the user group. A full functional analysis is undertaken and – according to the results – is revised and re-evaluated before a range is selected for production. Graded patterns, garments and final costings will then be produced and evaluated.

Production

In preparation for production the following will be provided: labour and material costings, together with appropriate markers; full base materials and component specifications; garment size and manufacturing specifications, with directions for the colour and size apportionments; labelling and packaging details. Those specifications will be used to monitor the quality of the merchandise throughout the production process, (cutting, manufacturing, finishing, pressing and packaging) in accordance with the schedule.

Communication and consumption

It is important that evaluation continues as the product becomes available to the user group. Observation, monitoring and analysis of the users' responses, preferences and feedback is essential to continue the fine tuning of the designs, including evaluation of the performance of individual products within the entire clothing system, their durability and any unforeseen problems in performance. This information can be fed incrementally into the renewal programme, so that enhancement is continuous. Some of the products designed for protective clothing may enter the mainstream market and evaluation can be conducted within the commercial retail environment at various market levels.

3.2.4   Tree of functional design requirements

To aid navigation through the complexity of choices and decisions to be made, the parameters of the design process can be represented as a 'tree' of requirements, which demonstrates the primary and secondary factors for consideration together with an increasing number of detailed sub-choices and issues which need to be taken into account, as outlined in the previous section and in Fig. 3.2. This method is adapted from McCann (1999). The primary factors are

- identification of users' needs and context
- overall product concept – clothing system: collective operation, number of layers, equipment and weight
- protective functions
- form and style
- performance and cost.

These primary factors can be further subdivided as:

- protective function – demands of the activity/needs of the body/specific protection required

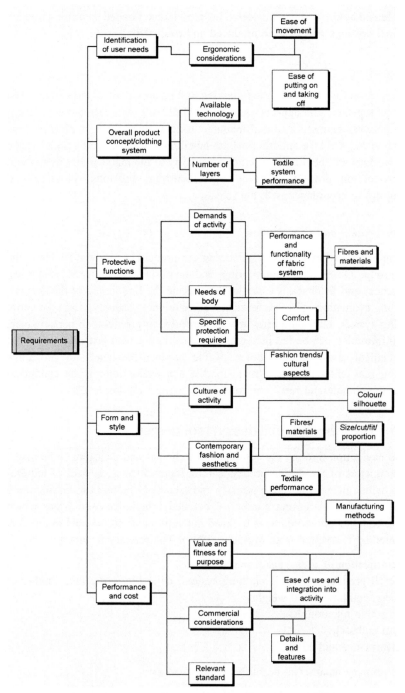

3.2 Tree of requirements in the design development of protective clothing (Source: S Black and V Kapsali).

- form and style – culture of the activity/contemporary fashion and aesthetics
- performance and cost – relevant standards/commercial considerations and value/fit for purpose.

Taking the contemporary fashion and aesthetics requirement further will include aspects of fibres, materials, colour, functionality, fabric performance, size, silhouette, cut, fit, proportion, comfort, manufacturing methods, in addition to the fashion trends and cultural factors. As progress is made, this branch begins to overlap and converge with aspects of other branches of the tree, such as performance and ergonomics and the needs of the body. Some of the considerations and choices which have to be made include

- garment system – form of overall collective, base, mid or outer layer?
- fabric system performance and functionality – e.g., strength/water permeable or barrier/ insulating or conductive/weight to bulk, i.e., density/stretch or rigid/wicking properties
- fibre and fabric construction: natural/synthetic/blends; woven/knit/non-woven/composite/single or bicomponent fibres/single or layered fabric elements
- aesthetic elements: colour/appearance/handle/touch/comfort on skin
- manufacturing processes: stitching/ bonding/moulding/welding/taping
- ease of putting on and taking off: openings and fastening systems/speed of operation
- details and features: pockets/attachments/accessories
- style/cut/fit/ease of movement.

Despite the complexity of the above requirements, the successful design and product development solution balances the primary factors through consideration of the sub-factors in such a manner as to arrive at a solution which meets the design brief to the required cost parameters and satisfies the key physical, social, psychological and aesthetic needs to combine protection, performance and fashion. However, one of the key inhibiting factors to the ideal design solution is often cost, therefore maximising the effectiveness of a clothing system for the available budget is a key skill practised by designers.

3.2.5   Selection of fabrics for protection

There is a bewildering array of possibilities when selecting textiles for protection, with many new and established fabrics available, some generic and others commercially branded. The mapping in Fig. 3.3 illustrates some of the fabric performance categories and examples of commercial products which are currently on the market. This map is indicative and not exhaustive, it illustrates the process of fabric selection from the initial stages of the design development where the requirements are being formulated to the final choice of fabric, which

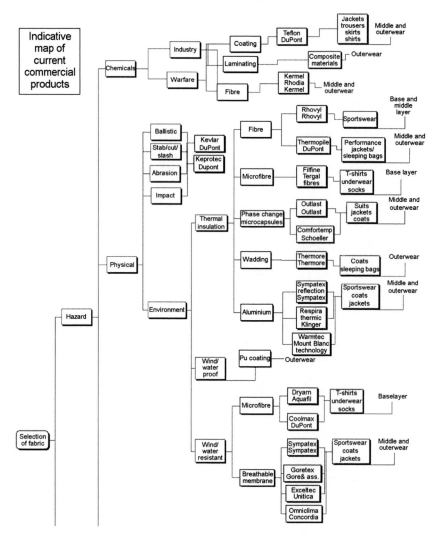

3.3 Selection of fabric functionalities (Source: V Kapsali).

may be an existing product or a tailored functionality. Many of the readily available protective textiles, such as those sourced at fashion industry trade fairs (e.g. Premiere Vision in Paris) are those with functionalities that protect the user from physiological hazards, such as UV rays and the build up of bacteria within a clothing system. Other fabrics are sourced from specific specialist suppliers. Each function incorporated into a clothing system is in danger of being disabled unless careful consideration is given to the compatibility of each component at the beginning of the process.

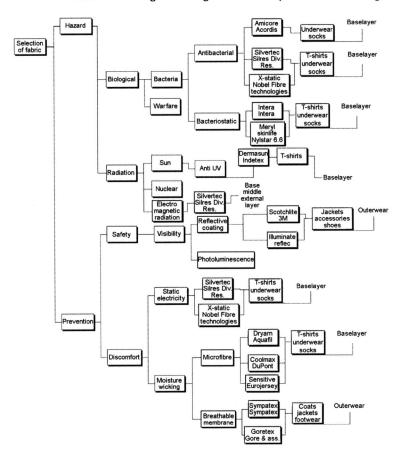

3.3 Continued

3.3   Clothing systems and functionality

3.3.1   Clothing systems and protection

In a sociological reading, clothing is 'a system of signs that derives meaning from its context while enabling us to carry on our activities' (Joseph, 1986). The process of dressing is a ritual activity executed at various levels of consciousness and performed daily by each individual in preparation for social engagement. Items of apparel are selected and combined to create a 'system of clothing' generally composed of base, middle, and outer garment layers. This complex architecture of materials and the pockets of air contained within both the fabrics and garment structures extends from the skin's surface through all levels to the outer face of the fabric comprising the external layer of clothing. This system possesses a dynamic microclimate resulting from a variety of external factors such as the activity of the individual (which generates heat and moisture), and the climate of the immediate environment, and from internal factors such as the types of fibres, the structure of the textiles, the properties of the materials employed as well as the design of each garment. It is therefore evident that every component operates within a dynamic system of inter-relationships and the properties of any individual textile or functional design feature are subsumed and may be lost within the complexity of the collective clothing system, unless it is designed to work in an integrated manner (Renbourn, 1971).

In the field of protective clothing, systems are engineered to shield the wearer from hazards such as extreme environmental conditions, chemicals, fire, etc. Their aim is to enable the user to perform tasks in these conditions without endangering the wearer's health or well-being. Successful design is paramount and can affect the survival of the wearer in extreme conditions. Although protective clothing is designed to enhance the user's comfort and safety, if the system is not carefully engineered it can have a negative effect on the wearer's performance by causing heat stress and discomfort, reducing task efficiency and restricting the range of motion. These factors may cause the user to reject the protective clothing and thus increasing the risk of injury or disease (Adams et al., 1994).

The success or failure of each clothing system cannot be measured by the performance of each of its parts individually but in the efficiency in which the components operate within a collective. In *A quest for thermophysiological comfort*, Brownless et al. reviewed the concept of physiological comfort in terms of technological efforts to improve the comfort sensation by developing wicking and insulation properties in textiles. The key observation made was that previous research in the area had misconstrued the concept of comfort in fabric requirements, resulting in the development of highly wicking and insulating textiles, which in extreme cases can lead to dehydration. The authors' findings highlight that 'a comfortable fabric is one which is not necessarily highly wicking or strongly thermally insulating, but one which has these two factors finely balanced in order to aid thermoregulation' (Brownless et al., 1995).

Firefighters' uniforms must act as a barrier between the user and radiant heat from fire; they therefore need to insulate the wearer from intense heat. In 2003 Bristol Uniforms, supplier of protective clothing systems to the fire-fighting industry, commissioned physiology consultancy Human Vertex to investigate the effects of heat stress in existing uniforms and used serving firefighters as volunteers. The results highlighted that existing test methods used for the assessment of ergonomic and thermal effects of the clothing systems actually underestimated the impact of heat stress on the volunteers. Firefighters on duty mostly operate below their individual anaerobic threshold and the assumption has been that they should be able to perform in these circumstances for extended periods ('Out of the hot ashes', *Company Clothing*, July 2004, p28). Because the firefighters' garment systems are designed to have high levels of insulation, on one hand the user is protected from the external heat generated by flames, however, the internal heat generated from the movements in addition to being in a very hot environment remains within the system of clothing, this increases the intensity of the exercise, resulting in the firefighter becoming exhausted more quickly and not being able to perform for extended periods. (See also chapter 22 by Mäkinen.)

Other factors that affect the success of a protective clothing system are users' attitudes and beliefs. In a study of Alberta farmers focusing on their attitude towards the use of disposable protective coveralls during exposure to pesticides, it was found that the system was being rejected on the premise that it was perceived to be costly; further analysis revealed that the users had misconceptions about the necessity of the level of protection, and placed comfort and convenience at a higher priority (Perkins *et al.*, 1992). When the functionality of the garment system is not transparent to the target user, such as in the case of the Alberta farmers, the users can be deterred from integrating the protective clothing system as a necessary tool into their particular industry.

In industries where wearers are reliant on their clothing systems, users can become attached to certain products and tend to reject new developments. For example, a clothing system for cold-storage workers recently underwent a change ('Glacial shift', *Company Clothing*, July 2004). In this particular situation, thinner insulation was introduced into the system replacing the thicker counterpart, although the new insulation is equally as effective yet lighter, enhancing mobility of the user. It appeared that the target user in this situation failed to accept the new insulation because of their attitudes and beliefs; they were not convinced that the new types of insulation had equal performance although less bulk, and therefore rejected the new system. A designer must ensure the traditions and culture of the relevant industry or occupation are taken into account before proposing radical design solutions, and consider the information which must accompany any major change.

Aesthetic design can affect the success or failure of a clothing system through the way it makes the user feel, allows for personal expression, and generally enables the psychological functions of clothing, but there is very little research on this particular area. However, evidence is growing that fashionability affects the way protective clothing is perceived. For example, health care workers in Belgium and Holland found the garments they used for work boring and basic, so Belgium healthcare supplier Sacro introduced 'denim look' textiles into the garments in their range, which are now being used as part of the clothing systems in major hospitals in Antwerp, Brussels and Ghent ('Denim look hits healthcare', *Company Clothing*, July 2004, p32). Although there is no substantial evidence due to lack of research in the area, this article suggest that design and aesthetics can influence the attitude and possibly the performance of the user, through increased self-esteem. (Similarly, in a medical context, when researcher Rebecca Earley from Chelsea College of Art and Design recently designed a range of printed medical gowns for patients following operations, this was seen to enhance their self-esteem and aided recovery.) This is indicative of a growing trend towards both higher aesthetic and performance characteristics in everyday clothing, which has changed at a faster pace than the equivalents in corporate and protective clothing, creating a gap of expectations and usage between workers' on- and off-duty clothes. Off-duty clothes are more fashion conscious and can provide a higher level of social acceptability than many occupational working clothes.

Performance clothing is a testing ground for new innovations due to the demanding nature of the functionalities required from the systems; often technology transfer from other fields such as space exploration and aviation is introduced. However, innovation and practicality do not always go hand in hand. For example, electrically heated gloves were introduced to cold-storage workers who rejected them because they were expensive and the battery pack necessary for powering the function was too heavy ('Glacial shift', *Company Clothing*, July 2004).

3.3.2 System anatomy

The composition of a system of clothing can vary from a single garment or layer to multiple layers expanding outward from the skin. There are however three fundamental strata: a base layer, middle and external layer. Each stratum performs a range of basic operations; in protective garment systems additional functionality is usually incorporated at the middle and external level. However, a system of clothing does not operate as an individual entity, but in a dynamic relationship with the wearer, activity and environment. There may be choice in whether certain layers are worn, or different ways of configuring the system according to particular circumstances and need.

Base layer

The base layer lies on the surface of the skin and remains in direct contact with the body, therefore the fibres used in the garments need to be soft and smooth, while the textile system needs to be able to wick moisture away and insulate the body without restricting the wearer's movement. This can be engineered through combinations of fibre choice with woven or knitted structures that channel moisture away from the skin and towards the outer surface for evaporation. Base layer items are intimate underwear, vests, t-shirts, socks, sportswear or swimming costumes, this being a vital layer for comfort and insulation. The traditional 'string vest' was designed to trap as much air as possible, but is now considered old fashioned. Knitted textile structures are almost universally used in base layer garments for their inherent stretch properties ensuring the wearer ease of movement and flexibility without chafing. Additional functionalities such as antibacterial properties can also be introduced into base layer garments through specialised treatment of the fibres and textiles; this technology prevents the development of bacteria in base layer garments, thus managing the development of body odours, particularly used in socks.

Physiological comfort is possibly the most important functionality for this layer to operate successfully within a system. Physiological comfort from clothing can be highly subjective, and is in fact the attainment of a neutral sensation, which can be described as a state in which we are physiologically and psychologically unaware of the clothing we are wearing (Smith, 1993). In contrast, any sensations actually experienced indicate various levels of discomfort.

One of the parameters that influences discomfort is the system's ability to manage the air enclosed within its constructs, in particular the ability to wick moisture vapour through its strata. It is when the moisture emitted from the skin begins to saturate the base layer garments that the wearer begins to experience discomfort. Traditionally, cotton fibres have been employed, as they are naturally soft and absorbent, but cotton fibres have poor wicking ability, whereas historically underwear used woollen knit fabrics which provided absorbency and wicking, and wool is currently being re-introduced in yarn blends.

Cotton is widely used in base layer garments, although it is not the most functional of the natural fibres for thermal protection. During the 2004 annual Survival Conference, (Leeds UK) a new concept for a base layer garment was presented (Ellis and Brook, 2004). Manchester firefighters were experiencing problems during the summer months, when extinguishing wild land fires. Firefighters would not use any of the upper middle and external layers of their clothing systems, because of the heat, and would extinguish the fires dressed only in their cotton t-shirts. (See Fig. 3.4: Manchester Fire Brigade uniform.) The problem that emerged was that the hot embers produced by the wild land fires would burn holes in the firefighters' t-shirts. The collaboration between the Manchester Fire Service, Leeds University and Bolton Institute resulted in the development of a double jersey knitted garment using a 20% wool and 80% polyester blend yarn named Sportwool® by The Woolmark Company. T-shirts made from Sportwool, utilising the natural absorbency and vapour transportation of the wool component, were shown to have a range of functionalities superior to their cotton or 100% polyester counterparts. During the testing and wearer trials, Sportwool t-shirts proved to have an increased wicking ability, resistance to ember burns (due to the natural flame resistance of the wool and polyester fibres), and were considered more comfortable by the firefighters. Unsolicited testimonials confirm this fibre blend's enhanced comfort in other circumstances such as cycling (see, for example, www. teamestrogen).

Single-layer clothing systems such as swimsuits belong to the base layer category, the system being comprised of the swimmer, the suit, and the water. During the activity of swimming, the suit needs to stay firmly in place next to the body but when out of the water the suit needs to dry quickly in order for the system to be successful and not cause the user discomfort. Early swimming

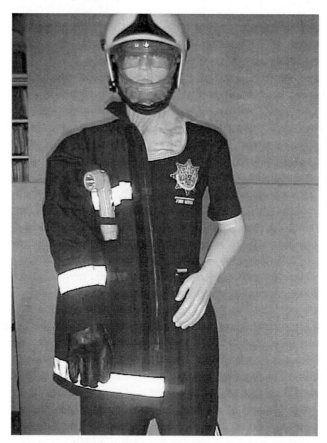

3.4 Layers of firefighter's uniform, Greater Manchester Fire service (Source: K Whitehead).

costumes were made of knitted wool, to take advantage of the latent heat emitted on evaporation of water which reduced the wind chill effect, but manufacturers failed to recognise that they would also sag due to the weight of water absorbed. Speedo were the first to introduce the 'racing back' cutaway design in 1927 to reduce drag in the water (McCann 1999), a functionality which they recently introduced into the intrinsic structure of the textile employed by the system. Of biomimetic inspiration, this textile was initially designed for Olympic competitors, to imitate the structure of sharkskin, which has remarkable ability to minimise drag when swimming in the water. The interpretation has been achieved through the engineering of dermal denticles which are small hydrofoils having V-shaped ridges which decrease drag and turbulence. A more recent development is Aquablade® which uses stripes to create a channelling effect to move water across the body (source http://www.speedo.com/). Similarly, Jetconcept® by Adidas uses V-shaped ridges to reduce drag in the water and was worn by Ian Thorp in the 2004 Athens Olympics ('Jet Stream', *Textile Horizons*, p17, Sept–Oct 2004).

Mid-layer

The middle layer is positioned above the base layer but can also have areas that are in direct contact with the skin. Middle layer garments need to be durable, wicking, provide insulation, and ease of movement in a similar manner to the base layer. Additional protective functions can also be introduced at this level ranging from stain resistance to stab and ballistic protection, however this stratum begins to carry aesthetic values that communicate and socialise the body. Garments at this level can be upper bodywear items such as shirts, blouses, sweaters, fleeces, and lower bodywear items such as skirts and trousers and whole body protective coveralls.

Insulation and ventilation are important functions at this mid-level and are most often achieved through trapping and releasing air respectively. For thermal stability, pockets of still air need to be kept within the system long enough for the air to be warmed from the heat generated by the body. However, this will cause an increase in internal humidity and lead to the saturation of the underlying layers unless the system's air is renewed frequently or the moisture is drawn out of the system (Sari and Berger, 2000). Structurally, air can be trapped or released most easily round the neck, wrists and waist areas through the use of a convertible collar that can be worn open or closed, a blouson style top or draw string trousers, and by using long sleeves with elasticated or turn back cuffs. Insulation properties can be engineered into the clothing system through use of both textiles and structural design. In fabrics, air pockets are created through constructions such as foam, cellular knits, or weaves, napping, pile, double woven or knitted structures, or batts of loose fibre wadding quilted together between two layers of fabric.

In addition to the management of heat and moisture generated from the body, the middle layer in the system can also carry additional protective properties such as ballistic protection, which is introduced into the system through specialist textile technology. Often protective functionality is achieved through rigid plate technology almost welded into the garment system and the textiles used lack flexibility both structurally and aesthetically. This is the point where protective functionality collides with design aesthetic. The textile system providing the function can dictate the overall shape and look of a garment, as rigid protective systems tend to be stiff and bulky which limits aesthetic design possibilities; in addition, surface manipulation processes such as printing and embroidery can damage the functionality. Waterproof/breathable membranes lose their water resistance when stitched, although this can be remedied by taping the seams, but breathability can be compromised by printing pastes or repellent finishes.

In the Metropolitan Police, body armour is an aesthetic issue for women. To accommodate the contours of the female, shaped ceramic body armour panels are used, which are technically very difficult to construct, in order to pass all the required tests; shaped panels which were intended to increase comfort and reduce weight on the breasts, were rejected by some women because of image. 'The "Madonna look" was ridiculed and women felt their authority was undermined' (authors' interview R von Szaley).

The innovative concept behind the Arctic Heat vest (also a collaboration with Sportwool®) is the encapsulation into a body protector of hydrophilic crystals which form a gel when immersed in water and can maintain the core body temperature, either hot or cold, for a considerable time. This vest can be worn directly over a garment and provide instant relief from heat stress but is a rigid and solid structure that limits the design possibilities. The aesthetic aspects have hardly been developed in the highly functional but unsympathetic styling. Another system that dictates the look of the garment is Airvantage by Gore and Associates which consist of inflatable tubes inside the vest. The technology provides adaptive insulation but is stiff and does not leave much scope for design. Further innovative design solutions are still needed in these areas.

External layer

The external layer is the final stratum; garments in this layer cover areas of the mid-level clothing and can also have direct contact with the skin at the extremities, usually at the hands, neck or on the head. This is perhaps the most complex of layers as it meets multiple requirements. It may need to be wicking, insulating, provide ease of movement, be wind and water proof and carry most of the protective functions that shield the user from the hazards of their particular external environment, in addition to aesthetic aspects. Garments in this group can be coats, jackets, heavy fleeces, whole body coverings, footwear, hats and gloves and can employ woven, knitted, non-woven and composite fabric structures.

One of the core requirements this layer must achieve is the management of heat and humidity generated by the body. Heat loss can be controlled structurally by the use of zippers and openings at strategic places, i.e., sleeves, back of neck, irrespective of the degree of moisture permeability particular to the material used in the external layer (Ruckman and Murray, 1999). However, this particular method requires the user to manage the renewal of air manually. This may not always be practical as the wearer's professional practice may require his full attention, such as in the situation of a firefighter. Any distraction could prove lethal.

Breathable membranes, for example Gore-Tex®, are introduced to the external strata through laminating methods that bond the membrane to one or more surfaces of a textile. This makes the garment resistant to drops of water from the external environment but permeable to moisture vapour generated from the body. Evidently, the individual properties of a textile can be lost within a system, and the success of the clothing system depends on the relationship between the different components. Therefore, if one of the items has low permeability to water vapour, for instance, the wicking performance of the entire system deteriorates. In addition, attention must be paid to co-ordination of different layers of the system. If both an outer and a mid-layer garment have impermeable membranes, then their effectiveness may be cancelled out. The procurement procedures for protective clothing must therefore also be co-ordinated. A designer should be sure to consider the wider context even if only commissioned to design certain elements of a clothing system. (See also Chapter 10 by Rossi.)

3.4   Reconciling fashion and function

3.4.1   Marketing and evaluation

Where the design of new products is carried out as part of a major research and development project, testing and evaluation naturally forms part of the standard procedures built into the development process outlined in section 3.2 above. However, trials are necessarily limited and often take place in fairly unrealistic laboratory circumstances. There is no substitute for wearer trials carried out over a period of time by the real users of the clothing systems, where feasible. The product development process therefore never concludes, but continues in response to user feedback, with adaptations being made to the products whenever possible as renewal and replacement occurs. The development of highly sophisticated responsive manikins for use in thermal comfort, fire, pressure, and crash testing has provided enhanced data in areas which are difficult to test safely with humans, e.g., military, fire and space applications (SATRA, 2003). (See also Chapter 9 by McCullough.)

Many claims are made by clothing manufacturers in promotional material concerning the performance and longevity of their products, especially in sports clothing which is not governed by CE marking regulations, and which are endorsed by well-known personalities. In the early development of sportswear and outdoor wear some of the prominent sportsmen and women within the activities themselves became consultants in the research and development process and helped to refine existing designs and develop new products for protective and performance clothing. These have developed into some of the most successful sportswear and outdoor brands such as Helly Hansen® and O'Neill®.

Branding and marketing is increasingly important to convey the technical message of invisible performance factors inherent in modern clothing. Wool fibre has recently undergone such a re-branding to attempt to rid itself of its persistent but somewhat anachronistic image of the past, and to position its natural properties and structural complexity within a technical and performance framework as 'nature's first smart fibre'. This was branded Woolscience in 2003, in order to appeal to a more masculine and younger user, reposition wool against its synthetic competitors, and to reverse the recent decline in its market share. Included in these recent developments are Deolaine®, a bacterial inhibitor for odours, and Arcana®, a descaled wool aimed to give a softer feel, similar to cashmere. The dual message is performance and comfort engendering wellbeing.

3.4.2   Case studies

Police

Within different contexts, such as the police, the protective aspect of clothing has to be balanced with its dual function as a recognisable uniform, serving as a form of corporate identity for the public. Following complaints from officers about the performance of their clothing, police forces across Scotland have recently collaborated to redesign their uniforms which have ranged from the formality of the original tunic first developed in the late 19th century, to casual clothing such as anoraks, 'old-fashioned woven jerkins' and sweaters. According to one reporter 'The new look finally sweeps away the impractical and old-fashioned tunic and replaces it with clothing that suits the realities of modern-day policing' (Qureshi, 2004). The fabrics were chosen from those developed for outdoor sports and aimed at preventing an officer from sweating profusely under the increasing burden of equipment, now weighing several kilograms. The design solution for the clothing system now has a more military ambience, and is currently being trialled. It includes a high visibility/reflective jacket, waistcoat comprising a polyurethane coated nylon material designed to protect officers from blood-borne viruses which may be transmitted at accidents or during violent incidents, and a shirt made from fabric which wicks away moisture. To aid ventilation, officers will be allowed to remove ties, although other forces have abandoned the shirt and tie in favour of t-shirts. There is also a lightweight fleece jacket which can be zipped into a waterproof outer jacket and 'cargo' trousers with multiple pockets (Qureshi, 2004). With the increasing impact of equal opportunities legislation, more consideration is now being given to the female form and fit, (so-called 'lady fit') in both military and civilian uniforms. In this case, the newly designed lighter weight body armour will be available in 'sculpted' form for female officers. However, carrying bulky equipment in trouser pockets is not a flattering female look, and could meet resistance for aesthetic reasons (see also Chapter 23 by Fenne).

Firefighters

In a similar manner to the Scottish police mentioned above, the diverse clothing practices of the fire services across the UK are being investigated within the

Integrated Clothing Project which has a remit to look at garment performance and make improvements to personal protective equipment and clothing by 2007. It is to be noted that there is no specific design input planned into this project, which will be tendered directly to manufacturers. The Manchester Fire Brigade currently uses a typical fire tunic comprising three strata of fabric: an outer shell made of Nomex® aramid fibre for flame resistance, with a fluorocarbon finish which gives an 80% fluid run-off and chemical protection; inside the shell a Gore-tex® membrane for sweat management, which allows moisture from the inside out, but is waterproof; an insulating layer of synthetic wadding by Duflot; and finally an inner lining of fire-retardant rayon. A cotton t-shirt is typically worn under the tunic with cotton/polyester trousers. (See Fig. 3.4.)

In the development of new base layer garments outlined in section 3.3.2, inspiration was taken from ski-wear, the most fashion-conscious of the active sports, to move to polyester fabrics. Whilst Coolmax® polyester t-shirts were popular and comfortable, they did not perform well in burn tests, and the Sportwool® solution identified had the best performance for the available budget. In addition, the durability of the new garment was three times that of a cotton t-shirt which balanced out the increased costs. Styling changes were made from inset sleeve to raglan sleeve and the use of a scooped back in order to emulate the fit of sports garments, and create a more inclusive fit for women. Certain clothing introduced for climatic protection such as anoraks had not been taken up by the firefighters, as they were deemed unfashionable, but crossovers from sportswear including the use of brand names such as Polartec® fabric encouraged a new outer garment to be accepted and worn. A similar effect was observed when safety goggles which were both unattractive and uncomfortable were replaced by more fashionable ski-type goggles with added features such as anti-misting.

One brigade recently completely redesigned their fire-fighting clothing, and developed a system which, contrary to the traditional loose and long fire tunic, comprised salopette style trousers and a short blouson jacket in a silhouette and fit that was much closer to the body. However, the change in the amount of air trapped within the clothing system caused in some cases much greater heat stress, resulting in an unacceptable increase of 2.5 °C in core body temperature (authors' interview K Whitehead, Greater Manchester Fire Service). Whether fashion or corporate identity played a significant part in any of these decisions remains unknown, however it appears that the design and product development process broke down in certain areas such as preliminary research and user testing.

Safety clothing and accessories

Personal safety is a functional prerequisite and preventive aspect of protective clothing which can be designed into garment systems, most obviously through the use of colour, especially fluorescent colours and retro-reflective strips for high visibility when working in hazardous environments (railway workers, road workers, mountain climbers, etc.). However, ironically, the ubiquitous nature of fluorescent clothing and trimmings, and its use in consumer goods, has now begun to reduce its effectiveness in visibility in some urban contexts, in so far as work-based clothing is often used as a criminal disguise, and drivers tend not to notice cyclists and motorcyclists, as accident statistics have recently shown (Ronson, 2005). These 'invisible' fluorescents indicate that there is a need for constant vigilance, updating and reinvention, and how attention must be paid to wider cultural aspects in the context of design.

Accessories (bags, hats, gloves, etc.) and footwear have gained in importance in commercial fashion over the last decades as a vital part of daily apparel, as is evident in the market for sports and non-sports trainers. They offer many opportunities for both functional and innovative solutions particularly to the exigencies of the urban working environment, where an increase in use of portable electronic items is most marked, together with a perceived increase in thefts. Designers and urban style leaders Vexed Generation are credited as the originators of the one-shouldered bag with integrated front pocket for a mobile phone, an accessory which left hands free for cycling for example, but allowed for instant communication (Bolton, 2002: 44). Some highly functional protective items become design classics (for example, the Dr Martens boots or Levi 501 jeans) and as such do not change radically with fashion. Footwear designer Nick O'Rorke cites comfort and performance as the key factors in his Tsubo range (named from the Japanese for pressure point). Taking the protective aspects of rock-climbing shoes into fashion, they incorporate an air bubble outsole for cushioning, with a double-density polyurethane moulded midsole that allows air flow, and also have an antibacterial sock liner. ('Tsubo charged' *Footwear News* 23.8.04.)

Protective clothing has recently begun to merge with fashion clothing in what Bolton (2002) articulates as the 'supermodern wardrobe', garments specifically designed for mobility and protection in urban environments by leading designers, using high-technology technical fabrics and integrating many functionalities. Bolton states 'Often, the metropolitan experience is characterised by insecurity and paranoia. Indeed, the contemporary urban landscape is shaped by our preoccupation with fear.' An initiative set up in 2000 by Dr Lorraine Gamman named 'Design against Crime' has taken personal protection into the design education arena, and aims to promote design as a tool for crime prevention and personal safety, particularly in the realm of clothing and accessories. Supported, amongst others, by the Home Office and the Design Council, the philosophy is to anticipate likely crime opportunities and correct them in advance rather than apply 'retro-fit' solutions to existing products. Accordingly, a range of prototypes was developed, in co-operation with military and emergency service bag specialists H Fine and Co, of designs for handbags and computer bags, both high risk for theft and mugging. Some concepts were commercialised in the Karrysafe® range by Vexed Generation, which included the 'Screamer' laptop bag which triggers a 138 decibel alarm if snatched, and the Bodysafe® fabric belt worn under jeans for personal credit cards, etc. (Gamman and Hughes 2003; www.karrysafe.com.)

3.5   Future trends

This is an exciting phase of experimental research and development within the textile industry, as the universality of textiles within our most immediate and intimate surroundings (the 'portable environment' posited by Watkins (1984)) and in our daily working and living spaces (interior textiles) provides a perfect interface for integrated and enhanced functionality. Several products are beginning to enter the most avant-garde areas of the fashion industry, which are traditionally the first to experiment with radical new ideas whether derived from new technologies or conceptual frameworks. The continued development of textiles with integrated electronics, phase changing or other physical properties, which respond to external environmental changes or internal body conditions, will eventually lead to more effective functional clothing for protection. Some technologies are at an advanced stage of research with existing prototypes, others are still very much at the concept stage. The impact of nanotechnology can already be found in coatings for textiles which allow stain and liquid repellency (marketed rather confusingly as self-cleaning) and the use of nanoparticles of silver in yarns to create anti-bacterial woven or knitted fabrics, currently being used in socks (e.g. SoleFresh®) and other applications.

Much research and development is being focused on the concept of clothes which morph and change to suit the user's individual needs. Shape memory polymers have begun to be utilised for cloth, and Italian manufacturers Corpe Nova's R&D unit Grado Zero Espace have created a prototype metallic polymer shirt whose sleeves they claim will shorten with heat from the sun, and lengthen again on cooling. The same shirt, it is claimed, can be 'ironed' by directing warm air over the crumpled surface whilst being worn (Battrick, 2003).

Over the past ten years, there has been widespread interest and research into the concept of 'wearable computers'. The first wave of research produced clothing prototypes with integrated computers wired into garments, and distributed the functions of the computer –processor, keyboard and monitor – into different parts of the clothing or accessories. Portable electronic devices for communication, location and entertainment have also been incorporated. A limited number of commercial applications resulted, such as the 2001 Cagoon jacket produced as a collaboration between Levi's and Philips, but they often experienced difficulties with aesthetics, bulky power sources, washing or cleaning procedures and with the fact that the requirements of the user will change constantly with fashion and lifestyle. Dent (2004) put it succinctly: 'Consider the Mithril® shirt: is this really a "portable computer" or simply the hardware of a computer strapped to a vest. If these devices are to become wearable, they must become a seamless and unobtrusive part of the clothing.' However, the 'mp3blue' jacket by Rosner and Infineon Technology, launched onto the market in 2004, incorporated removable or encapsulated electronic components designed to withstand washing. Softswitch and Eleksen pioneered soft keyboards and interfaces using different technologies, and further commercial products are in preparation.

The next generation of research has focused on the embedding of electrical functions directly into textiles using conductive polymer yarns or metal coated yarns introduced into woven or knitted structures, and using soft textiles sensors and actuators to provide functionality and monitoring. Medical research is now a key driver for functional integration for remote monitoring of patients and for wellbeing during sports or remote and hazardous activities (for example, Wealthy, Eleksen, De-tect). Further developments have seen conductive circuits printed onto textiles, and the incorporation of optical fibres and photonics into woven structures, which have the potential to change colour or act as a flexible display (see also Chapter 7). The key factor still to be resolved is the robustness of the systems, before ubiquitous applications can be developed. When this transpires, there will be a significant impact on both protective and everyday clothing.

A further factor which will have a major impact on the industry is the paradigm shift from mass production to mass customisation and personalisation, particularly relevant wherever military and civilian uniforms have to be adjusted for individuals. Three-dimensional body scanning and visualisation technologies provide an opportunity and the means for customised design and fitting of many types of apparel – shoes, jeans, suits, shirts, etc. Examples can be seen in both American and European retail and web-based environments, such as Brooks Brothers of New York, Odemark in Germany, and the customised 'virtual try-on' system for jeans established in 2002 by Bodymetrics at Selfridges in London.

These developments, together with continued research into new materials and manufacturing processes, and technology transfer are being applied to a number of processes such as knitting, weaving and rapid prototyping technologies for the production of clothing and accessories, which will integrate fashion design, materials science and micro computing. One project undertaken by footwear company Prior2Lever, in conjunction with researchers at London College of Fashion, is enhancing the performance and comfort of football boots, by creating a customised pair of boots on demand from an individual foot scan, which also combines a customised sole directly manufactured from powdered polymer material, using stereolithography to produce an individual 3D form to exactly fit the footballer's feet, improving fit and reducing injury.

One of the key drivers for new advances in technology for textiles and clothing comes from the military sector, which seeks to equip troops with the highest specification personal protective equipment, and therefore sponsor specific research programmes. The US military has recently embarked on the Rapid Fielding Initiative, the 'largest single project to properly protect and outfit the modern soldier', sourcing mainly from America. The majority of the technology that appears in the final product will be introduced at the fibre and textile level, which is increasingly where real innovation lies. Future operational equipment and clothing are envisaged to be lightweight multifunctional garments that integrate:

> intelligent textiles serving as the backbone for warfighters' electronics, optics and sensor suites
> lightweight ballistic protection
> improved camouflage and signature management
> self-deactivating chemical/biological protective membrane
> antimicrobial protection
> improved environmental protection
> reduced weight and bulk with improved fit, comfort and durability.
>
> (Kinney, 2004)

Other NATO Nations including the UK, France, Holland, Denmark, and Germany also have research and development programmes with similar objectives. This menu of functionalities may eventually be incorporated into multi-layered textile structures with external camouflage layer, internal membranes and embedded sensors, data and power systems. Enabling technologies are nano- and microtechnology, electro-textiles, flame-retardant chemistry in fibres and finishes, and composite structures, but power sources, textile materials and weight continue to be barriers to achieving the goal. Total integration may be achieved only when multiple protective capabilities can be enabled in a single textile without increasing weight, bulk and heat stress. This type of military research push certainly points the way for further integration of functionalities at the textile level in future civilian protective clothing, and so eventually to both functional and fashionable everyday wear.

Clothing is the interface between the wearer and their environment, within which will be embedded an increasing range of functionalities, within the materials and textiles structures and in the design and realisation of the garments, working to complement each other. Research already under way will result in direct customised manufacturing using both traditional and novel methods, and technologies will emerge which allow clothing to be made in entirely new ways. There will continue to be higher demands and expectations of functionality and performance from clothing, and clothing systems for protection will move towards lighter weight but effective solutions. Multi-functionality and versatility in fibre and textile systems will mean a reduction in the number of individual layers of clothing required. All the indications are that current research will deliver, in the relatively near future, garments and clothing systems which will be more protective, but also more adaptable in a multitude of ways: sensory, emotive, responsive, receptive, transmissive, regenerative, and aesthetic. In the meantime, fashion is playing an increasingly important role in achieving positive acceptance of personal protective clothing in today's design-conscious environment.

3.6   Sources of further information

Standards

See British Standards Institute websites www.bsonline.techindex.co.uk and www.bsi-global.com
American Society for the Testing of Materials (ASTM)

Journals

Clothing and Textiles Research Journal, International Textile and Apparel Association
The Journal of Clothing Technology and Management
International Journal of clothing science and technology
Company Clothing

Website

www.director-e.com Corporate Clothing Directory

Organisations

The Textile Institute
British Textile Technology Group www.bttg.co.uk
SATRA Technology Centre www.satra.co.uk

3.7   References

Adams SA, Slocum AC, Keyserling WM (1994) A model for protective clothing effects on performance. *International Journal of Clothing Science and Technology*, Vol. 6, No. 4, p. 6, MCB University Press.
Battrick B (ed.) (2003) *A thread from space to your body*, ESA Technology Transfer Programme, Noordwijk, ESA Publications.
Bolton A (2002) *The Supermodern Wardrobe,* London, V&A Publications.
Brownless NJ, Anand SC, Holmes DA, Rowe T, Silva AD (1995) 'The quest for thermo physiological comfort', *The Journal of Clothing Technology and Management*, Vol. 12, No. 2, pp. 13–23
Carr H and Pomeroy J (1992) *Fashion design and product development*, Oxford,

Blackwell.

Dent A (2004) 'Wearables', *www.toy-tia.org/access/development/d-articlearchive12.html* (accessed 1.7.04).

Ellis A, Brook D (2004) 'Moisture Management Base Layer Project', paper presented at Survival 2004 Conference in Leeds, UK (Sept).

Entwistle J (2000) *The Fashioned Body*, Cambridge, Polity Press.

Entwistle J (2001) 'The Dressed Body', in Entwistle J and Wilson E, *Body Dressing*, Oxford, Berg.

Gamman L, Hughes B (2003) 'Thinking Thief' *Ingenia*, London, RSA Publications.

Joseph N (1986) *Uniforms and Nonuniforms: Communication through Clothing* Contributions in Sociology no 61, New York, Greenwood Press.

Kinney RF (2004) 'Future DOD directions and its impact on textiles', Techtextil Symposium North America (May).

Lamb JM, Kallal M J (1992) 'A conceptual framework for apparel design', *Clothing and Textiles Research Journal*, Vol. 10, No. 2, pp. 42–47.

Maslow A H (1954, 1970) *Motivation and Personality*, London, Harper Row.

McCann J (1999) Establishing the requirements for the design development of performance sportswear. Unpublished MPhil thesis, University of Derby.

Perkins HM, Crown EM, Rigakis KB, Eggertson BS (1992) 'Attitudes and behavioural intentions of agricultural workers toward disposable protective coveralls', *Clothing and Textile Research Journal*, Vol. 11, No. 1, Fall.

Pitimaneeyakul U, LaBat KL, DeLong MR (2004) 'Knitwear product development process: a case study', *Clothing and Textile Research Journal*, Vol. 22, No. 3, pp. 113–121.

Qureshi Y (2004) 'The new black for the thin blue line', *Scotland on Sunday* 18.04.04.

Raheel M (ed.) (1994) *Protective clothing systems and materials*. New York, Marcel Dekker.

Renbourn ET (1971) *The psychology of clothing with materials in mind*, Third Shirley International Seminar 'Textiles for comfort' New Century Hall, Manchester.

Ronson J (2005) 'Now you see them ', *The Guardian Weekend* supplement 15.01, pp. 14–19.

Ruckman JE, Murray R (1999) 'Engineering of clothing systems for improved thermophysiological comfort', *International Journal of Clothing Science and Technology*, Vol. 11, No. 1, pp. 37–52, MBC University Press.

Sari H, Berger X (2000) 'A new dynamic clothing model. Part 2: Parameters of the underclothing microclimate', *International Journal of Thermal Science*, Vol. 39, pp. 646–654, Elsevier.

SATRA Spotlight (2003) *Manikin testing* (protective clothing; test methods; burns) p. 15.

Smith EJ (1993) 'The comfort of clothing', *Textiles,* Issue No. 1, pp. 18–22.

Sweetman P (2001) 'Shop-Window Dummies? Fashion, the Body, and Emergent Socialities', in Entwistle J and Wilson E, *Body Dressing*, Oxford, Berg.

Watkins SM (1995, 1984) *Clothing the portable environment*, Iowa, Iowa State University Press.

Steps in the selection of protective
clothing materials

A S H A W , University of Maryland Eastern Shore, USA

4.1   Introduction

Technological advances in the field of technical or high performance textiles have made it possible to engineer materials designed to meet specific needs. However, there is no 'ideal' fabric that will provide protection against all hazards. Careful selection of appropriate textiles is crucial for the performance, use, care, and maintenance of protective clothing. Broadly, a clear understanding of the work environment and performance requirements as well as knowledge of the textiles, is essential for the decision making process. According to the OSHA Technical Manual, 'In general, the greater the level of chemical protective clothing, the greater the associated risks. For any given situation, equipment and clothing should be selected that provide an adequate level of protection. Overprotection as well as under-protection can be hazardous and should be avoided'.[1]

The above statements are also applicable to other types of protective clothing. Heat, physical and psychological stress, as well as reduced dexterity and mobility, are examples of additional hazards that may be a result of the use of protective clothing. In addition to protection, the selection of textiles takes into consideration factors such as the impact of protective clothing on job performance; comfort; durability; availability and cost; use, care, and maintenance, and cultural factors. To select protective clothing materials, relevant factors have to be identified and prioritized in a multi-step process. It is imperative that such a holistic approach be taken while selecting materials for protective clothing. Basing the decision on just the performance properties of the material may result in low user acceptance. For example, a glove that protects a surgeon may be rejected if it hinders the surgeon's ability to operate effectively.

This chapter presents a model of the decision-making process and gives some examples of the level of detail necessary in each step. In addition, it describes an online system as a case study to illustrate how technology can be used to assist in the selection of textile materials for protective clothing. The scope of this chapter is limited to the selection of protective clothing materials and not the entire assembly. It is important to note that while careful selection of suitable materials is crucial for the performance of the clothing ensemble, testing of just the materials is not sufficient to determine the performance of the ensemble, especially for multi-component and fully encapsulating suits. Part II includes the performance of clothing ensembles for selected end uses.

The selection of textile materials requires a step-by-step approach in which the potential hazard is clearly defined, and the material is selected in accordance with the existing standards or guidelines. If no standards or guidelines are available, relevant test methods must be identified and used. This initial screening, based on protection provided by the material against potential hazard, is used to narrow the choices. Factors other than protection are then considered to make the final selection. Figure 4.1 provides a model of the selection process discussed in this chapter.

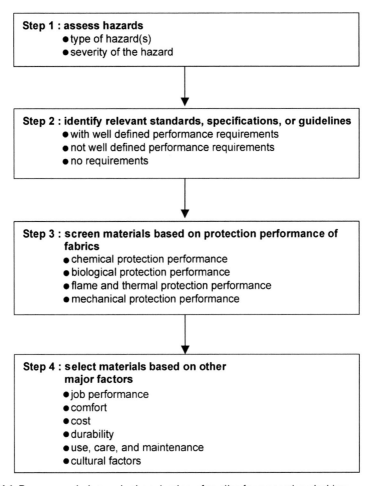

4.1 Recommended steps in the selection of textiles for protective clothing.

4.2   Assess hazards

The hazards, as well as a scenario that includes the work environment, are used to define potential risks to the individual. Hazards and scenarios to assist in the selection of materials are very well defined in some sectors, but in other sectors there may not be such well defined scenarios. In general, well defined risk scenarios and performance specifications based on the risks are available for sectors where the use of protective clothing is crucial for the safety of the individuals. For example, selection of protective clothing for firefighters and police is based upon well developed scenarios and requirements. With increased risk of terror attacks, the work scenarios for police and firefighters have changed in countries such as the United States, and the new risks impact the guidelines, standards, and selection of protective clothing materials. These new risks may also require the development of new materials that provide additional protection to such individuals. With the change in work scenario, firefighters and other professionals who are now categorized as 'first responders' need protection against other hazards. The United States Department of Homeland Security has adopted personal protective equipment (PPE) standards developed by the National Institute for Occupational Safety and Health (NIOSH) and the National Fire Protection Association (NFPA) to protect first responders against chemical, biological, radiological, nuclear, and explosive (CBRNE) threats.

In case of serious new threats, decisions regarding protective clothing may have to be made in a very short time frame. For example, decisions had to be made immediately regarding selection of appropriate personal protective equipment for protection against viruses such as severe acute respiratory syndrome (SARS) and anthrax. In those two scenarios, assessing the hazards was the crucial first step in the selection of the protective clothing.

4.2.1   Type of hazard(s)

The types of hazards are broadly classified as chemical, biological, physical/ mechanical, radiological, and flame and thermal. This classification is used by national and international standards organizations as a framework for organizing their committees and for related publications regarding terminology, standards, and performance specifications.[2–6] Limited information is available regarding protective clothing against radiological hazards due to the nature of the hazard. The current approach is to incorporate detecting devices to warn individuals of the hazard and thus prevent them from being exposed. The ASTM F23 Committee on Protective Clothing has recently established the F23.70 Radiological Hazards subcommittee to address issues related to radiological hazards.

Factors such as composition and physical form of the hazard are used to further categorize each type of hazard. Given below are the common sub-categories for each of the main types of hazards. Chemical hazards are categorized by the chemical composition, physical form, and toxicity. They are commonly identified by the common name, the Chemical Abstract Registry Service (CAS) number, and the chemical class to which they belong. The CAS numbers are numerical unique identifiers assigned to each substance. Chemicals from the same class, in general, have similar chemical composition. Thus, some commonly used chemicals from the various classes are used to determine the performance of materials against a broad range of chemicals. In cases such as pesticides where mixtures are used, it is recommended that the performance be measured against a representative mixture, rather than determining the performance based on a broad range of chemicals. The chemicals have different physical forms – solids, liquids, aerosols, and gases. The tests used to determine the performance are often determined by the physical form and the severity of the hazard. Additional information is included in section 4.4.1.

Biological hazards are divided by the type of microorganism and the mode of contamination. In the medical field, the primary potential risk to the individual is exposure to hazardous bacteria and viruses through contact with contaminated blood and other potentially hazardous body fluids. Surrogate microbes are commonly used to assess the penetration of extremely hazardous substances such as Hepatitis (B and C) and Human Immunodeficiency Viruses. The other risk is exposure to hazardous airborne pathogens such as influenza and the SARS viruses. Airborne pathogens may pose a risk to the general public, in addition to medical personnel. The hazard may be due to natural causes or as a result of bio-terrorism. As the use of biological agents poses a threat in wars, protective clothing materials are also used to protect military personnel.

Physical hazards, categorized by the type of contact of the object with the material, can be broadly divided into cut, ballistic, puncture, and abrasion. Cut injuries occur in work environments where individuals work with sharp, high-speed objects such as chainsaws. In most cases, resistant materials are used to protect the hands and legs. Ballistic protection is required to protect individuals from bullets. Bullets may also cause blunt trauma injury by deforming the body armor. Thus, body armor designed to protect individuals is evaluated for ballistic protection as well as the ability to prevent blunt trauma injury. Puncture or stab resistance is required to protect individuals from sharp objects such as knives, as well as comparatively blunt objects commonly used by inmates in correctional facilities. In addition to the above mentioned hazards, protection against abrasion is also used as a criterion for selecting the material. Protection against multiple physical hazards is commonly required for individuals in the military, police force or correctional institutions. Body armor is commonly used to provide protection against physical hazards. Puncture resistance tests are conducted on numerous fabrics that provide protection from hazards such as radiological and biological protection, as the individuals may be exposed to other hazards, if the material is punctured by a sharp object.

Flame and thermal hazards are grouped by the source of flame or heat. The most common types of hazards are exposure to open flame, radiant heat, and molten material. Individuals such as firefighters are susceptible to the potential of burn as a result of exposure to open flame and radiant heat. Individuals working in industries where they are exposed to molten metal need materials which reduce burn injuries by shedding molten metal into the floor.

4.2.2   Severity of the hazard

The toxicity of the hazardous material, duration of exposure, and level of exposure are used to determine the severity of the hazard. All three factors are interrelated and should be considered as a set. For all categories, toxicity of the materials and the by-products are the primary concern. Highly toxic substances or hazardous conditions have to be handled with extreme care in all categories, regardless of the duration and level of exposure. The level of exposure and the duration of exposure are considered with reference to the toxicity. For example, exposure to even a small amount of a chemical agent for a short duration may be far more harmful than exposure to a Class IV pesticide while spraying for an entire work day.

For many end uses, PPE is rated based on the severity of the hazard. For example, body armor classification developed by the National Institute of Justice (NIJ) is used for the selection of the body armor based on the potential workplace hazard. European Union certification for PPE uses Type I-VI categories that are based on very broad hazard scenarios. Although there are ratings or classifications for the various types of hazards, there is no universal system to rate the severity of all hazards. In the future, there may be a system to assess the severity and rate it regardless of the type of hazard. For example, the extremely hazardous category may be defined to include all hazard scenarios that are life threatening in very small amounts. Thus, the extremely hazardous category may include SARS and anthrax from the biological category, nerve gas from chemical, ballistic from physical, and radioactive materials from radiological hazards. This type of rating would help in the testing and selection of PPE materials. For example, when the severity of hazard is extreme, stringent rules would apply for the handling and testing of fabric, and thorough testing of the whole garment ensemble would be necessary.

4.3   Identify relevant standards, specifications or guidelines

Regulatory standards, performance specifications, guidelines, and test standards established by national and international standards organizations, governmental agencies, as well as associations for various professions are available to determine performance and assist in the selection of the textile materials.[2–6]

Regulatory standards are often mandated by governmental agencies as well as other organizations that have jurisdiction for a particular group. Knowledge of the performance specifications and standards is essential. Often companies that manufacture materials and protective clothing may have to comply with more than one set of requirements, as these may vary by geographic locations. For example, due to differences in minimum specifications as well as the required test methods, garments may meet the requirements in the United States but not in the European Union or vice versa. Thus, it is not unusual to find different products being marketed by the same manufacturer in different countries.

The specificity of the regulations and standards varies considerably. They range from very clearly defined selection, use, care, and maintenance criteria for personal protective equipment (PPE) to requirements in which the employer is responsible for providing 'appropriate protective clothing.' In some cases there is either no performance standard or guideline, or they are in the process of being established. Given below are examples of standards and specifications for different scenarios and the impact they have on the selection process for the particular hazard.

4.3.1 Standards with well defined tests standards and performance specifications

Protection of Industrial Personnel against Flash Fire: The NFPA 2112 Standard for Flame-Resistant Garments for Protection of Industrial Personnel against Flash Fire and NFPA 2113 Standard on Selection, Care, Use, and Maintenance of Flame-Resistant Garments for Protection of Industrial Personnel against Flash Fire are standards established by NFPA International.[7] These standards have clearly defined test methods and minimum requirements that assist in the selection of protective clothing materials. The manufacturers of the materials often provide data to guide product selection. Well defined standards, performance specifications, and guidelines that specify minimum performance requirements are very beneficial in the selection process, as they allow the individual responsible for selection to be able to compare the performance characteristics as well as other factors that play an important role in choosing the best material for the end use.

4.3.2 Standards with no tests standard and performance specification defined

The PPE standard mandated by the U.S. Occupational Safety and Health Administration (OSHA) for occupational exposure to blood and other potentially infectious materials is applicable to a wide range of scenarios. According to the standard, the employer is required to provide 'appropriate' personal protective equipment (PPE) for the employees.[8] However, the test method and the minimum requirements are not specified. It is thus the responsibility of the employer to determine the hazard and select appropriate PPE to protect the employees at work. As the regulations do not provide guidelines on the types of test or material to be used, it is up to the manufacturer to determine the test method to evaluate the performance of their materials and up to the employers to select the materials suitable for their end uses. Standards with no test method and minimum requirements pose a major problem in the selection of suitable materials, often resulting in over or under protection of the individuals.

4.3.3   Standard with recommendations based on garment design and not performance

Requirements for PPE for pesticide applicators in the United States are currently based on garment design and not on the performance of the garments. Test results show that the penetration of pesticides through garment materials in the same category varies considerably. Thus, recommendations such as use of long sleeve shirts and pants, commonly seen as part of the safety instruction on a pesticide label, are not sufficient.[9] Basing the recommendations on just garment design may result in over or under protection of the individuals. A task force has been formed as part of ASTM-F23 Committee for Protective Clothing to develop performance specifications for protective clothing during pesticide application.

4.4   Screen materials based on protection performance

The specificity of the standards, performance specifications, and guidelines determines the tests required to assess protection performance of the materials. For end uses where the hazard and minimum requirements are clearly defined, test data is usually available through fabric and garment manufacturers. Compliance of the material and garments with relevant standards is also commonly posted on the manufacturer's website.[10–13] In situations where regulations do not specify the test methods, or there is no performance specification, the individual responsible for selecting the protective clothing has the added responsibility of determining the scenario and test method suitable for assessing the performance of the materials. Often the appropriate test method will have to be selected. A clear understanding of the exposure scenario that includes type, level, and duration of exposure, and environmental conditions is essential. The information should be used to identify the standard that represents the scenario of the proposed end use. Often the test methods allow for testing at different levels of severity and/or duration of exposure. Defining the potential scenario is helpful in selecting the appropriate parameters for testing. In some cases, the method may have to be modified if the test parameters and procedures are not adequate to test for the potential scenarios.

As there are many probable scenarios, several test standards have been developed for each of the broad hazard categories. It is not unusual to find two or more test methods developed by different groups to measure the same end use performance. The results from the test methods may be similar or may vary considerably depending on the scenario or the basis for the test method. For example, three national and/or international standards are used to measure the penetration of liquid pesticides. Comparison of the three methods showed major differences in the pesticide penetration values for the same materials.[14] This was due to the fact that the scenario or the basis of the test method ranged from simulation of accidental spill to fine spray.

Given below is information regarding selection of materials based on protection performance against the broad categories of hazards. The categories and examples used are not all inclusive as there are numerous scenarios within each of the categories. The purpose of this detail is to illuminate the level of complex analysis needed.

4.4.1 Chemical protection performance

The route of entry into the body determines the type of protection required. For example, face masks or respirators with different filters are used to protect individuals against exposure due to inhalation. Chemical protective clothing is used to protect against dermal exposure. Chemical protective clothing materials are broadly divided into woven; nonwoven; laminated to microporous or hydrophilic membranes; coated, bonded, or laminated with plastic films or rubber; and films, sheets, or moulded plastic or rubber. The performance of the chemical protective materials varies considerably based on the air permeability, chemical composition, and material characteristics. Penetration, permeation, and/or degradation of the materials are measured to determine the chemical resistance. In general, the physical form of the chemical that poses a potential risk and the type of material determine whether penetration, permeation, and/or degradation tests have to be conducted. Permeation is 'the process by which a chemical moves through a protective clothing material on a molecular level. Permeation involves (i) sorption of molecules of the chemical into the contacted (challenge side) surface of the material; (ii) diffusion of the sorbed molecules in the material, and (iii) desorption of the molecules from the opposite surface of the material'.[15]

The permeation rate and breakthrough time are important for measuring permeation. Permeation tests are conducted to determine the performance of non-porous materials such as monolithic films against volatile and non-volatile liquids and gases. It is not used to measure the performance of woven or knitted fabrics. For such porous materials, penetration tests are used to measure performance. Penetration is defined as the 'process by which a solid, liquid, or gas moves through closures, seams, interstices, and pinholes or other imperfections on a non-molecular level, in a protective clothing material or item.'[15] Penetration through a material can be measured by several national and international standards. The selection of a suitable standard depends on the exposure scenario. For example, there are three national and international test methods to measure the penetration of pesticide through textiles.[14] Degradation is defined as 'a deleterious change in one or more properties of a material'.[15]

Materials that have degraded due to chemical exposure, use or exposure to environmental conditions such as sunlight should be discarded, as these affect the fabric performance.

Given below are the ASTM, ISO and EN test standards used to evaluate performance of materials against chemical hazards. In addition to the test methods stated below, national and international standards are also used to measure the performance of whole body ensemble.

- ASTM F1186-03 Standard Classification System for Chemicals According to Functional Groups.
- ASTM F1001-99a Standard Guide for Selection of Chemicals to Evaluate Protective Clothing Materials.
- ASTM F739-99a Standard Test Method for Resistance of Protective Clothing Materials to Permeation by Liquids or Gases Under Conditions of Continuous Contact.
- ASTM F1383-99a Standard Test Method for Resistance of Protective Clothing Materials to Permeation by Liquids or Gases Under Conditions of Intermittent Contact.
- ASTM F1407-99a Standard Test Method for Resistance of Chemical Protective Clothing Materials to Liquid Permeation-Permeation Cup Method.
- ASTM F1194-99 Standard Guide for Documenting the Results of Chemical Permeation Testing of Materials Used in Protective Clothing.
- ASTM F903-03 (2004) Standard Test Method for Resistance of Materials Used in Protective Clothing to Penetration by Liquids.
- ASTM F2053-00 Standard Guide for Documenting the Results of Airborne Particle Penetration Testing of Protective Clothing Materials.
- ASTM F2130-01 Standard Test Method for Measuring Repellency, Retention, and Penetration of Liquid Pesticide Formulation through Protective Clothing Materials.
- EN 943-1:2002 Protective clothing against liquid and gaseous chemicals, including liquid aerosols and solid particles – Part 1: Performance requirements.
- EN 467:1995 Protective Clothing – Protection against liquid chemicals – Performance requirements for garments providing protection to parts of the body.
- EN374-3:1994 Protective gloves against chemicals and micro-organisms – Part 3: Determination of resistance to permeation by chemicals.

- EN ISO 6529 Protective clothing – Protection against chemicals – Determination of resistance of protective clothing materials to permeation by liquids and gases.
- ISO 6530 (1990) Protective clothing – Protection against liquid chemicals – Determination of resistance of materials to penetration by liquids.
- ISO 13994 (1998) Clothing for protection against liquid chemicals – Determination of the resistance of protective clothing materials to penetration by liquids under pressure.
- ISO 17491(2002) Protective clothing – Protection against gaseous and liquid chemicals – Determination of resistance of protective clothing to penetration by liquids and gases.
- ISO 22608 (2004) Protective clothing – Protection against liquid chemicals – Measurement of repellency, retention, and penetration of liquid pesticide formulations through protective clothing materials.

Permeation data is generally available for commonly used chemical protective clothing materials. Breakthrough time, the time taken by the chemical to pass through the material, is commonly used to measure permeation. In addition, permeation rate, which is the rate at which the chemical moves through the material, is also recorded. Chemical resistance data can be obtained from published guidelines as well as from manufacturers of the chemical protective clothing. 'Guidelines for the Selection of Chemical Protective Clothing' and 'Quick Selection Guide to Chemical Protective Clothing'[15,16] include color-coded recommendations for sixteen commonly used barrier materials. The color codes and corresponding breakthrough times used in the guidelines are green for breakthrough detection time of >4 hours of continuous contact (a >8 is used when the breakthrough is greater than eight hours); yellow with breakthrough detection time between 1–4 hours; red with breakthrough times of less than one hour. White is used when no data is available for the material. Testing the material against the challenge liquid is recommended prior to use. Contact between the material and the chemical may result in degradation of the material or the breakthrough may be a result of the degradation of the material. It is important to note that the data reported in the guidelines is a compilation of published and unpublished permeation test data obtained from various sources. The authors of the publication state that at least 90% of the tests were conducted using ASTM F739-99, Standard Test Method for Resistance of Protective Clothing Materials to Permeation by Liquids or Gases Under Conditions of Continuous Contact.[15] The majority of the data for generic materials is a summary of results from different sources. According to the OSHA technical manual, 'The major limitation for these guidelines are their presentation of recommendations by generic material class. Numerous test results have shown that similar materials from different manufacturers may give widely different performance. That is to say manufacturer A's butyl rubber glove may protect against chemical X, but a butyl glove made by manufacturer B may not'.[1]

A majority of manufacturers test the materials against a battery of liquid and gaseous chemicals, which represent a wide range of commonly used chemicals in different classes. Often the recommended chemicals published by national and international standards organizations are used by the manufacturers as a common group of chemicals. The recommended list of challenge chemicals as well as the chemical class that is published by ASTM International is included in Table 4.1.[17] The performance data, based on tests conducted by independent laboratories, is publicly available through a manufacturer's website. This data is specific to the material manufactured by the company, and so may be different from that obtained from the above mentioned publications. Often information regarding compliance with minimum requirements is also provided by the manufacturers.

However, judicious use of such data is important, as there are numerous protective clothing and chemical combinations. The data from the recommended list should not be a basis for selecting material for protection against chemicals not represented on the list or against a mixture of chemicals. To the extent

Table 4.1 List of Recommended Gaseous and Liquid Test Chemicals (extracted with permission, from F 1001-99a Standard Guide for Selection of Chemicals to Evaluate Protective Clothing Materials, copyright ASTM International, 100 Barr Harbor Drive, West Conshohocken, PA 19462.)

Chemical	Class
Gaseous test chemicals	
Ammonia	Strong base
1,3-Butadiene	Olefin
Chlorine	Inorganic gas
Ethyl oxide	Oxygen heterocyclic gas
Hydrogen chloride	Acid gas
Methyl chloride	Chlorinated hydrocarbon
Liquid test chemicals	
Acetone	Ketone
Acetonitrile	Nitrile
Carbon disulfide	Sulfur-containing organic
Dichloromethane	Chlorinated hydrocarbon
Diethylamine	Amine
Dimethylformamide	Amide
Ethyl acetate	Ester
n-Hexane	Aliphatic hydrocarbon
Methanol	Alcohol
Nitrobenzene	Nitrogen-containing organic
Sodium hydroxide	Inorganic base
Sulfuric acid	Inorganic acid
Tetrachloroethylene	Chlorinated hydrocarbon
Tetrahydrofuran	Oxygen heterocyclic
Toluene	Aromatic hydrocarbon

possible, the selection of textile materials should be based on test data for the potential challenge. Materials manufactured to provide protection against specific chemical hazards require the use of challenge liquids that represent the potential risk. For example, a pharmacist working with drugs for chemotherapy needs protective clothing and accessories that protect the individual from those drugs. A garment that provides general protection against commonly used chemicals may not be recommended for this use. Special testing would have to be conducted for the proposed end use.

4.4.2 Biological protection performance

The selection of protective clothing that provides protection against biological hazards depends on the proposed end use of the clothing. Some of the common end uses are to protect the patient as well as medical professionals who are exposed to blood-borne and other pathogens. Until recently, the majority of biological hazard test methods and guidelines focused on applications in the medical field and for military personnel. With the current bio-terrorism threats, biological protection is important for 'first responders', workers with the potential of exposure to hazardous materials, and the general public. Biological hazards are commonly bacteria, viruses, or toxins that exist as very fine air or liquid-borne particles.

Given below are examples of ASTM International and ISO standards that illustrate the specificity of the garments.

- F1670-03 Standard Test Method for Resistance of Materials Used in Protective Clothing to Penetration by Synthetic Blood.
- F1671-03 Standard Test Method for Resistance of Materials Used in Protective Clothing to Penetration by Blood-Borne Pathogens Using Phi-X174 Bacteriophage Penetration as a Test System.
- F1819-04 Standard Test Method for Resistance of Materials Used in Protective Clothing to Penetration by Synthetic Blood Using a Mechanical Pressure Technique.
- F1862-00a Standard Test Method for Resistance of Medical Face Masks to Penetration by Synthetic Blood (Horizontal Projection of Fixed Volume at a Known Velocity).
- F2100-04 Standard Specification for Performance of Materials Used in Medical Face Masks.
- F2101-01 Standard Test Method for Evaluating the Bacterial Filtration Efficiency (BFE) of Medical Face Mask Materials, Using a Biological Aerosol of *Staphylococcus aureus*.
- F2299-03 Standard Test Method for Determining the Initial Efficiency of Materials Used in Medical Face Masks to Penetration by Particulates Using Latex Spheres.

- F1868-02 Standard Test Method for Thermal and Evaporative Resistance of Clothing Materials Using a Sweating Hot Plate.
- ISO 16603:2004 Clothing for protection against contact with blood and body fluids – Determination of the resistance of protective clothing materials to penetration by blood and body fluids – Test method using synthetic blood.
- ISO 16604:2004 Clothing for protection against contact with blood and body fluids – Determination of resistance of protective clothing materials to penetration by blood-borne pathogens – Test method using Phi-X 174 bacteriophage.

4.4.3   Flame and thermal protection performance

The selection of protective clothing that provides protection against heat and burning objects depends on the proposed end use of the clothing. Some of the common applications are to protect individuals from open flames (including flash fires), molten materials, and radiant heat. In general, materials that have inherent flame-resistant properties are used for flame and thermal protection. Often thermal protective apparel requires the use of multi-layered fabric construction, with each layer performing a specific function. Thus performance is measured for individual layers as well as for the whole garment assembly. Protective apparel is worn by firefighters, wildland firefighters, electrical linesmen, racing- and rally-car drivers, emergency services personnel, and many other groups. Standards and performance specifications on selection, use, care and maintenance of flame and thermal clothing, including fabric, are available through national and international standardization bodies such as NFPA International, ISO, OSHA, and ASTM, CEN/TC 162 (European Technical Committee for Protective Clothing). Due to the very specific requirements, there are many specifications for protective clothing used for protection against thermal and flame hazards. Given below are examples of ASTM, EN, and NFPA standards that illustrate the specificity of the garments.

- F955-03 Standard Test Method for Evaluating Heat Transfer through Materials for Protective Clothing upon Contact with Molten Substances.
- F1002-96 Standard Performance Specification for Protective Clothing for Use by Workers Exposed to Specific Molten Substances and Related Thermal Hazards.
- F1060-01 Standard Test Method for Thermal Protective Performance of Materials for Protective Clothing for Hot Surface Contact.
- F1358-00 Standard Test Method for Effects of Flame Impingement on Materials Used in Protective Clothing Not Designated Primarily for Flame Resistance.
- F1930-00 Standard Test Method for Evaluation of Flame Resistant Clothing for Protection Against Flash Fire Simulations Using an Instrumented Manikin.

- F1939-99a Standard Test Method for Radiant Protective Performance of Flame Resistant Clothing Materials.
- NFPA 70E Standard for Electrical Safety Requirements for Employee Workplaces.
- NFPA 2113 Selection, Care, Use and Maintenance of Flame-Resistant Garments for Protection of Industrial Personnel against Flash Fire.
- NFPA 1971, Standard on Protective Ensemble for Structural Fire Fighting, 2000 edition.
- NFPA 1976, Standard on Protective Ensemble for Proximity Fire Fighting, 2000 edition.
- NFPA 1977, Standard on Protective Clothing and Equipment for Wildland Fire Fighting, 1998 edition.
- NFPA 1991, Standard on Vapor-Protective Ensembles for Hazardous Materials Emergencies, 2000 edition.
- NFPA 1992, Standard on Liquid Splash-Protective Ensembles and Clothing for Hazardous Materials Emergencies, 2000 edition.
- NFPA 1999, Standard on Protective Clothing for Emergency Medical Operations, 1997 edition.
- NFPA 2113, Standard on Selection, Care, Use, and Maintenance of Flame-Resistant Garments for Protection of Industrial Personnel Against Flash Fire, 2001 edition.
- EN 469: European standard for fire fighters' personal protective equipment.
- EN 531: European standard for heat and flame protective clothing for industrial workers.
- EN 659: European standard for fire fighters' gloves.
- ISO 17492:2003 Clothing for protection against heat and flame – Determination of heat transmission on exposure to both flame and radiant heat.
- ISO 2801:1998 Clothing for protection against heat and flame – General recommendations for selection, care and use of protective clothing.
- ISO 6942:2002 Protective clothing – Protection against heat and fire – Method of test: Evaluation of materials and material assemblies when exposed to a source of radiant heat.
- ISO 8194:1987 Radiation protection – Clothing for protection against radioactive contamination – Design, selection, testing and use.
- ISO 9150:1988 Protective clothing – Determination of behaviour of materials on impact of small splashes of molten metal.
- ISO 9151:1995 Protective clothing against heat and flame – Determination of heat transmission on exposure to flame.
- ISO 9185:1990 Protective clothing – Assessment of resistance of materials to molten metal splash.
- ISO 11612:1998 Clothing for protection against heat and flame – Test methods and performance requirements for heat-protective clothing.

- ISO 11613:1999 Protective clothing for firefighters – Laboratory test methods and performance requirements.
- ISO 12127:1996 Clothing for protection against heat and flame – Determination of contact heat transmission through protective clothing or constituent materials.
- ISO 14460:1999 Protective clothing for automobile racing drivers – Protection against heat and flame – Performance requirements and test methods.
- ISO 15025:2000 Protective clothing – Protection against heat and flame – Method of test for limited flame spread.
- ISO 15383:2001 Protective gloves for firefighters – Laboratory test methods and performance requirements.
- ISO 15384:2003 Protective clothing for firefighters – Laboratory test methods and performance requirements for wildland firefighting clothing.
- ISO 15538:2001 Protective clothing for firefighters – Laboratory test methods and performance requirements for protective clothing with a reflective outer surface.

4.4.4 Mechanical protection performance

Protective clothing designed to provide protection from mechanical hazards ranges from cut protection for chainsaw workers to body amour designed for bullet, stab, puncture, or impact protection for police and military personnel. Ballistics tests and performance standards, developed by the National Institute of Justice (NIJ), are examples of the specialized ballistic tests that are used to determine the performance of the materials (6). Numerous standards are available to measure the cut resistance of materials used to protect individuals from chainsaw cuts. Given below is a list of test standards and performance specifications developed by ASTM International for cut resistance.

- F1790-04 Standard Test Method for Measuring Cut Resistance of Materials Used in Protective Clothing.
- F1414-99 Standard Test Method for Measurement of Cut Resistance to Chain Saw in Lower Body (Legs) Protective Clothing.
- F1458-98 Standard Test Method for Measurement of Cut Resistance to Chain Saw of Foot Protective Devices.
- F1818-97(2003) Standard Specification for Foot Protection for Chain Saw Users.
- F1897-98 Standard Specification for Leg Protection for Chain Saw Users.
- F1342-91(1996)e2 Standard Test Method for Protective Clothing Material Resistance to Puncture.
- F1414-99 Standard Test Method for Measurement of Cut Resistance to Chain Saw in Lower Body (Legs) Protective Clothing.

- ISO 11393-1(1998) Protective clothing for users of hand-held chain-saws – Part 1: Test rig driven by a flywheel for testing resistance to cutting by a chain-saw.
- ISO 11393-2 (1999) Protective clothing for users of hand-held chain-saws – Part 2: Test methods and performance requirements for leg protectors.
- ISO 11393-5(2001) Protective clothing for users of hand-held chain-saws – Part 5: Test methods and performance requirements for protective gaiters.
- ISO 13995:2000 Protective clothing - Mechanical properties – Test method for the determination of the resistance to puncture and dynamic tearing of materials.
- ISO 13996:1999 Protective clothing – Mechanical properties – Determination of resistance to puncture.
- ISO 13997:1999 Protective clothing – Mechanical properties – Determination of resistance to cutting by sharp objects.
- ISO 13998:2003 Protective clothing – Aprons, trousers and vests protecting against cuts and stabs by hand knives.
- ISO 14877:2002 Protective clothing for abrasive blasting operations using granular abrasives.
- NIJ Standard-0101.03 – Ballistic Resistance of Police Body Armor.
- NIJ Standard-0101.04 – Ballistic Resistance of Personal Body Armor.
- NIJ Standard-0115.00 – Stab Resistance of Personal Body Armor.

Knowledge of how the fabric provides the protection as well as the limitations of the material is important. Materials may have similar barrier performance; however, the manner in which the protection is provided may vary considerably. One may provide protection due to the inherent fiber properties, whereas another from a finish applied to the fabric. It would be important to obtain additional information on the durability of the finish to determine the level of protection that will be provided over the life of the garment. The durability of the finish may be a limitation for the material.

The above discussion provides a sense of the level of complex analysis that must be applied for each type of hazard (chemical, biological, physical, radiological, or thermal).

After such an analysis, a list is prepared of materials that meet the minimum protection requirements. This list is then used to select potential materials based on other major factors discussed in the following section. Some manufacturers provide online interactive software to assist in the selection of suitable clothing. DuPont™ SafeSPEC™ [10] is an example of an interactive tool that assists the user with hazard assessment and selection of apparel. As stated in the software, the responses are based on the information provided by the user. Thus, it is crucial for the individual entering the information to provide accurate information. With systems that are manufacturer specific, it is difficult to compare products from different manufacturers.

4.5   Selection of materials based on other major factors

For protective clothing, protection provided by the material is the primary factor in its selection. In addition, there are several other factors that have to be considered while selecting the materials. As with any other selection processes, the various factors have to be weighted in order to select the appropriate material. Impact of the use of PPE on job performance is an important factor to be considered. In addition, factors such as comfort, durability, maintenance, cost, and design considerations may apply. The major selection categories with applicable sub-categories are discussed below.

4.5.1   Job performance

Problems with dexterity, added weight and bulk of the PPE, and heat stress are examples of factors that may affect job performance. The type of burden the use of material will pose is dependent on the part of the body that is covered with the textile material as well as the type of job. For example, the weight of the protective gear may be a factor for firefighters fighting wildland fires; whereas, dexterity may be crucial for protective gloves worn by surgeons and other health professionals. Dexterity is an important factor while selecting glove materials for various end uses. Test standards have been developed to measure the dexterity of gloves. F2010-00 Standard Test Method for Evaluation of Glove Effects on Wearer Hand Dexterity Using a Modified Pegboard Test is an example of a test standard used to measure dexterity of the glove. The suitability of standards to assess dexterity of the material for a specific job should be considered carefully as the type of hand movements required to do the job may not be similar to those used in the standard.

In situations where the garment has to be worn on a regular basis for an extended period of time, bulk and weight often become major factors in the selection of protective clothing materials. The problem is compounded by the use of multiple layer clothing and additional equipment that has to be carried to perform the job. Protective clothing materials for military personnel and firefighters fighting wildland fires are examples of end uses where weight and bulk of protective clothing and equipment would be a factor in selection. Appropriate body armor worn by law enforcement personnel is an excellent example of an end use where bulk, weight, and flexibility of the body armor are very important to material selection. Given below is information on selection of body armor published as part of the body armor standard[18] (see also Chapter 19 on ballistic protection):

> Type I body armor, which was issued during the NIJ demonstration project, is the minimum level of protection that any officer should have, and is totally suitable for full-time wear. A number of departments desiring more than

minimum protection wear type II-A armor, which has been found sufficiently comfortable for full-time wear where the threat warrants it, particularly for those departments that use lower velocity 357 Magnum service weapons. Type II armor, heavier and more bulky than type II-A, is worn full time by some departments, but may not be considered suitable for full-time use in hot, humid climates. Type III-A armor, which provided the highest level of protection available as soft body armor, is generally considered to be unsuitable for routine wear, however, individuals confronted with a terrorist threat may be willing to tolerate the weight and bulk of such armor while on duty. Type III and IV armor are clearly intended for use only in tactical situations when the threat warrants such protection.

Job performance and comfort are closely related. Often discomfort affects the ability of the individual to perform his/her job efficiently and effectively. Comfort-related examples are included in the next section.

4.5.2 Comfort

In many cases the use of PPE may negatively impact the comfort of the user. Reduced comfort results in lower user compliance and potential for heat stress injuries. Environmental or climatic conditions as well as the type and level of activity are major factors that contribute to the potential for heat stress. Thus, adequate comfort is a primary concern in hot and humid climates, as well as in environments where the individual is exposed to a heat source. According to the guidelines published by NIOSH:[19]

> ... the frequency of accidents, in general appears to be higher in hot environments than in more moderate environmental conditions. One reason is that working in a hot environment lowers the mental alertness and physical performance of an individual. Increased body temperature and physical discomfort promote irritability, anger, and other emotional states which sometimes cause workers to overlook safety procedures or to divert attention from hazardous tasks.

To the extent possible, material should be carefully selected to balance the protection and comfort properties. In end uses where suitable protective clothing materials cannot provide the required protection and comfort, solutions such as the use of cooling devices, or use of work and rest cycles reduce the risk of heat stress. For example, cooling devices are used to reduce the potential for heat stress for firefighters. They are generally designed to provide the required comfort for a limited period of time. Willingness to use protective clothing materials that provide protection but are uncomfortable depends on factors such as awareness of risk and consequences for non-compliance. An individual may be willing to bear the discomfort in order to remain protected or comply with the requirements if he/she sees the benefit in wearing the protective clothing.

The measurement of comfort is very complex as many factors contribute to the comfort provided by a garment to an individual. A variety of tests are conducted to measure these properties. Material characteristics such as air permeability and moisture vapor transmission are used to quantify comfort characteristics. However, these do not take into consideration factors such as heat generated as a result of physical activity. Human subject studies that include physiological monitoring as well as self-reported questionnaires are also used to measure comfort. Factors such as perceived comfort, physical condition of the individuals, and duration of the test are some of the factors that affect the results of human subject studies. A newer method measures the micro-environment that is produced between the garment and the skin as an indicator of comfort. Thermal, sweating, and movable manikins are also being used to measure comfort.

4.5.3   Cost

Cost is often an important concern for the individuals procuring the garments. It is not unusual for the procurement department, especially government departments, to purchase PPE based on the cost, using competitive tendering processes. The type of garment plays a major role in determining the cost of the garment. For some end uses, limited use or reusable PPE may be selected. In these situations the added cost of cleaning, maintaining and disposing of PPE should be considered while selecting the materials. Often the initial cost of a limited use garment may be fairly low, but the added cost of purchasing new garments and disposal of the contaminated garments might outweigh the initial cost benefit. On the other hand, the cleaning cost of reusable garments may justify the use of limited-use garments.

4.5.4   Durability

Durability of the material determines the performance or the wear life of the garment. The extent to which durability is a concern is dependent on the proposed end use. The material should be able to withstand the normal wear and tear expected for the job that is being performed. Physical characteristics that are commonly measured and reported for protective clothing materials are tensile strength, tearing strength, bursting strength, abrasion resistance, puncture resistance, and cut resistance. In addition fabric weight, flexibility, flammability, and degradation due to environmental conditions may also be measured. Often the performance specifications include minimum require-ments for limited use and reusable materials. For materials in which a finish is applied to provide protection, determination of the durability of the finish is very important. The protective properties and thus the wear life of these materials may be reduced.

4.5.5   Use, care, and maintenance

Intended use, decontamination, maintenance, storage, and disposal factors should be considered prior to selecting a material. The impact the material will have on the use, care, and maintenance of the garment needs to be considered. The compatibility of the various materials in the manufacture of multi-component materials should also be considered.

4.5.6   Cultural factors

In addition to the above factors, selection of protective clothing materials may be affected by cultural factors, especially for comparatively lower risk applications. Thus, it is difficult to develop protective clothing materials that are accepted globally. It is also important for the individual responsible for making decisions to have an understanding of the culture. Decisions made by individuals solely on performance specifications have resulted in non-compliance due to low user acceptance of the recommended PPE for cultural reasons. User acceptance of protective clothing worn by pesticide applicators is used as an example of how factors such as color and materials can be important in the selection of protective clothing. A report published on the Safe Use project in Kenya explains:[20]

> Persuading farmers to use protective clothing was one of the biggest challenges. They see it as uncomfortable and expensive. Locally designed and manufactured clothing is addressing these problems. More specifically, women in Africa have a cultural aversion to being seen in trousers. Kenyan women were therefore asked to design their own protective clothing and the results have been widely publicised.

4.6   Future trends

The current trend of standards organizations focusing on the development of performance based specifications will continue. Comprehensive specifications will include information on selection, use, care, and maintenance (SUCM). The development of these specifications will assist in the selection of appropriate protective clothing materials. The need to protect individuals from multiple hazards such as chemical, biological, and thermal will require streamlining of the selection process. The use of online systems, databases, and knowledge bases will continue to expand. In the future online systems pertaining to protective clothing will, in all probability, include global mega-knowledge bases with built-in information on compliance requirements. Selection of protective clothing materials based on performance based specifications will be an integral part of the system. In building these systems, it will be important to ensure that the system is used to assist but not replace the decision makers. The individual selecting the materials(s) should be involved with each step of the selection process. An online system entitled 'Work and Protective Clothing for Agricultural Workers' is being used as an example to illustrate the use of databases and the Internet as tools for the selection of protective clothing materials.[21] A brief description of this system is included in section 4.6.1.

4.6.1   Sample online system

The system provides a structured systematic process for data collection, analysis, and dissemination. Online data entry forms are used to enter the raw data and material information. The raw data is used to calculate the values in accordance with applicable standards. These are stored for access through easy to use dropdown menus.

4.6.2   Data entry and calculations

The first step in the data entry process is the auto generation of a fabric code based on information regarding fiber content, country, and fabric construction (Fig. 4.2). The next step is entry of information regarding type of garment, cost, source, etc. (Fig. 4.3). Then the fabric is tested and the raw data entered by the operator using online forms. The hard copies of the raw data are retained for verification. The system uses the raw data to calculate results such as mean and standard deviation. These results are stored in a table for later use.

4.6.3   Statistical model

Data from over sixty fabrics in the database was used to develop a statistical model to predict percent pesticide penetration through the material. Data was analyzed to identify the key factors that affect pesticide penetration. Currently

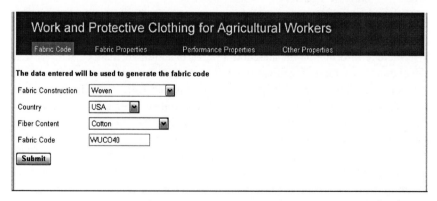

4.2 Screen to auto generate fabric codes (Source: Work and Protective Clothing for Agricultural Workers, University of Maryland Eastern Shore).

Fabric Weight Data

Enter the following information:

	Unit	Specimen #1	Specimen #2	Specimen #3
Width of the fabric	mm ▼			
Length of the specimen	mm ▼			
Width of the specimen	mm ▼			
Mass of specimen	gms ▼			

[Submit]

4.3 Data entry screen for fabric weight (Source: Work and Protective Clothing for Agricultural Workers, University of Maryland Eastern Shore).

the scope of the statistical model is limited to estimating pesticide penetration of three formulations through woven fabrics (Fig. 4.4).

4.6.4 Data dissemination

Easy to use dropdown menus are used to disseminate textile material and penetration data sorted into various categories (Fig. 4.5). Individuals can also use the system to customize their search by defining the search criteria (Fig. 4.6). Once performance specifications have been established, individuals will be able to select materials that meet the performance specifications.

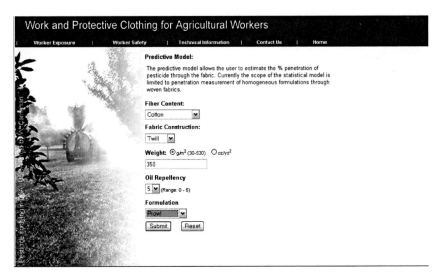

4.4 Predictive model screen (Source: Work and Protective Clothing for Agricultural Workers, University of Maryland Eastern Shore).

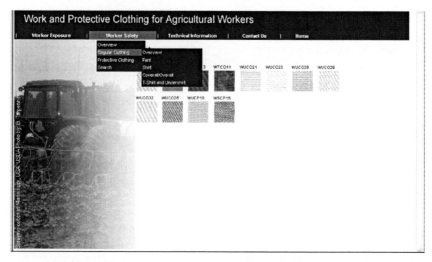

4.5 Dropdown menu for accessing information (Source: Work and Protective Clothing for Agricultural Workers, University of Maryland Eastern Shore).

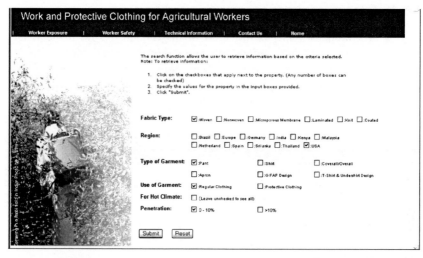

4.6 Screen to enter search criteria (Source: Work and Protective Clothing for Agricultural Workers, University of Maryland Eastern Shore).

4.6.5   Suggestions model for future systems

The flowchart that outlines the steps involved in the selection process is given in Fig. 4.7. The work scenario(s) and hazard(s) will be entered by the user through the online input screen. Based on the information entered, the system will list relevant standards. If relevant standards are available, the user will select the

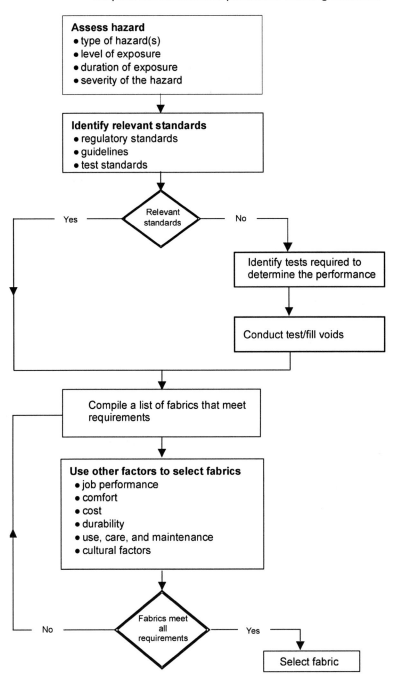

4.7 Recommended steps in the selection of textiles for protective clothing.

standard and proceed to the next step. If no relevant standards are available, the user will have the option to use the system to select the test methods, and/or enter data, and then move to the next step (see Fig. 4.7). In the next step, the user will enter the information regarding other factors such as durability, cost, etc., that are important in the decision making process. This information will be used to identify the fabrics that meet all requirements. If fabrics meet the requirements, final selection will be made. If fabrics do not meet all requirements, the requirements will have to be adjusted prior to final selection.

4.7   Sources of further information and advice

Book of Papers, 2nd European Conference on Protective Clothing (ECPC) and NOKOBETEF 7, Challenges for Protective Clothing, Montreux, Switzerland, 21–24 May 2003.

Guidance Manual for Selecting Protective Clothing for Agricultural Pesticides Operations, Risk Reduction Engineering Laboratory, Office of Research and Development, U.S. Environmental Protection Agency, Cincinnati, Ohio 45268, USA, 1991.

Kohloff, F. H., 1999, PPE Guidelines for Melting and Pouring Operations, *Modern Casting* 89, (5) pp 57–59.

Protective Clothing Systems and Materials, M. Raheel (ed.), Marcel Dekker 1994.

STP 1237 – *Performance of Protective Clothing*, Fifth Volume, J. S. Johnson and S. Z. Mansdorf (eds), ASTM, 100 Barr Harbor Drive, West Conshohocken, PA 19428, USA.

STP 1273 – *Performance of Protective Clothing*, Sixth Volume, J.O. Stull and A.D. Schwope (eds), ASTM, 100 Barr Harbor Drive, West Conshohocken, PA 19428, USA.

STP 1386 – *Performance of Protective Clothing: Issues and Priorities for the 21st Century*, Seventh Volume, C. N. Nelson and N.W. Henry (eds), ASTM, 100 Barr Harbor Drive, West Conshohocken, PA 19428, USA.

Stull, J.O., 2004, A Suggested Approach to the Selection of Chemical and Biological Protective Clothing – Meeting Industry Needs for Protection Against a Variety of Hazards, *International Journal of Occupational Safety and Ergonomics*, Vol. 10, No. 3, pp. 271–290.

Sun, G., H.S. Yoo, X.S. Zhang, and N. Pan, 2000, Radiant protective and transport properties of fabrics used by wildland firefighters, *Textile Research Journal*, Vol. 70, No. 7, pp. 567–573.

Walsh, D. L., M. R. Schwerin, R. W. Kisielewski, R. M. Kotz, M. P. Chaput, G. W. Varney, and T. M , 2004, Abrasion Resistance of Medical Glove Materials, *Journal of Biomedical Materials Research*, Vol. 68B, No. 1, pp. 81–87.

Whitford, F., Stone, J., and MacMillian, T., Chapter 11 – Personal Protective Equipment: Selection, Care, and Use, Whitford, F. (ed.), *The Complete Book of Pesticide Management*, John Wiley & Sons, Inc., 2001.

4.8   References

1. Protective Clothing Selection Factors, Section VIII: Chapter 1, Chemical Protective Clothing, OSHA Technical Manual, Occupational Safety & Health Administration, U.S. Department of Labor, www.osha.gov.
2. ASTM Standards Vol. 11.03, Standards on Protective Clothing, ASTM International. 100 Barr Harbor Drive, West Conshohocken, PA 19428, USA, www.astm.org.
3. ISO – International Standards Organization, Case Postale 56, CH-1211, Geneva 20, Switzerland.
4. CEN – European Committee for Standardization, Central Secretariat, rue de Strassart 36, B-1050 Brussels, Belgium.
5. NFPA (National Fire Protection Association), 1 Batterymarch Park, P.O. Box 9101, Quincy, MA 02269-9101, USA.
6. Technology Assessment Program, National Institute of Justice, U.S. Department of Justice, Washington, DC 20531. Publications available through National Criminal Justice Reference Service.
7. NFPA – National Fire Protection Association, NFPA 2112 (2001), Standard on Flame-Resistant Garments for Protection of Industrial Personnel Against Flash Fire; NFPA 2113 (2001), Standard on Selection, Care, Use, and Maintenance of Flame-Resistant Garments for Protection of Industrial Personnel Against Flash Fire, 1 Batterymarch Park, P.O. Box 9101, Quincy, MA 02269-9101.
8. Standard # 1910.1030 (d)(3) Personal Protective Equipment – Bloodborne pathogens, Toxic and Hazardous Substances, Occupational Safety and Health Standards, OSHA Occupational Safety & Health Administration, U.S. Department of Labor.
9. Shaw, A. and Abbi, R (2002), Performance of Work and Protective Clothing Worn by Agricultural Workers in Different Countries, XVIth World Congress on Safety and Health at Work, Vienna, Austria, May 2002.
10. DuPont Nonwovens, http://www.personalprotection.dupont.com/
11. Kimberly-Clark Corp, http://www.kimberly-clark.com/
12. Kappler Safety Group, http://www.kappler.com/home.asp
13. Ansell Edmont Industrial, Inc., http://www.ansell-edmont.com/
14. Shaw, A., Cohen, E., Hinz T., and Herzig B. (2003), Laboratory Test Methods to Measure Repellency, Retention and Penetration of Liquid Pesticides through Protective Clothing: Part I Comparison of Three Test Methods, *Textile Research Journal*.
15. Forsberg, K. and Mansdorf, S.Z. (1989), *Quick Selection Guide to Chemical Protective Clothing*. Van Nostrand-Reinhold, New York.
16. Forsberg, K. and Keith, L.H. (1989), *Chemical Protective Clothing Performance Index Book*. John Wiley & Sons, New York.

17. ASTM, F1001-99a *Standard Guide for Selection of Chemicals to Evaluate Protective Clothing Materials*, Vol. 11.03, ASTM International (*www.astm.org*).
18. NIJ Standard 0101.03 Ballistic Resistance of Police Body Armor Appendix A – Selection Body Armor Selection, National Institute of Justice Technology Assessment Program, U.S. Department of Justice.
19. *Working in Hot Environments*, DHHS (NIOSH) Publication No. 86-112, National Institute for Occupational Safety and Health, 4676 Columbia Parkway, Cincinnati, Ohio 45226.
20. Deasy, N. and Rigby, H. (eds) (1998), *Safe Use Pilot Project: Guatemala, Kenya, Thailand*, Croplife International publication, Avenue Louise 143, B1050, Brussels, Belgium. *www.croplife.org/library/attachments/9e4382cc-3d6d-4dd8-b9f5-83579b991a90/9/Safe_Use_pilot_projects.pdf*
21. Work and Protective Clothing for Agricultural Workers, www.umes.edu/ppe.

Fibres and fabrics for protective textiles

J W S H E A R L E, Consultant, UK

5.1   Introduction

5.1.1   Review of fibre types

A full classification of nearly 100 textile fibre types can be found in *Textile terms and definitions* (Denton and Daniels, 2002). With a few exceptions, they can be roughly divided into six major categories in order of mechanical strength:

- highly extensible, elastomeric fibres, e.g. Lycra
- very brittle fibres, e.g., rock wool
- widely used, natural and regenerated fibres, e.g., cotton, wool, flax, rayon
- tough fibres, moderately strong and extensible, e.g., silk, nylon, polyester
- moderately strong, inextensible fibres, e.g., glass, ceramic fibres
- high-modulus, high-tenacity fibres, e.g., carbon, aramid, High Modulus Poly Ethylene (HMPE).

The step change in mechanical properties from first- to second-generation synthetic fibres is illustrated in Fig. 5.1.

At some time or another, almost every fibre type has been used for some form of protection – though, for the first two categories, this may be stretching the definition of protection. Elastomeric fibres protect by firmly holding the body shape; they might also be incorporated in some protective clothing. The weak mineral fibres are used for thermal insulation. Protection from the elements, sun, rain and snow, was the first use of natural fibres over 10,000 years ago. Cotton and wool, together with rayon, nylon, polyester and other general textile fibres, continue to be the fibres used for protection from the weather and other environmental stresses. Particular surface treatments may be needed to make them effective. Brief accounts of these fibres are given in the next section.

In ancient times, a variety of natural fibres were used in textile armour. Today, the main use of the weaker textile fibres for this type of protection is to provide comfort in a blend with high-performance fibres. For example, *Canesis* has recently developed a low-cost, comfortable vest, consisting of wool and a

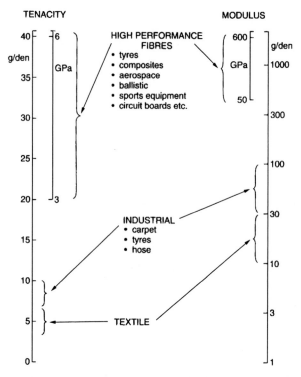

5.1 Step change in strength and stiffness from first-generation to second-generation manufactured fibres. The natural fibres fall in the textile group. (After Mukhopadhyay, 1993.)

high-performance filament, to protect prison officers against attacks with sharpened tooth-brushes (Anon., 2004). The tough fibres give protection through high energy absorption. In the context of this book, the major emphasis will be on the high-performance fibres developed in the latter part of the 20th century to give high strength and stiffness or to give high chemical or thermal protection. Other aspects of protection depend on fibre size and the interactions of fibre surfaces. Micro- and nano-fibres can form structures with small pore sizes to stop particle penetration and high surface areas for absorption of chemicals.

There are recent and continuing developments of 'smart' fibres, with a variety of special functions (Hongu *et al.*, 2004; Tao, 2001, 2004). Some change with a change in the environment. Others have particular chemical, physical, biological, or medical properties. Where relevant, such specialised fibres will be mentioned in other chapters. For electrostatic protection, fine metal wires, similar in diameter to textile fibres, may be incorporated in fabrics. Alternatively, manufactured textile fibres can be loaded with carbon black or other conducting material. Carbon fibres are the only conducting fibres with the high mechanical performance that justifies coverage in more detail (see section 5.3). Test methods for fibres are covered by Saville (1999). Other information on fibres relevant to protective textiles is given by Adanur (1995) and Horrocks and Anand (2000).

5.1.2   Units for fineness and mechanical properties

The fineness of fibres is most conveniently expressed in terms of their linear density (mass/length) and the preferred unit is tex, which is g/km (10^6 times the strict SI unit of kg/m). Area of cross-section is less easy to measure and in fibre assemblies is ill-defined because of the spaces between fibres. It is therefore better to express mechanical properties on a mass basis and not on the area basis commonly used in physics and engineering. The unit for stiffness (modulus) and strength (tenacity) is N/tex, which equals kJ/g and $(km/s)^2$. The latter two units are interesting because they relate to two factors that give good ballistic impact resistance: the specific work of rupture (energy to break) is expressed in kJ/g; the wave velocity is the square root of modulus in N/tex or $(km/s)^2$. Conversion to conventional units of stress is through multiplication by density: GPa = (N/tex) × (g/cm^3). Details of the many alternative units are given in an Appendix to *High-performance fibres* (Hearle, 2000). Two convenient bench-marks are that high-tenacity forms of nylon and polyester fibres approach 1 N/tex in strength and that the high-modulus variants of aramid fibres have a stiffness of around 100 N/tex.

5.1.3   Cellulosic, protein and synthetic 'textile' fibres

Cotton and other fibres from plants are composed of cellulose, which is laid down in helical forms in plant cells in the seed coats, stems or leaves of plants. The helix angle of 21°, the helix reversals and the convolutions in cotton give it a break extension of about 7% and a tenacity of 0.1 to 0.5 N/tex. Other cellulose fibres, which stiffen the leaves or stems of plants, have a lower helix angle and are stronger and less extensible.

Rayon and acetate fibres (see Woodings, 2001) can be manufactured by extruding solutions of cellulose or cellulose derivatives and solidifying in liquid baths (wet-spinning) or by evaporation (dry-spinning), with stretch orienting the long-chain molecules. The commercial fibres have strengths similar to cotton, but lower modulus and higher break extensions. Cellulose will form liquid-crystals in solution in phosphoric acid, and following aramid production technology, a cellulose fibre with a tenacity of 1.1 N/tex and a modulus of 30 N/tex has been produced (see Hearle, 2000, p. 3). Higher values would be expected if this fibre was commercialized, but because the ring structures have large side groups and are joined by single >C-O-C< bonds, the cellulose molecule is not ideal for making high-performance fibres.

Wool and hair fibres have an extremely complicated chemical and physical structure (see Simpson and Crawshaw, 2002). They are formed from a collection of protein molecules with many different side-groups, which are packed in assemblies with features at a number of structural levels. Wool is comparatively weak, highly extensible and crimped, so that it makes soft and bulky fabrics. Silk (from silk-worms) has a simpler structure and is a tough fibre comparable in properties to nylon. The versions of nylon, polyester and polypropylene fibres used in clothing and related products have strengths of about 0.5 N/tex or less and break extensions of 20% or more. Information on the more highly oriented types used in technical applications is given later.

5.1.4   Spider silk

Spiders produce a range of filaments from aqueous solution with properties adapted to different functions. The drag-lines of some spiders have strengths comparable to the synthetic high-performance fibres, but much greater break extensions. The resulting extremely high energy-to-break would make them ideal fibres for ballistic protection. Unfortunately, spider silk cannot be farmed. This has led to attempts to mimic its production artificially.

There are two problems. The first is to produce the spider-silk protein. It is reported that genetic engineering has achieved this through bacteria or goat's milk. The second step, mistakenly thought to be easier, is to extrude fibres with the right physical fine structure. All attempts to mimic spiders have failed. The extrusion methods that have been tried have produced weak fibres.

5.1.5 Comparative properties

Table 5.1 lists properties of fibres. The emphasis is on the fibres described in sections 5.2–5.7, for which mechanical properties are most relevant. The entries for ceramic, chemically resistant and thermally resistant fibres cover a variety of chemical types and necessarily have wide ranges of values. Figure 5.2 shows the superior strength and stiffness and low break extension of the high-performance fibres.

5.2   More extensible fibres

5.2.1   Energy absorption

There are situations in which protection from large energy shocks is required. An example is the safety of a falling rock-climber. Energy-to-break, or work of rupture, is approximately given by $\frac{1}{2}$(strength $\times$ break extension). However, the product is generally lower for the high-tenacity fibres than for the more extensible fibres. In addition, the forces developed must not be too large, if the

Table 5.1 Comparative fibre properties, from Hearle (2000), TTI and Noble Denton (1999), Morton and Hearle (1993), and Ford (1966). Note that these are typical values and variant types may have different values. Where ranges are given for linear polymer fibres, high tenacity and modulus goes with low break extension

Fibre	Density g/cm³	Moisture % at 65% rh	Melting point °C	Tenacity N/tex	Modulus N/tex	Work to break J/g	Break ext %
Cotton	1.52	7	185*	0.2–0.45	4–7.5	5–15	6–7
Flax	1.52	7	185*	0.54	18	8	3
Wool	1.31	15	100**/300*	0.1–0.15	2–3	25–40	30–40
Silk	1.34	10	175*	0.38	7.5	60	23
Rayon	1.49	13	185*	0.2–0.4	5–13	10–30	7–30
Nylon	1.14	4	260***	0.35–0.8	1–5	60–100	15–25
Polyester	1.39	0.4	258	0.45–0.8	7–13	20–120	9–13
Polypropylene	0.91	0	165	0.6	6	70	17
Para-aramid	1.44	5	550*	1.7–2.3	50–115	10–40	1.5–4.5
Meta-aramid	1.46	5	415*	0.49	7.5	85	35
TLCP (*Vectran*)	1.4	<0.1	330	2–2.5	45–60	15	3.5
HMPE	0.97	0	150	2.5–3.7	75–120	45–70	2.9–3.8
PBO	1.56	0	650*	3.8–4.8	180	30–90	1.5–3.7
Carbon	1.8–2.1	0	>2500	0.4–3.9	20–370	4–70	0.2–2.1
Glass	2.5	0	1000–12000*****	1–2.5	50–60	10–70	1.8–5.4
Ceramic	2.4–4.1	0	>1000	0.3–0.95	55–100	0.5–9	0.3–1.5
Chemical res'tant	1.3–1.6	0–0.5	170–375*****	0–0.65	0.5–5	15–80	15–35
Thermal res'tant	1.25–1.45	5–15	200–500*****	0.1–1.3	2.5–9.5	10–45	8–50

* decomposes (chars); ** softens; *** for nylon 66, nylon6 at 216 °C; **** liquidus temperature; ***** various limiting temperatures

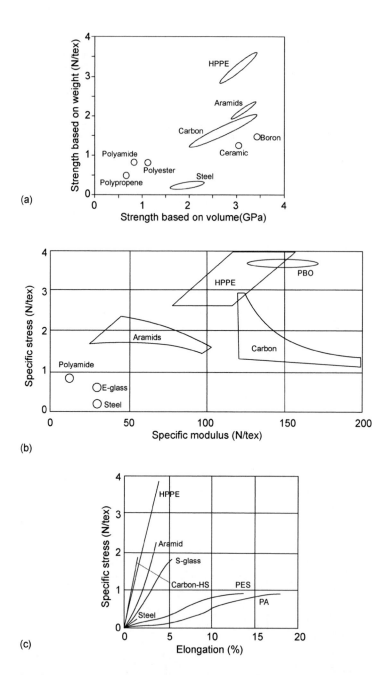

5.2 Comparative mechanical properties, from Van Dingenen (2000). (a) Strength on a weight and a volume basis. (b) Strength and stiffness on a weight basis. (c) Stress-strain curves. PA is polyamide (nylon); PES is polyester.

climber is not to be hurt by arresting the fall. Consequently, the best available combination is moderately high strength, low modulus and high break extension. This requirement is met by industrial forms of nylon, which has a lower modulus, and polyester.

A related application is in deepwater oil-rigs, providing protection against loss of the mooring. The mechanics is mainly extension driven due to the changing height of waves, and polyester ropes have more suitable properties than the higher-strength fibres.

5.2.2 Nylons (polyamides)

Nylon fibres, which are produced by melt spinning, are polyamides, with a molecular alternation of $-CH_2-$ groups (5 and 5 in nylon 6; 4 and 6 in nylon 66) and -CO.NH- groups. The $-CH_2-$ sequences provide flexibility, and the -CO.NH-groups form hydrogen bonds, which hold the long molecules together in a mixture of crystalline and amorphous regions. Following polymerisation, the melt is extruded and then drawn to orient the molecules. An increase of orientation increases strength and modulus and decreases break extension. Most filament yarns are produced in the range of 10 to 1,000 filaments; for staple fibre production, large tows are extruded and later cut into short fibres. For general textile purposes, a lower modulus is preferred. For industrial yarns, drawing and heat treatments are selected to maximise strength and modulus. Finishes will be applied to fit the particular end-uses. For example, some applications need high friction and others need low friction; this choice is also influenced by whether fibre-to-fibre contact or fibre-to-metal is relevant and by the speed of sliding. The major market for industrial yarns is in tyre cords, but similar fibres can be produced for protective applications.

Properties of nylon yarns are listed in Table 5.1 and a typical stress-strain curve is shown in Fig. 5.2(c). Nylon fibres absorb water and this reduces strength and modulus. Nylon 66 melts at 260 °C (nylon 6 at 220 °C), but the mechanical properties deteriorate above the glass-transition temperature, which is around 100 °C dry and about 20 °C wet. Nylon 66 fibres start to stick together at about 220 °C. In addition to its high work of rupture, a major advantage of nylon is its good elastic recovery from high strains. This makes it the fibre of choice for climbing ropes.

5.2.3 Polyesters

The common polyester, which is now the dominant general-purpose fibre, is poly[ethylene terephthalate] (PET or 2GT). Chemically the polymer molecules are an alternation of $-CO-O-CH_2-CH_2-O-CO-$ groups and benzene rings. As in nylon the aliphatic sequence gives flexibility, but the stiffness and phenolic interactions of the rings gives a higher modulus and higher glass transition temperature than nylon. The melting point is similar to nylon 66. The small moisture absorption of polyester fibres is on the surface and the properties are little affected. The manufacture of polyester fibres is by melt spinning and drawing. Except when the modulus and other properties match the specific engineering demands, as in ropes for deep-water moorings, polyester fibres may only be a choice for protective applications when low cost and durability are more important criteria than maximum performance.

Other polyester fibres are available. Poly[trimethylene-terephthalate] (3GT) and poly[butylene-terephthalate] (4GT) contain three and four -CH_2- groups in the aliphatic sequence. This lowers the modulus and gives mechanical properties closer to nylon. Poly[ethylene-naphthalate] (PEN) is of more potential interest for mechanical protection. The double naphthalate ring increases the modulus and the melting temperature.

5.3   Carbon fibres

5.3.1 'Graphitic' fibres

When suitable organic fibres are heated to high temperatures, the residual carbon atoms form hexagonally bonded sheets. The extreme form of this structure consists of the perfectly formed and regularly packed sheets of graphite. However such a structure would be useless in fibres, because the separate planes would fall apart transversely. Fortunately, the fibre processing gives an imperfect graphitic structure. The locally parallel planes are irregularly stacked as turbostratic carbon. Over larger distances, they are bent and interlocked; molecular defects may link one plane to another. There are many variant possibilities for these partially ordered structures, which have been represented by many speculative drawings in the literature. The structure and hence the properties depend on the starting material and the details of the thermal processing. Currently, the two starting materials for commercial production are polyacrylonitrile fibres [PAN; -CH_2-CH(CN)-] and pitch.

The PAN fibres that are preferred for carbon fibre production are produced by wet-spinning from a suitable solvent into a coagulating bath, and are optimised in molecular weight, orientation, purity, and comonomer content. Moderate heating causes the pendant nitrile groups to cyclize and then added oxygen renders the fibres infusible. Increasing the temperature to over 1000 °C drives off the non-carbon atoms and leads to the formation of aromatic sheets. Final heating in an inert atmosphere, usually at 2,000–2,500 °C, increases the structural perfection. The processes are slow, so that large ovens and long residence times are needed.

Low-strength carbon fibres can be made from isotropic pitch. High-strength fibres are made from mesophase pitch, produced from certain petroleum or coaltar fractions. Polymerised molecules consist of linked hexagonal rings, and, as heating progresses, large, liquid-crystalline sheets are formed. Manufacture of these pitch-based carbon fibres is by melt-spinning of mesophase pitch, followed by drawing and heat treatments, which eliminate hydrogen.

Structural irregularity and the presence of voids leads to densities of $1.8\,g/cm^3$ for PAN-based fibres and $2.1\,g/cm^3$ for pitch-based, compared to $2.28\,g/cm^3$ for pure graphite. Figure 5.3 shows the range of strength, modulus and break elongation of carbon fibres, obtained by varying the starting material and processing conditions. There is correlation with other properties. As modulus increases from 200 to 900 GPa, the electrical conductivity increases from 500 to $2 \times 10^4\,\Omega^{-1}\,cm^{-1}$ and the thermal conductivity increases from 4 to $900\,Wm^{-1}\,K^{-1}$. Both PAN-based and pitch-based carbon fibres are not subject to creep or fatigue failure. More information on carbon fibres is given by Lavin (2000) and Donnet *et al.* (1998).

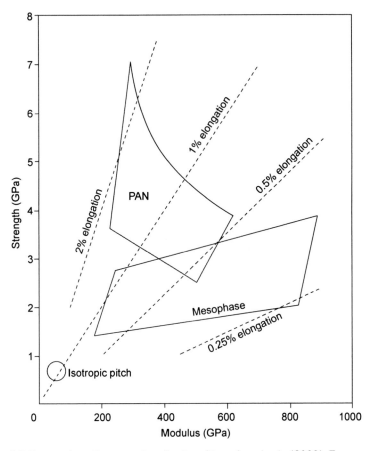

5.3 Range of tensile properties of carbon fibres, from Lavin (2000). To convert to N/tex divide by density of $1.8\,g/cm^3$ for PAN-based fibres and $2.1\,g/cm^3$ for pitch-based.

5.3.2 Other carbon fibres

Vapour-grown fibres are produced by a catalytic process. One process produces long fibres that are tangled in a ball; another produces short straight fibres. The fibres can be made in nanometre sizes and surface modifications lead to excellent absorbency, which may make them suitable for some protective uses. Carbon nano-tubes are covered in section 5.9.2. Semi-carbonised fibres, which are thermally resistant, are included in section 5.8.2.

5.4   Aramid and related fibres

5.4.1 General features of HM-HT polymer fibres

Although Staudinger had postulated an ideal structure in 1932, it was not until the 1970s that high-modulus, high-tenacity (HM-HT) fibres, which were the second generation of synthetic polymer fibres, became available. The requirements are high-molecular weight, with fully extended polymer molecules highly oriented along the fibre axis. Usually this means high crystallinity. It is also desirable that side-groups, which contribute to the weight but not to the strength, should not be substantial. There are two routes to such a structure. The first method, which led to the aramid HM-HT fibres, uses stiff, interactive molecules, which in solution form liquid-crystal or other organised structures without molecular folds, and which become highly oriented through the spinning process. The second method, used for high-modulus polyethylene fibres, is at the opposite extreme of polymeric nature: flexible, inert molecules, which can be highly drawn after extrusion to extend and orient the molecules. Unfortunately, no way has been found to produce the required structure from polymers of intermediate character, such as nylon or polyester.

The manufacturing processes for HM-HT fibres are more complicated and expensive than the melt-spinning used for polyester and nylon fibres. In addition to their high strength and stiffness, the HM-HT linear-polymer fibres differ from the inorganic fibres, including carbon, in characteristic ways. The polymer fibres melt or degrade between 150 and 500 °C, whereas even glass does not soften until about 800 °C, and the other inorganic fibres can stand higher temperatures. Another difference is that, although the tensile properties may be comparable, the compressive strength of linear-polymer fibres is much lower, whereas for inorganic fibres the compressive strength is similar to or better than the tensile strength. The polymer fibres have a low compressive yield stress, which allows for a large compressive deformation. The difference is a result of the fact that an oriented assembly of linear molecules can buckle internally with the formation of kink-bands, but an assembly bonded in three dimensions cannot (carbon with a two-dimensional form is in an intermediate position). The most important manifestation of this difference is in bending. Inorganic fibres follow the classical mechanics: there is a neutral plane at the centre and the fibre fails when the tensile strain on the outside of the bend reaches the low tensile break extension. The low compressive yield stress of linear-polymer fibres means that deformation is much easier on the inside than the outside of the bend. The neutral plane moves outwards and the limiting tensile strain is never reached. The fibres can be fully bent back on themselves without breaking; the tensile strain on the classical model would be 100%. There is some internal damage to the fibres, but the loss of strength is small; application of tension after a single bending pulls out the kink-bands, but repeated flexing eventually leads to fatigue failure.

5.4.2   Para-aramids (*Kevlar* and *Twaron*)

Kevlar (from DuPont) was the first of this class of fibre; it was followed by *Twaron* (from Akzo-Nobel, now with Teijin). Aramids are polyamides, defined as having at least 85% of the amide groups linked to two aromatic rings. In the para-aramids, which give HM-HT fibres, the links are to opposite corners of the rings. The commercial fibres, *Kevlar* and *Twaron*, are made of poly[*p*-phenylene terephthalamide] (PPTA):

Following polymerisation, PPTA is dissolved in concentrated sulphuric acid. Spinning is by dry-jet wet-spinning. The solution passes through an air-gap before entering the coagulating bath. Orientation of the liquid-crystal solution is achieved partly by shear in the spinning holes and partly by elongation in the air-gap.

The crystallinity and orientation of PPTA fibres, although very high, is not perfect. As spun, the fibres are highly crystalline, but have some fibrillar

Table 5.2 Aramid types and properties. From Rebouillat (2000)

Type	Tenacity N/tex	Modulus N/tex	Break ext. %
Kevlar 29	2.03	49	3.6
Kevlar 49	2.08	78	2.4
Kevlar 149	1.68	115	1.3
Nomex	0.485	7.5	35
Twaron	2.10	75	2.5
Twaron high modulus	2.10	75	2.5
Technora	2.20	50	4.4

disorder. The structure is pleated to give an orientation angle of about 12°. The pleated structure pulls out under tension and leads to the initial stiffening curvature of the stress-strain curve in Fig. 5.2(c). It also results in creep at low stresses, but this is not the continuing creep at high stress, which leads to fibre breakage. A brief heat treatment under tension increases orientation and gives higher modulus fibres.

Table 5.2 lists various types of aramid fibres with their mechanical properties. Aramid fibres do not melt, but decompose above 430 °C. They absorb a small amount of water. Further information on aramid fibres is given in Chapter 19 on ballistic protection, and by Rebouillat (2000) and Yang (1993).

5.4.3   A copolymer (*Technora*)

After a study of the physics of fibre formation, Teijin introduced a copolymer fibre, *Technora*, with the formula:

There are more than 85% PPTA groups, so this is counted as an aramid. The molecules do not form liquid-crystals in solution. They are polymerized in an organic solvent to give an isotropic solution without molecular entanglement. The solution can be spun into an aqueous coagulating bath. Orientation results from drawing and heat treatment. The manufacturing process is simpler than for *Kevlar* and *Twaron*, and, while broadly similar, *Technora* has a good balance of properties.

5.4.4   Meta-aramids (*Nomex*)

The first aramid to be commercialised was the meta-aramid *Nomex* (from DuPont). Poly [*m*-phenylene isophthalamide] has links that are to next-but-one atoms on the benzene ring. As can be seen from Table 5.2, the mechanical properties of *Nomex* are comparable to general-purpose textile fibres. Its application is in thermal resistance. More information on meta-aramids and related fibres will be given in section 5.8.2.

5.4.5   Other aromatic HM-HT fibres

Another route to HM-HT fibres is through fully aromatic polyesters, which, when coplymerised, can be spun from a thermotropic liquid-crystal melt. The commercially available fibre is *Vectran* (from Celanese), which has the formula:

The polymer has a market as a resin, which makes it economical to produce the fibre. Although melt-spinning is an inexpensive process, cost is added by a long subsequent heat-treatment, which is needed to increase the molecular weight. The tensile properties are broadly similar to p-aramids, but the fibres do not creep and have better abrasion and flex resistance. The combination of properties gives *Vectran* a niche market. Further information is given by Beers (2000). In addition to PPTA fibres, heterocyclic para-polyamide and para-copolyamide fibres have been produced in Russia, as described by Perepelkin (2000). They are claimed to have superior properties.

5.5   High-modulus polyethylene

5.5.1   Gel-spun fibres

In a suitably concentrated hydrocarbon solution, polyethylene[1] (-CH$_2$-)$_n$ forms a gel. Following extrusion this can be drawn to a high degree in order to fully extend and orient the molecules. As a starting material, ultra-high molecular weight polyethylene (UHMWPE) is used for the gel-spinning of the high-modulus or high-performance polyethylene fibres (HMPE or HPPE): *Spectra* (from Allied-Signal now Honeywell) and *Dyneema* (from DSM and Toyobo).

A typical stress-strain curve for HMPE fibre is shown in Fig. 5.2(c). Table 5.3 lists the various types with their mechanical properties. Gel-spun HMPE fibres combine high modulus and high tenacity, which makes them right for ballistic impact resistance. The fibres melt at 150 °C and their properties deteriorate as temperature rises above room temperature. Under high stresses, the fibres creep extensively and can break after short times under load. A secondary slow heating under tension close to the melting point increases modulus and reduces creep. In applications in which creep is a potential problem, the manufacturer's advice should be sought, because of the large difference of creep behaviour of different HMPE fibres. HMPE fibres are extremely resistant to chemical and biological attack and have better abrasion and fatigue reistance than aramid fibres. Further information on gel-spun HMPE fibres is given by Van Dingenen (2000).

---

[1] *Note on polyethylene terminology.* The first polyethylene was *low-density polyethylene*, with branched molecules. *High-density polyethylene*, made with Ziegler catalysts, has linear molecules. *Ultra-high molecular weight polyethylene* (UHMWPE) is a form of high-density polyethylene, used in superior engineering plastics. *HMPE* results from the special manufacturing processes of UHMWPE fibres.

Table 5.3 HMPE types and properties. From Van Dingenen (2000)

Type	Filament denier	Tenacity N/tex	Modulus N/tex	Break ext. %
Dyneema SK60	1	2.8	91	3.5
Dyneema SK65	1	3.1	97	3.6
Dyneema SK71	1	3.5	122	3.7
Dyneema SK75	2	3.7	120	3.8
Dyneema SK60	2	3.7	120	3.8
Spectra 900	10	2.6	75	3.6
Spectra 2000	5	3.2	110	3.3
Spectra 2000	3.5	3.4	120	2.9

5.5.2   Other high-modulus polyethylenes

Melt-spun polyethylene fibres can be highly drawn to give a high modulus in a process developed by Ward at the University of Leeds. This cheaper route was commercialised as *Certran* fibres by Hoechst-Celanese, but they no longer make the fibre. The market was limited because the strength is only about half that of gel-spun fibres.

Another way of making a high-modulus polyethylene fibre, *Tensylon*, described by Weedon (2000), is by solid-state extrusion of UHMWPE with low chain entanglement. The three basic operations are powder compaction, rolling and ultra-drawing. The resulting tapes can be split into fibrillated yarns. The modulus of *Tensylon* is similar to that of gel-spun fibres, but the strength is only half that of gel-spun fibres. However, the inherent costs of the process are also about half that of gel-spun fibres, which provides a market opportunity for *Tensylon*.

5.6   PBO and M5

5.6.1   Polybenzoxazole (PBO)

In a search for fibres with properties superior to *Kevlar*, the US Air Force had a research programme on polymers with more complicated ring structures. Finally, this led to the commercial production of fibres of poly[*p*-phenylene benzbisoxazole], often called polybenzoxazole (PBO), by Toyobo. The formula of PBO is:

The five-membered rings on either side of the benzene ring give a stiffer chain molecule. Fibre production is by dry-jet wet-spinning from a solution in phosphoric acid. The character of PBO fibres is generally similar to that of aramids, but, as shown in Table 5.1 and Fig. 5.2(b), the modulus and strength are about twice as high. One problem, which has recently come to light, is that the fibre degrades by hydrolysis in warm moist conditions, as experienced by body armour. Further information on PBO is given by Young and So (2000). See also Chapter 19 on ballistic protection.

5.6.2   PIPD or M5

The HM-HT fibres so far described have excellent tensile properties, but their transverse compression and shear properties are poor, because of the weak intermolecular forces, mostly van der Waals forces. The aramids do have hydrogen bonding between (-CO-NH-) groups, but these are all in one plane, so that there is weakness in the other direction. In the 1990s, Sikkema (2000) synthesised a polymer with a ring structure similar to PBO, but with hydroxyl groups, -OH, hydrogen-bonding in all transverse directions. The resulting M5 fibres have greatly improved transverse properties. The polymer is poly[2,6-di-imidazo[4,5-b4',5'-e]pyridinylene-1,4-(2,5-dihydroxy)phenylene] (PIPD) with the formula:

The modulus of laboratory-spun samples of M5 fibre is slightly higher than that of PBO and the tenacity is slightly lower, but it is expected that industrially optimised production will increase these values. M5 fibres from a pilot plant are expected on the commercial market in 2005.

5.7   Inorganic fibres

5.7.1   General features of inorganic fibres

As discussed in section 5.4.1, inorganic fibres may have similar tensile properties to linear-polymer HM-HT fibres, but differ in that their compressive strength is similar to or better than the tensile strength. Eventual shearing in compression leads to a low limiting compressive strain. Inorganic fibres are brittle in bending, which makes textile processing difficult, though, with care, many types of inorganic fibre yarns can be woven or knitted. The high modulus makes inorganic fibres suitable for composite reinforcement, and with the ceramic fibres in appropriate matrices this gives high-temperature composites. Applications in protective textiles are more limited, but could include heat shields and particulate protection at high temperatures.

Table 5.4 Ceramic fibres and properties. From Bunsell and Berger (2000)

Type	Manufacturer	Name	Composition: weight %	Diameter μm	Density g/cm³	Strength GPa	Modulus GPa	Break ext. %
Si–C based	Nippon Carbide	Nicalon NLM 202	56.6 Si/31.7 C/11.7 O	14	2.55	2.0	190	1.05
		Hi-Nicalon	62.4 Si/37.1 C/0.5 O	14	2.74	2.6	263	1.0
	Ube Industries	Tyranno Lox-M	54.0 Si/31.6 C/12.4 O/2.0 Ti	8.5	2.37	2.5	180	1.4
		Tyranno Lox-E	54.8 Si/37.5 C/5.8 O/1.9 Ti	11	2.39	2.9	199	1.45
Near stoichiometric SiC	Nippon Carbide	Hi-Nicalon type-S	68.9 Si/30.9 C/0.2 O	13	3.0	2.5	375	0.65
	Ube Industries	Tyranno SA1	SiC/Al <1	10	3.0	2.6	330	0.75
		SA3	small amounts of C and O	7.5	3.1	2.9	340	0.8
	Dow Corning	Sylramic	96 SiC/3.0 TiB$_2$/1.0 C/0.3 O	10	3.1	3.0	390	0.75
α-Al$_2$O$_3$	Mitsui Mining	Almax	99.9 Al$_2$O$_3$	10	3.6	1.02	344	0.3
	3M	Nextel	99 Al$_2$O$_3$/0.2–0.3 SiO$_2$/0.4–0.7Fe$_2$O$_3$	10–12	3.75	1.9	370	0.5
Alumina-silica	ICI	Saffil	95 Al$_2$O$_3$/5 SiO$_2$ (short fibres)	1–5	3.2	2.0	300	0.67
	Sumitomo	Altex	85 Al$_2$O$_3$/15 SiO$_2$	15	3.2	1.8	210	0.8
	3M	Nextel 312	62 Al$_2$O$_3$/24 SiO$_2$/14 B$_2$O$_3$	10–12 or 8–9	2.7	1.7	152	1.12
		Nextel 440	70 Al$_2$O$_3$/28 SiO$_2$/2 B$_2$O$_3$	10–12	3.05	2.1	190	1.11
		Nextel 720	85 Al$_2$O$_3$/15 SiO$_2$	12	3.4	2.1	260	0.81
Al'a-zirc'a	3M	Nextel 720	89 Al$_2$O$_3$/10 ZrO$_2$/2 Y$_2$O$_3$	11	4.1	2.5	360	0.7

5.7.2 Glass fibres

Low-cost staple glass fibres are made and laid down in mats by processes such as steam-blowing, or centrifugal spinning. They enter the protective market as thermal and sound insulation. Glass yarns for reinforcement of composites are made by melt-spinning under gravity, but drawing is not needed. The application of finish adapted to the end-use is an important feature of glass fibre production. Jones (2000) describes the composition, manufacture and properties of glass fibres. Silica, SiO_2, is the major component of glass, but varying amounts of about a dozen metal oxides are included in compositions intended for different uses. On solidification, silica forms an amorphous network with entrapped metal atoms. When expressed in GPa, glass fibres have a stiffness and strength comparable to aramid fibres, but the density is almost twice as high, so that the values in N/tex on a weight basis are much lower than for aramids. Glass softens at around $800\,°C$.

5.7.3 Ceramic fibres

The considerable variety of ceramic fibres are described by Bunsell and Berger (1999, 2000). They can be used at temperatures of $1{,}000\,°C$ or higher. Inorganic 'fibres' can be made by vapour deposition on a substrate, but these have diameters greater than $100\,\mu m$ and, in a textile context, would not be regarded as fibres. Similarly large single-crystal 'fibres' of alumina have also been made. Finer commercial ceramic fibres are made by melt-spinning or solution-spinning of precursor fibres, which are then heat-treated to form the ceramic. These fibres are either silicon or alumina based, although other compositions are being developed in laboratory or pilot-plant conditions for particular end-uses.

Silicon carbide, SiC, fibres are spun from a melt of an organico-silicon polymer, followed by cross-linking to allow pyrolysis above $1{,}200\,°C$. Sintering is employed in the production of *Tyranno SA, Syramic* and *Hi-Nicalon Type S*. The details of the processing influence the resulting polycrystalline microstructure. Alumina fibres are spun from solutions of aluminium salts with polymeric additives that cause a gel to form after extrusion. Progessive heat treatment to above $1{,}000\,°C$ gives the ceramic fibre. Other alumina-based fibres may be from dispersions of aluminium compounds in spinnable solutions. Table 5.4 lists ceramic fibres with their composition and mechanical properties.

5.8   Resistant polymer fibres

5.8.1 General comment

Some of the high-performance fibres already described have good chemical or thermal resistance. Ceramic fibres and carbon fibres (if oxidation is prevented) are stable to high temperatures and chemically resistant. The HM-HT polymer fibres made from stiff polymer molecules (described in sections 5.4 and 5.6 and excluding HMPE) will withstand fairly high temperatures; in particular PBO is reported to be stable up to 650 °C in air and over 700 °C in an inert atmosphere. HMPE fibres (described in section 6.5) have good chemical resistance at ambient temperatures, but this deteriorates with a moderate increase of temperature. Generally these expensive fibres will only be used for chemically or thermally resistant applications where their particular combinations of properties make them specially suitable.

The emphasis in this section is on chemically and thermally resistant fibres that have good textile properties. They need only moderate strength, but should be extensible in order to be suitable for protective clothing. The mechanical demands for protective filters and similar uses may be less stringent. In addition to performance requirements, economic considerations influence the choice of fibre for given end-uses. In general terms, greater chemical and thermal resistance, particularly when these are combined, requires more complicated chemistry and spinning into fibres is more difficult; these factors increase the manufacturing cost. More information on chemically resistant fibres is given by Horrocks and McIntosh (2000) and on thermally resistant fibres by Horrocks *et al.* (2000). Fire retardant materials are described by Horrocks and Price (2001), and in Chapter 15.

5.8.2   Chemically resistant fibres

Polyethylene (PE) fibres would have good chemical resistance, but due to their low strength (except for HMPE fibres), little or no PE fibres are currently in production. Polypropylene fibres (PP), produced by melt-spinning and drawing, are manufactured as a cheap general-purpose fibre; their properties are somewhat similar to nylon and polyester, but melt at a lower temperature and have poor elastic recovery. PP fibres have good chemical resistance, but this deteriorates above about 50 °C, which limits their applicability. This section deals with speciality fibres that have good chemical resistance up to moderately high temperatures.

The first category of chemically resistant fibres are halogenated derivatives of polyethylene. Poly[vinylidene chloride] ($-CH_2-CCl_2-$, PVDC) has good chemical resistance but is difficult to process, so that copolymers with about 15% of other vinyl or acrylic monomers are used. Saran is a copolymer with vinyl chloride, which can be melt-spun at about 180 °C. Fluorinated fibres are more expensive, but have extreme chemical inertness. *Teflon* (from DuPont), poly[tetrafluorethylene] ($-CF_2-CF_2-$, PTFE), has to be spun from suspension in a viscose dope followed by high-temperature sintering. Alternative PTFE fibres can be made by powder processing (W L Gore) or split-peeling from a cylindrical billet (Lenzing). PTFE fibres will stand temperatures up to nearly 300 °C. Poly[vinyl fluoride] ($-CH_2-CHF-$, PVF), poly[vinylidene fluoride] ($-CH_2-CF_2-$, PVDF), and other fluorinated ethylene polymers and copolymers (known as FEP) have good chemical resistance. Their lower melting points make processing easier, but limit performance.

Polyetherketones, which are produced for engineering plastics, can be formed into fibres by high-temperature melt-spinning. Although other variants have been produced by pilot plants, the only commercial success has been poly[etheretherketone] (PEEK), with the formula:

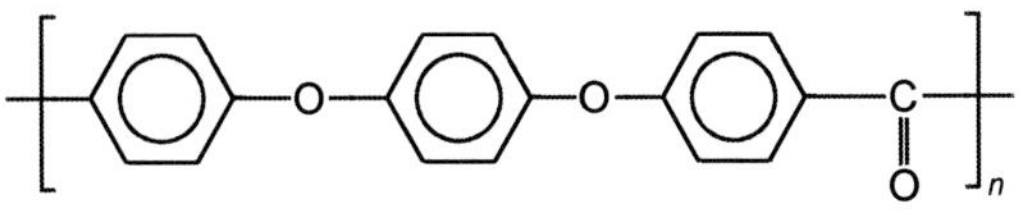

PEEK fibres can be used up to 260 °C, with short excursions to 300 °C, and are resistant to high-temperature steam and most chemicals, excepting strong oxidising acids.

Poly[phenylene sulphide] (PPS) can also be melt-spun. Its chemical formula is:

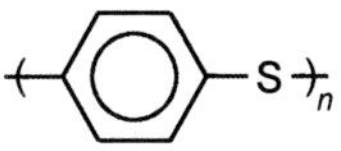

PPS fibres have excellent chemical resistance, but a second-order transition temperature of 93 °C limits use above 100 °C under high stress.

Another engineering plastic with good chemical resistance, poly[ether-imide] (PEI) has slightly inferior temperature resistance to PEEK, but is cheaper. PEI is an amorphous polymer with a formula reported as:

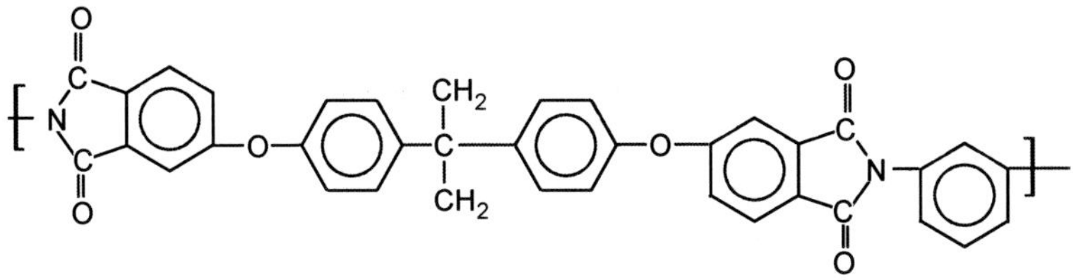

Some fibres described in the next section have good oxidation resistance at high temperatures, but may be less resistant to other chemicals.

5.8.3 Thermally resistant fibres

This section mainly covers fibres that withstand high temperatures, but do not have high mechanical performance. The dominant heat- and flame-resistant fibres are the aramids and arimids. The widely used fibre, with good textile properties, is the meta-aramid, poly[m-phenylene isophthalamide], commercialised in 1967 by *DuPont* as *Nomex*. Meta-aramid fibres of the same or variant compositions are also now made by other companies. The formula of this meta-aramid group, which must account for at least 85% of the structure, is:

Recent versions of these fibres can be dyed in a range of colours, which makes them attractive for protective clothing. They have a second-order transition at about 275 °C and start to melt and degrade at 375 °C. However, the resulting char is tough and remains as a thermally protective layer.

The para-aramids, (*Kevlar, Twaron, Technora*), which were described in sections 5.4.2 and 5.4.3, have higher degradation temperatures as well as high strength and stiffness. They may be selected for thermal protection when the demands are severe, or incorporated in blends with meta-aramids to give added protection. Blends of aramids with viscose rayon are also used. Due to the stiffer chains, PBO (section 5.6.1) is reported to be 'the most thermally stable and flame resistant of all organic polymer fibres commercially available at the present time' (Horrocks *et al.*, 2000).

A number of arimid fibres were investigated in 1960-70. *P84* (from *Lenzing*, now produced by *Inspec Fibres*) is made from poly[4.4'-diphenylmethane-*co*-tolylene benzophenotetra-carboxylic-imide]:

where R = $C_6H_4.CH_2$ or $C_6H_4.CH_2.C_6H_4$. *P84* can be crimped to provide increased thermal insulation in protective clothing and will withstand higher temperatures and has better chemical resistance than the meta-aramids.

Kermel (from *Rhodia Performance Fibres*) is a poly[amide-imide] with the reported formula:

The properties of *Kermel* are similar to the meta-aramids.

Poly[benzimidazole] (PBI) is reported by Horrocks *et al.* (2000) to be 'the premium product for many performance-based applications' because of 'its exceptional thermal stability and chemical resistance ... along with its excellent textile processing characteristics'. The full chemical name of the commercially available PBI fibre is poly[2,2'-(*m*-phenylene)-5,5'-dibenzimidazole]:

Most PBI fibre is treated with sulphuric acid to increase flame stability. For more demanding applications phosphonated PBI may be used.

Whereas most textile fibres are made from linear polymers, which are thermoplastic unless degradation occurs before melting, thermally resistant fibres can be made from thermosetting polymers. Uncured or partially cured resin can be extruded and then fully cross-linked during or after extrusion. An early example is the novoloid fibre, *Kynol*, a phenyl formaldehyde fibre with the structure:

Kynol has many of the properties of phenolic resins; it does not melt; has good chemical resistance and low smoke generation on heating. It will withstand temperatures of 2,500 °C for 12 seconds or more, but the limit for long-term use is 150 °C in air or 250 °C in absence of oxygen. *Basofil* is a melamine-formaldehyde fibre, which is somewhat stronger than *Kynol*; its long-term use temperature is 190 °C. Both of these fibres can be incorporated in comfortable protective clothing.

The black, partially carbonised or oxidised acrylic fibres, such as *Panox*, have a cross-linked structure, which gives thermal resistance while retaining some of the properties of the acrylic precursors. Their production follows the initial stages of carbon fibre production described in section 5.3.1. *Panox* fibres are oxidatively degraded at long times over 210 °C. For *Curlon* fibres (*Orcon Corporation*), additional heat treatment is reported to give improved properties.

5.9   Nano-fibres

5.9.1   Fibre fineness

Most textile fibres are in the range of 2 to 20 dtex, with diameters, depending on density, between about 10 and 50 μm. Polyester microfibres have been developed for apparel textiles, including leather-like products. With suitable precautions, direct melt-spinning can be taken down to 0.4 dtex (6 μm);

conjugate melt-spinning followed by splitting to 0.1 dtex (3 μm); and islands-in-a-sea spinning, in which an enveloping material is dissolved away, to 10^{-4} dtex (0.1 μm or 100 nm). The last of these could be called a nano-fibre, but the development of electro-spinning makes finer fibres possible. As a goal, 1 nm would correspond to about 10^{-8} dtex and would have few molecules in its cross-section

The fineness of fibres is a determinant of barrier performance. A closely packed assembly of fibres will have small spaces[2] between fibres, which are less than the fibre diameter, thus inhibiting particulate entry. The large surface area is effective in absorbing chemicals: 1 kg of a 100 nm diameter fibre would have a surface area of about 30,000 m^2 (200 $\times$ 150 m); at 1 nm the figure would be 3 $\times$ 10^6 m^2 (2 $\times$ 1.5 km).

5.9.2   Electro-spinning

The production of nano-fibres has been revolutionised by the development of electro-spinning. If a liquid is extruded through a nozzle with a high potential between the tube and the collector plate, a matt of extremely fine fibres is laid down. It was thought that the fineness was due to splitting of the extruded filament, but it is now known that it is due to an intense whirling action, which results in an extremely high stretching and attenuation of the single filament.

Electro-spinning has two great advantages. It is versatile, since any solution or melt can be electro-spun, and the equipment needed is of a small scale. The process is an old one, which was neglected until the work of Reneker in the 1990s (Fong and Reneker, 2000). Now many laboratories have bench-size equipment. This makes it easy to explore new fibre types and economic to make small quantities, which would be sufficient for specialised protective purposes. If the demand becomes greater, more productive multi-head machinery will be developed. Fibre diameters are typically in the range of 10 to 100 nm. With current procedures a matt of fibres is laid down on a moving belt. This non-woven fabric would be suitable for many protective applications. If woven or knitted fabrics were needed, it can be expected that ways will be found to collect the random fibre assembly and orient the fibres into a yarn.

5.9.3   Carbon nanotubes

In contrast to the versatility of electrospinning, another newly emerging form of nanofibres is highly specific, since they consist of pure carbon. The development starts with the discovery of C$_{60}$, in which carbon atoms are covalently bonded in

---

[2] Commonly referred to as 'pore size', though the paths between the fibres are not tubular pores, but are a branched network.

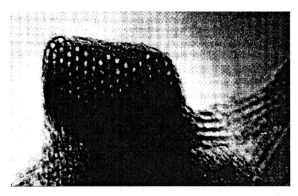

5.4 Ropes of single-walled carbon nanotubes. Reprinted with permission from Thess *et al.*, *Science* 273: 483-87. Copyright 1996, AAAS.

a spherical network called a buckyball. In another idealised form, two hemispherical ends could be linked by a cylindrical tube with a similar covalent network. Structures of this general form can be made by carbon-arc discharges or laser ablation. In 1993, it was discovered that long single-walled carbon nanotubes with diameters of about 1.5 nm were produced when transition metal catalysts were added to the carbon arc (Bethune *et al.*, 1993; Ijima, 1993). Guo *et al.* (1995) produced similar fibres by laser ablation. Single-walled nanotubes can be collected in ropes, Fig. 5.4 (Thess *et al.*, 1996). Previously, short (lengths $\sim \mu$m) multi-walled carbon nanotubes with diameters of 4 to 50 nm had been produced (Ijima, 1991). They have multiple layers of carbon atoms, rather like tree rings, Fig. 5.5. For over 100 years, it had been known that similar, but much coarser multiple-walled carbon fibres could be made by vapour deposition at around 600 °C. The term 'nanotubes' is now being applied to vapour-grown

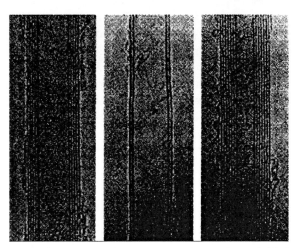

5.5 Multi-walled carbon nanotubes. From Ijima (1991). Copyright 1991, reproduced with permission from *Nature*.

fibres of less that 100 nm in diameter. Such nanotubes formed at high temperatures (~1,000 °C) are relatively free of defects and are similar to carbon-arc or laser-ablation fibres.

The early work on carbon nanotubes was suitable only for the production of small amounts for scientific study or for such uses as small, single-fibre, electron field-emitters. More recently, there have been accounts of ways of assembling carbon nanotubes into long lengths by various post-processing techniques (Li *et al.*, 2004). A more promising development in the Department of Materials Science of the University of Cambridge, which is moving into commercial operation, is direct spinning by chemical vapour deposition synthesis (Li *et al.*, 2004). The process has been colourfully described as winding up smoke (Windle, 2004). In one procedure, ethanol containing ferrocene and thiocene catalysts is injected into a hydrogen gas stream, which carries the solution into a furnace at 1,050–1,200 °C. Nanotubes are formed and associate into an aerogel, which is 'stretched by the gas flow into the form of a sock and then continuously drawn from the hot zone by winding it onto a rotating rod' in indefinitely long lengths. The terminology is slightly confusing: the product is referred to as a 'fibre', but in textile usage it would be regarded as a yarn composed of nanotube fibres. Variants in the carbon sources and the process conditions lead to a variety of forms and purities, made either of single-walled (SWNT) or multi-walled (MWNT) carbon nanotubes. MWNT had diameters of 30 nm and aspect ratios of ~1000; SWNT had diameters of 1.6–3.5 nm in bundles with a lateral size of 30 nm. Preliminary measurements showed strengths in the range of 0.05–0.5 N/tex and break strains of 100% or more; these properties must result from the packing within the 'fibres' and not the inherent properties of the nanotubes. Nonwoven assemblies can also be laid down.

An account of hand-spinning carbon nanotubes drawn from a MWNT 'forest' on a substrate into metre-length yarns is given by Zhang *et al.* (2004), who also discuss the role of twist and other factors according to the theories discussed by Hearle *et al.* (1969). As produced the yarns have break strains of up to 13%, but this is accompanied by a large, irreversible lateral contraction. Two-ply yarns reached strengths of 460 MPa based on the initial yarn cross-section, with the true stress at break being 30% higher. The specific stress at break was estimated to be 575 MPa/g cm^{-3} (0.575 N/tex). Considerable optimisation is needed in the development of a continuous process in order to approach the fibre strengths.

5.10   Fibres to fabrics

5.10.1 Nonwovens

Nonwoven fabrics, more positively described by the German term *Fliesstoffe*, can be made from staple fibres by laying down, in-parallel or cross-laid, multiple layers of the web of fibres that comes from conventional textile carding, in which fibres are stripped from a cylinder covered with a clothing of projecting wires. Bonded fibre fabrics are produced by application of adhesive or by thermal treatment of thermoplastic fibres. Alternatively, the web can be laid down from an air-stream or, for short fibres, from a dispersion in water. Continuous filaments can be laid down from a bank of spinnerettes as they are produced, with the resulting web fed directly into a bonding unit to form spunlaid (spunbonded) fabrics. Flash-spun fabric is made by extrusion of a polymer solution into a temperature high enough to cause rapid evaporation, with the formation of a coherent web of fine fibrils; the best-known example is *Tyvek*. Melt-blown nonwovens are another type.

Needled fabrics, with an integrity resulting from fibre entanglement, are made by feeding a cross-laid web into a bank of reciprocating needles. Hydroentangled (spunlaced) fabrics are formed by passing a web, usually air-laid, under high-pressure water jets. Stitch-bonded fabrics are made by using yarns to stitch through webs in a warp-knitting operation. More information on nonwovens is given in Jirsak and Wadsworth (1999) and Russell (2005).

5.10.2 Yarns

Most textile fabrics are made by the interlacing of yarns, as described below. Practical information on yarn production is given by Lord (2003) and on theoretical aspects by Grosberg and Iype (1999). As already described, continuous filament yarns are directly produced by extrusion. They can be given coherence by inserting a low level of twist, or, now more commonly, by interlacing in an air-jet on the way to wind-up. Cords, which are mechanically more robust, are made by twisting (plying) a number of singles yarns together. Ropes are the extreme of one-dimensional textile usage, and are made by twisting in multiple levels of 'textile yarn', rope yarn, strand, and rope; alternatively ropes can be made by braiding (see below) or assembling yarns or sub-ropes within a jacket (McKenna *et al.*, 2004).

The fibres in continuous filament yarns, as made or plied, are tightly packed, which is good for most technical uses, but not usually wanted in clothing. Several methods have been used to give bulk, stretch and texture to continuous filament yarns (Hearle *et al.*, 2001). The most widely used is false-twist texturing, in which, in a continuous operation, the yarn is heated and cooled while twisted and then untwisted. A single-stage operation, usually with nylon, gives a stretch tarn. The addition of a second heating stage with a small overfeed gives a yarn, usually of polyester, with high bulk but low stretch. Air-jet texturing gives a yarn with loops protruding from a tightly packed core.

In the traditional method of making yarns from staple fibres, used for all natural fibres except silk (which is reeled with a small amount of twist), it is first necessary to go through a sequence of processes (opening, carding and drawing) to produce a coarse roving of oriented fibres. The roving is then attenuated and twisted as it is fed onto a rotating spindle. Ring-twisting became the dominant method. The take-up package (bobbin) is rotated at high speed to insert twist and a slight lag due to the traveller on the ring causes the yarn to be wound up. In open-end spinning (break spinning) the twist is inserted at a break at the point of yarn formation, thus avoiding the need to rotate a large package at high speed. In rotor-spinning, the fibres are fed into and withdrawn from a rotating drum; in friction spinning, the fibres are fed between a pair of rotating rollers and the yarn is withdrawn axially; air-jet spinning is another method. As well as having a hairy surface of projecting fibre ends, staple fibre yarns have a more open, less regular fibre packing than untextured, continuous filament yarns.

A variety of other methods of producing yarns, which may involve inter-lacing, wrapping or bonding as well as twisting, were developed in the second half of the 20th century, but the above ways are those of most current commer-cial importance. For fibres that are difficult to spin in other ways, hollow-spindle spinning may be possible. A strand of fibres is fed through a rotating hollow tube that carries a package of a fine binder yarn, such as monofilament nylon. The speciality fibres are then held in the yarns by the wrapping.

Versaspun, a new integrated process for making staple fibre yarns, shown by DuPont at ITMA 2003, may be useful for some protective textiles. A thick continuous filament yarn is fed into a stretch-break zone, which converts the filaments into long-staple fibres, and then to a spinning head and wind-up. One economic advantage is clear. Going in one step from extruded fibre to spun yarn eliminates much processing. Another is more subtle. Fine aramid yarns are much more expensive to make than coarse yarns. The drafting, which occurs with *Versaspun*, offers a way of converting a cheaper coarse yarn into a fine yarn by an inexpensive operation. Because of the length of filaments, the strength loss is small. *Versaspun* is also well adapted to making blended yarns.

5.10.3 Weaving

The oldest way of making fabrics is by weaving. The essential parts of a weaving machine (loom) are shown in Fig. 5.6(a). One set of parallel yarns (warp) is interlaced by another set (weft). In a plain weave, Fig. 5.6(b), there is an alternation of over and under at each crossing. In other single-layer weaves, there will be floats where one yarn will remain over or under two or more yarns in the other set, before interlacing.

The weft is traditionally inserted by banging a shuttle, containing a small pirn of yarn, to-and-fro across the loom. In the newer, shuttleless weaving machines, free lengths of weft yarn, which are withdrawn from a large stationary package and cut, are propelled across by a metal carrier projectile, a rapier, an air-jet, or a water-jet. In multi-phase weaving machines, several weft yarns cross the machine at the same time, to give a spread of interlacing points.

The warp is usually prepared by winding the required number of yarns from a creel onto a beam (a creel could also feed the loom). The warp yarns are then

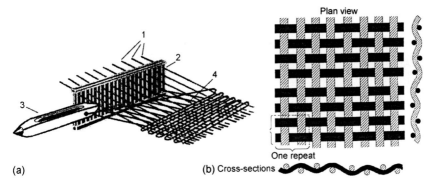

5.6 (a) Basic features of inserting weft by a shuttle. 1: Warp yarns. 2: The reed which beats up the weft yarns when the shuttle has passed through the loom. 3: The shuttle carrying the pirn of weft yarn. 4: Fell of cloth. Note that the healds, which raise or lower the warp yarns are not shown. Reproduced by kind permission of Sulzer Textile. (b) Plain weave. From Horrocks and Anand (2000).

threaded through the eyes of the healds, which are the devices allowing the yarns to be lifted above or below the crossing weft yarn, after the heald frame has been placed on the loom. The warp yarns pass through the reed, which is activated to beat up the yarns into the woven fabric. For a plain weave only two sets of healds mounted in frames are needed and these may be controlled by cranks. For slightly more complicated weaves, more frames are needed and they may be controlled by cams (tappets) or a dobby. For the most complicated weaves, each warp yarn can be raised or lowered individually by a jacquard, originally controlled by punched cards and now electronically from a computer.

Weaves may have more than one layer. For example, to make velvets two warp layers are linked by a third warp, which is subsequently cut to provide a pile. This leads on to 3D weaving, which produces either multi-layer solid structures or hollow fabrics with honeycomb-like structures. Although there were special machines made to produce 3D fabrics, it has been found that most forms can be made on conventional looms, if necessary with minor adaptations. Accounts of 3D fabrics, which are particularly important for composites, are given in Miravete (1999). Some examples are shown in Figs 5.7, 5.8 and 5.9. Clearly many variants are possible.

The ratio of yarn spacing to yarn diameter leads to a wide variety of woven fabrics. Open fabrics, with large pore sizes and low crimp, are formed when the spacing is large compared to the diameter. At an intermediate level, crimp interchange allows for substantial easy elongation as the fabric is pulled in the warp or weft directions to the limit when the yarns are straight or the structure jams. At the other extreme, the structure would be completely jammed in both directions, with minimal space between yarns.

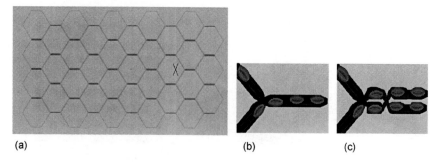

(a)   (b)   (c)

5.7 Hollow weave. (a) Honeycomb structure. (b), (c) Two ways in which the thicker section, where two weaves come together, can be constructed. Courtesy of Dr Xiaogang Chen, University of Manchester.

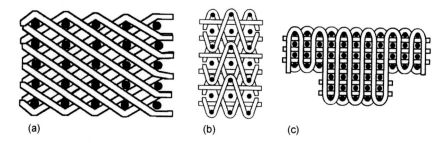

(a)   (b)   (c)

5.8 Examples of 3D solid weaves. From Hearle and Du (1990).

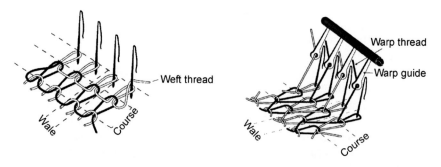

5.9 (a) Weft knitting. (b) Warp knitting. From Horrocks and Anand (2000).

The orthogonal structure of simple weaves leads to low shear stiffness up to the point of jamming, which is also shown as easy extension in the 45° (bias) direction. There has been some development of triaxial weaving to overcome this effect. More information on weaving is given by Ormerod and Sondhelm (1995), Adanur (2000) and Mohamed (2005).

5.10.4 Braiding

Braiding produces structures that are similar to weaves, but all the yarns run along the length of the braid in interlacing paths at an angle to the braid axis. For circular braids, the yarn packages are mounted on two sets of carriers that run in opposite senses round a circular track, crossing from inside to outside to provide the interlacing. For flat braids, the carriers reverse direction as they reach the ends of the tracks. Additional yarns may be laid in, to be trapped within the braid structure without interlacing, thus giving a tri-axial structure. As with weaving, there have been developments of 3D braids.

5.10.5 Knitting

In the simplest form of knitting, typified by hand-knitting, one yarn runs backwards and forwards across the growing fabric. The knitting needles cause the yarn to form loops with the preceding row (course), as shown in Fig. 5.9(a). Industrially, this mode of fabric formation is flat-bed, machine weft-knitting, which is now a high-speed operation. A great variety of detailed knit structures can be made, as illustrated by two examples in Fig. 5.10. The circular arc of the loop may be over or under the crossing yarn. Stitches can float past one or more stitches in the previous course or across rows along the wales (rows in the length of the fabric), before forming another loop. Stitches can be added or subtracted. Yarns can be hidden at the back of the fabric and then reintroduced to give surface patterns. Two layers form interlock fabrics. Cylindrical shapes can be made by knitting the yarn round one path and then back along another parallel path. In the past, flat fabrics had to be cut and sewn to make up garments or other products. Computer control now allows complicated 3D shapes to be knitted in a single operation, so that cutting is unnecessary and, if any sewing is needed, it is very limited.

An alternative form of weft knitting is circular knitting. In its simplest form, as in knitting socks where hand knitting shows how complex shapes can be made by following a knit pattern, one end of yarn follows round a continuous

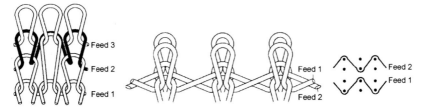

5.10 Two examples of weft-knit variants. (a) Tuck stitch. (b) Interlock. From Horrocks and Anand (2000).

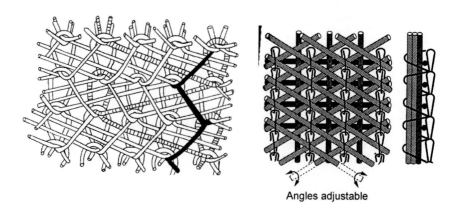

Angles adjustable

5.11 (a) A complicated warp-knit, from Horrocks and Anand (2000). (b) Warp-knit with laid-in yarns, from Hearle and Du (1990).

spiral path, forming loops with the previous course of the spiral. In commercial machines, the track is circular and multiple feeds can follow one another to give very high-speed fabric production. In warp knitting, a large number of yarns run along the length of the fabric, and form loops with their neighbours on either side, as shown in Fig. 5.9(b). The resulting structures can be complicated, as illustrated in Fig. 5.11(a). Warp knitting is growing in importance as a way of making technical fabrics.

Because of their loop structure, simple knit fabrics are easily deformed in all directions, though the presence of straight yarn lengths in more complicated weaves will tend to reduce this effect. A greater change can be made by laying in additional yarns, which are trapped by the knit structure but not incorporated in the loops. In weft knitting, laid-in yarns must be fed in together with the knit yarns; they may run along the courses or the wales or along both to stiffen the axial and transverse direction, but not the 45° direction. In warp knitting, a sheet of yarns, which is formed at the back of the machine, is fed into the warp knitting action. The yarns in the sheet can be in layers in any or all of the 0°, $+\theta°$, 90°, $-\theta°$ directions, Fig. 5.11(b). In the limit, the laid-in yarns dominate the fabric properties and the warp yarns merely hold the structure together. If $\theta° = 45°$, so that the directions are evenly spaced, and the yarn densities are all the same, the fabric will have isotropic properties. Warp-knits with laid-in yarns are of growing importance for technical fabrics. More information on knitting is given by Spencer (2001).

5.10.6 Fabric processing

After they have been formed, fabrics are commonly subject to further processing before they are made into products. It is beyond the scope of this chapter to do more than briefly indicate the nature of these operations. Where they have particular significance for protection, they will be covered in the relevant chapters. If colour has not been introduced through the yarns, it can be applied by dyeing or printing. Bleaching may also be needed. Information on coloration is given by Broadbent (2001).

Finishing is a more important factor for the properties of technical fabrics. Stentering, in which the fabric is gripped to control its width and length as it passes through a hot zone, controls the fabric dimensions and stabilises the fabric by heat setting. Chemical treatments, acting at the molecular level, can also stabilise or otherwise modify the fabric. Calendering between hot rollers will change the thickness and the surface character. Other processes also modify the physical nature of the surface. The chemical properties can be influenced by adding surface finishes – the most obvious example being water repellency without losing breathability. More information on fabric finishing is given by Heywood (2003) and Schindler and Hauser (2004).

Finishing leaves the fabric structure clearly apparent with little change in weight. Coating is a more extreme operation, in which a substantial amount of another material is added to the fabrics for particular functional reasons. Again foul weather garments are an obvious example, but the greater protection is at the expense of comfort. Coated textiles are described by Sen (2001) and Fung (2002).

5.10.7 Composites

Rigid composites, which may use textile preforms, are used for protective purposes, especially in helmets for impact and ballistic protection (see Chapter 19 on ballistic protection). The matrix turns them into solid, engineering materials. However, it is appropriate, in the fibre context, to mention hot compaction as a way of making novel composites. The process was invented at the University of Leeds (Hine *et al.*, 1993). If an assembly of thermoplastic fibres is put under pressure at a temperature just below the melting point, a small amount of the material melts and bonds the fibres together. The same chemical material provides both the matrix and the reinforcing fibres, though these are physically different phases. Hot compaction has been particularly useful with high-performance polyethylene fibres. In one set of tests, the tensile strength of hot compacted *Dyneema* was close to that of a conventional *Dyneema* composite, the modulus was somewhat lower, but the compressive strength was higher (Morye *et al.*, 1999).

5.11   References

Adanur S, ed. (1995), *Wellington Sears handbook of industrial textiles*, Cambridge, Woodhead Publishing.

Adanur S (2000) *Handbook of weaving*, Cambridge, Woodhead Publishing.

Anon. (2004), *New wool protective clothing*, The Network, Canesis Network Ltd, Christchurch, New Zealand, Issue 2.

Beers D (2000), 'Melt-spun wholly aromatic polyester', in Hearle (2000), 93–100.

Bethune D S, Kiang C H, de Vries M S, Gorman G, Savoy R and Vasquez J (1993), 'Cobalt-catalysed growth of carbon nanotubes with single-atomic-layer walls', *Nature*, **363**, 605.

Broadbent A D (2001), *Basic principles of textile coloration*, Cambridge, Woodhead Publishing.

Bunsell A R and Berger M-H (2000), 'Ceramic fibres', in Hearle (2000), 239–258.

Bunsell A R and Berger M-H, editors (1999), *Fine ceramic fibers*, New York, Marcel Dekker.

Denton M J and Daniels P N, editors (2002), *Textile terms and definitions*, Manchester, The Textile Institute.

Donnet J-B, Wang T K, Peng J C M and Rebouillat S, editors (1998), *Carbon Fibres*, 3rd edn, New York, Marcel Dekker.

Fong H and Reneker D H (2000), 'Electrospinning and the formation of nanofibers', in Salem D R, ed. (2000), *Structure formation in polymeric fibers*, Munich, Hanser.

Ford J E (1966), *Fibre data summaries*, Manchester, The cotton, silk and man-made fibres research association.

Fung W (2002), *Coated and laminated textiles*, Cambridge, Woodhead Publishing.

Grosberg P and Iype C (1999), *Yarn production: theoretical aspects*, Cambridge, Woodhead Publishing.

Guo T, Nikolaev P, Thess A, Colbert D T and Smalley R E (1995), 'Catalytic growth of single-walled nanotubes by laser vapourization', *Chem. Phys. Lett.*, **243**, 483–487.

Hearle J W S, ed. (2000), *High-performance fibres*, Cambridge, Woodhead Publishing.

Hearle J W S and Du Q W (1990), 'Forming rigid fibre assemblies: the interaction of textile technology and composites engineering', *J. Textile Inst.*, **81**, 360–383.

Hearle J W S, Grosberg P and Backer S (1969), *Structural mechanics of fibers, yarns and fabrics*, New York, Wiley-Interscience.

Hearle J W S, Hollick L and Wilson D K (2001), *Yarn texturing technology*, Cambridge, Woodhead Publishing.

Heywood D (2003), *Textile finishing*, Cambridge, Woodhead Publishing.

Hine P J, Ward I M, Olley R H and Bassett D C (1993), 'The hot compaction of high modulus melt-spun polyethylene fibres', *J. Materials Sci.*, **28**,316-324.

Hongu T, Phillips G O and Takigami M (2004), *New millennium fibers*, Cambridge, Woodhead Publishing.

Horrocks A R and Anand S C, eds (2000), *Handbook of technical textiles*, Cambridge, Woodhead Publishing.

Horrocks A R and McIntosh B (2000), *Chemically resistant fibres*, in Hearle (2000), 259-280.

Horrocks A R and Price D (2001), *Fire retardant materials*, Cambridge, Woodhead Publishing.

Horrocks A R, Eichhorn H, Schwaenke H, Saville N and Thomas C (2000), 'Thermally resistant fibres', in Hearle (2000), 281–324.

Ijima S (1991), 'Helical microtubules of graphite carbon', *Nature*, **354**, 56.

Ijima S (1993), 'Single-walled carbon nanotubes of 1 nm diameter', *Nature*, **363**, 363.

Jirsak O and Wadsworth L C (1999), *Nonwoven textiles*, Cambridge, Woodhead Publishing.

Jones F E (2000), 'Glass fibres', in Hearle (2000), 191–235.

Lavin J G (2000), 'Carbon fibres', in Hearle (2000), 156–190.

Li Y-L, Kinloch A and Windle A H (2004), 'Direct spinning of carbon nanotube fibers from chemical vapor deposition synthesis', *Science*, **304**, 276–278.

Lord P R (2003), *Handbook of yarn production*, Cambridge, Woodhead Publishing.

McKenna H A, Hearle J W S and O'Hear N (2004), *Handbook of fibre rope technology*, Cambridge, Woodhead Publishing.

Miravete A (1999), *3-D textile reinforcements in composite materials*, Cambridge, Woodhead Publishing.

Mohamed M (2005), *Weaving technology and woven fabrics*, Cambridge, Woodhead Publishing.

Morton W E and Hearle J W S (1993), *Physical properties of textile fibres*, 3rd edn, Manchester, The Textile Institute.

Morye S S, Hine P J, Duckett R A, Carr D J and Ward I M (1999) 'A comparison of the properties of hot compacted gel-spun polyethylene fibre composites with conventional gel-spun polyethylene fibre composites', *Composites: Part A*, **30**, 649–660.

Mukhopadhyay S K (1993), 'High performance fibres', *Textile Progress*, **25**, 1–85.

Ormerod A and Sondhelm W S (1995), *Weaving: technology and operations*, Cambridge, Woodhead Publishing.

Perepelkin K E (2000), 'Russian aromatic fibres', in Hearle (2000).

Rebouillat S (2000), 'Aramids', in Hearle (2000), 23–61.

Russell S, editor (2005), *Handbook of nonwovens*, Cambridge, Woodhead Publishing.

Saville B P (1999), *Physical testing of textiles*, Cambridge, Woodhead Publishing.

Schindler W D and Hauser P J (2004), *Chemical finishing of textiles*, Cambridge, Woodhead Publishing.

Sen A K (2001), *Coated textiles*, Cambridge, Woodhead Publishing.

Sikkema D J (2000), 'PIPD or M5 rigid-rod polymer', in Hearle (2000), 108–115.

Simpson W S and Crawshaw G H, eds (2002), *Wool: science and technology*, Cambridge, Woodhead Publishing.

Spencer D J (2001), *Knitting technology*, 3rd edn, Cambridge, Woodhead Publishing.

Tao X M, ed. (2001), *Smart fibres, fabrics and clothing*, Cambridge, Woodhead Publishing.

Tao X M, ed. (2004), *Wearable electronics and photonics*, Cambridge, Woodhead Publishing.

Thess A, Lee R, Nikolaev P, Dai H J, Petit P, Robert J, Xu C H, Lee Y H, Kim S G, Rinzler A G, Colbert D T, Scusetta G E, Tománek D, Fischer J E and Smalley R E (1996), 'Crystalline ropes of metallic carbon nanotubes', *Science*, **273**, 483 ff.

TTI and Noble Denton (1999), *Deepwater fibre moorings: an engineers' design guide*, Ledbury, Oilfield Publications Ltd.

Van Dingenen J L J (2000), 'Gel-spun high-performance polyethylene fibres', in Hearle (2000), 62–92.

Weedon G (2000), 'Solid-state extrusion high-molecular weight polyethylene fibres', in Hearle (2000), 132–144.

Windle A H (2004), *Carbon nanotube based continuous fibre*, presented at Polymer Fibres 2004 conference, Manchester.

Woodings C, ed. (2001), *Regenerated cellulose fibres*, Cambridge, Woodhead Publishing.

Yang H H (1993), *Kevlar aramid fiber*, New York, Wiley & Sons.

Young R J and So C L (2001), 'PBO and related polymers', in Hearle (2001), 101–108.

Zhang M, Atkinson K R and Baughman R H (2004) 'Multifunctional carbon nanotube yarns by downsizing an ancient technology', *Science*, **306**, 1358–1361.

Technical textiles for protection

P POTLURI and P NEEDHAM, University of Manchester, UK

6.1   Introduction

Protective textiles in the broadest sense can be defined as fibrous materials in the form of textiles used to enable an object to perform its function under hazardous conditions. This definition is so broad that it would include most functional textiles. For example, the object to be protected could be inanimate, packaging would then come under the definition, or the object could be the landscape so geotextiles would be included. However, this book concentrates on protective textiles that enable humans to perform their function or task under hazardous conditions.

Textiles have been used for human protection from time immemorial, primarily for protection from cold and rain. Primitive humans used hides and bark for protection. As civilisation advanced, wool felts in Northern Europe and woven cotton fabrics in India started to clothe the populations. While all the textiles have some protective function, protective textiles are those that are specifically designed for specific hazards such as fire, impact, chemical, etc.

6.2   Technical textiles

Technical textiles are defined as 'Textile materials and products manufactured primarily for their technical performance and functional properties rather than their aesthetic or decorative characteristics. A non-exhaustive list of end-uses includes aerospace, industrial, military, safety, transport textiles and geotextiles'.[1] From this definition, one can conclude that all the textile materials used in protective equipment may be considered as technical textiles. Protective equipment typically consists of both flexible and rigid components. Flexible components include textiles and films, either polymeric or metallic. Rigid components may include plastic, metallic, ceramic and fibre reinforced composites. Individual components are joined together by various techniques including stitching, lamination, and adhesive bonding.

Currently, around 200,000 tonnes of technical textiles with an estimated value of 2.5 billion dollars are used every year in protective clothing applications.[2] These figures would be substantially higher if we consider the protective textiles used in automotive (airbags, seat belts) and medical (anti-bacterial) applications. The protective textiles market is growing at an annual rate of 6.6%, and expected to grow faster in the post-9/11 scenario and also due to recent epidemics such as SARS and bird flu.

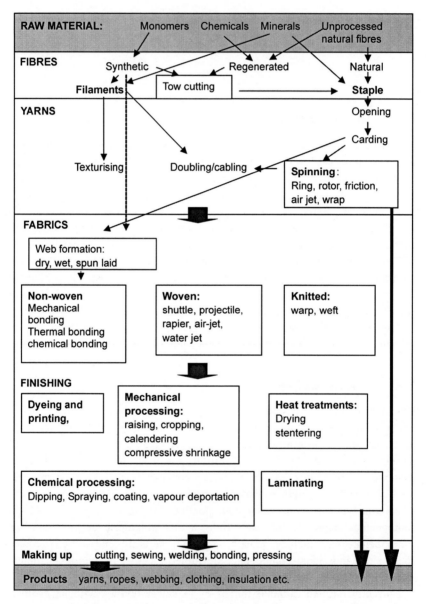

6.1 Summary of the hierarchy and principal processes of fibre assemblies.

6.2.1   Hierarchy of fibre assemblies

Textiles represent a complex hierarchy of fibres, yarns, fabrics and 3D fabric assemblies. This hierarchy and the principal processes involved are summarised in Fig. 6.1. To produce any protective textile takes many different manu-facturing processes. They can broadly be classified under processes to produce fibres, yarns, fabrics, finished fabrics and made-up textiles. If raw material is considered a separate category then the protective function of a textile product can be introduced or modified in any of the category stages. For example, nanomaterials can be used at the raw material stage to add to fibres to impart, flame retardance, chemical absorbance or electrical conductivity to a fibre. Selection of the type of fibre production process and fibre type can all impart protective properties.

The surface structure of fibres can be modified giving interesting properties; fibres can be crimped increasing their insulation ability. At the yarn processing stage, strength, stretch, bulk properties can be introduced and yarns can be blends of fibres or yarns. Strength, impact resistance, permeability, drape and stretch are some of the properties that can be introduced at the fabric production stage. At the finishing stage extra protective properties can be added. For example, waterproof finishes can be applied to give water protection. Fabrics can be laminated giving layered materials with differing properties on each side. Therefore at each stage protective properties can be added or modified to tailor the textile product for protection against a specific hazard. The following sections contain a brief description of the stages of textile processing, primarily aimed at non-specialist-textile readers.

6.2.2   Fibres

Fibres are the base units of any protective textile. Numerous books have been written about the fibres, a notable book on recent developments is by Hongu *et al.*[3] The fibre properties that are most often quoted are their dimensions because they are a major factor in determining how they are processed into a textile structure and the properties of the structure. Fibre lengths can be continuous (continuous filament) or of a known length distribution (staple fibres). The fibre fineness is quoted in terms of linear density, a standard unit is tex which is the weight in grammes of one kilometre of fibre or yarn. Decitex and Denier are commonly used units for fibres, where as Tex is used for yarns. Decitex is grams per 10 km whereas denier is grams per 9 km.

Fibres are often classified in terms of how they are produced: natural fibres, e.g., cotton, wool, silk; regenerated fibres, e.g., viscose, acetate, synthetic fibres, e.g., polyester, nylon. Synthetic fibres are the main types used in protective textiles, although natural fibres are used occasionally. The selection of a fibre for a particular application is determined initially by the fibre properties. In most cases a number of alternative fibres can be used so price, supply and tradition will determine the selection. The fibres commonly used are polyester/wool blends, nylon, wool, polyethylene, polypropylene, modacrylics, oxidised acrylics, aramids, glass and metallic fibres. More recently, Zylon (PBO fibre) has become popular for protective applications.[4]

6.2.3 Yarns

Individual micro-scale fibres, staple or continuous, are combined to form meso-scale structures called yarns. The yarns are typically fractions of a millimetre wide and several kilometres long and hence in a convenient form for fabric forming processes. Staple fibres are arranged in the form of a thick sliver, with fibres oriented along the length. The sliver is then drafted (stretched) into a finer cross-section, using a series of drafting rollers, before imparting twist.

A variety of spinning techniques, ring, rotor, friction, air-jet and wrap spinning, can be used to create inter-fibre cohesion through friction. The higher the twist level the higher the yarn strength up to a point; beyond this point strength drops due to increased off-axis orientation of the fibres. In wrap spinning, a fine continuous filament wraps fibres aligned in the axial direction.

Continuous filament fibres are commonly used in the form of 'roving' with little or no twist at all. However, twist is imparted for producing folded or ply yarn. Ply yarns have improved abrasion resistance and hence are commonly used as sewing thread. Yarns may be texturised to give three-dimensional crimp to the fibres thus improving bulk and elasticity; fabrics woven with textured yarns have improved comfort.

6.2.4 Woven textiles

Woven fabrics are the most common type of textile structures used for protective applications. They are constructed with two sets of yarn, warp and weft, interlaced at 90 degrees to each other. Several thousand-warp yarns, required for the fabric width, are assembled on a cylindrical drum, known as warp beam. The warp yarns are coated with a protective material (sizing) to minimise breakages during weaving. Once the warp beam is installed on a weaving machine, the process of weaving involves three primary motions – shedding, weft insertion and beat-up, and two secondary motions – let-off and take-up (Fig. 6.2). Shedding is the process of splitting the warp into two sheets, an upper and lower shed, to enable weft insertion. For example, even and odd numbered yarns are split to create a plain-woven fabric; other weaves require a more complex shedding pattern. A weft yarn is inserted through a shed using a variety of mechanisms including shuttle, projectile, rapier, air-jet and a water-jet. The newly inserted weft yarn is beaten-up into the cloth using a comb-like device, known as a reed. Let-off mechanisms release a required length of warp for each

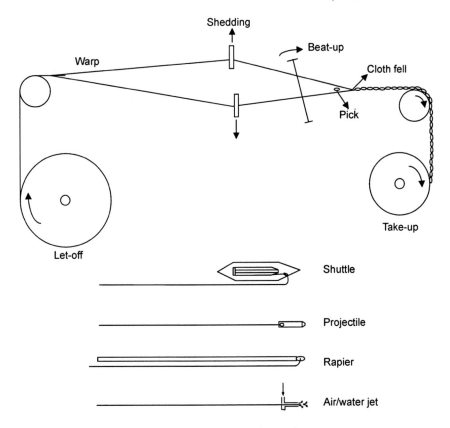

6.2 Weaving process and methods of weft insertion.

loom cycle while maintaining constant warp tension. The take-up mechanism winds the cloth at a uniform rate in order to maintain the weft spacing. Readers may refer to the excellent book by Ormerod and Sondhelm[5] for further details on the weaving processes.

Woven fabric structure

Depending on the degree of interlacement, basic weave structures may be classified as plain, matt, twill and satin/sateen with thousands of derived structures (Fig. 6.3). Plain weave has the simplest form of interlacement, one up and one down, denoted as 1/1. Matt weaves look similar to plain weave except that two or more warp yarns are lifted over two or more weft yarns; 2/2, 4/4, etc. Twill weaves are characterised by diagonal lines; the width and the angle of the twill lines depend on the interlacement 1/2, 1/3, 2/2, etc. Satins and sateens have long floats with occasional interlacement between warp and weft yarns. The type of woven structure has a significant influence on the physical, mechanical

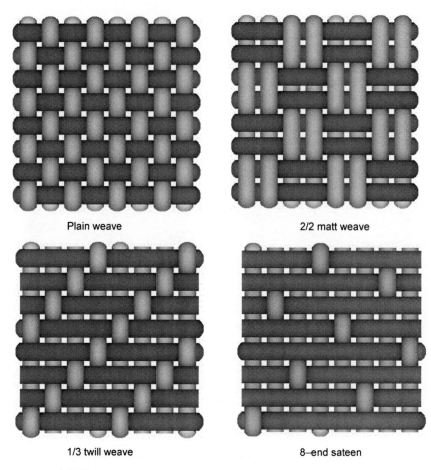

Plain weave   2/2 matt weave

1/3 twill weave   8–end sateen

6.3 Woven structures.

and fluid flow properties of fabrics. For example, a plain weave fabric has the highest degree of interlacement and hence offers highest resistance to inter-yarn slippage during penetration of a sharp object; at the same time, plain weave also has the highest degree of crimp and hence the lowest longitudinal wave speed.

Fabric specifications

- Fabric sett: fabric sett refers to number of warp yarns (ends/cm) and weft yarns (picks/cm) per unit length.
- Yarn diameter: for a given fibre type, yarn diameter is proportional to $\sqrt{T}$ where T is tex.
- Cover factor: cover factor indicates the proportion of the area covered by the yarns. Higher the cover factors lower the inter-yarn voids. Fabric cover factor is the sum of warp and weft cover factors.

Warp cover factor, $C_w = 0.1w\sqrt{T}$
Weft cover factor, $C_e = 0.1e\sqrt{T}$
where, w = warp density (ends/cm), e = weft density (picks/cm) and T = linear density in tex. Higher warp/weft densities should be used for finer yarns in comparison to coarser yarns in order to achieve a given cover factor. While fabrics are commonly woven with a cover factor of 24, most protective fabrics require higher cover factors, up to 32. This makes the weaving process much harder requiring heavier beat-up.

- Yarn crimp: due to interlacement, warp and weft yarns have a degree of waviness called crimp. Crimp level depends on cover factor and softness of the yarns and the weave type. Plain weave has the highest crimp while a satin weave has the lowest. Crimp level increases with cover factor. Softer yarns, i.e., yarns with little or no twist, spread out resulting in lower crimp in comparison to yarns with high twist level.
- Area density: as it is relatively difficult to estimate and measure fabric thickness, it is a common practice to use area density (g/m^2) instead of volume density.

6.2.5   Knitted textiles

Knitted fabrics essentially consist of a series of interlinked loops of yarn. Each loop along the width of a knitted fabric is produced by an individual needle. The fabrics are divided into two main categories; weft knitted and warp knitted. Weft knitting is where the loops are made horizontally across the fabric and they can be all produced by a single yarn. Figure 6.4(a) shows the simplest weft knitted structure 'plain weft knitting', this is where each row of loops are drawn through the previous row in the fabrics This structure can be modified to produce rib structures and by introducing special stitches, e.g., the miss stitch or tuck stitch, eyelet structures can be produced. The interested reader is referred to Spencer[6] for greater details of both warp and weft knitting fabrics and processors.

Warp knitting is where each needle is supplied by an individual yarn so that the loops formed by a yarn tend to run down the length of the fabric. This is illustrated in Fig. 6.4(b) showing the knitted structure of the face side of a warp knitted half tricot. This is not a stable fabric and so is unsuitable for clothing. To produce this fabric each yarn passes through a guide attached to a guide-bar. The fabric is produced by a combination of needle motion and sideways movement of the guide-bar. Multiple guide-bars are used to produce modifications on the basic half tricot giving for example locknit, satin, sharkskin fabrics.

Terms used to define a knitted fabric include:

- Wales per cm: number of loops per cm in the fabric width direction
- Courses per cm: number of loops per cm in the fabric length direction
- Stitch density: (wales per cm) × (courses per cm)

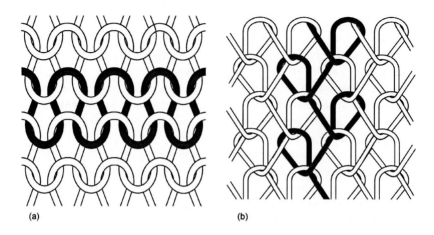

6.4 (a) Plain weft knitted structure – back side; (b) warp knitting – half tricot – face side.

- Stitch length: the length of yarn in one loop
- Tightness factor (originally called cover factor): $\sqrt{\{(\text{yarn tex})/(\text{stitch length})\}}$. For most weft knitted fabrics the tightness factor is in the range 1 to 2.
- Machine gauge: number of needles per cm or the distance in 0.1 mm units between the needles (metric system). The machine gauge limits the maximum yarn tex that can be knitted on a machine. The yarn tex is inversely proportionate to the square root of the machine gauge (in needles per cm units).

Weft knitted fabric can be produced on a number of different types of knitting machines. Circular or flat bar machines using a latch needle can produce both fabrics and knitted garments. Straight bar or circular machines using a bearded needle can produce shaped knitwear. Many machines can produce a double fabric structure with differing knitted structure on each fabric face. Spacer yarns can be inserted between the front and back fabric, thus creating a complex three layer structure. The properties of each layer are determined by the fibre, yarn properties and structure of that layer. These structures can be tailored for specific applications and are useful in the protective textile field.

Warp knitted fabrics can also be produced on a number of different types of knitting machine. Raschel machines using a latch or compound needle produce high pile upholstery fabrics, industrial furnishing fabrics and bags for vegetables. Tricot machines using bearded or compound needles produce lace, nets, outerwear fabrics. Weft insertion with, for example, elastic yarns or fleeces can produce directionally orientated fabrics. Warp knitted fabrics are commonly used in linings for protective clothing and laminated with polyurethane foams to provide a strong flexible base for the foam.

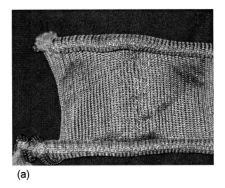

(a)   (b)

6.5 Slash-proof knitted fabrics produced with metal-cored yarns (courtesy A Rowe Ltd).

The main output of knitted textiles is for the apparel and household markets but they are being increasingly used in protective textiles. A major advantage is that seamless clothing and made-up textiles can be produced only by knitting. Knitted fabrics are highly elastic and less stable in comparison to woven fabrics – this is an advantage for applications such as cut, slash or abrasion resistance. Figure 6.5 illustrates a range of slash-proof knitted fabrics produced with metal core yarns.

6.2.6   Non-woven textiles

Non-wovens are a class of fabrics that are produced directly from fibres, and in some cases directly from polymers, thereby obviating a number of intermediate processes such as spinning, winding, warping, weaving/knitting. Hence non-wovens can be produced inexpensively for both single use and durable applications. Non-wovens are used extensively for chemical and biological protection. They are also commonly used as breathable, waterproof and windproof layers in skiwear and other protective clothing. Non-wovens are produced in two distinct steps:

1.   Web formation: arrangement of fibres into a 2D sheet, and
2.   Consolidation: bonding the fibres together to create a non-woven fabric.

Web formation methods

Web formation may be classified into dry laid, spun laid and wet laid processes.

Dry laid process

Dry textile fibres are carded, using a carding machine similar to the ones used in the spinning industry, to arrange the fibres in a 2D sheet with fibre orientations

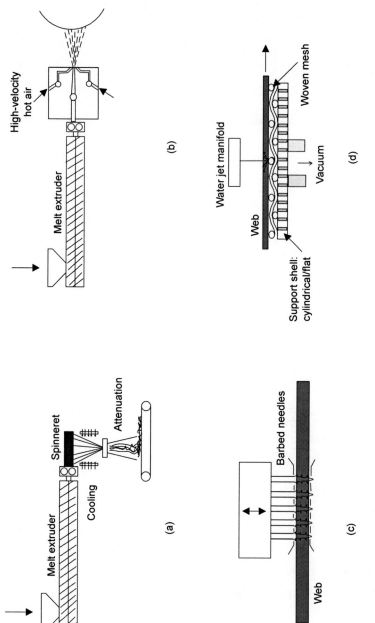

6.6 Non-woven processes: (a) spun-laid; (b) melt-blown; (c) needle punching; (d) hydro-entanglement.

predominantly in the machine direction. The web is subsequently folded using a cross-lapping machine to increase the web thickness and to achieve transverse fibre orientation. In some cases, conventional carded and cross-lapped webs are combined to produce a web with bi-directional fibre orientation. Alternatively, an aerodynamic system is used for creating a web with random fibre orientation.

Spun laid process

This is a method of producing fabrics directly from polymer chips, hence eliminating the entire textile supply chain. Fibres are extruded from a spinneret similar to conventional melt spinning process. These fibres are attenuated (stretched) using high-velocity air streams before depositing on a conveyer in a random manner (Fig. 6.6(a)). The spun laid process is the most commonly used method for producing both disposable and durable nonwovens for protective applications.

There are other related systems such as flash spinning, melt blowing and electro-spinning. Flash spinning involves extrusion of a polymer film dissolved in a solvent; subsequent evaporation of the solvent and mechanical stretching of the film results in a network of very fine fibres. These fibres are subsequently bonded to create a smooth, microporous textile structure. For example, Dupont's Tyvek[7] is produced using a flash spinning process, and is widely used in protective applications. The melt-blowing process produces microfibres by attenuating the polymer jet, coming out of the spinneret, using high-velocity air jet (Fig. 6.6(b)). Since the polymer is stretched in the molten state, extremely fine fibres can be produced. 3M Thinsulate[®8] is a very popular microfibre insulation material produced using melt-blowing technology. Because of the lack of molecular orientation, melt-blown fibres are weak and hence are generally used in conjunction with other type of non-wovens. For example, a composite non-woven consisting of melt-blown layer and a spun-bond layer is becoming popular for medical protective applications.

Electrospinning is a relatively new technique for producing nanofibres. Doshi et al.[9] described the electrospinning process: 'an electric field is used to create a charged jet of polymer solution. As the jet travels in air, the solvent evaporates leaving behind fibres that can be electrically deflected or collected on a screen'. Gibson et al.[10] explored the application of electrospun nanofibres for chemical protective clothing.

Wet laid process

Developed from the traditional paper-making process, relatively short textile and wood fibres are dispersed in large quantities of water before depositing on an inclined wire mesh. These materials find application in hospital drapes and filters.

Consolidation processes

Fibrous webs can be consolidated using a number of techniques depending on the area density and the desired properties. They can be classified into mechanical, chemical, thermal and stitch bonding processes.

Mechanical bonding

Needle-punching and hydro-entanglement are two complementary mechanical processes. Relatively thick webs ($150 \, \text{g/m}^2$ to $1,000 \, \text{g/m}^2$) are felted with the aid of oscillating barbed needles (Fig. 6.6(c)). The hydro-entanglement process (Fig. 6.6(d)) uses high-velocity water jets to consolidate relatively thin webs ($<140 \, \text{g/m}^2$). The resulting spunlaced fabrics are highly drapable and hence popular for medical protective clothing.

Chemical bonding

Fibres are bonded together with a suitable adhesive and subsequently cured under heat. Saturation bonding is seldom used for protective applications, as this process results in a relatively stiff non-porous material. Spray and print bonding instead of saturation bonding improves the flexibility and permeability.

Thermal bonding

Relatively thin webs are passed through a heated calender, resulting in partial melting and bonding of fibres. Thermal bonding is a high-speed process and hence commonly used in conjunction with spun laid webs.

Stitch bonding

Cross-laid webs are stitched together with a relatively large number of needles across the width. Alternately, stitch bonding is also used to bond a series of non-interlaced thread systems.

6.2.7 3D woven structures

Protective textiles, especially produced for ballistic applications, consist of a number of fabric layers stitched or quilted together. An alternative and cost effective method would be to weave all the layers together. Relatively thick fabrics consisting of a number of warp and weft layers can be produced on conventional and specialised 3D weaving machines. The warp and weft yarns are held together with interlacing z-yarns: orthogonal and angle-interlocked are the two prominent structures used. In these structures, most of the yarns remain non-crimped and hence these structures have high in-plane modulus and high longitudinal wave velocity. 3Tex developed a number of orthogonal weaves for the ballistic protection market.[11] They can typically weave up to 14 warp layers with a corresponding number of weft layers. Potentially, 3D weaves have a number of advantages over broad cloth:

- Fabrics can be woven with much higher cover factors since there are only small percentages of interlacing yarns.
- Warp and weft yarns have very little or no crimp at all in 3D weaves. Hence, coarser yarns can be used as opposed to fine yarns being used in 2D fabrics, to minimise the effect of crimp.
- Labour cost can be reduced as a result of using coarser yarns and eliminating subsequent stitching processes.

6.2.8   Non-interlaced structures

In 3D woven fabrics, a small percentage of binding yarns are used to hold a number of warp and weft layers together. Honeywell[12] developed unidirectional tape technology by binding a series of parallel yarns/fibres using a flexible resin. Two layers are then cross-plied at right-angles (0°/90°) and fused into a composite structure under heat and pressure. These non-interlaced structures are more efficient at dissipating shock waves from the point of impact.

6.2.9   Finishing

In many cases the functional requirements of a protective textile application cannot be met by the fibre and structure alone. Finishing is used to improve or add to the protective properties of the textile structure. For the purpose of this outline, dyeing and printing are being considered as part of finishing. There are a vast number of processes in finishing and they cannot hope to be covered in this brief outline. There are many textbooks and reviews on the subject.[13,14]

Finishing protective textiles can be divided into five main areas:

1. Dyeing and printing are a means of adding colour to a material. Colour can be chemically linked to the fibre (dyeing) or bonded by an adhesive (printing). The coloration process can be introduced at a number of stages in the protective textile manufacturing process. Pigments can be added directly into the synthetic fibre spinning process producing coloured fibres; bulk fibres, rovings and yarns. Fabrics can be dyed, printed or pigments added to coatings. Coloration is mainly used for aesthetic reasons even with protective textiles. It does have important applications in camouflage protective textiles and high visibility textiles.

2. Mechanical processing is the mechanical treatment of textiles fabrics to change their structure. The main processes are compression of fabrics

between heavy rollers (calendering), brushing the surface to produce a nap (raising), removing surface raised fibres (cropping), shrinkage of a textile structure to produce a denser structure (compressive shrinkage). Mechanical processing for protective textile applications tends to be used to aid further processing or to control the final thickness of the structure. It can also be used to control the textile structure pore size for filtration applications.

3. Heat treatments are used to dry a textile material and stabilise it at the required dimensions.

4. Chemical processing is the application of chemicals to a textile. The chemicals can be applied by spraying, dipping the textile in a bath of chemicals, coating, or vapour depositions. Chemical processing is widely used for protective textiles, for example, water-repellent finishes, oil-repellent finishes, flame-retardant finishes and antistatic finishes are all commonly used.

5. Lamination is the joining of two or more fabrics or a textile structure with a plastic sheet, film or foam. Examples include waterproof materials such as Goretex®.

6.2.10 Sewing and seams

There are a number of textbooks on the subject.[15,16] Sewing and seams are important for all made-up textiles but they have a particular importance in waterproof and chemical protective textiles. The seams need to be designed and manufactured in such a way that they do not allow water or chemicals to penetrate. Traditionally sewn seams or ultrasonic welded seams for PVC coated textiles have been used. The sewn seams are often coated with a polymer or tape stuck over the seams; so that the seams have the same protective properties as the bulk material. Depending on the application, the tape can be either impermeable or micro-porous. In recent years adhesive and bonded seams have been developed.[17] They have the advantage of producing a lighter product compared to a sewn seam and no further seam treatment is required. This is an area of potential growth with protective textile products.

6.3   Types of hazards

Protective textiles are used against a variety of hazards; these hazards may be classified as mechanical, environmental, pressure, thermal, fire, chemical, biological, electrical and radiation. Each hazard has its own functional behaviour leading to the development of specialised protective textiles for that hazard. In many cases protection against more than one hazard is a requirement, for example, the firefighter may require protection against fire and chemicals. Also the person must be able to function effectively when using protective textiles. Therefore the design of protective textiles is complex depending on numerous factors. In order to simplify the discussion, protective textiles will be specifically related to a particular hazard.

6.4   Mechanical hazards

Mechanical hazards are the most common and most diverse form of threat facing humans. The threat may be in the form of a bullet, knife, blunt object, high-speed rotating machinery, falling from heights, motor cycle or car accidents to name a few. In most cases, protective textiles absorb some of the kinetic energy from the impact and modify the nature of the impact and, by this process, minimise the trauma and protect the vital organs. The protective textile may be:

- worn on the body; bullet-proof vests and other personal protection equipment fall into this category
- attached to the body; parachutes, body harness attached to climbing ropes, seatbelts, etc.
- present in the immediate surroundings; air bags and deployable side curtains in cars.

6.4.1   Ballistic and knife protection

Ballistic armour is designed based on both the type and level of threat. For example, choice of the textile and its area density depends on type and velocity of the projectile as well as the expected degree of comfort and level of protection. The kinetic energy from the penetrating object must be absorbed by the fibrous network, and dispersed to prevent localised damage. Horsfall et al.[18] classified the threats in terms of kinetic energy density (KED) i.e., incident kinetic energy per unit area, as presented in Table 6.1.

Table 6.1 Threat classification based on KED[18]

Threat	Velocity (m/s)	Kinetic energy (J)	Presented area (mm^2)	KED (J/mm^2)	Typical armour type
Knife	10	43	2.5 (blunt) 0.2 (sharp)	17 210	Special textiles or plates
Handgun bullet (0.357″)	450	1,032	65 (initial) 254 (final)	16 4	Textiles
Assault rifle bullet (AK47)	720	2,050	45	45	Composites
High-velocity bullet (SA80)	940	1,805	24	75	Ceramic

Handgun bullets that deform under impact are stopped effectively by flexible textile structures. The bullet deforms under impact and is subsequently captured within the first few layers. Kinetic energy is absorbed by the fibres primarily as tensile energy and to a smaller extent bending and inter-fibre/yarn frictional energy. Energy is dissipated in the form of stress waves, initially along the yarns and then to the transverse yarns through the points of interlacement. High-performance fibres such as Kevlar, Spectra and Zylon are commonly used for ballistic armour; these fibres have high modulus so that stress waves are dissipated rapidly, and they have high strength and good transverse strength in order to avoid premature failure.

Weave selection is a compromise. For example, a plain weave is the most preferred structure as it has the highest interlacement that minimises yarn slippage, however, plain weave has highest crimp that minimises wave propagation speed. A recent trend is to use finer yarns and tighter weaves to improve the ballistic performance while keeping lower crimp levels. Traditional compromise between the degree of tightness and the crimp level is solved in Honeywell's Spectrashield®, which is a bidirectional non-crimp fabric bonded together with a flexible resin system. Orthogonal 3D woven fabrics from 3Tex have very low-crimp warp and weft yarns held in place by binding yarns. It is expected that these 3D fabrics will eventually replace conventional multilayer constructions in body armour.

For high-velocity bullets (e.g. AK 47 assault rifle) with kinetic energy density in excess of $30 \, J/mm^2$, shear failure occurs in the first few fabric layers.[18] As a result, conventional textile armour becomes very bulky for this application. As the assault rifle bullets do not deform upon impact, textile armour must be rigidised by creating a polymer matrix composite.

Military ballistic armour, designed for high-velocity projectiles with kinetic energy density in excess of $50 \, J/mm^2$ consists of both soft and hard parts. The hard part consists of a ceramic plate with a polymer composite backing. The polymer composite prevents the brittle fragmentation of the ceramic plate. The ceramic-composite insert is placed in a flexible armour.

Knives apply a relatively modest force over a very small contact area, hence highest energy density. Sharp knives and needles have a tendency to force through the weave; this can be resisted by finer and tighter weaves, especially plain weaves. Film lamination and abrasive coatings are commonly used for improving penetration resistance. However, protection against sharp needles is still a problem and open to new development in the area of protective textiles.

Kevlar 29® from Dupont was the first high-performance para-aramid fibre to replace ballistic nylon, to be followed by Kevlar 49®. Twaron is another para-aramid fibre developed by Akzo Nobel. A newly developed PBO fibre from Toyobo in Japan is Zylon®, consisting of rigid-rod chain molecules of poly(p-phenylene-2,6-benzobisoxazole). Zylon, similar in density to Kevlar, has exceptional ballistic properties, although a decrease in ballistic resistance occurs under heat and humidity. Therefore its mechanical properties degrade much more quickly than other ballistic fibres. Other common ballistic fibres are Dyneema® from DSM and Spectra® from Honeywell, both ultra-high-strength polyethylene fibres with a specific gravity less than one.

Although analytical models have been under development for a while, protective systems are designed using empirical methods and experience. In the early 1980s, Leech[19,20] analysed the mechanics of ballistic impact on textiles using variational principles. Leech[20] predicted that the transverse wave-front for an orthogonal fabric is a rhombus whose sides depend on longitudinal wave speeds. Gu[21] has recently extended this work with the help of material properties measured at high strain rates. Hetherington et al.[22] did extensive work on optimisation of two component composite armours. Naik et al.[23] compared the ballistic performance of carbon-epoxy and E glass-epoxy composites and found that E glass-epoxy composites are superior.

6.4.2   Blunt impact protection

Protection from blunt impact is another important area, e.g., injuries from vehicle accidents, falls and physical assaults with 'unconventional' weapons such as bats, bottles, planks and metal bars. Dupont has recently developed a multi-threat-protection (MTP) body armour for protection from a variety of threats including blunt trauma. Protection from blunt trauma requires a shock-absorbing cushion between the human body and the outer shield. In a number of situations, especially involving automobile accidents and falling from heights, deployable textile structures are popular. For example, parachutes are routinely used for sky jumping activities both in the military and sports area. Essentially, a parachute is a carefully folded textile structure that is deployed on activation – deployment is carried out manually in most cases or in some cases automatically. Parachutes have been used as part of aircraft seat ejection systems, and more recently on Mars missions. For example, Boeing[24] has recently been awarded a contract to develop advanced parachute systems with wind drift compensation for future Mars missions.

Airbags, invented by Allen Breed in 1968, have become almost mandatory for protecting vehicle passengers from collision and newer systems are being developed to protect the pedestrian. Automotive airbags are a huge market with roughly 100 million bags produced per year. Airbag technology uses a textile cushion, neatly folded and tucked away in the steering wheel. When a deceleration sensor detects a collision, it triggers a gas generator to inflate the textile cushion. Autoliv[25] introduced inflatable curtains in 1998 and has recently developed a pedestrian protection airbag (PPA) to be part of a car bonnet (Fig. 6.7). The cushion is a tightly woven fabric made with Nylon 6,6 multifilament yarns and subsequently coated with silicone to improve resistance to ageing and to reduce air permeability. Airbags are traditionally cut and sewn from

 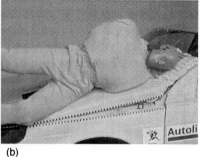

(a)   (b)

6.7 (a) Inflatable side curtains. (b) Pedestrian protection air bags (courtesy Autoliv).

broadcloth. However, Autoliv developed one-piece-weaving (OPW) technology, based on 3D Jacquard weaving, to eliminate the need for cutting and sewing.

Airbags are also used in Mars missions. Airbags must be strong enough to cushion the spacecraft if it lands on rocks or rough terrain and allow it to bounce across the surface of Mars at motorway speeds after landing. To add to the complexity, the airbags must be inflated seconds before touchdown and deflated once safely on the ground. Mars airbags are made with Vectran®, a copolymerised aromatic polyester fibre.

6.5   Pressure hazards

The pressure hazard can be divided into three categories: blast, atmospheric pressure and noise. Blast was dealt with in the mechanical hazard section.

6.5.1   Atmospheric pressure

This is a specialised hazard that relatively few people are likely to be exposed to. The oil industry with deep-sea work, working in space and military flights are where the hazards are experienced. All protective clothing designed to cope with this type of hazard are fully enclosed air impermeable suits with their own air supply. The suit is designed not to leak under a high pressure difference.

6.5.2   Noise

The hazard from noise has increased in importance over the past fifty years partly due to a clearer understanding of the damage noise can cause. This has led to the development of sound insulation materials for buildings and automotive industries. Polyurethane foams, non-woven or composite materials of wool, glass fibres and man-made fibres are typically used for these applications. The soundproofing characteristics of the material are a combination of the fibre properties, polymer properties for composite materials and the structure of the material.

6.6   Environmental and fire hazards

6.6.1   Environmental

If the temperature is outside the range 18 °C to 25 °C most people without clothing would begin to feel uncomfortable and if the temperature is outside the range 0 °C to 40 °C for any length of time, it can be life threatening. If water wets a person their body temperature drops making them feel uncomfortable and in extreme conditions this can also be life threatening. Therefore, the functions of thermal protective textiles are to protect against heat, cold and water. There has been extensive research and development in these three areas over the years. Recent developments in protection from heat and cold are in phase change materials. These materials are usually paraffinic hydrocarbons which are added to manmade fibres during spinning or encapsulated into microcapsules which are added to textiles during finishing. They work by changing phase at a set temperature; by tailoring the paraffinic hydrocarbon used the phase change temperature can be modified. If the temperature falls below the phase change temperature, heat is given off, likewise if the temperature rises above the phase change temperature heat is absorbed into the material. Outlast Technologies is the main company in this area, many outdoor pursuits clothing and footwear are beginning to use these materials. However, test methods for these materials are still being developed and the indications are that the materials can only give off heat or take in heat for a short time. This is likely to limit their use to specialised applications.

6.6.2   Fire

Wool and asbestos fibres, due to their inherent flame-retardant properties, were traditionally used in fire-protective textiles. Today, most fibres have a flame-retardant version, and fibres such as Nomex, Kevlar, glass, carbon or textile structures with flame-retardant finishes have replaced asbestos due to the health risk. The selection of fibres, textile structure and finish will depend largely on the application. Horrocks and Price[26] give a good review of the subject.

6.7   Chemical and biological hazards

6.7.1   Chemical

The chemical hazard is the most complex and varied of all the hazards. It is estimated[27] that more than 100,000 chemical products with very different toxicological properties are in use throughout the world. Chemical hazards are experienced in most industries, examples range from protection against particle contamination in a clean room, for electronic component manufacture, to protection against chemical warfare agents. There are various ways of defining hazardous chemicals. The level of risk is one method; the amount of chemical, exposure time, toxicity are all factors that need to be taken into account. Another common method is by classification into particles, liquids and gases. This is useful for research and development of protective textiles because each chemical phase will behave differently and so require different strategies for protection.

There are four possible interactions between a chemical and a chemical protective textile:[28]

1. Chemical degradation, which is a breakdown of the textile structure.
2. Chemical penetration, which is the chemical flow through the textile structure by wicking or pressure effects in air permeable structures, or through imperfections or seams and closures in impermeable structures.
3. Chemical permeation, which is the molecular flow of chemicals through the material of the structure.
4. The chemical may not interact but evaporate and the vapour will either go into the atmosphere or enter the garment.

Any combination of the four possibilities could occur. Most research has been reported on permeation. There is a vast amount of chemical permeation data from manufacturers and published papers, for example around a third of the papers in the *Performance of Protective Clothing* book of papers[29] are on permeation data. Fibre, fabric type, fabric finishes, coatings and lamination are factors that can have a major influence on the level of protection.

The chemical degradation of a material is determined by the reaction of the fibres, coating and lamination material to the chemical, which is determined by their chemical nature. For example, sulphuric acid at room temperature will attack cotton fabric, producing holes while nylon fabric is resistant. Chemical penetration is determined by the pore size and wickability of the material, which is determined by fibre linear density, yarn structure, fabric structure, fabric coating types and material surface energy. Examples of toxic chemical particles are carbon black, asbestos, dye powders, coal dust, beryllium particles and antineoplastic drugs. Lung damage can occur when particles around 1 μm are inhaled.[30] Particles that are in the size range 0.001 μm to 100 μm are considered an aerosol and when liquid droplets reach micron dimensions, their behaviour becomes similar to solid particles of the same size. If the particles are smaller than the textile structure pore size then other factors including Brownian diffusion and electrostatic capture will determine penetration.[31,32]

The ability of a chemical to be adsorbed in the fibres and in any fabric finishes and the diffusion of chemical molecules through them determines the chemical permeation properties of a textile structure. Thus chemical permeation of a material depends on the chemical nature and molecular structure of the fibre, coating and lamination material. Historically there have been two approaches in designing chemical protective textiles:

1. Impermeable structures. These structures are used for protection against all chemical product phases, mainly for liquid and gas protection. These are used for the majority of chemical protective clothing and the majority of research has been carried out in this area. The clothing is usually a single layer of textile material coated or laminated with thermoplastic polymers or synthetic rubbers; examples are PVC and butyl rubber. The main disadvantage is that because the structure is impermeable to water vapour and air the time a person can remain within the structure is limited. To overcome this, independent air supplies can be used, for example, spacesuits, and specialist firefighting suits.
2. Permeable structures. These structures are used for particle filtration, which was discussed above but are also used for gas permeable chemical protective clothing. Gas permeable structures are described as a discontinuous material that, by selective sorption, permits the transport of water vapour and air but limits the diffusion of hazardous chemicals. They are normally textile fabrics, single or multilayer and can be coated or laminated. Gas permeable chemical protective clothing is usually used for liquid splash, aerosol, dust, hazardous vapour or gas protection, particularly against chemical warfare agents.

To stop penetration of the hazardous gas the pore size of the structure must be less than the diameter of a single gas molecule. This would exclude air molecules thus making it impermeable. Therefore there must be penetration of hazardous gas into the structure. In order to give protection, the material contains substances which either neutralise the chemical or hold them in such a way that they cannot pass through to the inner face of the material. Brief details of the neutralising approach are given by Rajan et al.[33] where they describe studies on two catalysts that could be imbedded in the structure. The holding approach is usually where activated charcoal, in the form of powder or beads bonded by resin or an activated charcoal textile laminate, is used.[34] Most chemical protective clothing of this type has water, oil and flame-retardant finishes and many chemical protective clothing systems are multilayered. For example the UK MKIV NBC suit has two layers, an outer wicking layer which reduces the local concentration of toxic agent droplets on the suit and an inner layer of impregnated carbon to absorb the toxic agent. An example of how complex a system can be is the 'casualty care system'.[35] This is basically a protective bag that a casualty is placed in when under chemical or biological warfare threats. The bag is of multilayer construction consisting of a 50/50 cotton/nylon woven fabric with oil/water-repellent finishes, a non-woven containing activated charcoal in bead form to absorb chemicals, an electrostatically charged fine non-woven to filter out aerosols and particulates, and a final biologically reactive layer consisting of a knitted fabric with a chemical resin finish.

6.7.2   Biological

The biological hazard covers a variety of threats from biological warfare, insects, food contamination, and disease infection. The objective of protective textiles is to stop a person coming into contact with the hazard or to stop a person transmitting a potential infection. To achieve the objective two approaches are used and often a combination of the two. These are the barrier approach and the chemical release approach. The barrier approach is used extensively for biological warfare where NBC suits described above are used, for highly infectious agents and diseases fully encapsulated suits with their own air supplies are often used.

The earliest form of insect protection was the mosquito net, which utilises the filtration ability of woven and knitted fabrics. Insecticides and insect-repellent finishes are sometimes applied to the fabrics to improve protection. The barrier approach is often used for mattress covers to stop bed mites. In this case spun-bonded non-wovens or non-woven/plastic film laminates are often used. They may have a hydrophobic finish to give some bacterial protection. Disposable non-woven clothing for staff in operating theatres is commonly used to stop cross-infections. These are more generally used today due to the increase of MRSA (methicillin-resistant staphylococcus aureus) and other antibiotic-resistant bacteria in hospitals. The UK national audit office reports that hospital acquired infections kill at least 5,000 people in the UK each year and it has been reported that the bacteria can live on textile materials for longer than three months.[36]

The chemical release approach is where chemicals that stop or limit bacterial growth or even kill bacteria are applied to textiles. The application could be either in the form of a fabric finish or by putting the chemical directly into the fibre structure. Rajendran and Anand[37] give a good brief review of this approach.

6.8   Electrical and radiation hazards

6.8.1   Electrical

The electrical hazard can be divided into electrostatic charge, lightning strike and high-voltage electricity. Protective textiles for all these hazards incorporate electrical conductive materials. A proportion of conductive fibres, for example carbon, metal, synthetic fibres with a carbon core and conductive polymer fibres are commonly used. The other methods are to apply a conductive finish, coating or laminate the fabric to conductive polymer films.

6.8.2 Nuclear

Radiation hazards are a problem in the nuclear power industries and the health care services. Traditional protection is by shielding with materials containing heavy metals, for example, leaded aprons are used to protect against X-rays. Disposable non-woven clothing such as polyethylene spun-bonded materials are used in the nuclear power industries. They will not protect against gamma rays but stop radioactive dust and grease containing radioactive particles from coming in contact with the skin.

A recent development is a material called Demron[38] which is produced by laminating a polymer film between a woven and non-woven fabric. The polymer is a composite of polyurethane and polyvinylchloride that incorporates organic and inorganic salt particles that block X-rays, low-energy gamma, alpha and beta emissions. The polymer must be tailored for a specific threat. Suits are now being made of this material.

6.9   Future trends

The technical textiles industry is likely to continue to lose the old craft approach to product development, relying more on scientific understanding to tailor products and less of the 'feel' and experience of the technician. With the development of faster computers, process and quality control will continue to improve. As understanding of textile materials and their interactions increases it may be possible to use the engineering material section approach in designing protective clothing against particular threats. This approach has recently been partially used in the design of the spacesuit for the future Mars mission.[39]

Smart textiles are likely to play a large part in the development of protective textiles in the next decade. This will lead to the synthesis of parts of the technical textile industry with the electronic and materials industries. The ultimate aim of any protective textile would be to have the ability to recognise a hazard and then to change its properties to give complete protection against the hazard. This may not be achievable but with the recent developments in smart textiles we can begin to contemplate it.

6.10   References

1. The Textile Institute, *Textile Terms and Definitions*, 9th edn, Textile Institute, 1991.
2. A R Horrocks, S C Anand, *Handbook of Technical Textiles*, Woodhead Publishing, Cambridge, 2000.
3. T Hongu, G O Phillips, *New Fibres*, Ellis Horwood Ltd, UK, 1991.
4. PBO fiber Zylan technical information, Toyobo Co. Ltd, Japan, www.toyobo.co.jp
5. A Ormerod, W S Sondhelm, *Weaving – technology and operations*, Textile Institute, Manchester, 1999.
6. David J Spencer, *Knitting Technology*, Pergamon Press, Oxford, UK, 1986.

7. www.tyvek.com

8. 3M thinsulate website: http://cms.3m.com/cms/US/en/2-147/creFRFT/view.jhtml

9. J Doshi, D Reneker, Electrospinning process and applications of electrospun fibers, *Journal of Electrostatics*, 1995, 35, 151–160.

10. P Gibson, H Schreuder-Gibson, D Ravin, Transport properties of porous membranes based on electrospun nanofibers, *Colloids and Surfaces A*, 2001, 187, 469–481.

11. J N Singletary, A Bogdanovich, Orthogonal weaving for ballistic protection, *Technical Usage Textiles*, 2000, 37, pp 27–30

12. Honeywell's Spectra shield website: www.honeywell.com/sites/sm/afc/spectra_shield.htm

13. *Textile finishing*, edited by Derek Heywood, Society of Dyers & Colourists, 2003.

14. Rouette, Hans-Karl, *Encyclopedia of textile finishing*, vols 1–3, Springer, New York, 2000.

15. R N Laing, J Webster, *Stitches & Seams*, The Textile Institute, 1999.

16. H Carr, B Latham, D J Tyler, *Technology of Clothing Manufacture*, 3rd edn, Blackwell Science, UK, 2000.

17. K Desmarteau, Sew No More?: Adhesive seams gain momentum, *Apparel Magazine*, Vol. 45, Issue 6, 2004.

18. I Horsfall, C H Watson, Ballistic and stab protection, Short course in impact and explosion engineering, UMIST, Manchester, 25–28 March 2003.

19. C M Leech, 'Dynamics of flexible filament assemblies'. In Hearle J W S, Thwaites J J, Amirbayat J, eds. *Mechanics of flexible fibre assemblies*. The Netherlands, Sijthoff & Nooordhoff, 1980.

20. C M Leech, J W S Hearle, J Mansell, A variational model for the arrest of projectiles by woven fabrics and nets, *J Text Inst*, 1979, 70(11), 469–78.

21. B Gu, Analytical modelling for the ballistic perforation of a planar plain-woven fabric target by projectile, *Composites B*, 2003, 34, 361–371.

22. J G Hetherington, Optimisation of two component composite armours, *Int J Impact Eng*, 1993, 12(3), 409–14.

23. N K Naik, P Shrirao, composite structures under ballistic impact, *Composite Structures*, 2004, 66(1–4), 579–590.

24. www.boeing.com/news/releases/2004/q3/nr_040816n.html

25. www.autoliv.com

26. A R Horrocks, D Price, *Fire Retardant Materials*, Woodhead Publishing Ltd, UK, 2001.

27. M. Raheel, *Protective Clothing Systems and Materials*, Marcel Dekker Inc, 39, 1994.

28. S P Berardinelli, L Cottingham, Evaluation of Chemical Protective Garment Seams and Closures for Resistance to Liquid Penetration, *Performance of Protective Clothing*, ASTM, STP 900, 263–275, 1985.

29. *Performance of Protective Clothing*, ASTM, STP 900, 1986, STP 989, 1988, STP1273 vol. 6, 1996.

30. S Kim, S J Karrila, *Microhydrodynamics: Principles and Selected Applications*, Butterworth-Heinemann, 1991.

31. W C Hinds, *Aersol Technology: Properties, Behaviour and Measurement of Airborne Particles*, John Wiley, 1982.

32. S M Mainini, S P Hersh, P A Tucker, Barrier Fabrics for Protection against aerosols, *Textile Progress*, The Textile Institute, 26, 1, 1995.

33. K S Rajan, S Mainer, J E Walker, A Colunga, Exploration of a Catalysis Approach for Application to Chem-Protective Garments, Protection against CW Agents 5th Symposium Supplement, *UMEA*, 219, 1995.
34. S M Watkins, *Clothing: The Portable Environment*, Iowa State University Press, 2nd edn, 172, 1995.
35. M.Kelar, To serve and protect, *Industrial Fabric Products Review*, 89, 11, 2004.
36. BBC Web News Bulletin – http://news.bbc.co.uk/1/hi/health/653490.stm
37. S Rajendran, S C Anand, Developments in Medical Textiles, section 4.5.3; *Textile Progress*, 32, 4, The Textile Institute, 2002.
38. S Ashley, X-ray Proofing, *Scientific American*, May 2003.
39. J L Marcy, A C Shalanski, M A R Yarmuch, B M Patchett, Material Choices for Mars, *J of Material Engineering and Performance*, 13, 2, April 2004.

7
Intelligent textiles for protection

L VAN LANGENHOVE, R PUERS and D MATTHYS,
University of Ghent, Belgium

7.1   Introduction

The discoveries of shape memory materials in the 1960s and intelligent polymeric gels in the 1970s were generally accepted as the birth of smart materials. The concept 'smart material' as such was defined in Japan in 1989. It was not until the late 1990s that intelligent materials were introduced into textiles. The first textile material that, in retrospect, was labelled as a 'smart textile' was silk thread having a shape memory (by analogy with the better known 'shape memory alloys' which will be discussed later in this chapter). It is a new type of product that offers the same potential and interest as technical textiles.

What does the term, 'smart textiles' actually mean? Smart textiles can be described as textiles that are able to sense stimuli from the environment, to react to them and adapt to them by integration of functionalities in the textile structure. The stimulus as well as the response can have an electrical, thermal, chemical, magnetic or other origin. Advanced materials, such as vapour permeable barriers, fire-resistant or ultra-strong fabrics, are, according to this definition, not considered as intelligent, no matter how high-tech they might be. The extent of intelligence can be divided in three subgroups:[1]

- passive smart textiles can only sense the environment, they are sensors;
- active smart textiles can sense the stimuli from the environment and also react to them, besides the sensor function, they also have an actuator function;
- finally, very smart textiles take a step further, having the gift to adapt their behaviour to the circumstances.

Sometimes, the change in the material is clearly visible, but sometimes it takes place on a molecular level, completely invisible to the human eye.

The possible applications offered by these materials are limited only by human imagination. Although some products have been reported and presented, most of them are still in a development phase and far away from a true 100%

textile character. A first successful step towards wearability was the ICD+ garment line at the end of the 1990s, which was the result of co-operation between Levi Strauss and Philips electronics.[2] The garment's construction at that time required that all these components, including the wiring, were carefully removed from the coat before going into the washing machine. In the MP3 player system developed by Infineon,[3] robust and wash-proof packing protects the different components during laundry.

The Wearable Motherboard[4] is probably the first intelligent suit that can be used for medical purposes. The basic shirt included an optical wiring structure that could be equipped with conventional sensors to measure different body parameters. Therefore, smart clothing will find applications initially in specific areas where monitoring and actuation can save lives, and where professional and skilled people will use them in a controlled way. Professional environments like hospitals and industry are examples of such places. User friendliness and comfort are maybe less critical than safety. These types of textiles include, for example, wearable smart textiles (biomedical clothing), designed to fulfil certain functions, but apart from that are very straightforward. However, as experience and familiarity increase, breaking down barriers, the field of application will in the long term definitely widen to more routine applications such as sports and leisure, the work environment, and so on. More casual applications are possible which are expected to be functional as well as fashionable. It can also go as far as daily skin care, where the comfort factor is even more critical. Additionally, smart wound dressings, bandages and hygiene applications are envisaged. The main problems to be overcome in the initial phase of the development will be communication between textile engineers and medical people in order to be able to define and demonstrate the benefits of new applications.

7.1.1   Why smart textiles?

Textiles have several advantages and clothes are unique in several aspects. They are extremely versatile in product form as well as processes: various polymer types can be mixed at the level of fibre, yarn, fabric or clothing. High-tech materials, for instance, can be used in hazardous conditions involving heat, toxic or aggressive chemicals, or they provide extreme strength or impact resistance. Three-dimensional shapes can be made to fit for each individual. Specific treatments allow the creation of very special properties such as having a hydrophilic/ hydrophobic nature, selective permeability and being antimicrobial, etc.

Maintaining textiles is a daily practice; domestic as well as industrial laundries are well developed. Last but not least, textiles and clothes can be produced on fast and productive machinery at reasonable cost.

These characteristics will eventually open up a number of applications that were not possible before, especially in the area of monitoring and treatment, such as:

- long-term or permanent contact without skin irritation
- home applications
- applications for children in a discreet and fun-filled way
- applications for the elderly; discretion, comfort and aesthetics are important.

Full success in the long term however, will be achieved only when the sensors and all related components are entirely converted into 100% textile materials. This is a big challenge because, apart from technical considerations, concepts, materials, structures and treatments must focus on the appropriate use in or as a textile material. This includes criteria like flexibility, water or solvent (laundry) resistance, durability against deformation, radiation, etc. As for real devices, ultimately most signals will be transformed into electrical ones. Electro-conductive materials are consequently of utmost importance in the field of intelligent textiles.

7.2   Applications of smart textiles for protective purposes

Smart textiles can contribute to protection and safety in three ways:

- they are able to detect conditions that signal increased danger
- they prevent accidents by sending out a warning when hazardous conditions have been detected
- in the case of serious threats they can react by providing instantaneous protection.

Apart from obvious threats like heat, chemicals, gases etc., danger can also be caused by people themselves. Individuals can be threatened by an acute disease such as a heart attack, a stroke, or other physical conditions may cause them to be unable to perform their tasks safely. Personnel driving a machine or a vehicle for instance must maintain a high and continuous level of concentration and awareness. Fatigue, consumption of alcohol or medication can negatively affect such parameters, leading to an increased risk of accidents. Ultimately the suit may communicate to the machine that the driver is no longer able to continue his operations safely and the machine may be stopped. Sending out information can be achieved also at several levels. One of the simplest reactions is colour change, providing straightforward visual signals to both the wearer and the environment. More complex systems process the information and send out a warning to the wearer or to an external system, so that adequate measures can be taken in time; leaving the location, sending help or many other actions. Some examples of communication will be dealt with in a separate paragraph later.

An important benefit of smart textiles in protective applications is that the textile can react when necessary, in a passive way or by active control mechanisms. Passive protection systems that are in use today usually have an important negative impact on comfort and freedom of movement. A good example are firefighting suits where the insulation level is so high that the firemen suffer fatigue because of overheating caused by their own body heat, irrespective of the external fire. Bio-chemical protection requires a significant decrease in material porosity, leading to a strong reduction of ventilation capacity – a major factor influencing comfort. Hence, when such clothes have to be worn throughout the day, thermal comfort is limited.

In general, smart clothes offer the possibility of adaptation to the environment, providing protection only when required, for instance, when the temperature is too high, when harmful chemicals or micro-organisms have been detected and so on. When protection is only temporarily required, the balance between protection and comfort is totally different; higher levels of protection become possible.

Actively controlled smart textiles can even go one step further in this process. Not only can they adapt their level of protection as a function of instantaneous needs, additionally, they can neutralise the effect of the threatening factor; the suit can cool down the body when it gets excessively hot (for instance for firemen) or warm it up when it is cooling down (in case of people working in cold rooms or outdoors). Neutralising chemicals can be released as well as antibiotic products but only when necessary. Indirectly, smart clothes can contribute to safety by providing optimal working conditions so that fatigue and stress are minimised. Consequently, it will be possible for the worker to maintain a higher level of concentration during a longer period.

Optimal conditions may involve a large number of parameters:

- thermal comfort (definitely one of the main factors inducing stress), in itself includes:
  - temperature
  - humidity
  - air velocity
- illumination
- external noise
- odour.

Textile products are extremely flexible in design. Therefore a smart design may foresee storage facilities for tools, small devices and so on. This can also contribute to better and safer working conditions. A useful example in this respect is the Reima suit for snow-riding.[5] The ability of smart clothes to assess the level of comfort or stress in a person, their overall health or level of concentration opens up several other fields of application. It can be the basis for combining increased performance of employees with safer working conditions.

Apart from technical features, two of the main challenges are maintenance and durability. All functions must resist normal wear and cleaning processes.

Professional clothes are often cleaned in industrial processes with high-temperature, mechanical forces and so on. Functional failure may lead to injuries, so reliable products are required. Therefore, testing of functionality, and particularly self-testing capabilities of the textile product, are required. To fully achieve the capabilities mentioned above, real very smart materials have to be designed. Basically, five functions can be distinguished in a smart suit, namely:

- sensing
- data processing
- actuating
- storage
- communication.

The different components all have a clear role, although not all smart suits will contain all functions. The functions may be quite apparent, or may be an intrinsic property of the material or structure. They all require appropriate materials and structures, and they must be compatible with the function of clothing, being comfortable, durable, resistant to regular textile maintenance processes and so on.

7.3   Sensor function

The textile is in contact with the skin over a large body area. This means that monitoring can take place at several locations on the body. On the other hand, the clothing can also measure ambient parameters. Some types of parameters that are mentioned in literature are

- temperature
- electro-magnetic signals (biopotentials: cardiogram, electrostatic field, etc.)
- acoustic/ultrasound
- motion
- chemicals (liquids as well as gases)
- electrical properties of the skin
- mechanical parameters (pressure, stress/strain, shear, impact, acceleration)
- radiation (UV, IR, visible, radioactive, etc.)
- odour.

It will be clear to the reader that this list is not complete.

Sensors in general and textile sensors in particular struggle with the following problems:

- the flexibility and deformability required for comfort interfere with sensor stability
- signals tend to have relatively low amplitude (e.g. μV)
- long-term stability is affected by wear and laundry.

The first generation of intelligent clothing consisted of conventional components attached to the textile structure. Gradually they were being replaced by true textile materials. Several studies are taking place in the area of health monitoring, with various levels of transformation into full textile structures. Body parameters that are currently being measured using textile or textile compatible sensors are cardiograms, respiration rate, motion, temperature, blood pressure, acceleration (Mamagoose (B),[6] ANBRE (B),[6] Smart Shirt,[7] Life Shirt,[8] WEALTHY,[9] Intellitex,[10] VTAM[11]). They focus on medical, sports, space and military applications.

7.3.1   Heart signals[10]

Heart signals are one of the basic body parameters in health assessment. The heart is basically a muscle that is controlled by the brain through electric impulses. Electro-conductive textile structures are used to capture these signals while instruments are analysing the results, extracting the required parameters such as frequency, phases, etc. Electro-conductive gels are commonly used to improve the skin-electrode contact. Unfortunately, such gels cause skin irritation within a period of 24 hours. So when using textile electrodes during long-term monitoring electro-conductive gels should not be applied in order to avoid skin reactions. Unfortunately, this leads to a small increase in noise on the signal. This is illustrated by Fig. 7.1, where the signal originating from a conventional electrode (gel electrodes by 3M) and the textile electrodes were recorded at the same time.

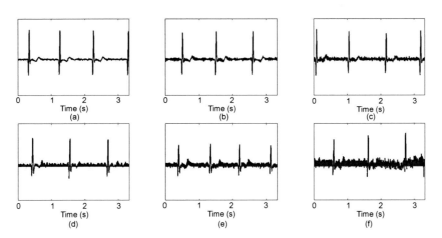

7.1 Conventional electrodes (a, b, c) versus textile electrodes (d, e, f) in three different configurations.

7.3.2   Strain sensors: respiration and motion

Textile materials composed of fibres form complex networks of conducting paths that make multiple contacts. During deformation a number of mechanisms take place:

- the number of contact points changes
- fibres are extended
- fibre cross-section is decreased.

The number of contact points changes drastically at low extension values, although fibre deformation takes place at higher levels of strain. An increase in the number of contact points reduces electrical resistance, whereas fibre extension and reduction of cross-section lead to an increase in electrical resistance. As a result, electrical resistance changes due to deformation in a way that depends mainly on the textile structure. This gives textile structures piezo-resistive properties enabling their use as strain or deformation sensors. From such signals, motion and even position can be extracted. Unfortunately, any action during use, such as laundry, extension, etc., may affect the structure of the textile material and consequently its piezo-resistive properties. In addition, one can consider the conductive textile belt placed around the chest, as a coil. The inductance of a coil varies with cross-section.

The Intellitex suit mentioned earlier combines heart and respiration rate measurements in one garment. The respiration sensor is a knitted belt called 'Respibelt'.[10] It is also made from stainless steel yarn. The accuracy and stability of this sensor are illustrated in Fig. 7.2. The signals show the inductance (Fig. 7.2(a)) with respect to resistivity (Fig. 7.2(b)) of the belt during breathing. Inductance analysis considers the belt as a circular circuit which reacts to changes in cross-section within the belt. Resistivity reacts to changes in circumference of the chest due to the breathing motion. Long-term stability turns out to be a worry for some applications where absolute deformations have to be recorded. One can expect that long-term use will require regular calibration procedures.

Information on body kinetics is important with respect to, for instance, detection of falls, identification of activity, rehabilitation, sports and dance, ergonomics. CEA-LETI has developed a 3D orientation tracker based on conventional accelerometers and magnetometers.[12] The system is based on a simplified skeleton model, which describes main bones and joints with rotational constraints. Human hands are able to perform very complex movements in a very accurate way. Reduced abilities to control the hand have a significant impact on a person's life. A smart glove could help to make a detailed analysis of a person's hand movements allowing identification of right and wrong movements. Lorussi et al.[13] have developed a sensorised glove and algorithms that enable the measurement of position and posture of the hand. The

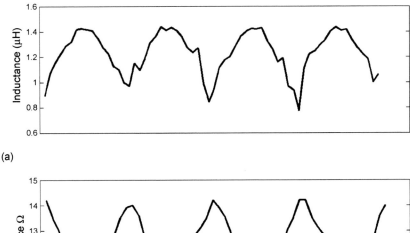

7.2 Respibelt signals.

basic sensor is a piezo-resistive sensor that has been developed at the University of Pisa and Smartex.[14] Threads or fabrics are given their sensing capacities by coating them with rubber loaded with a micro-disperse carbon phase.

7.3.3   Pressure sensors

Basically two principles are used in textile pressure sensors. A first example is a pressure-sensitive textile material that is already on the market, Softswitch[TM]. The Softswitch[TM] technology[15] uses a so-called 'quantum tunnelling composite (QTC).[16] This composite has the remarkable characteristic of being an isolator in its normal condition and changing to a metal-like conductor when pressure is applied to it. Depending on the application, the pressure sensitivity can be adjusted. Through the existing production methods, the active polymer layer can be applied to every textile structure, a knitted fabric, a woven fabric or a non-woven. The pressure-sensitive textile material can be connected to existing electronics.

A second type of material that has the property of being pressure sensitive, was developed by a team of the Design for Life Centre from Brunel University in Surrey. The English firm Eleksen[17] commercialised the sensory fabric under the name Elektex[TM]. The sensory fabric[18,19] consists of two layers of carbon-based electrically conductive textile, divided by a layer of non-conductive mesh.

Each conductive, or partially conductive fabric sheet, constitutes an electrical switch contact. The layer of non-conductive mesh in between serves to keep the two conductors apart. When the textile is pressurised, the top conductive sheet is pushed through the holes in the mesh to make electrical contact with the lower conductive sheet. The higher the applied force, the more conductive fibres will make contact and the higher the conductivity in the channel. The sensitivity depends upon the size of the meshes and the thickness of the insulating layer. In this way, the fabric can be adapted to its application. The basic material of the sensory fabric consists of a combination of conductive fibres and nylon. This combination results into a durable, reasonably priced, washable and even wearable 3D structure.

7.3.4   Optical fibres: a multifunctional tool

Fibre Bragg Grating (FBG) sensors are a type of optical sensors receiving a lot of attention lately. They are used for the monitoring of the structural condition of fibre-reinforced composites, concrete constructions or other construction materials. At the Hong Kong Polytechnic University, several important applications of optical fibres have been developed for the measurement of tension and temperature in composite materials and other textile structures.[20] FBG sensors look like normal optical fibres, but inside they contain at certain places specific diffraction grids that reflect particular wavelength bands, depending on the type and dimensions of the grid. Optical fibres can be used to measure a wide range of other parameters when used in the appropriate concepts. Temperature, for instance, leads to expansion that in its turn affects the characteristics of the signals inside. When covered with the right responsive coatings, the fibres could indicate the presence of chemicals or micro-organisms.

7.3.5   Colour change mechanisms: sensors we can see

The concept of producing textiles that readily vary in colour has long been anathema to the textile colourist, for whom achieving permanency of colour is a primary goal. During this search, particular colorants have shown sensitivities to light, pH, temperature, polarity of solvents, and many others.[21] Indicator colorants that change colour in various conditions have already been used for a long time in chemistry. Thermochromic dyes have increasingly been the subject of investigation over the past decades for use in producing novel coloration effects in textiles as well as other applications.[22] Such dyes can give a fast and straightforward indication of temperatures and temperature distribution. An analogous phenomenon is photochromism, where the colour will change under the influence of light. Photochromism is defined as the reversible change in colour of a chemical substance under the influence of electromagnetic radiation, such as UV light.

7.4   Data processing

Data processing is one of the components that are required only when active processing is necessary to realise a smart and adequate response. The first constraint at present is the interpretation of the data. Detection of problems can be fairly straightforward; temperature exceeding a certain level, presence of chemicals or micro-organisms, etc. The human body, on the contrary, is an extremely complex machine and behaviour or a feeling of well-being are very difficult to assess. Some projects in progress derive parameters like stress, based on the measurement of body parameters.[23] A second problem consists of feedback and control algorithms. Passive feedback consists of sending out a warning or giving an alarm. Active feedback requires control strategies. It is quite clear that the best response in a given situation will be highly dependent on the person and is time dependent. This will require smart algorithms that are capable of learning continuously about the immediate response of the body to a reaction of the textile.

Context awareness could considerably contribute to the level of intelligence of a smart suit.[24] It analyses information on the person and the environment provided by the sensors, and from this it derives the type of activity and the situation a person is involved in. This information is then used by the suit to decide on the most appropriate reaction. Apart from this, the textile material in itself does not have any computing power at all; electronics are still necessary to fulfil this task. However, they are available in miniaturised and even in a flexible form. They are embedded in waterproof materials but durability is still limited. The interconnections with the data-transmitting component are at present still a weak point. Research is going on to fix the active components on the fibres (Ficom project[25]). Many practical problems need to be overcome before real computing fibres will be on the market, e.g., fastness to washing, deformation, interconnections.

7.5   Actuators

Actuators respond to an impulse resulting from the sensor function, possibly after data processing. Actuators make things move, they release substances, make noise, heat up or cool down, and perform many more functions.

7.5.1   Mechanical actuators

Mechanical actuators make fibres move in the textile structure, changing properties like thermal insulation, permeability, etc. A more ambitious objective is to create fibres with muscle-like properties. When such fibres are integrated in clothes that fit as a second skin, the textile could provide considerable support or even take over body motion. Such fibres can be incorporated in the textile in any predetermined form so that any movement can be obtained. The fibres would need to be electro-active materials, capable of reaching a high level of contraction, high contraction forces, with a short reaction time and low voltage actuation. High-performance muscle-like materials, however, are not yet within reach.[26] They require either high voltage or a specific chemical environment that slows down reaction as diffusion mechanisms are involved. Recently some promising results have been presented that could open up this window.[27] Current available mechanical actuators provide a mechanical reaction in response to thermal, chemical or physical impulses. Some examples will be given below.

Gel-based actuators

Polymer gels differ in many ways from solid materials. Polymer chains in the gels are considered to be chemically or physically cross-linked and to form a three-dimensional network structure. For instance, polymer gel is usually a substance swollen with its appropriate solvent, and the characteristics range from a nearly solid polymer almost to a solution with very low polymer content, but still maintaining its shape by itself. From the standpoint of an actuator, the gel behaves like a conventional solid actuator or like a shapeless amoeba. The gels also have various actuating modes, symmetric or asymmetric deformation behaviour, depending on the structure in which it is used. There are a wide range of triggers that cause the actuating deformation. Chemical triggers are

- pH
- oxidation and reduction
- solvent exchange
- ionic strength change.

Physical triggers include

- light irradiation
- temperature change
- physical deformation
- magnetic field
- electric field
- microwave irradiation.

Gels incorporated into fibres or in an actual textile structure allow systems that open up or close down in response to one of the triggers mentioned above. Smart designs allow actuators of various types. An example of yarn consisting of a gel core wrapped with Z- and S-twisted filaments is derived from nature; worms contract their body by local expansion of body diameter. The result is an elementary version of an artificial muscle.[28]

pH triggered polymers

In 1950, two researchers (W. Kuhn and A. Katchalsky) developed fibres able to contract under the influence of a pH change. However, a disadvantage was the slow reaction time of the polymer; the change lasted for some minutes. Through the years, other fibres have been developed having much shorter reactions, from a few seconds to a few tenths of seconds.[54] These fibres also have a high tensile strength. The degree of contraction and the developed forces equal those of a human muscle.

At the University of New Mexico, Albuquerque, USA, research is taking place into the development of artificial muscles. To this end, a polyacrilonitrile fibre, Orlon[29] is used. A few years ago, researchers discovered that PAN fibres are able to contract when the degree of acidity in the surrounding changes. In two-tenths of a second, they can shrink 20%, which is almost as fast as a human muscle. Depending on the degree of acidity, PAN fibres can contract to half or a tenth of their original length. Moreover, the fibres are strong; they can bear up to four kilograms per square centimetre. This is more than a human muscle.

Shape memory materials

Shape memory materials return to a predetermined form above a given transition temperature. Shape memory materials exist in the form of metal alloys and polymers. Although shape memory polymers are cheaper, they exhibit lower performance in terms of magnitude of deformations, forces and durability. On the other hand, their reaction to temperature can be modified quite easily; with one set of monomers, it is possible to have a whole set of shape memory materials. Shape memory alloys (Nitinol, a nickel-titanium alloy) have been used for a long time as stents. They can be used to create materials with an adaptive level of insulation[39] (see section 7.8.2).

7.5.2   Chemical actuation

Chemical actuators release specific substances in predefined conditions. The substances can be stored in 'containers' or be chemically bound to the fibre polymer. The coat of the container or the chemical bonds steer the release rate.

Containers consist of the full fibres coated with smart coatings, micro and nano-capsules with well-designed shells or molecules like cyclodextrines. The latter have constant release rates.

Materials that release substances already have several commercial applications. They release fragrances, skin-care products, antimicrobial products, etc. However, actively controlled release is not obvious, although some basic research projects have started. Release could be triggered by temperature, pH, humidity, chemicals and many more parameters. Obviously, controlled release opens up a huge number of applications such as drug supply systems in intelligent suits that are also capable of making an adequate diagnosis.

7.6   Energy

Although usually not a goal as such, smart clothing often needs some storage capacity, as the suit must be able to function as a stand-alone unit. Sensing, data processing, actuation and communication usually need energy, mostly electrical power. Efficient power management will consist of an appropriate combination of energy supply and energy storage capacity in combination with an efficient distribution strategy of components and computing power. Sources of energy that are available to a garment are for instance body heat, mechanical motion (elastic from deformation of the fabrics, kinetic from body motion), radiation, etc.

Infineon[30] had the idea to transform the temperature difference between the human body and the environment into electrical energy by means of thermo-generators. This principle is known as the Seebeck effect. The prototype is a rigid, thin micro-module that is discreetly incorporated into the clothing. The module itself is not manufactured from textile materials. However, the possibility of using textiles exists. The use of solar energy for energy supply is also a possibility. At the University of California, Berkeley, a flexible solar cell has been developed which can be applied to any surface.[31]

At this moment, all energy transformation mechanisms are far too inefficient to provide a full power supply. Some microwatts can be gained, just enough to drive single low-power components like sensors. As mentioned before, energy supply must be combined with energy storage. When hearing this, one thinks of batteries. Batteries are becoming increasingly smaller and lighter. Even flexible versions are available, although their performance is poor. Currently, lithium-ion batteries are found in many applications. Fibre batteries made of super capacitor fibres based on carbon nanofibres are under development.

7.7   Communication

For intelligent textiles, communication has many faces. Communication may be required

- within one element of a suit
- between the individual elements within the suit
- from the wearer to the suit to pass instructions
- from the suit to the wearer or his environment to pass information or instructions.

Within the suit, communication is currently realised by either optical fibres,[32] either conductive yarns[33] or conventional electrical wires. They all clearly have a textile nature and can be built into the textile seamlessly.

Communication with the wearer is important as the wearer may want to be informed about some of the information an intelligent suit has gained. This could be achieved by different technologies. For the development of a flexible textile screen, the use of optical fibres is an obvious approach. France Telecom[34] has managed to realise some prototypes (a sweater and a backpack). At certain points, the light from the fibre can emerge and a pixel is formed on the textile surface. The textile screen can emit static and dynamic colour images. In order to increase the resolution, the concept will need to be reviewed, as currently one pixel requires several optical fibres. Nevertheless, in this way, these clothes could represent a first-generation attempt at graphical communication.

Pressure-sensitive textile materials,[14,35] as mentioned earlier, allow putting in information, provided a processing unit can interpret the meaning of pressing specific areas of a textile, i.e., it must understand the commands. This technology is already applied to make 'soft' telephones and a folding keyboard. The ElekTex™ fabric is also used in the development of car seats, with the aim of having an optimal weight distribution in order to increase seating comfort.

Communication with the wider environment is very important in medical applications. As the wearer is often in risky situations, help must be provided instantly in case of life-threatening events. The concept of a stand-alone suit does not allow direct contact, so wireless communication is required. This can be achieved by integrating an antenna. The step was also taken to manufacture this antenna in textile material.[36] The advantage of integrating antennas in clothing is that a large surface can be used without the user being aware of it. In the summer of 2002, a prototype was presented by Philips Research

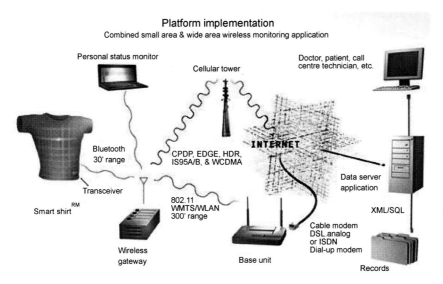

Platform implementation
Combined small area & wide area wireless monitoring application

Personal status monitor

Cellular tower

Doctor, patient, call centre technician, etc.

Bluetooth 30' range

CPDP, EDGE, HDR, IS95A/B, & WCDMA

INTERNET

Data server application

Transceiver

Smart shirt RM

802.11 WMTS/WLAN 300' range

XML/SQL

Wireless gateway

Base unit

Cable modem DSL.analog or ISDN Dial-up modem

Records

7.3 Communication with the Smart Shirt from Sensatex (http://www.sensatex.com). (Source: Sensatex).

Laboratories, UK, and Foster Miller, USA at the International Interactive Textiles for the Warrior Conference (Boston, USA). An example of a communication network linked to a smart suit for health monitoring is given in Fig. 7.3.

7.8   Thermal protection

Thermal actuators can have several levels of activity. Actual heating or cooling systems have the highest impact on thermal comfort. Adaptive insulation adjusts its level of thermal conductivity according to temperature. Materials with super-high thermal absorption capacity will support the wearer in maintaining its core temperature.

7.8.1   Active heating/cooling systems

Conductive materials can act as an electric resistance and can consequently be used as a heating element. Polartec has recently presented a heatable fleece.[38] On the other hand, cooling is a more complex process. D'Appolonia has presented a cooling shirt for F1 racers. Tiny cooling tubes are integrated in the jacket, a liquid that is cooled by a central Peltier cooling element is circulated through these tubes. Semi-active temperature regulation can be achieved by using micro capsules filled with waxes that have a melting point near the targeted temperature (see section 7.8.3 below).

7.8.2   Adaptive insulation

At the end of the 1990s, the DCTA R&TG (Defence Clothing & Textiles Agency), Colchester, UK[39,53] started research into the use of shape memory alloys for developing heat-protective clothing. The first application was to provide reactive protection against heat and flames. In the experiment, springs made of a shape memory alloy (Nitinol) are used. At room temperature the springs are in a flat state. At increasing temperatures the springs open up. The system consists of two separate layers in which cotton bands are introduced in order to incorporate the springs. The springs are fixed only at the outside layer, so that they are minimally obstructed when expanding. The springs are conic and have a 25 mm diameter. Air is a good insulator. Under the influence of the transformed springs, both textile layers will move away from each other as a result of which an insulating air layer is formed between them. The springs have only a one-way function; when cooling down, they will not regain their original shape unless a mechanical action is applied. Depending on the shape transitions, garments can be obtained with increasing levels of insulation at high or at low temperatures. An example of the design is given in Fig. 7.4.

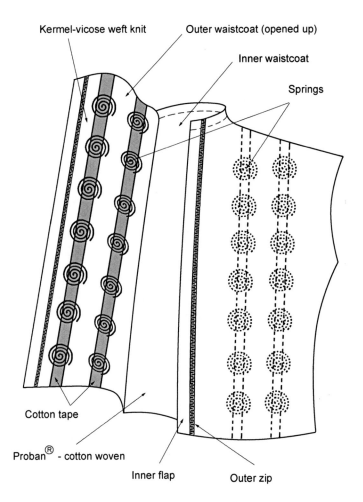

Kermel-vicose weft knit   Outer waistcoat (opened up)

Inner waistcoat

Springs

Cotton tape

Proban® - cotton woven

Inner flap   Outer zip

7.4 Protective waistcoat with incorporated springs made of a shape-memory alloy.

During the Gulf War the DCTA R&TG (Defence Clothing & Textiles Agency)[40] was searching for materials with adaptive ventilation characteristics. The purpose was to cope with the huge temperature differences that occur in the desert without additional weight for the soldiers. They came up with materials which mimicked the action of the pine cones that remain closed in wet conditions (e.g. because of sweating during the day) while opening up when dry (e.g. at night when feeling cold). The fabric consists of a knitted structure coated with a polymer having a high hygral expansion coefficient. U-shaped holes are punched in the fabric. The 'valves' obtained in this way are pushed open when the coating expands due to high humidity, and close down again when the coating regains its original shape when the internal humidity (sweating) stops.

7.8.3 Phase-change materials (PCMs)

The concept of micro-encapsulation of PCMs was developed by the NASA at the end of the 1970s and the beginning of the 1980s. NASA was looking for materials to protect its delicate instruments against the extreme temperature fluctuations prevalent in space. NASA had already published a *Phase Change Materials Handbook* in 1971, describing more than 500 of these substances, which distinguished themselves by phase-change temperatures and their ability to capture heat. Phase-change materials are materials whose phase changes within a predetermined and restricted temperature interval. The latent heat energy involved in changing phase between liquid and solid can be up to 200 times higher than the energy involved in heating up or cooling down an equal mass of material. The heat capacity of water and a wax used in PCM is shown in Fig. 7.5. The most famous example of such a material is water/ice, which at the freezing point changes from the fluid into the solid state. However, water is not an appropriate material for body temperature control, the phase transition temperature is 0 °C, which is far away from the targeted body temperature and water molecules are too small, they would slowly migrate through the shell of the capsule.

The phase change materials used in the textile sector are usually paraffin waxes in solid or liquid state.[42] These are hydrocarbons with a different chain length, such as heneicosane ($C_{21}H_{44}$), eicosane ($C_{20}H_{42}$), nonadecane ($C_{19}H_{40}$) and octadecane ($C_{18}H_{38}$). With these materials, the phase change occurs within a temperature range that lies in the vicinity of body temperature. The substances used in the textile sector are characterised by their capacity to capture or release large quantities of warmth without changing in temperature. The microcapsules are integrated in a textile material, in the actual fibre mass, in a coating or in a foam.

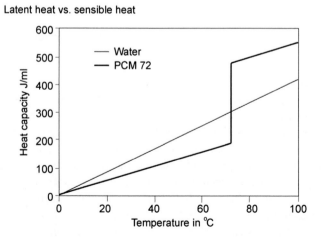

7.5 Melting process of a PCM.

7.9   Electric actuation

Electro-stimulation is the stimulation of muscles by electrical impulses. It is a complex process with many physiological effects. Electro-conductive textile structures can be used to take the electric impulses to any part of the body.[43,44] Research is going on focusing on the physiological effects of electro-stimulation such as strengthening of muscles,[23] skin sensorial performance and other.[46–49] Active control of muscles would enable control of movements. Of course this could be a life-saving tool in critical situations. Unfortunately, full remote control of this kind is a long way in the future. Detailed knowledge and understanding is required of several aspects of muscle control; which muscles are involved, time sequences, magnitude and position of signals, control models, feedback, and many more. Some initial research has started in such areas, for instance, on the analysis and modelling of myographical data.[50]

7.10   A story on impact protection

Impact can occur in many situations: impact by falling objects, bullets, fall of a person, etc. For each of these applications, different solutions are required. Bullet-proof vests are readily available. They provide passive protection. Real smart protective suits will detect a risk on impact and will react to it. Detection can be based on relatively simple acceleration measurements in the case of falling, however, approaching objects are more difficult to sense.

First the suit should try to prevent the impact, for instance, by warning the person to move, or by moving the person by using the built-in muscle-like fibres. Alternatively, the necessary parts of the suit must transform into protective zones, for instance, by built-in air bags or by local thickening and hardening. The suit detects that impact has occurred, analyses the magnitude and location of impact and again provides an adequate reaction, such as fixing ruptures, providing medication, calling for help should this be necessary. Smart protective vests with built-in air bags have been developed for motor cycle[51] and horse riders. They are a great example of providing comfort during normal use whilst protecting only when necessary.

7.11   References

1. X Zhang and X Tao, 'Passive smart', *Smart textiles*, June 2001, pp. 45–49, 'Active smart', *Smart textiles*, July 2001, pp. 49–52, 'Very smart', *Smart textiles*, August 2001, pp. 35–37, Textile Asia.
2. http://www.time.com/time/2002/greencentury/entechnology.html
3. http://www.wearable-electronics.de/intl/fotos_vorbereitungen.asp
4. Park S, Jayaraman S, 'The Wearable Motherboard: new class of adaptive and responsive textile structures'. International Interactive Textiles for the Warrior Conference, 9–11th July 2002.

5. http://www.reimasmart.com
6. http://www.verhaert.com
7. http://www.smartshirt.gatech.edu
8. http://www.vivometrics.com
9. http://smartex.it/uk/projects/physensor.htm
10. Van Langenhove L, Hertleer C, 'Smart Textiles for medical purposes'. MEDTEX 03. Int. Conference on Healthcare & Medical Textiles, July 7–9th 2003, Bolton, UK.
11. http://www.medes.fr/VTAMN.html
12. Guillemond R, Caritu Y, David D, Favre-Reguillon F, Fontaine D, Bonnet S. (2003) 'Body motion capture for activity monitoring'. *Smart Textiles*, New generation of wearable systems for e-health. Dec. 11–14th, Lucca (1), pp 209–215.
13. Lorussi F, Tognetti A, Tesconi M, Pastacaldi P, de Rossi D (2003) 'Strain sensing fabric for hand posture and gesture monitoring'. *Smart Textiles*, New generation of wearable systems for e-health. Dec. 11–14th, Lucca (1), pp 175–180.
14. Pacelli M *et al.* (2001) 'Sensing Threads and fabrics for monitoring body kinematics and vital signs'. Fibres & Textiles for the Future seminar. Tampere, Finland.
15. http://www.softswitch.co.uk
16. Donoso A, Martens C C (2001) 'Quantum Tunnelling using entangled classical trajectories'. *Phys. Rev. Lett.* 87, 223202, (26th November).
17. http://www.eleksen.com
18. Patent US 0119391 (2003) 'Conductive Pressure Sensitive Textile'. June 26th 2003.
19. Swallow S, Peta-Thompson A, (2001) 'Sensory Fabric for Ubiquitous Interfaces'. *Int. J. Human-Computer Interaction*, June, Vol. 13, issue 2, pp 147–159.
20. Tao X (2002) 'Sensors in garments'. *Textile Asia*, January 2002, pp 38-41.
21. Lewis D (2004) 'Smart Dyes'. Int. Conference on Intelligent Textiles, 25th June, Ghent, Belgium.
22. Anon. (1999) 'The heat is on for new colours'. *JSDC*, Vol. 15, July/Aug. 1999.
23. http://www.aubade-group.com
24. http://www.wearable.ethz.ch/fileadmin/pdf_files/pub/junker_ewsn04.pdf
25. http://www.fibercomputing.net
26. First World Congress on Biomimetics and Artificial Muscles, December 9–11, 2002, Albuquerque, USA.
27. De Rossi D, Lorussi F, Scilingo E P (2003) 'Artificial kinesthetic systems for telerehabilitation', *Smart textiles*, New generation of wearable systems for e-health, December 11–14, Lucca (I), pp. 129–133.
28. http://www.Reading.ac.uk
29. Orlon is an acryl fibre produced by the company Du Pont.
30. Lauterbach C *et al.*, 'Smart clothes selfpowered by body heat', *Avantex Proceedings*, 15th May 2002.
31. Chapman K (2002), 'High tech fabrics for smart garments', *Concept 2 Consumer*, September, pp 15–19.
32. Park S, Jayaraman S (2002), 'The wearable motherboard: the new class of adaptive and responsive textile structures', International Interactive Textiles for the Warrior Conference, 9–11 July 2002.
33. Van Langenhove L *et al.*, 'Intelligent textiles for children in a hospital environment', World Textile Conference Proceedings, 1–3 July 2002, pp 44–48.
34. Deflin E, Weill A, Koncar V, 'Communicating clothes: optical fiber fabric for a new flexible display', *Avantex Proceedings*, 13–15 May 2002.

35. http://www.tactex.com
36. Catrysse M, Puers R, Hertleer C et al., 'Towards the integration of textile sensors in a wireless monitoring suit', Sensors and Actuators A – Physical, 114 (2–3): 302–311, Sept 1 2004.
37. http://www.sensatex.com
38. Polartec, Innovation Award TechTextil, Frankfurt, April 2003.
39. Russell D A, Elton S F, Squire J, Staples R, Wilson N, Proffitt A D (2000), 'First experience with shape memory material in functional clothing', Proceedings Avantex Symposium, November 29, 2000.
40. Jeronimidis G (2000) 'Biomimetics: lessons from nature for engineering', G., 35th John Player Memorial Lecture, The Institution of Mechanical Engineers, 22nd March, 2000.
41. http://pb.merck.de/servlet/PB/show/1014580/pcm_br_eng.pdf
42. Cox R (1998), 'Synopsis of the new thermal regulating fibre outlast', Chem. Fibres Int. 48, 475–479.
43. Axelgaard J, Heard S, Medical method of manufacture, US patent 6.198.955 B1, 2001.
44. Kirstein T, Lawrence M, Tröster G, 'Functional electrical stimulation (FES with smart textile electrodes', Smart textiles, New generation of wearable systems for e-health, December 11–14, Lucca (I), 2003, pp. 201–208.
45. Siff M C (1990), 'Application of electrostimulation in physical conditioning', J of Applied Sports Science Res, 4(1): 20–26.
46. Hainaut K, Duchateau J (1992), 'Neuromuscular electrical stimulation and voluntary exercise', Sport Med, 14(2) 100–113.
47. Feedar J F, Kloth L C, Gentzkow G D (1991), 'Chronic dermal ulcer healing enhanced with monophasic pulsed electric stimulation', Phys Ther, 71, 639–649.
48. Mulder G D (1991), 'Treatment of open-skin wounds with electric stimulation', Arch Phys Med Rehabil, 1991, 72, 375–377.
49. Wood J M, Evans P E III, Schallrenter K U, et al. (1992), 'A multicenter study on the use of pulsed low intensity direct current for healing chronic stage II and III decubitus ulcers', J Invest Dermatol, 98: 4.
50. Triolo R J, Chae J, 'Preventing Pressure Sores with Neuromuscular Electrical Stimulation', http://feswww.fes.cwru.edu/projects/rjtscrf.htm, ATP production: http://www.cellstim.com/research.htm
51. Michael Shestakov, Russia; Alexey Averkin, 'Holonic multi-agent system for simulation of locomotor movements of a multijoint biomechanical system'. Eunite conference Aachen 2004.
52. http://www.dainese.com/eng/d-air.asp
53. Scott R A (2000), 'Fibres, Textiles and Fabrics for Future Military Protective Clothing'. In Ergonomics of Protective Clothing. Proceedings of NOKOBETEF Conference on Protective Clothing. Stockholm, May 7–10, pp 108–113.
54. Brock D. L., 'Dynamic Model and Control of an Artificial Muscle based on Contractile Polymers', Massachusetts Institute of Technology Artificial Intelligent Laboratory, A.I. Memo No. 1331, November 1991.

Surface treatments for protective textiles

R B U C K L E Y , Eastgate Consulting, UK

8.1   Introduction

Many textile apparel fabrics in themselves offer levels of protection that are quite adequate in certain environments and situations:

- Multi-layer clothing or garments made from thick woollen fabrics do not require special finishes or effects to offer good insulation against cold in dry environments; entrapped air space being the functional medium.
- For spraying of agrochemicals and pesticides in hot climates (IGNAMAP 1991), cotton or cotton/polyester coverall clothing provides a physical barrier to reduce skin contamination risk. Clothing must be washed if contaminated.
- Some technical fabrics depend on sophisticated construction for protection; chain- saw protective fabric contains threads that will readily slide out of the structure on contact with the saw to jam the chain before the blade can make contact with the body (Heathcoat & Co, 1999).
- Many ballistic-resistant materials rely on the energy absorbing and dissipating properties of multi-layer, high tenacity and high modulus yarn fabrics whereby designed yarn slippage and disintegration of fibres, as well as blunting of the projectile and impact area shockwave spreading are the important functions (Anderson *et al.* 1998).

8.2   Types of surface treatment

For sufficient protection against a whole range of agencies and situations that are life-threatening or cause health and comfort problems, applied surface treatments are necessary. Even taking all of the above listed examples into account, it is likely that the vast majority of protective textiles could be upgraded in their protective performance with the aid of applied products. If untreated, many standard non-high-performance fibre fabrics will burn in air in a fire situation, and offer little protection, or even aggravate flesh burn problems if thermoplastic. When considering protection against liquids, surface energy levels of natural fibres are such that they normally wet easily, and even the more hydrophobic synthetic fibres would not demonstrate sufficiently effective liquid repellency, but would, in some cases, encourage wicking. Proprietary applied products designed for the purpose will impart a surface-repelling barrier to the textile (as in the cases of oil, solvent, water and chemicals' repellency), a transmission-resistant barrier (as in the case of UV protection) and an interactive surface such as an anti-microbial finish, and gas-absorbing barrier layers of activated carbon. Reflective effects against radiant heat can aid thermal protection and provide visual/infra-red camouflage to military clothing. Intumescent FR finishes undergo a metamorphosis to form a heat barrier on the fabric. References are also made in this chapter to applied products that rely on absorption or diffusion of active products into the fibre polymer structure from the surface application. Examples include many FR (fire resistant) products in order to modify chemical changes during burning of the fibre. It is clear to see that there is a profusion of treatment products that provide great diversity in their effect on protective textiles.

   The chemistry and functional behaviour of products used in textile finishing treatments for protection are dealt with in this book and in other publications devoted to protection against each type of threat (Bajaj 2000, Textile Institute Proceedings 1999, Horrocks 2003, Holme 2003, Breunig *et al.* 1998) Coverage here is therefore more general in that respect. The author of this chapter, however, has endeavoured to provide an insight into the range of methods of treating textiles for protection, application aspects and best practice for finishing procedures which are important for performance. The important application considerations such as yarn and fabric preparation treatments, finishing treatments and effects of distribution of treatment products within the fibre, yarn and textile structure are addressed.

8.3   Early treatments for protective textiles

Skins and furs, worn to protect against the elements in pre-history, would not of course have been washed with detergent or alkali. They would therefore have had a long-lasting inherent degree of water and rain resistance by virtue of the natural oils and fats still present from the animals they were taken from, whilst being 'breathable', i.e., water vapour permeable. They would have been of sufficiently solid cover to offer a degree of protection to the body from sunlight, rain, cold winds, and partly from insects; the inherent natural fatty products would possibly have improved water repellency and prolonged the life of the 'garment' by virtue of some protection from bacterial and fungal degradation via reduced wetting. They were, presumably, not sufficiently tough to protect effectively at that time against a sabre-toothed tiger attack (unless of 'high performance' rhino-type hide) but the serious side to this is that the effectiveness of breathable rainwear has not advanced so dramatically over those millennia. Certainly tanned leather still finds protective apparel uses in addition to footwear, it also exhibits shower-proof effects in clothing, and an inherent degree of fire-retardant (FR) properties in firemens' tunics in some parts of the world.

In the middle ages, deliberate attempts were made to design clothing for protection on the battlefield. Sophisticated chain-mail armour provided lighter weight designs than metal breastplates and similar solid metal cover for other body parts. These chain-mail armours retained a useful effectiveness against some weapons of the day. This concept was to withstand the test of time due, in addition to its lighter weight, to being highly breathable and flexible for improved comfort and ergonomics; these are sought-after attributes in today's protective clothing. Their downside of course is lack of resistance to sharp point penetration.

As far as applied finishes of the past are concerned, oiling of cloths of woven linen or cotton was achieved with drying oils such as linseed (from the seeds of the flax plant, which oxidises to become more durable). With natural wax incorporated, this finish formed a good hydrophobic water-resistant base for foul-weather protection. With sufficiently tightly woven constructions for a high-cover factor, fabrics could be made usefully impervious to water. Water-soluble ammonium salts, sulphamic acid and borax/boric acid were impregnated in cotton fabrics to reduce flammability, but were (and are) of somewhat limited usefulness for personal protective garments, due to the significant loss of functional product on washing the garment.

8.4   Progression to modern treatments

For a non-durable (to washing) water-repellent finish a simple process of molten wax, applied via an engraved copper roller with doctor blade was used. Saturated paraffin wax from the higher distillation fractions of petroleum, with sufficiently high melting point, was heated with closed coil steam pipes in a trough to exceed its melting point. In this process the half-submerged, heated engraved copper roller revolved with the pressure roller nip (Fig. 8.1), transferring the wax to the fabric which simply cools to a dry finish. Drying oils (possibly synthetic or natural) for improved durability, and oil-soluble anti-microbial agents (such as pentachlorophenyl laurate – PCPL) have been used as compatible additives for such a system for application to rain-resistant cloaks or tentage. This might seem a very basic and antiquated arrangement, but its virtues are extremely low energy usage (no water or solvent to dry off) and being environmentally friendly, with no vapours or effluent to dispose of. To put this into some context, one has a fairly idyllic application system which transfers only the functional products to the textile. If one substituted modern auxiliaries in this application system, such as fusible epoxies (which cure in the drier at a higher temperatures than their melting points), synthetic drying oils and higher temperature-activated blocked isocyanates, scope would exist for maintaining the system's advantages, whilst improving performance and durability of a range of applied protective finishes.

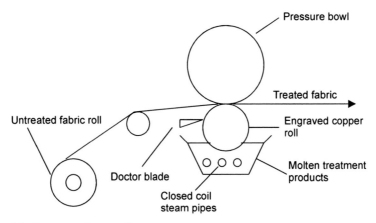

8.1 Molten product applicator.

Advances on the above state of the art that existed at the time were deemed especially necessary to achieve higher levels of effectiveness, high durability, and new effects to combat modern civilian working and living situations and military threats. An early synthetic innovation for water repellency was the ICI product Velan PF® that is stearamido-methyl pyridinium chloride; this relied on the residual reactive fatty part of the molecule, left after heat treatment, to produce a hydrophobic fabric surface. A range of fabrics bearing the name 'Ventile' were produced, initially for military pilots ditching in the sea. Pioneering attempts at designing further improvements in durability of water repellents are recorded by Garner (1966).

This area of technology is continuing to rapidly advance in sophistication to provide surface treatments, design and materials. Much of the research and development effort is being driven by the need to combat new and evolving military, environmental and social conditions and threats. Also modern trends in improved health and safety (especially in industry) and personal care, whereby the population is being constantly made aware of the need for extra protection are apparent. The use of insecticides, minimising dust generation, harmful ultraviolet light and hospital infections are examples where new legal and legislatory needs have to be met. R&D trends in finishing products, however, also have to consider health and safety in the workplace and VOC (volatile organic carbon) restrictions. This has resulted in an increase in the availability of aqueous-based treatment systems in place of solvent-based ones.

8.5   Choice of treatments in relation to fibre and fabric types

These are appropriate considerations because of the very different physical and chemical properties encountered across the range of textile fibres. Even within the natural fibres, wool burns and reacts to FR treatments in a very different way than does cotton (Horrocks 2003).

8.5.1   Treatment of natural and standard synthetic fibre fabrics

Modern FR treatments of natural fibre fabrics are effective and usefully durable, whilst largely retaining the fabric's inherent comfort and aesthetic properties. Different considerations have to be made in the application of Zirpro® (IWS – potassium hexafluorozirconate or titanate) to wool, which relies on the substantivity of the negatively charged product to the positively charged wool via an exhaustion process in acidic conditions. The ability of the finishing plant to cope with the effluent that contains residues of the product needs to be addressed by the finisher. During the drying and curing of Pyrovatex® (Ciba 1998) FR on cotton, acceptable local levels of atmospheric formaldehyde need to be controlled, although this is easier with the more recent Pyrovatex CP. For Proban® (Albright, Wilson) FR on cotton, there is the special need for dedicated ammonia after-treatment equipment.

The types of textile fabrics and fibres used in protective situations in recent years have had some influence on the range of treatments and applied products in use and under development. The establishment of nylon in the 1940s and of polyester fibre in the 1950s made way for stronger, hydrophobic materials, and continuous filament forms. They brought with them, however, related problems with new coloration processes, restrictions due to their thermoplastic behaviour, and, in many cases, greater difficulty in achieving durability of applied finishes compared to the existing natural fibres like cotton, wool, linen and silk. Besides their new chemical makeup, (a consideration of importance in flame-resistant finish design), such thermoplastic fibres have more hydrophobic surfaces and functional groups that are more difficult to access. However, for some functional protective effects such differences in properties are of benefit; one such property is a higher degree of inherent water resistance, or lower moisture regain, that reduces the water-holding capacity of the fabric compared to natural fibre fabrics. Paradoxically, it is probably relevant at this juncture to compare a natural fibre, whereby the high water absorption of cotton has actually been exploited in protection. Densely woven Ventile fabrics, of Oxford construction whereby two warp threads are woven as one, are very breathable (moisture permeable) in the dry state and become water- resistant when wet due to fibre swelling causing densification of the fabric. This system was developed by the Shirley Institute (now BTTG) for military pilots to survive in the event of the aircraft ditching into the sea. Velan PF®, referred to in section 8.4 above, was applied to ventile fabrics to impart some surface water resistance into the product.

Although there are more difficulties fixing surface finishes onto nylon and polyester fibres, the latter does allow organo-phosphorous compounds to be added in the fibre-manufacturing stage to produce fibre-inherent FR properties as in Trevira CS® (additive phosphinic acid derivative) and Trevira FR from Hoechst (Bajaj 2000). Very similar FR effects, including good wash durability, are achievable by pad/thermofixing similar organic phosphorous-based functional products into polyester fabrics as a finishing process. An example of one of the available products applied in this way is Flacavon AZ® from Schill and Seilacher; a final rinse is desirable to rid the fabric of unfixed product.

FR viscose is an attractive product in that it has some of the comfort factors of natural fibre, being regenerated cellulose, whilst possessing the ability to incorporate the functional products into the undrawn liquid state of the polymer. Sandoflam 5060® (Clariant), bis (2-thio-5,5 dimethyl-1,3,2-dioxaphosphorinyl) oxide has been a popular additive to the viscose dope and another well established version is Sateri's polysilicic acid-containing Visil® (Bajaj 2000). A popular application of viscose modified in this way is in blends with high-performance flame-resistant fibres such as meta-aramids; it does not possess sufficient ease of care, hard-wearing and strength properties on its own for most protective wear.

8.5.2   Treatment of high-performance fibre fabrics

More recently we have seen the introduction of high performance fibres (Hearle 2001, Lewis 1999) whose main attributes include high modulus, high strength, some with high strength to weight ratio, and many with inherent non-flammability or high FR properties. However, by virtue of their highly oriented molecular order, being less amorphous and more ordered and crystalline, such fibres possess reduced accessibility for applied finishes to penetrate or find sites in which to deposit or attach. Their low surface energies are particularly unhelpful in this respect, reducing useful adhesion and fixation levels with applied functional products (Lewis 1999). These high-tenacity fibres have not had an extensive effect on treatment capability, except possibly that reagents and finishing chemicals can be more potent without severely degrading the tensile strength of the fibres. The formation of fibrils on para-aramid fibre surface (and to some extent polyester) can be detrimental to durable bonding with other applied materials. Fibrillation – the splitting off of fine linear fragments parallel to the fibre axis – occurs when subjecting para-aramids to abrasion or other dynamic surface stresses; the weak radial tie-molecules in the fibres having low resistance to such shearing forces. The effectiveness of interfacial bonds with polymer or resin matrices could be compromised if fibrils split off within the matrix.

8.6   Treatment process fundamentals

There are various means of upgrading fibre and fabric surface properties to render them more receptive to protective treatments; physical, chemical and radiation pre-treatments are all capable of providing useful improvements in ultimate protective performance.

8.6.1   Physical pre-treatments

The ability of a textile fabric, of a particular fibre polymer type and fabric construction, to hold treatment liquor during the commonly used pad impregnation system depends in part on the fibre/filament configuration of the yarns. It is in the order for yarn types:

Spun staple > false twist textured > flat multifilament > monofilament.

This particular order of graduation also applies to the keying-in ability of deposited functional reagents and products, especially polymeric and coated forms; this particular behaviour can be enhanced by raising and exposing tiny loops in multifilament fabrics, or fibre ends by a sueding treatment with fine-gauge emery paper. Air-jet texturing generates loops along filaments, and fabrics with such configured yarns will have improved keying properties to applied products compared to flat filament fabrics. They avoid the inherent stretch exhibited by false-twist textured yarn fabrics, which can cause problems.

Thermoplastic fibre fabrics can be heat calendered to reduce their interstitial pore size to optimise resistance to penetration by harmful fluids. This process can be improved with moistened fabric, and friction calendering sometimes enhances the effect particularly on nylon, whilst minimising the loss of tensile strength.

8.6.2   Chemical preparation treatments

Pre-activation of some of the hydrophobic fibre surfaces, such as polyesters, and particularly in the more inert polyolefins (Brewis and Mathieson 2002), aramids and glass fibre (Chabert and Nemoz 1985), is done to aid wetting and fixation of applied finishes and matrices for protective products. Such preactivation treatments for aramids and industrial polyester are typically epoxy or blocked isocyanate linking/adhesive reagents, or a combination of the two. More common preparations for glass fibre are organo-silanes and Chabert and Nemoz (1985) have developed formulae showing how hydrogen bonding occurs with these coupling agents at the glass/fibre interface to improve the bond with an epoxy matrix. High-tenacity nylon yarns are more receptive to bonding agents anyway and adhesive preparations, often applied to the fabric by the finisher, consist of either resorcinol/formaldehyde/latex (RFL – mainly to bond to rubbers) blends (Wootton 2002) or polyurethanes.

It is important to set the appropriate pH of the fabric to suit the particular formulation being applied in finishing. The ionic nature of any residual reagents remaining on the fabric from earlier processes needs to be examined for possible incompatibility with any of the finishing formulation components in the treatment bath. Choice of other dyebath aids should be made with full knowledge of functional compatibility with the applied finish or intended use of the textile; for instance, residues of silicone dyebath antifoams on the fabric during fluorochemical oil or petrol repellent finishing could detract from optimum functional effect. They could also reduce bond strengths of coatings or laminations. Surface-active agents are dealt with in section 8.6.4.

8.6.3   Radiation and plasma surface treatments

It is invariably the case that oils, waxes, sizes, etc., that are used as fibre and yarn processing aids need to be efficiently cleansed from the fabric to allow good interfacial contact between fibre surface and deposited product to realise the desired protective performance and durability. The level of residual preparation aids remaining are normally quantified with Soxhlet solvent and aqueous extraction.

A particularly effective means of removal of lower levels of unwanted surface organic matter is low-pressure plasma treatment. The high-frequency forerunner to this process was corona treatment, but that is mainly confined to film treatments; the main difficulty is the long time lapse between corona treatment and applying the finish. Decay of the imparted surface energy which improves fibre wettability, can be too rapid. The modern plasma processes (Shishoo 1999, Gregor and Palmer 2001, Ohisson and Shishoo 2001) do not normally exhibit this problem and are capable of reducing the residues of contaminants on the fibre surface to insignificant levels, sometimes needed where durability of functional finishes is to be optimised. In addition to this extra-cleansing via vaporisation of residuals, the plasma process is capable of adding etched irregularities into the morphology of the fibre surface; this occurs due to a degree of stripping-off of localised parts of the surface molecular layers. The microscopically uneven nature and increased surface area generated improves keying-in of applied product.

A more sophisticated phenomenon possible with plasma treatment is the transformation of the chemical character of fibre surfaces. Depending on the gases introduced into the autoclave, which become ionised in the charged low-pressure environment, fibre surfaces can be improved by virtue of new functionalities; dramatic wettability improvement, via surface oxidation, is of particular benefit for aramid fibres and other oleophobic or low-energy surface synthetics. Examples of the introduction of new functional elements and groups are fluorine, for inherent oil and water resistance, and carboxyl groups that appear to be formed when incorporating carbon dioxide; the importance of this latter capability is in chemical linking of applied protective finishes onto otherwise fairly inert surfaces This might be by direct covalent bonding of a suitably designed finishing agent, or via a chemical cross-linking system to form a bridge between the introduced reactive surface radical and the applied product. Opportunities for hydrogen bonding are also generated.

High-performance polyolefin fibres, such as high-density polyethylene (HDPE), possess particularly low energy surfaces, and could also benefit from plasma pre-treatment with surface formation of oxidised polymer species; Lewis (1999) reports about a tenfold increase in lap shear bond strength compared to no pre-treatment of HDPE. Although they are lower melting point fibres they can easily withstand the lower temperatures used in the low-pressure plasma system.

A three-year EU funded programme (Shishoo 1999) which was completed in 1999, studied the application of fluorochemicals on textiles using plasma system for highly durable oil and water repellency. Also included in the programme were textile plasma treatments to improve adhesion of polyurethanes (PU) and finishes for improved soil release, antistatic and antibacterial effects. Gregor and Palmer (2001) addressed the environmental advantages of using oleophobic treatments, whilst Ohisson and Shishoo (2001) examined the plasma process for treatment with tetrafluoromethane.

Ultraviolet (UV) radiation treatment of polymer surfaces is another interesting emerging technology, since research with this process has recently examined its advantages in textile treatments of polymer films and membranes. Graft copolymerisation is one of the techniques used for polymer surface functional modification and is described by Cook and Smith (2003). Basically, the process involves extraction of hydrogen atoms by using UV radiation in combination with a photo initiator to generate free radicals. Monomers can then combine with the free radicals to form grafted polymer chains. The integrity of the base polymer is unaffected by the process and the grafts are chemically fixed.

8.6.4 Distribution of applied treatments

In pursuit of optimisation of effect and durability, the distribution of the deposited solids needs to be considered in any textile treatment system. From the textile fabric surface and internal space and yarn surfaces, down to the individual fibre coverage the location of active product and extent of encapsulation of the textile components are important to consider. For impregnation by padding or similar techniques, the inevitable movement on drying of solutes or dispersed or emulsified products should be of prime concern. Migration of successive layers of pad liquor into surface-dried layers will tend to occur on passage of the wet treated fabric through the drier. This movement of liquid vehicle will progressively carry more and more solids into the layers drying most quickly, resulting in uneven distribution of the functional product. Drying energy from the hot air moving over the wet fabric surface will have more influence on migration than purely high temperature levels. Even if there is good air speed (or pressure) balance above and below the fabric in the drier, migration to both outer surfaces can occur, leaving the inner layers of the fabric structure devoid of applied product. The penetrating radiation effect of infra-red pre-drying often promotes even drying, applied directly before the fabric enters the drier it imparts energy to all layers within the fabric. This encourages even-rate drying throughout the fabric.

One might consider surface-formed concentrations of product, due to migration, to be beneficial in some situations (e.g. for some types of transfer coating and lamination for waterproofing). As an example of a functional applied finish in its own right, a higher surface concentration of water or oil repellent would be effective in a low-intensity wetting situation, but with more prolonged exposure, greater pressure or with more dynamic energy in the impinging liquid, so-called back-wetting will occur once the surface resistance has been broken down. This results in liquid being absorbed into the inner layers of the fabric with some loss of liquid barrier effect.

Such inner layers-to-surface or face-to-face movement can be more visually demonstrated for a particular set of drying conditions by application of a coloured pigment dispersion on the fabric. However, the more representative approach to show distribution is to detect the position and coverage on the fibres of the active product. These include staining (coloured or fluorescent) with visual microscopy possibly with UV illumination, SEM (scanning electron microscope) photography, and SEM/XRF (X-ray fluorescence with a scanning electron microscope) elemental mapping.

Wetting and penetrating agents are often incorporated in the finishing bath, or left on the fabric after dyeing, to improve the efficiency of distribution of the solids throughout the fabric structure. The presence of surface-active agents has a deleterious effect on repellent treatments. Residual products from the scouring or dyebath might be in the form of detergents, emulsifiers, dispersing agents, levelling agents and softeners. If wetting/penetrating agents are considered necessary to improve ultimate dried solids distribution, they should normally be kept to a minimum, since there will be an inevitable monolayer barrier between fibre surface and functional product, whereas good fibre/product interfacial contact is needed for best function and durability. Fugitive wetting agents, often based on drying-stage volatile alcohols, are often a better proposition, although their contribution to factory VOC levels has often to be taken into account if scrubbers are not present on driers. The concentration of such vapours in a localised working area need to be considered in relation to tolerance levels and flammability hazards. There is also the possibility of preferential and uneven absorption of such surfactants and alcohols into the fabric during padding, resulting in reduced repellency, due to excess wetting agent or a reduced pick-up of functional product.

Fugitive surface active agents, such as a fatty acid (e.g. stearic) coupled with a volatile base, such as morpholine (1-oxa-4-azacyclohexane) (see Fig. 8.2) can

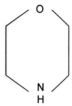

8.2 Structural formula of morpholine; 1-oxa-4-azacyclohexane.

be useful emulsifiers or wetting agents in water-resistant finishing of protective wear, but not for oil-repellent finishes, where the fatty acid residue would detrimentally effect performance. The dispersion of active products in a protective finish at the padding stage will invariably cause pad liquors to cling to hydrophobic fabrics at the boundary between two adjacent filaments, or at intersections of yarns in the fabric, leading to higher concentration of solids in those locations (Fig. 8.3). Deposits can also be spasmodically placed along the length of the filaments due to the low-energy fibre surface; a second impregnation treatment would be beneficial in rectifying such irregular deposition of solids. Migration inhibitors for aqueous treatment liquors consist of water-soluble polymers and gums that are of low viscosity; this allows a network of a sufficiently high concentration of polymer chains to inhibit the movement of dispersed solids (on drying) whilst retaining low viscosity, beneficial for good initial penetration and distribution of the liquor.

Similar restricted mobility effects are seen when using high-solids formulations, although this approach may cause initial penetration problems and other undesirable effects. In the thermofixation of disperse dyes, particle flocculation aids even dye distribution and in another thermofixation approach (Bajaj 2000) migration of an FR finish for polyester was shown to rely on a gel formation mechanism. Coagulation of dissolved or dispersed polymers, to fix applied products in the location they are deposited during treatment, is another

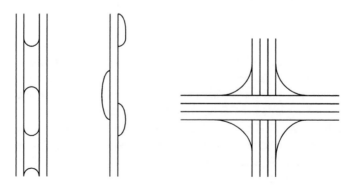

8.3 Common pad-liquor deposit patterns on hydrophobic filaments/yarns in fabric.

8.4 Structural formula of para toluene sulphonic acid (PTSA).

means of migration prevention. One coagulation approach is to incorporate acid donors such as volatile amine salts (e.g. amino methyl propanol) of para toluene sulphonic acid (Fig. 8.4) into the formulation. These would work where polymer emulsions depended on alkaline-based emulsifying agent which would be rendered inactive. Such deactivation and consequent conglomeration of the emulsified particles would ideally occur in the first stage of the drying process, or during infra-red pre-heating, where the latent acid or other emulsifier-deactivator would be rendered active via liberation of the amine. The employment of low-liquor application systems, referred to in 8.7, is another means of reducing the tendency of migration during drying.

8.7   Treatment application systems

Major factors that determine the choice of method to apply the protective treatment are end product use, the range of properties needed and the form of the products to be applied. The choice of application method will also be influenced by the type of textile material being modified; for instance, direct coating would not normally be considered for open or unstable structures such as some warp and weft knitteds, whereas transfer coating (Bobet 2002) or lamination (CS Interglas) might well be possible.

   With tighter legislation controls on solvent use, manufacturers have for some years paid more attention to aqueous-based solutions, dispersion and emulsion versions of functional products for application to textiles. However, application from organic solvent is still important to many leading coaters, helped by the practices of solvent recovery or evaporated solvent burning. Reduced energy is needed for drying of solvent, compared to water-based systems but more elaborate healthy working environmental control and sealed switchgear for flammable solvents are more demanding aspects. Advantages of products with solvent applied finishes include degreasing of the fibre surface (although the danger is in increasing contamination of the applied liquor throughout a finishing run), better wetting, penetration and interfacial contact with hydrophobic fibres, and improved film properties. Table 8.1 compares the major commercial continuous application systems in use, and the relevant applied product distribution, in the author's experience, is shown.

Table 8.1 Production application of finish

Application system	Machine/process	Treatment form	Distribution character of deposited solids
Conventional impregnation	Horizontal or vertical pad mangle	Dispersion, solution or emulsion	Fibre surface, fibre and yarn intersections. Migration onto surface on drying is possible
Foam low liquor add-on	Foam applicator	Dispersion, solution or emulsion; foamed	Low tendency to migrate; lower drying cost. Control of foam not easy
Spray low liquor add-on	Spray and rotating disc	Dispersion, solution or emulsion	Low tendency to migrate; even area distribution difficult
Molten products	Heated trough partly immersed, grained copper roller and doctor blade.	Liquid, above melting point, below cure temperature	Not continuous film. Porous
Sprinkle/scatter/ spread/ electro static	From hopper and lower rotating brush. Doctor blade spread is possible	Powder or small granules	Discontinuous deposits on surface unless larger quantity and fused
Knife/air coat	Doctor blade on tensioned fabric; scrapes surface. Sharper blade can reduce penetration	Thickened aqueous dispersion or solvent solution	Penetration into fabric
Knife/roller coat	Doctor blade over fabric supported by roller; set gap	Thickened aqueous dispersion or solvent solution	Less penetration, greater surface cover; better than k/a for proofing and breathable coating. Knife must lift at sewn joins, etc.
Knife/blanket coat	Blade contacts fabric above flexible part of continuous rubber blanket	Thickened aqueous dispersion or solvent solution	More forgiving than knife/roller, reduced tendency to trap sewn joins and surface irregularities

Table 8.1 Continued

Application system	Machine/process	Treatment form	Distribution character of deposited solids
Extrusion coat	Slot extruder with sophisticated slot gap control	Molten granules extruded as thin film	Most direct product to fabric and high speeds. No solvent to dry off
Transfer coat	Adhesive coat of film or fabric from paper; transfer by pressure and drying	Cast film on release paper	Coater can design own film, including surface grain via paper texture choice
Lamination	Adhesive application and (multi-ply) laminator range	Bought-in rolls of film	Less film design flexibility than transfer coat. Multi-layer composites possible though

8.8   Brief overview of finishes for protection

More detailed treatment of the applied product chemistry and the finish/fibre interaction mechanisms can be found in Part II of this book and elsewhere in the literature (Bajaj 2000, Horrocks 2003, Holme 2003). The following overview addresses the range of approaches and product types in current use for imparting protection to the most commonly encountered situations in work, civil and military environments.

8.8.1   Protection against the elements

Shower proofing

Paraffin wax, emulsified with a fugitive surfactant and in admixture with aluminium or zirconium salts, provides a semi-durable treatment for textiles; slightly higher wash durability being exhibited on cotton than synthetics due to ionic attraction of the heavy metal to the cellulose. Polysiloxane (silicone) based repellents (Breunig *et al.* 1998) offer good general durability with softness and drape but can often show an undesirable oily handle on synthetics, with greater tendency to attract oily stains than fluorochemical-based finishes. The latter type tend to provide a 'drier' handle and are also used as effective oil repellents. With well-designed formulations and finishing practices appropriate to the particular fibre and fabric type, very high levels of water shedding are achievable.

However, the fabric retains normal textile permeability and will not therefore be fully waterproof. Microporous or hydrophilic 'breathable' (i.e. maintaining a useful moisture transmission rate for comfort) membranes are available for lamination as drop-liners, or directly onto the main outer rainwear fabric for foul-weather wear; their high resistance to water penetration renders the garment largely waterproof and windproof. Useful levels of moisture transmission can also be maintained by using hydrophilic PU and polyester polymers for direct waterproof coating (Holme 2003, Holmes 2000).

UV screening effect fabrics

Textiles of sufficiently dense construction and in dark colours will offer improved shade barriers to harmful UV radiation wavelengths compared to open textile structures of pale shades. For garments to be especially protective against this form of radiation, proprietary UV absorbers are available (Clariant, L J Specialities, Ciba 1998) for application to textiles. T-shirts and other summer clothing with such treatments are particularly popular in Australia, and are in increasing demand with the growing awareness of the potential hazards of unprotected skin exposure to strong sunlight.

8.8.2   Industrial environment protection

The chemical industries are the businesses that have the greatest need for garments with protective treatments to repel both water and oil based chemicals. Legislatory controls, legal concerns, and increasing employer awareness of hazards to workers in these industries have increased the use of protective clothing possessing good liquid-repellent effects. Fabric treatments with fluorochemicals are particularly in demand, due to the wide range of repellent effects against aqueous and solvent-based chemicals. The chemical stability of the applied product itself also aids durability of protection. One should expect a higher concentration of fluorochemical applied, compared to that used, for example, in shower-proof rainwear, to ensure the required repellency performance and durability standards for industrial protective wear. Resistance to penetration, for any dynamic liquid contact or higher pressures would still depend on tightly woven constructions, proofed coatings, breathable laminates or use of these as drop liners in the garment. For protective situations where coveralls are deemed unnecessary, PVC or neoprene rubber-coated aprons and gloves are common.

8.8.3   Protection for emergency services

Petrol repellency

FR finished fabrics used in protective gear for riot police and some firemen's tunics calls for particularly high oil-repellency levels, in the order of 7 to 8 on

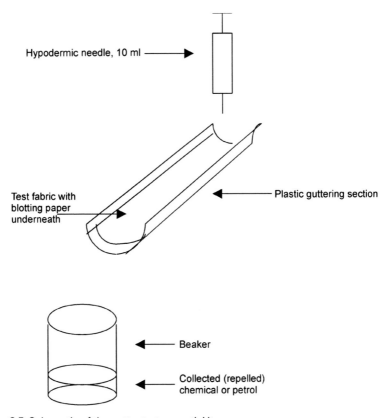

Hypodermic needle, 10 ml ⟶

Test fabric with blotting paper underneath

Plastic guttering section ⟵

Beaker ⟵

Collected (repelled) chemical or petrol ⟵

8.5 Schematic of the gutter test; essential items.

the 3Ms or AATCC Method 118 test, and pass the dynamic nature of the Gutter Test (Holmes 2000) (Fig. 8.5). Because such finishes are for garments employed in life-threatening situations it is also important to ensure very good durability to dry cleaning or laundering or possibly both, depending on the instructions on the garment label. In order to meet this high level of repellency and durability, carefully designed finishing formulae are necessary that normally call for further fluorochemical concentration increases over and above that used for chemical-resistant protective wear. Incorporation of either a blocked isocyanate, a formaldehyde condensation resin or both is required. The finite optimisation of such finishing treatments is often proprietary to the textile finisher (e.g. Heathcoat), but recommendations are available from the fluorochemical suppliers (Clariant, LJ Specialities, Ciba 1998). The optimised treatment task is more complicated, due to the inherently FR high-performance fabrics normally used in these end uses. Aramid fabrics and blends often do not lend themselves to good treatment coverage and fixation/durability. Loss of treatment product due to abrasion during laundering can be a contributory reason for insufficient wash durability. To resist penetration by petrol bombs or molotov cocktails, breathable back coating might be needed, although a drop liner has the thermal insulation advantage to protect against heat.

Fire, heat and flash protection

Papers by Horrocks (Textile Institute 1999, Horrocks 2003) and a chapter in this book provide detailed coverage of the science of FR treatments and mechanisms that include the chemistry of interactions between applied product and the fibre and flame environment. Many fire, police and some parts of the military services in the UK currently employ FR aramid or aramid/FR viscose blend yarns, i.e., inherently flame-resistant yarns, in robust fabrics for demanding operational protective outer-garments; such fabrics also have the additional advantage of high strength and durability. Anti-flash hoods, gloves and knitted underwear in Aramids for Naval personnel and tank crews are used by some national armed forces. An important requirement for garments in contact with the skin is non-thermoplasticity, especially so in some military environments. Unfortunately, applied FR finishes will not normally convert thermoplastic fibres into a non-melting form. The UK MOD limits polyester in blends to a 25–30% content for certain types of military clothing, especially for Naval Action Clothing and military pilots of fixed-wing and rotary-wing aircraft (Scott 2000); higher proportions of polyester in blends with cotton also present more difficulties with effective FR treatment.

In order to ensure maximum resistance to burning of these aramid and aramid blend fabrics there is a need to avoid application of products that might increase their flammability. It is wise to re-evaluate flammability level after any such changes in formulation. Appropriate fabric and multi-layer garment constructions will produce the required thermal insulation. For the most severe fire-entry situations aluminium reflective coatings provide a high specular reflection to radiant energy for some firefighters' tunics. These can be made by lamination (CS-Interglas 2004), transfer coating (Bobet 2002) or vacuum deposition. A novel effective treatment to reduce the effects of thermo-nuclear flash on military combat clothing is the use of a thermochromic green pigment from LJ Specialities. This particular thermochromic is based on encapsulated leuco dye in solid form that allows the colour developer to combine with the dye and produce full colour. The liquid state is formed within the capsule on heating, causing dissociation of developer and dye which results in its colourless form. Green fabric becomes white in this application at 60 °C.

Another thermochromic system uses liquid crystals in the form of thermally activated extending and contracting spiral molecules that alter the interaction of light, according to wavelength, in relation to angle of spirality. Such colour change systems have potential in other protective systems to visually indicate dangerous high temperature.

For FR treatment of cellulosic fabrics the Proban CC (Rhodia) product, based on tetrakis (hydroxymethyl) phosphonium chloride (THPC) appears to be favoured for military use, in the UK (Scott 2000) and is less expensive than high-performance fibres. It requires an ammonia cure after-treatment with urea and oxidation to form a cross-linked poly(phosphine) oxide polymer. Together with Pyrovatex (Ciba 1998) treated cotton, such FR fabrics form a large proportion of lower risk protective applications in Europe. Pyrovatex CP is also phosphorous based (N-methylol dimethylphosphonopropionamide) and uses a condensation process with trimethylol melamine to form a chemical link with the cellulose (Bajaj 2000, Horrocks 2003). The employment of antimony oxide and brominated organic admixtures appears to be currently more popular as FR back coatings for a range of fabric and fibre types.

Treatment of fabrics with intumescent compositions has continued to be developed and refined in recent years (Bajaj 2000, Textile Inst. 1999) and can provide effective flame and heat barriers, independent of chemical interaction with the fibre, when applied as textile coatings. The formation of a dry, largely carbon, foam on heating provides a non-flammable barrier that also has thermal insulation properties. This prevents heat penetrating to the fibres, thereby reducing emission rate of flammable volatiles, and reducing the rate of melting of thermoplastic fibres (Horrocks 2003). The major components of this catalysed coating composition are a high-carbon former (such as pentaerythritol), a non-flammable gas former (melamine; spumificant – to generate the foam) and an acrylic binder; the catalyst for the system that activates the spumificant can be phosphoric acid generated from incorporated ammonium phosphate. More recent work (Bajaj 2000, Horrocks 2003) has demonstrated the formation of a stronger carbon-foam structure that is more oxidation resistant, by more intimate coating within the fibre structure.

Stab, ballistic and abrasion protection

High-performance/high-modulus fibres with no functional finishes applied, satisfy a good proportion of these markets. For stab-proof fabrics, a highly dense construction from fine-denier yarns can achieve a useful resistance to penetration from sharp blades, spikes and even needles. However, an advanced example of recent developments has incorporated hard material deposits onto the surface of fabrics to improve performance, called Twaron (Teijin) SRM® (Bottger 1999). The protective coating on the product (Fig. 8.6) consists of silicone carbide particles deposited onto a special matrix composition that coats the Twaron aramid fabric substrate. A major function of the hard particles is to blunt sharp blades and points, adding to the energy dissipation and penetration resistance of the fabric.

Fabrics for motorcyclists' jackets that offer protection against abrasion of the body caused by sliding along road surfaces during accidents, need to be more able to withstand the very dynamic abrasive effects involved. Aramid fibre fabrics are used to absorb energy in such an event, but a novel, less expensive

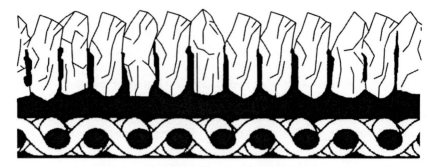

8.6 Cross-section of a schematic of Twaron SRM.

attempt has been developed whereby small domes of tough resin are bonded to the outer surface of the fabric (Vora 2002).

For bullet-proof vests, up to 36 layers of high-performance woven fabric are used but, where less bulky protection is demanded, one approach is to replace some layers with a fairly rigid composite plate. Such plates commonly consist of high-performance/high-modulus fabric composites in an epoxy resin matrix; such a component is designed to spread the area of impact and reduce soft tissue trauma. In order to improve the energy-dissipation effect, there is often a degree of relative movement between matrix and fibre surface by regulating the bond strength. The disadvantages of such plates are the rigidity and impervious nature, which reduces wearer tolerance level.

Impact resistance and heat stress

Protection from impact on the body by objects such as bricks (e.g. against police) and from falling over (the elderly, causing hip fractures) is now offered by a novel system of coiled plastic springs encapsulated into double-structure fabrics (Buckley 1996). A similar arrangement (Buckley 1996, Buckley *et al.* 2002) of springs forms tubular cavities that allow forced-air ventilation of the body. The evaporation of sweat by the moving cool air takes heat from the body via latent heat of evaporation, in heat stress situations. These systems can be incorporated as panels in garments, since they are very flexible and permeable. Air-cooling systems provide efficient, lightweight and simple garments compared with liquid cooling systems.

8.9   Future trends

Threats to military personnel are designed to be deliberate and effective, and these will continue to improve in effectiveness to counter the developments in protection measures. New types of bullets and anti-personnel devices are prime examples, and there will be a continual need for advances in technology to improve the effectiveness of protective systems. In civil and working environments, increasing demands for personal protection will no doubt come from stronger safety awareness, legal and legislatory considerations. Some or all of the following more recent emerging technologies for treatments will play their part in new and improved protective effects:

- Low-pressure, cold plasma treatment, vapour deposition and possibly UV treatment, for improved treatment distribution and fixation (e.g. via grafting or rendering cross linkable), especially onto higher-performance fabrics. (Shishoo 1999, Gregor and Palmer 2001, Ohisson and Shishoo 2001, Cook and Smith 2003).
- Shear thickening fluids for instantaneous reaction to applied forces like impact (Lee *et al.* 2003)
- Three-dimensional textile constructions designed to contain, and retain, larger quantities of functional products, such as phase change materials (www.outlast.com) and other flexible and shape memory materials for thermal and other types of protection (Heathcoat & Co. Ltd 1996).
- Microencapsulation of functional products to, e.g., allow protective effects to activate at the critical juncture and also to hold fluid forms of protective products.
- Nanotechnology – nanometer-sized particles and fibres for lighter-weight and stronger composite materials (Bourbigot *et al.* 2002).

8.10   References

Anderson M A, Brown G L, Wachter D R, Fabrics Having Improved Ballistic Performance and Processes for Making the Same. Affiliations – Clarke Scwebel Inc., Anderson, S.C., USA. Patent Number US 5788907 A 19980804.

Bajaj P, Heat and Flame Protection (Chapter 10); *Handbook of Technical Textiles*; ed. Anand S C and Horrocks A R.; Woodhead Publishing, 2000.

Bobet, High Technical Coated Fabrics, 5 bd Pierre Brossolette-B.P. 5-76126 Grand-Quevilly Cedex (France), 2002.

Bottger C, Twaron SRM – A novel type of stab resistant material. Sharp Weapons and Armour Systems Symposium 8–10 November 1999, The Royal Military College of Science.

Boubigot S, Devaux E, Flambard X, Flamability of polyamide-6/clay hybrid nanocomposite textiles. *Polymer Degradation and Stability* 75 (2002), 397–402.

Breunig S *et al.*, Poly-Functionalisation of Polysiloxanes: New Industrial Opportunities. *Journal of Coated Fabrics*, Volume 27, 1998.

Brewis D M, Mathieson, Adhesive Bonding to Polyolefins; *Rapra Review Reports* Volume 12, Number 11, 2002.

Buckley R W, inventor; John Heathcoat & Co. Ltd., assignee: EU patent app. No. 97308353.0, 1996.

Buckley R W, Daanen H, den Hartog, Tutton W, Efficiency of an air cooling system for helicopter pilots. ICEE Conference, Japan, 2002.

Chabert P, Nemoz G, Uni. Cl. Bernard, Lyon and Inst. Textile de France, Lyon resp.;

Influence of Fibre Surface Matrix Interface in Composite Materials. High Performance Textiles and Composites, UMIST 1985.

Ciba Speciality Chemicals; *www.cibasc.com*, 1998.

Cook J, Smith I, Surface Modification of Polymeric Films and Fabrics by UV Graft Copolymerisation; SciMAT Ltd. Dorcan, 200 Murdock Road, Swindon. Coating 2003.

CS-Interglas Ltd., Alpha Division GB, Sherborne, Dorset, 2004.

Clay F, Polymer composites: the story. *Materials Today*, Elsevier, November 2004.

Garner W, Textile Laboratory Manual, Volume 2, Resins and Finishes. Heywood Books/ Elsevier.

Gregor R, Palmer J (Europlasma, Belgium); Future Dry and Environmentally Clean Alternatives for the Current Wet Chemical Oleophobic Treatment by Means of Low Pressure Gas Plasma. World Textile Congress 2001 on High Performance Textiles, Bolton Institute.

Hearle J W S, *High-performance fibres*. The Textile Institute. Woodhead Publishing Ltd.

Heathcoat & Co. Ltd., Tiverton, Devon, 1996.

Holme I, Water repellency and water proofing (Ch. 5); *Textile Finishing*; ed. Heywood D; The Society of Dyers and Colourists, Bradford, 2003.

Holmes D A, Waterproof breathable fabrics (Ch. 16), *Handbook of technical textiles*. ed. Anand S C, Horrocks A R; Woodhead Publishing, 2000.

Horrocks A R, Flame-retardant finishing (Ch. 6); *Textile Finishing*; ed. Heywood D; The Society of Dyers and Colourists, Bradford, 2003.

IGNAMAP (International Group of National Associations of Manufacturers of Agrochemical Products); Protective Clothing for the Safe Use of Pesticides in Hot Climates. GIFAP 1991, ISO EN 368 ('92) under review.

Lee Y S *et al., Advanced body armour utilising sheer thickening fluids*. Centre for Composite Materials and Dept. of Chemical Engineering, U. of Delaware, Newark, DE19716, 2003.

Lewis P R, High Performance Polymer Fibres. *Rapra Review Reports* Volume 9, Number 11, 1999.

Ohisson J, Shishoo R, Use of plasma technology in textile processing – the achievements and challenges ahead. World Textile Congress 2001 on High Performance Textiles, Bolton Institute.

Scott R A, Textiles in Defence (Ch. 16), *Handbook of technical textiles*, ed. Anand S C, Horrocks A R, Woodhead Publishing, 2000.

Shishoo R, Final report – EU Brite Eu-Ram project PLASMATEX (Development of Plasma Technology for Continuous Processing of Textile Fabrics and Nonwovens), 1999,

Textile Institute; Textile Flamability: Current and Future Issues – Proceedings. 30–31 March 1999.

Vora R&D, Voravagen 31, FIN-66600 VORA, Finland, 2002.

Wootton D B, *Application of textiles in rubber*. Woodhead Publishing, 2002.

www.clariant-textiles.com

www.lj-specialities.com

www.outlast.com

Evaluation of protective clothing systems using manikins

E A McCULLOUGH, Kansas State University, USA

9.1 Introduction

Compared to typical apparel, protective garments are made with specialized materials, innovative fastening systems, and unique designs. Protective clothing systems are often complex and require user training to don and doff them correctly. The garments are often worn with auxiliary equipment also, such as a tool belt or respirator. It is important to evaluate the comfort and performance of the entire protective clothing system – not just the properties of the component materials. Therefore, life-size manikins are being used increasingly in standard test methods to evaluate the performance of protective clothing systems. Manikins have been used to measure the flame and heat protection provided by different types of protective clothing (ASTM, 2005; Crown et al., 1998; Lee et al., 2002; Prezant et al., 2000; Rossi and Bolli, 2000) and the integrity of chemical suits to leaks (ASTM, 2005). Manikins have also been used to evaluate and compare the effectiveness of personal cooling systems worn under protective garments (Teal, 1990). This chapter will discuss the use of manikins to evaluate the heat transfer (i.e., comfort) properties of protective clothing systems.

9.1.1 Protective clothing systems

Protective clothing systems are usually designed to protect the wearer from hazards in the environment. These include fire, extreme heat and cold, water, chemicals, particulates in the air, blood-borne pathogens and other biological agents, electrical power, radioactive materials, physical force or impact (e.g., bullets, bomb fragments, sports equipment, work equipment, falling debris, etc.), and ultraviolet light. Some clothing has been designed to protect the environment from body contaminants also (e.g., clean-room apparel used in the production of computer components). The development or selection of a particular type of protective clothing depends upon many factors, some of which may be affected by standards. Clothing factors include the materials, design features, construction techniques, and fastening systems used in the garments, and the compatibility of component garments with each other and with auxiliary equipment in the clothing system. Environmental factors include the work conditions (i.e., air temperature, relative humidity, etc.) and the type of hazards present. Human factors include the functional needs of the wearer, the potential risk and duration of exposure to hazards, the ease of donning and doffing garments, garment fit (related to size availability), clothing comfort, and the possible restriction of mobility, hearing, and vision. Other considerations include initial product costs, life cycle costs, and the ease of maintenance, storage, and disposability.

9.1.2   Comfort vs. safety

Protective clothing systems often contribute to heat stress when worn by people in hot environments, humid environments, and/or at high activity levels. The body gains heat by exposure to a hot ambient environment or a heat source (e.g., fire) and through the generation of metabolic heat (which increases with activity). The body loses heat through conductive, convective, and radiant heat exchange with the environment and by the evaporation of sweat. The heat is lost from the body surface and through respiration (convection and evaporation). Protective clothing systems can provide significant resistance to the heat exchange between the body and the environment.

As the level of protection provided by a clothing system increases, its comfort level generally decreases. This is particularly a problem for systems that are designed to protect against multiple threats (e.g., terrorism). Highly protective systems usually cover most of the body surface and consist of multiple material and/or garment layers. Therefore, the thickness and insulation value of the materials may increase, and the permeability of the materials to liquid, air, and moisture vapor may decrease. In addition, the size and number of garment openings for ventilation may be low or nonexistent. Protective garments and equipment are generally higher in weight and stiffness than typical clothing and have a poorer fit due to limited size availability, consequently restricting mobility and increasing the energy production of the wearer. By restricting the heat loss from the body to the environment and causing the wearer's energy expenditure to increase during movement, protective clothing may decrease the thermal comfort of the wearer and contribute to heat stress. When the environment and the clothing is uncomfortably hot, workers may need more rest breaks, have lower productivity, and have more accidents. Workers often try to make their clothing more comfortable by altering it, wearing it incorrectly, or refusing to wear it at all – thus decreasing the protection provided.

9.2   Thermal manikins

Thermal manikins have been used by many researchers to measure the thermal resistance (insulation) and evaporative resistance of clothing systems. These resistance values are used in biophysical models to predict the comfort and/or thermal stress associated with particular environmental conditions and the activity of the wearer. Wyon (1989) described the historical development of manikins, and Holmér (1999) continued this work, describing standard test methods for the use of thermal manikins and standards that require data from manikin tests. It is impossible to count the number of heated manikins in use because some laboratories have published numerous scientific articles on data collected with the same manikin, and researchers from labs without manikins have collaborated with researchers with manikins. However, manikins are in use in the United States, Canada, France, Sweden, Finland, Norway, Denmark, Germany, United Kingdom, Switzerland, Hungary, China, Korea, and Japan. A recent trend has been the development of specialized heated body parts such as a head, hand, and foot/calf so that the thermal effectiveness of the design and materials used in head gear, gloves/mittens, and footwear can be determined with more precision (Kuklane *et al.*, 1997).

It is important to remember that manikins do not simulate the human body physiologically. They are thermal measuring devices in the size and shape of a human being that are heated so that their surface temperatures simulate the local and/or mean skin temperatures of a human being. They do not respond to changes in the environment or clothing like the human body does.

9.2.1   Segmented thermal manikins

Most manikins are divided into body segments with independent temperature control and measurement (even though they are intended to be used to quantify the heat transfer characteristics of total body systems). The segments can all be controlled at the same temperature (i.e., 34 °C), or a skin temperature distribution where the extremities have lower temperatures than the head and trunk. These manikins can indicate the relative amounts of heat loss from different parts of the body under specific environmental conditions and/or measure the insulation value or evaporative resistance value of each segment. However, most segmented manikins have some internal heat transfer from one segment to another and the movement of heat within the clothing layers between segments – both of which will lead to inaccurate local results. In addition, the use of segmented manikins has led to the use of two different methods for calculating clothing resistance values: the parallel (total) method where all heat losses, temperatures (area-weighted), and areas are summed before the total resistance is calculated, and the serial (local) method where the individual resistances for each body segment are calculated and then summed (Nilsson, 1997). The serial calculation usually produces higher values and more variable results due to the uneven distribution of insulation over the body (Nilsson, 1997; McCullough *et al.*, 2002). When a body segment (e.g., abdomen) is well insulated relative to the others, its heat loss may be very low or zero, causing the measured insulation value of the segment to be too high, and the resulting serial calculation for the body to be high.

9.3   Measuring the insulation of protective clothing systems

The first thermal manikin was developed by military researchers in the United States in the 1940s (Belding, 1949). A manikin was needed to measure the insulation properties of protective clothing and sleeping-bag systems because measurements on pieces of fabric could not be related to whole-body systems with accuracy.

9.3.1   Standards

The first thermal manikin test method was developed in 1996: ASTM F 1291, Standard Test Method for Measuring the Thermal Insulation of Clothing Using a Heated Manikin (ASTM, 2005). Recently, ISO 15831, Clothing – Physiological Effects – Measurement of Thermal Insulation by Means of a Thermal Manikin was approved also (ISO, 2004).

9.3.2   Method

To measure the thermal resistance, a manikin is dressed in the clothing system and placed in a cool/cold environmental chamber. Then the amount of electrical power required to keep the manikin heated to a constant skin temperature (e.g., 33–35 °C) is measured under steady-state conditions. The power input is proportional to body heat loss (see Fig. 9.1). The total thermal insulation value (R_t) is the total resistance to dry heat loss from the body surface, which includes the resistance provided by the clothing and the air layer around the clothed body. R_t is measured directly with a manikin and is calculated by:

$$R_t = (T_s - T_a) \cdot A_s / H \qquad 9.1$$

where R_t = total thermal insulation of the clothing plus the boundary air layer $(m^2 \cdot °C/W)$, T_s = mean skin temperature (°C), T_a = ambient air temperature (°C), A_s = manikin surface area (m), and H = power input (W).

A clo unit is normally used for expressing clothing insulation since it is related to commonly worn ensembles. A warm business suit ensemble provides an average of approximately 1 clo of insulation for the whole body, where 1 clo of insulation is equal to 0.155 $m^2 \cdot °C/W$ (Gagge et al., 1941). When insulation values are reported in clo units, the symbol I is usually used instead of R. Therefore,

$$I_t = 6.45 \cdot (T_s - T_a) \cdot A_s / H \qquad 9.2$$

9.1 The thermal manikin at Kansas State University is dressed for the evaluation of sports apparel.

Because R_t or I_t includes the resistance at the surface of the clothed body, it is influenced by air velocity and temperature level (as it relates to incident radiation). These factors can be easily dealt with in physiological models which predict how much heat a person will lose or gain under a specific set of conditions. However, in some models and applications, it may be preferable to separate the resistance of the clothing from the resistance of the air layer.

Intrinsic clothing insulation (R_{cl}, I_{cl}) indicates the insulation provided by the clothing alone and does not include the insulation provided by the surface air layer around the clothed body. R_{cl} is defined by

$$R_{cl} = R_t - (R_a/f_{cl})$$ 9.3

and I_{cl} is defined by

$$I_{cl} = I_t - (I_a/f_{cl})$$ 9.4

where R_{cl} – intrinsic clothing insulation (m$^2 \cdot$°C/W), R_t = total thermal insulation of the clothing plus the boundary air layer (m$^2 \cdot$°C/W), R_a = resistance of the boundary air layer around the nude manikin (m$^2 \cdot$°C/W), f_{cl} = clothing area factor (unitless), I_{cl} = intrinsic clothing insulation (clo), I_t = total thermal insulation of the clothing plus the boundary air layer (clo), I_a = resistance of the boundary air layer around the nude manikin (clo).

9.3.3   Clothing area factor

The clothing area factor (f_{cl}) is the ratio of the projected area of the clothed body area to the projected area of the nude body. Photographs of the manikin or a person are taken from three azimuth angles: 0° front view, 45° angle view, and 90° side view and two altitude angles: 0° and 60°. The distance between the camera and the manikin and the focal length must stay the same for all photographs. The size of the projected areas of the nude manikin and the clothed manikin can be determined in several ways. A planimeter can be used to trace around each silhouette and determine the projected area for each view. Alternatively, the areas may be determined by cutting along the outside edges of the silhouette (i.e., projected area) in each photograph and weighing them (assuming the same type of paper was used for printing all photographs). The relative size of the areas may also be determined by manipulating the photographs using computer software such as Adobe Photoshop to blacken the silhouettes and compare the relative number of pixels used in each area. The sum of the projected areas of the clothed manikin is then divided by the sum of the projected areas of the nude manikin; the resulting f_{cl} is a dimensionless index greater than 1 (McCullough et al., 2005).

It is time-consuming to take all of the photographs and determine the relative size of the projected areas. Therefore, the clothing area factor of clothing ensembles can be estimated from tables in standards such as ISO 9920 and ASTM F 1291 (ISO, 1995; ASTM, 2005) and clothing databases (McCullough et al., 1985, 1989). The f_{cl} varies from 1.00 (nude) to a maximum of about 1.70 for some protective clothing ensembles.

The term (R_a/f_{cl} or I_a/f_{cl}) is the resistance provided by the air layer around the clothed body. It is smaller than the air layer resistance for the nude body because the clothing increases the surface area and thus provides a greater area for heat transfer. The effect of the clothing area factor on the calculation of intrinsic insulation is shown in Table 9.1. As the f_{cl} increases, the intrinsic clothing insulation also increases (for a given level of total insulation). The f_{cl} for summer clothing probably ranges from about 1.10 to 1.30, winter indoor

Table 9.1 The effect of using different clothing area factors on the calculation of intrinsic insulation

Summer indoor clothing (I_{cl} = 0.5 clo)	Winter indoor clothing (I_{cl} = 1.0 clo)	Cold weather protective clothing (I_{cl} = 2.5 clo)
1.11 − (0.7/1.00) = 0.41	1.54 − (0.7/1.00) = 0.84	2.97 − (0.7/1.00) = 2.27
1.11 − (0.7/1.05) = 0.44	1.54 − (0.7/1.05) = 0.87	2.97 − (0.7/1.05) = 2.30
1.11 − (0.7/1.10) = 0.47	1.54 − (0.7/1.10) = 0.90	2.97 − (0.7/1.10) = 2.33
1.11 − (0.7/1.15) = 0.50	1.54 − (0.7/1.15) = 0.93	2.97 − (0.7/1.15) = 2.36
1.11 − (0.7/1.20) = 0.53	1.54 − (0.7/1.20) = 0.96	2.97 − (0.7/1.20) = 2.39
1.11 − (0.7/1.25) = 0.55	1.54 − (0.7/1.25) = 0.98	2.97 − (0.7/1.25) = 2.41
1.11 − (0.7/1.30) = 0.57	**1.54 − (0.7/1.30) = 1.00**	2.97 − (0.7/1.30) = 2.43
1.11 − (0.7/1.35) = 0.59	1.54 − (0.7/1.35) = 1.02	2.97 − (0.7/1.35) = 2.45
1.11 − (0.7/1.40) = 0.61	1.54 − (0.7/1.40) = 1.04	2.97 − (0.7/1.40) = 2.47
1.11 − (0.7/1.45) = 0.63	1.54 − (0.7/1.45) = 1.06	2.97 − (0.7/1.45) = 2.49
1.11 − (0.7/1.50) = 0.64	1.54 − (0.7/1.50) = 1.07	**2.97 − (0.7/1.50) = 2.50**
		2.97 − (0.7/1.55) = 2.52
		2.97 − (0.7/1.60) = 2.53
		2.97 − (0.7/1.65) = 2.55
		2.97 − (0.7/1.70) = 2.56

Note: The examples in bold font are using the correct f_{cl} for the ensemble type. The other examples illustrate how errors in the measurement or estimation of f_{cl} can affect the calculation of intrinsic insulation from total insulation.

clothing from 1.25 to 1.45, and protective clothing from 1.30 to 1.70. Comfort models have shown that a 0.1 clo increase in insulation is equal to a 0.6 °C (1 °F) decrease in preferred temperature by people at low activity levels (ASHRAE, 1992). Therefore, the error introduced by using the wrong clothing area factor or none at all will lead to small errors in thermal modeling. It would be more accurate to estimate the f_{cl} and use it than to ignore it altogether.

9.4 Measuring the evaporative resistance of protective clothing systems

There are relatively few sweating manikins available for measuring the evaporative resistance or vapor permeability of clothing (McCullough *et al.*, 2002). Some manikins are covered with a cotton knit suit and wetted out with distilled water to create a saturated sweating skin. However, the skin will dry out over time unless tiny tubes are attached to the skin so that water can be supplied at a rate necessary to sustain saturation (McCullough *et al.*, 1989). Other manikins have sweat glands on different parts of the body (Holmér *et al.*, 1996). Water is supplied to each sweat gland from inside the manikin, and its supply rate can be varied. A new type of sweating manikin uses a waterproof, but moisture-permeable fabric skin, through which water vapor is transmitted from the inside of the body to the skin surface (Fan and Qian, 2004). Some manikins keep the clothing from getting wet by using a microporous membrane between the sweating surface and the clothing, but this configuration may increase the insulation value of the nude manikin.

9.4.1 Standards

ASTM Technical Committee F23 on Protective Clothing has recently approved a new standard ASTM F 2370, Measuring the Evaporative Resistance of Clothing Using a Sweating Manikin (ASTM, 2005). It specifies procedures for measuring the evaporative resistance of clothing systems under isothermal conditions – where the manikin's skin temperature is the same as the air temperature (i.e., there is no temperature gradient for dry heat loss). An alternative protocol in the standard allows the clothing ensemble to be tested under environmental conditions that simulate actual conditions of use; this is called the non-isothermal test. The same environmental conditions are used for the insulation test and the non-isothermal sweating manikin test. The air temperature is lower than the manikin's skin temperature, so dry heat loss is occurring simultaneously with evaporative heat loss, and condensation may develop in the clothing layers. The evaporative resistance determined under non-isothermal conditions is called the apparent evaporative resistance value. The apparent evaporative resistance values for ensembles can only be compared to those of other ensembles measured under the same environmental conditions.

9.4.2 Method

To conduct a sweating manikin test, the surface of the manikin is heated to skin temperature and saturated with water. The manikin is dressed in the clothing, and the evaporative resistance of the clothing system is determined by measuring the power consumption of the heated manikin. Even under isothermal conditions, it will take electrical power to keep the manikin heated because the process of evaporating moisture on the surface removes heat. More power is needed under non-isothermal conditions.

The equation for calculating the total resistance to evaporative heat transfer provided by the clothing is

$$R_{et} = (P_s - P_a) \cdot A_s / [H - ((T_s - T_a) \cdot A_s / R_t)] \qquad 9.5$$

where R_{et} = resistance to evaporative heat transfer provided by the clothing and the boundary air layer (m$^2 \cdot$ kPa/W), P_s = saturated water vapor pressure at the skin surface (kPa), and P_a = the water vapor pressure in the air (kPa), R_t = total thermal insulation of the clothing plus the boundary air layer (m$^2 \cdot$ °C/W), T_s = mean skin temperature (°C), T_a = ambient air temperature (°C), A_s = manikin surface area (m^2), and H = power input (W). The mean R_t value from the dry tests on the ensemble is needed to calculate R_{et}.

The standard also allows the evaporative resistance to be determined by weighing the sweating manikin to calculate the evaporation rate of the moisture leaving the manikin's surface. Then the following equation is used:

$$R_{et} = [(P - s - P_a)A_s]/\lambda(dm/dt) \tag{9.6}$$

where λ = heat of vaporization of water at the measured surface temperature (W) and dm/dt = evaporation rate of moisture leaving the manikin's sweating surface (g/min). The equation for calculating the intrinsic evaporative resistance provided by the clothing alone (R_{ecl}) is analogous to that for dry resistance:

$$R_{ecl} = R_{et} - (R_{ea}/f_{cl}) \tag{9.7}$$

where R_{ea} = the resistance to evaporative heat transfer for a still air layer ($m^2 \cdot kPa/W$).

9.4.3   Moisture permeability index

The permeability index (i_m) indicates the maximum evaporative heat transfer permitted by a clothing system as compared to ideal maximum from an uncovered surface (i.e., a slung psychrometer). It was defined by Woodcock (1962) as

$$i_m = (R_t/R_{et})/LR \tag{9.8}$$

where i_m = permeability index and LR = the Lewis relation, commonly given the value of 16.65 °C/kPa. The permeability index usually ranges from about 0.50 for a nude manikin to about 0.05 for an impermeable single-layer ensemble with a low thermal resistance and high evaporative resistance.

9.5   Ensemble data

Total resistance values vary widely from lab to lab because the air layer resistances measured in the nude manikin test vary due to differences in instrumentation and air velocity. Better interlaboratory agreement has been found when intrinsic values are compared (McCullough et al., 2002). Data for some typical ensembles and protective clothing ensembles are given in Table 9.2. Total resistance values are given, but intrinsic values can be calculated using the data for the air layer resistances (see table footnote) and the clothing area factor for each ensemble.

9.6   Moving manikins

Most of the time, manikins are used in the standing position, but more and more researchers are attaching their manikins to external locomotion devices and measuring clothing insulation with the manikin walking (McCullough and Hong, 1994; Kim and McCullough, 2000; Nilsson et al., 1992; Olesen et al.,

Table 9.2 Insulation data for clothing ensembles measured with a thermal manikin[a]

Ensemble	Total thermal resistance R_t (m²·°C/W) I_t (clo)	Clothing area factor f_{cl}	Total evaporative resistance R_{et} (m²·kPa/W)	Moisture permeability index i_m
1. Men's business suit (briefs, t-shirt, long-sleeve shirt, suit jacket, vest, trousers, belt, socks, shoes, necktie	0.262 1.69	1.32	0.044	0.37
2. Women's business suit: panties, half-slip, long-sleeve blouse, double-breasted suit jacket, knee-length skirt, pantyhose, shoes	0.248 1.60	1.30	0.039	0.40
3. Jeans and rugby shirt: briefs, long-sleeved shirt, jeans, socks, athletic shoes	1.197 1.27	1.22	0.031	0.40
4. Shorts and shirt: briefs, short-sleeve shirt, shorts, athletic socks, athletic shoes	0.158 1.02	1.10	0.023	0.42
5. Women's shorts and tank top: panties, sleeveless tank top, shorts, sandals/thongs	0.144 0.93	1.08	0.022	0.40
6. Sweat suit: panties, long-sleeve sweat shirt and pants, ankle socks, athletic shoes	0.209 1.35	1.19	0.029	0.45
7. Insulated coverall and thermal underwear: long underwear top and bottoms, socks, insulated coverall, work shoe/boots	0.302 1.94	1.26	0.048	0.39
8. Cleanroom coverall: briefs, short-sleeved shirt, long trousers, belt, socks, shoes, woven coverall	0.240 1.55	1.26	0.039	0.38

Ensemble description				
9. Tyvek® coverall: briefs, short-sleeved shirt, long trousers, belt, socks, shoes, nonwoven coverall	0.237 / 1.53	1.26	0.045	0.33
10. PTFE water-resistant suit: briefs, short-sleeved shirt, long trousers, belt, socks, shoes, PTFE jacket and trousers	0.268 / 1.73	1.28	0.044	0.38
11. Chemical protective suit: briefs, short-sleeved shirt, long trousers, belt, socks, shoes, PVC & Vinyl acid splash jacket with hood and overalls	0.262 / 1.69	1.28	0.126	0.13
12. Neoprene-coated nylon work suit: briefs, short-sleeved shirt, long trousers, belt, socks, shoes, neoprene/nylon jacket with hood and overalls	0.264 / 1.70	1.28	0.120	0.14
13. Shirt and trousers: briefs, short-sleeve shirt, long trousers, socks, athletic shoes	0.173 / 1.12	1.30	0.028	0.38
14. Cold weather ensemble: briefs, short-sleeve shirt, long trousers, socks, athletic shoes, fiberfill jacket, mittens, hat	0.339 / 2.19	1.40	0.053	0.39
15. Flame-resistant ensemble: briefs, short-sleeve shirt, long trousers, socks, athletic shoes, long-sleeve shirt/jacket, long trousers, gloves	0.248 / 1.60	1.35	0.037	0.41
16. Chemical protective clothing: briefs, short-sleeve shirt, long trousers, socks, athletic shoes, long-sleeve shirt/jacket, long trousers, gloves, hood	0.285 / 1.84	1.50	0.098	0.18
17. Cold weather wind-proof, waterproof ensemble: briefs, short-sleeve shirt, long trousers, socks, athletic shoes, long underwear top and bottoms, fleece top and bottoms, waterproof jacket and long trousers, fiberfill mittens, hood	0.407 / 2.63	1.45	0.099	0.25

[a] Air layer thermal resistance was 0.112 m$^2 \cdot$ C/W (0.72 clo) on the nude manikin. Air layer evaporative resistance was 0.014 m$^2 \cdot$ kPa/W on the nude manikin.

Source: #1–12 McCullough et al., 1989; #13–17 KSU data from McCullough et al., 2002.

Table 9.3 Static and dynamic insulation values for cold weather clothing ensembles[a]

Ensemble description	Clothing area factor f_{cl}	Intrinsic clothing insulation I_{cl} (clo)		
		Static	Dynamic	% Change
1. Extreme cold weather expedition suit with hood (down-filled, one piece suit), thermal long underwear top and bottoms, mittens with fleece liners, thick socks, insulated waterproof boots	1.50	3.67	3.21	12
2. One piece ski suit, thermal long underwear top and bottoms, knit head/ear band, goggles, insulated ski gloves, thin knee length ski socks, insulated waterproof boots	1.28	1.60	1.13	30
3. One piece fiberfill ski suit with hood, thermal long underwear top and bottoms, goggles, insulated ski gloves, thin knee length ski socks, insulated waterproof boots	1.27	1.97	1.53	22
4. Extreme cold weather down-filled parka with hood, shell pants, fiberfill pants liner, thermal long underwear top and bottoms, sweat shirt, mitten shell with inner fleece gloves, thick socks, insulated waterproof boots	1.47	3.28	2.53	23
5. Knee length down-filled coat, thermal long underwear bottoms, jeans, T-shirt, long-sleeve flannel shirt, hat with fleece liner and ear flaps, insulated ski gloves, thick socks, low cut leather work boots	1.52	2.45	1.50	39
6. Fiberfill jacket, jeans, T-shirt, thermal long underwear bottoms, long-sleeve flannel shirt, baseball cap, thick socks, low cut leather work boots	1.40	1.68	1.30	23
7. Fleece long-sleeve shirt, fleece pants, briefs, athletic socks, athletic shoes	1.29	1.19	0.86	28

[a]Air layer thermal resistance around the nude manikin was 0.68 clo while standing and 0.49 clo while walking. These values can be used with data in the table to calculate total insulation values (I_t).
Source: Kim and McCullough (2000).

1982). Body motion increases convective heat loss and decreases the insulation value of clothing. This value has been referred to as resultant or dynamic insulation. ISO 15831 (ISO, 2004) gives a protocol for using a walking manikin to measure resultant insulation. Few laboratories have used a sweating manikin while it was walking (Richards and Mattle, 2001). Examples of how the insulation provided by an ensemble decreases as a function of body movement is shown in Table 9.3.

9.7   Manikin tests vs. fabric tests

Sweating guarded hot plates can be used to measure and compare the thermal resistance and evaporative resistance of different types of fabrics and multi-component systems (McCullough et al., 2004). The most commonly used methods for this purpose include ASTM F 1868, Standard Test Method for Thermal and Evaporative Resistance of Clothing Materials Using a Sweating Hot Plate (ASTM, 2005) and ISO 11092, Textile B Physiological Effects B Measurement of Thermal and Water Vapor Resistance Under Steady State Conditions (Sweating Guarded Hot plate Test) (ISO, 1995). Fabric insulation is often expressed as a performance index, such as insulation per unit weight or insulation per unit thickness when comparing different materials.

Manikin measurements account for many additional factors that affect the heat exchange between the body and the environment than measurements on flat pieces of fabric do. These include:

- the amount of body surface area covered by textiles and the amount of exposed skin
- the distribution of textile layers and air layers over the body surface (i.e., non-uniform)
- looseness or tightness of fit
- the increase in surface area for heat loss (i.e., clothing area factor) due to the textiles around the body
- the effect of product design
- the adjustment of garment features (i.e., fasteners open, hood up, etc.)
- variation in the temperature (and heat flux) on different parts of the body
- the effect of body position (i.e., standing, sitting, lying down)
- the effect of body movement (i.e., walking, cycling).

Therefore, manikin measurements are realistic, in that they quantify the effect of a clothing system on the heat exchange between the whole body and the environment. Although models are now available for estimating the heat transfer properties of clothing systems from the component fabrics and garment geometry (McCullough et al., 1989), manikins are easier to use. However, manikins, environmental chambers, and computer control and data acquisition systems are expensive to acquire and complex to maintain.

9.8   Using manikins under transient conditions

Manikins are designed for steady-state measurements, so their control systems do not work well during transients (i.e., changing environmental conditions). The changes in heat loss over time do not simulate the thermal responses of human beings; they depend upon the power capacity and control system (i.e., time constants) of the manikin. The thermal capacitance of a manikin is different than the human body, so even if the manikin's program allows the skin temperature to change during a transient, the net effect on energy balance for the manikin is not the same as it is for the human body. However, manikins have been used on a limited basis to quantify and compare the impact of thermal changes in clothing during step changes in relative humidity and temperature (Shim *et al.*, 2001; McCullough, 1991). For example, the temperature of a garment may change temporarily as the result of a change in environmental conditions (i.e., heat generation due to moisture absorption or the incorporation of phase change materials).

9.9   Conclusions

Thermal manikins will continue to be used to measure the thermal resistance (i.e., insulation value) and evaporative resistance of clothing systems. They are a valuable tool in the development of protective clothing systems. They can be used to evaluate protective clothing under steady-state and transient conditions and predict the comfort (or thermal stress) of the wearer.

9.10   References

American Society for Testing and Materials (2005), *2005 Annual book of ASTM standards, Vol. 11.03,* American Society for Testing and Materials, West Conshohocken, PA.

ANSI/ASHRAE 55-1992 (1992), *Thermal environmental conditions for human occupancy.* ASHRAE, Atlanta.

Belding H S (1949), 'Protection against Cold'. In Newburgh L H, *Physiology of heat regulation and the Science of Clothing.* Philadelphia, Pa. Saunders 351–367.

Crown E M, Ackerman M Y, Dale J D and Tan Y (1998), 'Design and evaluation of thermal protective flightsuits. Part II: Instrumented mannequin evaluation', *Clothing and Textiles Research Journal,* 16, 79–87.

Fan J and Qian X (2004), 'New functions and applications of Walter, the sweating fabric manikin', *European Journal of Applied Physiology,* 92,641–644.

Gagge A P, Burton A C and Bazett H D (1941), 'A practical system of units for the description of heat exchange of man with his environment', *Science,* 94, 428–430.

Holmér I (1999), 'Thermal manikins in research and standards', 3rd int conf *Thermal Manikin Testing,* Sweden, 1–7.

Holmér I, Nilsson H and Meinander H (1996), 'Evaluation of clothing heat transfer by dry and sweating manikin measurements in performance of protective clothing', in

Johnson J S and Mansdorf S Z, *Performance of Protective Clothing: Fifth Volume, ASTM STP 1237,* West Conshohocken, PA, ASTM, 360–366.

International Organization for Standardization (1993), *ISO 11092, Textiles – physiological effects – measurement of thermal and water vapour resistance under steady-state conditions (sweating guarded hotplate test),* International Organization for Standardization, Geneva.

International Organization for Standardization (1995), *ISO 9920, Ergonomics of the thermal environment – estimation of the thermal insulation and evaporative resistance of a clothing ensemble,* International Organization for Standardization, Geneva.

International Organization for Standardization (2004), *ISO 15831, Clothing – physiological effects – measurement of thermal insulation by means of a thermal manikin,* International Organization for Standardization, Geneva.

Kim C S and McCullough E A (2000), 'Static and dynamic insulation values for cold weather protective clothing', in Nelson C N and Henry N W, *Performance of Protective Clothing: Issues and Priorities for the 21st Century: Seventh Volume, ASTM STP 1386,* West Conshohocken, PA, ASTM, 233–247.

Kuklane K, Nilsson H, Holmér I and Liu X (1997), 'Methods for handwear, footwear and headgear evaluation', European seminar *Thermal Manikin Testing,* Sweden, 23–29.

Lee C, Kim I Y and Wood A (2002), 'Investigation and correlation of manikin and bench-scale fire testing of clothing systems', *Fire and Materials, 26,* 269–278.

McCullough E A (1991), 'Transient thermal response of different types of clothing due to humidity step changes', 2nd int symp *Clothing Comfort Studies,* Mt. Fuji, Japan, 1–11.

McCullough E A and Hong S (1994), 'A data base for determining the decrease in clothing insulation due to body motion', *ASHRAE Transactions,* 100 (1), 765–775.

McCullough E A, Jones B W and Huck J (1985), 'A comprehensive data base for estimating clothing insulation', *ASHRAE Transactions,* 91 (2), 29–47.

McCullough E, Jones B and Tamura T (1989), 'A database for determining the evaporative resistance of clothing', *ASHRAE Transactions,* 95 (2), 316–328.

McCullough E A, Barker R, Giblo J, Higenbottam C, Meinander H, Shim H, and Tamura T (2002), 'Interlaboratory evaluation of sweating manikins', 10th int conf *Environmental Ergonomics,* Japan, 467–470.

McCullough E A, Huang J, and Kim, C S (2004), 'An explanation and comparison of sweating hot plate standards', *Journal of ASTM International,* 1, (7), (online journal at ASTM.org).

McCullough E A, Huang J, and Deaton, S (2005), 'Methods of measuring the clothing area factor', in Holmér I, Kuklane K and Gao C, *Environmental Ergonomics XI: Proceedings of the 11th International Conference,* Ystad, Sweden, 433–436.

Nilsson H (1997), 'Analysis of two methods of calculating the total insulation', European conf *Thermal Manikin Testing,* Sweden, 17–22.

Nilsson H, Gavhed D and Holmér I (1992), 'Effect of step rate on clothing insulation measurement with a moveable thermal manikin', 5th int conf *Environmental Ergonomics,* The Netherlands.

Olesen B, Sliwinska E, Madsen T and Fanger P (1982), 'Effect of body posture and activity on the thermal insulation of clothing: measurements by a moveable thermal manikin', *ASHRAE Transactions,* 88 (2), 791–805.

Prezant D J, Barker R L, Bender M and Kelly K J (2000), 'Predicting the impact of a design change from modern to modified modern firefighting uniforms on burn injuries using manikin fire tests', in Nelson C N and Henry N W, *Performance of Protective Clothing: Issues and Priorities for the 21st Century: Seventh Volume,* ASTM STP 1386, West Conshohocken, PA, ASTM, 224–233.

Richards M and Mattle N (2001), 'Development of a sweating agile thermal manikin (SAM)', 4th int conf *Thermal Manikins,* Switzerland.

Rossi R M and Bolli W P (2000), 'Assessment of radiant heat protection of firefighters' jackets with a manikin', in Nelson C N and Henry N W, *Performance of Protective Clothing: Issues and Priorities for the 21st Century: Seventh Volume, ASTM STP 1386,* West Conshohocken, PA, ASTM, 212–223.

Shim H, McCullough E and Jones B (2001), 'Using phase change materials in clothing', *Textile Research Journal,* 7 (6), 495–502.

Teal W (1990), 'A thermal manikin test method for evaluating the performance of liquid circulating cooling garments', in Shapiro Y, Moran D S and Epstein Y, *Environmental Ergonomics: Recent Progress and New Frontiers,* London, Freund Publishing House, Ltd., 355–358.

Woodcock A H (1962), 'Moisture transfer in textile systems, Part I', *Textile Research Journal,* August.

Wyon D P (1989), 'Use of thermal manikins in environmental ergonomics', *Scandinavian Journal of Work, Environment and Health,* 15 (1), 84–94.

Interactions between protection and thermal comfort

R R O S S I , EMPA, Switzerland

10.1   Introduction

Over a period of several decades, the development of protective clothing aimed at improving the barrier effect of the garment. The clothing was supposed ideally to shield the body from external influences. In the last years, ergonomics and physiological considerations have become more and more important, as it was realized that the acceptance of uncomfortable clothing by the users was sometimes very low and the protective clothing was therefore not worn in many cases. Working with uncomfortable and bulky clothing increases the heat stress of the wearer and, apart from heat-related diseases like cardiovascular problems, this may cause a reduction in cognitive and physical performance (Hancock and Vasmatzidis 2003). The demands of protection and comfort are therefore often contradictory and the goal of protective clothing is to offer the highest level of protection and the best possible comfort.

This chapter deals with these contradictory requirements of thermal comfort and protection of protective clothing. In the first section, the notion of comfort is explained and the complex problem of finding an optimum between thermal comfort and protection is described. The second section presents different, mainly standardized test methods to assess the different parameters influencing the thermal and moisture transfer through fabric layers. In the third section, methods are presented to measure the interactions between thermal and moisture transport and to evaluate the thermophysiological impact of the clothing on the human body. In the fourth section, the influence of moisture on the heat and mass transfer is discussed using two examples of firefighters' and impermeable protective clothing together with some possibilities to improve the thermal comfort are shown.

10.2   Definition of comfort

According to the dictionary, the definition of comfort is a condition or feeling of pleasurable ease, well-being, and contentment. Comfort is a very complex mix of different very subjective sensations. In simpler terms, comfort is sometimes defined as the absence of discomfort, i.e., the absence of any pain or disagreeable feeling. In the context of textiles and clothing, wear comfort is probably a state when we are unaware of the clothing we are wearing.

10.2.1   Four comfort types

Four different types of comfort may be defined: thermal or thermophysiological comfort, sensorial comfort, garment fit and psychological comfort. Thermophysiological comfort was defined as 'the condition of mind which expressed satisfaction with the thermal environment' (ISO 7730 1984), which is the case when we are neither feeling too cold nor too warm, and when the humidity (sweat) produced by the body can be evacuated to the environment. The factors affecting thermophysiological comfort are the loss (or gain) of heat by radiation, conduction and convection, the loss of heat by evaporation of sweat, the physical work being done by the person, and the environment (ambient temperature, air humidity and air movement).

Sensorial comfort is the sensation of how the fabric feels when it is worn near to the skin. This feeling addresses properties of the fabric like prickling, itching, stiffness or smoothness. It can also be related to the thermophysiological comfort, as a fabric wetted through with sweat will change its properties and may, for instance, cling to the skin. The sensorial comfort is very difficult to predict as it involves a large number of different factors. Different studies have been performed mostly with human subjects (Schneider *et al.* 1996, Garnsworthy *et al.* 1968, 1988, Li *et al.* 1988, 1991, Ajayi 1992, Elder *et al.* 1984, Behmann 1990, Sweeney and Branson 1990a,b, Demartino *et al.* 1984, Matsudaira *et al.* 1990a,b, Naylor and Phillips 1997, Wang *et al.* 2003) to try to understand the relationship between fabric properties (protruding fibres, fibre and yarn diameters, fabric thickness, stiffness, etc.) and sensorial feelings on the skin. There are only few objective methods to assess the sensorial properties of a textile. The most widely recognized and used around the world is probably the KES-F system developed by Kawabata and his co-workers to measure the fabric hand. The system consists of four different apparatus (tensile and shear, bending, compression and surface friction/roughness) (Kawabata 1980).

The garment fit considers the tightness of the garment, and its weight. Again, this type of comfort will affect the thermophysiological comfort, as a loose-fitting garment may be perceived as cool during summertime. On the other hand, in the context of protection, loose garments may represent a hazard of being caught by a surrounding object. The fourth type of comfort is psychological comfort, dealing with aesthetics (colour, garment construction, fashion, etc.) and the suitability of the clothing for the occasion. This last type of comfort may be very important in protective clothing if the wearer feels stressed only because he/she does not feel adequately protected.

In the field of protective clothing, the most important comfort parameters are probably thermophysiological comfort and garment fit. The sensorial comfort and the psychological comfort play a secondary role, as long as the fabrics do not cause excessive friction problems for the skin, or allergies. While the garment ease may be checked directly by the wearer, thermophysiological comfort is much more complex to analyse right away when purchasing the clothing. It is dependent on the interactions between the human body, the clothing and the environment.

10.2.2   Thermoregulation of the human body

Human beings are homoeothermic, which means that they have to maintain their core temperature within close limits around 37 °C. During every activity, the body produces a certain amount of heat lying between 80 W while sleeping and over 1,000 W during most strenuous efforts. The surplus energy can be transferred to the environment by three means: respiration and release of dry (radiation, convection and conduction) and evaporative heat through the skin. The total heat loss at moderate temperatures (around 20 °C) and 50% RH is approximately divided into 20% evaporation, 25% conduction, 45% radiation and 10% respiration (Aschoff *et al.* 1971). At low temperatures, respiration can account for over 30% of the heat loss. When the ambient temperature is over about 34 to 37 °C, evaporation is the only way to cool the body. Evaporative cooling is a very efficient means of heat dissipation, as one litre evaporated sweat removes 672 Wh from the body at a temperature of 35 °C. The body evaporates from at least 0.05 J/cm^{2}min (about 22 g/h *perspiratio insensibilis*), up to 4 l/h during short periods of time. During longer work periods, the amount of sweat produced is reduced (1 l/h for 6 hours work, 0.5 l/h for 12 hours) (Wenzel and Piekarski 1982). When heat stress sets in, the amount of sweat increases more in the trunk than in the other surface regions. The higher the core temperature, the lower the skin temperature when sweating starts.

The metabolic balance of the human body can be described as follows:

$$M - P_{ex} = H_{res} + H_c + H_e \pm \frac{\Delta D}{\Delta t} \qquad 10.1$$

with M = metabolic heat (in W), P_{ex} = muscular power, H_{res} = heat emission per time unit by respiration, H_c = heat emission per time unit by dry heat, H_e = heat emission per time unit by evaporation of sweat, $\Delta S/\Delta t$ = change in the body heat content. Ideally, ΔS should be zero which means that the body should be in equilibrium with the environment.

10.2.3   Factors influencing thermal comfort

In terms of thermal comfort, the function of clothing is to support the thermoregulation of the human body. If the environmental temperature is low, the clothing has to prevent too large a heat release and on the other hand, if the temperature is high, the heat and moisture transport through the garment should be as high as possible to avoid heat stress problems for the body. Different mechanisms affect the thermal and moisture transport through fabric layers:

- Dry (conductive, convective and radiant) heat transfer between the body, the environment and the atmosphere, which is dependent on many factors like the temperatures of the body and the environment, the thermal resistance of the fabric layers, the wind speed and possible internal convection between the layers (ventilation effects), etc.
- Thermal energy stored within the clothing: this quantity may be neglected in many cases but some types of protective clothing like heat protective clothing can store a large amount of heat. The temperature regulating function of phase change materials (PCM) also works on the basis of heat storage and release through changes of the aggregate state of the PCM.
- Diffusion of water vapour molecules through the pores of the textile: this process is often called 'breathability' and can be indicated as water vapour permeability through the layers or a water vapour resistance of the layers.
- Adsorption and migration of water vapour molecules and liquid water along the fibre surfaces, as well as transport of liquid through the capillaries between the fibres and the yarns.
- Absorption and desorption of water vapour, and transport of liquid water, in the interior of the fibres. This process is dependent on the hygroscopicity of the fibres.
- Evaporation of liquid with thermal energy consumption or condensation with thermal energy release.

10.2.4   The protection/comfort contradiction

Protective clothing protects the body from external influence like heat, chemicals, mechanical hazards, foul weather, etc. To achieve this goal, the clothing has to shield the human body from the environment. From a physiological point of view, the human body feels comfortable at about 29 °C in an unclothed state (Wenzel and Piekarski 1982) and at about 26 °C with a clothing insulation of 0.6 clo (1 clo = 0.155 m²K/W) (Olesen and Fanger 1973). However, protective clothing usually has a higher insulation. The bulkiness and the weight of the clothing lead to higher metabolic heat production. The British Standard (BS 7963 2000) provides estimations for the increase in metabolic rate due to the wearing of different types of protective clothing. This increase can be as high as 155 W/m² when wearing highly insulating firefighters' personal protective equipment (including helmet, clothing, gloves and boots). Furthermore, many items of protective clothing have to be watertight, and the ensuing moisture barrier will reduce the transfer of sweat to the environment. Therefore the protective function of protective clothing is only achieved with a certain discomfort for the wearer, and the best balance between protection and comfort has to be found for every type of protective clothing, depending on the foreseen metabolic heat production and climatic conditions.

This contradiction is probably the most obvious for heat protective clothing, as it should prevent the external heat from flowing towards the body, but on the other hand allow the metabolic heat to escape to the atmosphere. Working at higher temperatures quickly leads to heat stress and stress-related heart attacks are frequently reported: in the United States, about 50% of the lethal accidents of firefighters on duty are due to heat stress (Washburn *et al.* 1999, LeBlanc and Fahy 2004). But even at moderate temperatures, the burden of increased weight and reduced permeability of this type of clothing can lead to heat stress. The results can, however, greatly vary depending on the set parameters: measurements at low temperatures and/or low relative humidity (22 °C and 20% to 45% RH (Reischl and Stransky 1980); 20 °C and 50% RH (Bartels and Umbach 1997); 45/65 °C and 15% RH (Sköldström and Holmer 1983), 22 °C and 56% RH (Ftaiti *et al.* 2001)) resulted in great differences in the skin and core temperature increases for different jacket types (PVC, Neoprene or leather vs. breathable materials). As soon as the temperature or the humidity of the environment approaches the conditions near the skin, the differences become smaller. (Schopper-Jochum *et al.* 1997) stated that the increase of body core temperature in an environment of 30 °C and 50% RH was independent of the jacket type. Griefahn *et al.* (1996) and Rossi (2003) did not find any significant differences between different types of protective clothing (breathable vs. non breathable) during exercises at higher temperatures either.

The weight of the equipment represents an additional load for the firefighter. An equipment set of 24 kg, for instance, reduces the performance of the wearer by 25% (Louhevaara *et al.* 1995). The size of the clothing and the number of textile layers (Teitlebaum and Goldman 1972, Lotens 1983) also increase the energy consumption of the wearer and thus the required heat loss to maintain thermal comfort. With chemical protective clothing and other types of impermeable clothing, the comfort problems are mainly caused by the lack of water vapour permeability, as the protection against chemicals often imply that the materials used are totally liquid- and sometimes also vapour-tight. Different studies report heat stress as the primary limitation of use duration of chemical protective clothing (Santee and Wenger 1988, Veghte 1988, Ilmarinen *et al.* 2000, Töpfer and Stoll 2001). Problems of body dehydration can also occur, as the sweat production cannot be easily compensated due to the wearing of a gas mask or self-contained breathing apparatus (Melin *et al.* 1999). Therefore, new developments of such clothing often aim at increasing the thermophysiological comfort (Amos and Hansen 1997, Reneau *et al.* 1999), primarily by improving the water vapour permeability (Wilkinson *et al.* 1997). For totally impermeable protective clothing, technical aids like ice-, liquid-, or air-cooling can help to reduce the thermal strain of wearing such equipment. However, the additional weight causes additional metabolic heat production and lowers the benefits of such systems (Glitz and von Restorff 1999).

Foul weather and cold protective clothing often contains a waterproof, moisture permeable barrier. This barrier ensures the water tightness, but at the same time still hinders the free flow of part of the water vapour produced by the body to the environment, as even the most breathable membrane or coating adds a resistance to the vapour flow. If the outside temperature is low, there is a certain risk of water condensation within the clothing layers when the water vapour pressure increases beyond saturation. The presence of condensation can then change the thermal and moisture transport properties of the clothing (Rossi *et al.* 2004b, Fukazawa *et al.* 2003, Finn *et al.* 2000).

Ballistic protective clothing also has to be a compromise between comfort and protection. Because of weight and bulkiness problems, the users sometimes refuse to wear these garments continuously. Furthermore, ballistic protective clothing using aramid-based panels has a significantly lower ballistic resistance when water enters the structure. For this reason, the panels have to be packed into impermeable outer shells, which have unfavourable consequences on the clothing physiology of the garment (Reifler *et al.* 2003).

10.3   Test methods for heat and moisture transfer

Many of the test methods to measure thermophysiological comfort try to mimic the heat and mass transfer from the human skin to the environment through the textile layers. The Hohenstein institute in Germany developed a 5-level system (Fig. 10.1) for the physiological evaluation of clothing (Umbach 1983). Levels 1 and 2 assess physical properties of the fabric layers and the clothing whereas levels 3 to 5 also consider the physiological impact of the clothing on the human body. The intrinsic thermophysiological properties of the fabrics like the thermal resistance or the water vapour resistance are determined in level 1. As the wear comfort of clothing is also very much dependent on the air layers trapped between the fabric and the body, the garments must be assessed with human-shaped, life-sized manikins (level 2). The validation of these results is made with human subject tests, under well-controlled conditions in the laboratory (level 3). Levels 4 and 5 are usually restricted to a very small number of newly developed, optimized clothing systems.

One of the most widely used methods is the sweating guarded hot plate to measure the water vapour permeability or the thermal insulation of material samples, as it is probably the best standardized test method to simulate the heat and mass transfer conditions on a clothed body. There are, however, several other methods to determine these characteristics. It is difficult for consumers to interpret these figures, especially concerning the breathability of clothing, as the different methods do not always provide comparable results. The reason for this

Clothing physiology test methods

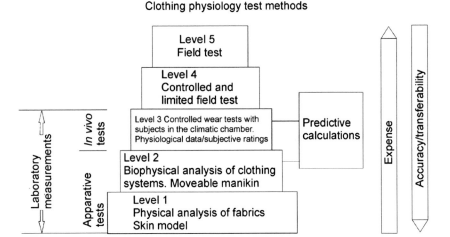

10.1 Five-level system of physiological evaluation of clothing (Umbach 1983).

is that these methods do not use the same test conditions. Additionally, the water vapour permeability of hydrophilic materials changes under different humidity conditions (Gretton *et al.* 1998, Farnworth *et al.* 1990, Osczevski and Dohlan 1989).

10.3.1   Sweating guarded hot plate (skin model)

The thermal insulation and the water vapour resistance of one or several textile layers can be assessed with the skin model (sweating guarded hot plate (ISO 11092 1993), and (EN 31092 1993)). The skin model consists of an electrically heated plate, which is located in a climatic chamber. Square samples are put onto the plate, and air at a defined temperature, relative humidity and velocity (1 m/s) is blown tangentially from a fan over the sample. The plate is heated to 35 °C and the measuring surface is surrounded by a guard that is heated to the same temperature in order to avoid any heat loss (Fig. 10.2).

The thermal resistance R_{ct} (m^{2}K/W) is assessed from the supplied steady-state heating power (Q), the temperature difference between the air in the wind channel ($T_a = 20$ °C) and the skin model (T_s) and the size of the measuring surface (A):

$$R_{ct} = A \cdot \frac{T_s - T_a}{Q}$$

10.2

The measurement of the water vapour resistance is made under isothermal conditions to avoid condensation effects in the samples that would influence the resistance. The plate is covered by a cellophane foil permeable only to water vapour to prevent a contact of the sample with the water on the plate. The heating

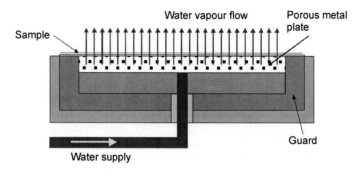

10.2 Scheme of the sweating guarded hot plate (skin model) according to ISO 11092.

power to compensate the evaporative cooling and thus maintain the plate at 35 °C is proportional to the water vapour permeability of the material. The water vapour resistance R_{et} (m²Pa/W) is determined by the supplied steady-state heating power (Q), the water vapour partial pressure difference between the air in the wind channel (p_a) and the skin model (p_s) and the size of the measuring surface (A):

$$R_{et} = A \cdot \frac{p_s - p_a}{Q} \qquad 10.3$$

The breathability of a fabric is often described with the water vapour transmission or permeability F:

$$F = \frac{1}{R_{et} \cdot \varphi} \qquad 10.4$$

with φ being the specific heat of evaporation of water at the test temperature ($\varphi = 0.672$ Wh/g at 35 °C). In eqn 10.4, F is given in $gm^{-2}h^{-1}mbar^{-1}$ but is usually expressed in $gm^{-2}h^{-1}$ by multiplying this figure by the water vapour pressure gradient.

10.3.2 Measurement of the thermal resistance and thermal transmission

The measurement of the thermal resistance with the sweating guarded hot plate according to ISO 11092 assesses the intrinsic resistance of the specimen plus a transition resistance from fabric to air. This transition resistance is dependent on the convective and the radiant heat loss from the surface of the fabric to the atmosphere. If a fabric is worn as an under-garment, there will usually be no convection and only limited radiation between the layers. Spencer-Smith (1977a) showed that internal convection between layers can be neglected if the air layer is smaller than 8 mm. Furthermore, as long as the fibre content in a fabric is higher than 9%, only thermal conduction needs to be considered in the fabric (Woo *et al.* 1994). It is therefore important only to assess the thermal conductivity of fabrics. ISO 5085-1 (1989) and ISO 5085-2 (1989) provide a method to determine the thermal resistance of fabrics. The specimen is placed onto a heating plate and covered by a cold plate with a defined pressure. The principle is that the temperature drop across the heating plate of known thermal resistance is assessed as well as the temperature on the surface of the cold plate. The thermal resistance of the specimen can then be calculated from these figures. The SI unit of the thermal resistance is m^2K/W, but a widely used unit is the 'tog' (1 tog $= 0.1$ m^2K/W). Another unit used to quantify the overall thermal insulation of garments is the 'clo' defined by Gagge *et al.* (1941). One 'clo' is equivalent to the insulation required to keep a seated subject comfortable at an air temperature of 21 °C with an air movement of 0.1 m/s, which corresponds to the insulation provided by an ordinary dress suit.

The new standard (ASTM D 7024 2004) describes a method to determine the overall (dynamic) thermal transmission coefficient of textile fabrics and measure the amount of latent energy in textiles. This method was developed to assess the thermal efficiency of phase change materials (PCM), which provide a temperature-regulating function by absorbing or releasing energy through aggregate state changes. It allows the measurement of the steady-state thermal resistance of a fabric as well as the determination of a 'temperature-regulating factor (TRF)'. The TRF is used to compare fabrics that store or release energy.

10.3.3   Cup methods

There are different methods to determine the water vapour transmission through fabrics. In the standard ASTM E 96 (2000), the desiccant method and the water method are described. Both methods use a test dish. The test specimen is sealed to the open mouth of the dish. The dish contains either a desiccant or distilled water, and is placed in a controlled atmosphere. By weighing the dish periodically, the rate of water vapour transfer can be determined. One problem of these two methods is that there is a still air layer in the dish between the fabric and the water/desiccant. This air layer has a high water vapour resistance that is often higher than the fabric itself. For waterproof materials, the dish may be inverted (ASTM E 96 Procedure BW) to avoid this air layer. The rest of the test protocol is identical to the upright dish test.

A similar method is described in ISO 15496 (2004) but instead of having an air layer between the specimen and the water/desiccant, the sample is placed between two semi-permeable membranes. The specimen is placed together with a waterproof but water vapour permeable, hydrophobic membrane on a ring holder, and then put in a water bath so that the membrane is in contact with the water. A cup containing a desiccant is covered with a second piece of the same membrane and then inverted above the specimen in the ring holder. The water vapour permeability of the specimen is determined by weighing the cup before and after 15 minutes in contact with the specimen.

Further dish methods are described in the standards BS 7209 (1990) and DIN 53122 (1974). Gibson (1993) compared the cup method ASTM E 96 with the sweating guarded hot plate ISO 11092 and found a correlation between both methods except for hydrophilic materials, which obtained much better water vapour transmission properties when measured according to ISO 11092. McCullough *et al.* (2003) also compared the different methods to measure water vapour permeability and found a high correlation between the sweating guarded hot plate ISO 11092 and the Japanese standard JIS L 1099 (1999), which was the basis for ISO 15496.

10.3.4   Dynamic moisture permeation cell

ASTM F2298 (2003) describes a method to determine the steady-state and transient water vapour permeation behaviour of fabrics. This method was developed by Gibson *et al.* (1997). A mixture of dry and water-saturated nitrogen streams are passed over the top and the bottom surfaces of the cell containing the sample. The relative humidity of the mixed streams is controlled by the proportion of dry and saturated components. Via measurement of temperature and relative humidity of the flows entering and leaving the cell, the flux of water vapour diffusing through the sample may be calculated. Gibson *et al.* (1997) compared this method with the sweating guarded hot plate ISO 11092 and found a very good correlation for different fabrics and microporous membrane laminates. There was also a correlation of this method with the inverted cup method (ASTM E 96 procedure BW).

10.3.5   Wicking test methods

During strenuous activities, the human body produces sweat in liquid form. In these cases, the wearing comfort of clothing worn next to the skin depends on its ability to absorb the liquid. The cooling of the body is dependent on the wicking effect of the fabric. The liquid can either be distributed in the fabric (lateral wicking effect) or be transported to the next clothing layer.

The wickability of a fabric can be determined with a vertical strip method as described in BS 3424 (1996) or DIN 53924 (1997). A test specimen is hung vertically and the lower end immersed in water. After a certain time (10, 30, 60 and 300 s for DIN 53924 and 24 hours for BS 3424), the height of liquid wicking above the water surface is measured. One of the drawbacks of these methods is that the vertical wicking is influenced by gravity. Furthermore, they are only qualitative tests that do not measure the amount of water absorbed.

Another very simple method to measure the hydrophilicity and the wickability of a fabric is the drop test, which consists of placing a drop onto the surface of the fabric and observing visually how the water is absorbed and distributed in the fabric. Van Langenhove and Kiekens (2001) proposed a more sophisticated version of this test by constantly supplying water from a water tank to the fabric, and measuring the electrical conductivity at a defined distance from the point of water supply. They showed that the horizontal displacement s of the liquid can be described as follows:

$$s = k_s \cdot t \qquad\qquad 10.5$$

with t the time and k_s the capillary transport constant. D'Silva *et al.* (2000) used a similar method to determine the absorption and the wickability of fabrics. They used a porous plate fed with water and put a test sample with a larger area onto it. A defined pressure was applied to the sample. The area of the plate was defined as absorption zone and the area around it as wicking zone. By measuring the weight loss of the water reservoir and the time during which the water was allowed to wick out of the absorption zone, the absorption and wickability of the fabric could be determined. A further method was described in Harnett and Mehta (1984) and Ghali *et al.* (1994), and consists of a strip of test fabric used as a siphon by immersing one end in a reservoir of water and allowing the liquid to drain from the other end, placed at a lower level, into a collecting beaker. The amount of liquid transferred is determined by weighing the beaker.

10.3.6   Measurement of water transport between layers

The transport of water between the different layers of a clothing combination was studied several times using gravimetric methods. Spencer-Smith (1977b) concluded that no liquid water transfer occurred between a wet and a dry fabric in contact until the regain of the wet fabric exceeded a critical value of about 80%. However, his testing method interrupted liquid transfer during measurements which made it difficult to obtain absolute values of the transfer wicking rate from fabric layer to layer. Adler and Walsh (1984) concluded that wicking did not begin until the moisture content was high, more than 30% above the regain for the woven fabrics. The knitted fabrics tested did not wick at all. Crow and Osczevski (1998) also reported that there had to be a threshold amount of water in the wet fabric layer before it wicked water to the second, initially dry layer. However, this amount varied considerably from fabric to fabric, depending on the pore sizes and their corresponding volume. Zhuang *et al.* (2002) studied the wicking transfer of knitted fabrics and reported that there was an optimum pressure for the maximum amount of liquid water transferred. With increasing external pressure, the number of contact points between the two layers increased, and therefore more liquid water was transferred. However, the void space in the dry layer of the fabric to accommodate the transferred water decreased when the fabrics were under a higher external pressure. Thus, the transfer wicking rate increased with increasing external pressure up to a point then decreased under a higher external pressure.

Gravimetric methods, which are based on weighing the different layers of a system, have the disadvantage that part of the moisture will evaporate during the weighing process. They also restrict the assessment of the dynamics of the wicking process, especially the space- and time-dependence of the distribution. Several methods using magnetic resonance imaging (Leisen and Beckham 2001), neutron radiography (Weder *et al.* 2004a) or micro-computer tomography (Weder *et al.* 2004b) have been proposed recently to measure liquid transfer between fabric layers in real time without altering the samples during the measurements.

10.4   Measurement of thermal comfort with practice-related tests – interactions between heat and mass transfer

Standardized test methods usually assess heat *or* moisture transport, but not both. In many cases, however, these two quantities interact with each other. As soon as the human body starts sweating in liquid form and evaporation occurs, or if moisture condenses within the fabric layers, the heat and mass transfers through the clothing have to be analysed simultaneously.

10.4.1   Cylindrical models

One of the easiest means to model the human body is the use of test apparatus with cylindrical shape. Lotens and Havenith (1991) proposed a model of the human body with 13 cylinders. The measurement of the heat and moisture loss from a cylinder will not give the same results as with a flat plate. The one-dimensional steady-state heat transfer Q through a cylindrical surface A_x is defined with Fourier's law and is dependent on the radius of the cylinder r and the thickness of the fabric x ($A_x = 2\Pi(r + x)L$):

$$Q = \frac{A \cdot \lambda(T_1 - T_2)}{r \cdot \ln\left(\dfrac{r_2}{r_1}\right)} \qquad\qquad 10.6$$

with A, r, L the area, radius and length of the cylindrical body, T_1, r_1 the temperature and radius on the inner surface of the fabric; and T_2, r_2 the temperature and radius on the outer surface of the fabric.

Different studies to determine the heat loss of cylindrical models were performed (Fonseca and Breckenridge Jr 1965a,b, Takeuchi *et al.* 1982, Kamata *et al.* 1988a,b, Lamb and Yoneda 1990, Anttonen and Oikarinen 1997). The heat loss from cylindrical bodies is usually higher than that from flat plates (Spencer-Smith 1977a). Meinander (1985) came to the same conclusion with experimental measurements on a sweating torso and supposed that this was due to a higher convective heat transfer due to the chimney effect on the vertical cylinder. The wind flow is a key factor in heat loss, as wind may penetrate the fabric on the windward side and move circumferentially towards the back of the cylinder. Kind and Broughton (2000) showed that the heat loss of multilayer, wind permeable clothing systems may be greatly reduced by introducing a layer with low resistance to airflow between the outer layer and the underlying layer. This layer acts as a bypass allowing air that has penetrated the outer layer to flow around the cylinder without penetrating the underlying layer.

Lamb (1992) analyzed water vapour transfer through fabrics with a cylinder model and concluded that a model of heat transfer can also be applied to the transfer of water vapour. Very few cylindrical apparati measure the heat and moisture transfer simultaneously. Meinander (1985) and Zimmerli and Weder (1997) made studies on a sweating cylinder and showed some influences of moisture in the clothing system on the heat transfer, especially the decrease of thermal resistance with increasing amount of moisture in the fabrics. Rossi (2000) used the latter system to characterize the simultaneous heat and moisture transport of multilayer combinations, especially the cooling effect of moisture evaporation and the influence of the hydrophilicity of fabrics on this cooling.

Cylinders constitute a good approximate geometry for a section of the human body. In particular, the convection effects around a clothed cylinder can be studied in a near-to-practice manner with such an apparatus. Another advantage of these models is that the fabrics can be stretched in a similar way as in practice (for instance underwear fabrics). This stretching process is also possible on flat plates, but the contact between the plate and the sample is never as close as on a cylindrically shaped apparatus. Sweating cylinder systems with sweating nozzles placed on defined spots around the surface of the cylinder allow the measurement of the influence of absorption and wicking of liquid sweat, as well as the evaporation of moisture, on the total heat loss in the presence of the clothing systems.

Cylinder models are more or less two-dimensional systems and represent a good step between the one-dimensional flat plate systems and manikins or human subject testing. The repeatability of the results is somewhat lower than that of flat plates but still remains quite high and cylinder systems may therefore be used for standardized assessments of the heat and moisture transfer of multilayer systems. On the other hand, these models are of limited use to study the effects of ventilation within the clothing. In sweating cylinders used in non-isothermal conditions, the different effects of dry heat loss, moisture driven heat loss, evaporative cooling and moisture transfer are superimposed, and it is therefore difficult to distinguish between the different contributions to total heat loss.

10.4.2   Thermal and sweating manikins

The topic of manikins will not be covered in detail in this chapter, as it is treated elsewhere in this book (see Chapter 9). However, as manikins are very important tools for the assessment of thermophysiological properties of the clothing, a short summary will be given here. Thermal manikins for the assessment of thermal insulation of garments have been used since the early 1940s and probably more than 100 are in use worldwide (Endrusick *et al.* 2001, Holmer 2003). The measurement of the thermal insulation of clothing is described in ISO 15831 (2003) as well as in ASTM F1291 (1999). The main advantage of manikin testing is probably the realistic simulation of the heat and moisture transfer from the body through the clothing to the environment. This method provides objective results and repeatability is usually fairly high. Manikins can assess the effects of air layers between the skin and the clothing and, if the limbs are movable, the pumping effects through the fabrics as well as ventilation effects though the garment openings. They are a valuable tool to assess the effects of the design of the clothing on heat and moisture transfer.

Some of these thermal manikins can additionally release moisture to measure the evaporative resistance of garments (Meinander 1992, Fan *et al.* 2001, Richards and Mattle 2001, Burke and McGuffin 2001) and ASTM is drafting a standard for sweating thermal manikins (Standard Test Method for Measuring the Evaporative Resistance of Clothing Using a Sweating Manikin, ASTM Committee F-23 on protective clothing). However, an interlaboratory study (McCullough *et al.* 2002, Richards and McCullough 2004) on different sweating manikins showed that the variability among the systems was relatively high, mostly because of the varied designs of the manikins.

10.4.3   Human trials

Apart from laboratory test methods, human trials are commonly used methods to evaluate the physiological impact of clothing on the human body. The repeatability and reproducibility of such measurements depend on the test conditions set. Measurements in a climatic chamber on treadmills, following a well-defined protocol, are usually used to validate the results obtained with small-scale test methods or manikins. The selection of the conditions is very important to be able to distinguish between different clothing systems since they can strongly influence the results. If, for instance, the physiological load of heat protective clothing is assessed at low or moderate temperatures and relative humidity (e.g. 20 °C and 65% RH), this will favour clothing types with good moisture permeability. On the other hand, if the environmental conditions are such that the heat and moisture transfer between the body and the environment is limited, a good moisture buffering effect of the clothing will become more important. In the case of higher temperature and water vapour partial pressure in the surroundings relative to the body, the insulating properties of the clothing will be determinant for the physiological comfort of the wearer. This example shows that an assessment of the conditions expected during actual use has to be made prior to the definition of the test parameters.

Human trials can also be performed in the field, either under controlled conditions, with a limited number of subjects, or as user trials on a large scale. These trials are time consuming and very expensive, and reproducibility decreases as the climatic conditions and the metabolic heat production of the subjects cannot be defined precisely. Several objective but also subjective parameters are usually assessed during human trials. The body core temperature can be monitored using different methods presented in ISO 9886 (1992). Three of the most widely used methods are the measurement of rectal temperature, intra-abdominal temperature (with a telemetry pill (O'Brien et al. 1998)) and tympanic temperature. The choice of the method will depend on the acceptability of the subjects, health and ethical considerations as well as the practicability of the method for the test. Tympanic temperatures, for instance, can be used only if the position of the sensor is maintained during the measurement, the air temperature is neither too high (<58 °C) nor too low (>18 °C) and the air velocity less than 1 m/s. Skin temperature can be measured by placing temperature sensors on the skin. Mean skin temperature is usually assessed as the weighted average measurements at specific points on the skin. The weighting system is often related to body surface area. Care has to be taken to ensure that the sensor remains in good contact with the skin during the whole test. However, the use of tape or foam patches may result in the measurement of higher skin temperatures (Buono and Ulrich 1998).

One very important parameter for the assessment of the thermal strain of the clothing is the body mass loss, which is mainly due to sweat loss. In order to measure the water vapour permeability and the moisture storage of the clothing system, all the pieces of the garment are also weighed before and after the test. Different other parameters like the heart rate, the metabolic rate or even specific hormone contents are sometimes also used to determine the thermal strain of the clothing for the wearer.

Subjective measurements use scales to determine how the subjects feel before, during, and after the trial (ISO 10551 1995). The subject has to judge his/her own thermal comfort feeling (on a scale between 'very hot' to 'very cold'). The questionnaire has to be prepared very carefully to avoid biases due to unclear or suggestive questions, or to training effects of the subjects (the subject may remember what he/she answered during a former test and adapt his/her answer in consequence). Some standards like prEN 469 (2004) try to incorporate human trials to assess the physiological load or the ergonomics of protective clothing. A possible method is presented in Havenith and Heus (2004). Several problems are, however, encountered in attempting to develop such a test method. Apart from the problems of choosing the right environmental conditions to avoid discrimination against certain products, the influence of the test subjects (different fitness, training effects, etc.) shall not be determinant for the result and, on the other hand, the tests shall not be too complex to maintain the costs for the certification of the clothing within reasonable limits.

10.5   Moisture storage and influences on protection

10.5.1   Transport of heat in a moist state

The thermal conductivity determines the transport of heat within a textile. It has been already shown with test subjects (Tokura and Midorikawatsurutani 1985) that increased thermal conductivity in the wet state can lead to a reduction in the quantity of sweat produced. According to Spencer-Smith (1977b), there is a linear dependency of the thermal conductivity with the moisture regain. This statement, however, is only partly valid. If the thermal conductivity through the air and water vapour ($\lambda = 0.019$ W/mK at 20 °C) is ignored, the complete thermal conductivity is dependent only on the conduction of heat through the fibres and the stored water:

$$\lambda_T = \frac{\lambda_{T,0} \cdot V_T + \lambda_W \cdot V_W}{V_T + V_W + V_L} \qquad 10.7$$

with λ_T = heat conductivity of the wet textile (in W/mK), $\lambda_{T,0}$ = heat conductivity of the dry textile (in W/mK), λ_W = heat conductivity of the water (0.598 W/mK at 20 °C), V_T = part by volume of the textile, V_W = part by volume of the water, V_L = part by volume of the air.

The volume can be described on the basis of the mass and density of the material ($M = \rho \cdot V$) and thus with the moisture regain R ($= M_W / M_T$ 100%) in the textile.

$$\lambda_T = \frac{\lambda_{T,0} + \lambda_W \cdot \dfrac{\rho_T}{\rho_W} \cdot \dfrac{R}{100}}{1 + \delta} \qquad 10.8$$

whereby $\delta = (V_W + V_L)/V_T$. Measurements of heat conductivity on eight different materials according to DIN 52612 confirm the formula presented in eqn 10.8 (Fig. 10.3), whereby the factor δ had to be determined numerically because the porosities of the individual materials were not known. The increase in the heat transfer coefficient could not be attributed to the different types of fibre because the storage of moisture is very dependent on the porosity and size of the capillaries between the fibres, and thus the type of yarn. In the case of the three polyester specimens, for example, the increase varied greatly.

10.5.2   Firefighters' protective clothing

The contradiction between protection and comfort is probably the most pronounced for firefighters' protective clothing. During firefighting activities, the temperatures and the water vapour pressure outside are often higher than those near the body. On the other hand, the tasks performed by the firefighter lead to high metabolic heat production lying often in the range of 300 to 500 W. The sweat production may reach 1 litre or higher in 20 minutes duty. As the

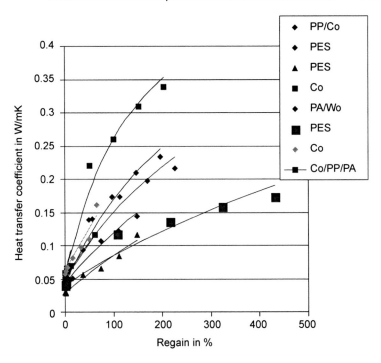

10.3 Heat transfer coefficient as a function of the moisture in the textile.

temperature and the water vapour pressure gradients are directed towards the body in these conditions, the body will not be able to dissipate this heat and moisture to the atmosphere. This will quickly lead to overheating and heat stress problems. The sweat released will have to be absorbed by the clothing layers. Additionally, water from external sources like rain or fire hoses may also be partly absorbed by the clothing. This moisture in the clothing will have different effects on heat transfer as the thermal conductivity and the heat capacity of the clothing are changed. When the temperature in the clothing increases, the moisture present may vaporize and lead to steam injuries. Mäkinen *et al.* (1988) suggested that many of the burns incurred by firefighters on duty may be due to steam. Brans *et al.* (1994) performed some tests on pig skin and found that part of the steam was absorbed by the skin. They suggested that steam burns may therefore be more severe than dry burns as the hot steam was partly transferred to deeper skin layers.

The influence of moisture on heat transfer has already been analyzed several times but the results were sometimes contradictory. Lee and Barker (1986) analyzed single-layer systems exposed to high radiant heat intensity ($80 \, kW/m^2$), and found that high moisture content reduced the heat protection. On the other hand, the influence was positive when the radiant heat intensity was lower ($20 \, kW/m^2$). Mäkinen *et al.* (1988) came to opposing conclusions; combinations with wet underwear exposed to a radiant heat source of $20\,kW/m^2$ gave a reduced protection. The study by Veghte (1986) showed that impermeable clothing combinations exposed to a mixed radiant and convective heat source ($84\,kW/m^2$) had a greatly reduced heat protection when wet. Benisek *et al.* (1986) found that the protection of an impermeable combination exposed to a convective heat source ($80\,kW/m^2$) was reduced when wet but that a permeable combination with a wet wool lining offered a higher protection. Rossi and Zimmerli (1996) obtained comparable results; breathable combinations with a wet lining offered better protection against a high radiant heat source ($80\,kW/m^2$), but the protection of an impermeable combination with a wet lining was reduced. On the other hand, if the heat exposure was low ($5\,kW/m^2$), or if all the layers of the combination were wet, protection was always reduced. Lawson *et al.* (2004) studied several outerwear/underwear combinations and found that under high heat flux flame exposure ($83\,kW/m^2$), external moisture tended to decrease heat transfer through the fabric systems while internal moisture had the opposite effect. However, under low radiant heat flux exposures ($10\,kW/m^2$), internal moisture also decreased heat transfer. Contact heat protection, however, seems to diminish in any case when the textile layers are wet (Veghte 1984, Zimmerli 1992).

The use of breathable, moisture-permeable barriers is certainly beneficial for the comfort of firefighters, as about 80 to 90% of their work does not involve fire. In a firefighting situation, however, as the water vapour pressure may be higher outside than on the body, moisture or steam may flow towards the body. Desruelle *et al.* (2001) and Rossi *et al.* (2004a) showed that the protection against hot steam of combinations with impermeable vapour barriers is generally higher than the one for combinations with semi-permeable barriers.

Considering all these results, about the only conclusion that can be drawn is that moisture has an effect on heat protection, but this effect will depend on the material type (how much moisture can be stored, location of the moisture storage (within the fibres, in the capillaries, along the fibres), permeability of the combination, etc.), the amount and the location of the water stored, the type of heat flux (radiant, convective or conductive), as well as the intensity and the duration of exposure. Therefore, the mechanisms of how moisture influences heat transfer are still not known for the moment, and further studies will be necessary to understand these processes.

10.5.3   Impermeable protective clothing

Impermeable protective clothing such as several types of chemical protective clothing cannot release sweat in liquid or vapour form to the environment. In this case, it may appear that the production of sweat is ineffective for cooling the body, as the sweat cannot leave the clothing. However, phenomena of internal evaporation and re-condensation can provide cooling to the human body; when liquid sweat evaporates near the skin, it will remove heat from the body. As long as the external garment layers have a lower water vapour pressure than that near to the skin, water vapour can flow to the outside and may condense on the inside of the impermeable layer. The heat released by the condensation process will partly hinder thermal transfer from the body to the environment, but on the other hand, most of this heat will be directly released to the atmosphere. Therefore, a 'moisture assisted thermal transfer' can occur. This cooling effect is obviously dependent on the moisture absorption capacity of the outer layers.

Lotens *et al.* (1995) were probably the first to verify this theory experimentally with human subjects. The subjects had to wear impermeable garments and, in order to prevent sweat evaporation, their skin was wrapped with an impermeable plastic. They compared data of the subjects with and without plastic wrap and reported that the heat flow through the clothing without plastic was about twice that of the condition with plastic. The outside temperature of impermeable clothing was higher than a comparable semi-permeable garment, which was attributed to heat of condensation.

For impermeable protective clothing that only covers parts of the body, other solutions are possible to improve the thermal comfort of the wearer. The correct design of the clothing is much more important for impermeable than for permeable clothing to optimize moisture transport through the garment openings (Weder *et al.* 1996, den Hartog 2000). In clothing tightly worn around the body, e.g., ballistic clothing (bullet-proof vests or protectors against cuts, etc.), the use of 3D-structures with very low densities that allow convective thermal and moisture transport within the material can increase the evacuation of surplus heat and sweat. The underwear must be chosen very carefully when wearing such protective clothing; it should wick the liquid sweat and transport it laterally towards regions of the body not covered by the protective clothing and thus compensate the lack of permeability in the regions covered.

The integration of phase change materials in such clothing was proposed to provide an additional cooling to the body (Wittmers *et al.* 1999, Pause 2000) but their effect is limited (Shim *et al.* 2001). Different active cooling systems were also studied as possible solutions to reduce heat stress (Heled *et al.* 2004, McLellan and Bell 1999). However, these systems have certain limitations due to the required additional weight, which increases metabolic heat production, and their bulkiness can reduce their practical use. The use of ice- or liquid-cooling systems will reduce the skin temperature and therefore also reduce dry heat dissipation. The use of air ventilating systems is thus probably more suitable to support natural thermoregulation of the human body.

10.6   Future trends

The European Council Directive 89/686/EEC, also known as the PPE-Directive (PPE: personal protective equipment), defines the basic health and safety requirements that have to be fulfilled by the PPE sold in Europe. Some requirements concern the ergonomics, the innocuousness and the comfort of the clothing. The development of new European standards thus considers these requirements in more detail than in the past. Nowadays, the main goal for the development of new materials for protective clothing is usually the improvement of ergonomics and thermal comfort while maintaining the protection level. For these different reasons, the need for ergonomic or physiological test methods will continue to grow in the future.

The classical approach of clothing physiology often considered the transport of heat and water vapour through the clothing as separate processes. Recently, different research groups have started to study the influence of liquid sweat and changes in the aggregate state of the water (evaporation and condensation). This is especially important for the characterization of protective clothing where liquid sweat is often a problem. Some more complex models of the coupled heat and mass transfer through fabric layers were developed to try to understand these mechanisms. Many processes occurring in the textiles are, however, still not completely understood and it can be expected that intensive research will be done on this topic in the future. New experimental methods for the real-time assessment of moisture distribution in multilayer fabric systems will help to understand these mechanisms. The trend towards more realistic methods implies that the thermal comfort of protective clothing is assessed in near-to-practice conditions. Cold protective clothing, for instance, should not be characterized at 20 °C but at or below 0 °C. These test conditions should be set after a thorough assessment of how and under what conditions the clothing will be used.

The development of new test methods for the assessment of thermo-physiological parameters will proceed in several directions; industry needs very simple, inexpensive test methods for quality control and marketing purposes. These methods should, however, be able to rank different products with regard to their thermal comfort performance and therefore one of the main tasks in the development of such tests will be their link to practice. For research purposes, the trend goes towards more and more complex systems that aim at simulating the human body as close to reality as possible. Sweating manikins should, for instance, replicate the sweat gland distribution of the human body and be able to sweat in liquid or vapour form. The movements of the limbs should reproduce typical activities like walking or cycling to be able to realistically evaluate the effects of ventilation of the clothing. One very ambitious task in the future will be to try to mimic the thermoregulation of the human body by regulating the temperature of the limbs in accordance to the chosen activity, the insulation of the clothing and the environmental conditions. The temperature of the skin is very seldom maintained at a constant temperature and can vary greatly locally. This will have an important effect on heat transfer between the skin and the environment. Therefore, a thermoregulation model of the human body should be implemented into a manikin system to take these effects into account.

10.7   References

Adler M M and Walsh W K (1984), 'Mechanisms of Transient Moisture Transport between Fabrics', *Text. Res. J.*, 54, 334–343.

Ajayi J O (1992), 'Fabric Smoothness, Friction, and Handle', *Textile Research Journal*, 62, 52–59.

Amos D and Hansen R (1997), 'The physiological strain induced by a new low burden chemical protective ensemble', *Aviat. Space Environ. Med.*, 68, 126–131.

Anttonen H E H and Oikarinen A (1997), 'The effect of ventilation on thermal insulation and sweating in cylinder and test person measurements', in Nielsen R and Borg C, *Proceedings of the Fifth Scandinavian Symposium on Protective Clothing (NOKOBETEF V)*, Elsinore, Denmark, 48–52.

Aschoff J, Günther B and Kramer K (1971), *Energiehaushalt und Temperaturregulation*, München, Germany, Urban & Schwarzenberg.

ASTM F1291 (1999), Standard Test Method for Measuring the Thermal Insulation of Clothing Using a Heated Manikin, ASTM, West Conshohocken, PA.

ASTM E 96 (2000), Standard Test Methods for Water Vapor Transmission of Materials, ASTM, West Conshohocken, PA.

ASTM F2298 (2003), Standard Test Methods for Water Vapor Diffusion Resistance and Air Flow Resistance of Clothing Materials Using the Dynamic Moisture Permeation Cell, ASTM, West Conshohocken, PA.

ASTM D 7024 (2004), Standard Test Method for Steady State and Dynamic Thermal Performance of Textile Materials, ASTM, West Conshohocken, PA.

Bartels V and Umbach K-H (1997), 'Die Bedeutung der physiologischen Funktion von Schutzkleidung für die Leistungsfähigkeit und Gesundheit des Trägers – Feuerwehr-, Krankenhaus-, und Wetterschutzbekleidung mit optimaler Gebrauchsfunktion', *Proceedings of the 36th International Man-Made Fibers Conference*, Dornbirn, Austria.

Behmann F W (1990), 'Versuche über die Rauhigkeit von Textiloberflächen', *Melliand Textilberichte*, 438–440.

Benisek L, Edondson G K, Mehta P N and Philips W A (1986), 'The Contribution of Wool to Improving the Safety of Workers Against Flames and Molten Metal Hazards', in Barker R L and Coletta G C, *Performance of Protective Clothing, ASTM STP 900*, ASTM, Philadelphia, 405–420.

Braus T A, Dutrieux R P, Hoekstra M J, Kreis R W and du Pont J S (1994), 'Histopathological evaluation of scalds and contact burns in the pig model', *Burns*, 20, 48–51.

BS 7209 (1990), Specification for Water Vapour Permeable Apparel Fabrics, British Standards Institution, London.

BS 3424 (1996), Part 19, Methods for the Determination of Resistance to Wicking and Lateral Leakage, British Standards Institution, London.

BS 7963 (2000), Ergonomics of the thermal environment – Guide to the assessment of heat strain in workers wearing personal protective equipment, British Standards Institution, London.

Buono M J and Ulrich R L (1998), 'Comparison of mean skin temperature using "covered" versus "uncovered" contact thermistors', *Physiol. Meas.*, 19, 297–300.

Burke R and McGuffin R (2001), 'Development of an advanced thermal manikin for vehicle climate evaluation', *Proceedings of the fourth International Meeting on Thermal Manikins*, St. Gallen, Switzerland.

Crow R M and Osczevski R J (1998), 'The interaction of water with fabrics', *Text. Res. J.*, 68, 280–288.

Demartino R N, Yoon H N and Buckley A (1984), 'Improved Comfort Polyester .5. Results from 2 Subjective Wearer Trials and Their Correlation with Laboratory Tests', *Text. Res. J.*, 54, 602–613.

den Hartog E A (2000), 'Effects of clothing design on ventilation and evaporation of sweat', *Proceedings of NOKOBETEF 6 and 1st European Conference on Protective Clothing*, Stockholm, Sweden, 281–284.

Desruelle A V, Schmid B and Montmayeur A (2001), 'Thermal protection against hot steam stress', *RTA/HFM Symposium 'Blowing hot and cold: Protecting against climatic extremes'*, Dresden, Germany.

DIN 53122 (1974), Bestimmung der Wasserdampfdurchlässigkeit – Gravimetrisches Verfahren (determination of water vapour permeability – gravimetric procedure), Deutsches Institut für Normung e.V., Berlin, Germany.

DIN 53924 (1997), Bestimmung der Sauggeschwindigkeit von textilen Flächengebilden gegenüber Wasser (velocity of soaking water of textile fabrics – method by determining the rising height), Deutsches Institut für Normung e.V., Berlin, Germany.

D'Silva A P, Greenwood C, Anand S C, Holmes D H and Whatmough N (2000), 'Concurrent Determination of Absorption and Wickability of Fabrics: A New Test Method', *J. Text. Inst.*, 91, 383–396.

Elder H M, Fisher S, Armstrong K and Hutchison G (1984), 'Fabric Softness, Handle, and Compression', *Journal of the Textile Institute*, 75, 37–46.

EN 31092 (1993), Textiles – Physiological effects – Measurements of thermal and water-vapour resistance under steady-state conditions (sweating guarded-hotplate test), CEN European Committee for Standardisation, Brussels, Belgium.

Endrusick T L, Stoschein L A and Gonzalez R R (2001), 'U.S. Military Use of Thermal Manikins in Protective Clothing Research', *RTA/HFM Symposium 'Blowing hot and cold: Protecting against climatic extremes'*, Dresden, Germany.

Fan J, Chen Y and Zhang W (2001), 'A perspiring fabric thermal manikin: its development and use', *Fourth International Meeting on Thermal Manikins*, St. Gallen, Switzerland.

Farnworth B, Lotens W A and Wittgen P (1990), 'Variation of Water-Vapor Resistance of Microporous and Hydrophilic Films with Relative-Humidity', *Text. Res. J.*, 60, 50–53.

Finn J T, Sagar A J G and Mukhopadhyay S K (2000), 'Effects of imposing a temperature gradient on moisture vapor transfer through water resistant breathable fabrics', *Text. Res. J.*, 70, 460–466.

Fonseca G F and Breckenridge Jr (1965a), 'Wind Penetration through Fabric Systems .3', *Text. Res. J.*, 35, 221–27.

Fonseca G F and Breckenridge Jr (1965b), 'Wind Penetration through Fabric Systems .I', *Text. Res. J.*, 35, 95–103.

Ftaiti F, Duflot J C, Nicol C and Grelot L (2001), 'Tympanic temperature and heart rate changes in firefighters during treadmill runs performed with different fireproof jackets', *Ergonomics*, 44, 502–512.

Fukazawa T, Kawamura H, Tochihara Y and Tamura T (2003), 'Water vapor transport through textiles and condensation in clothes at high altitudes-combined influence of temperature and pressure simulating altitude', *Text. Res. J.*, 73, 657–663.

Gagge A P, Burton A C and Bazett H C (1941), 'A practical system of units for the description of the heat exchange of man with his environment', *Science,* 94, 428–430.

Garnsworthy R K, Gully R L, Kandiah R P, Kenins P, Mayfield R J and Westerman R A (1968), *Understanding the causes of prickle and itch from the skin contact of fabrics,* CSIRO – Textile and Fibre Technology.

Garnsworthy R K, Gully R L, Kenins P, Mayfield R J and Westerman R A (1988), 'Identification of the Physical Stimulus and the Neural Basis of Fabric-Evoked Prickle', *J. Neurophysiol.,* 59, 1083–1097.

Ghali K, Jones B and Tracy J (1994), 'Experimental Techniques for Measuring Parameters Describing Wetting and Wicking in Fabrics', *Text. Res. J.,* 64, 106–111.

Gibson P W (1993), 'Factors Influencing Steady-State Heat and Water-Vapor Transfer Measurements for Clothing Materials', *Text. Res. J.,* 63, 749–764.

Gibson P W, Kendrick C E, Rivin D and Charmchi M (1997), 'An Automated Dynamic Water Vapor Permeation Test Method', in Stull J O and Schwope A D, *Performance of Protective Clothing: Sixth Volume, ASTM STP 1273,* ASTM, West Conshohocken, PA, 93–107.

Glitz K J and von Restorff W (1999), 'Reduction of heat stress in chemical protective suits', in Hodgdon J A, Heaney J H and Buono M J, *Envionmental Ergonomics VIII,* San Diego, USA, 191–194.

Gretton J C, Brook D B, Dyson H M and Harlock S C (1998), 'Moisture vapor transport through waterproof breathable fabrics and clothing systems under a temperature gradient', *Text. Res. J.,* 68, 936–941.

Griefahn B, Ilmarinen R, Louhevaara V, Mäkinen H and Künemund C (1996), 'Arbeitszeit und Pausen im simulierten Einsatz der Feuerwehr', *Zeitschrift für Arbeitswissenschaft,* 50, 89–95.

Hancock P A and Vasmatzidis I (2003), 'Effects of heat stress on cognitive performance: the current state of knowledge', *Int. J. Hyperthermia,* 19, 355–372.

Harnett P R and Mehta P N (1984), 'A Survey and Comparison of Laboratory Test Methods for Measuring Wicking', *Text. Res. J.,* 54, 471–478.

Havenith G and Heus R (2004), 'A test battery related to ergonomics of protective clothing', *Appl. Ergon.,* 35, 3–20.

Heled Y, Epstein Y and Moran D S (2004), 'Heat strain attenuation while wearing NBC clothing: Dry-ice vest compared to water spray', *Aviat. Space Environ. Med.,* 75, 391–396.

Holmer I (2003), 'Manikin history and applications', in Candas V, *The 5th International Meeting on Manikins and Modelling, 513M,* CEPA, Strasbourg, France.

Ilmarinen R, Lindholm H, Koivistoinen K and Helisten P (2000), 'Physiological strain and wear comfort while wearing a chemical protective suit with breathing apparatus inside and outside the suit in summer and in winter', in Kuklane K and Holmer I, *Proceedings of NOKOBETEF 6 and 1st European Conference on Protective Clothing,* Stockholm, Sweden, 235–238.

ISO 7730 (1984), Moderate Thermal Environments – Determination of the PMV and PPD indices and specification of the conditions for thermal comfort, International Organization for Standardization, Geneva, Switzerland.

ISO 5085-1 (1989), Textiles – Determination of thermal resistance – Part 1: low thermal resistance, International Organization for Standardization, Geneva, Switzerland.

ISO 5085-2 (1989), Textiles – Determination of thermal resistance – Part 2: high thermal resistance, International Organization for Standardization, Geneva, Switzerland.

ISO 9886 (1992), Evaluation of thermal strain by physiological measurements, International Organization for Standardization, Geneva, Switzerland.

ISO 11092 (1993), Textiles – Physiological effects – Measurements of thermal and water-vapour resistance under steady-state conditions (sweating guarded-hotplate test), International Organization for Standardization, Geneva, Switzerland.

ISO 10551 (1995), Ergonomics of the thermal environment – Assessment of the influence of the thermal environment using subjective judgement scales, International Organization for Standardization, Geneva, Switzerland.

ISO 15831 (2003), Clothing – Physiological effects – Measurement of thermal insulation by means of a thermal manikin, International Organization for Standardization, Geneva, Switzerland.

ISO 15496 (2004), Textiles – Measurement of water vapour permeability of textiles for the purpose of quality control, International Organization for Standardization, Geneva, Switzerland.

JIS L 1099 (1999), Testing Methods for Water Vapour Permeability of Clothes, Japanese Standards Association, Tokyo.

Kamata Y, Kato T, Ito A and Yahata N (1988a), 'Convective Heat Transfer from Human Body (Part 1. A Simulation by Vertical Cylinder)', *Sen-I Gakkaishi Transactions*, 42, 57–63.

Kamata Y, Kato T, Ito A and Yahata N (1988b), 'Convective Heat Transfer from Human Body (Part 2. Effect of Fabric on Heat Transfer)', *Sen-I Gakkaishi Transactions*, 44, 64–73.

Kawabata S (1980), *The standardization and analysis of hand evaluation*, Osaka, The Textile Machinery Society of Japan.

Kind R J and Broughton C A (2000), 'Reducing wind-induced heat loss through multilayer clothing systems by means of a bypass layer', *Text. Res. J.*, 70, 171–176.

Lamb G E R (1992), 'Heat and Water-Vapor Transport in Fabrics under Ventilated Conditions', *Text. Res. J.*, 62, 387–392.

Lamb G E R and Yoneda M (1990), 'Heat-Loss from a Ventilated Clothed Body', *Text. Res. J.*, 60, 378–383.

Lawson L K, Crown E M, Ackerman M Y and Dale J D (2004), 'Moisture Effects in Heat Transfer Through Clothing Systems for Wildland Firefighters', *International Journal of Occupational Safety and Ergonomics*, 10, (in press).

LeBlanc P R and Fahy R F (2004), *Firefighter Fatalities in the United States – 2003*, Quincy, MA, National Fire Protection Association.

Lee Y M and Barker R L (1986), 'Effect of Moisture on the Thermal Protective Performance of Heat-Resistanc Fabrics', *Journal of Fire Sciences*, 4, 315–331.

Leisen J and Beckham H W (2001), 'Quantitative magnetic resonance imaging of fluid distribution and movement in textiles', *Text. Res. J.*, 71, 1033–1045.

Li Y, Keighley J H and Hampton I F G (1988), 'Physiological-Responses and Psychological Sensations in Wearer Trials with Knitted Sportswear', *Ergonomics*, 31, 1709–1721.

Li Y, Keighley J H, McIntyre J E and Hampton I F G (1991), 'Predictability between Objective Physical Factors of Fabrics and Subjective Preference Votes for Derived Garments', *Journal of the Textile Institute*, 82, 277–284.

Lotens W A (1983), 'Clothing, physical load and military performance', *Aspects médicaux et biophysiques des vêtements de protection*, Centre de Recherches du

Service de Santé des Armées, Lyon, France, 268–279.

Lotens W A and Havenith G (1991), 'Calculation of Clothing Insulation and Vapor Resistance', *Ergonomics, 34*, 233–254.

Lotens W A, Vandelinde F J G and Havenith G (1995), 'Effects of Condensation in Clothing on Heat-Transfer', *Ergonomics, 38*, 1114–1131.

Louhevaara V, Ilmarinen R, Griefahn B, Kunemund C and Makinen H (1995), 'Maximal Physical Work Performance with European Standard Based Fire-Protective Clothing System and Equipment in Relation to Individual Characteristics', *Eur. J. Appl. Physiol. Occup. Physiol., 71*, 223–229.

Mäkinen H, Smolander J and Vuorinen H (1988), 'Simulation of the Effect of Moisture Content in Underwear and on the Skin Surface on Steam Burns of Fire Fighters', in Mansdorf S Z, Sager R and Nielsen A P, *Performance of Protective Clothing: Second Volume, ASTM STP 989*, ASTM, Philadelphia, 415–421.

Matsudaira M, Watt J D and Carnaby G A (1990a), 'Measurement of the Surface Prickle of Fabrics. 1. The Evaluation of Potential Objective Methods', *Journal of the Textile Institute,* 81, 288–299.

Matsudaira M, Watt J D and Carnaby G A (1990b), 'Measurement of the Surface Prickle of Fabrics. 2. Some Effects of Finishing on Fabric Prickle', *Journal of the Textile Institute,* 81, 300–309.

McCullough E, Barker R, Giblo J, Higenbottam C, Meinander H, Shim H and Tamura T (2002), 'Interlaboratory Evaluation of Sweating Manikins', *The 10th International Conference on Environmental Ergonomics*, Fukuoka, Japan, 467–470.

McCullough E A, Kwon M and Shim H (2003), 'A comparison of standard methods for measuring water vapour permeability of fabrics', *Meas. Sci. Technol., 14*, 1402–1408.

McLellan T M and Bell D G (1999), 'Efficacy of air and liquid cooling during light and heavy exercise while wearing NBC clothing', *Aviat. Space Environ. Med., 70*, 802–811.

Meinander H (1985), *Introduction of a new test method for measuring heat and moisture transmission through clothing materials and its application on winter work wear*, Technical Research Centre of Finland VTT.

Meinander H (1992), 'Coppelius – a sweating thermal manikin for the assessment of functional clothing', *Proceedings of the Fourth Scandinavian Symposium on Protective Clothing (NOKOBETEF IV)*, Kittilä, Finland, 157–161.

Melin B, Etienne S, Pelicand J Y, Charpenet A and Warmé-Janville B (1999), 'Light NBC protective combat suits and body hydration during physical activities under tropical climate', in Hodgdon J A, Heaney J H and Buono M J, *Envionmental Ergonomics VIII*, San Diego, USA, 113–115.

Naylor G R S and Phillips D G (1997), 'Fabric-evoked prickle in worsted spun single jersey fabrics. 3. Wear trial studies of absolute fabric acceptability', *Text. Res. J., 67*, 413–416.

O'Brien C, Hoyt R W, Buller M J, Castellani J W and Young A J (1998), 'Telemetry pill measurement of core temperature in humans during active heating and cooling', *Med. Sci. Sports Exerc., 30*, 468–472.

Olesen B W and Fanger P O (1973), 'The skin temperature distribution for resting man in comfort', *Arch. Sci. Physiol.*, 385–393.

Osczevski R J and Dohlan P A (1989), 'Anomalous Diffusion in a Water Vapour Permeable, Waterproof Coating', *J. Coated Fabrics, 18*, 255–258.

Pause B H (2000), 'New Heat Protective Garments with Phase Change Material', in Nelson C N and Henry N W, *Performance of Protective Clothing: Issues and Priorities for the 21st Century: Seventh Volume, ASTM STP 1386*, ASTM, West Conshohocken, PA, 3–13.

prEN 469 (2004), Protective clothing for firefighters – Performance requirements for protective clothing for firefighting, CEN European Committee for Standardisation, Brussels, Belgium.

Reifler F A, Lehmann E, Frei G and May H (2003), 'Water distribution and movement in wet aramid-based ballistic body armor panels detected with neutron radiography', *2nd European Conference on Protective Clothing and NOKOBETEF 7*, Montreux, Switzerland.

Reischl U and Stransky A (1980), 'Comparative-Assessment of Goretex and Neoprene Vapor Barriers in a Firefighter Turn-out Coat', *Text. Res. J.,* 50, 643–647.

Reneau P D, Bishop P A and Ashley C D (1999), 'A comparison of physiological responses to two types of particle barrier, vapor permeable clothing ensembles', *Am. Ind. Hyg. Assoc. J.,* 60, 495–501.

Richards M and Mattle N (2001), 'Development of a sweating agile thermal manikin (SAM)', *Fourth International Meeting on Thermal Manikins*, St. Gallen, Switzerland.

Richards M G M and McCullough E A (2004), 'Revised Interlaboratory Study of Wettable Thermal Manikins including Results from the Sweating Agile Thermal Manikin', *Eighth Symposium on Performance fo Protective Clothing: Global Needs and Emerging Markets*, ASTM, Tampa, Florida, (in press).

Rossi R (2000), 'Sweat management - Optimum Moisture and Heat Transport Properties of Textile Layers', *39th International Man-Made Fibres Congress*, Dornbirn, Austria.

Rossi R (2003), 'Fire fighting and its influence on the body', *Ergonomics,* 46, 1017–1033.

Rossi R M and Zimmerli T (1996), 'Influence of Humidity on the Radiant, Convective and Contact Heat Transmission through Protective Clothing Materials', in Johnson J S and Mansdorf S Z, *Performance of Protective Clothing, Fifth Volume, ASTM STP 1237*, ASTM, Philadelphia, 269–280.

Rossi R, Indelicato E and Bolli W (2004a), 'Hot steam transfer through heat protective clothing layers', *International Journal of Occupational Safety and Ergonomics,* 10, (in press).

Rossi R M, Gross R and May H (2004b), 'Water vapor transfer and condensation effects in multilayer textile combinations', *Text. Res. J.,* 74, 1–6.

Santee W R and Wenger C B (1988), 'Comparative heat stress of four chemical protective suits', in Perkins J and Stull J O, *Chemical Protective Clothing in Chemical Emergency Response, ASTM STP 1037*, ASTM, Philadelphia, 41–50.

Schneider A M, Holcombe B V and Stephens L G (1996), 'Enhancement of coolness to the touch by hygroscopic fibers. 1. Subjective trials', *Text. Res. J.,* 66, 515–520.

Schopper-Jochum S, Schubert W and Hocke M (1997), 'Vergleichende Bewertung des Trageverhaltens von Feuerwehreinsatzjacken (Phase I)', *Arbeitsmed. Sozialmed. Umweltmed.,* 32, 138–144.

Shim H, McCullough E A and Jones B W (2001), 'Using phase change materials in clothing', *Text. Res. J.,* 71, 495–502.

Sköldström B and Holmer I (1983), 'A protective garment for hot environments with improved evaporative heat transfer capacity', *Aspects médicaux et biophysiques des vêtements de protection*, Centre de Recherches du Service de Santé des Armées, Lyon, France, 289–294.

Spencer-Smith J L (1977a), 'The Physical Basis of Clothing Comfort, Part 2: Heat Transfer through Dry Clothing Assemblies', *Clothing Res. J.*, 5, 3–17.

Spencer-Smith J L (1977b), 'The Physical Basis of Clothing Comfort, Part 4: The Passage of Heat and Water through Damp Clothing Assemblies', *Clothing Res. J.*, 5, 116–128.

Sweeney M M and Branson D H (1990a), 'Sensory Comfort .1. A Psychophysical Method for Assessing Moisture Sensation in Clothing', *Text. Res. J.*, 60, 371–377.

Sweeney M M and Branson D H (1990b), 'Sensory Comfort .2. A Magnitude Estimation Approach for Assessing Moisture Sensation', *Text. Res. J.*, 60, 447–452.

Takeuchi M, Isshiki N and Ishibashi Y (1982), 'Heat-Transfer on Cylinder Covered with Close-Fitting Fabrics .1. Wind Penetration through Fabrics', *Bulletin of the JSME – Japan Society of Mechanical Engineers*, 25, 1406–1411.

Teitlebaum A and Goldman R F (1972), 'Increased Energy Cost with Multiple Clothing Layers', *J. Appl. Physiol.*, 32, 743–744.

Tokura H and Midorikawatsurutani T (1985), 'Effects of Hygroscopically Treated Polyester Blouses on Sweating Rates of Sedentary Women at 33-Degrees-C', *Text. Res. J.*, 55, 178–180.

Töpfer H-J and Stoll T P (2001), 'Physiological Assessments of Permeable NBC Protective Clothing for Hot Climate Conditions', *RTA/HFM Symposium 'Blowing hot and cold: Protecting against climatic extremes'*, Dresden, Germany.

Umbach K-H (1983), 'Biophysical evaluation of protective clothing by use of laboratory measurements and predictive models', *Aspects médicaux et biophysiques des vêtements de protection*, Centre de Recherches du Service de Santé des Armées, Lyon, France, 226–237.

Van Langenhove L and Kiekens P (2001), 'Textiles and the transport of moisture', *Textile Asia*, 32–34.

Veghte J H (1984), 'Effect of Moisture on the Burn Potential in Fire Fighters' Gloves', *Fire Technology*, 23, 313–322.

Veghte J H (1986), 'Functional Integration of Fire Fighters' Protective Clothing', in Barker R L and Coletta G C, *Performance of Protective Clothing, ASTM STP 900*, ASTM, Philadelphia, 487–496.

Veghte J H (1988), 'The physiological strain imposed by wearing fully encapsulated chemical protective clothing', in Perkins J and Stull J O, *Chemical Protective Clothing in Chemical Emergency Response, ASTM STP 1037*, ASTM, Philadelphia, 51–64.

Wang G, Zhang W, Postle R and Phillips D (2003), 'Evaluating wool shirt comfort with wear trials and the forearm test', *Text. Res. J.*, 73, 113–119.

Washburn A E, LeBlanc P R and Fahy R F (1999), *Firefighter Fatalities*, Quincy, MA, National Fire Protection Association.

Weder M S, Zimmerli T and Rossi R M (1996), 'A Sweating and Moving Arm for the Measurement of Thermal Insulation and Water Vapour Resistance of Clothing', in Johnson J S and Mansdorf S Z, *Performance of Protective Clothing: Fifth Volume, ASTM STP 1237*, ASTM, Philadelphia, 257–268.

Weder M S, Frei G, Brühwiler P A, Herzig U, Huber R and Lehmann E (2004a), 'Neutron radiography measurements of the moisture distribution in multilayer clothing systems', *Text. Res. J.*, (in press).

Weder M S, Laib A and Brühwiler P A (2004b), 'Computerized Tomography Measurements of the Moisture Distribution in Multilayered Clothing Systems', *Text. Res. J.*, (submitted).

Wenzel H G and Piekarski C (1982), *Klima und Arbeit*, München, Germany, Bayerisches Staatministerium für Arbeit und Sozialordnung.

Wilkinson M C, Scott R A, Williams J T and Lovell K V (1997), 'Considerations in the Design and Construction of Novel Integrated NBC Ensembles', in Nielsen R and Borg C, *Proceedings of the Fifth Scandinavian Symposium on Protective Clothing (NOKOBETEF V)*, Elsinore, Denmark, 121–126.

Wittmers L, Tabor C, Canine K and Hodgdon J (1999), 'Use of encapsulated phase change material (EPCM) as a cooling agent in microclimate cooling garments', in Hodgdon J A, Heaney J H and Buono M J, *Envionmental Ergonomics VIII*, San Diego, USA, 231–235.

Woo S S, Shalev I and Barker R L (1994), 'Heat and Moisture Transfer through Nonwoven Fabrics .1. Heat- Transfer', *Text. Res. J.*, 64, 149–162.

Zhuang Q, Harlock S C and Brook D B (2002), 'Transfer wicking mechanisms of knitted fabrics used as undergarments for outdoor activities', *Text. Res. J.*, 72, 727–734.

Zimmerli T (1992), 'Contact heat testing of dry and wet fire-fighters gloves', in Mäkinen H, *Fourth Scandinavian Symposium on Protective Clothing (NOKOBETEF IV)*, Kittilä, Finland, 297–303.

Zimmerli T and Weder M S (1997), 'Protection and Comfort – A Sweating Torso for the Simultaneous Measurement of Protective and Comfort Properties of PPE', in Stull J O and Schwope A D, *Performance of Protective Clothing: Sixth Volume, ASTM STP 1273*, ASTM, Philadelphia, 271–280.

11

Modeling thermal burn injury protection

G S O N G, University of Alberta, Canada

11.1 Introduction

In spite of high-performance fibers, fabrics, and advanced test methods, much remains to be learned to enhance the technical basis for improving thermal protective performance of materials and clothing to protect against burn injuries when subjected to thermal exposures. In order to understand the mechanism of heat transfer in protective clothing systems, some analytical and numerical models have been developed for these materials and tests in different conditions.[1-4] These models, which are based on some bench scale tests and manikin tests, once validated by experimental tests, can provide extra information in improving protective clothing performance.

11.1.1 Thermal protection and lab measurement

Three types of heat transfer exist in a firefighting field or thermal environment that could cause burns: conduction, convection, and thermal radiation. Conduction is the direct transfer of heat through contact with the hot object. Convection is the transfer of heat through a medium, for example, air. Thermal radiation is the transfer of heat in the form of light energy. Firefighters experience all three types of heat in a fire. Wet or compressed protective clothing increases the risk of being burned or scalded by conductive heat. Convected heat travels through the air, even if there is no immediate appearance of fire, or the fire seems distant. Convected heat can elevate the temperature of protective clothing to a point at which conductive heat burns can easily occur, particularly if the protective clothing is wet or damp. Thermal radiation is the transfer of heat in the form of light energy into a material, directly from flames or reflected from hot objects. Factors that affect the speed of radiant heat transfer include the temperature difference between two surfaces, their distance from each other, and the reflectiveness of each surface. Radiant heat becomes stored thermal energy as it strikes the surfaces of protective clothing. As it grows in intensity, it transfers heat inward from the outer (shell) layer and causes burn injuries.

Several methods have been developed to measure the insulative properties of fabrics against high-intensity thermal energy. Heat sources range from radiant panels to gas burners or a combination of the two, depending on the type of fire simulated. For most tests the magnitude of heat source intensity is in the range of 21 to $84 \, kW/m^2$. These test methods used a copper calorimeter (test sensor) placed behind the fabric. The calorimeter is of specified size, mass and mounting. It is used to measure energy transferred through the fabric or fabric layers being tested. The exposure time is limited to that required for the temperature of the copper calorimeter to reach some end point. The end point may be related to the time it takes for the onset of second-degree burning in human skin (blistering) as in the NFPA 1971 TPP test, ASTM F 1939, or for the calorimeter to reach a specified temperature rise, as in ISO 9151 and ISO 6942.

Depending on the test method, the fabric may be horizontal or vertical, in contact with or spaced away from the calorimeter. Commonly used methods in bench scale test include TPP in NFPA (the former ASTM D 4108, Standard Test Method for Thermal Protective Performance of Materials for Clothing by Open-Flame Method), the ASTM F 1939 (Standard Test Method for Radiant Protective Performance of Flame Resistant Clothing Materials tests) and ASTM F 1060 (Standard Test Method for Thermal Protective Performance of Materials for Protective Clothing for Hot Surface Contact). Similar test methods can be found in International Standards, such as ISO 9151 (Protective Clothing against Heat and Flame – Determination of Heat Transmission on Exposure to Flame), ISO 6942 (Protective Clothing Protection against Heat and Fire – Method of test: Evaluation of materials and material assemblies when exposed to a source of radiant heat) and ISO 12127 (Clothing for protection against heat and flame – Determination of contact heat transmission through protective clothing or constituent materials). All these methods are designed to evaluate the performance of thermal protective fabrics or assemblies exposed to different thermal hazards.

11.1.2   Manikin fire test

Currently, the most realistic laboratory assessment of the thermal protective performance of clothing is provided by instrumented manikin fire testing (ASTM F 1930, Standard Test Method for Evaluation of Flame Resistance Clothing for Protection Against Flash Fire Simulation Using an Instrumented Manikin). The manikin protective clothing evaluation system consists of an adult size male manikin with more than 100 heat sensors over the manikin body, a flash fire exposure system, and a data acquisition unit as well as a computer software package that controls the system operation and calculates the tissue burn injury. In addition, to measure the properties of fabric materials or their composites, the manikin fire test provides information on the effects of garment design, fit, and its components on clothing protective performance. It is the most complicated and powerful evaluation system for thermal protective clothing.

11.1.3   Key issues in thermal protection evaluation

Current standard tests for performance evaluation provide only a basis for measuring the thermal insulation of clothing materials. These methods focus on intense thermal environments in dry condition and consider no stored thermal energy effects. Data produced by decades of fire research on structural fires shows that most burn injuries sustained by firefighters occurred in the low-level thermal environments.[5,6] This thermal environment could be outside the flaming envelope, post-flashover or pre-flashover fires. In addition, these tests are conducted only under dry conditions before and after multiple cycles of laundering. Of particular concern is the effect of moisture in the clothing system on its protective performance. This moisture can be introduced both externally by hose spray and internally by sweating.

Water is a very poor insulator. It can create a conductive bond between surfaces that might not otherwise touch. This increases the potential of heat conduction by displacing the insulating air between and within the layers of clothing. Water conducts heat with dangerous and unpredictable efficiency. Conductive heat transfer burns can be caused by contacting heated surfaces or objects. Serious conductive heat burns can result by compressing parts of protective clothing and exposing to too much heat. Compression brings surfaces closer together and displaces air. This results in the transfer of heat between outside surfaces and clothing layers. An example of this type of injury is the blistering that occurs on knees while crawling on hot surfaces, or where the Breathing Apparatus (SCBA) straps have squeezed the surrounding fabric against the skin. Another common compression injury occurs (even without contacting a hot object) when the firefighter's forearm is extended toward the heat source while holding a hose.

Because moisture, present in protective clothing systems, has a complex influence on heat transmission and potential for skin burn injuries, there is significant interest in developing laboratory thermal protective performance testing protocols that incorporate reliable and realistic moisture preconditioning procedures. A major obstacle in the development of such testing methodologies is the lack of basic understanding of how moisture is absorbed in turnout clothing systems when exposed, either to perspiration from sweating firefighter, or to water from a fire ground source. Additionally, protective clothing systems are constructed by multilayer fabrics (shell outer fabric, moisture barrier and thermal liner), When exposed to thermal hazards, the firefighter's protective clothing system can store a large amount of thermal energy. This energy that is stored within the garment during exposure may be discharged naturally or by compression of the garment in a cooling period (off-exposure). The discharge, when delivered after an exposure, along with transmitted heat during the exposure, can cause burn injuries.[7] Currently, no standardized laboratory test methods or fully developed performance criteria exist for evaluating the thermal protective performance of firefighter clothing with regard to the discharge of the stored thermal energy.

The sensor used for measuring temperature rise in these standard tests is the copper calorimeter. This kind of sensor is rugged, easily cleaned and can be rapidly cooled with an air jet. However, it absorbs heat at quite a different rate from human skin. This higher rate of energy absorbtion will keep the fabric layers at a lower temperature than if they were placed over human skin or a skin simulant.[8] In addition, for longer duration, the copper calorimeter is not able to provide accurate measurement.[9]

Two methods are used as the basis for burn protection in evaluating garment and fabric thermal protective performance, Stoll criteria and Henriques Burn Integral (HBI). Most of the bench scale tests use the Stoll criteria to predict time to second degree burn. The Stoll curves were established by converting the total amount of energy that must be absorbed by the skin to sustain second-degree burn or cause pain for a given length of time to a temperature rise of copper calorimeter. The Stoll method has the advantage of simplicity; it does not need sophisticated numerical calculation to estimate burn injury. However, Stoll and Chianta stated that in applying these data it is essential that the incident heat pulse must be rectangular, for any variation from this shape invalidates the data.[10]

11.2   Thermal hazard and lab simulation

The physical environment faced by firefighters and first responders represents one of the most complex set of exposure conditions in terms of the type, amount, and duration of heat exposure. It can be thermal radiation, flashover fire, hot surface and molten substances. The intensity ranges from low-level thermal radiation ($1–10 \, kW/m^2$) to flashover condition ($80–180 \, kW/m^2$).[20] Firefighters can receive burns in thermal exposures that are considerably lower than flashover conditions. These exposures are usually several minutes in duration, and the exposure levels are generally not sufficient to degrade the turnout shell fabric. In flashover conditions, however, the current turnout system can provide only 10 to 12 seconds to escape.[11] The range of thermal conditions and possible air temperatures that can occur to firefighters and first responders are illustrated in Fig. 11.1 in which the thermal hazards are categorized into three regions: routine, ordinary and emergency.[12]

The routine region features a very low thermal radiation range of 1.1 to 2.1 kW/m^2 and the air temperature is in the range of 20–60 °C. Routine conditions are virtually the same as those encountered on a hot summer day. Protective clothing for firefighters typically provides protection under these

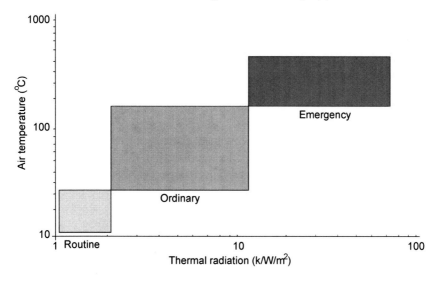

11.1 The range of thermal conditions in the firefighting field.

conditions, but it can generate a second-degree burn if exposed for excessively long times. In laboratory tests, these thermal conditions can be simulated by using a specially designed hot plate or quartz lamp.

The ordinary region describes thermal radiation in a range of 2.1 to 25 kW/m² representing air temperature range of 60–300 °C. Under these conditions, protective clothing may allow sufficient time to extinguish the fire or to fight the fire. However, the burn injury may occur as a result of long exposure or subsequent compression of the heated ensemble onto the body due to discharge of stored thermal energy. These ordinary thermal conditions can be simulated by RPP (radiant protective performance) testers or TPP (textile protective performance) testers using only quartz tubes.

The emergency region represents conditions in a severe and unusual exposure with intensity exceeding 25 kW/m², such as engulfment in a flash fire or next to a flame front. The protective clothing and equipment is simply to provide the short time needed for an escape without serious injury. Similarly, these thermal conditions can be simulated in the laboratory by RPP and TPP tests. The flashover thermal conditions can be obtained by typical TPP experimental arrangements using a methane gas flame in combination with a bank of quartz tubes to provide a 50% convective and 50% radiant heat source.

Schoppee *et al.* compared the behavior of quartz panels to a black body that, at maximum output, wavelengths coincide with temperatures above 1,023 K.[13] At lower temperatures, the emissive power of a quartz panel falls within the waveband containing 75% of the total emissive power of a black body at the same temperature. Gagnon compares quartz lamps and a cone heater having different spectral distributions of radiant energy.[14] Gagnon heated firefighter jacket materials, whose reflectivity and absorbtivity curve depend on the wavelength of the incident radiant energy, with the same thermal flux. These experiments show that the different temperature history occurred on the fabric, and a different prediction time to get second-degree burns was found. For these reasons, he concluded that a cone heater may be more representative of actual fires than quartz lamps.

Real fires produce a thermal environment characterized by turbulent buoyant diffusion flames.[15] The radiation characteristics of the fire depend on the degradation products of the combustion process. For this reason, fires have different characteristics depending on the type of fuel that is burned. Common combustible degradation products from polymers are methane, ethane, ethylene, formaldehyde, acetone and carbon monoxide.[16] Non-combustible products can include carbon dioxide, hydrogen chloride and water, or in the case of FR cotton, laevoglucose.[17] Water has strong absorption bands at 2.7 and 6.3 μm, CO_2 at 2.7 and 4.3 μm. Radiant intensity, which is wavelength dependent, determines the potential hazard of the fire exposure. Heat flux from a fire is similar to a black body at the fire temperature, usually in the range of 1500–1800 °C.[18] Wavelengths range 1–6 microns at heat flux levels above 84 kW/m^2, with a peak at about 2 microns. Holcombe and Hoschke measured heat fluxes of 130–330 kW/m^2 from simulated mine explosions, and 167–226 kW/m^2 for JP-4 fuel fires.[19] Krasny et al. reported estimates of 180 kW/m^2 in seven room fires from just below flashover, to flashover and severe postflashover fires.[20]

Lab-simulated flash within the manikin chamber fire is generated with liquid propane gas burned by several gas burners. If propane is assumed to react with stoichiometric air, then the chemical reaction for complete combustion can be written as[21]

$$C_3H_8 + 5O_2 + 5 \times 3.76N_2 \rightarrow 3 \times CO_2 + 4H_2O + 18.8N_2 \qquad 11.1$$

Adiabatic flame temperatures of about 2,400 K and 2,270 K were calculated using STANJAN® (software by Stanford University to calculate the adiabatic flame temperature) with and without dissociation, respectively. Adiabatic flame temperature is the maximum possible temperature for this flame. Actual flames are cooler due to heat transfer from the flame and incomplete combustion. Siegel and Howell report values of about 2,200 K for flame temperature in their experiments.[22] Maximum experimental values for laboratory burners using methane show flame temperature in the order of 2,000 K to 2,100 K.[23] Holcombe and Hoschke report that approximately 25% of the heat energy released by a Meker burner is thermal radiation.[19] Shalev found that a propane-burning Meker burner produced a heat flux which was approximately 70% convective and 30% radiative in nature.[24]

The characteristics of lab-simulated flash fires generated in a manikin chamber were fully investigated.[25] Flame temperatures measured in a Pyroman®

fire chamber ranged between 800–1400 °C in a four-second exposure to 84 kW/ m^2 heat intensity. The intensity of the incident thermal energy measured by 122 thermal sensors is normally distributed over the Pyroman® manikin. Variation in heat flux results from the three-dimensional shape of the manikin surface, and the complex and dynamic nature of the flame column surrounding the manikin. In order to estimate heat transfer from the fire to the manikin, heat transfer coefficients at each of the 122 thermal sensor locations were estimated. These experiments confirm that the flash fire generated by the torches produces a highly dynamic thermal environment. Measured heat transfer coefficients range from 80–140 W/m^2 °C for a four-second fire exposure. These studies show that the value of the heat transfer coefficient depends on the flame temperature and the specific location of sensors on the manikin. Locations that encounter higher flame temperatures produce high heat transfer coefficients. Heat transfer coefficients measured in the chest, back, and middle leg areas are higher compared to sensors located in the shoulder, thigh, and upper arms of the manikin body.

11.3   Modeling heat transfer in protective fabrics

Predicting thermal protective performance requires an ability to model heat transfer through protective clothing materials, through interfacial air gaps between the skin and clothing, and, finally, through the skin itself. Protective garments exposed to a fire hazard undergo three heating phases. During an initial warm-up phase, the temperature of the fibers in the fabric, plus the moisture retained within the fabric increases at a rate dictated by the system's thermal properties and by the intensity of the incident heat. Consequently, the amount of retained moisture and its thermal properties vary. For most fabric systems, the fiber content and its thermal properties remain constant during the initial heating phase. The second heating phase is marked by the onset of changes in the thermal properties of the fabric. Changes in the amount of retained moisture and its thermal properties continue to occur during this phase. Initially, most of these changes are due to the reaction of surface chemical treatments and finishes to heating, or to slight degradation of fiber surfaces.

Most of these changes begin on the exposed side of the fabric system and propagate toward the skin side of the protective material. If the fiber does not melt, or the transition temperature is not exceeded, the structural integrity of fabric system is maintained during this phase of heating. The end of the second heating phase is the temperature criterion below which thermally protective fabrics are designed to function. Protective fabrics exposed beyond the second phase lose their protective properties and, in some instances, become a source of harm to the wearer. In practice, the occurrence of the subsequent third phase occurs only when the protective clothing system is used beyond its intended limits of application. A third and final phase of the exposure is marked by chemical and structural degradation of the protective fabric. At this point, no moisture is retained by the fabric. This phase is followed by rapid fabric decomposition or combustion. At this point, the fabric itself becomes a source of off-gassing heat and flame.

11.4   Heat transfer models

Heat transfer models have been developed to characterize the behavior of protective fabrics in short-duration high heat flux exposures. Some models focus on specific mechanisms of heat transfer, while others provide a predictive model for a particular thermal test. Three models offer the most promising foundation for development into a complete generalized model. These models are fabric-air gap-test sensor model,[1] multiphase heat and mass transfer model,[2] and multi-layer composite model.[3]

11.4.1   Fabric-air gap-test sensor model

Torvi introduced a model to describe an experimental apparatus consisting of a fabric held horizontally over a Meker burner, with a copper calorimeter held over the fabric.[1] The model treats heat transfer in the vertical dimension only. It accounts for convection and radiation in the air gap between the burner and the fabric, conduction, absorbed radiation, and thermochemical reaction within the fabric, and conduction, convection, and radiation in the air gap between the fabric and the sensor. This model includes the most significant contributions to heat transfer from the burner, through the fabric to the sensor. It can be extended to treat heat transfer in multiple dimensions, and through multiple layers of fabric. Extensions of this model treat convective heat transfer in the fabric and heat conveyed by moisture within the fabric or in air gaps. The model accounts for convection and radiation on the outside of the material, exposed to the burner, and conduction/convection and radiation in the air gaps between fabric and skin. Radiation heat flux is expressed as the sum of blackbody components from hot gases, from the fabric to ambient air, and from the burner head to the fabric. Therefore,

$$q_{rad} = \sigma \epsilon_g T_g^4 - \sigma \epsilon_f F_a (1 - \epsilon_g)(T_f^4 - T_a^4)$$

$$+ \frac{\sigma F_b (1 - \epsilon_g)(T_b^4 - T_f^4)}{1 + F_b(1 - \epsilon_g)\left(\dfrac{1 - \epsilon_f}{\epsilon_f} + \dfrac{A_f}{A_b}\dfrac{1 - \epsilon_b}{\epsilon_b}\right)}, \qquad 11.2$$

where σ is the Stefan-Boltzmann constant, ϵ_g, ϵ_f, and ϵ_b are emissivities of the hot gases, the fabric, and the burner head, respectively, T_g, T_f, T_a, and T_b are the temperatures of the hot gases, the outside of the fabric, the ambient air, and the burner head, respectively, F_a and F_b are view factors accounting for the geometry of the fabric with respect to the ambient air and to the burner, respectively, and A_f and A_b are the surface areas of the fabric and the burner head, respectively.

Radiation heat flux on the inside is

$$q_{rad} = \frac{\sigma(T_f^4 - T_s^4)}{\dfrac{1 - \epsilon_s}{\epsilon_s} + \dfrac{A_s}{A_f}\left(\dfrac{1}{F_s} + \dfrac{1 - \epsilon_f}{\epsilon_f}\right)}, \qquad 11.3$$

where T_s, ϵ_s, and A_s are the temperature, emissivity, and surface area, respectively, of a test sensor taking the place of skin, and F_s accounts for the geometry of the fabric with respect to the sensor. The model accounts for conduction, thermochemical reaction, and absorption of incident radiation that occurs with the fabric. The resulting energy balance equation is written as

$$C^A(T)\frac{\partial T}{\partial t} = \frac{\partial}{\partial x}\left(k(T)\frac{\partial T}{\partial x}\right) + \gamma\, q_{rad}e^{-\gamma x}, \qquad 11.4$$

where T is the temperature, C^A is a temperature-dependent 'apparent' specific heat, which incorporates latent heat associated with thermochemical reaction, k is a temperature-dependent thermal conductivity, γ is the extinction coefficient of the fabric, and q_{rad} is the incident radiation heat flux.

11.4.2   Multiphase heat and mass transfer model

Gibson built on Whitaker's theory of coupled heat and mass transfer through porous media to derive a set of equations modeling heat and mass transfer through textile materials as hygroscopic porous media.[2,26] Gibson applies continuity, linear momentum conservation, and energy conservation equations to the fabric as a three-phase system consisting of a solid phase with a concentration of bound water, a free liquid water phase, and a gas phase of water vapor in air (Fig. 11.2). This model treats heat transfer in three dimensions, accounting for conduction by all phases, convection by the gas and liquid phases, and transformations among the phases. Gibson writes the thermal energy balance equation as:

$$\langle\rho\rangle c_p \frac{\partial\langle T\rangle}{\partial t} + \left(\sum_j (c_p)_j\langle\rho_j\mathbf{v}_j\rangle + \rho_\beta(c_p)_\beta\langle\mathbf{v}_\beta\rangle + \sum_i (c_p)_i\langle\rho_i\mathbf{v}_i\rangle\right)\cdot\nabla\langle T\rangle$$

$$+ \Delta h_{vap}\langle\dot{m}_{lv}\rangle + Q_l\langle\dot{m}_{sl}\rangle + (Q_l + \Delta h_{vap})\langle\dot{m}_{sv}\rangle$$

$$= \nabla\cdot(k_{eff}^T\cdot\nabla\langle T\rangle). \qquad 11.5$$

If we number each distinct species, (1) for water, (2) for dry solid, and (3) for inert air, the thermal energy balance equation can be written as:

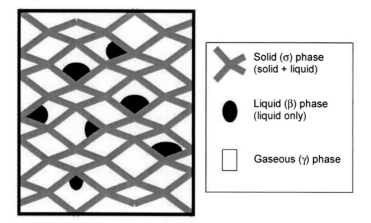

11.2 Three phases present in hygroscopic porous media.

$$\langle\rho\rangle C_p \frac{\partial T}{\partial t} + \left(\begin{array}{c} (c_p)_1(\rho_\beta\langle\mathbf{v}_\beta\rangle + \epsilon_\sigma\langle\rho_1\mathbf{v}_1\rangle^\sigma + \epsilon_\gamma\langle\rho_1\mathbf{v}_1\rangle^\gamma) \\ +(c_p)_2\epsilon_\sigma\langle\rho_2\mathbf{v}_2\rangle^\sigma + (c_p)_3\epsilon_\gamma\langle\rho_3\mathbf{v}_3\rangle^\gamma \end{array}\right) \cdot \nabla T$$

$$+ \Delta h_{vap}\langle\dot{m}_{lv}\rangle + Q_l\langle\dot{m}_{sl}\rangle + (Q_l + \Delta h_{vap})\langle\dot{m}_{sv}\rangle$$

$$= \nabla \cdot (\mathbf{K}_{eff}^T \cdot \nabla T), \qquad\qquad 11.6$$

where

$$\langle\rho\rangle C_p = \langle\rho_1\rangle(c_p)_1 + \langle\rho_2\rangle(c_p)_2 + \langle\rho_3\rangle(c_p)_3,$$

The bracketed terms $\langle\ \rangle$ denote a volume average over all phases, or over the single phase as indicated by a superscript. The term ϵ denotes the volume fraction of a single phase, $\mathbf{v}$ denotes velocity, $\mathbf{K}_{eff}$ is the effective thermal conductivity tensor, Δh_{vap} is the heat of vaporization of the liquid phase, Q_l is the heat of desorption from the solid phase, and $\dot{m}_{lv}$, $\dot{m}_{sl}$, and $\dot{m}_{sv}$ denote the mass flux desorbing from the solid to the liquid, desorbing from the solid to the gas, and evaporating from the liquid, respectively. Gibson and Charmchi extended the model to include heat transfer to skin in contact with the fabric.[27] Further extensions can be made to include air and moisture mass transfer in multiple layers and air gaps, as well as radiative transfer. Their model thoroughly treats convection heat transfer in fabrics, provided the heat flux is not too high. It further treats heat transfer in three dimensions and can be extended to treat high heat fluxes, the absorption of radiation, and to model thermochemical processes in fabrics.

11.4.3 Multi-layer composite model

Mell and Lawson use a model similar to Torvi's to treat heat transfer in a firefighter's turnout coat, or a multi-layer composite consisting of a shell layer, including trim material, moisture barrier, and thermal liner.[3] Like Torvi, Mell and Lawson account for conduction and absorption of incident radiation in the material layers. In addition, Mell and Lawson advance a forward-reverse radiation model, defining incident fluxes on both sides of each layer of material to account for interlayer flux due to reflected radiation between fabric layers. Their model is an example of extending the Torvi model to treat a multi-layer fabric assembles. It could be extended to treat heat transfer in three dimensions, to model convective heat transfer and heat conveyed by moisture within the air gaps.

11.5   Modeling thermal degradation in fabrics

The combustion and thermal degradation of polymeric fabrics are complicated processes involving physical and chemical phenomena that are only partially understood. A number of different approaches for modeling this problem have been suggested in the literature. Some researchers suggest modeling thermal degradation of polymers as a Stefan problem, where the degradation of the solid is assumed to occur infinitely rapidly once a critical temperature reached.[28-31] Other researchers model solid-phase degradation using limited global in-depth reactions.[32-34] Kashiwagi reviews physical and chemical phenomena involved in polymer combustion and highlights the complexity of this process.[35] Staggs suggests a heat transfer model that incorporated a general single-step solid-phase reaction for thermal degradation of polymer material.[36] Staggs' model does not account for heat transport by gaseous products escaping from the solid. The critical-temperature approach has been utilized by some researchers to derive simplified models of thermal degradation. This approach is overly simplified, because it is assumed that the polymer can volatilize only at surfaces exposed to heat. The thermal degradation of Staggs' model utilizing the following kinetic rate law is:[31]

$$\frac{D\mu}{Dt} = -f(\mu, T),$$ 11.7

where μ is a scalar quantity representing the progress of the reaction (the ratio of the mass of the material element to its initial mass), t is the time, and f is a function determined by the rate of degradation.

Different forms for the f function can be used to model specific decomposition mechanisms characteristic of different types of polymers or other chemicals used in thermally protective clothing. For the most polymeric fabrics, the use of an nth order Arrhenius reaction provides sufficient agreement with experimental data; so that:

$$f = A\mu^n \exp\left(-T_A/T\right),$$ 11.8

where T_A is the activation temperature, A is the empirically derived pre-exponential factor, and n is the order of the degradation reaction.

Accounting for degradation in the solid material produces an additional term in the energy conservation equation. If m_0 is the initial mass of the material element, then its mass at time t is $\mu(t)m_0$. The rate of heat consumption for the vaporization of polymer material in this element during its degradation can be expressed as $\Delta H m_0(D\mu/Dt)$, where ΔH is the heat of vaporization. Recasting this relation for a unit volume results in the following term for the energy equation of the fabric layer:

$$q''' = \frac{\rho \Delta H}{\mu} \frac{D\mu}{Dt} = -\frac{\rho \Delta H}{\mu} A\mu^n \exp(-T_A/T).$$

11.9

Staggs shows that the nth order Arrhenius reaction is used to model thermal degradation, the predicted surface temperature increases slowly during the mass loss period. Surface temperature is not determined solely by the properties of the polymer material, such as the specific chemical reaction that describes its degradation, but by interaction between reaction kinetics and the rate of heat loss.[31] This finding explains why the critical-temperature ablation models of polymer degradation do not provide sufficiently accurate data to predict heat transfer through fabrics exposed to intense heat.

As previously noted, Staggs' model does not consider transport of gaseous products from the degrading polymer material. This model can be supplemented with a model that describes transport of these products through the fabric layer, modeled as a thermally degrading porous medium. One of the issues to be addressed is the change of permeability of the porous medium as a result of its thermal degradation. It is expected that degradation will produce a decrease of permeability in the fabric and thus worsening its mass transport characteristics. This will hinder not only transport of the gaseous products of polymer degradation, but also moisture transport through the degrading fabric. This, in turn, can lead to accumulation of heat moisture in the fabric because the moisture that results from sweating will not be removed. Moisture accumulation in fabrics has been proven to have a profound influence on heat transfer and thermal protective performance.

11.6   Skin heat transfer model and burn evaluation

Burns, the result of thermal attack on human skin, are some of the worst injuries that can happen to human beings. Burn injuries require a long time to heal and are sometimes difficult to treat clinically. Burn injuries, which are time and temperature dependent, have been classified as first, second, third, or fourth degree burns.[37] In addition to burn injury, systemic heat trauma due to the thermal stress and the inflammatory mediators occurring within the body are released to the circulatory system. Most of these traumas result from the altered condition of skin due to intense heat exposure. Traumatic effects include the shock of fluid loss, decrease in cardiac output, and injuries to the respiratory system. An increase in body metabolic rate can occur to compensate for the large losses from outer evaporation from injured areas, as well as complications related to nutritional defects and altered immune function.

11.6.1   Bioheat transfer models

Pennes proposed a transfer equation to describe heat transfer in human tissues:[38]

$$\rho c \frac{\partial T}{\partial t} = k\nabla^2 T - G(\rho c)_b (T - T_c) \qquad 11.10$$

Pennes' model assumes that skin tissue above an isothermal core is maintained at a constant body temperature. The resulting simplified bioheat equation is based on the following specific assumptions:[39,40] heat is linearly conducted within tissues; tissue thermal properties are constant in each layer but may vary from layer to layer; blood temperature is constant and equal to body core temperature; negligible blood effect between the large blood vessels (arteries and veins) and the tissue; the local blood flow rate is constant. In long duration, low intensity heat exposure, the rate of metabolic energy production is included in the above equation. In a case of high intensity exposures ($84\,kW/m^2$), metabolic energy production can be assumed to be negligible.

Several investigators have questioned the validity of the assumptions underlying Pennes' model. Wulff claims that the blood flow contribution to heat transfer in tissue must be modeled as a directional term of the form $(\rho c)_b\, \bar{\mu} \cdot \nabla T$, rather than the scalar perfusion term suggested by Pennes.[41] Klinger points out that Pennes' equation does not include heat transfer in the vicinity of large blood vessels.[42] Deficiencies in Pennes' model result from the fact that the thermal equilibrium process occurs, not in blood capillaries, as he assumes, but rather in pre- and post-capillary vessels. Nor does Pennes' model account for convective heat transfer due to the blood flow, or for heat exchange between the small and closely spaced vessels.[42]

Chen and Holmes' model[43] groups blood vessels into two categories: large vessels, each treated separately, and small vessels, treated as part of the continuum that includes the skin tissue. In their model, heat transfer between small blood vessels and tissue is separated into three modes. The first, perfusion mode, considers equilibration of blood and tissue temperature. The thermal contribution is described by a term, q_p, similar to the perfusion term in Pennes' equation:

$$q_p = \omega^*(\rho c)_b (T_a^* - T), \qquad 11.11$$

where ρ, c, and T are defined as in Pennes' equation, is the perfusion parameter that reflects blood flow within vessels in the control volume, and T_a^* represents the temperature of the blood within the largest vessel in the control volume. The second, convective mode, deals with blood vessels that are already thermally equilibrated. This model represents the part of heat transfer that occurs when the flowing blood convects heat against a tissue temperature gradient. For this mode of heat transfer, blood is assumed to be in thermal equilibrium with the tissue at a temperature T; the heat transfer contribution can be estimated as:

$$q_c = -(\rho c)_b \, \bar{\mu} \cdot \nabla T, \qquad\qquad 11.12$$

where $\bar{\mu}$ is the net volume flux vector permeating a unit area of the control surface. A third mode describes thermal conduction due to small temperature fluctuations that occur in the blood along the tissue temperature gradient:

$$q_{pc} = \nabla k_p \nabla T, \qquad\qquad 11.13$$

where k_p is a perfusion conductivity tensor that depends on local blood flow velocity within the vessel, the relative angle between the directions of the blood vessel, the local tissue temperature gradient and the number of vessels in the control volume.

The Chen and Holmes model accounts for all three modes of heat transfer between blood and tissue, so that

$$\rho c \frac{\partial T}{\partial t} = \nabla \cdot (k \nabla T) + (\rho c)_b \omega^* (T_a^* - T) - (\rho c)_b \bar{\mu} \cdot \nabla T$$
$$+ \nabla \cdot k_p \nabla T + q_m. \qquad\qquad 11.14$$

The Chen and Holmes model has been applied to different biothermal situations in Xu and in Xu et al.[44,45]

Based on anatomical observations in peripheral tissue, Weinbaum et al. and Jiji et al. conclude that the main contribution of local blood perfusion to heat transfer in tissue is associated with incomplete countercurrent heat exchange between pairs of arteries and veins, not with heat exchange at the capillary level.[46,47] They propose a model that consists of three coupled thermal energy equations to describe heat transfer involving arterial and venous blood and skin tissue. Their models are stated as:

$$(\rho c)_b \pi r_b^2 \bar{V} \cdot \frac{dT_a}{ds} = -q_a$$

$$(\rho c)_b \pi r_b^2 \bar{V} \cdot \frac{dT_v}{ds} = -q_v$$

$$\rho c \frac{\partial T}{\partial t} = \nabla \cdot (k \nabla T) + ng(\rho c)_b \cdot (T_a - T_v)$$
$$- n \pi r_b^2 (\rho c)_b \bar{V} \cdot \frac{d(T_a - T_v)}{ds} \qquad\qquad 11.15$$

where g is the volumetric rate of the bleed-off blood flow (the flow out of or into the blood vessel via the connecting capillaries), n is the vessel number density, r_b is the vessel radius, and $\bar{V}$ is the blood velocity within the vessel. Applications of this model to a variety of biothermal situations are presented in Dagan et al., and in Song et al.[48–50]

11.6.2   Skin burn evaluation

Henriques and Moritz claim that skin burn damage can be represented as a chemical rate process, so that a first-order Arrhenius rate equation can be used to estimate the rate of tissue damage as:[51]

$$\frac{d\Omega}{dt} = P \exp\left(-\frac{\Delta E}{RT}\right), \qquad\qquad 11.16$$

where Ω is a quantitative measure of burn damage at the basal layer or at any depth in the dermis, P is frequency factor, S^{-1}, E is the activation energy for skin, J/mol, R is the universal gas constant, 8.315 J/kmol.K, T is the absolute temperature at the basal layer or at any depth in the dermis, K, and t is the total time for which T is above 44 °C (317.15 K). Integration of this equation yields:

$$\Omega = \int_0^t p \exp\left(-\frac{\Delta E}{RT}\right) dt. \qquad\qquad 11.17$$

Integration is performed for a time when the temperature of the basal layer of the skin, T, exceeds or equals to 44 °C. Henriques found that if Ω is less than, or equal to, 0.5 no damage will occur at the basal layer. If Ω is between 0.5 and 1.0, first-degree burns will occur, whereas if $\Omega > 1.0$, second-degree burns will result. The damage criteria can be applied to any depth of skin provided the appropriate values of P and ΔE are used. Mathematically, a second-degree burn injury has been defined as an $\Omega > 1.0$ at the epidermis/dermis interface and a third-degree burn injury as an $\Omega > 1.0$ at the dermis/subcutaneous tissue interface. This method is used in conjunction with temperature-time data from a skin model to predict times to second- and third-degree burns. This method is valid for any heat flux conditions.

Buetter also uses the Henriques burn integral. However, much of his work involves determining the threshold of unbearable pain when non-penetrating infra-red radiation was used to heat the skin of human volunteers.[52] Stoll published extensively on determining the skin pain threshold temperatures, and a constant used in the Henriques burn integral and the thermal protection of fabrics.[53] Mehta and Wong used the Henriques burn integral to predict skin burns.[54] However, they conclude that the Henriques equation is valid only for superficial (epidermal) burns. They point out that the temperature used to calculate the pre-exponential factor and activation energy had not been accurately measured. They express doubt as to whether data from low intensity, long duration tests can be used in high intensity, short duration tests. Mehta and Wong model skin as a finite solid with different layers, each with different properties. They alter the upper time limit in the Henriques integral to include cooling time. Takata used a large number of anaesthetized pigs exposed to JP-4 liquid jet fuel fires to analyze the skin data.[55] Data observed by these researchers are summarized in Table 11.1.

Table 11.1 Values of P and ΔE used in burn integral calculations

Source	P (Hz) for epidermis at temperature		ΔE (J/kmole) for dermis at temperature	
	$\geq 50\,C$	$< 50\,C$	$\geq 50\,C$	$< 50\,C$
Reference 53	2.18×10^{124}	1.82×10^{51}	7.784×10^{8}	3.222×10^{8}
Reference 54	4.32×10^{64}	9.39×10^{104}	4.143×10^{8}	6.654×10^{8}
Reference 55	1.43×10^{72}	2.86×10^{69}	4.604×10^{8}	4.604×10^{8}

11.7   Manikin fire heat test model development

Song developed a numerical model to predict skin burn injury resulting from heat transfer through a protective garment worn by an instrumented manikin exposed to laboratory-controlled flash fires.[4] This model incorporates characteristics of the simulated flash fire generated in the chamber and the heat-induced changes in fabric thermo-physical properties. The model also accounts for clothing air layers between the garment and the manikin. The model is validated using an instrumented manikin fire test system. Results from the numerical model helped to develop a better understanding of the heat transfer process in protective garments exposed to intense flash fires, and to establish systematic methods for the engineering of materials and garments to produce optimum thermal protective performance.

The basic elements of manikin fire testing and burn evaluation processes are illustrated in Fig. 11.3. Lab flash fire is generated in the burn chamber by eight propane burning torches. The heat from the fire transfers through the garment and air layer between the protective garment and manikin body. Subsequently,

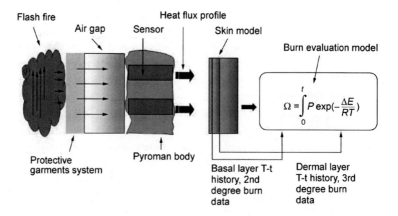

11.3 Heat transfer and burn evaluation process in the instrumented manikin system.

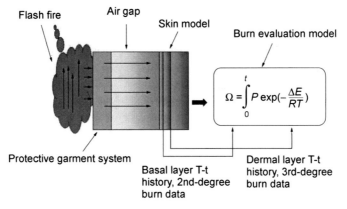

11.4 Elements of fabric air-layer and burn evaluation model.

the temperature changes registered by one hundred and twenty-two thermal sensors on the manikin body are translated into heat flux readings and predictions of skin burn injury are made using Henriques Burn Integral based on a multi-layer skin heat transfer model.

The model developed by this research considers the entire manikin burning process (Fig. 11.4). Heat transfer through the fabric and air layers is computed in conduction, radiation and convection modes. A heat transfer equation is applied in conjunction with a skin model to estimate the temperature profile in the basal and dermal layers of the skin. The schematic (Fig. 11.5) establishes how energy

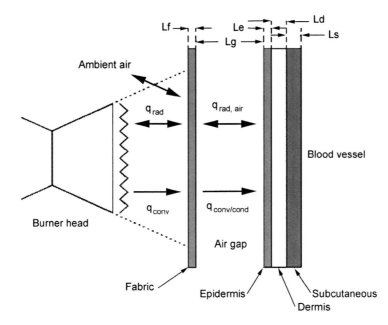

11.5 Schematic for a one-dimensional heat transfer model.

is transferred to the protective fabric in both the radiative and convective modes. Energy is conducted through the fabric simultaneously by conduction and radiation. At the skin surface, thermal energy is transferred through radiation from the protective garment and conduction/convection from the trapped air. The energy is then conducted through three layers of human skin to an isothermal boundary at the base of the subcutaneous layer.

The heat transfer deferential equation can be written as:

$$\rho_f(T)c_{p,f}(T)T\frac{\partial T}{\partial t} = \frac{\partial}{\partial x}\left(k_f(T)\frac{\partial T}{\partial x}\right) + \gamma \cdot q_{rad}e^{-\gamma x}, \qquad 11.18$$

where T is fabric temperature and ρ_f, $c_{p,f}$ and k_f are the temperature-dependent density, specific heat, and thermal conductivity of the fabric, respectively. The term $\gamma \cdot q_{rad}e^{-\gamma x}$ represents internal heat generated by thermal radiation transferred to the internal region of the fabric by the transmissibility of the fabric τ. γ is evaluated as:

$$\gamma = -\ln(\tau)/L_f, \qquad 11.19$$

where L_f is the thickness of the fabric. The boundary conditions at the external surface of the protective fabric ($x = 0$) are:

$$q_{rad} = \sigma\epsilon_g(T_g^4 - T_f^4) - \sigma\epsilon_f F_{fab-amb}(1 - \epsilon_g)(T_f^4 - T_{amb}^4)$$

and   11.20

$$q_{rad} = h_s(T_g - T_f) \qquad x = 0, t > 0$$

where σ is the Stefan-Boltzmann constant; ϵ_g and ϵ_f are the emissivities of the hot gases and the fabric, T_g is the gas temperature, T_f is the fabric surface temperature, h_s is the heat transfer coefficient, T_{amb} is the ambient air temperature, and $F_{fab-amb}$ is a view factor. The radiation boundary condition, however, is nonlinear and complicates the solution process.

To simplify this boundary condition, the total amount of energy transferred to the garment, both radiant and convected, was estimated using nude burn test results from the actual manikin system. The resulting boundary condition at the external surface of the garment is reduced to

$$(q_{conv} + q_{rad}) = h_{fl}(T_g - T_f), \qquad 11.21$$

where h_{fl} is the empirically estimated total heat transfer coefficient between the flame and the garment. Heat transfer by conduction/convection through the trapped air located between the internal surface of the fabric and human skin was modeled as a resistance to heat flow between the two surfaces. A value for the heat transfer coefficient between the fabric and skin was modeled as a function of the size of the air gap and the temperature of the trapped air as:

$$h_{gap}(L_{airgap}, T) = Nu\frac{k_{air}(T)}{L_{airgap}}, \qquad 11.22$$

where Nu is the Nusselt number, $k_{air}(T)$ is the thermal conductivity of the air and L_{airgap} is the size of the air gap. L_{airgap} is taken from garment air gap measurements described later in the chapter. The boundary at the interface between the fabric and skin can be expressed as:

$$q_{conv} = h_{gap}(T_f - T_s) \qquad x = L_f, t > 0 \qquad 11.23$$

Radiation heat transfer from the fabric to human skin was also considered. The radiation couple is modeled simply, for the one-dimensional case, as radiation exchange between two infinitely tall parallel plates. Taking the emissivities of the fabric and human skin to be the average values of 0.9 and 0.94 respectively,[24,1] the radiation boundary condition between the fabric and skin can be expressed as:

$$q_{rad} = 0.85\sigma(T_{fabric}^4(t) - T_{skin}^4(t)) \qquad x = L_f, t > 0 \qquad 11.24$$

This research adopted Pennes' approach[38] to model heat conduction in human skin. The model consists of three tissue layers: the epidermis, dermis, and subcutaneous layers which account for cooling blood perfusion in the subcutaneous layer. Therefore,

$$\rho_s c_{p_s} \frac{\partial T}{\partial t} = \nabla \cdot (k_s \nabla T) + (\rho c_p)_b \omega_b (T_a - T) \qquad x > L_f, t > 0 \qquad 11.25$$

In eqn 11.25, ρ_s, $c_{p,s}$ and k_s are the density, specific heat and thermal conductivity of human tissue and ρ_b and $c_{p,b}$ are the density and the specific heat of blood. ω_b is the rate of blood perfusion and is taken to be $0.00125 \, \text{m}^3/\text{s}/\text{m}^3$. The boundary condition at the base of the subcutaneous layer is set at a constant basal temperature of 37 °C. As an initial condition, linear distribution of skin temperature is assumed between a surface temperature of 34 °C and the basal temperature of 37 °C.

To estimate heat transfer from the fire to the manikin, it is necessary to determine heat transfer coefficients at each of the 122 thermal sensor locations. To estimate the heat transfer coefficient between the flames of the simulated flash fire and the clothed manikin, a series of experiments were conducted using the manikin test system. These experiments were conducted on a nude manikin to facilitate instrumentation. For these experiments, a special sensor was constructed that simultaneously measured surface temperature, surface heat flux, and gas temperature at a location 1 cm above the surface of the manikin body. The sensor was placed in each of the 122 manikin sensor locations and exposed to a nominal $2.00 \, \text{cal/cm}^2\text{sec}$ ($84 \, \text{kW/m}^2$) simulated flash fire. Using this experimental data for total heat flux, the surface temperature of the manikin, and gas temperature, the following relationship[56] can be used to estimate the total heat transfer coefficient at each thermal sensor location:

$$\hat{h}_M = \frac{\hat{q}_M}{T_f(T) - 0.5(\hat{T}_{OM} + \hat{T}_{OM-1})}, \qquad 11.26$$

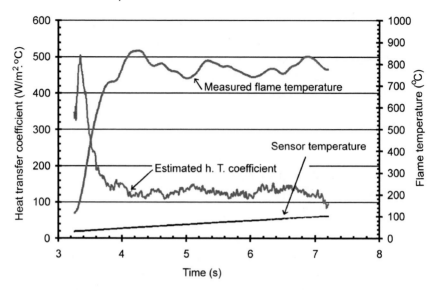

11.6 Estimated heat transfer coefficient using measured flame temperature and Pyrocal® sensor.

where $\hat{h}_M$ is estimated heat transfer coefficient, $\hat{q}_M$ is calculated heat flux, T_f is flame temperature and $\hat{T}_{OM}$ is estimated surface temperature at time t_M. An example of typical results, for a specific sensor location, is shown in Fig. 11.6. These experiments confirm that the flash fire generated by the torches produces a highly dynamic fire environment. Measured heat transfer coefficients ranged from 80–140 W/m² °C for a four-second fire exposure.

Fabric thermal conductivity and heat capacity are the main factors controlling heat transfer in fabrics. A parameter estimation approach was used to quantify changes in fabric properties that occur as a result of intense heat exposure.[57] Experiments were conducted to estimate heat induced changes in the thermal conductivity and volumetric heat capacity of Kevlar®/PBI and Nomex®IIIA fabrics in short duration (four-second) intense fire exposures. The results show that both thermal conductivity and volumetric heat capacity decrease during the exposure.[58]

Three-dimensional body scanning technology was used to measure air gap sizes and distribution between the manikin body and different sized protective garments for both pre- and post-exposure conditions. The process used data taken from a dressed manikin superimposed on data taken from a nude manikin. Figure 11.7 presents quantitative results for the various sizes of Kevlar®/PBI protective coveralls measured prior to exposure. The data show that air layers are not similarly sized or evenly distributed over the manikin body. In some locations, specifically the shoulder, knee and upper back areas, the protective garment is close to the body; while in other locations, such as the waist and thigh areas, a larger insulating air space is present. Areas of the legs exhibit the largest

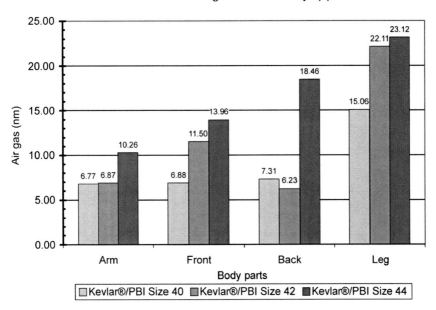

11.7 Different sized garments and their average air gap distribution in Pyroman® body.

air layers while the least air space between the manikin and protective garment is found in the arm and back. The size of the protective garment affects the distribution of the air layers over the surface of the manikin. Garment drapability and stiffness also affect clothing air layers, since more flexible materials tend to conform to body contours. Examining garments which have been exposed to simulated flash fire shows thermally induced fabric shrinkage plays a significant role in determining air gap dimensions. Nomex®IIIA coveralls (size 42, 'deluxe' style, 203 g/m^2) shrink significantly when subjected to a four-second, 2.00 cal/cm^2 sec exposure, thereby reducing the insulating air layer by 50% on average, and as much as 90% in areas around the legs of the manikin as shown in Fig. 11.8.

The numerical model calculates heat transfer at each of one hundred and twenty-two heat sensor locations distributed over the instrumented Pyroman® body. A finite difference method was used to solve the differential equations that describe heat transfer through the fabric, air gap, and skin layers. Due to the nonlinear terms of absorption of incident radiation, the Gauss-Seidel point-by-point iterative scheme was used to solve these equations. To avoid divergence associated with iterative methods, an underrelaxation process was utilized. In addition, the Crank-Nicholson implicit scheme was used to solve the resulting ordinary differential equations in time. The program was written using Microsoft FORTRAN PowerStation. Details of these numerical calculation procedures can be found in Song.[58]

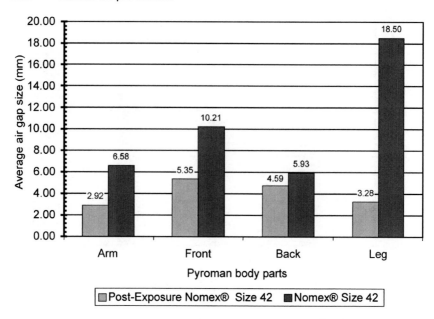

11.8 Comparisons of air gaps for NomexIIIA (203 g/m^2) coverall before and after a four-second exposure.

To validate the numerical model more than 40 experimental tests on the Pyroman® manikin were performed for Kevlar®/PBI and Nomex®IIIA coveralls covering exposure times from three to four seconds. Results from the numerical model successfully correlate with experimental Pyroman® results for one-layer Kevlar®/PBI and Nomex®IIIA protective coveralls exposed for short intervals. Figure 11.9 shows an example of comparison of Kevlar®/PBI coverall protective performance predicted by numerical model and manikin tests at three second exposure. Additionally, the manikin simulation model reproduced surface heat flux and skin damage results very well for the single-layer protective garments considered. Garment shrinkage during exposure can be taken into account by considering the changes of air gap size between the garment and manikin body. Based on these results, it can be concluded that the numerical model is capable of predicting heat transfer through one layer of protective clothing and the resulting human skin burn damage in flash fire conditions with good accuracy. The simulation model does not include many variables considered essential to fully simulate actual use conditions, such as multiple-layer garments and moisture contained in the fabric. It should be noted that as an extensive amount of garment-specific experimental data was required in order to execute the program, the model can prove to be a powerful tool for engineering more protective, efficient garments.

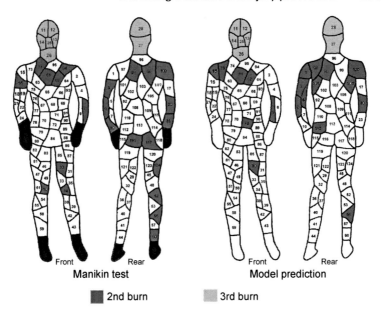

Front   Rear   Front   Rear
Manikin test   Model prediction

■ 2nd burn   ▧ 3rd burn

11.9 A comparison of burn distribution from manikin test and model prediction for three-second exposures.

11.8   Model predictions

11.8.1   Effect of lab-simulated flash fire on protective performance evaluation

In the manikin test, an average heat flux of $2.0\,cal/cm^2\,sec$ measured from 122 sensors of the manikin is used to simulate flash fire conditions, and the standard deviation of these heat flux distributions is between 0.25 and 0.5 and shows a bell shaped distribution. In order to further access this distribution normality, a normal scores plot is performed which demonstrated that the heat flux distribution with an average of $2.0\,cal/cm^2\,sec$ during a four-second exposure exhibits an approximate normal distribution. Variations in heat fluxes are expected due to the three-dimensional shape of the manikin surface, and because of the complex and dynamic nature of the flame column surrounding the manikin. Therefore, depending on the location on the manikin body (e.g. arms, legs, shoulders), sensors located in different positions will have different heat flux values with respect to the flash fire. Figure 11.10 shows the bell-shaped distributions with different standard deviations. The larger standard deviation indicates more extreme high and low flux values occurring during exposure. In order to examine the effects of different fire distributions, a series of heat flux data with different distributions were generated statistically to simulate the different fire distributions. These data, as a model fire boundary input, were

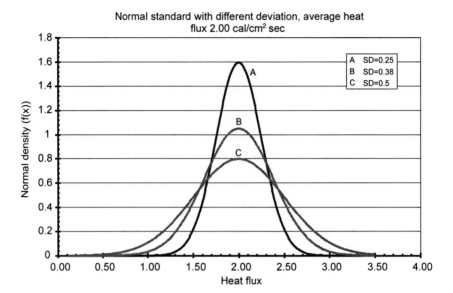

11.10 Heat flux distribution with different standard deviation.

predicted with single-layer garments using the numerical model. The model is based on single-layer Kevlar®/PBI coverall.

The standard deviation of 122 sensors' heat flux of Pyroman® predicted by the numerical model does not show a significant influence on third-degree burn prediction. The second-degree burn prediction, however, decreases as the standard deviation increases as illustrated in Fig. 11.11. This is expected as the standard deviation increases; more low heat flux values over the manikin body are generated. These lowered heat flux values could help reduce the burn predictions. For a three-second exposure, no significant difference was predicted on the results of second-degree burns.

11.8.2   Effects of variation in skin heat transfer model

The variations in skin model were analyzed using the manikin numerical model based on a single layer coverall. One of the functions of human skin is to help regulate the body's core temperature. The core temperature of the body must be maintained within a small range around 37 °C in order to keep biochemical reactions proceeding at required rates. Under normal ambient conditions, the skin surface temperature is about 32.5–34 °C. Some skin heat transfer models proposed constant initial skin temperatures of 32.5 °C, 34 °C, or 37 °C. Some other models suggest linear initial or higher-order initial temperature distribution. For a given incident heat flux history transmitted through the garment under flash fire conditions, different initial temperature distribution in the skin model could affect the temperature rising rate at specific depth in the

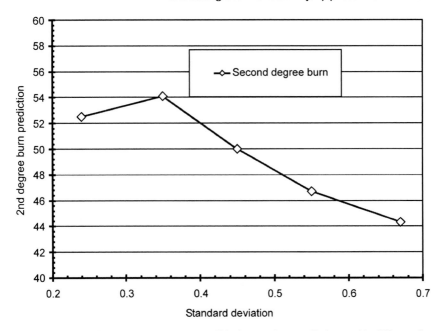

11.11 Single layer coverall manikin fire testing predictions with different fire distribution (overall average heat flux is 2.0 cal/cm² sec).

skin. In the three-layer skin model, the predictions are very sensitive to initial temperature distributions. The linear and quadratic temperature distributions show almost 10% difference on burn predictions, while the constant temperature (37 °C) distribution compared to the linear and quadratic indicates a large influence on second-degree body burn predictions.[25] The changes of these temperature distributions show little influence on third-degree burns. This result is expected because the temperature difference of these three distributions is very small at the interface of dermis and subcutaneous. In the one-layer skin model, however, initial temperature distribution shows little effect on burn predictions. The model simulation results demonstrate that the burn predictions on initial temperature distribution in the one-layer skin model are not as sensitive as in the three-layer skin model. This result is attributable to the relatively large thermal conductivity used in the one-layer skin model.

In summary, the skin model that assumes blood perfusion has minimal effect on predicted body burns in these exposures. The model assuming blood perfusion slightly increases the time to second- or third-degree burn; especially when the time to get second- or third-degree burns exceeds five seconds. Different temperature distributions in the skin models demonstrate a large effect on burn predictions in the three-layer skin model. The effect is pronounced in the constant temperature distribution compared to the linear and quadratic. This study suggests that the use of different skin models and their initial temperature distributions can greatly affect burn predictions of garment testing in an instrumented manikin. Therefore, a precise skin model selection and its standardization would be beneficial for manikin testing in thermal protective performance evaluation.

11.8.3 The role of air gaps in protective performance

Air entrapped between a protective garment and the human body is a major factor determining the garment's thermal protective insulation. Due to the geometry of the human body the size and the distribution of air gap layers are not evenly distributed. Thermally induced garment shrinkage during intense heat exposure can significantly reduce clothing air layers and therefore increase heat transfer to the skin. Under flash fire conditions, the size of air layer between the heated garment and the skin/manikin body affects the energy transfer in the air layer. The mode of heat transfer between a heated fabric and the skin can be radiation, conduction or convection. These modes of heat transfer in the air layers depend on air gap layer size and fire boundary conditions. The stagnant air can be a good insulator and if air is stagnant its insulating value will increase as the width of the air space increases. However, if the air space becomes wide enough, natural convection may occur, which will increase the heat transfer across the air space, and may decrease its value as an insulator. The gap at which convection is initiated is termed the critical or optimal air gap. In order to investigate the effect of the air gap size on burn predictions and forecast the dimensions of skin-clothing air gap for optimum thermal protection, the numerical model of manikin fire testing is used. The air gap size and distribution of protective coveralls dressed in a thermal manikin were determined using a 3D body-scanning technique. Based on the fire characteristics of lab simulation in the manikin chamber, the critical air gap, which could provide the maximum insulation under flash fire conditions, was obtained using the model analysis.

Under lab-simulated flash fire conditions, the model predicts the optimum air gap size is around 7–8 mm for a one-layer protective garment.[59] Figure 11.12 shows that the amount of total energy transferred to the skin (heat flux profile on skin surface) decreased significantly as the air gap size increased from 0.01 mm to 7.0 mm. Beyond this range, however, increasing the air gap size shows no significant effect on the amount of total energy transferred to the skin. This is due to the fact that convection was initiated. When the air gap size is beyond the range of 7–8 mm, the model predicted that convection currents occur, which could increase heat transfer across the air space. Therefore increasing the air gap size from the 7–8 mm range will not help to provide increased insulation value. As a matter of fact, the amount of energy transferred in different modes may change with different air spaces. At smaller air gaps (smaller than 3 mm), the predominant heat transfer mode is conduction. At larger air gaps, (greater than 7.5 mm) convection initiates, and the predominant form of energy transfer is

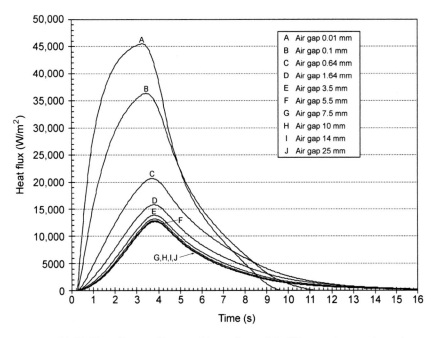

11.12 Heat flux profiles on skin surface with different air gaps for a three-second exposure.

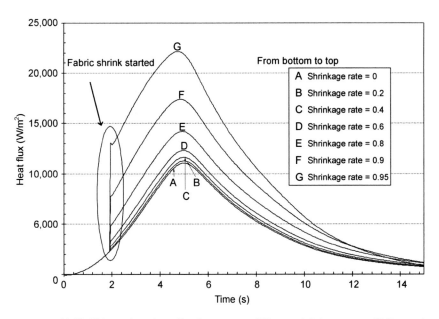

11.13 Skin surface heat flux increase at different shrinkage rates (7.5 mm air gap).

radiation heat transfer. Additionally, the model predictions demonstrated that thermal shrinkage as the result of exposure significantly increased heat transfer to the skin (Fig. 11.13).

Garment shrinkage during exposure could reduce the air gap layers, and therefore heat transfer between the heated fabric and skin could be enhanced. Figure 11.14 examines the shrinkage effect on thermal protection performance using model predictions. The model computes two predictions, with no shrinkage and with some shrinkage of Nomex®IIIA coveralls (size 42, 203 g/m²) for three- and four-second flash fire exposure (2.0 cal/cm² sec). The results demonstrated that shrinkage during exposure could increase burn injuries greatly for both three- and four-second exposure.

In summary, the existence of air gaps in a protective garment plays a vital role in providing thermal insulation in flash fire conditions. Manikin fire tests demonstrated that burn predictions occur in smaller air gap areas (locations). Under lab-simulated flash fire conditions, the model predicts the optimum air gap size to be around 7–8 mm for a one-layer protective garment. The insulating value will increase as the air gap size increases within optimum air gap size. Beyond this optimum air gap size, however, increasing the air gap will not help to provide an increased insulation value. In addition, garment shrinkage during exposure could tremendously lower the performance of thermal protective clothing.

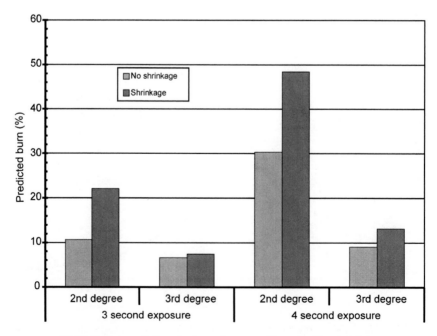

11.14 Model predictions for burn damage with and without shrinkage.

11.9   References

1. Torvi, D. A., Ph.D. Thesis, University of Alberta, 1997.
2. Gibson, P. 'Governing Equations for Multiphase Heat and Mass Transfer in Hygroscopic Porous Media with Applications to Clothing Materials,' Technical Report Natick/TR-95/004, 1994.
3. Mell, W. E. and Lawson, J. R., 'A Heat Transfer Model for Fire Fighter's Protective Clothing,' National Institute of Standards and Technology, NISTIR 6299, January 1999.
4. Song, G. and Barker, R., 'Numerical Analysis of Thermal Protective Garments under Simulated Flash Fire Exposures,' *Proceeding of 83rd World Conference: Quality Textiles for Quality Life*, Shanghai, China, May 23–27, 2004.
5. Abbott, N. J. and Schulman, S., 'Protection from the Fire: Nonflammable Fabrics and Coating,' *Journal of Coated Fabrics*, Vol. 6, July 1976, pp. 48–64.
6. Utech, H. P., 'High Temperatures vs. Fire Equipment,' *International Fire Chief*, Vol. 39, 1973, pp. 26–27.
7. Song, G. and Barker, R., 'Analyzing Thermal Store Energy and Clothing Thermal Protective Performance' *Proceeding of 4th International Conference on Safety & Protective Fabrics,* David L. Lawrence Convention Center, Pittsburgh, PA USA, Oct. 26–27, 2004.
8. Dale, J. D., Crown, E. M., 'Fundamental Issues in Bench Scale Testing of Fabrics for Thermal Protection,' *Proceeding of 2nd European Conference on Protective Clothing (ECPC),* Montreux, Switzerland, May 21–24, 2003.
9. Vettori, R. L., Twilley, W. H. and Stroup, D. W., 'Measurement Techniques for Low Heat Flux Exposures to Fire Fighters Protective Clothing,' National Institute of Standards and Technology, NISTIR 6750, June 2001.
10. Stoll, A. M. and Chianta, M. A., 'Method and Rating System for Evaluation of Thermal Protection,' *Aerospace Medication*, Vol. 40, 1969, pp. 1232–1238.
11. Krasney, J. F., Rockeet, J. A. and Huang, D., 'Protecting Fire Fighters Exposed in Room Fires: Comparison of Results of Bench Scale Tests for Thermal Protection and Conditions During Room Flashover,' *Fire Technology*, Vol. 24, February 1998, pp. 5–19.
12. Abbott, N. J. and Schulman, S., 'Protection from the Fire: Nonflammable Fabrics and Coating,' *Journal of Coated Fabrics*, Vol. 6, July 1976, pp. 48–64.
13. Schoppee, M. M., Skelton, J., Abbott, N. J., and Donovan, J. G., 'The Transient thermomechanical response of Protective Fabrics to Radiant Heat,' Technical Report, AFML TR-77-72, Air Force Materials Laboratory (1976).
14. Brian David Gagnon, 'Evaluation of New Test Methods for Fire Fighting Clothing,' A thesis of Master of Science of Worcester Polytechnic Institute, 2000.
15. Afgan, N. H. and Reer, J. M., *Heat Transfer in Flames*, Wiley and Sons, New York, NY, 1974.
16. Hilado, C. J., *Flammability Handbook for Plastics*, Technomic, Westport, RI, 1982.
17. Remy, D. E., Sousa, J. A., Caldarella, G. J. and Levasseur, L. A., 'Thermal Degradation Products of THPOH-NH₃ Treated Cottons,' Technical Report, Natick/TR-80/003, CEMEL-203 1979.
18. Lewis, B. and von Elbe, G., *Combustion, Flames, and Explosions of Gases*, 3rd edn, Academic Press, Inc., Orlando, 1987.
19. Holcombe, B. V. and Hoschke, B. N., 'Do Test Methods Yield Meaningful Performance Specifications?,' *Performance of Protective Clothing: First Volume*

ASTM STP 900, R. L. Barker and G. C. Coletta, eds, American Society for Testing and Materials, West Conshohocken, PA, 1986, pp. 327–339.

20. Krasny, J., Rockett, J. A. and Huang, D., 'Protecting Fire Fighters Exposed in Room Fires: Comparison of Results of Bench Scale Test for Thermal Protection and Conditions During Room Flashover,' *Fire Technology*, Vol. 24, 1998, pp. 5–19.

21. Drysdale, D., *An Introduction to Fire Dynamics*, John Wiley and Sons, Chichester, 1985.

22. Stoll, A. M., Chianta, M. A. and Munroe, L. R., 'Flame Contact Studies,' *Transactions of the ASME, Journal of Heat Transfer*, Vol. 86, 1964, pp. 449–456.

23. Lewis, B. and von Elbe, G., *Combustion, Flames, and Explosions of Gases*, 3rd edn, Academic Press, Inc., Orlando, 1987.

24. Shalev, I. 'Transient Thermophysical properties of Thermally Degrading Fabrics and Their Effect on Thermal Protection,' Ph.D. thesis, North Carolina State University, Raleigh, NC, 1984.

25. Song, G. and Barker, R., Effects of Simulated Flash Fire and Variations in Skin Model on Manikin Fire Test, *Journal of ASTM International*, Vol. 1, No. 7, July/August, 2004.

26. Whitaker, S., 'A Theory of Drying in Porous Media,' *Advances in Heat Transfer 13*, Academic Press, New York, 1977.

27. Gibson, P. W. and Charmchi, M., 'Integration of a Human Thermal Physiology Control model with a Numerical Model for Coupled Heat and Mass Transfer through Hygroscopic Porous Textiles,' presented at the 1996 ASME International Mechanical Engineering Congress & Exhibition, Atlanta, GA, Nov. 17–22, 1996.

28. Ricci, R., 'Traveling Wave Solutions of the Stefan and the Ablation Problems,' *SIAM J. Math. Anal.*, Vol. 21, pp. 1386–1393, 1990.

29. Whiting, P., Dowden, J. M., Kapadia, P. D. and Davis, M. P., 'A One-Dimensional Mathematical Model of Laser Induced Thermal Ablation of Biological Tissue,' *Lasers in Med. Sci.*, Vol. 7, pp. 357–368, 1992.

30. Delichatsios, M. A. and Chen, Y., 'Asymptotic, Approximate and Numerical Solutions for the Heatup and Pyrolisis of Materials Including Reradiation Losses,' *Comb. and Flame*, Vol. 92, pp. 292–307, 1993.

31. Staggs, J. E. J., 'A Discussion of Modeling Idealized Ablative Materials with Particular Reference to Fire Testing,' *Fire Safety Journal*, Vol. 28, pp. 47–66, 1997a.

32. Vovelle, C., Delfau, J. and Reuillon, M., 'Experimental and Numerical Study of Thermal Degradation of PMMA,' *Comb. Sci. and Tech.*, Vol. 53, pp. 187–201, 1987.

33. Wichman, I. S. and Atreya, A., 'A Simplified Model for the Pyrolisis of Charring Materials,' *Comb. and Flame*, Vol. 68, pp. 231–247, 1987.

34. Di Blasi, C. and Wichman, I. S., 'Effects of Solid-Phase Properties on Flames Spreading over Composite Materials,' *Comb. and Flame*, Vol. 102, pp. 229–240, 1995.

35. Kashiwagi, T., 'Polymer Combustion and Flammability – Role of the Condensed Phase,' *Proc. 25th Int. Symp. on Combustion*, The Combustion Institute, pp. 1423–1431, 1994.

36. Staggs, J. E. J., 'A Theoretical Investigation into Modeling Thermal Degradation of Solids Incorporating Finite-Rate Kinetics,' *Combust. Sci. and Tech.*, Vol. 123, pp. 261–285, 1997b.

37. Anthony, C. P. and Thibodeau, G. A., *Textbook of Anatomy and Physiology*, 10th

edn, The C. V. Mosby Co., St. Louis, 1979, pp. 70–73, 290–294, 530–533.

38. Pennes, H. H., 'Analysis of Tissue and Arterial Blood Temperatures in Resting Human Forearm,' *Journal of Applied Physiology*, Vol. 1, 1948, pp. 93–122.

39. Mehta, A. K. and Wong, F., 'Measurement of Flammability and Burn Potential of Fabrics,' Full Report from Fuels Research Laboratory, Massachusetts Institute of Technology, Cambridge, Massachusetts, 1973.

40. Hodson, D. A., Eason, G. and Barbennel, J. C., 'Modeling Transient Heat Transfer Through the Skin and Superficial Tissues – I: Surface Insulation,' *Transactions of the ASME, Journal of Biomechanical Engineering*, Vol. 108, 1986, pp. 183–188.

41. Wulff, W., 'The Energy Conservation Equation for Living Tissue,' *IEEE Transactions of Biomedical Engineering*, Vol. BME-21, pp. 494–495, 1974.

42. Klinger, H. G., 'Heat Transfer in Perfused Biological Tissue – I: General Theory,' *Bulletin of Mathematical Biology*, Vol. 36, pp. 403–415, 1974.

43. Chen, M. M. and Holmes, K. R., 'Microvascular Contributions in Tissue Heat Transfer,' *Annals of the New York Academy of Sciences*, Vol. 335, pp. 137–150, 1980.

44. Xu, L. X., 'The Evaluation of Heat Transfer Equations in the Pig Renal Cortex,' Ph.D. Dissertation, Univ. of Illinois at Urbana-Champaign, August, 1991.

45. Xu, L. X., Chen, M. M., Holmes, K. R. and Arkin, H., 'The Evaluation of the Pennes, the Chen-Holmes, the Weinbaum-Jiji Bioheat Transfer Models in the Pig Cortex,' *ASME WAM, HTD*, Vol. 189, pp. 15–21, 1991.

46. Weinbaum, S., Jiji, L. M. and Lemos, D. E., 'Theory and Experiment for the Effect of Vascular Temperature on Surface Tissue Heat Transfer – Part I: Anatomical Foundation and Model Conceptualization,' *ASME Journal of Biomedical Engineering*, Vol. 106, pp. 321–330, 1984.

47. Jiji, L. M., Weinbaum, S. and Lemos, D. E., 'Theory and Experiment for the Effect of Vascular Temperature on Surface Tissue Heat Transfer – Part II: Model Formulation and Solution,' *ASME Journal of Biomedical Engineering*, Vol. 106, pp. 331–341, 1984.

48. Dagan, Z., Weinbaum, S. and Jiji, L. M., 'Parametric Study of the Three-Layer Microcirculatory Model for Surface Tissue Energy Exchange,' *ASME Journal of Biomedical Engineering*, Vol. 108, pp. 89–96, 1986.

49. Song, W. J., Weinbaum, S. and Jiji, L. M., 'A Theoretical Model for Peripheral Heat Transfer Using the Bioheat Equation of Weinbaum and Jiji,' *ASME Journal of Biomedical Engineering*, Vol. 109, pp. 72–78, 1987.

50. Song, W. J., Weinbaum, S. and Jiji, L. M., 'A Combined Macro and Microvascular Model for Whole Limb Heat Transfer,' *ASME Journal of Biomedical Engineering*, Vol. 110, pp. 259–267, 1988.

51. Henriques, F. C., Jr., 'Studies of Thermal Injuries V. The Predictability and the Significance of Thermally Induced Rate Processes Leading to Irreversible Epidermal Injury,' *Archives of Pathology*, Vol. 43, 1947, pp. 489–502.

52. Buetter, K., 'Effects of Extreme Heat and Cold on Human Skin II. Surface Temperature, Pain and Heat Conductivity in Experiments with Radiant Heat,' *Journal of Applied Physiology*, Vol. 3, 1951, pp. 703–713.

53. Stoll, A. M. and Greene, L. C., 'Relationship Between Pain and Tissue Damage Due to Thermal Radiation,' *Journal of Applied Physiology*, Vol. 14, 1959, pp. 373–382.

54. Mehta, A. K. and Wong, F., 'Measurement of Flammability and Burn Potential of Fabrics,' full report from Fuels Research Laboratory, Massachusetts Institute of

Technology, Cambridge, Massachusetts, 1973.

55. Takata, A. N., Rouse, J. and Stanley, T., Thermal Analysis Program, I.I.T. Research Institute Report, IITRI-J6286, Chicago, 1973.

56. Beck, J. V., Blackwell, B. and St. Clair, C. R. Jr., *Inverse Heat Conduction: Ill-Posed Problems*, John Wiley & Sons, Inc., 1985.

57. Beck, J. V. and Arnold, K., *Parameter Estimation in Engineering and Science*, Wiley, 1977.

58. Song, G., Modeling Thermal Protection Outfits for Fire Exposures, Doctoral Dissertation, North Carolina State University, 2002.

59. Song, G. and Barker, R., 'Analyzing the Role of Air Layers and Shrinkage in Thermal Protective Garments,' *Proceeding of Autex 2004 Conference*, Roubaix, France, June 22–24, 2004.

Part II

General protection requirements and applications

Civilian protection and protection of industrial workers from chemicals

J O S T U L L , International Personal Protection Inc., USA

12.1   Introduction

The evaluation of chemical protective clothing (CPC) designs, design features, performance, and applications requires an understanding of the types of chemical protective clothing available for protection. The types of chemical protective clothing available in the marketplace and thus the choices available to the end user have changed dramatically over the past two decades and continue to increase in diversity. Chemical protective clothing exists in a variety of designs, materials, and methods of construction, each having advantages and disadvantages for specific protection applications. End users should have an understanding of the different types of chemical protective clothing and their features in order to make appropriate selections. It is important to realize that chemical protective clothing that appears to be similarly designed may offer significantly different levels of performance. Thus chemical protective clothing performance must be carefully scrutinized in addition to design and features. Furthermore, CPC must be properly sized to provide adequate protection. Improperly sized or ill-fitting chemical protective clothing may reduce or eliminate protective qualities of CPC.

Chemical protective clothing can be selected properly only when performance data indicate that resistance to chemicals lasts for the duration of anticipated, worst-case exposures. Relating chemical performance directly with possible exposures mandates a need for rigorously tested protective clothing. This clothing must also demonstrate acceptable integrity for overall protection and provide sufficient strength, durability, and physical hazard resistance. In addition, chemical protective clothing must be functional allowing the wearer to safely perform the required tasks at an acceptable level of comfort.

12.2   Classification of chemical protective clothing

Chemical protective clothing may be classified by its design, performance, and intended service life. These three characteristics of chemical protective clothing will permit the end user to understand the type of CPC item being considered or used, as well as indicate its potential limitations.[1] The following sections describe how CPC can be 'type' classified in these three ways.

12.2.1   Classification by design

Classification of CPC by its design usually reflects how the item is configured or the part of the body area or body systems that it protects. For example, a hood by design provides protection to the wearer's head. While there are a variety of different forms of clothing and equipment that are used for protection from industrial chemicals, the discussion in this chapter is limited to various types of textile-based full body or partial body garments. Gloves, footwear, and face/eye protection are not covered, though many of the same principles apply to these items. Table 12.1 shows the different CPC designs associated with different areas of body protection.

Classification of chemical protective clothing by design may also provide an indication of specific design features that differentiate CPC items of the same type. For example, totally encapsulating chemical protective suits are configured with significant design differences when compared to splash suits. Some designs of chemical protective clothing may offer varying protection against hazards in different parts of the CPC item. For example, the use of coated textile materials in the front portion of an apron may offer liquid chemical protection to wearer's

Table 12.1 Chemical protective clothing design types by body area

Body area(s)	Type
Entire body	Totally encapsulating suit
	2-piece suit (hooded jacket with visor with pants or overalls)
Torso, head, arms, and legs (excluding hands, feet, and face)	Hooded coveralls
	2-piece 'splash' suit (hooded jacket and pants or overalls)
Torso, arms and legs	Coveralls
	2-piece 'splash' suit (jacket and pants or overalls)
	Smock
Top torso and arms	Coat or jacket
	Lab coat
Bottom torso and legs	Pants
Torso (front) and arms	Sleeved apron
Torso (front)	Apron
Head and face	Hood with visor
Head	Hood
Foot	Booties
	Boot or shoe covers

front torso, but non-coated textile portions of the same clothing may offer no protection from chemicals. Therefore, it is important to realize that even apparent CPC coverage of a specific body area, in and of itself, does not guarantee protection of that body area.

A further means of distinguishing chemical protective clothing by design is to indicate the materials used in the CPC construction. Materials will possess different characteristics that impact both the performance and wearability of the CPC item. For example, rubber coated textiles behave much differently than plastic laminates. Owing to their elasticity, rubber materials are more likely to offer form fitting designs but may weigh more than comparable plastic laminates, which tend to offer better chemical resistance. Since material choices vary with the type of item, a more detailed discussion is provided in the respective section on the availability and characteristics of respective materials for different types of chemical protective clothing.

12.2.2   Classification by performance

Classification of chemical protective clothing by performance indicates the actual level of performance to be provided by the item of CPC. This may include a general area of performance or a more specific area of performance. For example, while two items of CPC might be considered to be chemical protective clothing, one item may provide an effective barrier to liquids but not to vapors, while the other item provides an effective barrier to both liquids and vapors. Classification of CPC by performance is best demonstrated by actual testing or evaluations of the chemical protective clothing with a standard test that relates to the type of desired protection (see sections 12.4 through 12.6). These tests can then be used as demonstrations of protection against the anticipated hazards and often become the basis of claims by the manufacturer for their products. However, intended or manufacturer-claimed performance does not always match actual performance. Furthermore, performance claims should be uniformly applied to all parts of the CPC item, i.e., the seam should offer the same performance as the material, otherwise the performance classification should be limited by the weakest element of the CPC design. The specific classification of chemical protective clothing will be related to the types of protection offered against chemical hazards. Table 12.2 describes a hierarchy of chemical barrier performance.

In general, chemical protective clothing that protects against gases and vapors, will also protect against liquids and particulates, and CPC effective against liquids will also prevent penetration of particulates. However, there are exceptions. For example, some chemical protective clothing based on adsorptive materials (such as those using air-purifying respirator cartridges) may prevent penetration of gases and chemical vapors, but not when exposed to liquids if the CPC is splashed by or immersed in liquid chemical. Other performance features

Table 12.2 Hierarchy of chemical barrier performance

Type of chemical barrier	Protection offered*
Permeation-resistant	Prevents or limits any contact with chemicals in the form of gases/vapors, liquids, or particulates (solid)
Vapor penetration-resistant	Prevents wearer contact with atmospheric vapors or gases
Liquid penetration-resistant**	Prevents wearer contact with liquids
Particulate penetration-resistant	Prevents wearer contact with particles

* Protection offered is specific to chemical or chemical characteristics.
** Liquid penetration resistance is sometimes subdivided into two categories where chemical protective clothing may be resistant to liquid penetration under pressure as might be associated with spraying liquid (e.g., from a burst pipe) versus CPC that may limit penetration under conditions of a light splash or mist.

may be related to the non-chemical hazards in the workplace, the durability of the CPC in different use environments, or the impact of the CPC on the wearer. The respective performance of CPC against these properties can similarly be ranked or rated, but may create tradeoffs between desired characteristics.

12.2.3   Classification by service life

The classification of chemical protective clothing by expected service life is based on the useful life of the CPC item. Thus, service life reflects the longevity of the product and how it relates to the user's expectations. The service life of chemical protective clothing generally fits into three classes:

- Disposable after a single use – CPC products that are relatively inexpensive, which cannot be adequately, cleaned, reserviced, or maintained after use, or it is easier to dispose of and replace the CPC rather than provide care or maintenance.
- Limited use – CPC where some cleaning, care, and maintenance is possible, but the CPC may not be reusable under rigorous physical conditions, or CPC is eventually degraded by use and maintenance processes.
- Reusable – CPC that can be readily cleaned and maintained, and still continue to provide acceptable performance.

Unfortunately, chemical protective clothing manufacturers do not always specify the service life of their products and may also not indicate conditions for retiring of CPC. Some chemical protective clothing may also have limited 'shelf life' (i.e., time in storage before use) because of material degradation that can take place in storage due to heat, ozone, or material self-degradation. Of course any item of chemical protective clothing can be rendered unusable if irreversibly contaminated or damaged in use.

Chemical protective clothing service life is a function of three factors:

1. Durability – how CPC maintains its performance with use.
2. Ease of serviceability – the user's ability to care for, maintain, and repair CPC so that it remains functional for further use.
3. Life cycle cost – the total costs for purchase, using, and maintaining an item of CPC.

Durability

Chemical protective clothing durability is demonstrated by the length of time that the CPC item provides acceptable performance given the range of use conditions, care, and maintenance. Unacceptable performance may be evident through physical changes in the chemical protective clothing item such as:

- rips, tears, or separation of materials and seams
- thin spots or cracks in coated materials or protruding fibers
- unexplained material discoloration
- diminished functionality of CPC component parts

Unacceptable performance may also not be readily evident unless products are carefully examined by product manufacturers or subject to destructive testing.[2] Estimates of product durability can be made through product testing for simulating product wear but most often are derived from field experience involving actual product use. Some products are expected to lose some performance in certain property areas; the acceptability of any drops in performance related to protection must be examined by the end user.

Ease of re-servicing

Some types of chemical protective clothing can be serviced or repaired to extend service life. This servicing and repair is considered part of a regular care and maintenance program to allow CPC to meet its expected service life. For example, NASA has an elaborate capability for cleaning, decontaminating, repairing, and testing propellent handlers' ensembles used for protecting launch site personnel during hypergolic fuel operations.[3] Other types of chemical protective clothing may not be easily repaired or cannot be repaired without manufacturer or special assistance.

The most significant aspect of reservicing chemical protective clothing is the ease of decontamination. This is an important issue because the end user must have some confidence that the CPC item is contamination free. One of the reasons for the increasing popularity of disposable clothing is because its use obviates the difficult decision of determining if the clothing is clean enough for reuse. However, in many work environments, the levels of contamination may be low or the type of contamination may be readily removed by standard decontamination practices. Otherwise, destructive testing of clothing may be needed to determine if chemical contaminants have been adequately removed by the selected decontamination process.[4] If an item of CPC cannot be reserviced to bring it to an acceptable level of performance, then it cannot be reused. In addition, if the costs to repair or maintain the chemical protective item represent a significant proportion of the item's original cost, then the CPC item will probably not be reused.

Life-cycle cost

The life-cycle of chemical protective clothing includes all aspects of its selection, use, care, and maintenance until its ultimate disposal. Therefore, the life-cycle cost of chemical protective clothing is the sum of all costs for associated with an item of CPC. To compare different products on the same basis, life-cycle cost is usually represented as the cost per use for a CPC item.

The following costs should be considered in determining the life-cycle cost:

• purchase cost
• labor cost for selection/procurement of CPC
• labor cost for inspecting CPC
• labor and facility costs for storing CPC
• labor and materials costs for cleaning, decontaminating, maintaining, and repairing CPC
• labor and fees for retirement and disposal of used CPC.

The total life-cycle cost is determined by adding the separate costs involved in the chemical protective clothing life cycle and dividing by the number of CPC items and number of uses per item. A detailed method for estimating chemical protective clothing life-cycle cost is provided by Schwope and Renard.[5]

12.2.4 Overall type classification of chemical protective clothing

The overall representation of chemical protective clothing should be made with respect to the three classification systems:

• Design classification is needed to establish which parts of the body will be protected.
• Performance classification will indicate the type of protection that will be offered by the CPC.
• Service life classification will establish the expected longevity of the CPC.

The following specific example illustrates the application of this type classification approach. The chemical protective clothing item pictured in Figure 12.1 is a hooded coverall constructed of a coated non-woven material that provides liquid

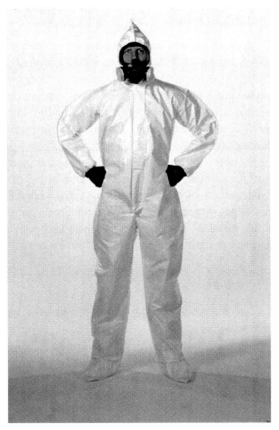

12.1 Example of a disposal coverall providing full-body protection against liquid splashes.

splash protection to the wearer's torso, arms, legs, and head (excluding the face). The coverall is considered disposable because the material has low durability, cannot be easily decontaminated if splashed and has a relatively low cost for a single use. If this CPC item is combined with gloves, boots, and full facepiece respirator, then protection is afforded to the wearer's hands, feet, and face. However, the interfaces between the clothing items (e.g., garment sleeve end to glove) and the integrity of clothing seams and closure will affect the level of overall performance of the ensemble.

12.3   Garment types, materials, design features and sizing

Further understanding the type classification of chemical protective clothing requires examining the specific designs, design features, materials, and other attributes associated with specific CPC garments.

12.3.1   Garment material types

Protection of the body from chemicals may be provided from either full-body garments or partial body garments. Full-body garments are designed to provide protection to the wearer's upper and lower torso, arms, and legs. Full-body garments may also provide protection to the wearer's hands, feet, and head when auxiliary CPC is integrated with the garment to form a suit. Full-body garments may be single or multi-piece clothing items. Partial body garments provide protection to only a limited area of the wearer's body including the upper torso, lower torso, arms, legs, neck, or head. Table 12.3 indicates some of the characteristics and features associated with each of these types of chemical protective garments. The extent of body protection varies with the garment design. Many garment designs do not provide uniform protection for all areas of the body covered by the garment. The specific materials and design features associated with the garment design will also influence protection.

12.3.2   Garment materials

Materials used in chemical protective clothing include seven basic types:

1. textiles
2. unsupported rubber or plastics
3. microporous film fabrics
4. adsorbent-based fabrics
5. coated fabrics
6. plastic laminates
7. combination or specialized materials.

Table 12.4 provides some characteristics and examples of these materials.

Textile materials

Ordinary textile materials are generally not considered suitable for protection against chemicals, however special non-coated textile materials are used for a variety of applications involving particulates and light liquid spray from relatively non-hazardous chemicals. Though woven textiles are not often found in chemical protective clothing, very tightly woven, repellent-treated fabrics can provide some very low minimum protection against liquid exposure.[6] More common are nonwoven fabrics that have demonstrated barrier performance against particles and repellency of liquids. Two predominant examples of non-woven fabrics are flashspun polyethylene (Tyvek®) and spunbond/ meltbown/ spunbond (SMS) polypropylene (Kleenguard®). These textiles are used because of their relatively low cost and because the materials provide a structure of microfibers that filter out dry particulates and many water-based liquids.

Table 12.3 Design characteristics and features associated with chemical protective garments

General garment type	Specific garment type	Characteristics and features
Full-body garments	Full-body suits	• One-piece garments which may offer a variety of entry options depending on the type and placement of the closure • Generally known as totally encapsulating suits, that encapsulate the wearer and other protective equipment such as the respirator (some designs may permit a respirator facepiece to integrate with a suit hood) • Includes other CPC attached to the suit such as gloves and footwear
	Jacket and pants combinations	• Mimic normal wearing apparel • Head protection provided by hood, usually without a visor • Generally provide some overlap of the waist portion of trousers with or without a collar, hood, or wrist protection • Generally use front closure • Pants usually rely on a zipper, snaps, or other front fly closure or drawstring/elastic waist • May have openings with closures at the foot end to allow entry while wearing footwear
	Jacket and overall combinations	• Similar to jacket and pants combination • Distinguished by a higher 'bib' style pants (overalls) that permits a shorter jacket • Some jacket combinations designed as pullover with hood • Overalls often use straps or gusset with snaps for adjusting garment on lower torso • Bib overalls usually provided with suspenders
	Coveralls	• One-piece garments, usually with a front closure, and have options for attachment of hoods, type of sleeve end (open or elastic), type of pant cuff end (open, elastic, or bootie)
Partial body garments	Hoods	• Cover the wearer's head and include either a face/eye opening or may be provided with an integrated faceshield or visor • Usually pullover design • Length affects integration with upper torso garments • Type of visor and size may vary (if present) • Size and type of face opening accommodates respirator or eye/face protection • Face opening closure options (elastic, pliable/stretchable material, drawstrings, ties, snaps, or hook and loop closures) • Bulk of the hood on the wearer's crown may affect the fit of helmets and respirator straps

Table 12.3 Continued

General garment type	Specific garment type	Characteristics and features
Partial body garments	Head covers and bouffants	• Provide protection to the wearer's upper head, but are most often used to contain or cover the wearer's hair • Bouffant style head covers are secured to the wearer's head by elastic around the periphery • Head covers may also be used as helmet covers as an aid to preventing contamination
	Aprons	• Consists simply of a flat piece of material contoured to the front of the body for providing lower or lower/upper torso protection • Lower part extends around the wearer's sides and ties in the back. • Aprons covering the upper torso are designated 'bib' aprons with a strap around the wearer's neck that holds up the 'bib' • Generally available in more than one size, designated by unisex rectangular dimensions, and are adjusted by the tie straps at the top or sides
	Lab coats	• Provide torso, arm, and upper leg protection • Of varying lengths and are generally offered in alphabetic sizing • Use front closures usually with snaps or buttons • May have open neck area or collar • Type and location of pockets are usually a design option
	Smocks	• Provide front torso, arm, and upper leg protection • Of varying lengths and are generally offered in alphabetic sizing • Use back closures with ties • May have open neck area or collar
	Shirts	• Provide protection to upper torso and arms (long-sleeved) • Mimic regular wearing apparel with differences in collar and type of closure
	Pants	• Provide protection to lower torso and legs • Mimic regular wearing apparel with differences in waist (fly or elastic)
	Sleeve protectors	• Provide protection from the wearer's hand to the shoulder area • Usually secured to the wearer's arm by several means or by elastic ends
	Chaps	• Partial pants which are open at the sides/back and are intended to provide wearer front leg protection (rarely used for chemical protection) • May incorporate belt or straps and may be flared at bottom to cover top of wearer's footwear
	Leggings	• Protect the wearer's lower leg (generally knee height) • Have elastic top or adjustable tops with snaps or tie closures • May have a flare for protecting the top of wearer's footwear

Table 12.4 Type of chemical protective garment materials

Material class	Characteristics	Examples	Type of protection
Textiles	Primarily nonwoven	Flashspun polyethylene (Tyvek) SMS Polypropylene (Kleenguard)	Dry particulate Light spray from water-based chemicals
Unsupported rubber or plastics	Thick rubber or plastic film	Polyvinyl chloride, PVC Chlorinated polyethylene, CPE (Chemturion)	Liquid splash, liquid or gas/vapor permeation
Microporous-film based	Polymer film with microscopic pores laminated to fabric	PTFE (Goretex) Polypropylene (NexGen, Kleenguard Ultra)	Liquid splash
Adsorbent-based	Material incorporating sorbent layer	Fabric/carbon/fabric (Lifetex, Saratoga)	Gas/vapor penetration
Coated fabrics	Woven fabric coated with rubber or plastic on one or both sides	PVC/nylon Polyurethane/nylon Neoprene/Nomex Butyl/Nylon/Butyl	Liquid splash, liquid or gas/vapor permeation
Plastic laminates	Plastic film laminated to one or both side of nonwoven fabric	Polyethylene/Tyvek* Tychem SL, BR, TK CPF I, II, III, IV Responder, Responder Plus	Liquid splash, liquid or gas/vapor permeation
Combination or specialized	Combines one or more of above material technologies or uses unique materials	Neoprene/nylon/neoprene/plastic laminate (VPS) Teflon/fiberglass/Teflon (Challenge)	Liquid splash, liquid or gas/vapor permeation

* Fabric is actually coated with polyethylene

Unsupported rubber and plastic materials

Normally, chemical protective clothing materials include supporting textile fabrics to provide strength. However, there are some CPC materials that do not include a fabric substrate. The rubber material or plastic is thick enough to provide sufficient strength for clothing use. Examples of polymers used in these materials are polyvinyl chloride and chlorinated polyethylene. Because the materials are continuous, they offer a barrier to liquids and can be used in the construction of CPC intended for protection against liquids and gases.

Microporous film-based materials

A relatively new class of CPC materials uses microporous films. As the name implies, microporous films have millions of microscopic pores per square inch of the film structure. In most cases, the pores are irregularly shaped with tortuous paths through the film. These pores are small enough to prevent the passage of most liquids, but still allow vapors and gases to pass through the material (see Fig. 12.2). This material feature makes the film 'breathable' and this feature is considered desirable for specifically allowing moisture vapor to transfer through chemical protective clothing to lessen the effect of wearer heat stress. The microporous films are generally glued or laminated to woven or nonwoven fabrics for physical support. Owing to their physical structure, these

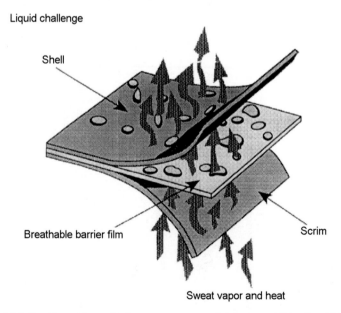

12.2 Configuration of microporous material for providing liquid barrier performance and breathability.

fabrics provide barrier performance against liquids but not gases. The type of liquids held out by microporous materials will depend on the surface tension, as lower surface tension liquids penetrate easier than higher surface tension liquids.

Adsorbent-based materials

One class of CPC materials uses adsorbents added to textile layers for providing chemical protection against hazardous vapors or aerosols. These engineered materials include adsorbents, such as activated charcoal or other sorbent materials. The principle of material operation is similar to cartridges used in air-purifying respirators. The outer and inner layers of these fabrics are treated with a liquid repellent to limit liquid contact that can saturate the adsorbent layer. For the most part, these materials have been used in military applications for protection against chemical warfare agents. In general, these materials can be optimized for adsorbing different chemical classes, but generally are most effective for large-molecule chemicals.[7]

Coated fabrics

Up until the mid-1980s, the majority of chemical protective clothing used coated fabrics. Woven fabrics such as cotton, nylon, and polyester are coated with a polymer such as butyl rubber, neoprene, polyvinyl chloride, or polyurethane to provide a continuous coating over the fabric substrate. Coating thickness is important for barrier performance. Coatings that are too thin may be prone to pinholes and other defects. For some materials, coating may be applied to both sides to create a thicker and more chemically resistant material. Materials using this construction tend to be heavy, but relatively rugged and are used in a variety of chemical protective clothing. Today, the most common clothing using coated fabrics are splash suits, but some higher-end totally encapsulating suits may be made from coated fabrics. Reusable splash clothing is similar to rainwear.

Plastic laminates

The majority of chemical protective clothing today uses plastic laminates. These materials combine various plastic polymer films with nonwoven substrate fabrics resulting in chemical-resistant, lightweight, relatively inexpensive materials. The majority of substrates are based on polypropylene and films are based on polyethylene and polyvinyl alcohol polymers, but more sophisticated films may employ a variety of different plastic films in a single laminate. Like coated fabrics, the plastic layer may be applied to one or both sides of the fabric substrate. Materials with plastic film on both sides tend to provide greater chemical resistance at the expense of increased stiffness.

Combination or specialized materials

This category includes chemical protective clothing materials not fitting into the classes above. These fabrics generally attempt to combine the best attributes of each class. For example, coated fabrics have been laminated with a plastic film on one side to provide the flexible features associated with rubber-based fabrics and the high levels of chemical resistance. Microporous films can be combined with adsorbent-based materials to provide a film with liquid chemical resistance that adsorbs large molecular weight chemical vapors. One unique special material developed was based on applying Teflon to woven or nonwoven fiberglass substrate fabrics.

12.3.3   Garment design features

In addition to material choices, features affecting the design of garments include:

- the type and location of seams
- the type, length, and location of the closure system(s)
- the type and characteristics of visors or faceshields, if integrated into garments
- the design of interface areas with other chemical protective clothing or equipment
- the types, function, and location of hardware.

Seams

Seams play a critical role in protective clothing because they directly affect the integrity of the CPC in providing protection against specific chemical challenges. General types of seams used in chemical protective clothing include:

- sewn
- glued
- sealed.

Seams may also be bound or reinforced with other material pieces such as additional fabric or tape, or even covered with a top coating of a polymer.
    The type of seam and how it is applied is usually affected by the material used in the garment construction and the intended integrity or performance of the chemical protective clothing:

- Sewing can be applied to any textile-based material.
- Gluing is usually performed for coated fabrics and unsupported rubber materials.
- Sealing is applied to materials where the surface can be melted to create a bond with the surface of another material. This seaming approach is used for

plastic laminates, thermoplastics, and other film-based materials. Sealing can be accomplished by using heat, ultrasonic radiation, or other means.

Different seam constructions can be used in protective clothing. A serged seam construction is popular for many styles of dry particulate-protective clothing because of its simplicity. However, lap or fell seams are also used in the fabrication of some CPC. Many CPC seams are sewn to provide garment strength against the stresses of wear and use. However, sewing by itself produces stitching holes that can provide a pathway for chemical penetration. Therefore, some seams will be bound for limiting these pathways for chemical penetration. Further protection of a sewn seam can occur when strips of tape directly cover the holes. The tape can be applied by either glues or by sealing the extra material over the seam. In some cases, seams are taped on both sides to provide increased chemical resistance.

The location of seams is also important. Not only do a large number of seams increase the garment manufacturing cost but also more seams can create potential exposure to the wearer in the event of seam failure. Therefore, some clothing designs are created to limit the number of seams and seams that are on the front of the garment where most chemical exposures occur.

Closure systems

Closures are typically the 'weak' link in the chemical protective clothing barrier. Closures are necessary to allow people to don and doff CPC, but should also not lower the integrity provided by the clothing. The simplest closures are zippers or a series of snaps. Because of the open construction of these closures and their needed placement often on the front of the garment, pathways for chemical penetration can be created. For this reason, many chemical protective clothing designs use storm flaps to cover the zipper or snaps. An extra storm flap on the interior side of the zipper can provide additional liquid protection. On many disposable or limited-use CPC, adhesive strips may be incorporated into the storm flap to keep it in place. Reusable CPC sometimes uses hook and loop closure tape to secure storm flaps or as the principal closure system.

Closure systems offering increased integrity against chemical penetration include liquid-repellent zippers, two-track extruded closures and special pressure-sealing zippers. Liquid-repellent zippers are conventional zippers that use rubber or plastic coated tape instead of woven cloth on the sides of the zipper and that have a special chain (teeth) that are coated to limit liquid penetration.

Two-track closures involve two extruded pieces of plastic that fit together to provide a seal much like many plastic sandwich bags. These closures offer good integrity against liquids (and vapors) but can be difficult to seal over a long length as might be required in a full-body garment. Some two-track closures use a zipper-like pull to seal the length shut but this design still leaves part of the end open. In addition, the types of plastics that work well for these closures are also limited and may not have chemical resistance compatible with the garment material.

The pressure-sealing zipper uses a compressible rubber or plastic material with a zipper chain to push the two sides of the closure material together for creating a seal. These zippers tend to be bulky and expensive but are generally required when total encapsulation of the wearer is required. As with seams, the location of the closure will be important. In some cases, the conventional design will feature the closure to be in the front of the garment. Some items such as totally encapsulating suits can have closures on the side or rear of the clothing item.

Garment visors

Visors are generally incorporated into separate hoods, coveralls, or suits to offer chemical barrier protection for the head and face area. In general, the visor is constructed from a material that provides clear undistorted vision as well as chemical resistance. While having some physical integrity, visors generally do not offer the same physical impact resistance provided by face shields and other primary face protection.

Garment visors vary in their size, stiffness, materials, and method of integration with the garment.

- The size of the visor affects the wearer's peripheral vision. In some garment designs, such as totally encapsulating suits, visor size has a significant effect on the wearer's field of vision since the suit visor does not move with the wearer's head. Consequently, more recent designs of these garments provide extra-large visors. This is less of a problem for hoods, which will rotate when the wearer turns his or her head.
- Visors can be flexible or rigid. Reusable suits and garments tend to use stiffer visors since overall weight may be less of a consideration, while many disposable garments incorporate relatively flexible visors to maintain low weight.
- The principal materials used in visor construction are polyvinyl chloride, polycarbonate, and polymethacrylate. Since the visor material must provide optical qualities, only transparent materials can be chosen as visors. When greater chemical resistance is required, the visor made be made of a composite material that includes fluorinated ethylene propylene (FEP) laminated to PVC. The FEP layer is kept thin because the material becomes opaque at large thicknesses, while the PVC provides support and scratch resistance.
- The joint between the visor and the garment is important for maintaining the integrity of the garment item. In some cases, the visor material is directly sealed to the garment material. In other cases, the visor may be perforated and

sewn in place, and then the seam is covered with tape. Some more robust designs for reusable clothing may use a gasket material to seal a rigid visor into a frame.

The visors used in suits, including the seams for integrating the visor in the garment, should be evaluated for the same properties as garments. These properties are important in addition to optical properties (such as visor clarity, light transmittance, and haze) that are specific to the visor. However, industry practice has typically not included chemical resistance of visors. At an incident in Benicia, California, United States in the mid 1980s, the failure of the visor in a totally encapsulating suit illustrated the potential life-threatening consequences of a visor failure. In particular, a leaking tank car of anhydrous dimethylamine required the response of a hazardous-materials team for incident evaluation and mitigation. Though the suit was recommended as being compatible with the leaking chemical, the visor of one responder's broke open during the incident. Fortunately, the self-contained breathing apparatus (SCBA) inside the suit protected the man's respiratory tract, but his unprotected skin was exposed to the dimethylamine. An analysis of the incident by the National Transportation Safety Board found that the polycarbonate visor material was unsuitable for dimethylamine exposure, though the suit itself was recommended.

Interface areas

The interfaces between the chemical protective clothing item and other CPC or equipment can vary with the design. Principal CPC interfaces include:

- amount of overlap for multi-piece garments
- upper torso garment sleeve to glove
- lower torso pant cuff to footwear
- upper torso garment collar to hood
- hood to respirator.

If different pieces of chemical protective clothing are worn, the items should provide sufficient overlap to cover the wearer's skin particularly during reaching or bending over. This is particularly an issue for jacket and pants sets of CPC.

The sleeve ends for garment-to-glove interfaces include a number of designs:

- For garments where there are no integrity issues, the sleeve end may be open. For a reusable garment, the sleeve end is hemmed; a disposable garment sleeve may be unfinished. If liquid leakage in the sleeve to glove area is a concern, duct tape is used but not recommended for providing integrity against liquid penetration.
- Elasticized sleeve ends provide somewhat better conformance of the sleeve with the glove. This design feature is particularly common for disposable and limited use clothing. However, chemical penetration may still occur at the

interface. The application of tape provides only a questionable improvement of integrity for chemical liquids or vapors.

- For suits requiring a higher level of protection, the gloves are attached to the garment sleeve end either permanently or in a manner in which they can be detached. These designs often include the incorporation of a hard ring into the sleeve end. An additional section of material at the sleeve end may be formed over the glove to act as a 'splash' guard.

Similar approaches are used for the pant cuff bottom in the interface area with footwear:

- A straight cuff is used. The cuff is generally pulled over the boot to keep liquid from entering the top of the boot or footwear. Like gloves, tape is sometimes used to keep the cuff in place.
- An elasticized cuff is used on some garment leg ends. This helps the pant leg stay in place on top of the boots. Tape may also be used, but only as a temporary measure for keeping the pant leg over the footwear.
- Some lower torso garment designs may use a covered zipper at the bottom of the leg so that the wearer can more easily insert his or her foot with attached boot or shoe through the pant leg.
- The garment material is fashioned into a bootie that is attached to the bottom of the garment leg and is worn inside the footwear. Garment booties are then worn inside an outer boot. This design may then include a splash guard mounted at the bottom of the garment legs that pull over the top of the boots.

Hoods can involve multiple interface areas. If a separate item, the bottom of the hood may be required to interface with the top of an upper torso garment. Most often, the length and flare of the hood control this interface. The bottom of the hood may also have slits to accommodate the wearer's shoulders. If without a visor, the hood will create the interface with the wearer's respirator facepiece or eye/face protection. This interface will provide limited protection if without some device or aid to close the hood face opening around the device. For this reason, hoods often incorporate drawstrings or elastic, or may have some other feature to cinch the hood opening around the respirator facepiece. Hoods may also need to accommodate hard hats or other head protection and hearing protection.

12.3.4 Garment sizing

There are few uniform sizing practices for the design of chemical protective garments. The availability of sizing often depends on the specific type of the garment and the relative volume of garments sold by the manufacturer. Sizing may be based on individual measurements for custom sizing, numerical sizing for regular wearing apparel (chest or waist size), or alphabetic sizing (e.g., small, medium, and large). Partial body garments are likely to be offered in fewer sizing choices. For example, sleeve protectors may vary in three lengths, while two sizes of aprons will be available, and only one size of hood is provided. It is important to realize that garment sizing must often take into account that the garment will be worn over regular work clothing and may need to accommodate different kinds of equipment, such as respirators, hard hats, and other devices.

Some sizing systems for garments usually use two or more wearer dimensions, such as height and weight, or height and chest circumference. This practice allows the wearer to determine the correct size by using their body dimensions. Unfortunately, sizing between manufacturers is often inconsistent. Even a standardized sizing scheme such as the one provided in ANSI/ISEA 101 for disposable or limited use coveralls has not found widespread acceptance among manufacturers. In addition, the sizing of protective garments often does not address the needs of women and special worker populations.[8]

12.4   Garment material chemical resistance testing

Chemical resistance is the principal basis on which CPC performance is based. Material test approaches can be classified into three types that describe how chemicals may interact with materials: degradation resistance; penetration resistance, and permeation resistance. A number of different procedures exist for the measurement of each type of performance depending on the type of chemical challenge and the level of sophistication for performing the tests. Of the material testing approaches, both penetration and permeation resistance testing allow assessment of the barrier qualities of a protective clothing material, whereas degradation resistance does not. Penetration testing may involve chemical particles, liquids, or vapors (gases). The individual procedures available for measuring chemical degradation, penetration, and permeation resistance are described in the subsections below.

12.4.1   Chemical degradation resistance

Degradation is defined as the 'change in a material's physical properties as the result of chemical exposure.' Physical properties may include material weight, dimensions, tensile strength, hardness, or any characteristic that relates to a material's performance when used in a particular application. As such, the test is used to determine the effects of specific chemicals on materials. In some cases chemical effects may be dramatic showing clear incompatibility of the material with the chemical. In other cases, chemical degradation effects may be very subtle.

Various groups have examined different approaches for measuring the chemical degradation resistance of barrier materials, but no single generalized test method has been developed by consensus organizations within the United States, Europe, or internationally that can apply to all protective clothing materials.[9] Nevertheless, a variety of techniques are commonly used for rubber and plastic materials within different barrier material industries. These procedures and their utility in evaluating chemical barrier materials are discussed below.

Degradation tests using immersion-based techniques

ASTM D 471 and D 543 establish standardized procedures for measuring specific properties of material specimens before and after immersion in the selected liquid(s) for a specified period of time at a particular test temperature. Test results are reported as the percentage change in the property of interest. ASTM D 471 provides techniques for comparing the effect of selected chemicals on rubber or rubber-like materials, and is also intended for use with coated fabrics. ASTM D 543 covers testing of plastic materials, including cast and laminated products, and sheet materials for resistance to chemical reagents. In each test, a minimum of three specimens is used whose shape and size are dependent on the form of the material being evaluated and the tests to be performed. An appropriately sized vessel, usually glass, is used for immersing the material specimens in the selected chemical(s). Testing with volatile chemicals typically requires either replenishment of liquid or a reflux chamber above the vessel to prevent evaporation.

Both test methods indicate that the selected exposure conditions and physical properties measured should be representative of the material's use. For CPC material testing, this will usually mean specifying significantly shorter test periods and ambient temperature exposures. Since the methods are intended for comparing materials against similar chemical challenges, no criteria are given for determining acceptable performance.

Degradation tests using one-sided exposure techniques

Section 12 of ASTM D 471 provides a procedure for evaluating the effects of chemicals when the exposure is one sided. This technique is particularly useful for the evaluation of protective clothing materials, particularly those involving coated fabrics, laminates, and any non-homogeneous material. In this procedure, the material specimen is clamped into a test cell that allows liquid chemical contact on its normal external (outer) surface. Usually, changes in mass are measured for this testing approach since the size of the material specimens is limited.

ASTM F 1407, while intended for measuring chemical permeation resistance, also serves as a useful technique for evaluating chemical degradation resistance of protective clothing materials.[10,11] This test employs a lightweight test cell in which the material specimen is clamped between a Teflon-coated metal cup filled with the selected chemical and a metal ring (flange). The entire cup assembly is inverted and allowed to rest on protruding metal pins that hold the test cell off the table surface. In the permeation testing mode, the weight of the entire assembly is monitored; however, for use as degradation test, the test cell serves as a convenient means for evaluating changes in material mass and thickness. Visible observations are also recorded as part of the testing protocol.

The International Safety Equipment Association undertook the development of a degradation resistance test method for providing a means to evaluate and classify glove performance in a new standard specific for hand protection (ANSI/ISEA 105). The test is based on the principle of one-sided exposure, employing a small flask containing the test chemical. The CPC specimen is sealed to the top of the flask and the flask is inverted to allow the test chemical to contact the specimen for a period of one hour. The flask is then positioned upright and mounted in a tensile testing machine outfitted with a compression cell and puncture probe. This test apparatus is used to measure the force required to puncture the specimen. The puncture resistance of exposed samples is then compared to the puncture resistance of pristine specimens. The percentage change in puncture resistance can thus be used as a measure of glove material degradation resistance to specific chemicals.

Application of test data

Chemical degradation by itself cannot fully demonstrate product barrier performance against chemicals. While a material that shows substantial effects when exposed to a chemical can be ruled out as a protective material, it remains uncertain whether materials that show no observable or measurable effect act as a barrier against the test chemical. For this reason, chemical degradation data are typically used as a screening technique to eliminate a material from consideration for further chemical resistance testing (i.e., penetration or permeation resistance).[12] Some sample degradation resistance data are shown in Table 12.5 comparing three different materials and three different methods of discerning degradation effects.

The majority of chemical degradation resistance data are reported in the glove industry. This is because most gloves are made from elastomeric materials. As a class of materials, elastomers when compared to plastics, show greater affinity for chemical adsorption and swelling.[13] Therefore, elastomeric materials are generally more susceptible to measurable chemical effects. This is particularly true today, because the majority of garment materials are composed of different plastic layers that have few observable degradation effects.

12.4.2 Chemical penetration resistance

Penetration is defined as 'the flow of chemical through closures, porous materials, seams, and pinholes and other imperfection in a protective clothing material on a non-molecular level.' This definition is intended to accommodate

Table 12.5 One-sided immersion degradation resistance data for selected materials and chemicals

Chemical	Viton/chlorobutyl laminate			Chlorinated polyethylene			FEP/surlyn laminate		
	Percent weight change	Percent elong.[a] change	Visual obs.	Percent weight change	Percent elong. change	Visual obs.	Percent weight change	Percent elong. change	Visual obs.
Acetaldehyde	10	0		24	11		0	-5	
Acrylonitrile	9	0	delam.	35	failed		-1	0	
Benzene	2	0		60	failed		0	0	
Chloroform	4	0		72	failed		0	0	
Dichloropropane	3	0		120	failed		-2	0	
Ethyl acrylate	17	0	curled	160	failed		0	0	
Ethylene oxide	2	0		13	11		0	0	
Hydrogen fluoride	4	0	discol.	2	11		4	0	
Nitric acid	9	0	discol.	8	-6	discol.	-1	0	

Abbreviations: FEP – Fluorinated ethylene propylene; elong. – tensile elongation; obs. – observations; delam. – delaminated; discol. – discolored.
[a] Percent elongation based on elongation measured using ASTM D412 for exposed and unexposed samples, 'failed' results indicate materials not tested due to weight changes over 25%.
Source: adapted from ref. 12.

both liquids and gases, but all U.S., European and International test methods focus on liquid penetration. Liquid suspended in air as aerosols and solid particles can also penetration protective clothing materials, but the discussion of penetration resistance in this section relates to liquids exclusively because particle-based test methods for CPC are still under development.

Liquid repellency and penetration resistance are related since wettability of the fabric affects the ability of the liquid to penetrate. For porous fabrics, a liquid of surface tension, γ, will penetrate given sufficient applied pressure, p, when its pores are of diameter, D, according the relationship known as Darcy's Law:

$$D = k \frac{4\gamma \cos \theta}{p} \qquad\qquad 12.1$$

where: θ = contact angle of liquid with the material, and k = shape factor for the material pores. For non-porous fabrics, particularly coated fabrics or laminate materials, liquid penetration may still take place as the result of degradation. Given a sufficient period of contact, chemicals may cause deterioration of the barrier film to allow pathways for liquid to penetrate. In this sense, penetration testing allows both an assessment of material barrier performance to liquid chemicals and chemical degradation resistance. There are two fundamentally different approaches used in liquid penetration resistance test methodologies, runoff-based methods, and hydrostatic-based methods. Table 12.6 compares the different characteristics of these test methods.

Runoff-based penetration tests

Runoff-based techniques involve contact of the liquid chemicals with the material by the force of gravity over a specified distance. The driving force for penetration is the weight of the liquid and the length of contact with the material specimen. Usually, the material specimen is supported at an incline, allowing the chemical to run off, hence the name for this class of penetration tests. Runoff-based tests are characterized by three features:

- impact of the liquid from a stationary source onto a material specimen
- orientation of the material specimen at an incline with respect to the point of liquid contact
- use of a blotter material underneath the material specimen to absorb penetrating liquid.

Runoff-based tests differ in the distance separating the liquid source from the point of contact with the material specimen, the type of nozzle through which liquid is delivered, the amount of liquid, the rate at which the liquid is delivered, the angle of the incline, and the type of test measurements made. The majority of these methods are intended for use with water as the liquid challenge only. Physically, many of the methods are suitable for testing with other liquids;

Table 12.6 Characteristics of runoff-based and hydrostatic-based penetration resistance tests

Test method	Type of delivery	Liquid amount and rate	Sample orientation	Measurements
Runoff-based				
AATCC 42	Specified nozzle above specimen	500 ml delivered at distance of 0.6 m	45 degrees, over blotter	Weight gain of blotter
ASTM F 2130	Applied through pipette	0.1 or 0.2 ml	Horizontal, over blotter	Weight gain of blotter
EN 368	Single, 0.8 mm bore hypodermic needle 100 mm above sample	10 ml at 1 ml/sec	45 degrees, over blotter in semicircular 'gutter'	Index of repellency; Index of penetration
ISO 6530	Same as EN 368			
ISO 22608	Same as ASTM F 2302			
Hydrostatic-based				
AATCC 127	Water pressurized above specimen at constant rate	Water pressure increased at rate of 10 mm H$_2$O/sec	Horizontal, clamped in test apparatus	Pressure at which water droplets appear in three separate specimen locations
ASTM D 751, Method A	Water pressurized above specimen at constant rate	Water pressure increased at rate of 1.4 ml/sec	Horizontal, clamped in test apparatus	Pressure at which water droplet appears
ASTM F 903	Liquid pressurized against specimen in test cell	60 ml liquid chemical, different pressurization options	Vertical, clamped in test cell	Visual observation of penetration
ISO 811	Water pressurized above specimen at constant rate	Water pressure increased at rate of 10 cm or 60 cm H$_2$O/sec	Horizontal, clamped in test apparatus	Pressure at which water droplets appear in three separate specimen locations
ISO 1420	Water pressurized underneath specimen at constant rate	Water pressure increase rate to be specified by lab	Horizontal, clamped in test apparatus	Pressure at which water droplet appears
ISO 13994	Same as ASTM F 903			

however, the containment aspects of these test methods vary and some are clearly inappropriate for use with hazardous chemicals. Some of the tests involve delivering relatively large quantities of water onto a sample and measuring the amount of water absorbed in a blotter paper placed underneath the material specimen. This approach is characteristic of AATCC 42. The large quantities of water specified, and the lack of containment in the design of the apparatus make these test methods unsuitable for other liquids.

Two of the test methods are essentially identical and are designated for use with various liquid chemicals. Both EN 368 and ISO 6530, the so-called 'gutter test,' use a system where the liquid chemical is delivered by a single, small-bore nozzle onto the material specimen at a distance of 100 mm. The material is supported in a rigid transparent gutter which is covered with a protective film and blotter material set at a 45 degree angle with respect to the horizontal plane. A small beaker is used to collect liquid running off the sample. The two results reported in these tests are the indices of penetration and repellency. The index of penetration is the proportion of liquid deposited in the blotter paper:

$$\text{Index of penetration } (P) = \frac{M_p \times 100}{M_t} \qquad \text{12.2}$$

where M_p is the mass of test liquid deposited on the absorbent paper/protective film combination and M_t is the mass of the test liquid discharged onto the test material specimen.

The index of repellency is the proportion of liquid deposited in the blotter paper:

$$\text{Index of repellency } (R) = \frac{M_r \times 100}{M_t} \qquad \text{12.3}$$

where M_r is the mass of test liquid collected in the beaker. A mass balance of the liquid also allows calculation for liquid retained in the material specimen. Variations of the runoff tests applied to the pesticide formulations are found in ASTM F 2130 and ISO 22608. Not all of the tests described above can be considered 'true' liquid penetration tests. Penetration with these procedures can only be characterized when some assessment or measurement of liquid passing through the material specimen is made. Typically this is done by examination of the blotter material, either visually or gravimetrically.

Runoff tests are generally used on textile materials which have surface finishes designed to prevent penetration of liquid splashes. Many of these tests easily accommodate uncoated or non-laminate materials, since the driving force for liquid penetration is relative low (when compared to hydrostatic-based test methods). As a consequence, runoff tests may be infrequently specified for chemical barrier-based clothing. The relatively small amount of liquid involved in some runoff tests is not considered a strong challenge. ISO 6530 recommends that the test be used only when the clothing item's overall integrity for preventing liquid penetration has been demonstrated. A discussion of liquid-based integrity methods appears later in this chapter.

Hydrostatic penetration testing

Hydrostatic-based techniques involve the pressurization of liquid behind or underneath the material specimen. It is this hydrostatic force which is the principal driver for liquid penetration. In this testing approach, liquid is contacted with the material specimen, with at least some portion of the test period having the liquid under pressure. Different devices or test cells are available for providing this type of liquid contact with the material specimen, in essence representing the differences among representative test methods. Like runoff-based test methods, the majority of the industry tests are designed for use with water. Many of the devices described below cannot be used with other liquids or may even be damage if anything but water is used in the respective tests.

Two different types of testing machines prevail for measuring hydrostatic resistance. AATCC 127 and ISO 811 use similar devices, where water is introduced above the clamped material specimen at a pressure controlled by water in a rising column. A mirror is affixed below the specimen to allow the test operator to view the underside of the specimen for the appearance of water droplets. Both the pressure and length of exposure are to be specified for the particular application. AATCC 127 defines water penetration as the pressure when a drop or drops appear at three different places of the test area (on the specimen). When a specific hydrostatic head is specified, test results are reported as pass or fail. ASTM D 751 and ISO 1420 use a motor-driven hydrostatic tester. Water contacts the underside of the material specimen that is clamped into a circular opening. Increasing hydraulic pressure is applied to the clamped material specimen at a specified rate until leakage occurs. The pressure at which this leakage occurs is noted and reported as the test result.

Of the listed tests above, only ASTM F 903 and ISO 13994 were developed for testing liquids other than water.[14] In these test methods, a 70 mm square material specimen is exposed on one side to the test chemical for a specified period of time using a special penetration test cell (see Fig. 12.3). The test cell is positioned vertically to allow easy viewing of the material specimen. During the chemical exposure, a pressure head may be applied to the liquid for part of the test period. Penetration is detected visually and sometimes with the aid of dyes or fluorescent light. The test is generally pass/fail, i.e., if penetration is detected within the test period, the material fails. Observations of material condition following chemical exposure are also usually provided. Different test specifications exist for the amount of chemical contact time and level of pressurization.

Penetration resistance using a hydrostatic-based test methods can accommodate a variety of different material types and clothing test specimens,

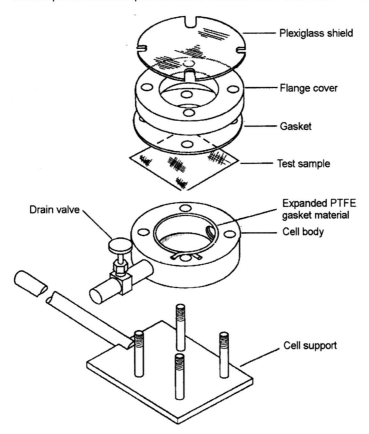

12.3 Exploded view of ASTM F 903 penetration test cell.

including CPC seams and closures. For these types of material specimens there are different modes of failure. Continuous film or film coated fabrics generally fail only due to imperfections in the material, such as cuts or pinholes or deterioration (degradation) of the film providing an avenue for liquid penetration. The latter type of failure often depends on the thickness of the film or coating as well as the contact time and amount of pressure applied to the specimen.

Textiles and microporous film products provide another set of possible failure mechanisms. Textiles may be considered as a liquid barrier when they have been treated with water/chemical repellent finishes. The ease of liquid penetration is therefore more a function of repellent finish quality and the surface tension of the liquid being tested. Also, penetration may still be the result of material degradation while in contact with the chemical. Microporous film products represent a unique test material since by design they afford transmission of vapors but prevent liquid penetration. These materials therefore require careful observations since significant vapor penetration may occur. Like textiles, surface tension may be a factor, though most microporous films have pore sizes that preclude penetration of most common liquids at relative low pressures (less than 12 kPa).

The integrity of seams, closures, and other clothing material interfaces are easily evaluated using penetration resistance testing.[15,16] Their uneven sample profiles must be accommodated through special gaskets or sealing techniques. For zipper closures, a groove has been cut in the test cell to provide a better seal on the protruding teeth portion of the zipper. In assessing penetration resistance for these items, failures may occur because of

- penetration of liquid through stitching holes in seams
- solvation of seam adhesives
- degradation of seam tapes or other seam components
- degradation of materials joining seam causing lifting of seam tapes or destruction of seam integrity
- physical leakage of closures.

Berardinelli and Cottingham[15] demonstrated the utility of this test on a number of material, seam, and closure samples. Understanding how clothing specimens may fail provides insight into identifying protection offered by the overall clothing item.

Specific use of penetration resistance tests

Penetration testing as per ASTM F 903 (and ISO 13994) provides a test for assessing the barrier performance of materials against liquid chemicals.[16] Though measuring specimen weight change is not required, this testing can also serve to measure material degradation since visual observations are required. In turn, degradation of the material may be a primary route for penetration by some chemicals. The difficulty in penetration testing lies in making a clear cut determination of liquid penetration. Many high vapor pressure, low surface tension solvents spread thinly over the material and evaporate quickly. Therefore, actual liquid penetration may be difficult to observe even when enhanced by using dyes. Still, the test serves as a good indicator of material performance against liquid contact or splashes. Since test length and pressurization periods depend on the selected procedure within the method, pass/fail determinations are clear-cut and easily define acceptable material-chemical combinations.

ASTM F 903 and ISO 13994 incorporate four types of contact time and pressure exposure formats (see Table 12.7). The original protocol consisted of exposing the material to the liquid for a five-minute period at ambient pressure, followed by a ten-minute period at 13.8 kPa. This exposure condition was selected as a test pressure to simulate the force on a protective garment of a liquid coming out of a burst pipe at an approximate distance of 3 m. A lower

Table 12.7 ASTM F 903/ISO 13994 penetration test variations

Procedure	Initial contact period (minutes at 0 kPa)	Pressurization period (minutes/pressure)	Subsequent contact period (minutes at 0 kPa)
A	5	10 min/7 kPa	None
B	5	10 min/14 kPa	None
C	5	1 min/7 kPa	54
D	60	None	None

pressure of 6.9 kPa was adopted later on because many materials would 'balloon' away from the test cell as pressure was applied. Some differences in material performance due to degradation effects have been noted as shown in Table 12.8.[14] The additional contact time/pressure formats in ASTM F903 were included to accommodate practices being used by the National Fire Protection Association in their requirements for chemical protective suit material and component penetration resistance.[17] Many unsupported film samples cannot be tested at high pressure since they burst when the pressure is applied. In these cases, the true barrier properties of the material to liquid penetration are not tested. For this reason, the optional use of a screen having more than 50% open area is specified in ASTM F 903 and ISO 13994 to prevent over-expansion of the clothing material specimen.

12.4.3   Permeation resistance

Permeation is a process in which chemicals move through a material at a molecular level. Material permeation resistance is generally characterized using two test results: breakthrough time and permeation rate. Breakthrough time is

Table 12.8 Effect of contact time and pressure on penetration of selected material-chemical combinations

Material	Chemical	Penetration test exposure protocol	Penetration time (minutes)
PVC/Nylon	Dimethylformamide	A	None
		B	None
		C	40
		D	50
Microporous film/ nonwoven laminate	Hexane	A	None
		B	5
		C	5
		D	None

Source: Adapted from ref. 16.

the time that chemical is first detected on the 'interior' side of the material. As discussed below, its determination is strongly dependent on how the test is configured and the sensitivity of the detector. Permeation rate is a measure of the mass flux through a unit area of material for a unit time. Permeation rate is most commonly expressed in units of micrograms per square centimeter per minute ($\mu g/cm^2 min$). For a given material-chemical combination, the steady-state or maximum observed permeation rates are reported. The measurement of chemical permeation resistance is specified in different standard test methods offered by ASTM, CEN, and ISO.

These permeation tests involve either liquid or gaseous chemical contact with the material and assessment of permeation as affected by both chemical solubility and diffusion through the test material. ASTM F 739 was first established in 1981 as the first test standard test method for measuring material permeation resistance to liquid chemicals.[18] ASTM F 739, EN 369, and ISO 6529 provide standardized procedures for measuring the resistance of protective clothing to permeation by chemicals using continuous contact of the chemical with the material's exterior surface. ASTM F 1383 is a variation of ASTM F 739 that involves testing under conditions of intermittent chemical contact. ASTM F 1407 represents a simplified form of testing where permeation is determined gravimetrically. Based on its limited sensitivity, this method is primarily used as a screening test or field method. In each of the tests (except ASTM F1407), a similarly designed test cell is used for mounting the material specimen (Figure 12.4). The test cell consists of two hemispherical halves divided by the material specimen. One half of the test cell serves as the 'challenge' side where chemical is placed for contacting the material chamber. The other half is used as the 'collection' side that is sampled for the presence of chemical permeating through the material specimen.

The basic procedure in each test is to charge chemical into the challenge side of the test cell and to measure the concentration of test chemical in the test cell as a function of time. Of principal interest in permeation testing are the elapsed time from the beginning of the chemical exposure to the first detection of the chemical (i.e., the so-called breakthrough time), the permeation rate, and the cumulative amount of chemical permeated. The results reported are dependent on the test method chosen:

- ASTM F739 requires reporting of breakthrough time and maximum or steady state permeation rate.
- ASTM F1383 specifies reporting breakthrough time and cumulative permeation.
- ASTM F1407 permits reporting either cumulative permeation or breakthrough time and permeation rate (maximum or steady state).
- EN 369 requires reporting of breakthrough time with the total cumulative mass permeated at 30 and 60 minutes.

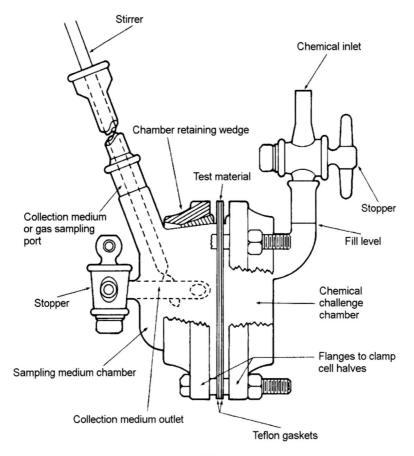

12.4 Specification for ASTM F 739 standard permeation test cell.

ISO 6529 combined ASTM and CEN approaches into its procedures. Other significant differences exist between the different listed test methods as described below. Table 12.9 provides a comparison of key characteristics for each of the different permeation test methods.

Parameters affecting permeation resistance testing

Although the permeation test procedure is simple in concept and generalized procedures are specified by each of the test methods above, a number of significant variations exist in the manner in which permeation testing can be conducted. These variables include:

- the general configuration of the test apparatus
- how the chemical contacts the material specimen in the test cell
- the type of collection medium used and frequency of sampling

Table 12.9 Differences between permeation test methods

Test method	Chemicals permitted	Type of contact	Collection medium flowrate(s)	Minimum test sensitivity ($\mu g/cm^2 min$)	Test results reported
ASTM F 739	Liquids and gases	Continuous	50 to 150 ml/min	0.1	Breakthrough time Permeation rate
ASTM F 1383	Liquids and gases	Intermittent	50 to 150 ml/min	0.1	Breakthrough time Cumulative permeation
ASTM F 1407	Liquids only	Continuous	Not applicable	≈ 20.0*	Breakthrough time Permeation rate Cumulative permeation
EN 369	Liquids only	Continuous	520 ml/min (gas) 260 ml/min (liq.)	1.0	Breakthrough time Cumulative permeation at 30 and 60 minutes
ISO 6529	Liquids and gases	Continuous or intermittent	Not specified	0.1 or 1.0	Breakthrough time Permeation rate Cumulative permeation

* Depends on analytical balance, exposed specimen surface area, and time interval between measurements.

- the type of detector and detection strategy used
- the test temperature
- the effect of multicomponent solutions.

The variety of available test techniques and conditions allows several different approaches for conducting permeation testing and can provide different results for testing the same material and chemical combination. This apparatus should be configured to meet testing needs and accommodate the characteristics of the chemical(s) being tested. Test cells are generally specified by the test method, but alternative designs are available and may be necessary for testing with specific chemicals or chemical mixtures. Likewise, the chemical delivery and collection/detection systems are dependent on the nature of the chemical and the requirements for running the test. The way that each part of the apparatus is operated comprises the test apparatus configuration. There are two basic modes for configuring permeation test systems, closed-loop or open-loop.

In closed-loop permeation systems, the volume of collection fluid is maintained throughout the test. This volume may be contained fully within the collection chamber or it may be circulated through the chamber, into a non-intrusive detector, and back into the chamber. Since the total volume of collection medium remains constant, permeating chemical accumulates within the collection medium. In this system, the permeation rate must account for this accumulation of permeant as follows:

$$\text{Rate} = \frac{(C_n - C_{n-1})}{(t_n - t_{n-1})} \times \frac{V}{A} \qquad 12.4$$

where $(C_n - C_{n-1})$ is the change in concentration of the challenge chemical between sampling periods, $(t_n - t_{n-1})$ is the time between sampling periods, V is the volume of collection medium, and A is the exposed area of the material specimen.

In the open-loop permeation systems, a gas or liquid collection medium is passed through the collection side out of the test cell to the detection system. This collection medium stream can be evaluated discretely or continuously depending on the detector selected. Therefore, collection of permeant is specific to the sample taken (over a discrete time period) and permeation rates can be directly calculated as a factor of the collection medium permeant concentration (C) and flow rate (F):

$$\text{Rate} = \frac{C \times F}{A} \qquad 12.5$$

The choice of a closed- or open-loop system is most often determined by the properties of the chemical and the available detector. Some chemicals such as inorganic substances often require closed-loop systems, particularly if ion-specific electrodes are used which have recovery time constraints. Open-loop testing is preferred for many volatile organic chemicals because these systems can be easily automated.

Permeation test methods are generally applied with neat chemicals under conditions of continuous exposure. In liquid exposures, the chemical or chemical mixture of interest is placed directly in the challenge portion of the test cell and left in contact with the material specimen for the selected duration of the test. ASTM F 739 and ISO 6529 permit testing with gases, using the modifications to the test cell. Special considerations are needed for testing of gases to ensure integrity of the test cell and proper disposal of the effluent challenge gas.[19,20] Permeation testing may also be conducted against vapors of liquid chemicals per ASTM F 739. These tests require a high level of temperature control to achieve consistency in the vapor concentration of the chemical and a different orientation of the test cell[19] and have shown ways of discriminating material performance as shown in Table 12.10. Some research has also been reported for conducting permeation tests with solids. Lara and Drolet[21] describe a modified test cell where a gel-containing nitroglycerin was placed on the surface of the material's external surface for permeation testing.

Intermittent forms of chemical contact akin to splash-like exposures are prescribed in ASTM F 1383 and ISO 6529. In these test methods, the time of material specimen exposure to chemical is varied in a periodic fashion. Chemical is charged into the challenge side of the test cell and then removed after a specified time. This type of exposure may be repeated in a cyclic fashion. The use of intermittent exposure conditions gives rise to permeation curves with a cyclic appearance (see Fig. 12.5). As a consequence, breakthrough time with cumulative permeation is reported in lieu of permeation rate for these tests. Schwope et al.[22] illustrated this behavior for a number of material-chemical combinations and found the cumulative permeation to be proportional to the relative exposure time. Man et al.[23] compared permeation breakthrough times of protective clothing materials against specific chemicals using liquid contact, liquid splashes, and vapors. Their findings showed significant differences between the different exposure conditions for some combinations of materials and chemicals, but lesser changes in breakthrough time for other material-chemical sets.

Table 12.10 Permeation data for chemical vapors for selected material-chemical combinations

Material-chemical combination	Chemical challenge	Breakthrough time (minutes)	Permeation rate ($\mu g/cm^2$ min)
Ethylene dichloride against PVC glove	Saturated vapor @ 27 °C 10 ppm in nitrogen	4 4	>25,000 350
Dichloromethane against Viton-Butyl suit material	Liquid Saturated vapor @ 27 °C 100 ppm in air	16 28 No BT	470 280 Not applicable

Abbreviation: No BT – no breakthrough observed in three-hour period for testing per ASTM F739.
Source: adapted from ref. 19.

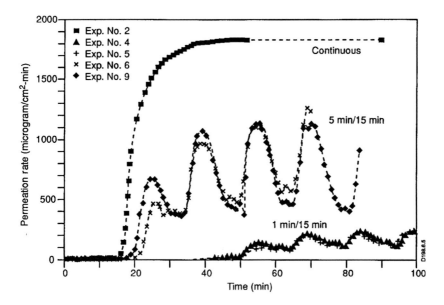

12.5 Cyclic permeation observed during permeation test involving intermittent contact (from ref. 23).

The choice of collection medium, sampling frequency and detector are determined by the chemicals to be evaluated. For example, the collection medium must have a high capacity for the permeating chemical(s), allow ready mixing, be readily analyzed for the chemical(s) of interest, and have no effect on the clothing material being tested.[24] Air, nitrogen, helium, and water are common collection media. In general, these collection media have no effect on the clothing material and are amenable to most analytical techniques. In cases where the test chemical has a relatively low vapor pressure, other approaches must be used. One approach for conducting permeation tests with these chemicals has been to use solid collection media.[22,25] This technique involves placing a solid, highly absorbent film directly against the material specimen. Ehntholt[25] designed a special test cell successfully using a silicone rubber material for collection of pesticides. An alternative approach advocated by Pinnette, Stull *et al.*[24] and Swearengen *et al.*[26] has been the use of a liquid splash collection. In these approaches, a solvent media is briefly contacted with the material specimen on the collection side and the extract evaluated for the chemical(s) of interest. The design of test system will determine whether permeant can be analyzed continuously or at some frequency of sampling. Since breakthrough time is totally dependent on the sensitivity of the detector, a detector must be used that can provide determination of concentrations that yield sufficiently low permeation rates.

Effects of temperature

Spence[27] first showed significant changes in the permeation resistance of protective clothing materials with increasing temperature as evidenced by shorter breakthrough times and larger permeation rates. Changes in temperature may have an influence on permeation by several mechanisms. Increased temperatures may increase the concentration of the challenge chemical adsorbed onto the material surface by increasing the solubility of the material-chemical matrix or by increasing the vapor pressure of the chemical.[28] The rate of diffusion step in the permeation process may also increase with temperature following an Arrhenius equation type of relationship.[27–31] Temperature, therefore, exhibits its effect on breakthrough time and permeation rate through the diffusion coefficient (D) and solubility (S). The expected effect manifests itself in a logarithmic-like relationship between permeation rate and temperature. Figure 12.6 shows this relationship for several material-chemical pairs and temperatures. Even small differences in temperature have been shown to significantly affect permeation breakthrough times as shown in Table 12.11. As a consequence, permeation testing must be performed under tightly controlled temperature conditions.

Effect of testing multi-chemical challenges

When permeation tests involve multi-chemical challenges, test configurations must employ detection techniques which permit the identification of each chemical in the permeating mixture. A number of researchers have investigated the effects of multicomponent chemical mixture permeation through barrier materials. Stampher et al.[32] investigated the permeation of PCB/paraffin oil and 1,2,4 trichlorobenzene mixtures through protective clothing. They used a small amount of isooctane in the collection medium to capture permeating PCBs. Schwope et al.[33] performed extensive testing with pesticides using different active ingredients and carrier solvents. Their tests demonstrated different breakthrough times and proportions of permeating chemicals between pesticide and carrier solvent. Bentz and Man[28] identified a case involving acetone/hexane

Table 12.11 Temperature effects on breakthrough time

Test material	Temperature (°C)	Acetone breakthrough time (minutes)
Viton/chlorobutyl laminate	20	95–98
	26.5	43–53
Chlorinated polyethylene	22	32–35
	24.5	27–31

Source: ref. 23.

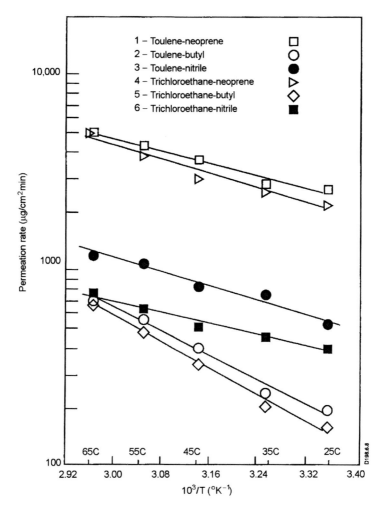

1 – Toulene-neoprene □
2 – Toulene-butyl ○
3 – Toulene-nitrile ●
4 – Trichloroethane-neoprene ▷
5 – Trichloroethane-butyl ◇
6 – Trichloroethane-nitrile ■

12.6 Plot showing effect of temperature on permeation rate for selected material-chemical combinations (from ref. 31).

mixtures where the mixture permeated a dual elastomer coated material at shorter breakthrough times than either of the pure components. This testing illustrated the potential synergistic permeation of mixtures. Mickelsen, Roder, and Berardinelli[34] evaluated elastomeric glove materials against three different binary mixtures and found similar permeation behavior where mixture permeation could not be predicted on the basis the individual mixture components. Ridge and Perkins[35] attempted to model mixture permeation using solubility parameters and found the technique to be only partially successful. Goydan *et al.*[36] were able to predict the mixture using a series of empirical rules when applied to a particular fluoropolymer laminate material.

Use and interpretation of permeation testing

As originally indicated, breakthrough time and steady-state or maximum permeation rate are typically provided as permeation test data. ASTM F 739 as well as CEN and other test methods also require reporting of key test parameters. In general, these include a complete description of the test material, test chemical, and test system configuration. Table 12.12 lists test parameters that should be reported with each test. ASTM F 1194 provides a more extensive list of testing reporting requirements

Since sensitivity significantly affects breakthrough times, ASTM, CEN, and ISO have adopted reporting requirements which are intended to normalize the effect of test system parameters on this measurement. Currently, ASTM F 739 and ASTM F 1383 specify reporting of the 'normalized' breakthrough time in additional to actual breakthrough time. Normalized breakthrough time is defined as the time when the permeation rate is equal to $0.10 \, \mu g/cm^2 min$. European test methods specify reporting breakthrough times at the time the rate equal $1.0 \, \mu g/cm^2 min$, a single order of magnitude difference with U.S. test methods. ISO 6529 permits both approaches. Therefore, it is important that permeation breakthrough time data be compared only when the respective sensitivities of the test laboratories are the same or if data is normalized on the same basis.[37]

Permeation resistance testing is the appropriate test when vapor protection is required. This does not mean that the test can only be applied for gas or vapor challenges, but rather that the test discriminates among chemical hazards at a

Table 12.12 Permeation test report parameters

Area	Reporting requirements
Test chemical	Components
	Concentration
	Source
Test material	Identification
	Source
	Condition at time of testing
	Thickness
	Unit area weight
Test system	Overall configuration (open or closed loop)
	Type of test cell
	Type of challenge (continuous or intermittent)
	Collection medium
	Collection medium flow rate
	Detector or analytical technique
Test results	Breakthrough time
	Normalized breakthrough time
	Test system sensitivity
	Steady-state or maximum permeation rate
	Cumulative permeation

molecular level owing to the sensitivity for detecting a permeating chemical in its vapor form (as opposed to liquids or solids). As such, permeation testing represents the most rigorous of chemical resistance test approaches. Table 12.13 shows representative permeation and penetration resistance data for selected materials and chemicals.

Within the protective clothing industry, many end users judge the acceptability of a material on the basis of how its breakthrough time relates to the expected period of exposure. Reporting of permeation rate offers a more consistent and reproducible means of representing material permeation. The inherent variability and test system dependence on breakthrough times make these data a less than satisfactory choice for characterizing material performance. Permeation rate data can be used to show subtle changes in material characteristics and determine cumulative (total) permeation when acceptable 'dose' levels of the test chemical can be determined. On the other hand some material-chemical systems take a long time to reach steady-state or exceed the capacity of the detector. In addition, the lack of widespread data on acceptable dermal exposure levels for most chemicals leads many specifiers to rely on breakthrough times exclusively.

The majority of permeation tests in the protective clothing industry are conducted using neat chemical continuously contacting pristine material at room temperature for a period of eight hours. General testing on common chemicals, such as those listed in ASTM F 1001 and ISO 6529, are generally performed for comparing material permeation performance (see chemicals listed in Table 12.14). Test sensitivities are at $0.10\ \mu g/cm^2 min$ or better but may be higher for difficult-to-evaluate chemicals. Other barrier materials are generally evaluated against chemicals for longer periods of time at slightly elevated temperatures for examining steady state permeation rates and cumulative permeation. These test conditions are considered worst case, because constant contact of the material with the chemical is maintained which may or may not be representative of actual use. When specific barrier product applications are identified, it is best to model the conditions of use through the selection of test parameters. If general performance is to be determined, using industry practices for test set up are preferred so that material performance may be compared against other available data.

12.5   Overall CPC integrity performance

Product integrity refers to the ability of entire clothing systems to prevent inward leakage of chemicals, whether in gaseous, liquid, or particulate forms. Tests for measuring product integrity complement material barrier tests because the clothing design has a significant effect on the overall protection to the end user. A material with good barrier characteristics against chemical vapors in clothing with a poor design (in terms of vapor or gas-tight integrity) will still result in exposure. Thus, product integrity tests evaluate the other parts of clothing such as:

Table 12.13 Comparison of penetration and permeation resistance for representative liquid splash-protective barrier materials

Chemical	PVC/nylon			Saranex/Tyvek laminate			Microporous film/nonwoven laminate		
	F903(C) Result[a]	F739 B.T.[b]	F739 P.R.[c]	F903(C) Result	F739 B.T.	F739 P.R.	F903(C) Result	F739 B.T.	F739 P.R.
Acetone	Pass	8	>50	Pass	28	3.4	Pass	<4	>50
Acetonitrile	Pass	12	25	Pass	88	0.27	Pass	<4	>50
Carbon disulfide	Fail (6)	4	>50	Pass	4	>50	Pass	<4	>50
Dichloromethane	Fail (6)	4	>50	Pass	4	>50	Pass	<4	>50
Diethylamine	Fail (20)	8	>50	Pass	20	20	Pass	<4	>50
Dimethylformamide	Fail (40)	28	>50	Pass	72	1.8	Pass	<4	>50
Ethyl acetate	Pass	8	>50	Pass	20	1.5	Pass	<4	>50
Hexane	Fail (40)	20	8	Pass	None	N/A	Pass	<4	>50
Methanol	Fail (55)	16	13	Pass	None	N/A	Pass	<4	>50
Nitrobenzene	Pass	32	>50	Pass	120	6.0	Pass	<4	>50
Sodium hydroxide	Pass	None	N/A	Pass	None	N/A	Pass	<4	>50
Sulfuric acid	Pass	120	6	Pass	None	N/A	Pass	<4	>50
Tetrachloroethylene	Fail (30)	16	>50	Pass	128	1.3	Pass	<4	>50
Tetrahydrofuran	Pass	8	>50	Pass	4	>50	Pass	<4	>50
Toluene	Fail (25)	12	>50	Pass	24	40	Pass	<4	>50

Abbreviations: B.T. – breakthrough time in minutes, P.R. – permeation rate in ig/cm²min, N/A – not applicable. Source: ref. 16.
Penetration results provide a pass or fail with penetration in parentheses; permeation tests per ASTM F739 at ambient temperature for three hours.

Table 12.14 ASTM F 1001 chemicals and key properties

Chemical	Class	Molecular weight	Vapor pressure (mm Hg)	Molar volume (cm³/mol)	Specific gravity
Acetone	Ketone	58	266	74.0	0.791
Acetonitrile	Nitrile	41	73	53.0	0.787
Ammonia	Inorganic gas	17	>760	—	N/A
1,3-Butadiene	Alkene	54	910	87.0	N/A
Carbon disulfide	Sulfur compound	76	300	62.0	1.260
Chlorine	Inorganic gas	70	>760	—	N/A
Dichloromethane	Halogen compound	85	350	63.9	1.336
Dimethylformamide	Amide	73	2.7	77.0	0.949
Ethyl acetate	Ester	88	76	99.0	0.920
Ethylene oxide	Heterocyclic	44	>760	—	N/A
Hexane	Aliphatic	86	124	131.6	0.659
Hydrogen chloride	Inorganic gas	37	>760	—	N/A
Methanol	Alcohol	32	97	41.0	1.329
Methyl chloride	Halogen compound	51	>760	—	N/A
Nitrobenzene	Nitro compound	123	≪1	102.7	1.203
Sodium hydroxide	Inorganic base	40	0	N/A	2.130
Sulfuric acid	Inorganic acid	98	<0.001	N/A	1.841
Tetrachloroethylene	Halogen compound	166	14	101.1	1.631
Tetrahydrofuran	Ether	72	145	81.7	0.888
Toluene	Aromatic	92	22	106.8	0.866

- seams
- closures
- interfaces with other clothing (such as sleeve ends with gloves and trouser cuffs with footwear)
- interfaces with other equipment (such as with a respirator).

As with material chemical resistance, there are three principal types of overall product integrity testing:

1. particulate-tight integrity
2. liquid-tight integrity
3. gas-tight integrity.

12.5.1   Particulate-tight integrity

Particle-tight integrity tests determine if particles enter whole items of CPC. In this testing, protective clothing samples are either placed on a manikin or worn by a test subject. The manikin or test subject is then exposed to a particulate atmosphere, usually an aerosol formed by a non-toxic, easily detectable liquid. Human subjects usually perform a series of exercises to put stress on the garment. The particulate-contaminated atmosphere and atmosphere inside the protective clothing are sampled either to determine the levels of particle intrusion inside the protective clothing, or measure rates of particle release from the protective clothing when worn inside a 'clean' chamber.

A number of informal test methods have been established in the United States that use corn oil or other aerosols in a chamber to measure intrusion into the PPE by sampling air. In Europe and internationally, ISO 13982-2 has been adopted as a test method, which uses a sodium chloride aerosol that is introduced into a chamber with a test subject wearing the protective clothing to be evaluated. Sampling of air to determine salt concentration both inside and outside the clothing is used to determine clothing effectiveness against particles. Garment performance is usually characterized in terms of 'intrusion coefficient' which is the ratio of the outside contaminant concentration to the concentration of contaminant measured on the inside. Higher intrusion coefficients indicate garments with better resistance to penetration by particles.

12.5.2   Liquid-tight integrity

Liquid integrity tests determine if liquid enters to the interior side of the CPC or onto wearer underclothing. Methods for measuring liquid-tight integrity involve spraying protective clothing on a manikin with a liquid and observing penetration of liquid onto an inner liquid absorptive garment or the interior of the clothing. Water is often treated with either surfactant to lower surface tension and to better simulate organic liquids, and/or dyes to enhance detection of penetrating liquid. In general, tests may be conducted at periods longer than expected exposure to assist in observing leakage. Several techniques can be used to quantify leakage such as in the use of spectrophometry on dye-based liquid challenges and electroconductivity for salt-based liquid challenges. However, the majority of testing is performed with visual detection of liquid leakage. Most results are therefore reported as detected penetration in terms of passing or failing performance. Observations of areas where leakage occurred may also be reported to help determine problems in the product that limit the clothing item's integrity. Table 12.15 provides a comparison of these methods.

The available test methods provide slightly different test approaches for measuring liquid-tight integrity:

- ASTM F 1359 involves placing clothing on a manikin and spraying the manikin with surfactant-treated water from five different nozzles. The volumetric flow out of each nozzle is specified. In various applications of this test method, the liquid is sprayed at the manikin from 4 minutes to one hour. During the exposure period, the manikin is rotated through four orientations to completely challenge the garment. An inner liquid-absorptive garment worn on the manikin underneath the clothing is used for detecting liquid penetration. The garment interior can also be examined. Test results are recorded as 'pass' or 'fail'.
- EN 468 is similar to ASTM F 1359, in that it involves surfactant treated water that is sprayed at the protective clothing in a specified pattern over a specified duration. However, this test involves a human test subject (who wears a liquid-absorptive garment and performs a series of exercises) and requires the use of a dye. Performance of the garment is determined by measuring the total area of dye staining on the liquid-absorptive garment and relating that area to the area of a 'calibrated' stain representing a specific volume of penetrating liquid.
- In EN 463, an aqueous jet, containing a fluorescent or visible dye tracer, is directed under controlled conditions at chemical protective clothing worn by a test manikin or human test subject. Inspection of the inside surface of the clothing and the outside surface of absorbent clothing worn underneath allows any points of inward leakage to be identified.

ISO 17491 covers all three approaches and provides criteria for applying each test:

- Method C specifies a method for determining the resistance of chemical protective clothing to penetration by jets of liquid chemicals (analogous to EN 463). This procedure is applicable to clothing which may comprise one or more items and which is intended to be worn where there is a risk of exposure to a forceful projection of a liquid chemical. It is also applicable to clothing which is intended to be resistant to penetration under conditions which

Table 12.15 Characteristics of liquid-tight integrity tests for complete chemical protective clothing

Test method	Clothing placement	Test liquid	Liquid application	Method of detection
ASTM F 1359 (also ISO 17491, Procedure E)	Clothing placed on manikin that is dressed in liquid-absorptive garment	Water treated with surfactant to 32 dynes/cm	Liquid sprayed from 5 nozzles positioned above and to sides of test clothing at 3.0 L/min from each nozzle for 60 minutes; manikin is rotated through 4 positions during test	Interior of garment and liquid-absorptive garment is examined for signs of liquid penetration; results are pass/fail
EN 463 (also ISO 17491, Procedure C)	Human subject wears test clothing over liquid-absorptive garment; alternatively clothing may be placed on manikin	Water treated with fluorescent dye	Liquid directed at specific locations on test clothing at pressure of 180 kPa at distance of 1 m for 5 seconds; if human subject is used, subject remains stationary	Staining of liquid-absorptive garment is compared with calibrated stain (0.1 ml); approximate area of staining is reported
EN 468 (also ISO 17491, Procedure D)	Human subject wears test clothing over liquid-absorptive garment; subject stands on turntable	Water treated with surfactant to 30 dynes/cm and fluorescent dye	Liquid sprayed from 4 nozzles positioned vertically on test stand beside test subject (at distance of 1.5 m); liquid is sprayed at 1.14 L/min from each nozzle for 1 minute; during spray, subject performs stationary exercises	Staining of liquid-absorptive garment is compared with calibrated stain (0.1 ml); approximate area of staining is reported
ISO 17491 Procedure F	Human subject wears test clothing over liquid-absorptive garment; subject stands on turn-table	Water treated with surfactant to 52.5 dynes/cm an fluorescent dye	Same as EN 468 (or Procedure D of ISO 17941) except that flow rate is at 0.47 L/min	Staining of liquid-absorptive garment is compared with calibrated stain (0.1 ml); approximate area of staining is reported

require total body surface cover but do not demand the wearing of gas-tight clothing.

- Method D specifies a method for determining the resistance of chemical protective clothing to penetration by sprays of liquid chemicals (based on EN 468). This procedure applies to protective clothing which may comprise one or more items and which is intended to be worn when there is a risk of exposure to slight splashes of a liquid chemical or to spray particles that coalesce and run off the surface of the garment.
- Method E specifies an alternative method for determining the resistance of chemical protective clothing to penetration by sprays of liquid chemicals (based on ASTM F 1359). This procedure applies to protective clothing which may comprise one or more items and which is intended to be worn when there is a risk of exposure to slight splashes of a liquid chemical or to spray particles that coalesce and run off the surface of the garment.
- Method F is a modification of Method D where the spray has been modified to light spray or mist by use of different nozzles and spray conditions. This method is intended for partial body protective clothing where the likelihood of splash exposure is low.

12.5.3   Gas-tight integrity

Gas-tight integrity tests determine if gas or vapor can penetrate protective clothing. Gas-tight integrity testing can only be performed on items that can be sealed, including totally encapsulating suits. The most common approach for testing gas-tight integrity of protective clothing is to inflate the item to a specified pressure and then observe for any change in pressure within the item after several minutes. Using a soapy water solution on exterior of the clothing can then identify the location of leaks. Alternative approaches involve placing PPE on a test subject in closed environment containing a gas (such as ammonia) and measuring concentration of challenge agent inside suit. In some rare cases, full-scale testing of suits against actual hazardous materials has been performed in special projects.

Both ASTM F 1052 and EN 464 involve similar testing approaches where the protective clothing is inflated and left at a specific pressure. ASTM F 1052 uses a lower test pressure than EN 464, but the results are similar. Extensive work by Carroll[38] has demonstrated that a test pressure of 1 kPa over a four-minute period is sensitive enough to evaluate most encapsulating clothing for very small leaks that could contribute to poor integrity. ISO 17491 involves a challenge environment where a test subject wears the clothing inside a closed chamber in contact with a test agent. The subject performs a series of exercises while wearing the clothing. The interior of the clothing is either measured during or after the exposure to determine the ratio of the concentration on the outside to the ratio of the concentration of test agent on the inside. This ratio, known as the intrusion coefficient, is used to measure clothing performance similar to the way protection factors are used to judge the performance of respirators.

12.6   Other CPC performance properties

A number of other methods for evaluating CPC garments and materials are often needed for assessing performance properties other than chemical resistance and integrity. These properties relate to material physical properties and ergonomic properties.

12.6.1   CPC material physical properties

A number of other properties are important for judging the effective of chemical protective clothing selections. Physical properties assess or determine:

- weight and thickness of clothing and materials
- strength of clothing
- resistance to specific physical hazards
- product durability.

In general, more than one test method may be available for a performance property based on the type of material or product. In addition, many material physical property results are reported in different material directions. Results that are parallel to the direction the material is fabricated or comes off the roll are referred to as warp direction for woven textile fabrics, course direction for knit textile fabrics, and machine direction for nonwoven textile, rubber, and plastic materials. Results that are perpendicular to the direction the material is fabricated or comes off the roll are referred to as the fill or weft direction for woven textile fabrics, wales or weft direction for knit textile fabrics, and cross-machine direction for nonwoven textile, rubber, and plastic materials.

The multitude of different physical properties prevents detailed coverage of this performance area in this chapter. Typical physical properties applied to chemical protective clothing include:

- Weight is reported as the weight of a material per unit area.
- Thickness is reported as the nominal or average thickness of a material in mils (thousands of an inch) or millimeters.
- Breaking strength measures the force required to break clothing materials or components when items are pulled along one direction; breaking strength tests can also be used to measure the strength of seams and closures.
- Burst strength measures the force or pressure required to rupture clothing materials or components when a force is directed perpendicular to the item; this property may be related to ability of PPE materials to prevent items from protruding through garments.

- Tear or snag resistance measures the force required to continue a tear in a clothing material once initiated or the resistance of a material in preventing a tear or snag from occurring.
- Abrasion resistance measures the ability of clothing surfaces or materials to resist wearing away when rubbed against other surfaces.
- Cut resistance measures the ability of clothing items or materials to resist cutting through by a sharp-edged objects or machinery.
- Puncture resistance measures the ability of clothing items or materials to resist penetration by a slow-moving, pointed object.
- Flex fatigue resistance measures the ability of clothing items or materials to resist wear or other damage when repeatedly flexed.
- Flame resistance determines if material will ignite and continue to burn after exposed to flame.

Table 12.16 provides representative test methods for many of the above listed physical properties. For each physical property reviewed, it is important to understand how the test method related to the intended use of the protective clothing and how to interpret the measurement. For some properties, increasing values generally mean better performance. But physical properties that are too

Table 12.16 List of representative physical performance test methods

Physical property	Type of test/material application	Available test methods
Weight	Woven and knit textiles	ASTM D 3776
		ISO 3801
	Nonwoven textiles	ASTM D 1117
		ISO 9073-1
	Coated fabrics	ASTM D 751
		ISO 2286
Thickness	Woven and knit textiles	ASTM D 1777
		ISO 5084
	Nonwoven textiles	ASTM D 5736
		ISO 9073-2
	Coated fabrics	ASTM D 751
		ISO 2286
Breaking strength	Grab method – textiles	ASTM D 5034
		EN ISO 13934-2
	Strip method – textiles	ASTM D 5035
		EN ISO 13934-1
	Nonwoven textiles	ISO 9073-3
	Rubber/coated fabrics	ASTM D 412
		ASTM D 751
		ISO 1421
	Plastic	ASTM D 638
		ASTM D 882
		ISO 1421

Table 12.16 Continued

Physical property	Type of test/material application	Available test methods
	Seam strength	ASTM D 751
		ASTM D 1683
		EN ISO 13934-1
Burst strength	Ball method – textiles	ASTM D 3787
		ASTM D 3940
	Ball method – rubber or plastic	ISO 3303
	Diaphragm method – textiles	ASTM D 3786
	Diaphragm method – rubber or plastic	ISO 2960
	Coated fabrics	ASTM D 751
Tear resistance	Elmendorf method – textiles	ASTM D 1424
		ISO 13937
	Elmendorf method – plastics	ASTM D 1922
	Tongue tear – textiles	ASTM D 2261
		ISO 13937
	Tongue tear – nonwoven textiles	ASTM D 5735
	Tongue tear – plastics	ASTM D 1938
	Trapezoidal method – textiles	ASTM D 5733
		ISO 9073-4
	Tear resistance – plastics	ASTM D 1004
		ISO 4674
	Tear resistance – coated fabrics, rubber	ASTM D 751
		ISO 4676
	Puncture propagation tear resistance	ASTM D 2582
		ISO 13995
Abrasion resistance	Taber method – textiles	ASTM D 3884
	Taber method – coated fabrics	ASTM D 3389
		ISO 5470-1
	Taber method – plastics	ASTM D 1044
	Wyzenbeek method – textiles	ASTM D 4157
	Martindale method – clothing materials	ASTM D 4966
		EN 530
		ISO 12974
	Flexing/abrading method – textiles	ASTM D 3885
	Flexing/abrading method – coated fabrics	ISO 5981
	Inflated diaphragm method – textiles	ASTM D 3886
	Uniform method – textiles	ASTM D 4158
Cut resistance	Clothing materials	ASTM F 1790
		ISO 13997
Puncture resistance	Clothing materials	ASTM F 1342
		EN 863
		ISO 13996
Flex fatigue	Flat material	ASTM F 392
		ISO 7854
Flammability	Clothing materials	ASTM F 1358
		EN 532
		ISO 15025

ASTM – American Society for Testing and Materials; EN – European Norm; ISO – International Standards Organization.

high may also create a tradeoff for such properties as stiffness or aspects of clothing related to use.

12.6.2   CPC ergonomic properties

Ergonomic properties describe how CPC affects the wearer in terms of functionality, fit, comfort, and overall well-being. Most ergonomic properties represent tradeoffs with protection, for example, barriers to chemicals versus thermal comfort. Common ergonomic properties for CPC include:

- Material compatibility evaluates the potential for skin irritation or adverse reactions due to contact with certain substances which may be present in or on clothing.
- Thermal insulation and breathability evaluate the ability of CPC materials to allow the passage of air, moisture, and the heat associated with body evaporative cooling and environmental conditions.
- Mobility and range of motion evaluate the effects of CPC on wearer function in performing work tasks.
- Clarity and field of vision evaluate the ability of an individual to see through a visor or a faceshield; field of vision testing evaluates peripheral vision for an individual wearing the visor or faceshield.
- Ease of communication evaluates the ability of CPC to allow intelligible (understood) communication of the wearer with other persons.
- Sizing and fit determines how well clothing fits the individual wearer.
- Donning and doffing ease evaluates how easily or how quickly individuals can put on and take off protective clothing.

Most ergonomic properties are difficult to measure and must employ human subjects. For this reason, most human-factor evaluations are conducted as wear trials where end users can rate or rank different properties related to function and comfort for different clothing items. In some cases, standardized test methods exist for measuring human factors. Representative standard test methods for ergonomic properties described in this subsection are summarized in Table 12.17.

12.7   CPC specification and classification standards

A number of overall standard specifications or classifications for CPC have been developed. Specifications generally establish minimum performance criteria while classifications provide a means for ranking products based on standard test methods. In the United States, the primary protective clothing standards that have been developed are related to emergency response applications. These include NFPA 1991 and NFPA 1992. Both standards cover ensemble that are the combination of suits or garments, with gloves and footwear for complete body protection. Performance criteria are provided in terms of:

Table 12.17 List of representative human factors test methods

Human factor	Type of test/application	Available test methods
Material biocompatibility	Medical device (gloves, gowns) biocompatibility tests	ISO 10993
	Natural rubber latex proteins	ASTM D 5712
Thermal insulation and breathability	Air permeability – materials	ASTM D 737 ISO 9237
	Moisture vapor transmission – materials	ASTM E 96 CAN/CGSB-4.2 No. 49 ISO 15106
	Thermal resistance – materials	ASTM D 1518 ISO 5085
	Thermal resistance – garments	ASTM F 1291
	Total heat loss – materials	ASTM F 1868 ISO 11092
Mobility and range of motion	Overall product evaluation	ASTM F 1154 NFPA 1991, Section 8.4 EN 943-1, Clause 7.2
Clarity and field of vision	Clarity/field of vision	NFPA 1991, Section 8.4 EN 943-1, Clause 7.2
	Haze and transmittance	ASTM D 1003 ISO 14782
Sizing and fit	Garment sizing/dynamic fit test	ASTM F 1731 ANSI/ISEA 101 EN 340 ISO 13688
Donning and doffing ease	Full body suits	NFPA 1991, Section 8.4

ANSI – American National Standards Institute; ASTM – American Society for Testing and Materials; EN – European Norm; CAN/CGSB – Canadian General Standards Board; ISO – International Standards Organization; NFPA – National Fire Protection Association.

- overall ensemble integrity
- material chemical resistance
- material flame resistance
- material physical properties
- clothing human factors
- component function (e.g., airline pass-throughs for respiratory equipment and exhaust valves).

Two different levels are established by these standards based on overall ensemble integrity and material chemical resistance. This scheme associated clothing gas-tight integrity and material permeation resistance with vapor protection and clothing liquid-tight integrity, and material penetration resistance with liquid splash protection. Using this hierarchy, the NFPA attempted to define protective clothing types on the basis of needed performance as demonstrated by test methods designed to measure the type of protection provided. In addition to performance criteria, these standards establish requirements for third-party certification and labeling of suits and clothing for demonstrating compliance with these standards. Manufacturers must also meet certain design criteria and provide documentation to the end user such as instructions and technical data supporting claims against the standard.

In Europe, several standards have been developed related to chemical protective clothing and equipment. These include:

- EN 465 (Type 4: spray-tight chemical protective clothing)
- EN 466 (Type 3: liquid-tight chemical protective clothing)
- EN 467 (partial body chemical protective clothing)
- EN 943-1 (Types 1 and 2: gas-tight chemical protective clothing)
- EN 943-2 (Type 1-ET: gas-tight chemical protective clothing for emergency teams)
- prEN 13034 (Type 6: chemical protective clothing with limited liquid protection)
- EN 13982-2 (Type 5: particulate-tight chemical protective clothing)
- prEN 14605 (Type 3, 4, and partial body chemical protective clothing) to replace EN 465, 466, and 467.

Each standard defines a specific type of chemical protective clothing with differences in the intended protection, based on general clothing design, clothing integrity and the type of chemical resistance provided. In each of the standards (with the exception of EN 943-2), clothing performance is classified in terms of material performance for chemical resistance, physical properties, and other properties. Each standard also specifies certain minimum design and integrity performance for the respective type of clothing.

Under the International Standards Organization, an effort to harmonize requirements for chemical protective clothing between Europe and North American has resulted in a proposed standard, ISO 16602. In essence, this standard consolidates many of the clothing types and performance classifications into one standard.

12.8 Summary

A wide variety of clothing and equipment is available in the workplace for industrial chemical protection. These items can be 'type' classified by their general design, performance, and service life. Specific details related to the choice of materials, design features, construction methods, and performance features further serve to differentiate among the type of chemical protective clothing. An understanding of performance tests affecting both materials and the overall product are especially important for determining the appropriateness of different CPC items for specific applications.

12.9   References

1. Stull J O, 'Selecting chemical protective clothing,' *Occup Health & Safety*, December 1995, 20–24.

2. Bray A R and Stull J O, 'A new nondestructive inspection method to determine fatigue in chemical protective suit and shelter materials,' *Performance of Protective Clothing: Fifth Volume, ASTM STP 1237* (J S Johnson and S Z Mansdorf, eds.), American Society for Testing and Materials, Philadelphia, 1996, 281–95.

3. Dudzinski D J, 'Kennedy space center maintenance program for propellant handlers ensembles,' *Performance of Protective Clothing: Second Symposium, ASTM STP 989* (S Z Mansdorf, R G Sager, and A P Nielson, eds.), American Society for Testing and Materials, Philadelphia, PA, 1988, 492–500.

4. Garland C E and Torrence A M, 'Protective clothing materials: Chemical contamination and decontamination concerns and possible solutions,' *Performance of Protective Clothing: Second Symposium, ASTM STP 989* (S Z Mansdorf, R G Sager, and A P Nielson, eds.), American Society for Testing and Materials, Philadelphia, PA, 1988, 368–75.

5. Schwope A D and Renard E, 'Estimation of the cost of using chemical protective clothing,' *Performance of Protective Clothing: Fourth Volume, ASTM STP 1133*, J P McBriarty and N W Henry, eds., ASTM, Philadelphia, 1992, 972–81.

6. Fraser A J and Keeble V B, 'Factors influencing design of protective clothing for pesticide application,' *Performance of Protective Clothing: Second Symposium, ASTM STP 989* (S Z Mansdorf, R G Sager, and A P Nielson, eds.), American Society for Testing and Materials, Philadelphia, PA, 1988, 565–72.

7. Baars D M, Eagles D B and Edmond J A, 'Test method for evaluating adsorptive Fabrics,' *Performance of Protective Clothing, ASTM STP 900* (R L Barker and G C Coletta, eds.), American Society for Testing and Materials, Philadelphia, PA, 1986, 39–50.

8. Keeble V B, Prevatt M B and Mellian S A, 'An evaluation of fit of protective coveralls manufactured to a proposed revision of ANSI/ISEA 101,' *Performance of Protective Clothing: Fourth Volume, ASTM STP 1133* (J P McBriarity and N W Henry, eds.), American Society for Testing and Materials, Philadelphia, 1992, 675–90.

9. Stull J O, 'Performance standards and testing of chemical protective clothing,' *Proceedings for Clemson University's Sixth Annual Conference on Protective Clothing*, Greenville, South Carolina, 1992.

10. Coletta G C, Mansdorf S Z and Berardinelli S P, 'Chemical protective clothing standard test method development: Part II. Degradation test method,' *Am Ind Hyg Assoc J, 1988,* **49,** 26–33.

11. Schwope A D, Carroll T R. Haung R and Royer M D, 'Test kit for evaluation of the chemical resistance of protective clothing,' *Performance of Protective Clothing: Second Symposium, ASTM STP 989* (S Z Mansdorf, R Sager, and A P Nielsen, eds.), American Society for Testing and Materials, Philadelphia, 1988, p. 314–30.

12. Stull J O, 'Early development of a hazardous chemical protective ensemble,' Final Report CG-D-24-86, AD A174, 885, Washington, D.C., 1986.

13. Van Amerongen G J, 'Diffusion in elastomers,' *Rubber Chem Technol,* 1964, **37,** 1065–81.

14. Mansdorf S Z and Berardinelli S P, 'Chemical protective clothing standard test method development, Part 1. Penetration method,' *Am Ind Hyg Assoc J,* 1988, **49,** 21–5.

15. Berardinelli S P and Cottingham L, 'Evaluation of chemical protective garment seams and closures for resistance to liquid penetration,' *Performance of Protective Clothing, ASTM STP 900* (R L Barker and G C Coletta, eds.), American Society for Testing and Materials, Philadelphia, 1986, 263–75.

16. Stull J O, White D F and Greimel T C., 'A comparison of the liquid penetration test with other chemical resistance tests and its application in determining the performance of protective clothing,' *Performance of Protective Clothing: Fourth Volume*, ASTM STP 1133 (J P McBriarity and N W Henry, eds.), American Society for Testing and Materials, Philadelphia, 1992, pp. 123–38.

17. Stull J O., 'Performance standards for improving chemical protective suits,' *Chemical Protective Clothing Performance in Chemical Emergency Response, ASTM STP 1037* (J L Perkins and J O Stull, eds.), American Society for Testing and Materials, Philadelphia, 1989, 245–64.

18. Henry N W and Schlatter C N, 'The development of a standard method for evaluating chemical protective clothing to permeation by liquids,' *Am Ind Hyg Assoc J*, 1981, **42**, 202–7.

19. Stull J O and Pinette, M F S, 'Special considerations for conducting permeation testing of protective clothing materials with gases and chemical vapors,' *Am Ind Hyg Assoc J*, 1900, **51**, 378–85.

20. Stull J O, Pinette M F S, and Green R, 'Permeation of ethylene oxide through protective clothing materials for chemical emergency response,' *App Ind Hyg*, 1990, **5** 448–62.

21. Lara J and Drolet D, 'Testing the resistance of protective clothing materials to nitroglycerin and ethylene glycol dinitrate,' *Performance of Protective Clothing: Fourth Volume, ASTM STP 1133* (J P McBriarity and N W Henry, eds.), American Society for Testing and Materials, Philadelphia, 1992, pp. 153–61.

22. Schwope, A D, Goydan R, Carroll T R, and Royer, M D, 'Test methods development for assessing the barrier effectiveness of protective clothing materials,' Proceedings of the Third Scandinavian Symposium on Protective Clothing Against Chemicals and Other Heath Risks, Norwegian Defence Research Establishment, Gausdal, Norway, September 1989.

23. Man, V L, Bastecki V, Vandal G, and Bentz A P, 'Permeation of protective clothing materials: Comparison of liquid contact, liquid splashes, and vapors on breakthrough times,' *Am Ind Hyg Assoc J*, 1987, **48**, 551–62.

24. Pinette M F S, Stull J O, Dodgen C R, and Morley M G, 'A new permeation testing collection method for non-volatile, non-water soluble chemical challenges of protective clothing,' *Performance of Protective Clothing: Fourth Volume, ASTM STP 1133* (J P McBriarity and N W Henry, eds.), American Society for Testing and Materials, Philadelphia, 1992, 339–49.

25. Ehntholt D J, Cerundolo D L, Dodek I, Schwope A D, Royer M D, and Nielsen A P, 'A test method for the evaluation of protective glove materials used in agricultural pesticide operations,' *Am Ind Hyg Assoc J*, 1990 **51**, 462–68.

26. Swearengen P M, Johnson J S, and Priante S, 'A modified method for fabric permeation testing,' Paper presented at 1991 Am Ind Hyg Conf, Salt Lake City, Utah, 1991.

27. Spence M W, 'Chemical permeation through protective clothing materials: An evaluation of several critical variables,' Presented at Am. Ind. Hyg. Conference, Portland, Oregon, 1981.

28. Bentz A P and Man V L, 'Critical variables regarding permeability of materials for

totally encapsulating suits,' Proceedings of the 1st Scandinavian symposium on protective clothing against chemicals, Copenhagen, Denmark, 1984.

29. Nelson G O, Lum B Y, and Carlson G J, 'Glove permeation by organic solvents,' *Am Ind Hyg Assoc J*, 1981, **42**, 217–24.

30. Huang R Y and Lin V J C, *J App Polymer Sci*, 1968, **12**, 2615–23.

31. Vahdat N and Bush M, 'Influence of temperature on the permeation properties of protective clothing materials,' *Chemical Protective Clothing Performance in Chemical Emergency Response, ASTM STP 1037*, (J L Perkins and J O Stull, eds.), American Society for Testing and Materials, Philadelphia, 1989, 132–45.

32. Stampher J F, McLeod M J, Bettis M R, Martinez A M, and Berardinelli S P, 'Permeation of polychlorinated biphenyls and solutions of these substances through selected protective clothing materials,' *Am Ind Hyg Assoc J*, 1984, **45**, 634–41.

33. Schwope A D, Goydan R, Ehntholt D J, Frank U, and Nielsen A P, 'Permeation resistance of glove materials to agricultural pesticides,' *Performance of Protective Clothing: Fourth Volume, ASTM STP 1133* (J P McBriarity and N W Henry, eds.), American Society for Testing and Materials, Philadelphia, 1992, 198–209.

34. Mickelsen R L, Roder M M, and Berardinelli S P, 'Permeation of chemical protective clothing by three binary solvent mixtures,' *Am Ind Hyg Assoc J*, 1986, **47**, 236–40.

35. Ridge M C and Perkins J L, 'Permeation of Solvent mixtures through protective clothing elastomers,' *Chemical Protective Clothing Performance in Chemical Emergency Response, ASTM STP 1037* (J L Perkins and J O Stull, eds.), American Society for Testing and Materials, Philadelphia, 1989, 113–31.

36. Goydan R, Powell J R, Bentz A P, and Billing C B, 'A computer system to predict chemical permeation through fluoropolymer-based protective clothing,' *Performance of Protective Clothing: Fourth Volume, ASTM STP 1133* (J P McBriarity and N W Henry, eds.), American Society for Testing and Materials, Philadelphia, 1992, 956–71.

37. Jamke R A, 'Understanding and using chemical permeation data in the selection of chemical protective clothing,' *Chemical Protective Clothing Performance in Chemical Emergency Response, ASTM STP 1037* (J L Perkins and J O Stull, eds.), American Society for Testing and Materials, Philadelphia, 1989, 11–22.

38. Carroll T R, Resha C J, Vencill C T, and Langley J D, 'Determining the sensitivity of international test methods design to assess the gas-tight integrity of fully encapsulating garments,' *Performance of Protective Clothing, Sixth Volume*, ASTM STP 1273 (J O Stull and A D Schwope, eds.), American Society for Testing and Materials, 1997, 3–15.

12.10   Appendix: list of referenced standards

American Association of Textile Chemist and Colorist (AATCC) Test Methods

AATCC 42, Water resistance: Impact penetration test.
AATCC 127, Water resistance: Hydrostatic pressure test.

American Society for Testing and Materials (ASTM)

ASTM D 412, Test methods for vulcanized Rubber and thermoplastic rubbers and thermoplastic elastomers – Tension.

ASTM D 471, Test method for rubber property – Effect of liquids.

ASTM D 543, Practices for evaluating the resistance of plastics to chemical reagents.

ASTM D 638, Test method for tensile properties of plastics.

ASTM D 737, Test method for air permeability of textile fabrics.

ASTM D 751, Methods for testing coated fabrics.

ASTM D 882, Test methods for tensile properties of thin plastic sheeting.

ASTM D 1003, Standard test method for haze and luminous transmittance of transparent plastics.

ASTM D 1004, Test method for initial tear resistance of plastic film and sheeting.

ASTM D 1044, Test Method for resistance of transparent Plastics to surface abrasion.

ASTM D 1117, Methods of testing nonwoven fabrics.

ASTM D 1242, Test methods for resistance of plastic materials to abrasion.

ASTM D 1424, Test method for tear resistance of woven fabrics by falling pendulum (Elmendorf) apparatus.

ASTM D 1518, Test method for thermal transmittance of textile material.

ASTM D 1683, Test method for failure in sewn seams of woven fabrics.

ASTM D 1777, Method for measuring thickness of textile materials.

ASTM D 1922, Test method for propagation tear resistance of plastic film and thin sheeting by pendulum method.

ASTM D 1938, Test method for tear propagation resistance of plastic film and thin sheeting by a single-tear method.

ASTM D 2261, Test method for tearing strength of woven fabrics by the tongue (single rip) method (constant-rate-of-extension tensile testing machine).

ASTM D 2582, Test method for puncture-propagation tear resistance of plastic film and thin sheeting.

ASTM D 3776, Test methods for mass per unit area (weight) of woven fabric.

ASTM D 3786, Test method for hydraulic bursting strength of knitted goods and nonwoven fabrics: Diaphragm bursting strength tester method.

ASTM D 3787, Test method for bursting strength of knitted goods: Constant-rate-of traverse (CRT), ball burst test.

ASTM D 3884, Test method for abrasion resistance of textile fabrics (rotary platform, double-head method).

ASTM D 3885, Test method for abrasion resistance of textile fabrics (flexing and abrasion method).

ASTM D 3886, Test method for abrasion resistance of textile fabrics (inflated diaphragm method).

ASTM D 3940, Test method for bursting strength (load) and elongation of sewn seams of knit or woven stretch textile fabrics.

ASTM D 4157, Test method for abrasion resistance of textile fabrics (oscillatory cylinder method).

ASTM D 4158, Test method for abrasion resistance of textile fabrics (uniform abrasion method).

ASTM D 4966, Test Method for abrasion resistance of textile fabrics (martindale abrasion tester method).

ASTM D 5034, Test method for breaking force and elongation of textile fabrics (grab test).

ASTM D 5035, Test method for breaking force and elongation of textile fabrics (strip test).

ASTM D 5712, Test method for analysis of proteins in natural rubber and its products.

ASTM D 5733, Test method for tearing strength of nonwoven fabrics by the trapezoid procedure.

ASTM D 5735, Test method for tearing strength of nonwoven fabrics by the tongue (single rip) procedure (constant-rate-of-extension tensile testing machine).

ASTM D 5736, Test method for thickness of highloft nonwoven fabrics.

ASTM E 96, Test methods for water vapor transmission of materials.

ASTM F 392, Test method for flex durability of flexible barrier materials.

ASTM F 739, Test method for resistance of protective clothing materials to permeation by liquids and gases under conditions of continuous contact.

ASTM F 903, Test method for resistance of protective clothing materials to penetration by liquids.

ASTM F 1052, Pressure testing of gas-tight totally encapsulating chemical protective suits.

ASTM F 1154, Practices for qualitatively evaluating the comfort, fit, function, and integrity of chemical protective suit ensembles.

ASTM F 1194, Guide for documenting the results of chemical permeation testing on protective clothing materials.

ASTM F 1291, Test method for measuring the thermal insulation of clothing using a heated manikin.

ASTM F 1342, Test method for protective clothing material resistance to puncture.

ASTM F 1358, Test method for effects of flame impingement on materials used in protective clothing not designated primarily for flame resistance.

ASTM F 1359, Liquid penetration resistance of protective clothing or protective ensembles under a shower spray while on a manikin.

ASTM F 1383, Test method for resistance of protective clothing materials to permeation by liquids and gases under conditions of intermittent contact.

ASTM F 1407, Test method for resistance of chemical protective materials to liquid permeation - permeation cup method.

ASTM F 1731, Practice for body measurements and sizing of fire and rescue services uniforms and other thermal hazard protective clothing.

ASTM F 1790, Test method for measuring cut resistance of materials used in protective clothing.

ASTM F1868, Test method for thermal and evaporative resistance of clothing materials using a sweating hot plate.

ASTM F 2130, Test method for measuring repellency, retention, and penetration of liquid pesticide formulation through protective clothing materials.

Canadian General Standards Board (CGSB)

CAN/CGSB-4.2 No. 49, Resistance of materials to water vapour diffusion.

European Committee on Standardization (CEN)

EN 368, Protective clothing – Protection against liquid chemicals – Test Method: Resistance of materials to penetration by liquids.

EN 369, Protective clothing – Protection against liquid chemicals – Test method: Resistance of materials to permeation by liquids.

EN 340, Protective clothing – General requirements.

EN 463, Protective clothing – Protection against liquid chemicals – Test method: Determination of resistance to penetration by a jet of liquid.

EN 464, Protective clothing – Protection against liquid and gaseous chemicals, including liquid aerosols and solid particles – Test method: Determination of leak-tightness of gas-tight suits.

EN 465, Protective clothing – Protection against liquid chemicals – Performance requirements for chemical protective clothing with spray-tight connections between different parts of the clothing (Type 4 Equipment).

EN 466, Protective clothing – Protection against liquid chemicals – Performance requirements for chemical protective clothing with liquid-tight connections between different parts of the clothing (Type 3 Equipment).

EN 467, Protective clothing – Protection against liquid chemicals – Performance requirements for garments providing chemical protection to parts of the body.

EN 468, Protective clothing – Protection against liquid chemicals – Test method: Determination of resistance to penetration by spray.

EN 530, Abrasion resistance of protective clothing material – Test method.

EN 532, Protective clothing – Protection against heat and flame – Method of test for limited flame spread.

EN 863, Protective clothing – Mechanical properties – Test method: Puncture resistance, 1995.

EN 943-1, Protective clothing against liquid and gaseous chemicals, including aerosols and solid particles – Part 1: Performance requirements for ventilated and non-ventilated 'gas tight' (Type 1) protective clothing and 'non-gas-tight' (Type 2) protective clothing.

EN 943-2, Protective clothing against liquid and gaseous chemicals, including aerosols and solid particles – Part 2: Performance requirements for 'gas tight'

(Type 1) protective clothing for emergency teams (ET).

prEN 13034, Protective clothing against liquid chemicals – Performance requirements for chemical protective suits offering limited protective performance against liquid chemicals (Type 6 equipment).

prEN 14605, Protective clothing against liquid chemicals – Performance requirements for protective clothing.

prEN 14325, Protective clothing against liquid chemicals – Material testing and classification.

International Safety Equipment Association (ISEA)

ANSI/ISEA 101, American National Standard for Limited-Use and Disposable Coveralls – Size and Labeling Requirements.

ANSI/ISEA 105, American National Standard for Hand Protection Selection Criteria.

International Standards Organization (ISO)

ISO 811, Textile fabrics – Determination of resistance to water penetration – Hydrostatic pressure test.

ISO 1420, Rubber- or plastics-coated fabrics – Determination of resistance to penetration by water.

ISO 1421, Fabrics coated with rubber or plastics – Determination of breaking strength and elongation at break.

ISO 2286, Rubber- or plastics-coated fabrics – Determination of roll characteristics.

ISO 2960, Textiles – Determination of bursting strength and bursting distension – Diaphragm method.

ISO 3303, Rubber- or plastics-coated fabrics – Determination of bursting strength.

ISO 3801, Textiles – Woven fabrics – Determination of mass per unit length and mass per unit area.

ISO 4674, Fabrics coated with rubber or plastics – Determination of tear resistance.

ISO 5084, Textiles – Determination of thickness of textiles and textile products.

ISO 5085, Textiles – Determination of thermal resistance Part 2; High thermal resistance.

ISO 5470-1, Rubber- or plastics-coated fabrics – Determination of abrasion resistance – Part 1: Taber abrader.

ISO 5981, Rubber- or plastics-coated fabrics – Determination of flex abrasion.

ISO 6529, Protective Clothing – Protection against chemicals – Determination of resistance of protective clothing materials to permeation by liquids and gases.

ISO 6530, Protective clothing – Protection against liquid chemicals – Determination of resistance of materials to penetration by liquids.

ISO 7854, Rubber- or plastics-coated fabrics – Determination of resistance to damage by flexing, 1995.

ISO 9073-1, Textiles – Test methods for nonwovens – Part 1: Determination of mass per unit area.

ISO 9073-2, Textiles – Test methods for nonwovens – Part 2: Determination of thickness.

ISO 9073-3, Textiles – Test methods for nonwovens – Part 3: Determination of tensile strength and elongation.

ISO 9073-4, Textiles – Test methods for nonwovens – Part 4: Determination of tear resistance.

ISO 9237, Textiles – Determination of the permeability of fabrics to air.

ISO 10993-1, Biological evaluation of medical devices.

ISO 11092, Textiles-Physiological effects – Measurement of thermal and water-vapour resistance under steady-state conditions (sweating guarded-hotplate test).

ISO 12974, Textiles – Determination of abrasion resistance of fabrics by the Martindale method.

ISO 13688, Protective clothing – General requirements.

ISO 13934-1, Textiles – Tensile properties of fabrics – Part 1: Determination of maximum force using the strip method.

ISO 13934-2, Textiles – Tensile properties of fabrics – Part 2: Determination of maximum force using the grab method.

ISO 13937, Textiles – Determination of Resistance to Tear for Woven Fabrics.

ISO 13982-1, Chemical protective clothing – Protection against dust – Part 1: Requirements.

ISO 13982-2, Protective clothing for use against solid particulate chemicals – Part 2: Test method of determination of inward leakage of aerosols of fine particles into suits.

ISO 13994, Clothing for protection against liquid chemicals – Determination of resistance of protective clothing materials to penetration by liquids under pressure.

ISO 13995, Protective clothing – Mechanical properties – Determination of dynamic puncture-tear propagation.

ISO 13996, Protective clothing – Mechanical properties – Determination of resistance to puncture.

ISO 13997, Protective clothing – Mechanical properties – Determination of resistance to cutting by sharp objects.

ISO 14782, Plastics – Determination of haze for transparent materials.

ISO 15025, Clothing for protection against heat and flame – Test methods for limited flame spread materials.

ISO 15106, Plastics – Film and sheeting – Determination of water vapour transmission rate.

ISO 16602, Clothing for protection against chemicals – Classification and performance requirements (proposed).

ISO 17491, Protective clothing – Protection against gaseous and liquid chemicals – Determination of integrity of protective clothing to penetration by liquids and gases.

ISO 22608, Protective clothing – Protection against liquid chemicals – Measurement of repellency, retention, and penetration of liquid pesticide formulations of protective clothing materials.

National Fire Protection Association (NFPA)

NFPA 1991, Standard on Vapor-Protective Ensembles for Hazardous Chemical Emergencies.

NFPA 1992, Standard on Liquid Splash-Protective Ensembles and Clothing for Hazardous Chemical Emergencies, 1994.

13

Textiles for UV protection

A K S A R K A R , Colorado State University, USA

13.1 Introduction

The past decade has witnessed an alarming increase in the incidence of skin cancer worldwide. A primary reason for the increased incidence of skin cancers is attributed to stratospheric ozone depletion. Because ozone is a very effective UV-absorber each one percent decrease in ozone concentration is predicted to increase the rate of skin cancer by two percent to five percent.[1] It is estimated by the United States Environmental Protection Agency that ozone depletion will lead to between three and fifteen million new cases of skin cancer in the United States by the year 2075.[1] Other reasons for the skin cancer epidemic can be traced to lifestyle changes such as excessive exposure to sunlight during leisure activities. These activities include playing outdoors and swimming in the case of children, and golfing and fishing in the case of adults. In the case of agricultural and other outdoor workers, exposure to the sun is an occupational hazard as they have no choice about the duration of their exposure to the sun.

Each year over 1000 Australians die from skin cancer while two-thirds of the Australian population will develop some form of skin cancer at some point in their lives.[2] Other countries with a high melanoma rate are Canada, New Zealand and Norway. The American Cancer Society estimates that more than one million new skin cancer cases are diagnosed each year in the United States. As frightening as the one million count is an even more devastating statistic is that an estimated 54,200 of those will be diagnosed with melanoma, the deadliest form of skin cancer.[3] In response, governments around the world have begun to initiate action to educate their populations with respect to protection from solar ultraviolet radiation (UVR) exposure. For example, the National Toxicology Program, US Department of Health and Human Services has classified ultraviolet radiation as a known human carcinogen.[4] The state of Hawaii, USA developed a SunSmart skin cancer prevention program. New Hampshire, USA advocated avoiding the sun between 11 a.m. and 3 p.m. through its SunSafe program.[5] The Swiss Cancer League sponsors an annual awareness campaign, 'How sunproof are your clothes?'.[6] In Australia state cancer councils run educational campaigns such as 'SunSmart' and an intervention program named 'Kidskin'.[7,8] Awareness promotion efforts are particularly directed towards reducing sun exposure in young children since it is well established that skin cancer is related to sun exposure early in life.[9]

13.2   Ultraviolet radiation

The ultraviolet radiation band consists of three regions: UVA (320 to 400 nm), UVB (290 to 320 nm), and UVC (200 to 290 nm). UVC is totally absorbed by the atmosphere and does not reach the earth. UVA causes little visible reaction on the skin but has been shown to decrease the immunological response of skin cells.[1,9] UVB is most responsible for the development of skin cancers. The most common types of skin cancer are squamous cell and basal cell carcinoma, both of which can be cured either by excision or topical treatments. Malignant melanoma however, can result in death if not detected and treated in its early stages.[1,10]

13.3   Assessment of ultraviolet protection of textiles

Quantitative tests to assess the ability of a textile to protect against ultraviolet radiation can be performed either through laboratory testing *in vivo* or instrumental measurement *in vitro*. Accordingly, two terms are used; sun protection factor (SPF) for *in vivo* testing and ultraviolet protection factor (UPF) for instrumental evaluation *in vitro*.

13.3.1   *In vitro*

The term, UPF has been widely adopted by the textile and clothing industry worldwide to denote the protective ability of a textile based on instrumental measurements and is defined in Australian/New Zealand standard AS/NZS 4399:1996.[11] UPF is the ratio of the average effective ultraviolet radiation (UVR) irradiance calculated for unprotected skin to the average effective UVR irradiance calculated for skin protected by the test fabric.[12] Effective UVR irradiance is the product of relative erythemal spectral effectiveness and the relative energy value of solar irradiance reaching skin. Though the erythemal effectiveness is obtained by irradiating human test subjects with monochromatic UV-radiation of various wavelengths it is not necessary for the individual researcher to measure the erythemal effectiveness. Instead, data of relative spectral erythemal effectiveness from 290 to 400 nm with 5 nm intervals are provided by the CIE International Commission on Illumination. The solar spectral distribution is usually representative of a 'noonday' solar spectrum. For example, measured sunlight spectra for midsummer in Albuquerque, New Mexico, USA and Melbourne, Australia are available in literature.[13–17]

To measure the *in vitro* UPF an effective UVR dose (ED) for unprotected skin is calculated by convolving the incident solar spectral power distribution with the relative spectral effectiveness function and summing over the wavelength range 290–400 nm. The calculation is repeated with the spectral transmission of the fabric as an additional weighting to get the effective dose (ED_m) for the skin when it is protected. Thus UPF defined as the ratio of ED to ED_m can be expressed by the following equation:[13–17]

$$UPF = \frac{ED}{ED_m} = \frac{\sum\limits_{290\ nm}^{400\ nm} E_\lambda S_\lambda \Delta\lambda}{\sum\limits_{290\ nm}^{400\ nm} E_\lambda S_\lambda T_\lambda \Delta\lambda} \qquad 13.1$$

where E_λ = CIE erythemal spectral effectiveness, S_λ = solar spectral irradiance in $Wm^{-2}nm^{-1}$, T_λ = spectral transmittance of fabric, $\Delta\lambda$ = the bandwidth in nm, and λ = the wavelength in nm.

Though the above equation is based on percent UVR transmission, the determination of UPF is weighted most heavily by a textile sample transmittance in the UVB band of the spectrum, with maximum weighting near 305 nm. The scheme is logical because UVB radiation is responsible for more biological damage than UVA radiation.[13–17] Thus, UPF is an elegant way of ranking the sun protective abilities of a textile. The higher the UPF of a textile, the better is its ability to protect the skin it covers. UPF indicates how much longer a person can stay in the sun with the textile protecting the skin, as compared with unprotected skin to obtain the same erythemal response. For example, wearing a textile with a UPF rating of 15 will minimize exposure to UV radiation by a factor of 15, or put another way a textile with a UPF rating of 15 will allow only one-fifteenth of the UV radiation to pass through it.[18]

There are two *in vitro* quantitative measurement techniques to test UVR transmission through textiles or measure UPF. One is radiometry where the total transmission of UVR through a fabric is measured using a real or simulated solar spectrum. The other is spectrophotometry where the transmission of UVR through a fabric is measured as a function of wavelength from which UPF is then calculated. For both radiometry and spectrophotometry the primary requirement is an ultraviolet radiation source that includes both the UVA and UVB radiation. Total diffuse radiation transmitted through fabric is measured by both techniques because it simulates radiation hitting the skin beneath the textile in a real-life scenario.

Radiometry

Radiometric UV transmission techniques use a broadband UV light source filtered for UVB or combined UVA and UVB bands to illuminate a fabric sample. The total UV transmission through a textile is measured by a radiometer. The radiometer produces an output reading corresponding to the total radiant energy passing through the textile and falling on the detector surface.[1] The radiometric technique was used by Bech-Thomsen et al.[19] in a study. In their study radiation from a solar stimulator lamp was filtered to a band in the region of 298 to 329 nm with a peak at 313 nm. A radiometer with a diffuser was used to measure the transmitted radiation in the presence and absence of a textile material. To ensure that the scattered light transmitted by the fabric was also measured by the detector the textile material was wrapped around the detector. The protection factor was determined by taking the ratio of the measured power in the absence of the textile material to the measured power in the presence of the textile material. It should be pointed out though that radiometric measurements do not yield a definitive value for the protection factor of a given textile. A probable reason could be that the absorption of the textile is not independent of the wavelength in the UVA and UVB bands of interest.[1] Nevertheless, the technique is useful when for example only a relative variation in UPF needs to be measured such as the variation in UPF from site to site within a textile or the effect of stretching the textile on the UPF.[14]

Spectrophotometry

The spectrophotometric technique relies on the collection of transmitted and scattered radiation with the aid of an integrating sphere positioned behind the textile material. The AS/NZS and European standards for measuring and labeling UPF of textiles suggests equipping the spectrophotometer with a UV radiation transmitting filter for wavelengths of less than 400 nm to minimize errors caused by fluorescence from optical brightening agents.[15] Spectrophotometric measurements are generally done in 5 nm or less steps in the wavelength range from 290 to 400 nm. Step size is important since it has been shown that using a step size of 1 nm results in the most accurate determination of UPF; using a 5 nm step size produces results within 0.5% while with 10 nm steps the UPF values are within 1.5 to 2.5%.[14] A minimum of four samples must be measured from a textile material with two in the machine direction and two in the cross-machine direction.[15]

One of the first attempts to measure spectral transmission of ultraviolet radiation through textiles by a spectroradiometer was reported by Robson and Diffey.[20] In their method the spectral irradiance from an unfiltered 75 watt xenon arc lamp was measured in the range from 290 to 400 nm in 5 nm steps using a spectororadiometer. Subsequently, a 3 × 3 cm fabric material was placed over the quartz input optics of the spectroradiometer and the spectral irradiance measurement was repeated. As in the radiometric technique, the wrapping of the fabric material directly around the detector ensured that the scattered radiation transmitted by the fabric was also measured by the detector. The monochromatic protection factor [PF(λ)] at wavelength λ nm was defined as the ratio of the spectral irradiance of the unfiltered radiation to that of the radiation transmitted through the fabric. The UPF values measured by this technique had a range from UPF 2 for a polyester blouse to more than 1,000 for cotton twill jeans.

Subsequently, many other researchers have used the spectrophotometric technique and interlaboratory testing has proven spectrophotometry to be an accurate and reproducible method of measuring UPF. It is noted, however, that spectrophotometric measurements are made with collimated radiation incident at right-angles to the fabric and thus represent a worst-case scenario. In practice, the actual protection afforded by a textile might well be higher than that predicted by an *in vitro* measurement since in the real world UVR would be incident on a textile at angles other than normal incidence and in addition there would be the presence of diffuse UV radiation from the sky. Indeed, Ravishankar and Diffey[21] reported that laboratory determination of UPF might underestimate protection. Their results are discussed in a later section of this chapter.

13.3.2 *In vivo*

The simplest way of *in vivo* testing is by attaching rectangular pieces of fabric to the back of a human subject and determining the minimum erythemal dose (MED) of the unprotected and protected skin. MED is defined as the minimum quantity of radiant energy required to produce first detectable reddening of the skin, 22 ± 2 hours after exposure. MED for unprotected skin is determined first using incremental UVB doses. Subsequently, MED for protected skin is determined by a series of incremental and decremental UVB doses centered at the estimated SPF of the given fabric as estimated from UPF values determined *in vitro*. The dose that results in a minimal erythema extending to the borders of irradiation is then used to calculate SPF as shown in eqn 13.2. The higher the SPF value, the better the fabric's ability to protect against sunburn.[1,10,15,17]

$$\text{SPF} = \frac{\text{MED(protected skin)}}{\text{MED(unprotected skin)}} \qquad 13.2$$

In vivo testing suffers from some disadvantages. Chief among the drawbacks is the use of filtered xenon arc solar simulators to reproduce the solar spectrum, which in essence means that the solar spectrum reproduced, is only for one particular latitude, atmospheric ozone concentration, and time period in the year. As reported by Hilfiker *et al.*[13] and Robson and Diffey[20] a slight change of the spectrum can have a large effect on SPF, especially for fabrics containing polyester and silk which show a large variation in absorption in the 290–400 nm range.[1] The second drawback is the subjective determination of the presence of erythema. Cost and impracticability of the use of human subjects including factors such as unevenness of human skin are other disadvantages.[15] Given its limitations, however, *in vivo* testing is a useful tool to measure the response of the human body to UV radiation and is especially useful for confirming the UPF values measured *in vitro*.

13.3.3   Comparison of *in vitro* measurement (UPF/spectrophotometric) and *in vivo* testing (SPF/human skin)

Menzies *et al.*[22] were among the early researchers to conduct a comparative study of fabric protection using both spectrophotometric and human skin measurements. Six textiles, woven and knitted and of various fiber content were examined. For five of the six textiles human skin measurements were significantly lower than *in vitro* UPF. Interestingly, there was a better agreement between SPF and UPF when human testing was done 2 mm off the skin. Further, when thin-film metal meshes that simulated idealized fabrics were used the agreement between fabric UPF and SPF were good. The metal meshes were superior to the textiles in the sense that they did not suffer from nonuniformity in weave structure. The researchers therefore theorized that the lack of agreement between textile fabric UPF and SPF values were most probably due to the nonuniform weave structure of a textile's surface that provided a clear path for the UVR to impinge upon the skin in some areas while blocking the UVR in other areas, the so-called 'hole effect'.

Greenoak and Pailthorpe[23] reported similar results in that the values obtained *in vivo* were lower than *in vitro* UPF values. Gambichler *et al.*[24] were also in agreement in their study of thirty summer textiles where they obtained statistically lower SPFs than UPFs. However, Ravishankar and Diffey[21] who detailed the findings of an *in vivo* study of diffuse UV transmission through various T-shirts at various anatomical sites of a life-size manikin and compared the SPFs with UPF values obtained *in vitro* reported contradictory results. They observed that there was a variation in SPF by a factor of two or more at different anatomical sites for a given T-shirt. Lower SPFs were observed at sites where the fabric was stretched compared to where the fabric was off the skin. Further, at every site and for each T-shirt the SPF was higher than UPF. As discussed previously the reason for this discrepancy was attributed to the use of collimated radiation in UPF determination. Ravishankar and Diffey[21] confirmed their theory by measuring UPF as a function of angle of incidence of the radiation beam by rotating the lamp in an arc around the sensor. Accordingly, an increase in path length through the fabric and scattering of radiation on account of increased angle of incidence from the normal resulted in higher values of UPF. At 45 degrees from the normal UPF was higher by a factor of three than when determined at normal incidence. Therefore, they estimated that the protection afforded by fabrics is about 50% higher than predicted by *in vitro* determination of UPF.

On the other hand, Gies *et al.*[25] compared the UPFs for sixteen fabrics against SPFs and found no statistically significant differences between the two sets of results. Adding to the confusion is a study by Hoffmann *et al.*[26] who compared UPF and SPF values for seven different viscose fabrics and seven different cotton fabrics. There was no agreement between *in vitro* UPF and *in vivo* SPF values. For viscose fabrics containing a delustering pigment and for the cotton fabrics the UPF values were lower than the SPF values. In the case of viscose fabrics without a delustering agent the UPF values were higher than the SPF values.

It is evident from the above conflicting reports that additional studies are needed to establish a reliable and valid relationship between *in vitro* UPF and *in vivo* SPF. Opportunities might also exist for developing completely new measurement methods as well. An effort towards new directions is described in a method by Knittel *et al.*[27] who used a biological UV-detector film. In their method cells were immobilized onto a polymer foil and were sealed in a small case. During measurement, DNA, bound into dehydrated spores, was exposed to transmitted UVR. The incident radiation dose was also stored. Subsequently, any defect on the DNA caused by exposure to UVR was quantified. It is claimed that the sensitivity of the UV-detector film to solar wavelengths is identical to that of the human skin.

13.4   Standards for UV protective textiles

AS/NZS 4399 (Australia/New Zealand),[11] American Association of Textile Chemists and Colorists (AATCC) Test Method 183 (USA),[28] BS 7914 (British Standards Institution, Great Britain),[29] and European standard EN 13758-1[30] are the standard test methods presently in use. All of these methods employ eqn 13.1 to calculate the ultraviolet protection factor of a textile. Differences among the test methods exist particularly with regard to scanning intervals, positioning of the textile in the instruments and the erythemal action spectrum designated.

13.4.1   Australia/New Zealand

The AS/NZS 4399:1996,[11] Sun Protective Clothing-Evaluation and Classification standard includes a definition for UV protective clothing; a detailed procedure for determining the UV transmittance of fabric; the formulae required to calculate UPF and percent of blocking from the UV transmittance data; and directions for taking those UPF or percent block numbers and determining the singular UPF or singular percent block value to appear on a label. This number is called the label UPF. The calculated label UPF is the mean sample UPF minus the standard error of the sample UPF, the result of which should be rounded down to the nearest multiple of five. When the calculated label UPF value is greater than the lowest UPF value for an individual specimen, then the UPF value to be placed on a textile should be the UPF value of the specimen with the lowest UPF value, which is then rounded down to the nearest multiple of five. No textile is given a UPF rating greater than 50. The standard also classifies fabrics in three categories of protection. They are excellent, very good and good. Fabrics with calculated label UPF values of 40 to 50+ are classified in the Excellent UV protection category (97.5% or more UVR blocked); those with UPF values greater than 25 but less than 40 in the Very Good UV protection category (96.0–97.4% UVR blocked), and those with calculated label UPF values higher than 15 but less than 25 in the Good UV protection category (93.3–95.9% UVR blocked). A textile must have a minimum UPF of 15 to be rated as UV protective.

The Australian Radiation Protection and Nuclear Safety Agency (ARPANSA)[18] issues a certification label to manufacturers who test and label their fabrics according to AS/NZS 4399. It should be noted, however, that this standard does not take into account the effects of stretch (tension state), wetness (moisture content), wear, and use. Also, the design of the garment as it affects protection is not addressed. As explained subsequently in this chapter, moisture content can have a significant influence on UV transmittance through a textile especially for hydrophilic cellulosic textiles. Stretch is important in the case of textiles containing an elastane fiber such as spandex. Use and wear are significant since the UV protective ability of a textile may decrease during its lifetime due to exposure to environmental elements and wear due to care procedures.

13.4.2   United States of America

The US standard definition for a UV protective textile is any textile (fabric or product made from fabric) 'whose manufacturer and/or seller claims that it protects consumers from ultraviolet radiation, claims the reduction of risk of skin injury associated with UV exposure, and/or uses a rating system that quantifies the amount of sun protection afforded'.[31] In the United States, AATCC Test Method 183,[28] 'Transmittance or Blocking of Erythemally Weighted Ultraviolet Radiation through Fabric' describes a procedure for determining the transmittance of UV radiation through fabric and subsequent calculation of UPF. In addition, American Society of Testing and Materials International standard method ASTM D 6544,[32] Preparation of Textiles Prior to UV transmittance Testing details fabric sample preparation prior to UV transmittance testing primarily with an eye towards labeling requirements. The ASTM standard specifies that fabric must be laundered 50 times and exposed to 100 AATCC fading units of simulated light. One hundred AATCC fading units is roughly equivalent to the amount of sunlight during a two-year period. As well, swimwear fabrics must be exposed to chlorinated pool water. The hope is that a UPF value determined after laundering and exposure to light will reflect the lowest protection of the fabric and in essence will have already taken into account the decreased ability of the fabric's protective ability due to environmental stresses and care procedures. ASTM D6603,[33] Standard Guide for Labeling of UV-Protective Textiles instructs that a label in a UV protective textile must contain a UPF value; a classification category similar to the AS/NZS 4399 standard; and a statement that the UV-protective textile product has been labeled according to this ASTM standard guide. Regulation of UV protective textiles is under the auspices of the Consumer Product Safety Commission and enforced by the Federal Trade Commission.[33–35]

13.4.3   Europe

The European standard EN 13758-1 strives to overcome many of the drawbacks of the AS/NZS 4399 standard by recommending that only textiles with UPFs 30+ be labeled as sun protective clothing. It is reasoned that a UPF of 30+ will be robust to accommodate the effects of stretch, wetness, wear and use. The European standard also includes requirements for the design of garments. Clothing designed to offer UV protection to the upper body must provide at least coverage from the base of the neck down to the hip and across the shoulders down to three-quarters of the upper arm. Clothing designed to offer protection of the lower body (from the waist to below the patellae) must similarly provide complete coverage. Ultraviolet protective clothing complying with this standard must be permanently marked with the number of the European standard and with the designation 'UPF+'.[36]

13.5   Textiles as protection from ultraviolet radiation

Apart from drastically reducing exposure to the sun, the most frequently recommended form of UV protection is the use of sunscreens, hats, and proper selection of clothing. Because fabric is composed of fibers that can absorb, reflect or scatter radiant energy, it has the ability to absorb and/or block most of the incident radiant energy and prevent it from reaching the skin.[1] Figure 13.1 is a schematic representation of the different ways a fabric can prevent UV radiation from coming into contact with the skin. However, a fabric's ability to block UVR is dependent on several parameters. Principal parameters include fiber chemistry; fabric construction, particularly porosity, thickness and weight; moisture content and wet-processing history of the fabric such as dye concentration, fluorescent whitening agents, UV-absorbers and other finishing chemicals that may have been applied to the textile material. Since a plethora of factors establish a fabrics protective ability, mere visual examination by holding up a textile material to sunlight to determine how susceptible a textile is to UV rays is of little merit. Even textiles which seem to be non-light-transmitting may

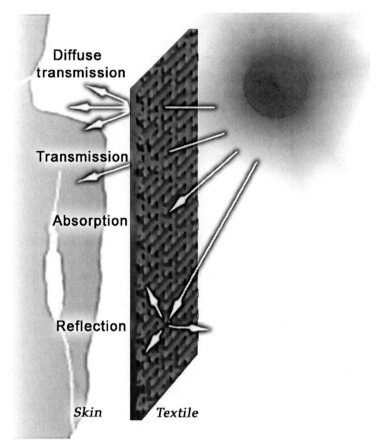

13.1 Schematic representation of a textile as a barrier to UV radiation. Illustration by Julia DiVerdi.

pass significant amounts of erythema-inducing UV irradiation.[37] Therefore, knowledge of the factors that contribute to the UV-protective abilities of textiles is vital as is appropriate testing and evaluation using the various test methods developed to gauge protection.

13.5.1   Fiber chemistry and UV protection

Fiber chemistry has a significant effect on UV protection as evidenced from data reported in numerous studies. However, it is advisable to interpret the data with caution particularly because many researchers neglected to take into account the processing history of the fibers. Such history might include addition of delusterants or UV stabilizers during fiber manufacture and/or the wet-processing history such as preparation, dyeing and finishing. In such cases the values reported reflect the 'fiber-dye-finish' combination rather than of the fiber itself.[15] It should also be noted that fiber chemistry is but one of several factors that influence the UPF of a fabric. Nevertheless, by analyzing the data with prudence broad generalizations can be made.

In a study of thirty commercial summer textiles by Gambichler et al.[24] it was found that white cotton, linen and viscose rayon afforded very little protection from UV exposure. These results are in agreement with earlier studies by Curiskis and Pailthorpe[38] who found that bleached cotton was transparent to UV radiation presumably because bleaching removes the natural pigments and lignins that can act as UV absorbers. Algaba et al.[39] further found that even bleached cotton fabrics with very compact fabric structures were ineffective in minimizing UV transmission radiation. The influence of bleaching in the form of increased UVR transmission was also reported by Crews et al.[16] in the case of silk fabrics. Bleached silk fabrics were found to have four times higher UVR transmission than comparable unbleached silk fabrics. Reinert et al.[40] investigated the UPF of wool fabrics and found that apart from wool muslin which has high porosity, wool fiber possesses excellent UV-protecting properties.

Davis et al.[41] conducted a comparison of fabrics of different fiber types but similar construction, count and weight and found nylon, acrylic and acetate to be poor inhibitors of UVR. All studies were unanimous in the assertion that polyester, by virtue of its large conjugated aromatic polymer system, was very effective in blocking UVB radiation. Polyester was less effective against UVA radiation because its UVR transmission increased significantly at 313 nm which is close to the boundary between the UVB and UVA spectral regions. However, this drawback is easily rectified by the addition of titanium dioxide delusterant that further improves the protective ability of polyester by blocking UVA radiation as well. In summary, it appears therefore that polyester and blends of polyester with other fibers may be the most suitable fabric in terms of UV protection particularly for white and undyed fabrics.

13.5.2   Fabric construction and UV protection

Porosity, weight, and thickness are the most important parameters under the fabric construction category. Numerous studies have concluded that fabric porosity is the single best predictor of UVR transmission through white and undyed fabrics with fiber chemistry coming in second.[1,2,15,16] Fabric porosity sometimes referred to as fabric openness is a measure of tightness of weave. According to Pailthorpe,[2] a Utopian fabric for sun protection is one in which the yarns are completely opaque to UVR, and the openings or pores in the fabric are sufficiently small to block the transmission of UV radiation. Pailthorpe[2] related UVR transmission through a fabric to its 'cover factor', the term used to describe a fabric's porosity. Percentage UVR transmission and UPF can then be calculated using the following equations:[2,16]

$$\%UVR \text{ transmission} = 100 - \text{cover factor} \qquad\qquad 13.3$$

$$UPF = \frac{100}{100 - \text{cover factor}} \qquad\qquad 13.4$$

Application of eqns 13.3 and 13.4 show that the cover factor of a fabric must be greater than 93% to achieve a minimum UPF rating of 15.[2] Also, once the cover factor exceeds 95%, very small increases in cover factor lead to dramatic improvements in the protective ability of the fabric. It should be noted, however, that fabrics with the same cover factor can have widely different UPF ratings, particularly if their fiber chemistries are different.

In general, knit fabrics have a lower cover factor than woven textiles because of their open structure. Among woven textiles, plain weaves have a higher cover factor than other weaves. This is because the interlacement of yarns in a plain weave creates double layers of fibrous material. In a knit fabric, loops are pulled through previously formed loops and double layers occur to a lesser extent.[1] Cover factor can also be modified by over-feeding on the stenter during heat-setting causing the fabric to shrink and hence increase the cover factor. On the contrary, under-feeding will cause the fabric to stretch decreasing the cover factor.[2]

Porosity can be quantified by several methods. Perhaps the simplest method is by calculating cloth cover. Cloth cover is a measure of the fraction of area covered by both the warp and weft threads in a given fabric and calculated simply by using thread count and yarn number according to an equation by Booth and Pollitt.[42]

$$\text{Cloth cover} = \text{Cover factor}_{warp} + \text{Cover factor}_{weft} - [(\text{Cf}_{warp} * \text{Cf}_{weft}/28]$$
$$\qquad\qquad 13.5$$

where Cover factor (Cf) = threads per inch/$\sqrt{}$yarn number (cotton count system). The above equation was developed for a plain weave fabric and the basic premise is that the cover factor would be 28 if all yarns just touched each other. In practice, cover factors can range from 8 to 28 for commercially available fabrics. However, Crews et al.[16] found that cloth cover was not a precise estimator of fabric porosity and therefore not a reliable predictor of UVR transmission through a fabric.

Hilfiker et al.[13] determined porosity by measuring the directed transmittance of fabrics. This method is based on the argument that only light passing through pores in a fabric is not scattered. All other light that passes through a fabric changes its direction before leaving the fabric. Therefore, if the measurement is done by placing the detector of a UV/VIS spectrophotometer far away from the sample only the transmitted light with the same direction as the incident light is detected.

Another method of quantifying porosity is determining percent cover by image analysis. Percent cover is defined as the percentage area occupied by warp and filling yarns in a given fabric area. The closer the weave or knitting, the more is the percentage area occupied by the yarns and more opaque is the fabric to UV radiation. This method reported by Crews *et al.*[16] involves magnifying a microscopic image of a fabric obtained using a light microscope and a video camera on a video monitor screen for a magnification of 130X. The monochrome image of the magnified fabric is then captured with a video capture card, converted to pixels on the computer screen, and analyzed with image analysis software. Each pixel represents a monochromatic value between 0 and 255. The area of fabric occupied by yarns is considered to be those areas represented by black pixels, i.e., monochromatic values between 0 and 75. Percent cover is then determined according to eqn 13.6.[16]

$$\text{Cover (\%)} = \frac{\text{Number of black pixels}}{\text{Total number of pixels}} \times 100 \qquad 13.6$$

Results using the image analysis showed that as percent cover increased, percent UVR transmission decreased. Interestingly, higher percent cover did not always translate into lower UVR transmission with correspondingly higher UPF values and vice versa. For example, among the fabrics studied by Crews *et al.*[16] polyester had a lower percent cover than rayon but a higher UPF than rayon. These results once again emphasize the importance of fiber chemistry in UVR transmission and the significance of a holistic approach with respect to studying textiles for UV protection.

Closely related to porosity is fabric weight. Fabric porosity is inversely proportional to weight per unit area. Heavier clothing minimizes UVR transmission by virtue of having smaller spaces between yarns thus blocking more radiation as reported by Wong *et al.*[43] Davis *et al.*[41] agreed with Wong *et al.*[43] but add that the relationship between UPF and fabric weight was not linear. The third parameter is fabric thickness. Thicker, denser fabrics tend to transmit less UV radiation as reported by Sliney *et al.*[44] However, Crews *et al.*[16] stated that thickness is useful in explaining differences in UVR transmission when differences in percent cover are also accounted for. In the study conducted by Crews *et al.*[16] the thinnest rayon and polyester lining fabrics did not have the highest UVR transmission. Instead, a slightly thicker nylon organza with a lower percent cover than the rayon and polyester exhibited the highest level of UVR transmission.

In summary, porosity is an excellent predictor of UVR transmission followed by fabric thickness and fabric weight. These three parameters along with fiber chemistry can be considered as the four pillars that determine the UV protective ability of white and undyed fabrics.

13.5.3   Colour and UV protection

The colour or shade of a fabric is a very significant factor in preventing UV radiation transmission through a textile. The colour of a dye is dependent on the absorbing properties of the dye in the visible band of the electromagnetic spectrum (380–770 nm). The absorption band for all dyes, however, extends into the UVR radiation band (290–400 nm) and hence dyes act as effective UVR absorbers. According to Pailthorpe,[2] the extinction coefficients of the dyes in the UVR spectral band determine their ability to increase fabric protectiveness against UV radiation. One of the first reports regarding the effect of colour on UV radiation transmission was by Gies et al.[45] who stated that for fabrics of identical weight and construction, darker coloured fabrics scored higher UPF ratings than lighter shades. In a subsequent report, the spectral transmission of fabrics dyed in different colours was illustrated (Fig. 13.2) to show that darker colours had a UPF rating of 50+.[8]

Davis et al.[41] conducted a study and found that white cotton denim showed a very low UPF rating whereas blue denim produced a very high UPF value. Since the white denim in their study differed only in the absence of colour from the blue denim the researchers theorized that the indigo dye in the blue denim must be a significant absorber of UV radiation. Their theory was confirmed after a dyeing experiment wherein cotton fabrics dyed with a blue reactive dye showed an improvement in UPF by more than a factor of three. Whether a different dyestuff would produce the same result was the question left to be answered. Pailthorpe[2] studied the UVR absorbing property of CI Reactive Red 6 and CI Acid Yellow 44 and concluded that each dyestuff was unique in its UVR absorbing property. CI Reactive Red 6 absorbed strongly in the UVR band whereas CI Acid Yellow 44 was ineffective as it absorbed at 310–330 nm.

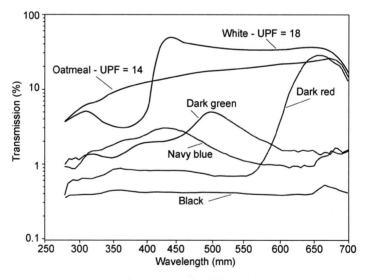

13.2 Effect of color on UPF rating. Reprinted from *Mutation Research*, Vol. 422, P.H. Gies *et al.*, Protection against solar ultraviolet radiation, 15–22, 1998, with permission from Elsevier.

Another study by Knittel *et al.*[27] examined the effect of CI Direct Green 26 on a cotton fabric. The resultant dyed cotton fabric exhibited a high UPF value. A more comprehensive study examining the ability of dyes to reduce UV transmission through fabrics based on dye concentration in the fabric and transmittance in solution was carried out by Srinivasan and Gatewood[46] who studied the effect of fourteen direct dyes on the UPF of a bleached print cloth. At 1.0% dye on weight of fabric (owf), all dyed fabrics had UPF values above 15 compared to an UPF value of 4.1 for the undyed fabric. The results also indicated that dyes applied at 1.0% owf gave higher UPF values than dyes applied at 0.5% owf concentration.

In agreement with the Pailthorpe[2] study referred to earlier, dyes with red hues (CI Direct Red 28 and CI Direct Red 24) provided greater UV protection than a dye with a yellow hue (CI Direct Yellow 28). CI Direct Black 38 provided greater protection than all other dyes. Additionally, the effective UVR transmittance of the dyes in solution was analogous to the UPF value of fabrics. Practically, this means that dyes can be screened for their ability to enhance UV protection prior to actually applying them to fabrics. Srinivasan and Gatewood[46] further noted that transmission/absorption characteristics of dyes in the UV band were a better predictor of UV protection than the colour of the dyestuff itself. A recent study by Gorensek and Sluga[47] studied the effect of orange, red, and blue disperse dyes on polyester fabric. The UV blocking ability of pale orange and blue dyed polyester fabrics increased with pale red dyed fabric coming in third. It is not necessary, however, that polyester be dyed to increase its UPF value because, as discussed previously, polyester exhibits good UV protection even in its undyed state.

In conclusion it must be remembered that while colour can dramatically increase a fabric's protective ability, the dyestuffs responsible for the protection must be colourfast to washing, perspiration, sunlight, crocking, and bleaching for the life of the fabric.

13.5.4   Additives and UV protection

It is evident from the preceding discussion that undyed fibres, be they natural fibres such as cotton, linen and silk or manufactured fibers such as viscose rayon, nylon and acrylic, offer little protection against UV radiation. It is necessary for these fibers to be treated with additives in order to improve their UV protection abilities. The simplest type of additive, albeit a very effective one, is the addition of the delusterant pigment TiO_2 which acts as an UV absorber. Since TiO_2 is incorporated during fiber manufacturing the effect is permanent. Treatment with ultraviolet absorbers (UVAs) is another efficient method of increasing UPF. UVAs are colorless compounds with chromophore systems that are effective absorbers in the UV band (290–400 nm). UVAs can be applied to fabrics by either exhaust, pad batch or continuous processes. UVAs

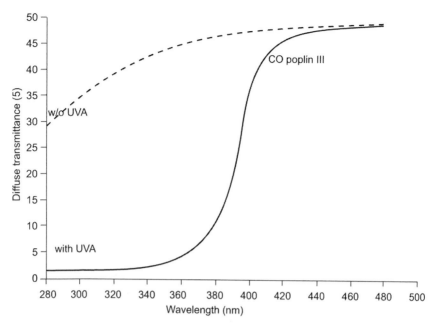

13.3 Effect of UV absorber on UV transmission. Originally published in *Textile Chemist and Colorist*, Vol. 29, No. 12, December 1997, p. 39; reprinted with permission from AATCC.

are available as a reactive water soluble oxaldianilide for cotton and cotton blend fibers; a monosulfonated benzotriazole derivative for wool, silk and nylon or as a dispersion of a benzotriazole derivative for polyester.[40]

Reinert *et al.*[40] reported the effect of dyeing and treatment with UV absorbers on cotton fabrics (Fig. 13.3). Two cotton fabrics differing in thickness and porosity were dyed to pale and deep shades with and without UV absorber. It was observed that the UV absorber was more effective in increasing the UPF of the cotton fabric with a lower percent porosity suggesting that a slightly porous fabric is probably the ideal candidate for UVA treatment. In other words, high UPF values post-UVA treatment can be expected only if the porosity of a fabric is low enough to begin with. Similar results were obtained for silk fabrics that showed dye and UV absorber improved UPF but fabric thickness and porosity were also important parameters.

Hilfiker *et al.*[13] have developed a model to calculate the UPF of fabrics as a function of fabric type, thickness, porosity and UVA type and concentration. In a recent paper Xin *et al.*[48] presented a new treatment for cotton fabrics using the sol-gel method. Sol-gel is a tool to create transparent titania films that adheres well to cotton. The high level of adhesion is attributed to a condensation reaction between the cellulosic hydroxyl groups of cotton and the uncondensed hydroxyl groups of titania. Treated cotton fabrics showed an excellent UPF rating of 50+. The excellent rating was maintained after 55 home launderings. The principle

behind this technology is the ability of titania films to absorb light energy that matches or exceeds their band gap energy and the band gap energy of titania lies in the UV band of the solar spectrum.

13.5.5   Stretch and UV protection

When a textile material is stretched, the porosity increases accompanied by a decrease in fabric thickness thereby allowing increased levels of UVR transmission. Gies et al.[45] were among the first researchers to report a decrease in UPF with stretch especially in the case of Lycra® spandex. Under significant stretch the decrease for Lycra® was an order of magnitude from UPF 200 to UPF 20. Osterwalder et al.[6] found the increase of UVR transmission to be almost linear with stretch. They calculated that choosing one size too small caused a 7% stretch around the chest. Moon and Pailthorpe[49] found that people tend to buy elastane type fabrics under size and stretch them by 15% during use thereby negating significant UV protective properties of the material. Kimlin et al.[50] measured the UPF of 50 denier stockings and found that their UPF decreased 868% when stretched 30% from their original size.

13.5.6   Moisture and UV protection

Many scientists have reported an increase in UVR transmission and a corresponding decrease in UPF when textiles are wet. According to Pailthorpe,[2] the effect is purely an optical one wherein due to refractive index differences the water filled interstices of the fabric causes less scattering than does air. The result is an increase in transmission of UV radiation through fabric. Gies et al.[45] observed an important relationship between water absorbed and decrease in UPF. In general, fabrics such as cotton that is more absorbent showed an increase of 50% UVR transmission and corresponding decrease in UPF when wet. For a cotton/elastane fabric the UPF decreased from 50+ in the dry condition to 32 when wet. However, Osterwalder et al.[6] argue that UV absorbers on a cotton fiber can minimize the effect of wetness on UPF. According to them, absorbance can be expected to be independent of environment and therefore treating cotton with a UV absorber that absorbs over both the UVB and UVA bands will afford complete protection when the fabric is wet.

A recent paper by Crews and Zhou[51] examined the effect of different water type on UVR transmission and UPF values. Thirteen undyed woven fabrics of different fiber contents were evaluated in their dry condition and after wetting with distilled water, sea water, or chlorinated pool water. Results showed that water type had no significant effect on the UPF, though in agreement with previous studies a majority of the fabrics in this study exhibited lower UPF values when wet.

13.5.7   Effects of laundering on UV protection

Sliney et al.[44] reported the first study on the effects of laundering. In that study various fabrics were laundered ten times at 60 °C. Post-laundering, a majority of the fabrics showed a decrease in UVR transmission with thicker fabrics exhibiting a higher decrease. The researchers surmised that the process of laundering led to compaction due to shrinkage presumably decreasing porosity and hence resulting in an improvement in UV protection. Stanford et al.[52] subjected five jersey-knit cotton T-shirts to a total of 36 washes. Fabric UPF was measured after one wash and after 36 washes. Mean fabric UPF approximately doubled for all T-shirts after the first wash and further additional change after another 35 washes was not significant. In practical terms, the results of this limited study meant that the protective ability of a new cotton T-shirt could be expected to last the lifetime of the T-shirt. The results of this study were also confirmed by a wear and wash trial using 20 human subjects. The subjects wore their T-shirts for 4–8 h per week and washed their T-shirt once per week. After ten wear and wash cycles, the rated UPF of the T-shirts increased to UPF 35 from UPF 15.

Zhou and Crews[53] did a more comprehensive study on the effect of repeated launderings on UVR transmission through fabrics. In their study, eight types of woven and knitted summer wear fabrics ranging from a 100% cotton sheeting to blends of cotton/polyester to a 100% nylon were subjected to 20 launderings using detergents with and without an optical brightening agent (OBAs). Optical brightening agents are additives found in home laundry detergents to enhance the whiteness of textiles. Since OBAs function by excitation in the UV band and re-emission in the visible blue band of the electromagnetic spectrum it stood to reason that fabrics laundered with a detergent containing an OBA would likely have enhanced sun protection. As expected, all the woven and knit cotton fabrics in the study showed an increase in UPF and decrease in percent UVR transmission after repeated launderings using a detergent with OBA. The positive results were attributed partly to the high absorbent properties of cotton fiber, which helps the buildup of OBA on cotton, and partly to the chemical affinity of the cotton fiber for the OBA in the detergent used in the study. Without an OBA in the detergent the results were different for the woven and knit cotton fabrics. Whereas the UPF of cotton knit doubled after 20 launderings changes in UPF for the woven cotton fabrics were small. Again, the results were not surprising since knit fabrics shrink to a larger extent on laundering which consequently leads to reduced porosity and increased UPF. Similar results were obtained for the polyester/cotton blends in the study.

For 100% polyester and 100% nylon delustered with TiO_2 the changes in UPF on laundering were not significant as they lacked affinity for the OBA and also because they did not shrink and compact to any appreciable degree. Results of the Zhou and Crews study[53] were confirmed by a real-life study conducted by

Osterwalder *et al.*[6] In this study approximately 100 cotton and viscose garments such as T-shirts, shorts and dresses were worn and washed in selected families over a whole summer. The garments were washed with or without optical brighteners and also with and without a fabric softener. It was found that the initial UVR transmission generally decreased or the UPF increased after one season of wear and wash. For fabrics with poor substantivity to OBAs, Osterwalder *et al.*[6] have suggested a detergent with an UV absorber additive. According to them, an UV absorber would exhaust very well from the wash liquor and stay on the fabric for several washes. In their test study they achieved an UPF of 30 on fabrics with initial UPFs of 3–5 using a detergent with an UV absorber additive.

Eckhardt and Rohwer[54] in their report agree with Osterwalder *et al.*[6] and further contend that an UV absorber additive in a detergent might be the better alternative to OBAs since OBAs exhibit a weakness in UV absorption around 308 nm which falls in the range of UV radiation that is most harmful to human skin. In addition fewer washes were required to provide good UPF values when using a detergent containing a UV absorber as compared to when using a detergent with an optical brightener.

13.6   Future trends

The preceding discussion outlines the need for UV protective textiles and the complexities associated with making a textile impervious to ultraviolet radiation. What is not certain, however, is the magnitude of the market for UV textiles. According to most industry experts, people have been led to believe over many decades that sunscreen lotions are an equivalent alternative to covering up with clothing. Thus currently the market is limited to audiences that are very aware of the risks of skin cancer due to UV exposure. It is apparent that campaigns to increase public awareness of the harmful risks of UV exposure and education regarding the beneficial effects of UV textiles have to be mounted to develop the UV textiles market. Nevertheless, there is guarded optimism among manufacturers that the need for UV protective fabrics will grow as people live longer lives and are more active outdoors. However, in the immediate future the most promising market for UV protective textiles appears to be children's wear since as alluded to previously, children are more vulnerable to the photo carcinogenetic effects of excessive sun exposure. The development of the UV protective children's apparel market will necessarily involve education of parents, caregivers, teachers, pediatricians, and public health officials in addition to textile and clothing manufacturers.[2,5,55]

In a survey by Black *et al.*[5] little selection of sun protective clothing was found in many retail stores which means that a niche market for UV protective children's apparel remains untapped. A related point that is also not known is whether apparel and textile designers are encouraged to design sun protective clothing for children. Design of UV protective clothing will have to take into consideration children and adolescent's preferences in design and fashion, a finding pointed out in a study by DeLong et al.[55] In that study educational intervention resulted in an increase in knowledge of sun protective factors in clothing. However, preference and intent to wear was based on style and fashion considerations. On a cautionary note, manufacturers who wish to exploit this market are advised to bear in mind that strict quality control measures and testing criteria will have to be instituted, since it has been reported that UPF values can vary by as much as 30 to 40% at different sites in the same batch of textile material.[2] A positive approach to expanding the UV textiles market is through building trust in UV labels and claims of UV protective properties in textile materials by way of stringent quality-control programs.

13.7   References

1. Capjack, L., Kerr, N., Davis, S., Fedosejevs, R., Hatch, K.L. and Markee, N.L. (1994). Protection of humans from ultraviolet radiation through the use of textiles: A Review. *Family and Consumer Sciences Research Journal*, *23*(2), 198–218.
2. Pailthrope, M. (1998). Apparel textiles and sun protection: a marketing opportunity or a quality control nightmare? *Mutation Research, 422*, 175–183.
3. Skin cancer facts [http://www.cancer.org/docroot/PED/content/ ped_7_1_What_You_Need_To_Know_About_Skin_Cancer.asp?sitearea=PED]. Retrieved 9/22/2004.
4. Carcinogens listed in the tenth report [http://ehp.niehs.nih.gov/roc/toc10.html]. Retrieved 9/22/04.
5. Black, C., Grise, K., Heitmeyer, J. and Readdick, C.A. (2001). Sun protection: knowledge, attitude, and perceived behavior of parents and observed dress of preschool children. *Family and Consumer Sciences Research Journal, 30*(1), 93–109.
6. Osterwalder, U., Schlenker, W., Rohwer, H., Martin, E. and Schuh, S. (2000). Facts and fiction on ultraviolet protection by clothing. *Radiation Protection Dosimetry, 91*(1–3), 255–260.
7. Milne, E., English, D.R., Corti, B., Cross, D., Borland, R., Gies, P., Costa, C. and Johnston, R. (1999). Direct measurement of sun protection in primary schools. *Preventive Medicine, 29*, 45–52.
8. Gies, P.H., Roy, C.R., Toomey, S. and McLennan, A. (1998). Protection against solar ultraviolet radiation. *Mutation Research, 422*, 15–22.
9. Truhan, A.P. (1991). Sun protection in childhood. *Clinical Pediatrics, 30*, 676–681.
10. Perenich, T.A. (1998). Textiles as preventive measures for skin cancer. *Colourage Annual*, 71–73.
11. Standards Australia/Standards New Zealand: AS/NZS 4399:1996; Sun-protective clothing-Evaluation and classification. Homebush, NSW, Australia/Wellington, New Zealand.
12. labsphere® (2000). Technical Notes. *SPF analysis of textiles*. labsphere®, North Sutton, NH 03260.
13. Hilfiker, R., Kaufman, W., Reinert, G. and Schmidt, E. (1996). Improving sun

protection factors of fabrics by applying UV-absorbers. *Textile Research Journal, 66*(2), 62–69.

14. Gies, H.P., Roy, C.R., McLennan, A., Diffey, B.L., Pailthorpe, M., Driscoll, C., Whillock, M., McKinlay, A.F., Grainger, K., Clark, I. and Sayre, R.M. (1997). UV protection by clothing: An intercomparison of measurements and methods. *Health Physics, 73*(3), 456–464.

15. Hoffman, K., Laperre, J., Avermaete, A., Altmeyer, P. and Gambichler, T. (2001). Defined UV protection by apparel textiles. *Archives of Dermatology, 137*(8), 1089–1094.

16. Crews, P.C., Kachman, S. and Beyer, A.G. (1999). Influences on UVR transmission of undyed woven fabrics. *Textile Chemist and Colorist, 31*(6), 17–26.

17. Menter, J.M. and Hatch, K.L. (2003). Clothing as solar radiation protection. *Current Problems in Dermatology, 31,* 50–63.

18. Australian Radiation Protection and Nuclear Safety Agency Fact Sheet 14. Clothing and solar UV protection [http://www.arpansa.gov.au]. Retrieved 9/22/2004.

19. Bech-Thomsen, N., Wulf, H.C. and Ullman, S. (1991). Xeroderma pigmentosum lesions relate to ultraviolet transmittance by clothes. *Journal of the American Academy of Dermatology, 24,* 365–368.

20. Robson, J. and Diffey, B.L. (1990). Textiles and sun protection. *Photodermatology, Photoimmunology and Photomedicine, 7,* 32–34.

21. Ravishankar, J. and Diffey, B. (1997). Laboratory testing of UV transmission through fabrics may underestimate protection. *Photodermatology, Photoimmunology and Photomedicine, 13,* 202–203.

22. Menzies, S.W., Lukins, P.B., Greenoak, G.E., Walker, P.J., Pailthorpe, M.T., Martin, David, S.K. and Georgouras, K.E. (1991). A comparative study of fabric protection against ultraviolet-induced erythema determined by spectrophotometric and human skin measurements. *Photodermatology, Photoimmunology and Photomedicine, 8* (4), 157–163.

23. Greenoak, G.E. and Pailthorpe, M.T. (1996). Skin protection by clothing from the damaging effects of sunlight. *Australasian Textiles,* Jan/Feb., 61.

24. Gambichler, T., Avermaete, A., Bader, A., Altmeyer, P. and Hoffman, K. (2001). Ultraviolet protection by summer textiles. Ultraviolet transmission measurements verified by determination of the minimal erythema dose with solar-simulated radiation. *British Journal of Dermatology, 144,* 484–489.

25. Gies, H.P., Roy, C.R. and Holmes, G. (2000). Ultraviolet radiation protection by clothing: Comparison of *in vivo* and *in vitro* measurements. *Radiation Protection Dosimetry, 91*(1–3), 247–250.

26. Hoffmann, K., Kaspar, K., Gambichler, T. and Altmeyer, P. (2000). *In vitro* and *in vivo* determination of the UV protection factor for lightweight cotton and viscose summer fabrics: A preliminary study. *Journal of the American Academy of Dermatology, 43*(6), 1009–1016.

27. Knittel, D., Schollmeyer, E., Holtschmidt, H. and Quintern, L. (1999). Measurements of UV transmission of textiles-Use of an *in vivo* simulating measuring method for UPF determination at various stages of cotton finishing. *Melliand Textilberichte (English), (4),* E73–75.

28. American Association of Textile Chemists and Colorists. AATCC Test Method 183-1998: Transmittance or blocking of erythemally weighted ultraviolet radiation through fabrics. AATCC Technical Manual, Research Triangle Park, NC, USA.

29. British Standards Institute: BS 7914: 1998. Method of test for penetration of erythemally weighted solar ultraviolet radiation through clothing fabrics. London, BSI.

30. CEN: The European Committee for Standardization 1999. Textiles-solar UV protective properties-methods of test for apparel fabrics. PrEN 13758. Stassart, Brussels.

31. Hatch, K.L. (2001). Fry not! *American Society for Testing and Materials Standarization News*, January, 18–21.

32. American Society for Testing and Materials (2000): ASTM D6544: Standard practice for the preparation of textiles prior to UV transmittance testing. Conshohocken, PA, USA.

33. American Society for Testing and Materials (2000): ASTM D6603: Standard guide for labeling UV protective textiles. Conshohocken, PA, USA.

34. Hatch, K.L. (2003). Making a claim that a garment is UV protective. *AATCC Review, 3*(12), 23–26.

35. Thiry, M.C. (2002). Here comes the sun. *AATCC Review, 2*(6), 13–16.

36. Gambichler, T., Rotterdam, S., Altmeyer, P. and Hoffman, K. (2001). Protection against ultraviolet radiation by commercial summer clothing: need for standarised testing and labeling. *BMC Dermatology, 1*:6. Available from [http://www.biomedcentral.com/1471-5945/1/6].

37. Rieker, J., Guschlbauer, T. and Rusmich, S. (2001). Scientific and practical assessment of UV protection. *Melliand Textilberichte (English), (7–8),* E155–156.

38. Curiskis, J. and Pailthorpe, M. (1996). Apparel Textiles and sun protection. *Textiles Magazine, 25*(4), 13.

39. Algaba, I., Riva, A. and Crews, P.C. (2004). Influence of fiber type and fabric porosity on the UPF of summer fabrics. *AATCC Review, 4*(2), 26–31.

40. Reinert, G., Fuso, F., Hilfiker, R. and Schmidt, E. (1997). UV-Protecting properties of textile fabrics and their improvement. *Textile Chemist and Colorist, 29*(12), 36–43.

41. Davis, S., Capjack, L., Kerr, N. and Fedosejevs, R. (1997). Clothing as protection from ultraviolet radiation: Which fabric is most effective? *International Journal of Dermatology, 36* (5), 374–379.

42. Booth, J.E. (1968). *An Introduction to Physical Methods of Testing Textile Fibers, Yarns and Fabrics*, 3rd edition, Boston: Newnes-Butterworths.

43. Wong, J.C., Cowling, I. and Parisi, A.V. (2000). Reducing human exposure to solar ultraviolet radiation. Available from [http://www.photobiology.com/v1/wong]. Retrieved 9/12/2000.

44. Sliney, D.H., Benton, R.E., Cole, H.M., Epstein, S.G. and Morin, C.J. (1987). Transmission of potentially hazardous actinic ultraviolet radiation through fabrics. *Applied Industrial Hygiene, 12*(1), 36–44.

45. Gies, H.P., Roy, C.R., Elliot, G. and Wang, Z. (1994). Ultraviolet radiation factors for clothing, *Health Physics, 67*(2), 131–139.

46. Srinivasan, M. and Gatewood, B.M. (2000). Relationship of dye characteristics to UV protection provided by cotton fabric. *Textile Chemist and Colorist, 32*(4), 36–43.

47. Gorenek, M. and Sluga, F. (2004). Modifying the UV blocking effect of polyester fabric. *Textile Research Journal, 74*(6), 469–474.

48. Xin, J.H., Daoud, W.A. and Kong, Y.Y. (2004). A new approach to UV-blocking treatment for cotton fabrics. *Textile Research Journal, 74*(2), 97–100.

49. Moon, R.L. and Pailthorpe, M.T. (1995). Effect of stretch and wetting on the UPF of elastane fabrics. *Australasian Textiles, 15*(5), 39–42.
50. Kimlin, M.G., Parisi, A.V. and Meldrum, L.R. (1999). Effect of stretch on the ultraviolet spectral transmission of one type of commonly used clothing. *Photodermatology, Photoimmunology and Photomedicine, 15,* 171–174.
51. Crews, P.C. and Zhou, Y. (2004). The effect of wetness on the UVR transmission of woven fabrics. *AATCC Review, 4*(8), 41–43
52. Stanford, D.G., Georgouras, K.E. and Pailthorpe, M.T. (1995). The effect of laundering on the sun protection afforded by a summer weight garment. *Journal of the European Academy of Dermatology and Venereology, 5,* 28–30.
53. Zhou, Y. and Crews, P.C. (1998). Effect of OBAs and repeated launderings on UVR transmission through fabrics. *Textile Chemist and Colorist, 30*(11), 19–24.
54. Eckhardt, C. and Rohwer, H. (2000). UV protector for cotton fabrics. *Textile Chemist and Colorist & American Dyestuff Reporter, 32*(4), 21–23.
55. DeLong, M., LaBat, K., Gahring, S., Nelson, N. and Leung, L. (1999). Implications of an educational intervention program designed to increase young adolescents' awareness of hats to sun protection. *Clothing and Textiles Research Journal, 17*(2), 73–83.

Textiles for protection against cold

I HOLMÉR, Lund Technical University, Sweden

14.1   Introduction

Cold is a hazard to human health. Cold environments may adversely affect physiological functions, work performance and wellbeing. The major threat is cooling, be it local skin cooling, extremity cooling or whole body cooling. Likewise the strategy for cold protection is prevention of cooling and maintenance of heat balance at acceptable temperature levels. This chapter deals with protection against cold. The cold environment and its climatic components are defined and described. The heat exchange between the human body and the environment is presented and the various forms of heat losses and methods for their determination are defined. A method for calculation of required clothing insulation is described. Thermal properties of clothing determine the various forms for heat losses. The essential factors are defined and methods for their determination are presented. The effects of fibres, fabrics, construction and clothing design are discussed. In a final section, the ultimate strategy for staying well protected against cold is presented.

14.2   The cold environment

A cold environment can be defined as an environment in which larger than normal heat losses can be expected. The climatic factors governing heat losses are

- air temperature
- mean radiant temperature
- air velocity
- humidity.

Factors like snow and rainfall affect heat exchange primarily by interaction with clothing heat transfer properties. Heat exchange between the body and the environment takes place at the skin surface by convection, radiation, conduction, evaporation and via the airways (respiration).

14.2.1   Convection

Air in contact with a warm surface warms up and becomes less dense. The warm air rises and causes a chimney effect close to the skin surface (natural convection). Wind strongly interferes with this process and increases convection. The following principal equation applies to convective heat exchange (C).

$$C = h_c \cdot (t_{sk} - t_a) \qquad 14.1$$

where h_c is the convective heat transfer coefficient in W/m²°C, t_{sk} is the mean skin temperature in °C and t_a is the air temperature in °C. The value of h_c depends primarily on wind for nude surfaces. In calm air a normal value for h_c is 3–4 W/m²°C. With clothed surfaces heat transfer becomes more complex. This is discussed in a following section.

14.2.2   Radiation

Heat is transported as electro-magnetic waves from a warm to a cold surface. This radiative heat exchange (R) is determined by the following equation.

$$R = h_r \cdot (t_{sk} - \bar{t}_r) \qquad 14.2$$

where h_r is the radiative heat transfer coefficient in W/m²°C and $\bar{t}_r$ is the mean radiant temperature in °C. The value of h_r is relatively constant about 4 W/m²°C and independent of, for example, wind. Clothing affects radiative heat transfer and will be discussed later.

14.2.3   Conduction

Heat is transmitted between two surfaces in contact with each other if there is a temperature difference. For a standing or moving person the body surface area in contact with another surface is negligible (foot soles). For a seated or reclining person larger areas contact other surfaces and determination of conductive heat exchange may be justified.

14.2.4   Evaporation

By producing sweat that evaporates on the skin surface the human body can get rid of significant amounts of heat. Sweating is a necessary and powerful mechanism for body cooling in response to high levels of body heat production and/or external heat loads. The formula for evaporative heat exchange (E) in air is defined by the following formula.

$$R = h_e \cdot (psk - p_a) = 16.6 \cdot h_c \cdot (p_{sk} - p_a) \qquad 14.3$$

where h_e is the evaporative heat transfer coefficient, p_{sk} is the water vapour pressure at the skin surface in kPa and p_a is the ambient water vapour pressure in kPa. For normal air a constant relation exists between convection and evaporation and can be expressed by $16.6 * h_c$ (Lewis relation).

In a cold environment the large temperature gradients between the skin surface and the environment is usually sufficient to allow control of heat balance by convection and radiation. Additional sweat evaporation may be required only at extremely high levels of metabolic heat production. For efficient cooling the evaporated sweat needs to be transported as water vapour through the clothing and air layers adjacent to the skin and/or by convection through openings in the clothing. The effects of clothing on evaporative heat transfer will be discussed later.

14.2.5   Airway heat exchange

Breathing air at low temperatures cools the airways of the respiratory system and adds to the skin heat losses. The cold air is warmed and saturated with water vapour in the lungs and airways. The amount of airway cooling increases with lowered air temperature. It increases with increased minute ventilation, but remains a relatively constant fraction of the metabolic heat production. The airway heat losses may amount to 15–20% of the total heat production of the body. Airway heat losses are not under any physiological control, but may be reduced by simple covers of mouth or by special masks for heat and moisture regain.

A classification of cold stress can be based on the auxiliary heat losses imposed by the gradually colder environment. Compared to conditions at +20 °C heat loss approximately doubles at +5 °C, increases three times at −10 °C and four times at −25 °C, everything else kept constant. Conditions at +20 °C represent indoor conditions with the body in good thermal balance. At the lower temperatures the body cannot preserve heat balance, tissues loose heat and temperatures drop.

14.3   Energy metabolism, heat production and physical work

Assessment of the protection requirements in a cold environment requires information about the energy metabolism of the individual. Metabolic rate is related to the intensity of physical work and can be easily determined from measurements of oxygen consumption. Tables are readily available that allows its estimation during different types of activity (ISO-8996, 2004). With few exceptions the values for metabolic rate also indicate the level of metabolic heat production. In most types of muscular work the mechanical efficiency is negligible. Table 14.1 can be used for a rough estimation of the metabolic rate and associated heat production.

Table 14.1 Examples of metabolic energy production associated with different types of work. Modified from ISO8996. Values refer to a standard man with $1.8\,m^2$ body surface area

Class	Average metabolic rate (Wm^2)	Examples
0 Resting	65	Resting
1 Low	100	Light manual work; hand and arm, work; arm and leg work; driving vehicle in normal conditions; casual walking (speed up to 3.5 km/h)
2 Moderate	165	Sustained hand and arm work; arm and leg work; arm and trunk work; walking at a speed of 3.5 km/h to 5.5 km/h
3 High	230	Intense arm and trunk work; carrying heavy material; walking at a speed of 5.5 km/h to 7 km/h
4 Very high	290	Very intense activity at fast pace; intense shovelling or digging; climbing stairs, ramp or ladder; running or walking at a speed of > 7 km/h
Very, very high (2 h)	400	Sustained rescue work; wild land fire fighting
Intensive work (15 min)	475	Structural fire fighting and rescue work
Exhaustive work (5 min)	600	Fire fighting and rescue work; climbing stairs; carrying persons

14.4   The human heat balance equation

Appropriate protection against cold is provided when the human body is in heat balance at acceptable levels of body temperatures (for example skin and core temperatures). This implies that heat losses are equal to metabolic heat production. The following equation describes the heat balance.

$$S = M - C - R - E - RES \qquad 14.4$$

where S is the rate of change in body heat content, M is the metabolic heat production, C is the convective heat exchange, R is the radiative heat exchange and RES is the airway heat loss, all in W/m^2. For simplicity, external mechanical work rate and conductive heat exchanges are neglected.

Equations for determination of C, R, E and RES have been described, but the role of clothing is yet to be defined. There are two principal thermal properties that determine clothing effects on heat exchanges by convection, radiation and evaporation.

1. Thermal insulation (I) defines the resistance to heat transfer by convection and radiation by clothing layers. It accounts for the resistance to heat exchange in all directions and over the whole body surface. It is an average of covered as well as uncovered body parts. This unique definition allows the introduction of clothing in the heat balance equation. The insulation of clothing and adjacent air layers is defined as the total insulation values (I_T) and is defined by the following equation.

$$I_T = \frac{t_{sk} - t_a}{R + C} \qquad\qquad 14.5$$

The value of I is given in m²°C/W or in clo-units (1 clo = 0.155 m²°C/W). This definition and the clo unit were introduced more than 60 years ago in order to facilitate understanding of human balance (Gagge *et al.*, 1941, Newburgh, 1949).

2. Evaporative resistance (R_e) defines the resistance to heat transfer by evaporation and vapour transfer through clothing layers. As for insulation, it also refers to the whole body surface. In reality, the property is a resistance to vapour transfer. Heat transfer takes place only when sweat evaporates at the skin and is transported to the environment by diffusion or convection. The evaporative resistance of clothing layers and adjacent air layers (R_{eT}) is defined by the following equation. The unit is Pa m²/W.

$$R_{eT} = \frac{p_{sk} - p_a}{E} \qquad\qquad 14.6$$

An alternative way of expressing clothing resistance to water vapour transfer is by using the moisture permeability index, i_m. The index provides a relation between evaporative and dry heat resistance of clothing systems. The following relation applies.

$$R_{eT} = \frac{I_T}{i_m \cdot L} = \frac{0.06}{i_m} \cdot \left(\frac{I_a}{f_{cl}} + I_{cl} \right) \qquad\qquad 14.7$$

where L is Lewis number (16.7 °C/kPa). The value of i_m varies between 0 for impermeable ensembles to 1 for a wet surface in strong wind. The value of i_m reduces with thickness of still air layers and number of fabric layers. Accordingly, cold protective clothing with several layers offers high resistance to evaporative heat transfer.

The human heat balance can now be written as follows. It can readily been seen how clothing affects heat exchange and the effect can be quantified.

$$S = M - \frac{t_{sk} - t_a}{I_T} - \frac{p_{sk} - p_a}{R_{eT}} - RES \qquad\qquad 14.8$$

Heat balance is achieved when the value of S is nil. This can occur for various combinations of the variables of the equation. However, only certain values for the physiological variables (M, t_{sk} and p_{sk}) are compatible with acceptable and tolerable conditions. These conditions can be analysed in terms of various scenarios for activity, climate and clothing.

14.5   Requirements for protection

Based on the human heat balance equation, the required values for I_t and R_{eT}, for combinations of activity and climate can be calculated. As previously mentioned, the appropriate strategy for efficient cold protection is to optimise clothing insulation, so as to avoid or minimise evaporative heat exchange. The value of p_{sk} in the equation becomes only slightly higher than ambient p_a and the evaporative heat loss will be small. Accordingly, the possible values of ambient t_a for which heat balance can be maintained is primarily determined by the clothing insulation value.

A method (and international standard) has been proposed that determines the required clothing insulation (IREQ) as a function of ambient climate and activity (ISO/DIS-11079, 2004). Figure 14.1 presents the required clothing insulation (IREQ-values) for combinations of activity and air temperature. This method of determining the insulation requirement implies that clothing must provide this final insulation level when used during the prevailing conditions. Also, the

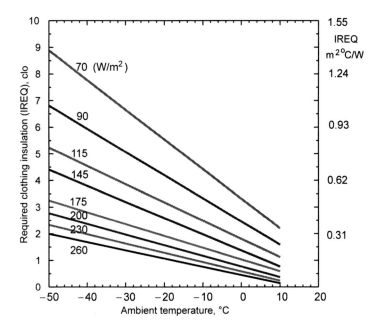

14.1 Required clothing insulation at various work intensities at low temperatures in still air (modified from ISO/DIS-11079). At very high activity levels (broken lines) the real insulation must be higher in order to prevent unacceptable local skin cooling.

IREQ-value specifies the requirement for the clothing layers only. The boundary air layer at the clothing surface is calculated separately. The justification for this is explained in the section on clothing measurements.

Insulation requirements increase steeply at low activity levels in the cold. The low levels of metabolic heat production require high thermal resistances to create heat balance. In contrast, high activity levels produce much heat that must be transferred instantly to the environment, in order to prevent overheating of the body. This is best done by reducing the insulative layer. Wind accelerates heat loss from a warm surface. Accordingly, surface layers of the body should provide high resistance to wind penetration in order to minimise microclimate cooling. The outer garment of the clothing ensemble, preferably, should be made of materials with low air permeability. Similarly, wet fabrics and layers reduce insulation and increase heat losses. In particular, the surface layer of the outer garment should be water repellent or water proof.

14.6   Measurements of clothing performance

The practical question now is how can actual clothing be tested in order to evaluate to what extent it meets the requirements for cold protective clothing. This will be answered in the following section.

14.6.1   Thermal insulation

Traditionally thermal properties of textiles are measured with a heated hot plate (ISO-5085-1, 1989, ISO-11092, 1993). Such information is of limited value for prediction of clothing properties. Heat flow through a clothing system is three-dimensional and passes through combinations of layers of fabrics and air that vary in thickness. Form, fit, design and coverage of the body are other factors modifying the insulation value. As previously mentioned the specific definition of clothing insulation requires the resistance to heat fluxes over the whole body surface area to be measured. For this purpose a thermal manikin is required (Fig. 14.2).

The use of thermal manikins in clothing research has a long history (Holmér, 2004). Today more than 100 thermal manikins are in use world-wide. Two international standards describe measurement of thermal insulation of clothing with a thermal manikin (ISO-9920, 1993, ISO-15831, 2003). In principle, a full-scale model of the human body is densely covered with resistance wires. Typically the body surface is divided into anatomical body parts that form independent zones. The surface temperature of each zone is measured and a regulation program supplies the electrical power required for maintaining a constant skin temperature (usually about 34 °C). The manikin is placed in a climatic chamber under defined conditions. From measurements of manikin skin temperature, ambient air temperature and power consumption during steady

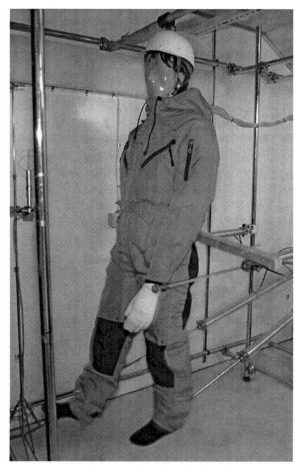

14.2 Walking thermal manikin for measurement of thermal insulation of clothing.

state conditions, the insulation of the clothing on the manikin can be determined (eqn 14.5).

The following insulation values apply. Total thermal insulation of clothing (I_T) defines insulation of all layers surrounding the body including adjacent air layers. Basic thermal insulation of clothing (I_{cl}) defines the insulation of the skin to clothing surface layers only. Air layer insulation (I_a) defines the insulation of the boundary air layer on the nude body surface.

The following relation applies:

$$I_T = I_{cl} + \frac{I_a}{f_{cl}}$$ 14.9

where f_{cl} is the clothing area factor. Total body surface area expands when clothing layers are put on the body. Typically f_{cl} varies from 1.0 to 1.5 for heavy and thick winter clothing. Normally I_a is measured with the nude thermal manikin and I_T measured with the test clothing. The I_{cl} value for the test clothing is calculated from eqn 14.9. The f_{cl} value is best determined with a photogrammetric method or with body scanning.

14.6.2   Evaporative resistance

Evaporative resistance of fabrics is measured with a heated water hotplate (ISO-5085-1, 1989, ISO-11092, 1993). Sweating manikins are available, but results show the relative effect of sweating on heat exchange rather than the actual evaporative resistance (McCullough, 2001; Meinander, 2000). A wetted cover on a dry, thermal manikin may provide reliable values for the permeability index (Breckenridge and Goldman, 1977). A combination of fabric and manikin measurements may allow the determination of clothing evaporative resistance (Umbach, 1992).

14.6.3   Wind resistance

The air permeability of fabrics can be measured using (ISO-EN-9237, 1995). The effects of form, fit, stiffness, layers and other factors are not accounted for.

14.6.4   Water resistance

Water resistance of fabrics is measured using (ISO-811, 1981). A new test is proposed that measures water resistance of full ensembles (ENV-14360, 2002).

14.6.5   Standards for protective clothing against cold and foul weather

A large number of international standards have been developed for various types of protective clothing. A few of them deal with protection against cold and foul weather.

ENV-342. Protective clothing – Ensembles and garments for protection against cold. 2002.

ENV-343. Protective clothing – Protection against foul weather. 1998.

ENV-14058. Protective clothing – Garments for protection against cool environments. 2002.

EN-511. Protective gloves against cold. 2002.

ENV-342 requires measurement of thermal insulation air permeability and evaporative resistance. The two latter properties are measured on fabrics. Thermal insulation is measured with a manikin either static or walking. Equations for wind correction are provided. Clothing performance is indicated by marking the label with the insulation value and the class of air permeability. Evaporative resistance is optional.

ENV-343 (ENV-343, 1998) requires measurement of water permeability and evaporative resistance of the fabric. Performance is indicated by marking with classes for the measured values.

ENV-14058 requires measurement of thermal insulation and air permeability of the fabric.

EN-511 (EN-511, 2004) requires measurement of thermal insulation of the complete glove with a thermal hand model. It also requires measurement of contact resistance measured with a hotplate on a sample of the glove from the palm side. Results are presented in one of four classes.

14.7   Performance of clothing for cold protection

14.7.1   Standard values for clothing insulation

The standard value for clothing insulation of an ensemble is the basic insulation value (I_{cl}). According to the standards (ISO-9920, 1993, ISO-11399, 1995), this value is measured with a standing, static manikin under still wind conditions. The basic insulation value is listed in most tables with information about clothing insulation, for example ISO-9920. Table 14.2 provides examples of clothing insulation values intended for cool to cold environments. Insulation increases with number of layers and thickness of clothing. Insulation is built up by air layers. Fabrics and textiles that trap air in layers and minimise internal convection insulate well. The chemical and physical properties of the fibres are less important, because the space they occupy in a fabric or textile is small. Also, in multi-layer ensembles the fabrics themselves may be less important then the intermediate air layers they create. A rough estimate is that fabrics on top of each other and battings provide about 1.5 clo/cm thickness, irrespective of fibre type.

In practical use the thermal properties of a clothing ensemble change dynamically as a result of the influence of for example body movements, wind and moisture accumulation. Such factors disturb the microclimate air layers and change the thermal properties. The basic insulation value, in practice, applies only to a standing person in still air. For use in evaluation of dynamic work conditions, corrections are required. This can be done in two ways.

1.   Additional measurements are carried out with a thermal manikin.
2.   Basic insulation value is corrected using empirical algorithms.

The first approach comprises measurements with a walking manikin under the influence of different air velocities. This provides realistic information about the performance of the tested ensemble, but only for the actual test conditions. The cost of testing quickly becomes enormous when number of conditions increases.

Table 14.2 Basic insulation values (I_{cl}) of selected garment ensembles measured with a thermal manikin (modifed from ISO/DIS-11079 (2004))

Clothing ensemble	I_{cl}m^2°C/W	clo
1. Briefs, shortsleeve shirt, fitted trousers, calf length socks, shoes	0.08	0.5
2. Underpants, shirt, fitted trousers, socks, shoes	0.10	0.6
3. Underpants, coverall, socks, shoes	0.11	0.7
4. Underpants, shirt, coverall, socks, shoes	0.13	0.8
5. Underpants, shirt, trousers, smock, socks, shoes	0.14	0.9
6. Briefs, undershirt, underpants, shirt, overalls, calf length socks, shoes	0.16	1.0
7. Underpants, undershirt, shirt, trousers, jacket, vest, socks, shoes	0.17	1.1
8. Underpants, shirt, trousers, jacket, coverall, socks, shoes	0.19	1.3
9. Undershirt, underpants, insulated trousers, insulated jacket, socks, shoes	0.22	1.4
10. Briefs, T-shirt, shirt, fitted trousers, insulated coveralls, calf length socks, shoes	0.23	1.5
11. Underpants, undershirt, shirt, trousers, jacket, overjacket, hat, gloves, socks, shoes	0.25	1.6
12. Underpants, undershirt, shirt, trousers, jacket, overjacket, overtrousers, socks, shoes	0.29	1.9
13. Underpants, undershirt, shirt, trousers, jacket, overjacket, overtrousers, socks, shoes, hat, gloves	0.31	2.0
14. Undershirt, underpants, insulated trousers, insulated jacket, overtrousers, overjacket, socks, shoes	0.34	2.2
15. Undershirt, underpants, insulated trousers, insulated jacket, overtrousers, overjacket, socks, shoes, hat, gloves	0.40	2.6
16. Arctic clothing systems	0.46–0.70	3–4.5
17. Sleeping bags	0.46–1.4	3–9

This option is available in a European standard for cold protective clothing (ENV-342, 2003). The second approach requires one or more equations that correct the I_{cl} value for actual activity and climatic conditions (usually wind). This model is included in the computer programme for the revised IREQ-standard (ISO/DIS-11079, 2004).

14.7.2   Influence of walking and wind

Equations have been derived from experiments in different laboratories. Nilsson *et al.* (2000) investigated winter clothing ensembles in a climatic wind tunnel and proposed a correction equation taking into account walking speed, wind speed and air permeability of the outer layer fabric. Havenith and Nilsson (2004) added data from other sources and proposed a slightly modified equation.

$$I_{T,r} = \left\lfloor e^{[-0.0512 \cdot (v_{ar}-0.4)+0.794\cdot10^{-3} \cdot (v_{ar}-0.4)^2 -0.0639*w] \cdot p^{0.1434}} \right\rfloor \cdot I_{T,static} \quad 14.10$$

Low permeability

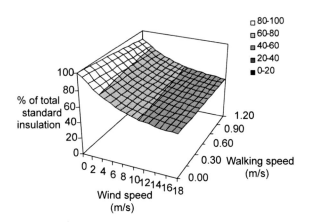

High permeability

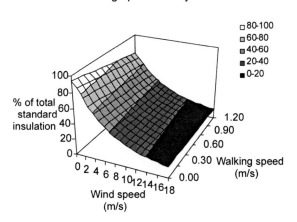

14.3 Effects of wind and walking speed on the thermal insulation of clothing. Effect is expressed as a percentage reduction of the values measured with a standing manikin in still air. Air permeability of the outer garment layer is 1 l m^{-2} s^{-1} (left panel) and 1,000 l m^{-2} s^{-1} (right panel), respectively.

where v_{ar} is the relative air velocity in m/s, w is the walking speed in m/s and p is the air permeability in l/(m^2 s).

The correction equation applies to the total insulation value and much of the effect on the boundary air layer is picked up and included. Figure 14.3 shows that also with an outer ensemble, which is highly wind proof, the reduction in insulation is more than 30–40% at high wind speeds. Much of this reduction is due to reduced boundary air layer insulation and to compression effects on the wind side of the body. The number and size of pores in the construction determine the air permeability of a textile. Textiles with no pores are impermeable and offer good protection against wind effects. But also materials with sufficiently small pores may in practice be almost impermeable to air penetration.

14.7.3   Influences of water and moisture

Clothing may become wet from precipitation or contacts from outside. Wetting from the outside should be prevented by selection of water-repellent or waterproof fabrics for the outermost clothing layer. A number of materials on the market are absolutely impermeable to water. Different treatments of fabrics render them more or less waterproof – a property, however, that may deteriorate with ageing and washing. Some microporous materials and even treatments of fabrics comprise high levels of waterproofness (and windproofness). In addition, they may allow the passage of water vapour ('breathability'). This combination of properties results in garments that are waterproof (and windproof), yet breathable in the sense that a limited amount of evaporated sweat may pass from the skin to ambient air. These features may be of relevance in temperate and warm climates. In the cold, however, a physical phenomenon limits the function.

In the cold there is a steep temperature gradient from skin, across clothing layers to ambient air. The dewpoint temperature and eventually also the freezing temperature may be reached inside the clothing by the moist air passing from the skin. Condensation occurs and moisture builds up in discrete layers. Wetting of clothing reduces the effective thermal insulation (Holmér, 1985, Meinander, 1994, Meinander and Subzerogroup, 2003). Four winter ensembles were measured during three hours with a sweating thermal manikin at two sweating rates, 100 and 200 g/(m^2∗h), respectively. The air temperatures varied from 0 to −40 °C. Table 14.3 shows the accumulated water during the three hours and the associated reduction in effective insulation (Meinander and Subzerogroup,

Table 14.3 Actual evaporation and difference in total thermal insulation as result of sweating and sweat accumulation

Ensemble	Total dry insulation, m^2°C W^{-1} (clo)	Sweating rate 100 g h^{-1} m^{-2}		Sweating rate 200 g h^{-1} m^{-2}	
		Evaporation %	Insulation reduction %	Evaporation %	Insulation reduction %
Ens. 1 at 0 °C	0.32 (2.1)	54	0	61	−7
Ens. 2 at −10 °C	0.46 (3.0)	48	−13	38	−21
Ens. 4 at −25 °C	0.65 (4.2)	17	−21	17	−28
Ens. 4 at −40 °C	0.65 (4.2)	21	−20	15	−21

2003). It is readily shown that sweat evaporation reduces significantly in the cold. At -10 and $-25\,°C$ about 40–60% of sweat evaporates. At lower temperatures evaporation is less than 20%.

14.7.4   Effects of solar radiation

Solar radiation in the cold may improve heat balance. Dark colours absorb more radiative heat of the visible spectrum than light colours (Nielsen, 1991). However, due to the many layers normally worn on a cold day the net effect of heat absorption on the outermost layer is likely to be small.

14.7.5   Effects of treatment

Many properties of textiles and clothing deteriorate with wear and washing. Surface treatments are likely to gradually disappear and must be repeated. This is particularly true for treatments against water permeation. Thermal insulation may reduce after repeated washing. The magnitude depends on many factors such as quality, type and properties of fibres and textiles, shrinking and type of construction. Thick, insulative garments and sleeping bags have been reported to lose up to 20% in thermal insulation after washing.

14.7.6   Prediction of protection

As previously mentioned, the conditions for heat balance can be described by a heat balance model (ISO/DIS-11079, 2004). The model allows the determination of

- required clothing insulation (IREQ) for given combinations of activity and climate
- exposure times for acceptable cooling for a defined clothing insulation for combinations of activity and climate
- wind and motion effects on thermal insulation and heat balance.

The input values for the model are ambient climatic conditions (air temperature, mean radiant temperature, wind speed and humidity) and activity level for IREQ. The other two analyses require a basic insulation value to be defined for the actual clothing ensemble. (ISO/DIS-11079, 2004) lists a computer program that performs the necessary calculations. The program is available online (ThEL-Lund, 2004).

   Figure 14.4 shows the protective value of clothing ensembles at nominal insulation values of 1–4 clo at a light work intensity corresponding to a metabolic rate of $110\,W/m^2$. The clo-values are basic values that can be taken from, for example, Table 14.2. When available insulation value is insufficient a recommended exposure time is calculated. Values in Fig. 14.4 apply to still air.

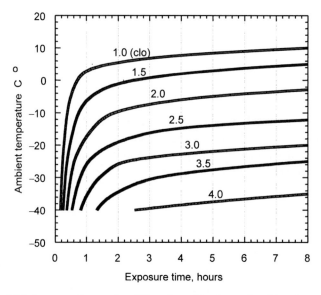

14.4 Exposure time and at different combinations of ambient temperature (still air) and clothing insulation for light activity (110 W m^{-2}). Clothing insulation is given as the basic insulation value (see text and Table 14.2). The value is automatically corrected for wind and walking speed in the computer program (ISO/DIS 11079).

All curves will shift to the left (shorter exposure times) with increasing wind and walking speed (cf. Fig. 14.3).

14.8   Specific materials and textiles for cold protection

14.8.1   The multi-layer principle

As previously mentioned, protection against cold is determined by the thickness of still air layers covering almost all parts of the body surface. Accordingly, all kinds of textiles and materials may be used provided this principle is adhered to. Typically, a cold protective ensemble is built up by several layers of garments on top of each other. The layers, apart from contributing to the total insulation, serve specific purposes.

The inner layer is worn directly on top of the skin and controls the microclimate temperature and humidity. With low activity the layer must reduce air movements. With high activity heat and moisture should be transported from the layer to cool the skin. Moisture control can be performed by absorption, by transportation to next layers or by ventilation. Absorption reduces skin humidity and retains relative comfort, but moisture remains in the clothing system and may be detrimental to heat balance at a later stage. Using hydrophobic textiles next to skin quickly increases humidity and moisture is transferred to outer layers. The advantage is an increased awareness of heat imbalance (discomfort), but moisture remains in the clothing system. The ventilation principle requires a vapour barrier worn next to skin. Microclimate humidity quickly rises with sweating, but water vapour cannot escape to outer layers. The humid microclimate forces the wearer to open up the clothing and ventilate the microclimate. This principle is preferably used in resting and low-activity conditions.

The absorbing technique may be useful for low to moderate activities with limited sweating. The transporting principle may be applied to all kind of activities, but in particular for high activity with sweating. They are most suitable for sports events and long-term exposures. Another important factor for cold protection is moisture control. This means that wetting of layers must be avoided at all time (from inside by sweating or from outside by rain or snow). If this is not possible the consequences of moisture accumulation must be controlled.

The middle layer provides most of the insulation. It comprises one or several garments of thicker material depending on the requirements. Choice of textiles is more or less arbitrary as long as good insulation is provided. Non-absorbing materials should be selected for long-term exposures with limited heating and drying opportunities (expeditions, etc.).

The outer layer must provide protection against environmental factors such as wind, rain, fire, tear and abrasion. In occupational contexts, protection may also be required against, for example, chemical and physical agents. This layer can also add to the total insulation by the provision of insulation liners and battings.

14.8.2   Natural and synthetic fibres

Textiles such as wool and wool blends possess a high absorbing capacity and can handle smaller amounts of moisture without losing their insulation properties. Wool can be used as a next-to-skin fabric and may keep the skin relatively dry. When the fabric becomes saturated, however, moisture control is reduced. Cotton is absorbing as well, but clings to skin when wet and should not be close to skin in cold environments. Many synthetic textiles are hydrophobic and the moist air moves from skin through the fabric to the next layer. The skin microclimate quickly becomes humid during sweating. This humidity is uncomfortable and causes the wearer to take action for appropriate adjustments. This is also the main effect and purpose of using vapour barrier fabrics next to skin.

Moisture absorbed in garments, in addition to causing discomfort at some stage, adds to the weight carried by the person. In addition it gradually reduces thermal insulation of that particular layer. When activity drops and sweating ceases, the drying of wet clothing layers may deprive the body of more heat than is generated by metabolic rate. The result is a post-chilling effect that may endanger heat balance and result in hypothermia. As long as one can stay dry, the choice of material (natural or synthetic) for the various layers is a matter of taste or other preferences and requirements. With sweating and, in particular, with longer outdoor excursions, the advantage of lightweight, strong and hydrophobic materials as parts of the clothing ensemble should be recognised.

14.8.3   Improved insulation

Wool and down fabrics are highly insulative due to the very nature of the fibre materials. Modern synthetic textiles such as battings made of polyester hollow fibres or polyolefin micro-fibres resemble in a way the natural materials and provide good insulation per unit thickness. As a spin off to space technology, reflective materials (mostly aluminised fabrics or fibres) are used in garments and survival kits. The idea is that much of the heat the body loses through radiation will be reflected back to the skin. Such an ensemble will transmit less overall heat and the net insulation is higher compared to a similar one without a reflective layer. Practical tests, however, have shown that the net effect is small – in certain conditions negligible. The reasons are several. Radiation heat loss is only a minor part of the overall heat loss in the cold, in particular in the presence of wind and/or body movements (10–15%). Reflection of radiation requires spacing of layers, which is difficult to achieve and maintain. Most reflective fabrics are impermeable and interfere with moisture transfer. Aluminised insoles for shoes are common but provide no additional insulation compared to soles of similar thickness without aluminium. Gloves and socks with aluminium threads in the fabric do not give higher insulation compared to those without.

14.8.4   'Breathing fabrics'

In foul weather good protection against rain and snow is required. However, waterproof fabrics may interfere with evaporative heat exchange. During activity the person gets wet from inside instead of from outside. Micro-porous materials help to solve this problem to some extent. The small pores allow water vapour to pass but stops liquid water. This works reasonably well in temperate and warm climates. Due to the 'cold wall' principle it becomes less valuable in colder climates. During intermittent cold and warm exposures (in and out) it may allow absorbed moisture to escape in the warm conditions. Also textiles of this kind are often highly windproof, which is beneficial in cold environments.

14.8.5   Intelligent textiles

In recent years several new types of materials and fabrics have been put on the market that contain some active component. Examples are phase change materials (PCM), inflated tubings and electrical heating. Fabrics containing

PCM respond to cooling by releasing heat from a range of waxes in the fibre or fabric (fibre content changes from solid to liquid phase). Once solid it reacts to heating by absorbing heat (fibre content melts). By choosing a certain temperature for the phase change the fabric (in a garment) could assist the wearer's thermoregulatory adjustment to hot and cold environments. Although the principle is physically sound (compared with the industrial ice-vest), the PCM fabrics on the market contain insufficient amounts of the phase change material. The heat transfer involved is almost negligible, difficult to measure and almost impossible to perceive (Shim *et al.*, 2001, Weder and Hering, 2000, Ying *et al.*, 2003).

Inflatable fabrics, in principle, should allow the expansion of thickness of the ensemble, thereby increasing the effective insulation. Fabrics with a system of thin tubing can be inflated by the mouth. The effect is a thicker layer of that particular garment that should add insulation. Electrically heated elements incorporated into fabrics have been available for many years. The disadvantage, so far, has been the low capacity of portable batteries, and the durability of the wiring system. The rapid development of mobile phones and portable computers has resulted in the availability of powerful, long-lasting, portable batteries that can be used also for auxiliary heating. This concept is likely to be most beneficial to the heating of hands and feet, but garments are already available on the market that have built-in batteries and are charged from the mains supply.

14.9   Sources of further information

Several international conferences have focused on clothing, climate and protection.

- Problems with cold work (Holmér and Kuklane, 1998)
- Nokobetef 6 and 7 organised together with the 1st and 2nd European Conference for Protective Clothing (ECPC) (Kuklane and Holmér, 2000, Richards, 2003)
- Environmental Ergonomics Conferences (see Holmér *et al.*, 2005)

There are also several fibre and textile conferences that provide valuable information.

14.10   References

Breckenridge JR, Goldman RF. Effect of clothing on bodily resistance against meteorological stimuli. In: Tromp J, ed. *Progress in Human Biometeorology*. Amsterdam: Sweits & Zeitlinger, 1977: 194–208.
EN-511. Protective gloves against cold. Comité Européen de Normalisation, 2004.
ENV-342. Protective clothing against cold. Comité Européen de Normalisation, 2003.
ENV-343. Protective clothing – Protection against foul weather. Comité Européen de Normalisation, 1998.

ENV-14360. Protection against foul weather – Test method for the rain tightness of a garment – Impact from above with high energy droplets. Comité Européen de Normalisation, 2002.

Gagge AP, Burton AC, Bazett HC. A practical system of units for the description of the heat exchange of man with his environment. *Science* 1941; 94: 428–430.

Havenith G, Nilsson H. Clothing evaporative heat exchange in comfort standards. *European Journal of Applied Physiology* 2004; 92(6): 636–640.

Holmér I. Heat exchange and thermal insulation compared in woollen and nylon garments during wear trials. *Textile Research J.* 1985; 55: 511–518.

Holmér I. The history and use of thermal manikins. *European Journal of Applied Physiology* 2004; 96(6): 614–618.

Holmér I, Kuklane K, eds. *Problems with cold work*. Stockholm: Arbete & Hälsa, 1998: 265.

Holmér I, Kuklane K, Gao C. *Environmental Ergonomics XI*, Lund University, Lund, 663 pp, 2005.

ISO-5085-1. Textiles – Determination of thermal resistance – Part 1: Low thermal resistance. 1989.

ISO-8996. Ergonomics – Determination of metabolic heat production. International Standards Organisation, 2004.

ISO-9920. Ergonomics – Estimation of the thermal characteristics of a clothing ensemble. International Standards Organisation, 1993.

ISO-11092. Textiles – Physiological effects – Measurement of thermal and water-vapour resistance under steady-state conditions (sweating guarded hotplate test). International Standards Organisation, 1993.

ISO-11399. Ergonomics of the thermal environment – Principles and application of International Standards. International Standards Organisation, 1995.

ISO-15831. Thermal manikin for measuring the resultant basic thermal insulation. 2003.

ISO-811. Textiles – Determination of resistance to water penetration. Skip International Standards Organisation, 1981.

ISO-EN-9237. Textiles – Determination of permeability of fabrics to air. Skip International Standards Organisation, 1995.

ISO/DIS-11079. Ergonomics of the thermal environment – Determination and interpretation of cold stress when using required clothing insulation (IREQ) and local cooling effects (Technical Report). International Standards Organisation, 2004.

Kuklane K, Holmér I. Ergonomics of protective clothing. *Proceedings of NOKOBETEF 6 and 1st ECPC*. Solna, Sweden: Arbetslivsinstitutet, Stockholm, Sweden, 2000 Arbete & Hälsa; vol 2000: 8).

McCullough E. Interlaboratory study of sweating manikins. *5th European Conference on Protective Clothing and NOKOBETEF7*. Montreux, Switzerland: EMPA, 2001.

Meinander H. Influence of sweating and ambient temperature on the thermal properties of clothing. *Fifth International Symposium on Performance of Protective Clothing Improvement through Innovation*. San Francisco, 1994.

Meinander H. Extraction of data from sweating manikins tests. In: Nilsson H, Holmér I, eds. *Thermal manikin testing 3IMM*. Stockholm: National Institute for Working Life, 2000: 99–102.

Meinander H, Subzerogroup. *Thermal insulation measurements of cold protective clothing using thermal manikins*. Technical University, 2003.

Newburgh LH, ed. *Physiology of heat regulation and The science of clothing.*

Philadelphia London: W. B. Saunders Company, 1949: 457.

Nielsen B. Solar heat load an clothed subjects. *Proceedings of 2nd International Symposium on Clothing Comfort Studies* in Mt. Fuji. Fuji Institute of Education and Training, Japan: The Japan research association for textile end-uses, 1991: 243–255.

Nilsson HO, Anttonen H, Holmér I. New algorithms for prediction of wind effects on cold protective clothing. *NOKOBETEF 6, 1st ECPC*. Norra Latin, Stockholm, Sweden, 2000: 17–20.

Richards M. Challenges for Protective Clothing. Second Eurpean Conference on Protective Clothing. Montreux, Switzerland: *EMPA*, 2003: 4.

Shim H, McCullough E, Jones B. Using phase change materials in clothing. *Textile Res. J.* 2001; 71(6): 495–502.

ThEL-Lund. www.eat.lth.se/Research/Thermal, 2004.

Umbach KH. Prediction of the physiological performance of protective clothing based on laboratory tests. In: Mäkinen H, ed. *Nokobetef IV: Quality and usage of protective clothing*. Kittilä, Finland, 1992: 151–156.

Weder M, Hering A. How effective are PCM-materials? Experience from laboratory measurments and controlled human subject tests. *EMPA*, Switzerland, 2000.

Ying B, Kwok Y, Li Y, Yeung C, Song Q. Thermal regulating functional performance of PCM garments. *International Journal of Clothing Science and Technology* 2003; 16(1–2): 84–96.

15
Thermal (heat and fire) protection

A R HORROCKS, University of Bolton, UK

15.1 Introduction

That textiles on the one hand protect us from external agencies in our daily lives and yet pose a significant fire hazard may appear as a paradox to the average consumer, especially if, for example, he or she is to consider the requirements of a protective garment for an industrial worker in a hazardous environment or a firefighter. However, as demands for higher levels of protection for wearers in hazardous environments (usually associated with occupational and professional requirements) increase, the textile and apparel industries continue to overcome design and constructional challenges that offer the right levels of protection for defined external agencies that operate individually or in concert. In addition, having engineered into a textile the desired level, the product must still function as a textile in terms of its aesthetic, comfort, durability and aftercare expectations.

Given that protective textiles often have to protect against more than one agency and that technologies available may not only operate against each other and compromise aesthetics and comfort, the final solution is often a balance. In the case of heat and fire protection, addressing these hazards is the principal goal and other properties, such as breathability and comfort, may be less well satisfied. In fact, in the case of firefighters' garments, for example, the outer fire protective garment assembly, because of its weight and lack of moisture permeability, easily gives rise to heat stress or shock to wearers in typical fire surroundings. Therefore, conditions may have to applied to their use such as definition of maximum wearing times in real fire scenarios. In the case of barrier textiles designed to protect underlying and inanimate, but potentially flammable materials such as upholstery interior fillings, this is obviously not a problem. However, in commercial aircraft seating where fire protection levels have to be high, so also does comfort since passengers may be in contact with seating for up to 12 hours at a time; far more than is usually the case in a domestic environment.

In this chapter, major focus will obviously be on fire and heat protection and the means of conferring these features into textile products and assemblies and so other factors, such as comfort, ballistic resistance, chemical and biological agency resistance will only be commented on if particularly relevant to a particular thermal protective solution or system. Other chapters in this book consider related and often interacting issues such as multifunctional treatments (Chapter 8), comfort (Chapter 10) and biological and chemical (Chapters 12, 16 and 20) and ballistic (Chapter 19) resistance.

This chapter will concentrate on factors that assist us in understanding the nature of the thermal hazards as they apply to textiles and the means of addressing them. The latter will examine the available strategies at fibre, yarn, fabric, chemical finishing and textile assembly levels for introducing defined levels of thermal protection. Finally, the related testing and performance assessment methodologies that enable the quantification of defined levels and qualities of protection are discussed. Typical examples of thermally protective textiles include the following:

- protective clothing for workwear, hazardous industrial occupations, firefighters and defence personnel
- outer furnishing fabrics that protect underlying fillings in domestic, contract and transport applications
- barrier textiles that may prevent fire spread (e.g., fire curtains) or protect underlying surfaces from heat and fire spread
- extreme hazard protection, e.g., furnace operators' aprons to protect against hot metal splash, fire entry suits, racing car drivers' suits.

While there is a considerable literature in this area and this chapter is not meant to be a comprehensive literature review, a recent selection of relevant publications are those by Barker and Coletta (1986), Mansdorf *et al.* (1988), Raheel (1994), Johnson and Mansdorf (1996), Stull and Schwope (1997), Scott (2000), Bajaj (2000), Horrocks (2001) and Hearle (2001).

15.2   Fire science factors

Thermal protection relates to the ability of textiles to resist conductive, convective, radiant thermal energy or a combination of two or more. For example, a flame constitutes a convective oxidative chemical reaction zone in which the energy is contained within the extremely hot gas molecules and particulates, including smoke. Typical textile flame temperatures may range from about 600 to 1,000 °C (see Table 15.1) and when a flame impinges upon a textile surface, not only is it subjected to the high temperature of the flame but also the chemical intermediates of the reaction zone, which may increase the likelihood of its degradation and ignition. Thus the selection of textile outer fabrics for a firefighter's tunic or an industrial worker's coverall should take account of both the temperature and reactivity of an impinging flame source.

Radiant thermal energy, however, is electromagnetic radiation emitted by a hot surface or a flame and this infra-red is absorbed by the molecular structure of

Table 15.1 Thermal transitions and flame temperatures of the common fibres

Fibre	T_g, °C (softens)	T_m, °C (melts)	T_p, °C (pyrolysis)	T_c, °C (ignition)	T_f, °C*	LOI, %
Wool	—	—	245	570–600	680, 825 (v)	25
Cotton	—	—	350	350	974 (h)	18.4
Viscose	—	—	350	420	—	18.9
Nylon 6	50	215	431	450	—	20–21.5
Nylon 6.6	50	265	403	530	861 (h)	20–21.5
Polyester	80–90	255	420–447	480	649 (h), 820 (v)	20–21
Acrylic	100	>220	290 (with decomp)	>250	910 (h), 1050 (v)	18.2
Polypropylene	−20	165	470	550	—	18.6
Modacrylic	<80	>240	273	690	—	29–30
PVC	<80	>180	>180	450	—	37–39
Meta-aramid (e.g. Nomex)	275	375–430 (decomp)	425	>500	—	29–30
Para-aramid (e.g. Kevlar)	340	560 (decomp)	>590	>500	—	29

Note: * flame temperatures recorded only for fibres that burn in air; h is for horizontal and v is for vertically downward burning fabrics (Rebenfeld *et al.*, 1979).

the surface fibres and any coatings present, raising the temperature to several hundred degrees Celsius, which if high enough (typically >300 °C) promotes thermal degradation (or pyrolysis) and even ignition. In a real fire, it is usually the radiant heat that causes it to spread rapidly by heating up and igniting materials some distance away from the fire. Again, it is obvious that protective clothing should resist the effects of heat radiation if it is to be effective.

Finally, conductive heat is that which flows from a hotter surface to a cooler one, which if a textile may promote thermal degradation and ignition. Protective clothing for furnace personnel who risk contact with molten metal splashes must resist such threats. However, in many thermally hazardous environments, a combination of conduction, convection and radiation may be operating in concert, and usually the last two are associated with flame sources in particular.

15.2.1   Criteria that define heat and/or fire performance

Once heat is absorbed by a textile fibre, it may promote physical or chemical change or both. All organic fibre-forming polymers will eventually thermally degrade at or above a threshold temperature often defined as the pyrolysis temperature, T_p. This may be influenced by the presence of oxygen in the air and so T_p may vary whether the fibre is heated in nitrogen, air or vitiated air (often present in a typical fire). Typical values of T_p are listed in Table 15.1 for the more common fibres. Once heated above this temperature in air, ignition at a higher temperature, T_i, usually follows and these are also listed in Table 15.1. However, flame temperatures are usually higher still, and where available, are included for those fibres that burn in air.

While the above temperatures are associated with chemical change, some fibres undergo physical change as an initial softening followed by melting in some cases. The former is defined as a second-order temperature, T_g, and the latter as a melting temperature, T_m. Table 15.1 lists these, which for the so-called thermoplastic fibres, are less than T_p and T_i. For the conventional textile fibres like polypropylene, polyester and nylons (or polyamides) 6 and 6.6, the relatively low temperatures at which these physical transitions occur mean that textiles comprising them offer little of no protection against thermal energy. Not only will they shrink and deform when heated, but also complete melting and disintegration of any previous textile structure will occur with dire consequences for the wearer or underlying surface.

15.2.2   Levels of thermal resistance

Based on the above discussion, it is evident that different thermal threats will promote a number of often interrelated physical and chemical changes in an exposed textile substrate including any finishes or coatings that are present. While there is no accepted hierarchy of threat and hence levels of resistance and thermal protection required, some performance and standard test methods do recognise this. For example, BS5852:1979 Parts 1 and 2 and its variants since 1979 (see section 15.5) tests the ability of upholstered furnishing fabric internal filling composites to resist ignition and of the outer cover to protect the inner components. The related standard BS7176:1991 advises how the test may be used to assess ignition hazards relating to various application environments of differing hazard level (see section 15.5). This has led, of course, to a whole family of textile solutions that have different levels of thermal and fire protection.

Levels of thermal resistance may be simply defined with respect to maximum temperature of exposure criteria, although time of exposure is a significant parameter as well. For thermal environments where flame or spark sources of ignition are absent (e.g. hot surfaces, radiant energy, molten metal splash), temperatures may be significantly less than flame temperatures and so even thermoplastic fibres can resist some thermal hazards for significant periods of time. Once temperatures rise significantly above respective T_g values and approach melting temperatures, then most textiles comprising these fibres lose any protective value. In fact, they can prove to be a hazard through shrinkage onto underlying surfaces thereby removing any thermally insulating air layers. Fabrics designed for defence purposes often require only non-thermoplastic fibres to be present, especially for inner clothing items, and so remove the possibility of this often-termed 'shrink wrapping' effect (Scott, 2000). Since most conventional synthetic fibres undergo physical transitions no greater than about 265 °C (in the case of nylon 6.6 and polyester), temperatures of about 200 °C are seen as a maximum for their safe use, if the relatively more sensitive polyolefin fibres are ignored (see Table 15.1).

For thermal threats yielding temperatures above 200 °C, essentially non-thermoplastic fibres only should be used and we should differentiate between those fibres that are flame retardant and those that are heat resistant – the two are not necessarily related! The former will yield textiles that resist defined ignition sources and if they do ignite when the source is present, have reduced burning rates and times once the source has been removed. Ideally, if textiles comprising this type of fibre are to continue to offer thermal protection, the heated fibres should convert to char and so the textile now provides a carbonaceous char replica of itself that is extremely flame resistant (Horrocks, 1996). Most flame retardant fibres start to become effective above about 250 °C when component fibres start to thermally degrade (see Table 15.1) and the flame retardants present start to interact either independently or in concert with respective fibre chemistries (Horrocks and Price, 2001).

Heat resistant fibres, however, are those having chemical structures that are little changed by temperatures well above the 300 °C level and, in the case of

Table 15.2 Maximum service lifetimes for heat resistant fibres in thermally protective textiles (Horrocks *et al.*, 2001)

Fibre genus	Second-order temperature, °C	Melting temperature, °C	Onset of decomposition, °C	Maximum, continuous use temperature, °C	LOI, %
Melamine-formaldehyde	NA	NA	370	190	32
Novoloid	NA	NA	>150	150/air; 250/inert	30–34
m-Aramid	275	375–430 (decomp)	425	150–200	28–31
p-Aramid	340	560 (decomp)	>590	180–300	29–31
Arimid (P84)	315	—	450	260	36–38
Aramid-arimid	<315	—	380	NA	32
Semicarbon	NA	NA	NA	~200/air	55
PBI	>400	>NA	450/air; 1000/inert	~300 (est)	>41
PBO	—	—	650; >700/inert	200–250(est)	68

Notes: NA = not applicable; (decomp) = with decomposition; (est) = value estimated from data in cited references.

ceramic fibres, above 1,000 °C. For textiles used in high-temperature industrial processes, such as gas and liquid filtration, long-term exposures to temperatures of about 100 °C are often required, but not all these fibres are used in thermally protective applications (Horrocks *et al.*, 2001). However, in long-term exposure, thermally protective applications, we need to be able to define maximum service life temperatures and these are listed in Table 15.2 for selected heat resistant fibres (Horrocks *et al.*, 2001).

Heat resistant fibres are more often than not also flame resistant because they have chemical structures resistant to thermal degradation in the first place (see Table 15.1) and may also form carbonaceous chars and few flammable, fuel-forming volatiles (Horrocks, 1996). Flame resistance may be conveniently measured as a limiting oxygen index value (Horrocks *et al.*, 1987), which for textiles with values greater than 21 (the percentage volume concentration of oxygen in air), indicates less vigorous burning than if less than 21. Fibres and textiles having LOI values >26–28 tend to be flame retardant in air and will pass simple vertical fabric strip tests (see section 15.5), while those having values in excess of 30 are highly flame retardant. LOI values are also included in Table 15.1.

15.3   Flame retardant fibres and textiles

For most thermal protective applications, textiles comprising 100% synthetic and hence thermoplastic fibres must be avoided for reasons already stated. The only time 100% synthetic textiles may be used is if either they are part of a composite structure in which, for example, they are present as a component with a flame retardant/heat resistant majority fibre construction or a char-forming, flame retardant back-coating is present. Examples here are the 100% medium weight (e.g. 200–300 gsm) polypropylene contract seating fabrics to which a flame retardant back-coating (e.g. an acrylic based, antimony-bromine-containing system (Horrocks, 2003)) has been applied, often at high levels (e.g. typically >50% w/w with respect to fabric). Typical blend examples that include thermoplastic fibres are:

- flame retardant cotton-rich (e.g. 55%) – polyester (e.g. 45%) blends for protective workwear
- flame retardant wool (e.g. 90%) – nylon 6.6 (e.g. 10%) blends for aircraft seating outer fabrics.

The major flame retardant fibres used in flame retardant, protective textiles are listed in Table 15.3.

Within this group, the flame retardant properties may be conferred by chemical treatment or finishing of conventional fibres, use of inherently flame retardant fibres, blending of flame retardant and non-flame retardant fibres or by the application of a surface coating. Some flame retardant finishes are non-

Table 15.3 The major flame retardant fibres including thermoplastic variants used in protective fibre blends

Fibre	Flame retardant structural components	Mode of intro-duction
Natural		
Cotton	Organophosphorus and nitrogen-containing monomeric or reactive species, e.g. Proban CC (Rhodia), Pyrovatex CP (Ciba), Aflammit P and KWB (Thor), Flacavon WP (Schill & Seilacher)	F
	Antimony-organo-halogen systems, e.g. Flacavon F12/97 (Schill & Seilacher), Myflam (Noveon)	F
Wool	Zirconium hexafluoride complexes, e.g. Zirpro (IWS); Pyrovatex CP (Ciba), Aflammit ZR (Thor)	F
Regenerated		
Viscose	Organophosphorus and nitrogen/sulphur-containing species, e.g. Sandoflam 5060 (Clariant, formerly Sandoz) in FR Viscose (Lenzing)	A
	Polysilicic acid and complexes, e.g. Visil AP (Sateri)	A
Inherent synthetic		
Polyester	Organophosphorus species: Phosphinic acidic comonomer, e.g. Trevira CS (Trevira GmbH, formerly Hoechst), Avora CS (KoSA); phosphinate diester comonomer, Heim (Toyobo, Japan); phosphorus-containing additive, Fidion FR (Montefibre), Brilén FR (Brilén, Spain)	C/A
Acrylic (modacrylic)	Halogenated comonomer (35–50% w/w) plus antimony compounds, e.g. Velicren (Montefibre); Kanecaron including Protex (Kaneka Corp.)	C
Polypropylene	Halo-organic compounds usually as brominated derivatives, e.g. Sandoflam 5072 (Clariant, formerly Sandoz);	A
	Bromo-phosphorus compound: FR-370/372 (Dead Sea Bromine Group)	A
	Hindered amine: NOR116 (Ciba)	A
Polyhalo-alkenes	Polyvinyl chloride, e.g. Clevyl (Rhovyl SA), Fibravyl (Rhone-Poulenc)	H
	Polyvinylidene chloride, e.g. Saran (Saran Corp.)	H

Key
F: chemical finish
A: additive introduced during fibre production
C: copolymeric modifications
H: homopolymer

durable to laundering, others may be semi-durable and withstand a single water soak or dry cleaning process and fully durable ones may withstand many domestic or commercial wash cycles, usually in excess of 50. Durability requirements are usually determined by the application and specified by regulation or the customer. Since the accumulation of oily soils on textiles during use often negates the effect of flame retardant properties present, laundering and aftercare requirements may also be closely defined by regulation or need. For example, protective workwear may require laundering after one day's wearing and protective textiles in defence applications require military personnel to be aware of the aftercare requirements if performance is to be maintained (Scott, 2000).

The most commonly used, flame retarded protective textiles for the workwear markets are those comprising flame retardant cellulose fibres. They also find application in military and emergency applications as well as furnishings. Flame retardant cellulosic protective textiles generally fall into three groups based on fibre genus, namely flame retardant cotton, flame retardant viscose (or regenerated cellulose) and blends of flame retardant cellulosic fibres with other fibres, usually synthetic or chemical fibres. They all have the advantage of being durable to most aftercare treatments, they are very cost-effective and, most importantly, when exposed to a flame or heat source, convert to a carbonaceous char and so continue to offer protection to underlying surfaces.

15.3.1   Flame retardant cottons

All flame retardant cottons are usually produced by chemically after-treating fabrics as a textile finishing process which, depending on chemical character and cost, yield flame retardant properties having varying degrees of durability to various laundering processes. For extreme durability chemically reactive, usually functional finishes are required to give durable flame retardancy (e.g. alkylphosphonamide derivatives (Pyrovatex, Ciba; Aflammit KWB, Thor; Flacavon WP, Schill & Seilacher; Pekoflam DPM (Clariant)); tetrakis (hydroxymethyl) phosphonium (THP) salt condensates (Proban, Rhodia; Aflammit P, Thor). Generally, however, phosphorus levels of 1.5 to 4% (w/w) on fabric are used which can give finish add-ons in the range of 5 to 20% (w/w) depending on the finishing agent phosphorus content. The actual level used depends on both the fabric weight (typically 200–350 gsm for protective end-uses) and the level of protection required.

Most of these treatments have become well-established during the last thirty to forty years and few changes have been made to the basic chemistries since that time. The earlier review by Horrocks (1986) and the more recent update (Horrocks, 2003) fully discuss the relevant chemistries and technologies. During this same period, many other flame retardants based on phosphorus chemistry, have ceased to have any commercial acceptability for reasons which include toxicological properties during application or during end-use, antagonistic inter-actions with other acceptable textile properties and cost (Horrocks, 1986). The examples cited above may be considered to have stood the test of time in terms of their being able to satisfy technical performance demands and enable flammability regulatory requirements to be met, while having acceptable costs and meeting health and safety and environmental demands. It is probably true to say, that durably flame retardant industrial workwear commands a significant proportion (probably >90%) of this market where a combination of FR performance, comfort, launderability and cost are the determining factors. This is not to say that FR cotton does not find application in more exotic end-uses, however – its use in the US Space Shuttle programme for protective apparel was reported 25 years ago (Dawn and Morton, 1979).

15.3.2   Flame retardant viscose

These usually have flame retardant additives incorporated into the spinning dopes during their manufacture, which therefore yield durability and reduced levels of environmental hazard with respect to the removal of the need for a chemical flame retardant finishing process (see Table 15.3). The thermal performance of these fibres is very similar to those of the flame retarded cotton fibre examples above. The polysilicic acid-containing Visil (Sateri, Finland) flame retardant viscose fibre is particularly interesting in that not only is it phosphorus-free but on heating, both a carbonaceous and siliceous char is formed (Horrocks, 1996). The presence of the silica in the residue, ensures that thermally exposed fabrics revert to a ceramic char, thus affording high levels of protection to temperatures as high as 1,000 °C.

15.3.3   Flame retarded cellulosic blends

In principle, flame retardant cellulosic fibres may be blended with any other fibre, whether synthetic or natural. In practice, limitations are dictated by a number of technical factors including compatibility of fibres during spinning, fabric formation and chemical finishing. Of particular relevance to fire performance is the last factor since flame retardant treatments must not adversely influence the characteristics of the other fibres present in the blend during their chemical application. Furthermore, additivity and, preferably synergy, should exist in the flame retardant blend. It is important to note, however, that with some flame retardant blends, antagonism can occur and the properties of the blend may be significantly worse than either of the components alone (Horrocks, 1986). Consequently, the current 'rule of thumb' for the simple flame retardant finishing of blends is to apply flame retardant only to the majority fibre present and enable this component to confer flame retardant behaviour on the minority components.

The prevalence of polyester-cotton blends coupled with the apparent flammability-enhancing interaction in which both components participate (the so-called scaffolding effect, reviewed elsewhere (Horrocks *et al.*, 1987; Horrocks, 1986)) has promoted greater attention than any other blend. However, because of the observed interaction, only halogen-containing coatings and back-coatings find commercial application to blends which span the whole blend composition range (see below). Most durable finishes for protective cellulosic textiles function best on cellulosic-rich blends with polyester where the char-forming majority component not only supports and extinguishes the minor polyester present, but also prevents hole formation and disintegration of the exposed fabric. Thus, phosphorus-containing cellulose flame-retardants, such as the THP-based systems like Proban CC (Rhodia) on blends containing no less than 55% cotton offer a combination of flame retardation and acceptable handle and comfort, unlike polyester-rich blends. The THP condensate is substantive only on the cellulose content and so requires over 5% (w/w) phosphorus to be present on this component in order to confer acceptable flame retardancy to the whole blend. In order to achieve the high finish levels necessary, often a double pass pad (or foam)-dry stage is required before the THP-condensate-impregnated fabric is ammonia-cured in the normal way. However, high phosphorus and hence finish levels may lead to excessive surface deposits on fibres which often reduce durability to laundering and create unacceptable harshness of handle. Because such an application only works well on medium to heavy weight fabrics (> 200 gsm), they are particularly effective for protective clothing applications. The use of a cotton-rich blend here is also advantageous because the lower polyester content confers a generally lower thermoplastic character to the fabric with less tendency to produce an adhesive molten surface layer when exposed to a flame.

Less used in protective workwear, but often preferred in furnishings because of superior dyestuff and print compatibility, are the methylolated phosphonamide finishes (e.g. Pyrovatex CP, Ciba), but these are effective only on blends containing 70% or less cellulose content. This is because the phosphorus present is less effective on the polyester component than in THP-based finishes (Horrocks, 1986). The reasons for this are not clear but are thought to be associated with some vapour-phase activity of phosphorus in the latter finish on the polyester component.

If a blend component does have vapour phase activity then this may be transferred to the cellulosic component. This certainly is the case in blends of cotton and PVC and, more recently, with modacrylic fibres. The Protex M range of fabrics feature blends of the modacrylic Kanecaron Protex fibre (see Table 15.3) with cotton in almost equivalent fractions as a 55% modacylic/45% cotton composition. The fabrics are suitable for furnishings, bedding, protective clothing including welding and similar hazard protection.

15.3.4   Flame retarded wool and blends

Within the area of flammability of all so-called conventional fibres, wool has the highest inherent non-flammability and so is particularly attractive for protective textile end-uses such as uniforms, coveralls, transport seatings and domestic and contract furnishings, where heavier fabrics may be used and the aesthetic character of wool may be marketed. Table 15.1 above shows it to have a relatively high LOI value of about 25 and a low flame temperature of about 680 °C.

Once again, char-promoting flame retardants are preferred, although bromine-containing, vapour phase-active surface treatments are effective. Horrocks (1986) has comprehensively reviewed developments in flame retardants for wool up to 1986 and very little has changed since that time. However, although considerable research has been undertaken into the use of functional phosphorus-based finishes, including the effectiveness of methylolated phosphonomides (e.g. Pyrovatex CP) (Hall and Shah, 1991), and substantive halogenated species like chlorendic, tetrabromophthalic anhydride (TBPA) and dibromo-maleic anhydrides and brominated salicylic acid derivatives, the most commonly used durable flame retardants are probably those based on Benisek's Zirpro (IWS) system (Horrocks, 1986). This treatment is applicable from the dyebath and has no obvious associated discoloration or affect on wool aesthetics.

The Zirpro process is based upon the exhaustion of negatively charged complexes of zirconium or titanium, usually as potassium hexafluorozirconate (K_2ZrF_6) or a mixture of this and potassium hexafluorotitanate ($K_2Ti F_6$), onto positively charged wool fibres under acidic conditions at a relatively low temperature of 60 °C. Application to wool is possible at any processing stage from loose fibre to fabric using exhaustion techniques either during or after dyeing; these have been fully reviewed and described elsewhere (Horrocks, 1986, 2003). The relatively low treatment temperature is an advantage because this limits the felting of wool and by maintaining a low pH (3), penetration is maximised and wash-fastness to as many as 50 washes at 40 °C or 50 dry cleaning cycles in perchloroethylene is achieved. The general simplicity of the whole process enables it to be used either concurrently with 1:1 premetallised and acid levelling dyes or after dyeing when applying acid milling reactive 1:2 premetallised and chrome dyes. Furthermore, Zirpro treatments are compatible with shrink-resist, insect-resistant and easy-care finishes.

It has been proposed that the Zirpro treatment enhances intumescent char formation (Benisek, 1971) although this view is not universally held (Beck *et al.*, 1976). However, its effectiveness in protective fabrics to flame and heat at high heat fluxes, is associated with the intumescent fibrous char structures generated. Because of heavy metal effluent problems, there has been pressure to identify alternatives and limited interest remains in sulphation with ammonium sulphamate (Lewin and Mark, 1997) followed by curing at 180–200 °C in the presence of urea which can give a 50 hard water wash-durable finish for wool fabrics with little change in handle. While there has been concurrent interest in the use of intumescents (Horrocks and Davies, 2000) no commercial exploitation has taken place.

Wool blends pose different challenges, because of the complexity of wool on the one hand and the specificity of flame retardants with respect to each blend component fibre on the other. In the absence of any back-coating treatment, acceptable flame retardancy of Zirpro-treated blends is obtainable in 85/15 wool/polyester or polyamide combinations. Such fabrics are ideal for aircraft and other seatings as well as heavier fire protective clothing and uniforms. For lower wool contents in blends and without the possibility of using alternative FR treatments, flame retardancy can be maintained only if some of the Zirpro-treated wool is replaced by certain inherently flame retardant fibres, except for Trevira CS polyester where antagonistic effects have been noted (Benisek, 1981). Chlorine-containing fibres such as PVC and modacrylics are particularly effective in this respect.

15.3.5   Flame retarded and inherently flame retardant synthetic fibres

Notwithstanding their associated thermoplasticity and often fusible behaviour (see Table 15.1), these fibres may, however, be used as minor components in blends with char-forming fibres, particularly cellulosic and wool and these have been discussed above. Rarely when present as a minor component are these synthetic fibres individually treated (Horrocks, 2003) and they rely on the flame retarded majority component to reduce any flammable tendency. Their presence is often included to enhance tear strength and abrasion resistance as well as conferring some easy-care character but their generally non-char-forming behaviour remains a problem, unless heavily back-coated (see section 15.3.6).

However, some synthetic fibres modified during production by either incorporation of a flame retardant additive in the polymer melt or solution prior to extrusion or by copolymeric modification can yield inherent flame retardancy, although still thermoplasticity remains a challenge. Perhaps the oldest available fibres in this group (see Table 15.3) are the modacrylics which have been commercially available for forty years or so although at present few manufacturers, such as the Kaneka Corporation of Japan and Montefibre in Italy (see Table 15.3) continue to produce them. These fibres have some char-forming capacity because of the comonomeric chloro-species present. It is this char-forming capacity that makes their presence in the previously mentioned Protex M blends with cotton so effective along with the ability of the chlorine radicals released by the Kanecaron Protex component to flame retard adjacent cotton fibres. Not surprisingly, therefore, the chlorofibres such as PVC and PVDC themselves (see Table 15.3) are highly char-forming and also belong to this group. However, because both these fibre groups have poor thermal physical properties and release hydrogen chloride gas during burning, they will not find application in closed environments such as exist in aircraft and other transport systems.

On the other hand, one group which continues to be successful is FR polyester (see Table 15.3) typified by the well-established Trevira CS (Horrocks, 1996), which contains the phosphinic acid comonomer. Toyobo's latest version of its Heim FR polyester is also based on a phosphorus-containing, phosphinate diester and this has claimed superior hydrolytic stability (Weil and Levchik, 2004). Other flame retardant systems, believed to be based on phosphorus-containing additives, such as Fidion FR (Montefibre) are also commercially available. However, these FR polyester variants do not promote char but function mainly by reducing the flaming propensity of molten drips normally associated with unmodified polyester. Thus they are rarely used in protective applications and only if blended with compatible flame and/or heat resistant, char-forming fibres as the major element.

15.3.6   Coatings and back-coatings

Coatings are applied to both sides of a fabric or impregnate the whole structure. Their general impermeability restricts their use unless associated waterproof properties are desired. Obviously, the use of the moisture-permeable coatings, typified by the microporous, copolymeric acrylics and polyurethanes (Woodruff, 2003) will improve their aesthetic and handle. Protective applications will include waterproof mattress coverings and tickings for use in hospitals as well as seating fabrics for contract furnishing and transport applications where the underlying filling must be protected from an impinging ignition source. The flame retarding elements within these coatings may arise from the use of flame retardant copolymers, typically vinyl or vinylidene chloride and/or the presence of additives in common with back-coating formulations. Back-coatings, however, are specifically applied to create flame retardant barrier properties only and they describe a family of application methods where the flame retardant formulation is applied in a bonding resin to the reverse surface of an otherwise flammable fabric. In this way the aesthetic quality of the face of the fabric is maintained while the flame retardant property is present on the back or reverse face. While the flame retardant components present are the major additives, careful use of viscosity modifiers and other proprietary chemicals ensures that 'grin-through' is minimised and that fabric handle is not compromised. Application methods include doctor blade or knife-coating methods and the formulation is as a paste or foam. These processes and finishes are used on fabrics where aesthetics of the front face are of paramount importance, such as domestic and contract furnishing fabrics where, again, ignition protection of the underlying filling material is of paramount importance.

Generally, the major flame retardants used are combinations of brominated organic species in synergistic combinations with antimony III oxide (Dombrowksi, 1996). The vapour phase activity of the typical Sb-Br flame retardants present ensures effectiveness of the finish if the coating is truly on the rear face of the fabric and they work well on all single and multifibre-containing fabrics independently of area density or construction. Most typical brominated derivatives are decabromodiphenyl oxide (DBDPO), and hexabromocyclo-dodecane (HBCD) and while the former has just passed a European Risk Assessment and has been shown to be safe (Buszard, 2004), HBCD is still undergoing assessment. The present back-coating formulations based on these are very effective, can withstand defined cleansing processes such as the 30 min., 40 °C water soak defined in UK Furniture, and Furnishing (Fire) (Safety) Regulations (1988) (Consumer Protection Act, 1987) (see section 15.5) and are very cost effective. The recent environmental pressures to replace Sb-Br by less environmentally questionable retardants based on phosphorus (Horrocks, 2003) have not been successful to date and residual flame retardancy after a water soak treatment, often yields only marginal passes (Horrocks *et al.*, 2005). This is because not only do phosphorus-based retardants work in the condensed phase and so have difficulty transferring from the back to the front fabric face when exposed to a flame, but also they may be excessively water soluble.

Use of chlorine-containing resins, such as PVC-vinyl acetate and PVC-ethylene-vinyl acetate copolymers may decrease both the amounts of antimony-bromine additives as well as the less effective phosphorus-containing replacement retardants required and hence coating application levels themselves. For technical and economic viability, typical coating levels of 20–30% by weight are required and by careful formulating, passes to standard tests may be achieved.

15.4   Heat and fire resistant fibres and textiles

Heat resistant textiles comprise fibres that can withstand high working temperatures for reasonable service lives in applications like hot gas or fume filtration and where flammability is of secondary importance. Table 15.2 lists the more commonly available examples of these fibres with respective maximum service exposure temperatures. The limiting oxygen indices of these fibres are also included and it is evident that all are flame retardant to varying degrees with LOI values as high as 55 and 68 being possible with the semicarbon and poly-benzoxazoles (PBO) respective examples. It may thus be said that these organic heat and flame resistant fibres may offer varying levels of fire resistance subject to the limit that all carbon-containing fibres and textiles eventually burn. However, extreme thermal protection is a desirable feature in applications such as furnace linings or hot component insulation in car exhaust catalysts or around combustion chambers in jet engines where working temperatures and occasional flash temperatures are in excess of 500 °C and even 1,000 °C in extreme circumstances. These are indeed extreme cases and component fibres must be inorganic such as glass, silica and alumina for many such applications. While fire resistance is an intrinsic feature of these inorganic fibres and textiles, their poor aesthetics limits their use to these extreme technical applications. However, glass-cored, organic fibre-wrapped yarns may be used in more conventional textile roles and examples do exist (Tolbert *et al.*, 1989) where such yarns have been used for barrier fabrics in furnishing applications.

Selected examples of these heat and flame resistant fibres have been reviewed rigorously elsewhere (Horrocks *et al.*, 2001; Perepelkin, 2001), but their particular relevance to thermally protective textiles is more fully explored below. The main groupings of these fibres may be divided into the following groups, namely, the thermosets, the aramids and arimids, the polybenzazoles and the semicarbons. Not only are the generic chemistries similar within each grouping, but their properties and potential application suitabilities are similar.

15.4.1   Thermoset polymeric fibres

These are typified by the melamine-formaldehyde fibre Basofil (BASF) and the phenol-formaldehyde (or novoloid) fibre Kynol (Kynol GmbH) and have the common feature that when heated, they continue to polymerise, cross-link and thermally degrade to coherent char replicas. These chars have especially high flame and heat resistance as a consequence of their high carbon contents. In fact under controlled heating, novoloid-derived fibres can give rise to carbon fibres in their own right. While the parent polymers are cross-linked and not fibre-forming in the manner that linear polymers tend to be, they do give rise to fibres having acceptable textile properties. Their respective properties are listed in Table 15.4 from which it is seen that they have quite low strengths that prevent their being processed easily into yarns. Thus they are more often incorporated into nonwoven fabric structures. Furthermore, because each has an inherent colour, they are usually used as a barrier fabric and not in face fabrics. However, the melamine-formaldehyde structure in Basofil does allow the fibre to be dyed if small molecular disperse dyes are used under high-temperature applications similar to those used for polyester. Dyeing of Kynol is not possible. Both fibre

Table 15.4 Selected properties of melamine-formaldehyde (Basofil) and novoloid (Kynol) fibres

Property	Basofil (BASF)	Kynol (Kynol GmbH)
Tenacity, N/tex	0.2–0.4	0.12–0.16
Modulus, N/tex	6	2.6–3.5
Elongation-at-break, %	15–20	30–50
Moisture regain (at STP), %	5	6
Colour	Pale pink	Gold

types have relatively high moisture regains for synthetic fibres and so have desirable moisture transfer properties.

Respective thermal properties are listed in Table 15.2, and this indicates very similar properties with high levels of heat and flame resistance and complete freedom from any melting and dripping, unlike conventional synthetic fibres. Typical end use applications of both Basofil® and Kynol in thermal protection include fire blocking and heat insulating barriers or blockers and heat and flame protective apparel. Fibres may be blended typically with meta- and para-aramid fibres to improve tensile properties including strength and abrasion resistance in both nonwoven felts and fleeces for fire blocking aircraft seat fabrics and firefighters' clothing for example. For workwear, blending Basofil with inherently flame retardant viscose is also possible as this reduces cost, improves aesthetics and comfort and enables lighter fabrics to be constructed. However, Kynol is often favoured in heavier applications including outer and inner fabrics for firefighters' and racing drivers' clothing, hoods and gloves. Such fabrics may be aluminised to improve heat reflection and hence fire performance.

Both fibre types are reported to have low or even negligible emissions of smoke and toxic gases, such as carbon monoxide and hydrogen cyanide, when subjected to a flame and so find application in closed environments typically in transport, e.g., aircraft, cars, trains, ships, and even submarines. In public buildings they are also of use as fire blockers in furnishings, seat linings, smoke barriers and curtains.

15.4.2   The aramid and the aramid family

This group is perhaps the most well known and exploited of all the inherently heat and flame resistant fibres developed since 1960. They are typified by having aromatic repeat units bonded together by strong amide – CO.NH-, imide – CO.N< groups or both in alternating manner. In polyaramids the single C–N bond in the amide group determines thermal resistance, whereas in polyimides, this same C–N bond is strengthened by the presence of increased conjugation and so the former have inferior heat resistances to polyarimid analogues as will be shown below. However, all members of this group are typified by having thermal resistances in excess of 300 °C for short-term exposures and high levels of inherent flame resistance (see Table 15.2).

The most commonly used thermally resistant aramids are based on a meta-chain structure as typified by the original Nomex (Du Pont) fibre introduced over 40 years ago. Typically, these commercially available fibres are based on poly (meta-phenylene isophthalamide), with other brands being available, e.g., Conex (Teijin), Apyeil (Unitika) and Fenilon (former USSR), in addition to modifications having modifed tensile properties (e.g., Inconex, Teijin) and antistatic properties (Apyeil-α, Unitika). Over the years, these fibres have become available with improved dyeing properties and so are now available in full colour ranges. Their advantage over many other thermally resistant fibres is their acceptable tensile and physical properties that are very similar to those of conventional nylons 6 and 6.6, hence they are often claimed to have an overall more acceptable textile performance. Their rigid, all-meta aromatic polymer chains ensure that the fibres have minimal thermoplastic characteristics with second-order transition temperatures (T_g) of about 275 °C and an ill-defined melting point accompanied by thermal degradation starting at 375 °C (see Tables 15.1 and 15.2). These enable the fibres to be used in textiles where shrinkage should be minimal at temperatures of continuous use of 150–200 °C and to shorter exposures as high as 300 °C and so are ideal for use in protective clothing. LOI values are 28–31 and so they compete effectively with many other non-thermoplastic flame retardant fibres such as flame retarded cotton and wool, which have similar LOI values, although they are inferior to the highly cross-linked (e.g. Kynol), semicarbon and polybenzazole groups of fibres.

In order to improve the performance of the meta-aramids, Du Pont in particular, has introduced a number of variants over recent years. For example, blending with small amounts of para-aramid fibres (e.g. Nomex III contains 5% Kevlar) enables the thermal protective behaviour to increase primarily through increasing the char tensile strength. Nomex III is available in a limited range of spun-dyed colours and is recommended for applications where direct heat exposure is possible, e.g., single or multi-layer garments such as firefighters' station uniforms, coveralls, jackets, trousers, gloves, flight suits or tank crew coveralls. Nomex Comfort incorporates microfibre technology and has improved antistatic and moisture management properties and is recommended for coveralls, foul-weather gear, fleece, shirts, hoods and a range of knitwear fabrics including T-shirts and underwear. A third variant is Nomex Outershell, specifically designed for firefighters' clothing although it may be used in other suggested applications for Nomex III.

There have also been attempts to reduce costs by blending meta-aramids with lower cost flame retardant fibres like flame retardant viscose. Thus Du Pont introduced the fabric Karvin, comprising 30% Nomex, 5% Kevlar para-aramid and 65% Viscose FR (Lenzing). While having similar flame retardancy to their respective parent single aramid components, their char structures are weaker and so do not offer sustained fire protection at high heat fluxes and temperatures as the 100% meta-aramid fabrics do. However, recent work has indicated that use of aramid blends may offer overall advantageous properties and this is exemplified by research at ENSAIT in Lille, France which has shown that the fire performances of wool/para-aramid blends containing over 30% and up to 75% para-aramid are superior to that for 100% para-aramid measured in terms of reduced peak heat release rate from cone calorimetry (Flambard *et al.*, 2005).

The para-aramids are typified by Kevlar (DuPont) and Twaron (Teijin) and are based on poly (para-phenylene terephthalamide). While having enhanced tensile strengths and moduli as a consequence of the extreme symmetry of their polymer chains and hence order or crystallinity, they also have enhanced thermal performance. The increased structural chain rigidity and order raises the second-order transition temperature to about 340 °C and melting point to about 560 °C before decomposing above 590 °C (see Table 15.1). Furthermore, higher continuous working temperatures of 180 °C and above are possible with resistance to short-term exposures to temperatures as high as 450 °C being achievable (see Table 15.2). However, thermal degradation is similar to that occurring in the meta-aramids and so the LOI values are similar at 30–31%. However, the higher cost, poorer textile processing properties and higher modulus of para-aramid fibres ensure that their use in applications such as protective textiles are limited to 100% contents only when performance demands are exceptional. Thus, and as is more usually the case, additions of small amounts of para-aramid to blends with meta-aramid (e.g. 5% in Nomex III) and blending with other cheaper flame retardant fibres offer enhanced improved tensile and heat and flame resistance relative to the major fibre present.

With regard to arimid fibres, while a number have been reported and reviewed (Raheel et al., 1994; Horrocks et al., 2001), only the example P84 introduced by Lenzing during the mid-1980 period has been commercially exploited. P84 fibres are now produced by Inspec Fibres (USA). As Table 15.2 indicates, these fibres have superior thermal properties to aramid and so find use in protective applications. For example, outerwear, underwear and gloves may be made from 100% P84 or blended with lower cost fibres like flame retardant viscose. For instance, a 50/50 P84/Viscose FR (Lenzing) blend is used for knitted underwear with high moisture absorbency. Alternatively, blending with high tenacity polyaramids increases wear and tensile characteristics. Spun-dyeing of P84 fibres enables their natural gold colour to be replaced by those often demanded by customers who may require more appropriate bright safety colours.

The final member of this grouping is the poly (aramid-arimid) fibre, Kermel, which was produced initially by Rhone-Poulenc of France in 1971 and is now produced by Rhodia Performance Fibres. The chemical structure of Kermel is a combination of amide –CO.NH– and imide –CO.N< functional chain groups.

While its overall properties are very similar to those of the meta-aramids, it shares certain problems including difficult dyeability although in 1993, a so-called third-generation Kermel was announced which claimed to have superior colouration properties. Like the poly (meta-aramids), however, Kermel has poor UV stability, and so must be protected from intense radiation sources. Its second-order and decomposition temperatures fall between those of the para-aramid and the arimid fibre P84 (see Table 15.2) and this reflects the lower chain rigidity and the weaker polyamide bond structure. With an LOI value of about 32%, comparable to the meta-aramids, it competes in protective clothing markets where again it is used as 100% Kermel or as blends with other fibres, including FR viscose. Blends with wool have been shown to be ideal for uniforms, jerseys and pullovers. Production of composite yarns with high modulus aromatic fibres like the poly (para-aramids) has yielded the modification Kermel HTA, a yarn with a para-aramid core (35%) and a Kermel fibre wrapping (65%). Not only does this yarn enable the aesthetics and colour of the outer fibres to be prominent, but also like Nomex III discussed above, the para-aramid content enables the strength and abrasion resistance fabrics to be increased.

15.4.3   Polybenzazole group – polybenzimidazole and polybenzoxazole fibres

These fibre-formimg polymers are so-called 'ladder polymers' and essentially have wholly aromatic polymer chains. The two common examples available commercially are the polybenzimidazole PBI (Celanese) with the chemical structure:

and full chemical name poly (2,2'-(m-phenylene)-5,5'-bibenzimidazole) and the polybenzoxazole, Zylon (Toyobo) with the structure:

and full chemical name poly(para-phenylene benzobisoxazole) and generic acronym, PBO. The similarity in general polymer chain structures is apparent and it is the high degree of chain rigidity that gives both of these fibres their superior thermal properties as shown in Table 15.2. Both fibres have thermal degradation temperatures well in excess of 400 °C and can operate at continuous temperatures of well over 200 °C and demonstrate superior LOI values well over 40%.

PBI has been introduced to the commercial markets only during the last 20 years or so in spite of its development during the early 1960s. It was then that the US Air Force Materials Laboratory (AFML) contracted with the former Celanese Corporation to develop a novel high-temperature resistant fibre. In the aftermath of the 1967 Apollo spacecraft fire, AFML and NASA examined PBI fibre as a non-flammable material for flight suits that would afford maximum protection to astronauts in oxygen-rich environments. Subsequently, in 1983, Celanese built a full-scale manufacturing plant for the production of PBI polymer and fibre to develop civilian markets.

The current PBI fibre is a sulphonated version of the chemical formula above and this improves shrinkage resistance at high temperature. The fibre has a moderate strength of about 0.25–0.30 N/tex, a lowish modulus of about 3 N/tex and a remarkably high moisture regain of 15%. It is easily processible by normal textile methods and gives rise to very comfortable fabrics, although the inherent colour is bronze-like or even gold in some forms. The fibre cannot be dyed and it is often blended. One well known blend is PBI Gold in which a yarn spun with both PBI and Kevlar in a 40/60 blend that gives rise to gold-coloured fabrics with fire protective properties claimed to be superior even to those made from Nomex III. Presently, this blend is being introduced into firefighters' clothing both in the USA and UK. This includes outershells as well as underwear, hoods, socks and gloves. Other uses include industrial workwear, aluminised proximity clothing, military protective clothing and fire barrier/blocker applications. However, PBI is several times as expensive as the meta-aramids and so this superior performance comes at a price.

Zylon or PBO is a more recently developed fibre than PBI and has outstanding tensile properties as well as superior thermal and fire properties to any of the fibres mentioned in this section (see Table 15.2). Its tensile strength is well above 3 N/tex and modulus exceeds 150 N/tex, both values being about twice respective values for Kevlar. While there are at least two variants of fibre, Zylon-AS and Zylon-HM of which the latter has the higher modulus, they both have these same thermal and burning parameter values. The fibre is available in staple, filament and chopped forms and finds applications in the area of thermal and fire resistance products where a combination of these with high tenacity and modulus are required. Principal examples of thermally protective textiles include heat protective clothing and aircraft fragment/heat barriers. The price of this fibre is probably similar to that of PBI and so its use is restricted to applications where strength, modulus and fire resistance are at a premium.

15.4.4 Semicarbon

The semicarbon fibres include any in which the structure is essentially carbon while retaining acceptable textile properties, unlike true carbon fibres (Hearle, 2001). Within the group, the oxidised acrylics represent the sole commercial group and are produced following controlled high-temperature oxidation of acrylic fibres during the first stages of carbon fibre production. First reported about 1960 as 'Black Orlon' (Vosburgh, 1960), they became of commercial interest as potential high-temperature-resistant fibres during the early 1980s.

During this period a number of commercial versions were announced including Celiox (Celanese), Grafil O (Courtaulds), Pyron (Stackpole), Sigrafil O (Sigri Elektrographit, now SGL) and Panox (SGL UK Ltd., formerly R K Textiles). There are presently large production plants for oxidised acrylics in North America, UK, France, Germany, Hungary, Israel, Korea, Taiwan and Japan.

The fibres have only a low tenacity at about 0.15–0.22 N/tex, a moderate extensibility of 15–20% and a reasonable modulus of 5–8 N/tex. This creates the problem of ease of processability for these weak fibres although they can be spun into yarns by the woollen system. Thus they are produced as a continuous tow that is stretch-broken by conventional means for eventual conversion into coarse woollen-type yarns, although fine filaments of the order of 1.7–5 dtex are possible. The surprisingly high moisture regain of 10% enables fabrics to have an unexpected comfort. The limiting oxygen index is typically about 55% and so fabrics are extremely thermally resistant giving off negligible smoke and toxic gases when subjected to the even the most intense of flames. Unfortunately, of course, the fibres are black and so are rarely used alone except in military and police coverall clothing where the colour is a bonus. More usually, oxidised acrylic fibres are blended with other fibres, typically wool and aramid in order to dilute the colour and introduce other desirable textile properties. Because of their extreme fire resistance and lower cost than PBI and PBO, they find applications usually as blends in anti-riot suits, tank suits, FR underwear, fire blockers for aircraft seats and heat resistant felts (insulation), hoods and gloves. When aluminised, they are very effective in fire entry/fire proximity suits. More specialist end-uses include fabrics for protection against phosphorus and sodium splash and welding blankets. Usually, however, other finishes to reduce fabric permeability are essential if full protection, say, against petrol or molten liquids is to be afforded. Finally, fabrics may be overprinted to offset the underlying natural colour.

15.5   Design Issues

15.5.1   Maximising flame, heat and fire resistance

The factors that determine maximisation of fire resistance in a given textile product lie primarily in the properties of the fibre and fabric characteristics for each component. These include the thermal inertia of the fabric (Day and Sturgeon, 1987) and the abilities to reflect radiant thermal energy and to insulate the underlying surface (or wearer) from the incident fire or thermal source (Krasny, 1986). Heat reflective treatments such as aluminisation are use effectively in both fire protective clothing and fire blocker applications. Thermal inertia and insulative character are both related to fabric weight or area density, although heavier fabrics become more uncomfortable in hot environments. In addition, air permeability, moisture and construction can influence thermal behaviour with nonwovens tending to give superior properties to wovens or knitted structures, all other variables being equal (Lee and Barker, 1987). Baitinger (1979), however, suggests that only at low levels of heat exposure do physical variables like thickness correlate with insulative properties, whereas it is the fibre type and finish and respective degradative resistances that determine high heat exposure performance. Essentially, this means that it is the ability of a fabric assembly to retain its thickness when subjected to a heat flux, whether radiant, convective or both (Shalev and Barker, 1986).

Most thermally protective textiles have area densities in excess of 250 gsm or so with workwear and the lighter upholstery and barrier fabrics being in the 250–350 gsm range. However, it is the air trapped within the interstices of any fabric structure and between layers of fabrics within a garment assembly that provides the real thermal insulation. The need to have thermal and fire resistance in the component fibres ensures that these insulating air domains are maintained. Any slight tendency of a garment to shrink will reduce its insulative effectiveness, even if the fibres present remain undamaged. It goes without saying, therefore, that the presence of any thermoplastic fibre is to be avoided and in fact banned from use in certain applications. However, while the essentially non-thermoplastic fibres, both natural and synthetic, described above and which feature in Tables 15.1 and 15.2 are generally assumed to have negligible shrinkage at elevated temperatures, at very high temperatures even small levels may cause concern. Barker and Brewster (1982) have published shrinkage data at 300 °C where charring is minimal and at 400 °C where it may be considerable in some fibres. For instance, even FR cotton shows 25–29% shrinkage at 300 °C increasing to about 40% as it chars at 400 °C. Nomex aramid is shown to give 0% at 300 °C and yet 5–12% at 400 °C. Kevlar aramid shows 0% at both temperatures thus demonstrating why it is introduced into Nomex III. Even PBI shows a measurable shrinkage of 2% at 300 °C rising to 6% at 400 °C.

However, even with the most fire resistant fibres present, no single fabric can perform all the tasks required of it in a given product and many garments and clothing assemblies are multilayered and may consist of a set of components to form a system. While a protective fire resistant coverall or item of workwear may be a single-layered garment comprised of the most cost-effective fabric that has the desired level of protection, underneath may be the requirement for fire resistant linings and underwear, including socks and gloves in order to provide full and fail-safe protection. For example, a very recent study (Valentin, 2004) has compared the performance of three competing designs of firefighters' outer jackets comprising external and liner fabrics in the following combinations:

- meta-aramid/para-aramid blended external fabric plus knitted modacylic-polyurethane membrane-meta-aramid felt/FR viscose composite liner
- meta-aramid/para-aramid blend-polyurethane membrane-meta-aramid nonwoven external fabric composite plus a liner comprising a meta-aramid/FR viscose blend with para-aramid spacers

- meta-aramid/para-aramid nonwoven blend with para-aramid spacers/PTFE membrane external fabric assembly plus meta-aramid felt/FR viscose liner.

It is noteworthy that each of these has an outer fabric containing meta- and para-aramid blends for good fire protection and durability. However, a number of inner layers, whether joined to the outer or part of the liner comprising breathable polyurethane or PTFE membranes, are also a common feature. Thirdly, the use of thermally insulative nonwoven interlayers and comfortable meta-aramid/FR viscose innermost linings are also present. It is the ordering and exact interlayer compositions that appear to be the major differences while each design attempts to provide the common properties of fire resistance, thermal insulation, moisture transport and comfort.

Anticipated wearing times may also influence design because some coveralls or outershells may be required to be worn only as an emergency garment over normal working clothes which may be fire resistant or not, as the case may be. However, the more complex coveralls and jacket/trouser assemblies may be required to be worn for a full working period or, indeed, more extreme lengths of time in some firefighting, military and civil emergency environments. Furthermore, in addition to fire resistance may be the requirement of resistance to water (e.g. firefighters' outershells), petrol (e.g. police anti-riot gear) and other aggressive agencies (e.g. chemical and biological threats). In some cases, a multi-functional character will be an essential feature. However, in the main, fire resistance is the first concern that requires an effective heat and flame resistant base fabric to be designed to which surface finishes, coatings or other treatments may be applied to yield all the desirable properties.

The exposure time of a particular fire and related threat is of considerable importance in choosing the base fabric. For example, in workwear where fire resistance is a low to medium risk (e.g. chemical workers' overalls), flame retardant cotton or viscose fabrics are ideal compositions and exposure to a fire hazard will prevent outer garment ignition for a limited period. Furthermore, the char formed following exposure will provide a degree of further protection while the victim escapes from the fire hazard. However, if the fire risk is high and there is a need to provide an increased escape time, then fabrics will be constructed from the higher performance fibres such as the aramids, arimid, aramid-arimid, PBI, PBO, semicarbon, etc. Here charring occurs much more slowly and outer garments retain their integrity for periods of many minutes even in intense fire environments that firefighters and racing drivers experience, for example.

It is important to realise, however, that a working human cannot survive for very long without oxygen and so even the most fire-resistant garments will not protect for long unless suits such as fire proximity and racing car drivers suits are fitted with internal oxygen or air breathing supplies and smoke and heat masks. Even if the immediate threat is not so great as to need this requirement,

then the ability to work efficiently in a hot, high-risk environment for considerable periods of time is an important concern – this is the typical firefighters'
scenario.

Similar issues are important in fire blocking fabrics from simple furnishing
fabrics that in the UK must resist a simulated match and cigarette ignition, to
those used in aircraft seating that must resist the US Federal Aviation
Administration kerosene burner test FAR 25.853(c). In the latter a full seat
mock-up is exposed to a burner with a heat flux of $115 \, kW/m^2$ for 120s and
fabric/seat assemblies must suffer a weight loss of less than 10%, fall within
specified maximum and average burn length criteria as well as not sustaining
afterburning and smouldering for more than five minutes. Fire blockers that
enable passes to be achieved may have a variety of weights depending on
whether aluminised and the choice of fibres present. For instance, fabrics
comprising Zirpro wool, oxidised acrylics, aramid, arimid and glass in various
blends and in woven or nonwoven structures with area densities from 250–400
gsm are typically used.

One final thermal threat requiring specialist textile solutions is that of molten
metal splash where temperature, metal density, size of splash droplets and
droplet surface reactivity determine the solution to be adopted. These factors are
all different for molten iron or steel, copper, tin, lead, zinc or aluminium and so
protective aprons and overalls have to be tailored to fit the threat. Issues here are
whether or not:

• the temperature of the molten metal drop is sufficient to ignite the fabric
• the density and hence mass of a drop enables the drop to 'burrow' into the
  textile as it chars the underlying fibres
• the drop sticks or slides from the fabric surface.

Traditionally, furnace operators have always worn leather aprons which are
heavy and uncomfortable. Heavy metal processing, e.g., steel, iron, lead and
copper, requires heavier fabrics to offset the effect of burrowing and flame
retarded wool and cotton fabrics with area densities in the range 260–660 gsm
have been shown to be superior to 100% aramid, novoloid and glass fabrics.
Their effectiveness is partly related to the excellent char-forming abilities of
cotton and wool although the intumescent charring character if wool is generally
believed to explain its overall superior properties in this respect (Benisek and
Edmondson, 1981). For Zirpro-treated wool, it was noted that the degree of
damage decreased in the order: iron, copper > aluminium, zinc > lead, tin. The
advantageous effect of aluminised surfaces and use of ceramic fibres in
improving resistance to molten iron was noted by Barker and Yener (1981).
However, one real problem is that posed by aluminium, which, because of its
reactive surface, sticks to many flame retardant fabrics, including aramids and
glass, although it does not do so to Zirpro-treated wool and untreated cotton
(Benisek and Edmondson, 1981). This effect is compounded by the low density

of aluminium and so droplets do not have the mass to break away from the surface fibres on the fabric.

15.5.2   Maintaining comfort and aesthetics

Heat fatigue is as much a hazard to the firefighter as the fire itself to a large extent and is a major consideration in designing clothing assemblies (Holmér, 1988). Thus, while having a high level of protection afforded by his or her firefighting suit, the wearer must be able to wear it for long enough to do an effective job while ensuring that comfort is acceptable and that moisture loss is minimised to prevent dehydration. Thus, while offering fire and water protection, the outer shell should be breathable and the inner lining assembly should enable moisture to be wicked away from the body and underclothing. This same lining should have an appropriate level of fire resistance should the outer shell be damaged. However, for every slight increase in the clothing weight, so the risk of heat fatigue is increased and so a compromise situation is reached as the clothing system becomes more complex in structure. Underclothing should, therefore, enable both heat and moisture to be transmitted away from the body, air gaps should be maintained and there should be no chance of shrinkage of any part of the garment assembly.

In the case of fire blocker fabrics used in seating assemblies, the flexibility and aesthetics of the face fabrics should not be compromised by their presence. This is a particular issue in both contract furnishings in the UK that require ignition resistance to more intense sources than matches and cigarettes (see BS 7176:1995 in section 15.5) and in aircraft seating where passengers may be sitting continuously for considerable periods.

15.5.3   Combining other properties

The need for multifunctionality in thermally protective textiles is becoming increasingly a necessity for many applications. Table 15.5 lists typical requirements and possible solutions. Often when multifunctionality is required, competing effects and finishes may cancel unless carefully engineered. Furthermore, such treatments must not add to the potential flammability of the overall textile. For instance, water repelling characteristics demand hydrophobic coatings or finishes while oil and petrol require hydrophilic treatments. Luckily, with the advent of the very low surface energy fluorocarbons (Holme, 2003), both goals may be achieved providing fibre surfaces are clean and inherent flammability properties are low or negligible. Not only do these fluorocarbon finishes produce both water and oil repellency but also offer soil release characteristics that ensure fabrics remain cleaner for longer. Thus from conventional UK domestic flame retardant furnishings to the more sophisticated firefighters' outershell, the presence of fluorocarbon treatments is common.

Table 15.5 Typical multifunctional requirements and solutions for thermally protective textiles

Function	Solution	Reference
Moisture transport and management	Microporous membranes and coatings	Holmes, 2000; Holme, 2003; Woodruff, 2003
Oil and petrol repellency	Fluorocarbon finishes	Holme, 2003
Water repellency and proofing	Breathable coatings	Holme, 2003; Woodruff, 2003
Ballistic properties	Inclusion of high modulus fibres	Hearle, 2001; Scott, 2000
Chemical resistance	Impermeable barriers (e.g. PTFE, PVC, neoprene) Adsorptive barriers (e.g. active carbon)	Raheel, 1994
Microbiological resistance	As for chemical resistance plus use of antimicrobial fibres	Vigo, 1994
Radioactive particle resistance	As for chemical resistance	Raheel, 1994; Scott, 2000
Colour, camouflage, radar signature	Dyestuff and colourant selection	Scott, 2000

The need for cleanliness also influences the effect of other properties, not least flame retardancy when small amounts of oily soil may offset the flame retardant character of the underlying textile. While it is assumed that all flame retardant finishes used in protective textiles have the desired level of laundering durability (Horrocks, 1986, 2003) and those containing inherently FR fibres have very high levels, laundering itself can cause deposits of water hardness salts and detergent residues that may increase ignitability. Thus, sophisticated protective textiles from which sustained high levels of performance are demanded often have stringent aftercare instructions issued with them. This is the case with aircraft and transport seatings, firefighters' clothing assemblies and protective clothing for military personnel (Scott, 2000).

15.6   Testing and performance

The details of the many test methods available across the world are outside the scope of this chapter, although Eaton (2000) has recently reviewed those more relevant to the UK. Most test methods are defined by a number of national and international bodies such as air, land, and sea transport authorities, insurance organisations and governmental departments relating to industry, defence and health, in particular. While the reader is recommended to refer to the indexes of major standards institutions such as the British Standards Institution (BSI) and the American Society for Testing and Materials (ASTM), normalisation across the European Union during the last ten years or so has seen traditional 'BS' tests replaced by BS EN and BS EN ISO alternatives as methods of test become truly international. For a detailed examination of most of the world's test methods relating to textiles and plastics materials, the reader should consult Troitzsch (2004).

Because textiles are 'thermally thin' and have high specific volumes and hence accessibility to oxygen from the air and these properties are determined by variables in yarn, fabric and product geometries, textile tests usually focus only on ease of ignition and/or burning rate behaviour which can be easily quantified for fabrics and composites in varying geometries. Few, however, yield quantitative and fire science-related data unlike the often criticised oxygen index methods (Horrocks *et al.*, 1987).

15.6.1   Textile component testing

Most tests for protective textiles examine and test the potential ignition resistance and ability of single fabrics or eventual product components to protect underlying surfaces or materials. These tests may be undertaken on a single fabric or a composite that reflects a real application or product. Table 15.6 lists a selection of tests useful in the thermal and flammability assessment of protective textiles. In all flammability test procedures conditions should attempt to replicate real use and so while atmospheric conditions are specified in terms of relative humidity and temperature ranges allowable, fabrics should be tested after having been exposed to defined cleansing and aftercare processes. Table 15.6 includes BS 6561 and its CEN derivatives as being typical here and these standards define treatments from simple water soaking, through dry cleaning and domestic laundering to the more harsh commercial laundering processes used in commercial laundries and hospitals, for instance.

While simple vertical strip tests such as the now discontinued BS 6249:Part 1:1982 originally used for assessing flame retardant workwear performance are still well-established in other areas of textiles, they are embedded within the more recent European tests for protective clothing also listed in Table 15.6. Thus BS 6249, which uses the established standard test method BS 5438, has been replaced by the general performance standard BS EN 533:1997 in which BS EN 532:1994 is the test method (see Fig. 15.1).

Tests for furnishing fabrics and blockers as barriers

When assessing the ability of a furnishing fabric to resist ignition and protect the underlying filling, BS 5852 Parts 1 and 2:1979 and its subsequent EN and BS

Table 15.6 Selected test methods for protective textiles

Nature of test	Textile type	Standard	Ignition source
British Standard based vertical strip method BS 5438	Protective clothing (now withdrawn)	BS 6249:Part 1:1982 (replaced by BS EN 533:1997)	Small flame
ISO vertical strip similar to Tests 1 and 2 in BS 5438	Vertical fabrics	BS EN ISO 6940/1:2004	Small flame
Small-scale composite test for furnishing fabrics/fillings /bedding materials	Furnishing fabrics	BS 5852: Pts 1 and 2:1979 (retained pending changes in legislation)	Cigarette and simulated match flame (20 s ignition)
	Furnishing fabrics	BS 5852:1990(1998) replaces BS 5852: Pt 2	Small flames and wooden cribs applied to small and full-scale tests
		ISO 8191:Pts 1 and 2 (same as BS 5852:1990) BS EN 1021-1:1994 BS EN 1021-2:1994	Cigarette Simulated match flame (15 s ignition)
Cleansing and wetting procedures for use in flammability tests	All fabrics	BS 5651:1989	Not applicable but used on fabrics prior to submitting for standard ignition tests
	Commercial laundering Domestic laundering	BS EN ISO 10528:1995 BS EN ISO 12138:1997	

Table 15.6 *Continued*

Nature of test	Textile type	Standard	Ignition source
Protective clothing	General requirements	EN 340	
	Resistance to radiant heat	BS EN ISO 6942:2002 (formerly BS EN366:1993 which replaced BS 3791:1970)	Exposure to radiant source
	Resistance to convective heat (flame)	BS EN 367:1992	Determine heat transfer index
	Resistance to molten metal splash	BS EN 373:1993	Molten metal
	Gloves	BS EN 407:1994*	Composite standard (including firefighters' and welders' gloves)
	Firefighters' clothing	BS EN 469:1995*	Composite standard
	Welders' and allied industrial clothing	BS EN 470-1:1995/ISO 11611	Composite standard
	General flame spread	BS EN ISO 15025:2002 (formerly BS EN 532:1995 which replaced BS 5438)	Small flame
	Workers exposed to heat	BS EN 531:1995/ISO 11612	Composite standard
	General protection	BS EN 533:1997* (replaces BS 6249)	Small flame
	Contact heat transmission	BS EN 702:1995	Contact temps. 100–500 °C
	Firefighters' hoods	EN 131911	

Note: * Test methods presently under review.

(a)   (b)

15.1 Flame retarded cotton being tested to the BS EN 532 test method; (a) shows the flame application and (b) shows the char after igniting flame extinction (courtesy of Rhodia Consumer Specialities, Oldbury, UK).

EN editions combine both the test method and performance-related set of defining criteria. With the recognition of the hazards posed by upholstered fabrics, the development of the small-scale composite test BS 5852 in 1979 represented a milestone in the development of realistic model tests which cheaply and accurately indicate the ignition behaviour of full-scale products of complex structure. One major strength of this test is that it defines a set of different ignition sources from the cigarette, through a series of gas burners of increasing intensity to a series of wooden cribs in a total order of increasing thermal output to represent real ignition source intensities. In the UK, BS 7176:1995 specifies which source should be used for furnishing fabrics in environments of increasing ignition hazard and Table 15.7 summarises these. A similar advisory standard for testing bedding over a non-combustible fibre mat is defined in the sister standard BS 7175 and its related test methods for ignitability of mattresses of divans using sources in BS 5852 (BS 6807:1996 and ISO 12952-1/4:1998), but this is outside the present discussion for protective textiles.

Fire barrier or blocker fabrics are usually tested under conditions that realistically attempt to replicate their application thermal hazard. A prime example is the kerosene burner test used to test commercial aircraft seating, FAR 25.863(c), discussed above. In other cases, where a blocker is part of a larger system such as part of a building or transport system, then the test will be

Table 15.7 Ignition source/hazard combinations – BS7176:1995 (for full details, see the actual standard)

	Low hazard	Medium hazard	High hazard	Very high hazard
Requirements	Resistance to ignition source: smouldering cigarette of BS EN1021-1:1994 and the match flame of BS EN1021-2:1994	Resistance to ignition source: smouldering cigarette of BS EN 1021-1:1994 and the match flame of BS EN1021-2:1994 Resistance to ignition Source 5 in BS5852:1990	Resistance to ignition source: smouldering cigarette of BS EN 1021-1:1994 and the match flame of BS EN1021-2:1994 Resistance to ignition Source 7 in BS5852:1990	Resistance to ignition source: smouldering cigarette of BS EN 1021-1:1994 and the match flame of BS EN1021-2:1994 At discretion of the specifier but at least high hazard requirements
Typical examples	Offices Schools Colleges Universities Museums Exhibitions Day centres	Hotel bedrooms Public buildings Restaurants Places of public entertainment Public baths Public houses & bars Casinos Hospitals Hostels	Sleeping accommodation in certain hospital wards and hostels Off-shore installations	Prison cells

related to the related building, motor vehicle, rail vehicle, aircraft or ship fire testing regime (Troitsch, 2004). While building fire tests relate to national standards requirements, for aircraft and ships, there are overarching agencies responsible for these tests with international standing such as the US Federal Aviation Administration (FAA) and the International Maritime Organisation (IMO). For further details into this complex area, the reader is referred to Troitzsch (2004) and relevant organisational web-sites.

Protective clothing tests

Table 15.6 shows also a set of tests which has recently been developed across the EU to accommodate the different demands of varying types of protective clothing and hazards, whether open flame, hot surface, molten metal splash or indeed a combination. For test details, the author is referred to the specific method but it is significant to note that for a fabric to be used in a given defined protective clothing end-use, general ignition resistance and flame spread tendency (test method BS EN 532:1994 and performance requirement BS EN 533:1997; see Fig. 15.1) will complement more application-specific require-ments such as resistance to radiant (BS EN ISO 6942:2002) and convective (flame) (BS EN 367) heat or hot metal splash (BS EN 373:1993).

While it goes without saying that most protective clothing must pass a simple ignition and flame spread test such as BS EN 532, of particular relevance to thermally protective clothing and an area that has seen much research activity is that of heat transfer and the ability of a given fabric or layered assembly to minimise heat transmission to a wearer. Tests like BS EN 367 measure a heat transfer index (HTI) and this has developed from a history of similar measures of the ability of a fabric to insulate the wearer. These indices are usually measures of times to achieve a given burn level (Shirley Institute, 1982; Barker and Coletta, 1984). For example, Benisek and Phillips (1981) showed that the time taken for the temperature to reach a second-degree burn level behind a given protective textile fabric exposed to an open-flame, convective heat source was in the increasing order FR cotton (315 gsm) < Aramid (259 gsm) < Zirpro wool (290 gsm). Furthermore, an assembly of single-layered fabrics gave significantly improved times to reach this same level and was significantly longer than a single-layered fabric having the same total area density; this demonstrated the importance of entrapped air between adjacent layers in a garment assembly.

One early authoritative study of the thermal insulative properties of fabrics was that by Perkins (1979) who studied a large number of fabrics with an area density range of 85–740 gsm as single layers. He selected an incident radiant source intensity of up to $16.8\,kW/m^2$ and a convective flame source of $84\,kW/m^2$, which is considered to be commensurate with the exposure typically experienced by firefighters. Behind each fabric was a heat flux meter which enabled time versus heat flux to be determined. Using a standard burn-injury curve that relates delivered heat to incipient second-degree burn threshold level (Derksen *et al.*, 1961), performance of fabrics could be measured in terms of time to reach the latter. Their results may be summarised as follows:

- For radiant heat fluxes of 8.4 kW/m², fabric area density determines protection efficiency with air permeability also influencing performance – open constructions enhance heat flow. Differences in fibre type have little effect.
- At heat fluxes of 12.6 and 16.8 kW/m², the fibre properties become important and char-forming fibres like FR cotton and FR wool become superior.
- Exposures to the convective flame at 84 kW/m², shows the FR wool fabrics to yield significantly higher times than FR cotton and aramid fabrics of similar weight.

An alternative method is to determine the so-called thermal protective performance index as described in the ASTM Test Method for Thermal Protective Performance (TPP) of Materials for Clothing by Open-flame Method (D 4108-82, -87 and subsequent revisions) where TPP for a fabric assembly is the burn threshold time multiplied by the incident heat flux. This method has been used by a number of authors (Shalev and Barker, 1986; Day 1988; Barker *et al.*, 1996). At the same time, Benisek *et al.* (1979) published their study of the insulative behaviour of a range of fabrics subjected to a convective flame with a temperature of 1,050 °C, heat transfer from a radiant panel according to BS 3791:1970 and hot molten metal. For heat transfer from radiant heat, the time for a copper disc located behind the fabric to rise by 25 °C is defined as the Thermal Protective Index or TPI and this gives a simple measure of fabric thermal protection against first-degree burns. Table 15.8 lists TPI values determined by Benisek *et al.* (1979), which although determined over a quarter of a century ago, do indicate the effectiveness of aluminisation. Here aluminised polyester films are less reflective than aluminium spray coatings which are very much less effective than a bonded aluminium film to the fabric surfaces. It is important, however, that such fabrics must be kept clean if they are to remain efficient.

Table 15.8 Thermal protective indices of selected fabrics (Benisek *et al.*, 1979)

Fabric	Construction	Area density, gsm	Thermal protective index, s
100% Zirpro wool	Twill	260	8.8
100% Zirpro wool/aluminised polyester film	Twill	378	32.7
100% Zirpro wool/aluminised by spraying	Twill	297	172
85% Zirpro wool/15% glass	Twill	300	13.5
85% Zirpro wool/15% glass/ aluminised polyester film	Twill	448	34.3
85% Zirpro wool/15% glass/aluminised by spraying	Twill	351	143
100% aramid/bonded aluminium film	Twill	339	401

Their effectiveness under convective flame conditions will be compromised by both smoke deposition and film oxidation (Shalev and Barker, 1983).

The now-obsolete test BS 3791 used to generate the data in Table 15.8 is the basis of the current standard for protective clothing BS EN ISO 6942:2002 where a radiant source temperature 1,070 °C is used and times to achieve temperature rises behind a sample of 12 and 24 °C are recorded at any chosen incident heat flux level, Q_o, from the following ranges: Low – 5 and 10 kW/m²; Medium – 20 and 40 kW/m² and High – 80 kW/m². In parallel, transmitted heat flux density, Q_c, and the heat transmission factor, $TF(Q_o)$, may be calculated for a given assembly. The radiant heat transfer index (RHTI) at a given incident heat flux intensity, Q_o, is defined as the mean time taken, t_{24}, for the temperature taken in the calorimeter at the rear of the assembly to rise by 24 ± 0.2 °C.

BS EN 367 (see Table 15.6) defines the method for assessing convective heat transfer following exposure to a flame with a heat flux of 80 kW/m² and describes the means of calculating the heat transfer index or HTI which should not be confused with the TPI discussed above for radiant panel exposure or indeed, the TPP determined for a flame source according to ASTM D 4108-87 and subsequent revisions. The HTI, like RHTI above, is the mean time taken, t_{24}, for the calorimeter at the rear of the assembly away from the flame to rise by 24 ± 0.2 °C. This same standard offers a banding of fabric assembly performances from 1 to 5 that cover the HTI limits of 3–6 s for Band 1 (single layer fabrics) to over 31 s for Band 5. Fabric assemblies in this last category will be very thick and be for special applications whereas Band 3 covers HTI values from 13 to 20 s and spans the requirement for specialist firefighters' clothing of 16 s (see section 15.6.3).

Hot metal splash resistance is determined using BS EN 373:1993 and the challenges that different molten metals pose have been discussed in section 15.5.1 above, although setting standards for its determination and translation to actual performance are not simple (Proctor and Thompson, 1988). Work reported by Benisek and Edmondson (1981) assessed different metal splash-fabric combinations by determining the mass of a given metal required to damage a PVC skin-simulant film placed on the reverse of the fabric under test (Mehta and Willerton, 1977). Molten drops impinged upon upper fabric surfaces oriented at 45° to the horizontal so that droplets during impact had time to thermally degrade the fabric surface and either glance off or stick to and burrow into the fabric. In BS EN 373:1993, the underlying embossed PVC film (with an area density of 160 gsm) is designed to show loss of embossing and even small holes when exposed to excessive localised heat and hence exhibit damage commensurate to skin. In this test, shown in Fig. 15.2, 50 g molten metal (at about 50 °C higher than its respective melting point) is dropped onto a supported fabric. If no PVC damage is apparent, the test is repeated with fresh fabric and PVC samples but with an increasing incremental mass of molten metal (10 g) until damage is apparent. Conversely, if 50 g molten metal damages the PVC, incrementally decreasing masses are used until no damage is apparent. The molten mass index for a given fabric is the average of four highest masses that do not give rise to PVC damage.

15.2 A fabric being tested to BS EN 373:1993 (courtesy of BTTG Fire Testing Laboratory, Altrincham, UK).

15.6.2   Integrated component and product testing

As textile materials like protective clothing are used in more complex and demanding environments, so the associated test procedures become more complex. While the BS EN tests outlined above and listed in Table 15.6 relate to testing individual fabrics, there is an increasing need to test the ability of the final garment and even clothing system to protect the wearer during a defined thermal or fire environment. One of the first uses of a simulated human figure or manikin was reported by Finley and Carter in 1971 and in this work garments were ignited by a bunsen burner. Subsequently, the Du Pont 'Thermoman' (Chouinard *et al.*, 1973) provided the first attempt at recording the temperature profile and simulated burn damage sustained by the torso when clothed in defined garments (usually prototype protective garments) during exposure to an intense fire source. This latter is typically a series of gas burners yielding a heat flux of $80 \, \mathrm{kW \, m^{-2}}$ that impinge on the front and rear of the clothed manikin.

The use of manikins has increased since this time and reviews of their use have been compiled by Krasny (1982) and Norton et al. (1985). Over ten years ago Sorensen (1992) reviewed attempts to establish this and related manikin methods as a standard method and subsequently draft standard procedures have been produced (e.g. ISO/DIS 13506). While not a mandatory test, it is often specified by fire service purchasing authorities in order to select clothing systems and is now offered as an optional test in the internationally accepted standard for firefighters' clothing, BS EN ISO 469 (see Table 15.6). A major challenge with such a test is determining its sensitivity to garment fit since overly loose garments will enable flames to penetrate between overlapping layers (e.g. between jacket and trousers) and too tight a fit will reduce the thickness of underlying air layers. Furthermore, while testing the ability of a given clothing system to protect a wearer from receiving torso burns measured at first-, second- and third-degree levels, it also ignores the consequences of the effect of heat and flame on the exposed head of the wearer. Furthermore, it does not enable the heat fatigue experienced by wearers to be measured during use of garments that can 'pass' the test. However, since full-scale manikin tests offer a more realistic means of testing full clothing assemblies they continue to draw research interest (Bajaj, 2000). Notable among these more recent studies is the work of Crown, Dale and coworkers at the University of Alberta who have used an instrumented manikin to test garment performance in flash fires (Crown and Dale, 1993).More recent work by this group includes a comparison of results from bench scale test-derived data such as TPP with manikin results expressed as percentage of body suffering defined burn levels (Crown et al., 2002). The reader is referred to Chapter 9 for a more detailed evaluation of manikins in general.

15.6.3   Composite standards: firefighters' clothing*

In Table 15.6 are a number of composite standards which offer a series of tests for specific working or hazard environments such as clothing for welders (BS EN 470:1995) and workers exposed to heat (BS EN 531:1995). In addition, they exist for protective gloves (BS EN 420) and not least, for firefighters' clothing (BS EN 469:1995). These standards may be examined in more detail by direct reference to them and in summary form by referring to the review by Bajaj (2000). However, in order to illustrate their character and structure, it is worth examining BS EN 469:1995 for firefighters' clothing in greater detail. As part of the European Union Personal Protective Equipment Directive, the standard BS EN 469:1995 is an attempt to create a single, normalised standard method applicable to firefighters' clothing throughout the EU. Such a standard may be

---

* See also Chapter 22.

applied to any part of the full garment assembly from the outershell jacket and overtrousers (or a single outer coverall) to the underlying outer and underwear garments. Gloves and hoods are covered by other standards listed in Table 15.6. There are in the order of 2 million firefighters including part-time personnel in the EU and so not only is the market for clothing considerable, but use of a standard test should ensure that each member state is providing the same minimal levels of protection to its firefighting workforce. However BS EN ISO 469 is a composite test in which a portfolio of test methods and performance requirements are defined as shown in Table 15.9.

The performance specifications specifically relating to thermal protection include ignition and flame spread (BS EN 532) and heat transfer to radiant (BS EN ISO 6942) and convective (BS EN 367) heat. For instance, the radiant heat transfer index, RTHI, values for a multilayer assembly (to BS EN ISO 6942 Method B) at $40\,kW/m^2$ should be $\geq 13\,s$ and the RTI values for flame exposure $(80\,kW/m^2)$, $\geq 22\,s$. In addition are tests relevant to the overall performance of

Table 15.9 Test methods within the composite standard for firefighters' clothing, BS EN 469:1995

Property tested	Standard	Principal performance specifications
Flame spread	BS EN ISO 15025:2002 (replaces BS EN 532:1995 and BS 5438)	No flame extending to top or edge, no hole formation and afterflaming and afterglow times $\leq 2\,s$
Heat transfer (flame)	BS EN 367:1992 (ISO 9151)	$HTI_{24} \geq 13\,s$
Heat transfer (radiant)	BS EN ISO 6942:2002 at $40\,kW/m^2$	$RHTI_{24} \geq 22\,s$
Residual strength	ISO 5081 after BS EN ISO 6942 Method A at $10\,kW/m^2$	Tensile strength $\geq 450\,N$
Heat resistance	BS EN 469:1995 Annex A	No melting, dripping or ignition
Tensile strength	ISO 5081	Tensile strength $\geq 450\,N$
Tear strength	ISO 4674/A2	Tear strength $\geq 25\,N$
Surface wetting	BS EN 24920:1992 (ISO 4920:1981)	Spray rating ≥ 4
Dimensional change	ISO 5077	$\leq 3\%$
Penetration by liquid chemicals	BS EN 368:1993 (for 40% NaOH, 36% HCl and 36% H_2SO_4 at 20°C)	$\leq 80\%$ run off
Water resistance	BS EN 20811:1992 (ISO 811:1981)	Defined by manufacturer/ customer
Breathability	BS EN 31092:1994 (ISO 11092:1993)	Defined by manufacturer/ customer
Contact heat transfer	BS EN 702:1995 (ISO 12127)	Defined by manufacturer/ customer
Manikin (optional)	ISO/DIS 12127	Defined by manufacturer/ customer

15.3 The BTTG RALPH burner flames at 80 kWm^{-2} heat flux engulfing the instrumented manikin (courtesy of BTTG Fire Laboratory, Altrincham, UK).

firefighters' clothing, including tensile and tear strength, surface wettability and others. Again the reader is referred to the main standard and its component parts for full details of each test. Of interest is the optional inclusion of the manikin test ISO/DIS 13506 mentioned above in Section 15.5.2. In this, the manikin is that developed at British Textile Technology Group in the UK using the RALPH II (Research Aim Longer Protection Against Heat) manikin (Sorensen, 1992; Bajaj, 2000) in which a clothed manikin is subjected to a full flame exposure with gas burner flames of about 800 °C and heat flux, 80 kW/m^2 for an 8 s application time. The torso is instrumented to determine those areas which will suffer first- and second-degree burns and a map of burnt areas is produced along with respective percentage of body area burn values. Figure 15.3 shows the intensity of the impinging flames in the BTTG RALPH manikin; note that the whole torso is engulfed within the flames and only the boots are observable.

15.7   Future trends

With the recent normalisation of test methods across the EU and the generation of the individual and composite standards discussed above in section 15.6, it is probable that very little change in standard testing methodology will occur in the near future. The development of instrumented manikins and their acceptance as an optional test (e.g. ISO/DIS 13506) perhaps points the way toward their greater use on the one hand and their more general adoption for a wider range of thermally protective clothing performance requirements on the other. At the time of writing, EN 469:1995 is currently being updated and this optional test will be included as a new annex (Annex E). In addition, the heat and flame performance criteria defined in Table 15.9 are being modified to allow for differing levels of performance, level I being less than the present requirements and level II being similar. Thus the convective or flame heat performance criterion in Table 15.9 of a heat transfer index, $HTI_{24} \geq 13$ s will be replaced by two levels – level I with $HTI_{24} \geq 9$ s and level II with $HTI_{24} \geq 13$ s. Similarly the radiant heat transfer index values that define radiant heat performance will change from $RHTI_{24} \geq 22$ s to $RHTI_{24} \geq 10$ s for level I and to $RHTI_{24} \geq 22$ s for level II performance. Alongside these changes are the proposed requirement (instead of being optional in the present 1995 version) of comfort performance criteria in terms of quantifiable moisture transfer and ergonomic parameters.

While heat fatigue is about to be recognised in the proposed revision of EN 469, it will still remain a challenge associated with such clothing assemblies, and the next generation of instrumented manikins may include sensors that enable prediction of its onset and magnitude. This would be more likely to be the case, however, if a 'working manikin' could be developed for assessing fire resistance as well as comfort and eventual onset of heat fatigue by a typical wearer working in a typical fire or other thermally hazardous environment. While the published literature gives little detailed information regarding specific developments of thermal and fire protective clothing that fit within the 'smart' or 'sensored' textile fields, the concepts have been discussed by a number of authors (Gries (2003), Tao X (2002)) and indications of developments reported (Vogel (2002), Anon (2003), Butler (2003)).

With regard to possible new materials, it is doubtful whether any really new generic fibre groups exhibiting unusual thermal characteristics will appear in the next decade. This is because, as section 15.3 has shown, the thermal protective clothing designer has a fair range of fibres, with or without after-treatments, to choose from and which enable a balance of performance for a specific application and cost to be achieved. However, because of the need for protective textiles and garments to be more multifunctional and reactive to defined thermal threats, the next generation may also be required to determine the degree of a given threat and respond appropriately. This, of course, brings us into the realms of so-called 'smart' textiles as mentioned above in which both the use of embedded sensors and the means of increasing levels of protection within a given textile structure may be anticipated. At the moment, however, the development of such textiles is an aspiration that should provide direction for the present and future generation of thermally protective textiles.

15.8 References

Anon (2003), 'Uniforms of the future', *Textile Horizons,* July/August, 7.

Bajaj P (2000), 'Heat and flame protection', in *Handbook of Technical Textiles,* Horrocks A R and Anand S A (editors), Cambridge, Woodhead Publishing, pp 223–263.

Baitinger W F (1979), 'Product engineering of safety apparel fabrics: insulation characteristics of fire retardant cottons', *Text. Res. J.,* 49, 221.

Barker R and Brewster E P (1982), 'Evaluating the flammability and thermal shrinkage of some protective fabrics,' *J. Ind. Fabrics,* 1 (1), 7–17.

Barker R L and Coletta G C (1986), *Performance of Protective Clothing,* ASTM Special Technical Publication 900, Philadelphia, ASTM, Philadelphia.

Barker R and Yener M (1981), 'Evaluating the resistance of some protective fabrics to molten iron', *Text. Res. J.,* 51, 533.

Barker R, Geshury A J and Behnke W P (1996), 'The effect of Nomex®/Kevlar® fibre blend ratio and fabric weight on fabric performance in static and dynamic TPP tests', in *Performance of Protective Clothing, Fifth Volume,* Johnson S and Mansdorf S Z (eds), ASTM Special Technical Publication 1237, West Conshohocken, USA, ASTM.

Beck P P, Gordon P G and Ingham P E (1976), 'Thermogravimetric analysis of flame-retardant-treated wools', *Text. Res. J.,* 46, 478.

Benisek L (1971), 'Use of titanium complexes to improve the natural flame retardancy of wool', *J. Soc. Dyers. Col.,* 87, 277–278.

Benisek L (1981), 'Antagonisms and flame retardancy', *Text. Res. J.,* 51, 369.

Benisek L and Edmondson G K (1981), 'Protective clothing fabrics . Part I Against molten metal hazards', *Text. Res. J.,* 51, 182.

Benisek L and Phillips W A (1981), 'Protective clothing fabrics. Part II Against convective heat (open-flame) hazards', *Text. Res. J.,* 51, 191–196.

Benisek L, Edmondson G K and Phillips W A (1979), 'Protective clothing – evaluation of wool and other fabrics', *Text. Res. J.,* 49, 212–225.

Buszard D (2004), 'The present European position: textile flame retardants and risk assessment', in proceedings of the conference, *Ecotextile04, The Way Forward for Sustainable Development in Textiles*', 7-8 July 2004, Bolton, UK.

Butler N (2003), 'How to succeed in developing safety and protective fabrics', *Technical Textiles International,* 12(3), 23–26.

Chouinard M P, Knodel D C and Arnold H W (1973), 'Heat transfer from flammable fabrics', *Text.Res.J.,* 43, 166–175.

Consumer Protection Act (1987), the Furniture and Furnishings (Fire) (Safety) Regulations, 1988, SI1324 (1988), London, HMSO, 1988.

Crown E M and Dale J D (1993), 'Built for the hot seat', *Canadian Textile J.,*Mar., 16–19.

Crown E M, Dale J D and Bitner E (2002), 'A comparative analysis of protocols for measuring heat transmission through flame resistant materials: capturing the effects of thermal shrinkage', *Fire Mater.,* 26 (4–5), 207–214.

Dawn F S and Morton G P (1979), 'Cotton protective apparel in the Space Shuttle', *Text.*

Res. J., 49, 197–201.

Day M (1988), 'A comparative evaluation of test methods and materials for thermal protective performance', in *Performance of Protective Clothing, Volume 2,* Mansdorf S Z, Sager R and Neilsen A (editors), ASTM Special Technical Publication 989, Philadelphia, ASTM, pp 108–120.

Day M and Sturgeon P Z (1987), 'Thermal radiative protection of firefighters' protective clothing', *Fire Technol.,* 23 (1), 49–59.

Derksen W L, Monahan T I and DeLhery G P (1961), 'The temperatures associated with radiant energy skin burns', in *Temperature – Its Measurement and Control in Science and Industry, Vol.3, Part 3*, New York, Reinhold, pp 171–175.

Dombrowksi R (1996), 'Flame retardants for textile coatings', *J. Coated Fabrics*, 25, 224.

Eaton P (2000), 'Flame retardancy test methods for textiles', *Rev. Prog. Colouration,* 30, 51–62.

Finley E L and Carter W H (1971), 'Temperature and heat flux measurements on life-size garments ignited by flame contact', *J. Fire Flammability,* 2, 298–319.

Flambard X, Bourbigot S and Poutch F (2005), 'Fire reaction of seats: the wool/para-aramid blend', *Textiles & Usages Techniques*, 56, 23–25.

Gries T (2003), 'Smart textiles for technical applications', *Technische Textilien*, 46(2), E66.

Hall M E and Shah S (1991), 'The reaction of wool with N-hydroxymethyl dimethylphosphonopropionamide', *Polym. Degrad. Stab.,* 33, 207.

Hearle J W S (2001), *High Performance Fibres,* Cambridge, Woodhead Publishing.

Holme I (2003), 'Water repellency and waterproofing', in *Textile Finishing,* Heywood D (editor), Bradford, Society of Dyers and Colourists, pp 135–213.

Holmér, I (1988), 'Thermal properties of protective clothing and prediction of physiological strain', in *Performance of Protective Clothing, Volume 2,* Mansdorf S Z, Sager R and Neilsen A (eds), ASTM Special Technical Publication 989, Philadelphia, ASTM.

Holmes D A (2000), 'Waterproof breathable fabrics', in *Handbook of Technical Textiles,* Horrocks A R and Anand S A (editors), Cambridge, Woodhead Publishing, pp 282–315.

Horrocks A R (1986), 'Flame retardant finishing of textiles', *Prog. Rev. Colouration*, 16, 62–101.

Horrocks A R (1996), 'Developments in flame retardants for heat and fire resistant textiles – the role of char formation and intumescence', *Poly.Degrad. Stab.,* 54, 143–154.

Horrocks A R (2001), 'Textiles', in *Fire Retardant Materials,* Horrocks A R and Price D (eds), *Fire Retardant Materials,* Cambridge, Woodhead Publishing, pp 128–181.

Horrocks A R (2003), 'Flame retardant finishes and finishing', in *Textile Finishing,* Heywood D (ed.), Bradford, Society of Dyers and Colourists, 2003, pp 214–250.

Horrocks A R and Davies P J (2000), 'Char formation in flame-retarded wool fibres; Part 1 Effect of intumescents on thermogravimetric behaviour', *Fire Mater.,* 24, 151–157.

Horrocks A R and Price D (2001), *Fire Retardant Materials,* Cambridge, Woodhead Publishing.

Horrocks A R, Price D and Tunc M (1987), 'The burning behaviour of textiles and its assessment by oxygen index measurements', *Text. Prog.,* 18 (1-3), 1–205.

Horrocks A R, Eichhorn H, Schwaenke H, Saville N and Thomas C (2001), 'Thermally resistant fibres', in Hearle J W S, *High Performance Fibres,* Cambridge, Woodhead Publishing, pp 289–324.

Horrocks A R, Kandola B K, Davies P J, Zhang S and Padbury S A (2005), 'Developments in flame retardant textiles', *Polym. Degrad. Stab.*, 88, 3–12.

Johnson S and Mansdorf S Z (1996), *Performance of Protective Clothing, Fifth Volume,* ASTM Special Technical Publication 1237, West Conshohocken, USA, ASTM.

Krasny J (1982), 'Flammability evaluation methods for textiles', in *Flame Retardant Polymeric Materials, Volume 3,* Lewin M, Atlas S M and Pearce E M, (eds), New York, Plenum, pp 155–200.

Krasny J (1986), 'Some characteristics of fabrics for heat protective garments', in Barker R L and Coletta G C (eds), *Performance of Protective Clothing,* ASTM Special Technical Publication 900, Philadelphia, ASTM, pp 463–474.

Lee Y M and Barker R (1987), 'Thermal protective performance of heat-resistance fabrics in various high intensity heat exposures', *Text. Res. J.,* 57, 123.

Lewin M and Mark H F (1997), 'Flame retarding of polymers with sulfamates; Part 1 Sulfation of cotton and wool,' 8th annual conf *Recent Advances in Flame Retardancy of Polymer Materials*, Stamford, USA, Business Communications Company.

Mansdorf S Z, Sager R and Neilsen A (1988), *Performance of Protective Clothing, Volume 2,* ASTM Special Technical Publication 989, Philadelphia, ASTM.

Mehta P N and Willerton K (1977), 'Evaluation of clothing materials for protection against molten metal', *Text. Inst. Ind.,* 15, 334–337.

Norton R J, Kandolph S J, Johnson R F and Jordon K A (1985), 'Design, construction and use of Minnesota Woman, a thermally instrumented mannequin', *Text. Res. J.,* 55, 5–12.

Perepelkin K E (2001), 'Chemical fibers with specific properties for industrial application and personnel protection', *J. Ind. Text.,* 31 (2), 87–102.

Perkins R M (1979), 'Insulative values of single-layer fabrics for thermal protective clothing', *Text. Res. J.,* 49, 202–212.

Proctor T D and Thompson H (1988), 'Setting standards for the resistance of clothing to molten metal splashes', in *Performance of Protective Clothing, Volume 2,* Mansdorf S Z, Sager R and Neilsen A (editors), ASTM Special Technical Publication 989, Philadelphia, ASTM.

Raheel M (1994), 'Chemical protective clothing', in Raheel M (ed.), *Protective Clothing Systems and Materials,* New York, Marcel Dekker, pp 39–77.

Raheel M, Perenich T A and Kim C (1994), 'Heat- and fire-resistant fibers for protective clothing', in Raheel M (ed.), *Protective Clothing Systems and Materials,* New York, Marcel Dekker, pp 197–224.

Rebenfeld L, Miller B and Martin J R (1979), 'The thermal and flammability behaviour of textile materials', in *Applied Fibre Science, Vol 2,* ed. F Happey, London, Acad. Press, p 465.

Scott R A (2000), 'Textiles in defence', in *Handbook of Technical Textiles,* Horrocks A R and Anand S A (editors), Cambridge, Woodhead Publishing, pp 425–460.

Shalev I and Barker R (1983), 'Analysis of heat transfer characteristics of fabrics in an open flame exposure', *Text. Res. J.,* 53, 475.

Shalev I and Barker R (1986), 'Predicting the thermal protective performance of heat-protective fabrics from basic properties', in *Performance of Protective Clothing,* Barker R L and Coletta G C (eds), ASTM Special Technical Publication 900, Philadelphia, ASTM, Philadelphia, pp 358–375.

Shirley Institute (1982), *Protective Clothing,* Shirley Publication S45, Shirley Institute, Manchester.

Sorensen N (1992), 'A manikin for realistic testing of heat and flame protective clothing',

Technical Text. Int., 8–12.

Stull J O and Schwope A D (1997), *Performance of Protective Clothing, 6th Volume,* ASTM Special Technical Publication 1273, Philadelphia, ASTM.

Tao X (2002), 'Sensors in garments', *Textile Asia,* 33(10), 38–41.

Tolbert T W, Dugan J S, Jaco P and Hendrix J E, Springs Industries (1989), Fire Barrier Fabrics, US Patent Office, 333174, 4 April.

Troitzsch J (2004), *Plastics Flammability Handbook,* 3rd edn, Munich, Hanser.

Valentin N (2004), 'Thermal testing of firemen's clothing', *Textiles à Usage Techniques (TUT),* 3(53), 51–53.

Vigo T L (1994), 'Protective clothing effective against biohazards', in Raheel M (ed.), *Protective Clothing Systems and Materials,* New York, Marcel Dekker, pp 225–244.

Vogel C (2002), 'Firefighter turnout gear with integrated sensors', *Textiles a Usage Techniques,* 46, 38–41.

Vosburgh J (1960), 'The heat treatment of Orlon acrylic fibre to render it fireproof', *Text. Res J.,* 30, 882.

Weil E D and Levchik S V (2004), 'Commercial flame retardancy of thermoplastic polyesters. A review', *J. Fire Sci.,* 22(4), 339–350.

Woodruff F A (2003), 'Coating, laminating, bonding, flocking and prepregging', in *Textile Finishing,* Heywood D (ed.), Bradford, Society of Dyers and Colourists, pp 447–523.

15.8.1   Selected web-sites active at the time of publishing

Exemplar manufacturers of protective textiles

http://www.heathcoat.co.uk/military_products.htm
http://www.baltex.co.uk/
http://www.firegard.co.uk/SAFETY/Clothing/clothing.html
www.bristol-uniforms.com
www.bennettsafetywear.co.uk

Fibre producers

http://www.kermel.com/
http://www.toyobo.co.jp/e/seihin/kc/pbo/
http://www.teijin.co.jp/english/about/enterprise/index.html
http://www.roehm.de/en/performanceplastics?content=/en/performanceplastics/inspec.html
http://www.ibena.de/protect/english/Protective_fabric/Welding/welding.html
http://www.lenzing.com/lenzred/frameset.jsp
http://www1.dupont.com
http://www.personalprotection.dupont.com/
http://www.dupont.com/nomex/europe/protectiveapparel/index.html
http://www.pbigold.com/
http://www.celanese.com/index/about_index/innov-home/innovation-product-innovation/innov-prod/innov-pbi.htm
http://www.sglcarbon.com/sgl_t/fibers/panox.html

Microorganism protection

K K L E O N A S , University of Georgia, USA

16.1   Introduction

Among health care professionals today, there is increasing concern over exposure to, and transference of, various microorganisms that are commonly carried through bodily fluids. There is an undisputed need for effective barriers in the operating room that will eliminate or reduce the risk of infection (Abreu *et al.* 2003, Medical Nonwovens 1983). This concern is international in scope as demonstrated by the recent introduction of regulations, guidelines and test methods by organizations throughout the world. Gowns and drapes have been used in the operating theater since the late nineteenth century (Beck 1963, Belkin 1992). A review of the century's progress in surgical apparel was completed by Laufman, Belkin and Meyer (Laufman *et al.* 2000). Today, in addition to protecting the healthcare worker from patient microorganisms, gowns and drapes are used to protect the patient from microorganisms carried by the healthcare worker and patient-to-patient transference (Rutala *et al.* 2001, Woodhead *et al.* 2002).

The increase in the presence of resistant pathogens that can be transmitted is also of concern (Centers for Disease Control 1999). These issues have led to the production and development of innovative new products in the textile industry. From the time that Beck and Collette (1952) reported that surgeons' gowns and drapes of cotton lost their effectiveness as barriers to bacterial transmission when wet, the evaluation and development of fabrics with suitable barrier properties for operating room use has been under investigation. Textiles have the potential to be suitable barriers but they must prevent the transmission of bacteria and fluids found in the operating room. Protective surgical apparel can play an important role in minimizing disease transmission in the operating theater. Bacterial and viral diseases are spread through both airborne and blood borne pathways. Surgical apparel can reduce the transfer of microorganisms by creating a physical barrier between the infection source and a healthy individual. A surgical apparel is identified in 21 CFR, Part 878.4040, as a medical device intended to be worn by operating room personnel during surgical procedures to protect both the surgical patients and operating room personnel from transfer of microorganisms, body fluids and particulate material. It is agreed that although protective apparel for healthcare workers must meet a variety of needs, the one of greatest importance is to prevent the transmission of microorganisms (AWMF 1998, May-Plumlee et al. 2002, Whyte et al. 1983).

There are a number of factors that contribute to the occurrence of hospital-acquired infections. It is known that various protocols in the operating theater and routine care of patients have a significant impact on the healthcare associated infections. One factor that reduces these infections is the use of appropriate personal protective apparel that includes surgical gowns and drapes. Healthcare-associated infections result in substantial rates of morbidity and mortality. According to the CDC, nearly 2 million patients in the US get an infection as a result of receiving healthcare in a hospital (Centers for Disease Control and Prevention 1999). Of these, it is estimated that approximately 500,000 surgical site infections (SSIs) occur annually in the United States (Wong 1999). Although the cost can vary by location and surgery type, it is estimated additional care may result in $1–$10 billion in direct and indirect medical costs each year (Holtz et al. 1992, Wong 1999).

The CDC has estimated that infections cause 19,027 deaths and contribute to an additional 58,092 deaths each year. Surgical site infections (SSIs) have been estimated to cause 3,251 deaths directly and contribute to 9,726 deaths (Emori and Gaynes 1993). Multiple strategies have been recommended to reduce the incidence of SSIs, including the use of gowns and drapes to create a sterile field during invasive procedures (Perencevich et al. 2003). More recently, gowns have been evaluated as a means to reduce patient-to-patient transmission of resistance pathogens.

The CDC has identified three primary routes of contact between people in healthcare settings and microorganisms: (i) contact (direct and indirect), (ii) respiratory droplets, and (iii) airborne droplet nuclei (Centers for Disease Control and Prevention 2003). Contact transmission is considered the most common and direct contact occurs when microorganisms are transferred directly from one person to another. The precise definition of droplet transmission is under discussion but technically it is a form of contact transmission. It refers to respiratory droplets that are generated through coughs, sneezes or talking that are then propelled a short distance and deposited on another person. For both of these transfer mechanisms (direct contact and droplet), using appropriate protective apparel, it is possible to create a barrier to eliminate or reduce this contact, and therefore prevent the transfer of microorganisms between patients and healthcare workers.

In 1991 in the United States, the Occupational Safety and Health Administration (OSHA) implemented regulations designed to protect workers from exposure to blood-borne diseases if they are exposed to blood or other infectious body fluids in the course of their job. This regulation mandates the principles of universal precautions; mandates performance levels, and allows employers to specify what personal protective equipment is required and when it must be used. It also requires that employers supply personal protective equipment (PPE) including gowns, masks, caps, and shoe covers to employees if they are at risk of exposure to blood or infectious body fluids. The employers are also responsible for making sure the protective equipment is readily accessible, cleaned, laundered and/or disposed of after use, and ensuring its effectiveness. This regulation requires over five million health care employees to wear personal protective equipment. Perencevich *et al.* (2003) imply that surgical gowns provided barrier protection from microorganism transmission. Since the time of its implementation, researchers, textile producers and surgical gown manufacturers have been working to develop products that will provide ultimate protection against microorganism transmission while meeting the other critical needs of the workers (comfort, freedom of movement, flexibility, durability, flame resistance).

Protective surgical apparel can play an important role in minimizing disease transmission in the operating theater. Bacterial and viral diseases are spread through both airborne and blood borne pathways. Surgical apparel can reduce the transfer of microorganisms by creating a physical barrier between the infection source and a healthy individual (Abreu *et al.* 2003). In addition to the initial intent of the OSHA blood borne pathogen final rule to protect the healthcare worker from exposure to pathogens from the patient, the role of the gowns and drapes has been expanded to include protecting the patient from microorganisms the healthcare worker might be carrying, and prevent patient-to-patient microorganism transfer (Rutala *et al.* 2001, Woodhead *et al.* 2002).

This chapter is designed to examine those characteristics that govern the suitability of a fabric as a barrier to the transmission of bacteria. To do this it is necessary to have an understanding of the mechanism of bacterial transmission through fabrics and to be familiar with those fabric characteristics that influence this transmission. Additional factors necessary for a fabric to be acceptable for use in surgical gowns and drapes are low linting, no pilling or abrasion, resistance to liquid strike-through, high air porosity, sufficient strength and tear resistance, light weight and cloth-like aesthetic properties, and flexible enough to allow for the necessary movement of the part to be protected (Seaman undated and Mayberry 1989).

16.2   Bacterial and liquid transmission through fabrics

Fabrics are three-dimensional entities with void and non-void areas. The physical and chemical properties of the fabric, the shape and surface characteristics of the bacteria, and the characteristics of carriers contribute to control the movement of microorganisms through fabric structures. Bacteria are not thought to move from one location to another independently. Rather, they are transported by some other object such as dust, lint, skin particles or liquids (Beck and Carlson 1963, Dineen 1969). The carrier transports the bacteria as it moves, depositing the bacteria particles along the way, or at the point of destination. In medical procedures, fluids such as blood, perspiration and alcohol are commonly present in the environment and they can act as carriers transporting the bacteria through the fabric. Particles such as shedding skin cells, lint and dust, as well as respiratory droplets produced by coughs, sneezing or talking, are also a likely carrier of bacterial particles. Therefore, it is necessary to consider the interaction of the carrier and the barrier, as well as the effectiveness of the barrier to the movement of the bacterial particle.

The surgical team is the most important source of bacteria-causing infection (Whyte *et al.* 1983, Woodhead *et al.* 2002). Most wound infections are caused by germs originating from either the staff or the patient. While transmission can occur with or without liquids, the presence of liquids facilitates microbial transfer and therefore increases the probability of an infection (AWMF 1998). To examine the mechanism of bacterial transmission, one must consider the movement of the liquid through the fabric. Liquids and bacteria can be drawn through by capillary action enhanced by wicking (Woodhead *et al.* 2002). The yarns and fibers typically used in fabric construction are usually cylindrical and surrounded with narrow spaces (interstices) forming capillaries. The pressure drop across the curved surface is responsible for the rise or depression of a liquid in a capillary. Factors that play an important role in governing capillary absorption are identified by Gupta as (i) characteristics of the fluid (surface tension, viscosity and density), (ii) the nature of the surface (surface energy and surface morphology), (iii) interaction of the fluid with the surface (interfacial tension and contact angle), and (iv) pore characteristics (size, volume, geometry and orientation) (Gupta 1988).

It is the combination of these factors that determines the movement of a liquid through the fabric. The relationship between the surface tension of the fluid and the surface energy of the material governs adsorption, the first step in the transmission process. This relationship is measured as surface contact angle. Generally if the surface contact angle between the liquid and the surface is greater than 90 degrees the liquid will roll into a sphere and roll off the surface of the fabric. However, if the contact angle is less than 90 degrees, the liquid is likely to spread, adsorption of the liquid may occur and absorption of the liquid into the fabric and fibers will be enhanced. The contact angle is controlled by the difference between the surface tension of the liquid and the surface energy of the fabric. The greater this difference, the less likely adsorption will occur. The surface energy of the fabric is influenced by all components of the fabric. The determination of surface contact angle between fabrics and liquids is difficult, due to the irregularities of the fabric surface. Typically, contact angles are measured between the liquid in question and films with similar composition to that of the fibers in the fabric. These values are an inaccurate measure of the surface energy for a fabric because the roughness of the fabric surface is not accounted for. There are currently available measurement systems where it is possible to measure the contact angle between various liquids and fabrics.

Liquids typically found in the operating room have a wide range of surface tensions. Water and saline solutions have relatively high surface tensions (~70–72dyne/cm), blood is in the medium range (~42 dyne/cm) and isopropyl alcohol has a low surface tension (~22 dyne/cm). Therefore, fabrics used as barriers in healthcare facilities must be able to repel liquids of lower surface tensions than those used in other environments where water repellency is of primary concern.

If the liquid adheres to the surface, there is increased potential for adsorption and transmission to occur. When this occurs, or the carrier is a dry particle, the fabric must act as a filter to prevent the movement of the bacterial particles and their carriers through the fabric. The pore size, geometry and volume are critical in establishing the fabric as a filter media. These characteristics also influence the formation of capillaries in the fabric. If the movement of the liquid can be stopped, the movement of the bacteria will be inhibited. The pores must be smaller than the particle for effective filtering. The average particle diameter of *Staphylococcus aureus* is approximately one micron indicating that the pore size of the filtering media would need to be smaller than the size indicated. However, it is important to remember that fabrics are pliable and the pore size will change when the fabric is exposed to stress or pressure. Charnley and Eftekhar (1969) reported the transmission of bacteria containing dry particles through loosely woven fabric. Additionally, liquid carriers that aid in the movement of the particle also may act as a lubricant and/or energy provider. Therefore, even if the pore size of the fabric is smaller than the bacterial particle size, this does not ensure that the particle will be trapped within the fabric structure.

Microbial diversity can be seen in terms of variations in cell size and morphology, metabolic strategies, motility, cell division, development biology, adaptation to environmental extremes, and many other structural and functional aspects of the cell. A variety of microorganisms have been found in healthcare settings. The majority of these are bacteria, but viruses and some fungi are also of interest. Microorganisms in general are very small. Bacteria are typically smaller than fungi and are about 1–5 μm long and completely invisible to the naked eye. Viruses are smaller in size than bacteria; the reported size of the HIV virus is approximately 13 nanometers. The shape of microorganisms can also vary and this will influence their ability to move through a fabric structure. Research in the filtering area has shown that rod shaped bacteria (like *E. coli*) are more likely to be trapped than are bacteria of round shapes.

16.2.1   Influence of wicking

Microorganisms can be transferred through barrier materials by wicking of fluids and/or pressure or leaning on a flooded surface. Mechanical action such as pressure can result in both liquid and dry penetration of microbes if the pressure exceeds the maximum level of resistance that the material provides. It is recommended that the wearers' anticipated exposure to blood and body fluids be considered and a product of the appropriate barrier effectiveness be selected in accordance with the OSHA guidelines for use of personal protective equipment (Association of periOperative Registered Nurses 2003).

16.3   Fabrics used in gowns and drapes

Surgical gowns and drapes found in the marketplace today are produced from a variety of fabrics and a wide range of fibers (May-Plumlee *et al.* 2002). One method commonly used to classify these products is as 'disposable' or 'reusable'. Both disposable (single-use) and reusable surgical gowns are currently used for surgical wear; however, disposables comprise the majority of the market. Disposable products are typically made of nonwoven fabrics and are designed to be discarded after a single use. They can be produced from a variety of nonwoven fabrics with the most common being a spunbonded/meltblown/spunbonded polypropylene composite or spunlaced wood pulp/polyester. Reusable gowns are 'reprocessed' (laundered and sterilized) after each use. Common fabrics typically used in reusable gowns have a woven or trilaminate construction. Plain weave fabric is the most common fabric construction found in reusable surgical gowns and most are produced from cotton/polyester blends or polyester. Trilaminate fabrics are composed of a microporous membrane between layers of woven polyester and knitted polyester. Both woven and nonwoven products are some-times reinforced with additional layers to provide increased barrier effectiveness. The reinforcement of surgical gowns is discussed in detail in the gown design section of this chapter.

Regardless of whether the product is classified as disposable or reusable, or produced from nonwoven or woven fabrics, there are specific fabric charac-teristics that have been demonstrated to influence the barrier effectiveness of the product. These include repellency, pore size, thread count (woven), and thick-ness. In addition, hydrostatic resistance (impact water resistance) is commonly measured and used as an indicator for barrier performance. To achieve the properties necessary to be an effective barrier and meet the additional require-ments for surgical gowns and drapes, consideration of the fiber, yarns (when present), specific fabric construction characteristics, fabric finish and their interactions must be made. It is critical to understand that because there are complex interactions of these fabric components, and seldom is it an individual component or property that dictates the final performance, but rather the interactions of these components that control the final fabric properties. Following is a brief overview of the fabric components and critical properties.

16.3.1   Fibers

Fibers are the smallest unit of a textile fabric and their properties are dependent on their chemical and physical properties. Chemically, the absorbency of the fiber is important to the transmission mechanism. Highly absorbent fibers remove the liquid therefore halting the movement of the bacteria and trapping it within the fiber structure (Leonas 1997b). When fibers of low absorbency are present, the liquid will wick along the fiber surface enhancing capillary movement of bacterial containing liquid. Natural fibers, such as cotton and wool have higher absorbency capabilities than synthetic fibers including polyolefins, polyester and nylon. Rayon, a regenerated cellulosic fiber, also has higher absorbency that the synthetic fibers. It has been proposed that only a small percentage of the moisture held is actually held internally in the fiber (Gupta 1988). In addition it is important to remember that fibers with low absorbency produce a surface that is more likely to repel water-based liquids adding to the fabric's repellency.

The physical properties or the surface morphology of the fiber also influence the barrier effectiveness. Fibers with irregular cross-sections or irregular surfaces have the ability to trap particles as they travel inhibiting particle transmission. Cotton fibers have convolutions ideal for trapping the small particles. Rayon is characterized by striations, which provide some irregularity to the fibers surface. These striations are usually not pronounced and therefore are not as effective in trapping or inhibiting the movement of the bacteria as the convolutions found in cotton fibers on inhibiting particle movements. The synthetic fibers are inherently smooth with circular cross-sections that allow particles to easily slide by the fiber. This smooth circular cross-section also enhances the formation of capillary forces between fibers. However, synthetic fibers with irregular cross-sections can be produced using appropriate dies during extrusion or can be textured to impart irregularity to their inherently smooth surface. Recently, polyester fibers have been developed that have longitudinal channels, which enhance wicking. The channels, along with the inherently low moisture absorption of polyester (0.04%), contribute to the high wicking fibers. Fiber length also influences the barrier properties of the fabric. Shorter fibers are more effective in preventing the transmission of small particles than longer fibers.

A relatively new development in the area of fiber manufacturing is the development of micro fibers and ultra-fine fibers. Micro fibers were first commercially produced in 1989. A micro fiber is a fiber that has less than one denier per filament. Micro fiber is the thinnest, finest of all man-made fibers and fabrics made from micro-fibers can meet the high specifications of barrier fabrics (Bernstein 1996). Micro fibers are now used in some gown and drape fabrics.

16.3.2   Yarns

The second component of a fabric are yarns, which are produced from fibers. Traditionally yarns are found in woven fabrics but not in nonwoven fabrics. Twist is commonly used to hold the fibers together. The amount of twist significantly influences the yarn properties. Yarns produced from filament fibers usually contain less twist than yarns produced from staple fibers. The lower the twist, the bulkier the yarn with increased voids between the fibers. These interstitial spaces allow for the trapping of particles that move through the fabric. In studies evaluating the geometry of surgical gown fabrics, image analysis showed that pores existed both within the multifilament yarns and within the fabric structure (Aibibu *et al.* 2003). In the case of staple yarns, lower twist also results in protruding fibers ends at the surface of the yarn. These small projections are ideal for trapping small particles. Additionally they create an irregular surface, which disrupts the formation of capillary forces as well, inhibiting the liquid movement. Yarns are used to produce knit and woven fabrics but they are not commonly used in the production of nonwoven fabrics.

16.3.3   Fabric construction

Fabric characteristics of importance include pore size characteristics and surface characteristics. Pore size and geometry are determined by fabric construction characteristics. The three primary methods of fabric construction are woven, knit and nonwoven. This discussion will concentrate on woven and nonwoven fabrics as they are commonly found in surgical gowns and drapes. Woven fabrics are characterized by two or more sets of yarns interlacing at right-angles. Nonwoven fabrics are typically produced directly from fibers and involve web formation and bonding procedures.

The most common woven fabric used to produce surgical gowns and drapes is the plain weave. Fabrics of this weave have the simplest, most regular interlacing pattern. The interstices determine the size and geometry of the pores and also contribute to capillary formation. The plain weave with its simple and regular interlacing pattern is susceptible to the formation of capillary forces that enhance the movement of the liquid through the fabric.

Thread count is a parameter of woven fabrics and identifies the number of threads or yarns per inch. As the thread count increases, the yarns are closer together (more tightly packed) resulting in a smaller pore size but more effective capillaries. If the interlacing pattern is irregular (as in a twill weave), the orientation of the yarns or fiber to one another is disrupted, the capillaries are shorter and the ability to move the liquid though the fabric is reduced.

Nonwoven products are produced directly from fibers or fiber solutions and the process can be divided into two steps, web formation and bonding. The quality of the nonwoven product is largely determined by the quality of the web.

The fibers are converted directly into a web or sheet significantly faster than that of most woven fabrics. The elimination of the yarn production step and the increased production speeds result in lower production cost. However, it is also more difficult to maintain uniformity in the product (Shadduck *et al.* 1990). Nonwoven fabrics are characterized by fibers more randomly oriented than yarns found in woven fabrics. This successfully reduces the potential for the formation of capillaries thereby reducing liquid transmission. In addition, the random orientation provides a filtering media more effective in trapping bacteria and bacteria-carrying particles. In the past decade, three types of nonwoven fabrics, spunlaced, meltblown, and a three-layer composite (spunbonded/ meltblown/spunbonded) have been used frequently for the production of surgical gowns and drapes. These fabrics are characterized by their methods of production. Spunlaced fabrics are produced by hydro entanglement of the fibers. Meltblown fabrics are characterized by micro fibers; the web is produced by air layering and then bonded thermally to produce the desired fabric. The use of micro fibers results in excellent filtering properties and is effective in preventing the movement of the bacteria particles. The composite nonwoven fabrics commonly used in surgical gowns are three layers of nonwoven fabrics combined in a sandwich form. A composite of a meltblown fabric sandwiched between spunbonded fabrics has been shown to provide excellent protection against bacteria protection (Leonas and Miller 1990).

16.3.4   Finishes

The finish type most widely used in protective apparel for healthcare workers imparts repellency to the fabric. In addition to repellent finishes, there has been some interest in finishes that impart antibacterial activity to these products (Huang and Leonas 2000, Jinkins and Leonas 1994). Where repellency and repellent finishes are applied, the surface of the fabric is influenced by both physical and chemical properties of the fabric components. The physical properties of interest include the texture of the fabric, which is influenced by the fiber surface, yarn surface and fabric construction characteristics. Surfaces that are smooth will shed liquids more readily than rough surfaces. The chemical characteristics are related more directly to the chemical characteristics of the fiber. With all other factors equal, if the fibers in a fabric inherently repel water, then the fabric will repel water more readily than a fabric produced with absorbent fibers. The combination of these factors and other characteristics contribute to the surface energy of the fabric.

To enhance the repellency of a surface, the surface energy is reduced by treating (finishing) the fabric with chemicals to alter the surface energy. There are a number of chemical classes of repellent finishes used in the textile industry. Fluorocarbon-based finishes are most commonly used in protective apparel for healthcare workers. This class of repellent finishes are successful in reducing the surface energy of the fabric sufficiently to repel both water and oil-based liquids. They provide a fabric that is water resistant (will shed small amounts of water) but not waterproof, so they allow for comfort. However, they can be susceptible to penetration by liquids of low surface tension, such as isopropyl alcohol. As the number of fluorine molecules increases in the structure of the chemical finish, the surface energy decreases. Even if a fabric is effectively treated to improve repellency, once the barrier becomes wet, regardless of the wetting solution, it is no longer an effective barrier in the prevention of bacterial transmission. Studies have demonstrated that when repellent finishes are applied to fabric that have previously not prevented bacterial transmission, the barrier properties are improved (Laufman *et al.* 1975, Leonas and Miller 1990).

Antimicrobial agents are those that kill or inhibit the growth of micro-organisms. To effectively inhibit microorganism growth, the antimicrobial agent must interrupt the growth cycle. Some important targets in the growth cycle include the cell wall, cytoplasmic membrane, protein synthesis, and nucleic acid synthesis (Brock *et al.* 1994). Depending on those microorganisms the chemical agents are designed to attack, antimicrobial agents can be further classified as bactericides, fungicides, disinfectants, antiseptics, chemotherapeutic agents, and antibiotics. The sensitivity of microorganisms to antimicrobial agents varies. Gram positive bacteria are usually more sensitive to antibiotics than gram negative bacteria. A broad spectrum antibiotic will act on a wider range of microorganisms than just a single group. Some agents have an extremely limited spectrum of action, being effective for only one or a few species. Likewise, different antimicrobial agents vary in their selective toxicity. Some have similar effects on all types of cells, and others are more toxic to microorganisms than animal tissues. Antimicrobial compounds used to treat textiles include alcohols, oxidizing agents, heavy metals, acids, aldehydes, surfactants and antibiotics.

Broughton *et al.* (1999) provided an excellent review of the current chemicals used in textile applications. For the end-use discussed here, the antimicrobial would need to be selected to target suspect microorganisms and would need to have a quick effect (upon contact vs. hours of contact). The application of polyethylene glycol to surgical gown fabrics was studied and results showed that although there were some antibacterial properties imparted to fabrics at higher finishing add-ons, the treatment also increased the surface energy of the fabric that increased the absorbency and liquid transport, which is undesirable in this type of product (Jinkins and Leonas 1998, Leonas *et al.* 1996). Huang and Leonas (2000) reported the application of a combined repellent and antimicrobial treatment to nonwoven surgical gown fabrics to improve their performance. In these studies, durability was not of concern, as the fabrics were intended for disposable products. If the finish were to be applied to a reusable product, then it would need to maintain its effectiveness through the life of the garment.

In recent years there has been the development of antimicrobial systems that can be regenerated by laundering and exposure to ultraviolet light. In this system, active antimicrobials can be continually regenerated by laundering or exposure to ultraviolet light. Although the reservoir of the antimicrobial chemical is limited, the surface remains effective for long periods of time (Sun and Xu 1998).

16.4   Fabric properties that influence barrier properties

In this section, various fabric characteristics as they relate to fabrics typically used in surgical gowns and drapes have been presented. Several studies have identified those fabric properties that have an impact on the barrier effectiveness. These characteristics include repellency and pore size (Leonas and Jinkins 1998). Wicking has also been associated with transmission.

16.4.1   Repellency

The barrier effectiveness of the surgical gowns increases as the repellency of the fabric increases. The fabric must be repellent not only to water and saline, but also to lower surface energy liquids such as blood and alcohol. Studies have shown that once a fabric is wet out by a lower surface tension liquid, the barrier effectiveness is reduced (Blom *et al.* 2002, Flaherty and Wick 1993).

16.4.2   Pore size

To ensure barrier effectiveness, the pore size of the fabric must be smaller than that of the bacteria or its carrier. If the bacteria are carried by a liquid (such as perspiration or blood), then the fabric must filter the bacteria and the pore size must be smaller. If the bacteria are carried by shed skin cells, lint or dust, then the pore size must be smaller than the carrier. It has been shown that fabrics with smaller pore sizes, have improved barrier effectiveness to bacterial transmission (Leonas and Jinkins 1998). Conventional cotton fabrics that were used for many years had a pore size of at least 80 μm. This material can be penetrated by all bacteria and bacterial carrying particles smaller than 80 μm (AWMF 1998). Several studies have evaluated the fabric pore size of fabrics commonly used in surgical gowns and found there is a bimodal distribution of pores in woven gown fabrics which is not the case in nonwoven gown fabrics. This distribution is due to the pores within the fibers of the yarns and the interstices of the yarns in the fabric structure (Aibibu *et al.* 2003, Leonas 1997b, Leonas and Jinkins 1998).

16.5   Gown design

In addition to the selection of appropriate fabrics to reduce transmission, the design of the gown can also contribute to the barrier effectiveness. While it has been stated that the prevention of microbial transmission is of critical importance, surgical gowns must provide other advantages that allow the healthcare workers to perform their tasks (AWMF 1998).

May-Plumlee and Pittman (2002) completed a design analysis of existing products and in their research they reviewed some of the necessary design characteristics that are summarized below. The gowns must allow adequate freedom for the workers to move and designed to fit a diversity of body shapes and sizes as most healthcare facilities will stock only a limited number of sizes. The gowns are easy to put on and remove without contaminating the worker or the workplace. Although they must prevent the transmission of fluids, they must allow airflow to ventilate the workers body heat (Belkin 1993). Both woven and nonwoven products are sometimes reinforced in those areas that are exposed to the highest levels of blood exposure to aid in the prevention of transmission and improve their barrier quality. This is commonly accomplished by adding an additional layer of fabric and/or a microporous membrane in the areas most likely to be exposed to the blood and body fluids. Some nonwoven gowns are reinforced with a polyethylene film or micro porous film to increase barrier effectiveness (Association of periOperative Registered Nurses 2003, Leonas and Jinkins1998, Lickfield 1997, Rutala and Weber 2001).

Several studies have identified the areas of the gown that are considered critical as they receive high levels of blood exposure. The front of the torso is a critical area for barrier performance (Lickfield 1997). Additional in-use studies showed that the surgeon's forearm and abdomen were subject to the highest levels of blood exposure during general surgery (Closs and Tierney 1990, Quebbeman *et al.* 1992). Pissiotis *et al.* (1997) divided gowns into 11 areas and found the highest areas of contamination occurred in the areas of the forearm, chest and abdomen. However, in this study, the highest percentages of strikethrough occurred in the cuff, forearm and thigh areas. Design and performance characteristics vary as a result of trade-offs in cost, comfort and the amount of barrier protection provided (Association of periOperative Registered Nurses 2003).

16.6   Guidelines, recommended standards, practices and regulations

There has been an interest in the use of protective apparel in the healthcare industry for over a century (Belkin 2002, Rutala and Weber 2001). Currently a number of major organizations have published guidelines for healthcare workers to minimize risks of exposure, which include recommendations for personal protective equipment. These organizations include Centers for Disease Control and Prevention (CDC), Association of periOperative Registered Nurses (AORN), Occupational Safety and Health Administration (OSHA), The Operating Room Nurses Association of Canada (ORNAC), and the Association for the Advancement of Medical Instrumentation (AAMI).

Organizations in the US, Europe and Canada are preparing new performance standards for protective apparel, gowns, and drapes, which have either recently been published or are entering the final development stages. Both US and European performance standards will rely on quantitative test methods that should allow end users to make better decisions because they will be better able to compare products. This is unlike the current situation in which a variety of test methods are being used to characterize products (Bushman 2003). By specifying a consistent basis for testing and labeling protective apparel and drapes, and providing a common understanding of barrier properties, this new classification system should assist healthcare personal in making informed decisions (Cardinal Health 2004). Historically, the methods used by manufacturing companies to test the effectiveness of protective equipment have not been clinically based (Ahmad *et al.* 1998). Additional costs will be incurred for both reusable and disposable apparel to upgrade seams for higher level products (Bushman 2003). Following are brief summaries of the current guidelines, recommended practices and standards.

16.6.1 Occupational Safety and Health Administration – Final Rule on Occupational Exposure to Bloodborne Pathogen (OSHA)

In 1991, OSHA proposed the Bloodborne Pathogen: Final Rule – which is a standard designed to eliminate or minimize occupational exposure to hepatitis B virus (HBV), human immunodeficiency virus (HIV) and other blood borne pathogens and became effective in March 1992. This standard was developed in response to the determination that employees face a significant health risk as the result of occupational exposure to blood and other potentially infectious materials because they may contain blood borne pathogens. The Agency further concluded that this exposure could be minimized or eliminated using a combination of engineering and work practice, controls including the use of personal protective clothing. The OSHA Occupational Exposure to Blood borne Pathogens: Final Rule (1991) mandates the principles of universal precautions, mandates performance levels, and allows employers to specify what personal protective equipment is required and when it must be used. With the implementation of this regulation, employers are responsible for making sure the protective equipment is readily accessible, cleaned, laundered and/or disposed of after use and ensuring the effectiveness. This regulation requires over five million healthcare employees to wear personal protective equipment (Occupational Safety and Health Administration 1991).

16.6.2 Association of periOperative Registered Nurses (AORN)

In January 2003 the recommended practices for selection and use of surgical gowns and drapes developed by the AORN Practices Committee and approved by the Board of Directors were summarized and published. The practices provide guidelines for evaluation, selection and use of surgical gowns and drapes. Of the eight recommended practices reviewed in this publication, the following four apply directly to surgical gowns and drapes. The first two reflect the appropriate selection of the product with reference to its intended function. Recommended practice 1 covers the recommended practices for evaluation and selection of products and medical devices used in preoperative practice settings and states that the product should be selected based on the properties critical specific to its function and use. Recommended practice 2 concerns the materials and states that those used in surgical gowns and drapes should be resistant to penetration by blood and other body fluids as necessitated by their intended use. It is recommended that data be provided verifying that the materials used in gowns and drapes are protective barriers against the transfer of microorganisms, particulates, and fluids to minimize strikethrough, to address the issue of personal protection. The physical movements of the wearer must be considered, as pressure can result in both liquid and dry penetration of microbes if the pressure exceeds the maximum level of resistance that the material provides. Two of the recommended practices relate to the performance of the product over time with consideration give to normal use and care. Recommended practice 3 states that surgical gowns and drapes should maintain their integrity and be durable. Recommended practice 4 states that materials used for surgical gowns and drapes should be appropriate to the method(s) of sterilization. It is noted that tightly woven reusable textiles will lose their protective barrier quality after repeated processing (Association of periOperative Registered Nurses 2003). The complete recommendation is available in the 2004 Standards, Recommended Practices and Guidelines of AORN (2004).

16.6.3   Centers for Disease Control (CDC)

In 1985, the CDC recommended universal precautions for the care of all patients in which there is an increased risk of blood exposure and the infection status of the patient is unknown (Centers for Disease Control and Prevention 1985). In June of 2004 the CDC published a Draft Guideline for Isolation Precautions: Preventing Transmission of Infectious Agents in Healthcare Settings 2004 prepared by the Healthcare Infection Control Practices Advisor Committee (Centers for Disease Control and Prevention 2003). One purpose of the document is to provide infection control recommendations for use by infection control staff, healthcare epidemiologists, healthcare administrators, and other persons responsible for developing, implementing, and evaluating infection control programs for healthcare settings across the continuum of care. It updates and expands the 1996 Guideline for Isolation Precautions in Hospitals. The use of Personal Protective Equipment (PPE) is recommended in this draft and that the selection of protective apparel is based on the nature of the patient interaction and the anticipated degree of body contact with infectious material and level of need protection from fluid penetration (Centers for Disease Control and Prevention 2003).

16.6.4 American National Standards Institute (ANSI)/ Association for the Advancement of Medical Instrumentation (AAMI)

There are currently several AAMI standards that have been recently published or are under development that relate to protective apparel and drapes intended for use in healthcare facilities. A standard exists that establishes a classification system for protective apparel and drapes used in healthcare facilities based on their liquid barrier performance and specifies labeling requirements and test methods for determining their compliance. This standard, ANSI/AMMI PB7:2003 – Liquid barrier performance and classification of protective apparel and drapes intended for use in healthcare facilities, was approved in October 2003 and published in January 2004 (Association for the Advancement of Medical Instrumentation 2004). The standard includes four standard tests to evaluate the barrier effectiveness of surgical gowns and drapes. Standard test methods include AATCC 42-2000: Water Resistance Impact Penetration Test, AATCC 127-2000 – Water Resistance: Hydrostatic Pressure Test, ASTM 1670-03: Standard Test Method for Resistance of Materials Used in Protective Clothing to Penetration by Synthetic Blood and ASTM F 1671-03 Standard Test Method for Resistance of Materials Used in Protective Clothing to Penetration by Blood-Borne Pathogens Using Phi-X174 Bacteriophage Penetration as a Test System Bacteriophage Penetration Resistance. This standard allows for the classification of products at four levels based on their performance. Level 1 describes surgical gowns, other protective apparel, surgical drapes, and drape accessories that demonstrate the ability to resist liquid penetration when tested with laboratory test AATCC 42 (≤ 4.5 g). Levels 2 and 3 both describe surgical gowns, other protective apparel, surgical drapes, and drape accessories that demonstrate the ability to resist liquid penetration in two laboratory tests (AATCC 42 and AATCC 127), with level 3 requiring higher values than level 2 as follows – Level 2, AATCC 127 (≥ 20 cm) and AATCC 42 (≤ 1.0 g) and Level 3, AATCC 127 (≥ 30 cm) and AATCC 42 (≤ 1.0 g). Level 4 has two components, one for surgical gowns and protective apparel that have the ability to resist liquid and viral penetration in test ASTM F1671; the second component is intended for drapes and drape accessories with the ability to resist liquid penetration in test ASTM F1670.

A second standard is currently being developed and under review by the AAMI Protective Barriers Committee. This standard covers the selection and use of protective apparel and surgical drapes in healthcare facilities. Information on types of protective materials, safety and performance characteristics of protective materials, product evaluation and selection, choosing the level of barrier performance and care of protective apparel and drapes. The draft is currently titled AAMI/CDV-1 TIR11 – Selection and use of protective apparel and surgical drapes in health care facilities (Association for the Advancement of Medical Instrumentation 2004).

16.6.5   European Standards and Test Methods – European Committee for Standardisation (CEN)

The EU recently integrated the standards developed over the past decade in various European countries into a single medical device directive that begins to describe the necessary properties to be fulfilled for any product to be considered safe and fit. The European Committee for Standardisation (CEN) working group TC 205 WG 14 has written a New European Standard EN 13795; 'Surgical drapes, gowns and clean air suits used as medical devices for patients, clinical staff and equipment' (European Committee for Standardisation 2004). This document is being prepared in three parts. At the time of writing, Part I: General requirements for manufacturers, processors and products has been ratified. Part II: Test method to determine the resistance to wet bacterial penetration, is currently under approval and Part III, Performance requirements and performance levels, is under approval (http://www.cenorm.be/CENORM/BusinessDomains/Technical CommitteesWorkshops/CENTechnicalCommittees/WP.asp?param= 6186&title=CEN%2FTC+205). Table 16.1 provides a list of the characteristics and associated test methods included in this document. The new requirements devised by the CEN are intended to help prevent the spread of bacteria during operations and specify the adoption of highly impermeable fabrics, as well as strict laundering and reprocessing procedures. 'These standards will undoubtedly raise the safety performance of reusable products, although many will disappear unless they are substantially redesigned', says Jan Hoborn, Director of Medical Sciences

Table 16.1 A summary of European standards and test methods

Characteristic	Test method
Microbial penetration – dry	prEN 13795-3
Microbial penetration – wet	prEN 13795-4
Linting	ISO CD 9073-10
Hydrostatic resistance	EN 20811
Bursting strength – dry	prEN 13938
Bursting strength – wet	EN 13938
Tensile strength – dry	EN 29073-3
Tensile strength – wet	EN 29073-3
Cleanliness – microbial	EN 1174-2
Cleanliness – particulate	prEN 13795-2

at Mölnlycke Health Care, a leading international supplier of medical products. 'It will also make quality assurance easier for buying departments since every product on the market will have to meet the new requirements' (Molnlycke, 2003).

All medical textiles regarded as medical devices now placed in the market have to comply with the European Medical Devices Directive and bear the CE mark demonstrating that they have been declared fit for their intended purpose and meet the essential requirements of the Directive (Abreu *et al.* 2003).

16.6.6   Canadian Standards Association (CSA)

CSA Z314.10-03 'Selection, Use, Maintenance and Laundering or Reusable Textile Wrappers, Surgical Gowns, and Drapes for Health Care Facilities' (Canadian Standards Association 2003) is the standard that relates to surgical gowns and drapes (Table 16.2). This standard creates four classes of reusable textiles, as defined in this standard. They are described as follows: (i) Class A: a material that restricts airborne microbial penetration, is permeable to the sterilization of choice, and demonstrates little or no water resistance; (ii) Class B: a material that restricts water absorption, is permeable to the sterilization of choice and meets the requirements of surface wetting test; (iii) Class C: a material that restricts visible liquid penetration, is permeable to the sterilization of choice and meets the requirements of the synthetic blood penetration test in ASTM F 1670-03; and (iv) Class D: a material that restricts viral penetration, is permeable to the sterilization of choice and meets the requirements of the Phi-X174 bacteriophage penetration test in ASTM F 1671-03 (Canadian Standards Association 2003).

At the time of writing, everyone is carefully watching the standard development activity around the world. Textile and healthcare product manufacturers are actively participating on the committees along with those from the healthcare industry and regulatory agencies. There is ongoing work to develop products that will meet the barrier standards as well as the other critical needs of the healthcare industry (comfort, mobility, flammability) (Table 16.2).

Table 16.2 A summary of Canadian standards and test methods

Characteristics	Test method
Colorfastness	CGSB CAN2-4.2 No. 19.1
Spray resistance	CGDB CAN2-4.2 No. 26.3
High-pressure water resistance	CGSP CAN2-4.2 No. 26.5
Synthetic blood resistance	ASTM 1670-95
Bacteriophage (viral) resistance	ASTM 1671-95
Electrostatic propensity	ASTM 4238
Flammability	CAN/CGSB-4.162 *or equivalent*
Sterilizability	CAN/CSA-Z314.2 *(EtO)*
	CAN/CSA-Z314.3 *(Steam)*

16.7   Related studies

A number of studies have been completed that evaluated the effectiveness of the gown and/or drape as a barrier in the laboratory and the field. There has been limited study of particle and liquid transmission through fabrics.

16.7.1   Laboratory and field studies

During the past 25 years, there have been studies that successfully identified that surgical gowns on the market have a wide range of barrier effectiveness. In these studies, a wide variety of methods has been used to determine barrier effectiveness including visual penetration of blood and/or other fluids, monitoring the occurrence of hospital acquired infections (at the wound site (SSI)), attaching agar plates to the inside/outside of the gown/drape and then evaluating for the presence of microorganism growth as a result of transmission, and standardized and non-standardized laboratory tests. In many of the studies, specific fabric characteristics have not been identified but only general characteristics of the gowns are noted. There have been several excellent complete reviews of surgical gowns and drapes, their use and their development since the turn of the century (Belkin 2002, Rutala *et al.* 2001). Based on the high quality of these reviews, the review here will be brief.

There have been a number of clinical studies addressing the issue of barrier effectiveness of surgical apparel during the past 25 years (Bellchambers *et al.* 1999, Aibibu *et al.* 2003, Birebaum *et al.* 1990, Garibaldi *et al.* 1986, Lankester *et al.* 2002, Moylan *et al.* 1987, Moylan and Kennedy 1980, Pissiotis *et al.* 1997, Quebbeman *et al.* 1992, Renaud 1983, Rush *et al.* 1990, Shadduck *et al.* 1990, Smith and Nichols 1991, Telford and Quebberman 1993). There have also been laboratory studies evaluating the barrier effectiveness of the gowns under various conditions (Flaherty and Wick 1993, Leonas 1997a, Leonas and Jinkins 1998, McCullough 1993, McCullough and Shoenberger 1991). It is important to note that the majority of these studies were conducted over ten years ago and there are currently different products available in the marketplace. There have been several more recently published studies of clinical studies (Ahmad *et al.* 1998, Bellchambers *et al.* 1999, Lankester *et al.* 2002, Pissiotis *et al.* 1997) and evaluation of surgical gown materials in the laboratory (Leonas 1997b, Leonas and Jinkins 1998). Many of the initial studies have been reported in previous reviews of surgical gown performance (Belkin 2002, Rutala and Weber 2001). In general, many showed that differences could be attributed to the gown type by general classification (reusable vs. disposable or single layer vs. reinforced), but little technical information was provided on the specific fabric characteristics. One trend that was reported in several of the studies was the improved performance of disposable products when compared with reusable. However, there was one study that did not indicate a difference in the performance of these classes of gowns (Bellchambers *et al.* 1999). Also, in studies where reinforced gowns were compared with non-reinforced gowns, generally the reinforced gowns had enhanced performance.

Several of the most recent studies have evaluated penetration of blood or fluids through the gowns and results showed penetration occurs frequently. Ahmad *et al.* (1998) evaluated 241 obstetric procedures, involving 1,022 medical personnel. In the gowns examined, 44% of the cases had evidence of penetration. The frequency of gown failure varied with the type of surgical gown used. Quebbeman *et al.* (1992) found that many of the surgical gowns used in operating rooms are not impenetrable to blood or bodily fluids. They observed 234 operations and examined 535 gowns and noted gowns using a single layer or material allowed higher penetration rates than gowns reinforced with a second layer of material or coated with plastic. Pissiotis *et al.* (1997) reported on the blood strikethrough of different surgical gowns. Gowns were divided into 11 areas for assessment and the highest areas of contamination occurred in forearm, chest and abdomen areas. The highest percentages of strikethrough were in the cuff, forearm and thigh. In this study 250 general surgical operations were studied and nine different gowns were evaluated. One of the gowns was a reusable cotton gown, four were single-layer disposable gowns and four were disposable reinforced gowns. For those areas that were contaminated the percentage strike-through rates per gown group were 90% for reusable gowns, 11% for disposable single-layer gowns, and 3% for the disposable reinforced gown group. One recent study reported on the wound infections in coronary artery surgery; one variable was the use of reusable and disposable gowns and the researchers found no advantages in the reduction of postoperative infections when comparing the two systems (Bellchambers *et al.* 1999).

Several recent studies have reported the effects of pre-wetting of the fabric with blood or other liquids and the impact on the barrier properties of the fabric. Flaherty and Wick (1993) sampled the chest and forearm areas of three commercially available surgical gowns, one was classified as a reusable and two were classified as disposables. They reported the resistance to penetration after repeated exposures to ant coagulated or coagulating blood. The reusable gowns were tested as single layers and reinforced. Results showed that in some cases, prolonged contact with blood increases the amount of blood penetrations on application of an external pressure. The authors recommended that because of this, a criterion in the design and selection of gowns should be its ability to resist blood penetrations for prolonged periods of times. More recently, Blom *et al.* (2002) studied the effect of different wetting agents on the strikethrough rate of bacteria through reusable polyester/cotton surgical drapes. Within 30 minutes bacterial strikethrough of dry surgical drapes occurs. Wetting drapes with blood or normal saline enhances the strikethrough rate of bacteria. Wetting drapes with iodine or chlorhexidine diminishes, but does not stop bacterial strike-through.

In reviewing these studies, it is apparent that the conditions of use greatly influence the performance of the gown; however, there is limited information provided concerning specific gown characteristics therefore it is difficult to identify specific gown characteristics that relate to effective barriers. In future studies, additional information must be given on the specific characteristics of the fabric used in the gowns.

16.7.2   Mechanism of transmission

In the design and development of new products that have the potential to provide optimum barrier protection for healthcare workers while addressing the other necessary design features (comfort, movement ability, flame resistance, linting, resistance to tears), it is necessary to determine the transmission mechanism of small bacterial shaped particles through surgical gown fabrics. Where most studies have focused on laboratory evaluations and in-use performance, there have been several studies that have evaluated the transmission mechanism (Aibibu *et al.* 2003, Leonas 1997b, Leonas and Huang 1999). In these studies, surgical gown and drape fabrics have been exposed to solutions of either *E. coli* $Dh\alpha5$ or fluorescent latex microspheres (round in shape and with an average size of 1 μm similar to *S. aureus*) and then the fabrics examined using laser scanning confocal microscopy (LSCM). The solutions have included liquids of varying surface energies to represent those commonly encountered in operating theaters and have been applied at various pressures. The results indicated that neither the surface tension of the solution, nor the pressure influenced the transmission rate of the particle through the fabrics. Solutions of lower surface tensions (i.e. synthetic blood and alcohol) and exposures at higher pressures increased the rate of transmission. These studies also clearly indicated the point of failure in the fabrics when transmission occurred.

16.8   Critical issues today

There are a number of critical issues that face this area of study today; development and implementation of standards and regulations, development and selection of test methods that appropriately test the barrier effectiveness in the healthcare setting and the controversy of using disposable vs. reusable products. The primary purpose of this area of study is to improve safety for healthcare workers and patients who are at risk of exposure to pathogens. As this area of study progresses, the development and implementation of standards and regulations will have a significant impact on this area along with the implementation of new technology into the personal protective apparel market. Test methods used to evaluate the properties of interest is another point of controversy, which was not covered in this chapter. Although a variety of tests have been used to compare barrier qualities of fabrics, there is controversy about which test procedure most closely mimics actual conditions found in the surgical environment (Rutala *et al.* 2001). The controversy of using disposable vs. reusable products has been going on for a number of years. In addition to considering the barrier effectiveness of the product, environmental and cost per use issues are also concerns when the selection is made.

The desired outcome is a product that will provide optimum protection to the healthcare worker and patient in preventing the transmission of microorganisms during surgical procedures and stays in healthcare facilities that is cost effective and meets the additional criteria of the workers and patients. To effectively produce this product, a thorough understanding of many different areas including textiles, microbiology, rituals in the healthcare facility, their interactions and educating workers and patients are of critical importance.

16.9   References

Abreu, M.J., Silva, M.E., Schacher, L. and Adolphe, D. (2003) Designing surgical clothing according to the new technical standards. *International Journal of Clothing Science and Technology*, Vol. 15, No. 1, 69–74.

Ahmad, F.K., Sherman, S.J. and Hagglund, K.H. (1998) The use and failure rates of protective equipment to prevent blood and bodily fluid contamination in the obstetric health care worker. *Obstetrics and Gynecology*, Vol. 92, No. 1, 131–136.

Aibibu, D., Lehmann, B. and Offermann, P. (2003) Image analysis for testing and evaluation of the barrier effect of surgical gowns, *Journal of Textile and Apparel, Technology and Management*, Vol. 3, No. 2.

Association for the Advancement of Medical Instrumentaion. (2004) ANSI/AAMI PB70:2003 – Liquid barrier performance and classification of protective apparel and drapes intended for use in health care facilities. AAMI, Arlington, VA.

Association of periOperative Registered Nurses. (2004) Standards, Recommended Practices and Guidelines, 2004 edition, AORN, Denver CO.

Association of periOperative Registered Nurses. (2003) Recommended Practices for Selection and Use of surgical gowns and Drapes, *AORN Journal*, Vol 77, No. 1, 206–213.

AWMF (Arbeitsgemeinschaft der Wissenschaftlichen Medizinischen Fachgesellschaften) Online Recommendations (1998) OR Clothing and Patient Draping, Hospital Hygiene, mhp-Verlag, 2nd edn, Wiesbaden, p. 50.

Beck W.C. (1963) Justified faith in surgical drapes: A new and safe material for draping. *American Journal of Surgery*, Vol. 105, 560–562.

Beck, W.C. and Carlson, W.W. (1963) Aseptic Barriers. *Archives of Surgery*, Vol. 87, 288–292.

Beck, W.C. and Collette, T.S. (1952) False Faith in the Surgeon's Gown and Surgical Drapes. *American Journal of Surgery*, Vol. 83, 125–126

Belkin, N. (1992) Barrier materials: Their influence on surgical wound infections. *AORN Journal*, Vol. 55, 1521–1528.

Belkin, N. (1993) The challenge of defining the effectiveness of protective aseptic barrier, *Technical Textiles, International*, pp 22–24.

Belkin, N. (2002), Barrier surgical drapes vis à vis Protective Clothing, *The Guthrie Journal*, Vol. 71, No 3, 108–111.

Bellchambers, J., Harris, J.M., Cullinan, P. Gaya, H. and Pepper, J.R. (1999) A prospective study of wound infection incoronary artery surgery, *European Journal of Cardio-thoracic Surgery*, Vol. 15, 45–50.

Bernstein, U. (1996) Microfibre textiles for protective clothing with barrier properties, *Nonwoven Industrial Textiles*, March, pp 9–10, 12.

Birebaum, H.G., Gloriso L., Rosenberger, K.C., Arshad, C. and Edwards, K. (1990) Gowning on postpartum ward fails to decrease colonization in the new born infant. *American Journal of Diseases of Children*, Vol. 144, 1031–1033.

Blom, A.W., Gozzard, C. Heal J., Bowker, K. and Estela C.M. (2002) Bacterial strike-through of re-usable surgical drapes: the effect of different wetting agents. *Journal of Hospital Infection*, Vol 52, 52–55.

Brock, T.D., Madigan, M.T., Martinko, J.M. *et al.* (1994) *Biology of Microorganisms*, Seventh Edition. Prentice Hall, Englewood Cliffs NJ.

Broughton, R.M., Worley, S.D., Unchin, C. *et al.* (1999) Textiles with Antimicrobial Functionality. *Book of Papers, INDA Tec*, pp 20.1–20.12, Available from INDA, Cary, NC.

Bushman, B. (2003) Surgical Textiles: Battle Lines Being Drawn, Reusable Textiles. *Newsletter of the American Resusable Textile Association*, November/December 2003, p. 4.

Cardinal Health, Inc. (2004) Finally, an industry standard that eliminated the guesswork. Cardinal Health, Inc,. Lit. No. CDN01280 (0304/10M).

Centers for Disease Control and Prevention (1985) Recommendations for preventing transmission of infection with human T-Lymphotropic virus type III/ lymphadenopathy associated virus in the workplace. *MMWR Morbidity Mortality Weekly Report*, Vol. 34, 681–95.

Centers for Disease Control and Prevention (1999) Antimicrobial resistance: A growing threat to public health, Division of Healthcare Quality Promotion, Issues in Healthcare Settings, June 1999

Centers for Disease Control and Prevention (2003) Draft, Guideline for Isolation Pre-cautions: Preventing transmission of Infectious Agents in Healthcare Settings 2004, Recommendations of the Healthcare Infection Control Practices Advisory, 2004.

Canadian Standards Association (2003) CSA Z314.10-03, Selection, Use, Maintenance and Laundering or Reusable Textile Wrappers, Surgical Gowns, and Drapes for Health Care Facilities.

Charnley, J. and Eftekhar, N. (1969) Penetration of gown material by organisms by the surgeons body. *Lancet*, Vol. 1, 173.

Closs, S.J. and Tierney, A.J. (1990) Theater gowns: A survey of the extent of user protection (Short Report). *Journal of Hospital Infection*, Vol. 15, 375–378.

Dineen, P. (1969) Penetration of Surgical Draping Material. *Hospitals*, Vol. 43, No. 10, 82, 84–85.

Emori, T.G. and Gaynes, R.P. (1993) An overview of nosocomial infections, including the role of the microbiology laboratory. Clin. Microbial. Rev, Vol. 6, 428–442.

European Committee for Standardisation (2004) List of new approach mandated standarization projects, Directive Area: Medical Devices, CEN Technical Body: Non-active medical devices (CEN/TC 205), downloaded from CEN Management Centre database 7/6/2004.

Garibaldi, R.A., Maglio, S., Lerer, T., Becker, D. and Lyons, R. (1986) Comparison of nonwoven and woven gowns and drape fabric to prevent intraoperative wound contamination and postoperative infection. *American Journal Surgery*, Vol. 152, 505–509.

Flaherty, A.L. and Wick. T.M. (1993) Prolonged contact with blood alters surgical gown permeability, *American Journal of Infection Control*, October, Vol. 21, No. 5, 249–256.

Gupta, B.S. (1988) Effect of Structural Factors on Absorbent Characteristics of Nonwovens. *TAPPI Proceedings*, 1988 Nonwovens Conference, 195–202.

Holtz, T.H. and Wenzel, R.P. (1992) Postdischarge surveillance for nosocomial wound infection: a brief review and commentary. *American Journal of Infection Control*, Vol. 20, 206–213.

Huang, W. and Leonas, K.K. (2000) Evaluation of a one bath process for Imparting repellency and antimicrobial activity to nonwoven surgical gown fabrics, *Textile Research Journal*, Vol. 12, No. 4, 18–21.

Jinkins, R.S. and Leonas, K.K. (1998) Influence of a Polyethylene Glycol Treated Surface, Liquid Barrier and Antibacterial Properties. *Textile Chemist and Colorist*, Vol. 26, No. 12, 25–29.

Lankester, B.J.A., Bartlett, G.E., Garneti, N., Blom, A.W., Bowker, K.E. and Bannister, G.C. (2002) Direct measurement of bacterial penetration through surgical gowns: A new method. *Journal of Hospital Infection*, Vol. 50, 281–285.

Laufman, H., Eudy, W.W., Vandernoot, A.M., Liu, D. and Harris, C.A. (1975) Strikethrough of Moist Contamination by Woven and Nonwoven Surgical Materials. *Annals of Surgery*, Vol. 181, 857–862.

Laufman, H., Belkin, N.L. and Meyer, K.K. (2000) A critical review of a century's progress in surgical apparel: How far have we come?, *Journal of American College of Surgeons*, Vol. 191, No. 5, 554–568.

Leonas, K.K. (1997a) Effect of laundering on the barrier properties of reusable surgical gown fabrics. *American Journal of Infection Control*, 1998, Vol. 26, 495–501.

Leonas, K.K. (1997b) Using laser scanning confocal microscopy to evaluate microorganism transmission through surgical gown fabrics. *Medical Textiles '96 Conference Proceedings*, pp 60–65.

Leonas, K K. and Huang, W. (1999) Transmission of small particles through selected surgical gown fabrics. *International Nonwovens Journal*, Spring, Vol. 8, No, 1, 18–23.

Leonas, K.K. and Jinkins, R.S. (1998) The relationship of selected fabric characteristics and the barrier effectiveness of surgical gown fabrics. *American Journal of Infection Control*, Vol. 25 16–23.

Leonas, K.K. and Miller, E.M. (1990) Transmission of two bacterial species through selected fabrics-nonwoven and woven. *INDA Journal of Nonwovens Research*, Vol. 2, 29–32.

Leonas, K.K., Malkam, S., Huang, W. and Strawn, S. (1996) Study of one-sided PEG application to fabric. *American Dyestuff Reporter*, Vol. 85, No. 8, 41, 42, 44, 45.

Lickfield, D.K. (1997) Medical fabrics and future needs. *International Nonwovens Journal*, Vol. 6, No. 2, 37–41.

Mayberry, P. (1989) OSHA ruling to impact nonwovens. *Nonwovens Industry*, Vol. 20, 24–26.

May-Plumlee, T. and Pittman, A. (2002) Surgical gown requirements capture: A design analyiss case Study, *Journal of Textile and Apparel, Technology and Management*, Vol. 2, Issue 2.

McCullough, E.A. (1993) Methods for determining the barrier efficacy of surgical gowns. *American Journal of Infection Control*, Vol. 21, 368–374.

McCullough, E.A. and Shoenberger, L. (1991) Liquid barreir properties of nine surgical gown fabrics. *INDA Journal of Nonwovens Research*, Vol. 3, 14–20.

Medical Nonwoven Fabrics (1983) OR infection control discussed by American College of surgeons. *Medical Nonwoven Fabrics*, Vol. 4, 3–4.

Molnlycke Health Care Surgical Area, (2003) New EU standards to challenge operating theater dress codes. (2003-04-09).

Moylan, J.A., Fitzpatrick, K.T. and Davenport, K.E. (1987) Reducing wound infections. Improved gown and drape barrier performance. *Archieves of Surgery*, Vol. 122, No. 2.

Moylan, I.A. and Kennedy, B.V. (1980) The importance of gown and drape barriers in the prevention of wound infection. *Surgery, Gynecology and Obstetrics*, Vol. 151, 465–470.

Occupational Safety and Health Administration (1991) Occupational exposure to bloodborne pathogens: Final rule, Federal Register; **56**:64004-64182 (29 CFR Part 1910.1030).

Perencevich, E.N., Sands, K.E., Cosgrove, S.E., Guadagnoli, E., Meara, E. and Platt, R. (2003) Health and economic impact of surgical site infections diagnosed after hospital discharge. Emerg Infect Dis (serial online) 2003 Feb (*date cited*). Available from: URL: http://www.cdc.gov/ncidod/EID/vol9no2/02-0232.htm.

Pissiotis, C.A., Komborozos, V., Papoutsi, C. and Skrekas, G. (1997). Factors that influence the effectiveness of surgical gowns in the operating theater. *European Journal of Surgery*, Vol. 163, No. 8, 597–604.

Quebbeman, E.J., Telford, G.L., Hubbard, S., Wadsworth, K., Hardman, B. and Goodman, H. *et al.* (1992) In-use evalution of surgical gowns. *Surgery, Gynecology and Obstetrics*, Vol. 174, 369–75.

Renaud, M.T. (1983) Effects of discontinuing cover gowns on a postpartal ward upon cord colonization for the newborn. *JOGN Nurs*, Vol. 12, 399–401.

Rush, J., Fiorino-Chiovitti, R., Kaufman, K. and Mitchell, A. (1990) A randomized controlled trial of a nursery ritual, Wearing cover gowns to care for healthy newborns, *Birth*, Vol. 17, 25–30.

Rutala, W.A. and Weber, D.J. (2001) A review of single-use and resuable gowns and drapes in health care. *Infection Control and hospital Epidemiology*, Vol. 22, No. 4, 248–257.

Seaman, R.E. (undated) Properties and Performance of Operation Room Gowns. A report on file with Textile Fibers Department Spunbonded Products Division, E.I. duPont deNemours and Company, Inc., Wilmington, Delaware.

Shadduck, P.D., Tyler, D.S., Lyerly, H.X. *et al.* (1990) Commercially available surgical gowns do not prevent penetration by HIV-1. *Surgery Forum*, Vol. 41, 77–80.

Smith, J.W. and Nichols, R.L. (1991) Barrier efficiency of surgical gowns: Are we really protected from our patients pathogens? *Archives of Surgery*, Vol. 126, 756–763.

Sun, G. and Xu, X. (1998) *Durable and regenerable antibacterial finishing of fabric: Biocidal properties*, Vol. 30, No. 6, pp 26–30.

Telford, G.L. and Quebbeman, E.J. (1993) Assessing the risk of blood exposure in the operating room. *American Journal of Infection Control*, Vol. 21, 351–356.

Whyte, W., Bailey, P.V., Hamblen, D.L., Fisher, W.D. and Kelly, I.G. (1983) A bacteriologically occlusive clothing system for use in the operating room. *The Journal of Bone and Joint Surgery*, Vol. 65B, 502–506.

Wong, E.S. (1999) Surgical site infections. In Mayhall C.G., editor. *Hospital epidemiology and infection control*. 2nd edn. Philadelphia: Lippincott, pp. 189–210.

Woodhead, K., Taylor, E.W., Bannister, G., Chesworth, T., Hoffman, P. and Humphreys, H. (2002) Behaviours and rituals in the operating therater. *Journal of Hospital Infection*, Vol. 81, 241–255.

Textiles for respiratory protection

I KRUCIŃSKA, Technical University of Lodz, Poland

17.1 Introduction

Since prehistoric times, air pollution from dust has been an essential threat to human health. The presence of dust in the air is mainly the result of naturally occurring processes of which soil erosion is a good example. Together with the development of civilisation, the dust content in air can be increasingly attributed to human activity. A hypothesis can be put forward that as early as prehistoric times people protected themselves against breathing in dust. Davies in his work *Air Filtration* (1973) has presented an interesting review of the earliest literature considering problems connected with air pollution in the workplace.

Plinius the Elder, in his monumental work *Natural History*, written around the time of the Roman Empire, stated that the dust he observed had been produced by disintegrating lead carbonate (used for cosmetics) and oxides of lead applied for pigments. For centuries, miners have used special clothes to cover their nose and mouth and protect their respiratory system against dust. Bernardino Ramazzini, who lived around the turn of the 17th century, indicated in his work *De morbis artificum* the need for protection of the respiratory tracts against dust for workers labouring in various professions. In 1814 Brise Fradin developed the first device to provide durable protection of the respiratory tract. It was composed of a container filled with cotton fibres which was connected by a duct with the user's mouth. The first filtration respiratory mask was designed at the beginning of the 19th century with the aim of protecting the users against airborne diseases. At this time, firemen began to use specially designed masks. The first construction of such a mask was primitive: a hose was connected to a leather helmet and supplied with air from ground level. The construction was based on the observation that during fire, fewer toxic substances were found at ground level than at the level of the fireman's mouth. In addition, a layer of fibres protected the lower air inlet. John Tyndall in 1868 designed a mask which consisted of various layers of differentiated structure. A layer of clay separated the first two layers of dry cotton fibres. Between the two next cotton fibre layers was inserted charcoal, and the last two cotton fibre layers were separated by a layer of wool fibres saturated with glycerine. The history of the development of filtration materials over the 19th century has been described in a work elaborated by Feldhaus (1929).

Due to the use of toxic gases for the first time in the First World War the requirements of respiratory filtration changed over the twentieth century. For this reason, in 1914, the development of filtration materials was connected with the absorption of toxic substances, and filters were manufactured with the use of charcoal and fibrous materials. In 1930 there was another discovery, which changed the approach to the design of filtration materials. Hansen, in his filter, applied a mixture of fibres and resin as filtration materials. This caused an electrostatic field to be created inside the material. It was the action of electrostatic forces on dust particles that significantly increased the filtration efficiency of the materials manufactured by drawing and trapping the particles.

The brief historical sketch presented above highlights how textile fibres were one of the original material components used to protect the respiratory tract, and have been applied for as long as the need was recognised. From the beginning they had been used intuitively, without understanding the mechanism of filtration. The first attempts at scientific description of the filtration mechanism were presented by Albrecht (1931), and then by Kaufman (1936). But the beginning of modern filtration theory was documented in the works of Langmuir carried out during the Second World War (1942).

17.2   Filtration theories

The process of particle deposition of the dispersed phase, which takes place in a porous medium, is described by the notion 'filtration'. The dispersed particles may be solids or liquids, whereas the dispersing medium may be gas (mostly air) or liquid. In the case of protecting the respiratory tract, which is the subject of this work, we have to deal in the majority of cases with the problem of aerosol filtration where the solid body or liquid particles are dispersed in air. The porous media used in the filtration process are differentiated by structure, which can be fibrous, granular, and capillary. The considerations presented below are related to fibrous structures.

17.2.1   The velocity field in fibrous structures

Filtration theories are mainly based on considerations of the flow field around a single cylinder, which is placed at a right-angle to the aerosol flow (Fig. 17.1), and on determination of the filtration effectiveness of particles of different sizes on the cylinder. In this connection, a very important problem is the determination of the velocity field around the cylinder. The Navier-Stokes equations of gas motion with various simplifications is used for this calculation. The simplified Navier-Stokes equations may be used if a row of assumptions are

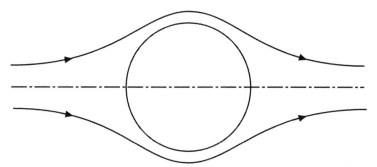

17.1 The flow of fluid around a fibre.

fulfilled. One of them is that the flowing medium is continuous. This assumption can be fulfilled if the Knudsen number expressed as Kn $= \lambda/R$ is less than 0.001; λ is the free path of molecules, and R is the radius of the obstacle. No slippage occurs in this case. This condition is fulfilled for fibres with a diameter equal to or greater than $10\,\mu$m. However, at present the ability exists to manufacture fibres with a diameter below one micrometre, and in this case the laws based on classical hydrodynamics can not be used. The reason for this is the phenomenon of gas particle slippage on the fibre's surface over the medium flow around the submicronal fibre, which causes a lack of continuity of the medium. This phenomenon also occurs during flow at very low pressures. In a case where the Knudsen number is in the range of 0.001–0.25, the calculation should be done according to the laws which arise from low-pressure hydrodynamics, whereas if Kn > 10 the calculation should be based on molecular flow. In this latter case, the obstacles are much less than the free path of fluid molecules, which for air equals $\lambda = 0.065\,\mu$m. The flows which are forced in filtration processes applied for the protection of respiratory tracts can be considered as Stokes' flows. This is characterised by a low Reynold's number; where flow inertia can be neglected and the flow pattern does not change with time. Therefore it has to deal with a steady flow, which means that the speed in each point of the flow field does not change with time.

In early theories of filtration developed by Albrecht (1931) and Kaufman (1936) the airflow around a transverse cylinder with the radius R was considered using equations of the ideal flow of an inviscid fluid. Langmuir presented a new concept of filtration in 1942, and also developed a theory called 'theory of isolated fibre'. In a method based on this theory, the velocity field around an isolated fibre placed at right-angles to the viscous flow of air is used for calculating the filter efficiency resulting from various mechanisms. The influence of adjacent fibres was expressed by introducing empirical coefficients to the equations. This method can be applied only to filters characterised by high porosity.

The relatively simple means of describing the two-dimensional steady-state flow results from a description of the stream function ψ. According to the

Value

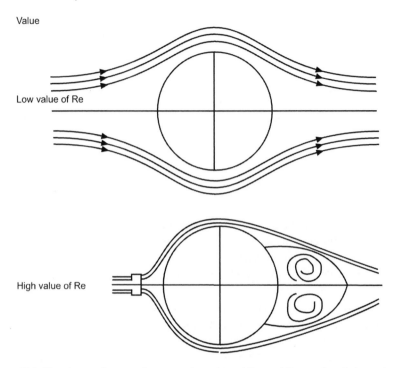

Low value of Re

High value of Re

17.2 The shape of streamlines as a function of Reynold's number (adapted from Kabsch (1992)).

definition given by Brown (1993) the value of ψ defined at any point, is a measure of air flowing between that point and some arbitrary origin where $\psi = 0$. Joining the points at which the stream function ψ takes the constant value creates the streamlines. There is no flow across them. Streamlines illustrate the flow pattern and they show both the directions of the flow and its speed. Closely spaced streamlines indicate rapid flow as shown in Fig. 17.2. The vector of velocity at any point of the flow field is tangential to the streamline at this point. The equations relating the stream function to velocity of fluid can be expressed in Cartesian co-ordinates as:

$$v_x = \frac{\partial \psi}{\partial y}; \quad v_y = -\frac{\partial \psi}{\partial x} \qquad 17.1$$

and in polar co-ordinates as:

$$v_r = \frac{1}{r}\frac{\partial \psi}{\partial \theta}; \quad v_\Theta = -\frac{\partial \psi}{\partial r} \qquad 17.2$$

The pattern of streamlines around the fibre depends on the Reynold's number (Kabsch, 1992, Gradoń and Majchrzycka, 2001). The Reynold's number can be calculated using the following equation:

Table 17.1 Reynold's number for various types of filtering materials

Type of filtering material	Diameter of filtering elements (μm)	Velocity of air (mm/s)	Reynold's number
Electrospun fibres	0.400–1	1–10	$0.264 \times 10^{-4} - 0.66 \times 10^{-3}$
Melt-blown fibres	1–5	1–10	$0.66 \times 10^{-4} - 3.3 \times 10^{-3}$
Classical fibres	10–100	1–10	$0.66 \times 10^{-3} - 0.66 \times 10^{-1}$

$$\text{Re} = \frac{2vR}{\eta} \rho \qquad 17.3$$

The analysis of eqn (17.3) indicates that the Reynold's number depends on the properties of the flowing fluid and also on the structure of the filter, mainly on the diameter of the obstacle. Examples of Reynold's number values, calculated taking that the density of air is approximately $1.20 \, \text{kg/m}^3$ and dynamic viscosity is $1.81 \times 10^{-5} \, \text{kg/ms}$, are given in Table 17.1.

In the literature various equations are given defining the stream function ψ for different filtering structures (Kabsch, 1992). A velocity field of Stokes' flow of viscous incompressible fluid round an isolated transverse cylinder, the model of isolated fibre, is described by Lamb equations as:

$$\psi = \frac{vr}{2(2 - \ln \text{Re})} \left[2\ln\left(\frac{r}{R}\right) - 1 + \left(\frac{R}{r}\right)^2 \right] \sin \theta$$

$$v_r = \frac{v}{2(2 - \ln \text{Re})} \left[-1 + \left(\frac{R}{r}\right)^2 + 2\ln\left(\frac{r}{R}\right) \right] \cos \theta \qquad 17.4$$

$$v_\theta = \frac{v}{2(2 - \ln \text{Re})} \left[2\ln\left(\frac{r}{R}\right) + 1 - \left(\frac{R}{r}\right)^2 \right] \sin \theta$$

Another way of expressing the boundary conditions and solving the Navier-Stokes equations for flow transverse to a set of parallel cylinders was used by Kuwabara (1959).

Kuwabara solved the flow equations of fluid mechanics in two dimensions to present the distribution of velocity around a cylindrical model of a fibre, placed transverse to the flow and surrounded by imaginary coaxial cylinders of radius b (see Fig. 17.3). The cylinder was one of a parallel set of randomly arranged cylinders, for which the volume fraction was c. If there are n parallel fibres per unit volume of the filter, the volume fraction or porosity is $c = n\pi R^2$. The following expressions result from Kuwabara's solution for the stream function ψ, and the radial and tangential velocities v_r and v_θ:

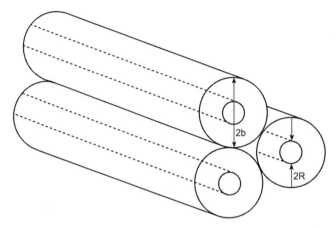

17.3 The Kuwabara concept of flow cells in a filter consisting of parallel fibres surrounded by imaginary coaxial cylinders of radius b, spaced randomly and transversal to the flow.

$$\psi = \frac{vr}{2\mathrm{Ku}}\left\{2\ln\frac{r}{R} - 1 + c + \frac{R^2}{r^2}\left(1 - \frac{c}{2}\right) - \frac{c}{2}\frac{r^2}{R^2}\right\}\sin\theta$$

$$\frac{1}{r}\frac{\partial\psi}{\partial\theta} = v_r = \frac{v}{2\mathrm{Ku}}\left\{2\ln\frac{r}{R} - 1 + c + \frac{R^2}{r^2}\left(1 - \frac{c}{2}\right) - \frac{c}{2}\frac{r^2}{R^2}\right\}\cos\theta \qquad 17.5$$

$$-\frac{\partial\psi}{\partial r} = v_\theta = \frac{v}{2\mathrm{Ku}}\left\{2\ln\frac{r}{R} + 1 + c - \frac{R^2}{r^2}\left(1 - \frac{c}{2}\right) - \frac{3c}{2}\frac{r^2}{R^2}\right\}\sin\theta$$

where $\mathrm{Ku} = -\frac{1}{2}\ln c - \frac{3}{4} + c - \frac{c^2}{4}$. These equations give $\psi = v_r = v_\theta = 0$ on the surface of the fibre where $r = R$.

Some limitation exists concerning the application of eqn 17.5 for the description of the velocity field of airflow. Kuwabara's theory is based on viscous flow hydrodynamics with fluid inertia negligible at all parts of the flow field. Therefore the Reynold's number of flow past the fibre should be small. According to Davies (1973), the Kuwabara theory can be used to describe the filter efficiency for materials composed of fibres which guarantee the fulfilment of the condition Re < 0.5 for each term.

17.2.2 Filtration efficiency

The single fibre is the basic element of fibrous materials assigned for filtration. A layer arrangement of these single fibres is used in the final product for respiratory protection. For this reason, the explanation of the filtration mechanism should begin with an analysis of the deposition of the particles of the dispersed phase on a single fibre. The efficiency of aerosol filtration by a

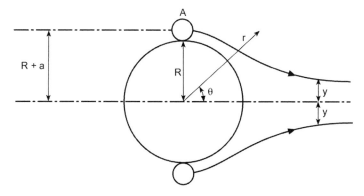

17.4 Definition of the single fibre efficiency (adapted from Davies (1973)).

single fibre is defined as the quotient of the cross-section area of this part of the medium stream from which all aerosol particles are separated when the fibre cross-section area is positioned perpendicular to the direction of flow (Davies, 1973). This stream from which all particles are separated by the fibre is limited by the boundary streamlines, from which one is shown in Fig. 17.4. The single fibre efficiency has a value determined by:

$$E_s = \frac{y}{R} \qquad\qquad 17.6$$

and after implementation the relation between the y component and the stream function far from an obstacle as $\psi = vy$, eqn 17.6 takes the form

$$E_s = \frac{\psi}{vR} \qquad\qquad 17.7$$

From the efficiency of aerosol deposition on a single fibre we may come to the efficiency of a layer with the thickness h, composed from parallel, cylindrical fibres in the following manner (Davies 1973). Consider an element δx of a filter of a unit cross-stream area at the distance x from the face of the filter as is shown in Fig. 17.5. The length of fibres in an element of thickness δx is equal to $\delta x L$, where L is the total length of all fibres composing the unite volume of a filter. The rate of a filtered amount of air per unit time is equal to V:

$$V = 2yvL\delta x \qquad\qquad 17.8$$

Mean velocity of fluid inside the filter v can be expressed as:

$$v = \frac{W}{1 - c} \qquad\qquad 17.9$$

where c is a volume fraction of fibres, $c = \pi R^2 L$ and $W = Q/A_f$ is the face velocity of fluid or flow per unit area of the filter. The changes of the particle concentration can be determined using the formula:

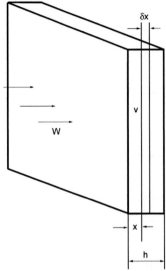

17.5 Element δx of a filter of thickness h. (adapted from Davies (1973)).

$$-\frac{dN}{N} = \frac{V}{W} = \frac{2yvL\delta x}{v(1-c)} \qquad 17.10$$

Introducing $L = c/(\pi R^2)$ and $y/R = E_s$, we have:

$$-\frac{dN}{N} = \frac{2y\delta xc}{\pi R^2(1-c)} = \frac{2E_s c\delta x}{\pi R(1-c)} \qquad 17.11$$

By integrating the last equation we get the value of penetration of particles of a filter of thickness h:

$$P = \frac{N}{N_0} = \exp\left[-\frac{2E_s ch}{\pi R(1-c)}\right] = \exp(-\alpha) \qquad 17.12$$

The filter efficiency is expressed as:

$$E = \frac{N_0 - N}{N_o} = 1 - \exp(-\alpha) \qquad 17.13$$

Equations 17.12 and 17.13 indicate that the most important issue in evaluation of the efficiency of filters is the determination of single fibre efficiency.

17.2.3   Mechanisms of single fibre filtration

Filtration phenomena result from the influence of the mechanical and electrical forces on airborne particles. Mechanical capture is realised without the influence of the attractive forces between the airborne particles and the filter fibres. Brown (1993) characterised the four basic physical phenomena of mechanical deposition in the following way:

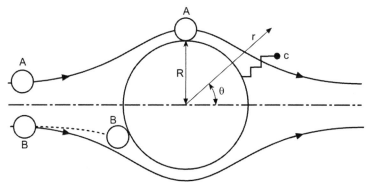

17.6 Particle capture mechanism: A – particle captured by interception; B – particle captured by inertial impaction; C – particle captured by diffusive deposition (adapted from Brown (1993)).

- direct interception occurs when a particle follows a streamline and is captured as a result of coming into contact with the fibre;
- inertial impaction is realised when the deposition is affected by the deviation of a particle from the streamline caused by its own inertia;
- in diffusive deposition, the combined action of airflow and Brownian motion brings a particle into contact with the fibre;
- gravitational settling resulting from gravitational forces.

Illustrations of the above mechanisms of filtration are presented in Fig. 17.6.

Particle capture by direct interception

According to Brown (1993) direct interception occurs when airborne particles are subject neither to inertial effects nor to diffusive motion, and they are not acted upon by any external forces, including gravity. In this situation, particles carried by air would follow a streamline as illustrated in Fig. 17.6. The efficiency of interception can be expressed by eqn 17.7. The value of ψ can be determined by choosing a streamline which approaches within a distance of the surface of the fibre, when $\theta = \pi/2$. Hence, substituting $r = R + a$ in eqn 17.5 we have:

$$\psi = \frac{vR}{2Ku}\left\{2\frac{a+R}{R}\ln\frac{a+R}{R} - \frac{a+R}{R}(1-c) + \frac{R}{a+R}\left(1-\frac{c}{2}\right) - \frac{c}{2}\left(\frac{a+R}{R}\right)^3\right\}$$

17.14

and the effectiveness can be expressed as:

$$E_R = \frac{1}{2Ku}\left\{2\frac{a+R}{R}\ln\frac{a+R}{R} - \frac{a+R}{R}(1-c) + \frac{R}{a+R}\left(1-\frac{c}{2}\right) - \frac{c}{2}\left(\frac{a+R}{R}\right)^3\right\}$$

17.15

Particle capture by inertial impaction

Any convergence, divergence or curvature of streamlines involves the acceleration of air, and due to this phenomenon a particle may not be able to follow the
airflow, as presented in Fig. 17.6. The lines which the particles follow depend
upon its mass or inertia, and upon the Stokes' drag exerted by air. The motion of
particles in two-dimensional space can be described by the following equation:

$$m\frac{du}{dt} = 3\pi\eta d_p(v - u) \qquad 17.16$$

The analytical solution of eqn 17.16 is difficult and very often numerical
methods are applied. During determination of the filtration efficiency of a single
fibre due to inertia, the Stokes' number is used expressing the ratio of the
particle inertia and the resistance force of air related to the diameter of d_f
according to the equation:

$$St = \frac{mv}{3\pi\eta d_p d_f} \qquad 17.17$$

Taking into consideration that $m = \rho_p\pi d_p^3/6$, eqn 17.17 can be expressed as:

$$St = \frac{\rho_p v d_p^2}{18\eta d_f} \qquad 17.18$$

The filter efficiency for a single fibre under the common influence of the
mechanism of direct interception and inertia was defined by Stechkina *et al.*
(1969) using the Kuwabara theory with the following formula:

$$E_{IR} = E_r + \frac{JSt}{2Ku} \qquad 17.19$$

The constant J, which is obtained by numerical integration, can be fitted by the
following analytical expression,

$$J = (29.6 - 28c^{0.62})N_R - 27.5N_R^{2.8} \qquad 17.20$$

provided that $0.01 < N_R < 0.4$ and $0.0035 < c < 0.111$, for the Kuwabara
model, where $N_R = d_p/d_f$.

Diffusive motion of particles

According to the theory delivered by Einstein (1905) the motion of particles can
be quantified by a coefficient of diffusion D. D is defined in terms of the average
displacement of a particle from its original position, say for x-direction,
according to the formula:

$$\bar{x^2} = 2Dt \qquad 17.21$$

Einstein found that the coefficient of diffusion of the particles is dependent on
their mobility according to the law:

$$D = \mu k_B T \qquad 17.22$$

where

$$\mu = \frac{C_n}{3\pi \eta d_p} \qquad 17.23$$

In the above equation C_n is the Cunningham factor resulting from the aerodynamic slip at the particle surface, k_B is Boltzmann's constant, and T is absolute temperature. The Cunningham factor for air is defined by:

$$C_n = 1 + \frac{2A\lambda}{d_P} + \frac{2Q_1\lambda}{d_p}\exp - \frac{Bd_p}{2\lambda} \qquad 17.24$$

where λ is the mean free path of molecules and the appropriate constants take the values as $A = 1.246$, $Q_1 = 0.42$, $B = 0.87$. The capture of particles by a fibre will depend on the relative magnitude of diffusive and convective motion of air past the fibre. The dimensionless parameter concerning these actions is the Peclet number Pe:

$$Pe = \frac{2vR}{D} \qquad 17.25$$

The Peclet number compares the transport of particles by fluid motions forced by convection with the transport by particle diffusion. Stechkina and Fuchs (1968) determined the filter efficiency for Kuwabara flow field due to diffusion mechanism according to eqn 17.26 and due to common influence of diffusion and interception in the form of eqn 17.27:

$$E_D = 2.9\mathrm{Ku}^{-1/3}Pe^{-2/3} + 0.62Pe^{-1} \qquad 17.26$$

$$E_{DR} = 1.24\mathrm{Ku}^{-1/2}Pe^{-1/2}\left(\frac{a}{R}\right)^{2/3} \qquad 17.27$$

According to Davies (1973) eqns 17.26 and 17.27 should be valid for $Pe > 100$. A limit of a/R must be also valid, and Stechkina and Fuchs (1968) defined it as less than 0.5.

Effect of gravitation

The gravitational settlement of particles upon the fibres of a filter was discussed by Stechkina *et al.* (1969). It was shown that the single fibre efficiencies due to gravitational deposition were:

$$E_G^\uparrow = -\left(\frac{a+R}{R}\right)\frac{v_g}{v} \qquad \text{for up-flow} \qquad 17.28$$

$$E_G^\downarrow = \left(\frac{a+R}{R}\right)\frac{v_g}{v} \qquad \text{for down-flow} \qquad 17.29$$

where v_g is the rate of fall of the particles due to gravity. The gravitational settlement is demonstrated at very low velocities of air flow.

Electrostatic interaction

The fibrous filtration material and the aerosol particles are often carriers of particular electric charges, which can cause an increase in filtration efficiency. As a result of mutual displacements of the particles of the dispersed phase in relation to the filtration medium, a frequent charge excitation occurs by means of induction, or as the result of action of the triboelectric effect. It was demonstrated that electrostatic charges of 1200 V can be created by the triboelectric effect when passing a gas stream, with a velocity in the range from 140 cm/s to 170 cm/s, through a nonwoven layer manufactured from fibres with high resistivity. In accordance with the Cohen law, if two different surfaces come into contact, the surface of the fibre with higher dielectrical permeability is charged positive, whereas the surface of the fibre with lower permeability receives a negative charge. In this case, while two surfaces of the same body come into contact, the charges are not generated so long as the surfaces are not displaced in relation to each other. During the mutual displacement of such surfaces, charges are generated as the result of friction. The solid aerosol particles, which are in the flowing air stream, may receive an electric charge as the result of diffusion charging, and charging in the electric field. This occurs as the result of aerosol particles colliding with ions which move under the action of electrostatic forces. During diffusion charging the ions, which are in the gas, depose themselves on aerosol particles by means of diffusion. Many approaches have been developed that determine a particle capture mechanism due to the action of electrostatic forces (Pitch, 1966; Stenhouse, 1974; Brown, 1993).

Two essential cases can be distinguished while considering the phenomena which precede electrostatic deposition of the dispersed phase particles on fibres. The first case takes place if the deposition occurs as the effect of interaction between a charged particle and a charged fibre. For this case the single fibre efficiency depends on the parameter defined by Brown (1993) as:

$$N_{qQ} = \frac{qQ_f}{3\pi^2 \eta d_p d_f v \epsilon_0} \qquad 17.30$$

The second case takes place when an electrical field developed by fibres influences a particle that has no charge of its own. In this case the material of the particle will be polarised by the electrical field. Efficiency of capture of particles by polarisation forces was defined by Brown (1993) as:

$$N_{Qo} = \left(\frac{D_1 - 1}{D_1 + 2}\right) \frac{d_p^2 Q_f^2}{3\pi^2 \eta \epsilon_o d_f^3 v} \qquad 17.31$$

Conclusions arising from the filtration theory review

The analysis of the equations determining filter efficiency indicates that the filtering properties of the material depend on the structure of the filtering material, properties of dispersed phase and properties of disperse medium. The following parameters belonging to these three specified groups have an important influence on the filtering efficiency:

- thickness of filter, diameter of fibres, porosity of filter and amount of electric charge carried by fibres
- dimensions of particles, amount of electrical charge carried by particles and dielectrical constant of particle
- velocity of air flow, temperature, viscosity of air.

Filter efficiency of textile materials does not depend on the area of filters characterised by uniform structure in the whole volume. Significant influence on the particle penetration is exerted by the filter thickness. According to eqn 17.12 which can be expressed as:

$$P = e^{-kh} \qquad\qquad 17.32$$

penetration decreases exponentially with the filter thickness.

The significance of the fibre diameter on filter efficiency results from the definition of such parameters as: E_D, E_R, E_{IR}, E_{DR}. It means that the mechanical capture of particles by direct interception, inertial impaction and diffusion are dependent on the fibre diameter; the filter efficiency governed by these mechanisms increases with the decrease of the fibre diameter. Identification of these phenomena was the basis for development of new technologies for manufacturing fibres characterised by a very small diameter. Additionally, elaborated equations also indicate that the filtering effect governed by direct interception and internal impaction is more efficient with a decrease in the porosity of the fibrous material.

Capture of particles passing the fibres also depends on the action of electrostatic forces. Therefore a useful way of improving filtration efficiency is the development of an internal electrostatic field. The analysis of eqns 17.30–17.31 indicates that the efficiency of filtration due to the action of electrostatic forces results from the densities of charges carried by fibres and particles and the dimensions of fibres and particles. Additionally, the distribution of charges inside the filtering material plays an important role. The monopolar distribution of charges is more effective than a dipolar distribution, due to the appearance of a greater gradient of the electrostatic field characterised by a higher range of action. In a case of interaction between charged particles and charged fibres the filter efficiency due to Coulombic attraction varies with the diameter of fibres as d_f^{-1} and for polarisation attraction varies as d_f^{-3}.

17.3   Theories describing the breathing resistance

Airflow resistance is the second essential parameter which characterises filtration materials suitable for respiratory protection. The airflow resistance through a filtration material is a parameter of the respiratory tract protection equipment that indicates the degree of disturbance of the user's breathing functions. This is the reason that close attention has been devoted to scientific investigations into the problem of fluid flow through porous materials. Model considerations, which are related to the description of gas flow through porous materials, are firstly based on the model of a capillary porous body. Next, the flow field around a single fibre is considered, followed by analysis of the whole layer represented by various geometric models. The basic equation of fluid flow through a porous medium was formulated by Darcy in 1856, and for one-directional laminar flow can be presented in the form:

$$v = K_1 \frac{\Delta P}{\eta h} \qquad\qquad 17.33$$

Davies (1973) proved that filters which obey Darcy's law must also obey the following equation:

$$\Delta P = \frac{4v\eta f(c)h}{d_f^2} \qquad\qquad 17.34$$

For fibrous beds the functions $f(c)$ were determined by Davies as:

$$f(c) = \kappa \frac{4c^2}{(1-c)^3} \qquad\qquad 17.35$$

where κ is called the Kozeny constant.

The law indicates that the tested filtration material will be characterised by higher air resistance for higher fibre packing in the filter and smaller diameter of the fibres composing the filter. On the other hand, considering the mechanism of the direct interception and inertial impaction, the capture efficiency of the particles depends significantly on the fibre diameter and porosity. The presented analysis indicates that the change of fibre diameter and porosity acts in opposition to both usability features, i.e., penetration and pressure drop of filters designed for protection of the upper respiratory tract. This is the reason that while designing filters the electrostatic mechanisms of filtration besides the mechanical mechanisms should be considered, as these could influence filtration efficiency advantageously, without increasing flow resistance.

17.4   Manufacturing methods of filtration materials used for respiratory tract protection

Nonwovens are the basic material used for purifying air from aerosol impurities in the form of solid or liquid particles. They fulfil a protective function of the respiratory tract thanks to their high filtration efficiency connected with small airflow resistance and large dust volumetric absorbability. Technological knowledge concerned with nonwoven manufacturing methods together with theoretical knowledge related to the phenomena which determine filtration are essential for the successful design of filtration systems.

17.4.1   Classification of nonwoven filtration materials

Nonwovens are the basic material used in the process of aerosol filtration. According to a definition by INDA (Jirsak and Wadsworth, 1999), 'a nonwoven is a sheet, web, or bat of natural and/or man-made fibres or filaments, excluding paper, that have not been converted into yarns, and that are bonded to each other by any of several means.' Nonwovens can be manufactured by means of different technologies, but the kind of raw material is the basic criterion for classification. Considering this criterion, nonwovens can be manufactured from fibres or directly from a polymer. The formation of a web is the first technological process for all nonwovens manufactured from fibres. The web composed of staple fibres can be manufactured by means of dry laying methods and wet laying methods. A further division of the technologies is based on the procedure of connecting fibres in the web. A detailed division of nonwovens manufactured from fibres according to the technologies used is presented in Fig. 17.7.

The procedure of polymer fluidisation is the first basis for dividing the manufacturing technologies of nonwovens obtained directly from a polymer. Polymer in the form of liquid can be obtained by melting or dissolution. Nonwovens can be obtained directly from a polymer fluid by means of various differentiated processes, or indirectly by film formation, which is then split into fibres and used for nonwoven processing. The procedure of nonwoven formation from a polymer fluid is the next dividing criterion for nonwovens applied in aerosol filtration processes. A detailed division of nonwovens directly manufactured from a polymer, and their connection with the technology used is presented in Fig. 17.8.

The nonwoven classification presented above indicates the broad range of possibilities for structure formation of the fibrous materials discussed and for modelling their properties. The need to protect the respiratory tract against particles of submicron sizes forces producers to apply multilayer systems, which can achieve high filtration efficiency, low airflow resistance, and high dust absorption capacity. The structure and design of the multilayer materials depend on their appropriation. Considering the general filter design, the following types can be distinguished:

- filters completed together with the face part (masks, half-masks, quarter-masks, and masks with a mouthpiece)
- filtration half-masks (Fig. 17.9).

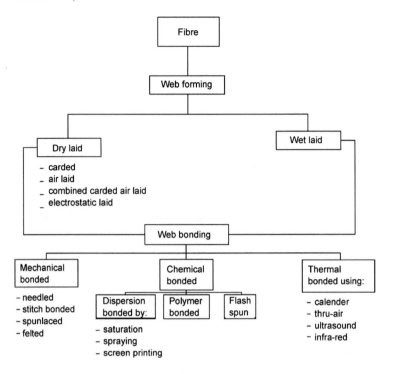

17.7 Classification of nonwovens manufactured from staple fibres.

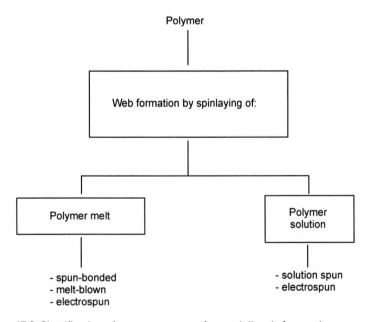

17.8 Classification of nonwovens manufactured directly from polymer.

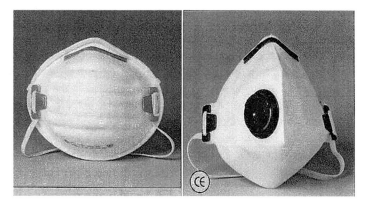

17.9 A view of two types of half mask.

Considering the design of the filter itself, the following filters can be distinguished:

• flat filters in the shape of some layers of filtration nonwovens connected together at the circumference by a welding process
• filters encapsulated in the form of a few layers of filtration nonwovens enclosed in a container with openings allowing for a free air flow
• filters encapsulated with a connector; the filters mainly in the form of a pleated nonwoven enclosed in a casing with an outlet from one side, and the connector from the other side.

The filtration half-masks consist of nonwoven filtration systems connected at the edges, and formed into the shape of a bowl. They shield the nose, mouth, and chin of the user. Often, an expiration valve is mounted with the aim of increasing the user's comfort, which can be seen in Fig. 17.9.

The succeeding nonwoven layers in the filter materials fulfil specified functions dependent on the type of filter. The following nonwovens are used most often:

• Nonwovens manufactured by needle punching which firstly are used as preliminary filtration layers absorbing the coarse-grained aerosol particles by the total layer volume; these nonwovens are applied in filters and half-masks.
• Nonwovens manufactured by the melt-blown technique are used mainly as the basic filtering layer in filters and filtering half-masks.
• Nonwovens manufactured by the paper-making technique which are the basic pleated filtration layer in filters with a connector.
• Nonwovens manufactured by the spun-bonded technique consist of the outer layer of filtration half-masks, which protect the internal layers against mechanical damage.

Recently it has been possible to observe new trends in the development of nonwoven technologies oriented for the manufacturing of the basic filtering layer made from nano-size fibres. Flashspinning and electrospinning are among these technologies. Overviews of flashspinning technologies are presented by Wehman (2004). Flashspun nonwovens made from fibres with very low denier can be obtained using as a raw material the splittable fibres for production of the conventional webs. Subsequently webs can be subjected to the classical needle punching or spunlace process during which sacrificial polymer is removed and the low denier fibres are obtained. The flashspinning process can be accomplished also using bicomponent melt-blown technology when two incompatible polymers are spun together forming the web which is then subjected to the splitting process.

Theoretical consideration also indicates that the activity range of fibres on the dust particles significantly increases if an electrostatic field is formed inside the nonwoven. This is the reason that nonwovens are additionally modified. The three following groups of fibrous electrostatic materials used can be distinguished, based upon their ability to generate an electrostatic charge:

1. Materials in which the charge is generated by corona discharge after fibre or web formation.
2. Materials in which the charge is generated by induction during spinning in an electrostatic field.
3. Materials in which the charge is induced as the result of the triboelectric effect.

17.4.2   Technologies for manufacturing nonwovens for respiratory tract protection

Technology of needled nonwovens

The technology of needle punched nonwovens can be traced as far back as the 1870s. This process was first used for the production of woollen felts. A review included in the work by Lunenschloss and Albrecht (1985) indicates that the first needling machine was manufactured by the Bywater Company in 1889. However, the needling technique was not commonly applied to the manufacture of synthetic and natural fibre nonwovens until the 1950s, when the production of synthetic felts commenced. The technology of needled nonwovens is based on web formation by carding machines and lappers, followed by connecting the fibres together by perforating the web with barbed needles. The use of the needling process results in intermingling the fibres and an increase in the inter-fibre friction caused by compression of the web. A schematic specification of the succeeding technological operations in the process of manufacturing needled nonwovens is presented in Fig. 17.10.

The first stage consists of web formation with the use of a carding machine. Revolving flat cards are used for cotton-type fibres, whereas roller cards are used for wool-type fibres. The web obtained from the carding machine is

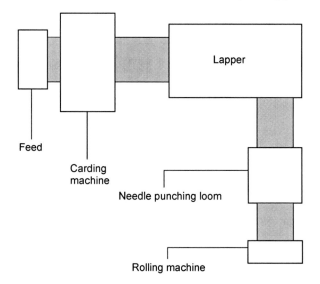

17.10 Scheme of a technological process for manufacturing needled nonwovens.

repeatedly folded in the second stage of the process. It is possible to obtain a web of a predominantly lengthwise, predominantly crosswise or combined arrangement of the fibres. The lengthwise arrangement is characterised by coinciding the majority of the fibres with the longitudinal web axis; crosswise by positioning the fibres at an angle α to this axis, and the combined arrangement is a combination of webs with lengthwise and crosswise arrangements, as illustrated in Fig. 17.11. The needling process is the final stage of nonwoven manufacturing, and it can be differentiated by the design of the needling machine. Needling can be one-sided or double-sided. The needles can penetrate the web or webs perpendicular to the web axis or at an angle of 45°, as illustrated in Fig. 17.12.

According to the theoretical considerations presented above, the following essential parameters influence filtration efficiency: fibre thickness, apparent density, thickness of the filtration layer, and the fibre blend's composition, which is requisite for creation of the phenomenon of electric charge generation as a result of the triboelectric effect. Electric charge generation by means of the triboelectric effect is a technological operation used only for manufacturing filtration materials. The oldest filter material charged by the triboelectrical charge exchange was made from a mixture of wool noils and particles of a natural resin. This material is called Hansen's filter (1931), and was described in detail by Feltham (1979). Wool fibres develop a positive charge during carding, and resin particles develop a negative charge. In the late 1980s, Brown and his co-workers developed a new class of filtering materials called 'blended fibre filter materials' (Brown, 1984, 1992; Smith *et al.*, 1988) which were related to

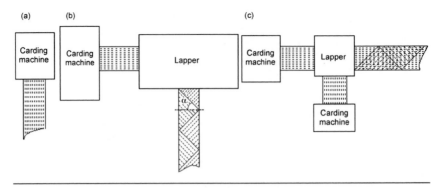

17.11 Scheme of obtaining nonwovens with differentiated kinds of fibre arrangement in the web: (a) lengthwise, (b) crosswise, (c) combined.

the concept of the Hansen's filter. These materials consist of a mixture of two synthetic fibres which are charged by the triboelectrical exchange during carding. One component of the mixture develops a positive charge and the other a negative charge. The mixture invented by Brown (1992) comprises a blend of clean polypropylene fibres, and fibres containing chloride and cyanide groups. The second known mixture developed by the Carl Freudenberg Company in

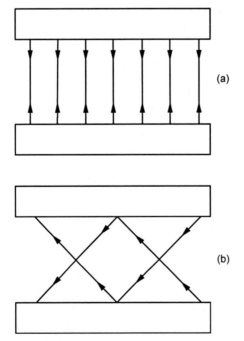

17.12 Arrangement of needles while needling (a) at an angle of 90°, (b) at an angle of 45°.

1988 consists of a multilayer structure of polypropylene and polycarbonate fibres. A successful solution was the application of a mixture of polyamide and polypropylene fibres, as described in the work of Kruciñska *et al.* (1995, 1997). A comparative analysis of filtration efficiency and airflow resistance for nonwovens of various fibre mixtures is presented in Krucinska *et al.* (1997) and the results of the analysis are shown in Table 17.2.

The results listed in the table show that nonwovens manufactured of PP and PA fibres are characterised by the smallest penetration values in the range of 1.19–2.22% (dependent on air humidity). This effect is maintained over at least five months after manufacturing by these nonwovens, which are obtained using the technique discussed. The air flow resistance for the above-mentioned nonwovens is not higher than 51 Pa.

Analysis of the influence of the atmosphere's relative humidity to which all the samples are exposed for 24 hours before testing shows an insignificant increase in sodium chloride penetration for the samples exposed to saturated air (100% r.h.). Precise analysis of the influence of technological parameters of needle punching was presented in a paper by Kruciñska (2002). To select the most important technological parameters influencing the efficiency of developed filter materials, a statistical analysis was completed using Anova and t-Student tests. The results indicate that manufacturing needled nonwoven fabrics using a mixture of fibres without antistatic spin finish results in a filtering material whose properties depend strongly on the correct settings of technological parameters. The most important parameters, which significantly influence the particle capture efficiency of the materials manufactured, are the linear density of both constituent fibres, i.e., PA and PP fibres, the gauge of needle, the depth of needle punching, the mass per unit area of a nonwoven sample, and the surface content of PA fibres. Another way to generate charges on the surface of nonwovens is application of the corona discharge method. A point electrode at high potential emits ions of its own sign, which are influenced by the electrical field drift to the surface of the nonwoven material. The first type of material from this group was made from a corona charged sheet of polypropylene fibres. This material was described in detail by (Turnhout *et al.*, 1976). The second type of electret material is also made from polypropylene fibres, but is charged by corona discharge in felted form (Baumgartner *et al.*, 1985). A corona discharge is often used as a means of applying electric charges to melt blown nonwovens, rather than by needle punching.

Technology of melt-blown nonwovens

Melt-blown nonwovens are one of the basic types of nonwovens assigned for filtration materials. The melt-blown method is based on extruding the elementary fibres, blowing them out of a jet by means of the kinetic force of an air stream, and depositing them after cooling on a perforated band, to form a

Table 17.2 Sodium particle penetration and breathing resistance of the filtering materials tested (source: Krucińska et al., 1997)

Type of mixture	Penetration at 0.1 m/s, %						Breathing resistance at 0.1 l/min, Pa					
	after fabrication			after five months			after fabrication			after five months		
	dry	65%	wet	dry	65%	wet	dry	65%	wet	dry	65%	wet
PP/PCV 30/70 494 g/m²	5.18	4.87	6.24	4.16	4.49	5.6	32	30	28	33	34	30
PP/PA 30/70 523 g/m²	1.19	1.20	1.67	0.79	0.81	1.9	38	38	34	38	39	35
PP/PA 50/50 554 g/m²	1.81	2.04	2.22	1.87	1.48	2.35	45	42	47	48	51	43
PP-PE/ PAN 30/70 546 g/m²	32.8	29.5	36.7	—	—	—	36	40	39	—	—	—

nonwoven composed of super-thin, non-continuous fibres. Fibres in the melt-blown nonwoven are characterised by random arrangement. The nonwoven stability is achieved mainly by the cohesion forces, as the fibres are mutually glued during solidification. Polypropylene (PP) has a low melting point and is the most popular polymer for melt-blown nonwovens. The following polymers can also be used: polyethylene terephthalate (PET), polyamide (PA) and its copolymers, polyethylene (PE), polycarbonates (PCs), simple and complex polyurethanes (PU).

The first nonwovens using melted organic polymers were manufactured in the 1950s, using a method similar to air-blowing the polymer melt. Application of this latter method enabled super-thin fibres to be obtained with a diameter smaller than 5 μm. The melt-blown technique of manufacturing nonwovens from super-fine fibres was developed at the Naval Research Laboratory in USA, and Wente (1956) was the author of the first designs. Buntin, a worker at Exxon Research and Engineering, introduced the melt-blown technique into industry for processing PP (Buntin, 1973). A technological scheme of an installation designed for manufacturing melt-blown nonwovens is presented in Fig. 17.13.

A head (6) with fibration jets (7) is the basic part of the equipment. The compressed air is supplied to the jets through a cut-off valve (10), a control valve (1), a rotameter (2), the valve (11), a cold air collector (3) and an air heater (4). The melted polymer is supplied to the head by the extruder (8). The fibres are blown out and deposited on the take-up equipment (9). Over the take-up equipment the corona discharge device can be installed (12). The compressed air

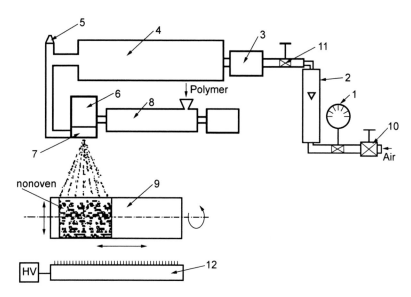

17.13 Scheme of an installation designed for manufacturing melt-blown nonwovens (adapted from Krzyżanowski (1986)).

is heated in the heater (4), and the temperature is maintained by the controller (5). The take-up equipment includes a cylinder, which rotates around its own axis and at the same time moves with a uniform reciprocating motion. The take-up cylinder can move vertically up and down, which enables the distance between the jet and the cylinder to be changed. The fibre stream comes into contact with the cylinder and is rolled on its surface in the form of a band. The degree of cover of one fibre layer by the next layer depends on the rotational speed of the take-up cylinder and its linear velocity. The fibration jet, applied the first time by Wente, had a milled channel through which the melted polymer was forced. Above and below a row of nozzles, which fed the melted polymer, slits were arranged at an angle of 60°, through which heated, compressed air was conducted to blow out the polymer melt into fibres. The fresh-formed fibres were torn off from the jet and were dispersed in the turbulent air stream that was separated from the fibre bundle after passing the net of the take-up equipment. The latter moved with controlled speed crosswise in relation to the fibre bundle. The fleece formed in this way was separated from the net and after bulking was wound onto reels.

The properties of nonwovens manufactured by the melt-blown technique depend on a set of technological parameters which can be related to the functional components which fulfil specified functions in the production process. A scheme showing these components is presented in Fig. 17.14. The components fulfil the following functions:

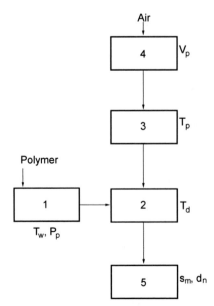

17.14 Block scheme of a nonwoven manufacturing installation by means of the melt-blown technique (adapted from Krzyżanowski (1986)).

1. Extruder – melting and homogenisation of the polymer, heating the melt to the required temperature T_w, and conveying it under control to the fibration jet, P_p
2. Fibration jet – heating the melt to the required temperature T_d and blowing it off into fibres
3. Heater – heating the compressed air to the required temperature T_p
4. Rotameter and manometer – measuring and controlling the output of the blown air, V_p
5. Take-up equipment – formation of a nonwoven with required area mass and compactness controlled by d_n – distance of the nonwoven take-up cylinder from jet and s_m – linear velocity of the take-up cylinder.

The extruding temperature depends on the kind of thermoplastic material used, the nonwoven's required area mass, the preceding thermal processing, and the dwell time of the material in the extruder. Temperatures in the range of 250–450 °C are used for PP. The amount of polymer mass fed by the extruder to the jets depends mainly on the number of polymer channel holes per jet length unit and the length of the jet. The polymer output is commonly chosen so that a flow through a single channel of 0.5–1.5 g/min is achieved. The polymer melt temperature, controlled by a heater placed in the head, is the main process parameter together with the air output. The melt temperature for polypropylene processing is in general controlled in the range of 300–400 °C. The temperature of the air blowing the polymer is generally maintained at the level of the polymer melt temperature. The air output is dependent on the required fibre thickness, mainly in the range from 2–8 m^3/h. The above-mentioned process parameters are connected with energetic factors and generally influence the thickness of the fibres formed. An increase in the polymer melt temperature causes a decrease in its viscosity, and at the same time facilitates the process of melt blowing into thin fibres. Similarly, an increase in the air output causes the lowering of the fibre diameter.

The influence of these parameters on the thickness of the fibres formed is presented in Fig. 17.15 which shows that with all other parameters constant, the increase in melt temperature, as well as the output of air blown out, causes a decrease in fibre diameters. The average fibre diameter can be changed in the range of 1–20 μm. The remaining nonwoven parameters, such as the area mass and the compactness, measured by the value of apparent density, are controlled by changing the working parameters of the take-up equipment. By changing the distance between the jet and the take-up cylinder surface the compactness of the nonwoven can be changed. At a take-up distance of less than 4 cm a nonwoven of the compact paper type is obtained, whereas at a distance of up to 50 cm a layer of loose connected fibres is formed. Naturally, further thickening of the nonwoven structure is possible using a calender. The nonwoven's area mass can be changed by changing the velocity of the take-up cylinder, or by repeatedly

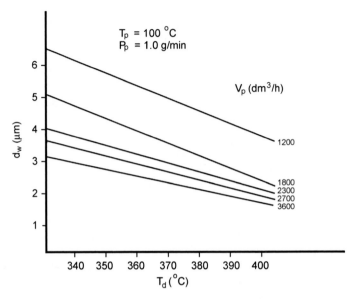

17.15 Dependence of fibre diameter on temperature T_d and output of blown air V_p (adapted from Krzyżanowski (1986)).

passing the take-up cylinder below the fibration jet. It is possible to obtain nonwovens with an area mass in the range from 5–200 g/m^2.

With the aim of increasing the nonwoven filtration efficiency, methods of inducing electrostatic forces between aerosol particles and the nonwoven have been developed similar to those for needled nonwovens. Methods for obtaining electret nonwovens are mostly derived from methods developed for obtaining electret films, and detailed descriptions are presented in the work of Hilczer and Małecki (1992). The technique of charge deposition by corona discharge is the basic activation method for melt-blown nonwovens. According to a method described in the Patent (Exxon Chem., 1989), a high potential point electrode emits ions with a sign characteristic of the electrode, which move under the action of the electrostatic field. If the ions meet an insulated surface they gather on it. Activation by means of corona discharge is realised mainly for a web positioned on the take-up beam. A device consisting of a head including electrodes with a width equal to the nonwoven take-up beam is located directly over the beam, as presented in Fig. 17.13. The effect of improving the filtration properties of melt-blown nonwovens manufactured from propylene after activation can be analysed in Table 17.3.

Technology of nonwovens made by electrospinning

Induction of electric charge is another mechanism used in filtering material technology. Induction consists of electric charge generation in a conductor

Table 17.3 Influence of the activation process on filtration properties of melt-blown nonwovens (source: Gradoń and Makhrzycka, 2001)

Type of nonwoven	Breathing resistance (Pa)	Penetration of sodium chloride (%)	Average area density of charge (nC/m^2)
Nonactivated nonwoven	52.8	45.5	24.6
Activated nonwoven	53.8	1.46	84.6

placed in an electric field. Therefore, fine fibres made from conductive solutions or melts charged during electrostatic extrusion belong to this group. Formation of nanofibres by the electrospinning method results from the reaction of a polymer solution drop subjected to an external electric field. This method enables the manufacture of fibres with transversal dimensions of nanometers. One of the first investigations into the phenomenon of interaction of an electric field with a polymer drop was carried out by Zeleny (1914), whereas Formhals (1934) patented the spinning process in an electrical field. Taylor (1964, 1969) proved that for a given type of liquid, a critical value of the applied voltage exists, at which the liquid drop flowing from the nozzle is transformed into a cone under the influence of the electric field. The electric charges which diffuse in the liquid, forced by the electric field, cause a strong deformation of the liquid surface in order to minimise the system's total energy. The electric forces exceed the forces of surface tension in the regions of the maximum field strength and charge density, and the liquid forms a cone at the nozzle outlet. A thin stream of liquid particles is torn off from the end of the cone.

Taylor's further investigations have been an inspiration for many researchers who carried out observations of the behaviour of different kinds of polymers in the electric field. These were the basis for the development of manufacturing technologies for a new generation of fibres with very small transversal dimensions. Schmidt (1980) demonstrated the possibility of the application of electrospun polycarbonate fibres to enhance dust filtration efficiency. In the 1980s Carl Freudenberg was the first company to use electrospinning technology commercially. Trouilhet (1981) and Weghmann (1982) presented the wide range of applications of electrospun webs especially in filtration area. In the 1980s the electrospinning method for manufacturing of filtering materials did not find common application. The revival of this technology has been observed for the last four or five years. In 2000 Donaldson Inc. USA realised dust filters with a thin layer of nanofibres.

A web of nanofibres can be manufactured using apparatus for the electrospinning technique presented in Fig. 17.16. Two electrodes make up the basic system elements. One is connected to a syringe for extruding the polymer, enabling the polymer drops to achieve a suitable electric potential. The second

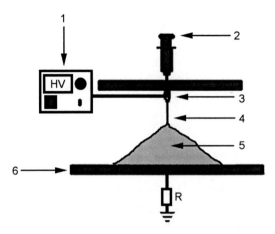

17.16 Scheme of a stand for electrospinning technique: 1 – high voltage generator, 2 – syringe, 3 – Taylor's cone, 4 – jet region, 5 – instability region, 6 – lower electrode.

electrode is the take-up electrode to which the electric potential of the polymer is applied, and on which the fibres are deposited during the manufacturing process. These electrodes can be of different shapes, although we proposed the shape of a flat plate. Both electrodes are mutually insulated. The whole system is insulated from external electric fields by a screen, which serves as a Faraday cage, and additionally isolates against turbulent air. A view of cross-sections of fibres and electrospun nonwovens obtained by the electrospinning method is illustrated in Figs 17.17 and 17.18. By using the electrospinning method it is possible to

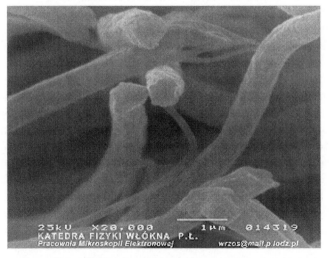

17.17 View of PAN fibres made by the electrospinning method (acc. to Klata *et al.* (2004)).

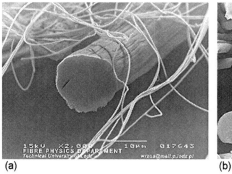 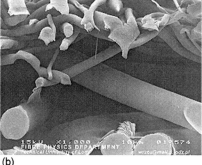

(a)   (b)

17.18 View of the two-layer filtering material; (a) needled punching/
electrospun, (b) needled punching/melt-blown.

Table 17.4 Properties of pleated filter material composed from melt-blown/
electrospun/spun-bonded nonwoven layers

Type of nonwoven	Breathing resistance (Pa)	Penetration of sodium chloride (%)
Filter composed of layer made by electrospinning method	12.75	15.05
Filter without layer made by electrospinning method	4.5	42.4

produce multilayer forms of filtering materials. Examples of filters manufactured by the combination of melt-blown, spun-bonded and electrospinning
techniques are presented in a paper by Hruza (2004). The great enhancement of
filter efficiency using this method is shown in Table 17.4 (Hruza, 2004).

17.5   Assessment of filter materials used for protection of the respiratory tract

The harmonised law governing consumer protection in workplace conditions
hazardous to human health is controlled by the New Approach Directive
introduced in1985 by the European Union. The 89/686/EEC Directive related to
a wide range of personal protective equipment (PPE) issued by the European
Council and dated 21st December 1989, was one of the first New Approach
Directives. The issue of such a document indicates that the unconstrained flow
of products destined for the protection of the respiratory tract against aerosols is
controlled by certification in the European Union by the decisions of the 89/686/
EEC Direction. The Direction dictates that certification for the above-mentioned
products must be achieved by meeting particular harmonised standards
established by CEN and CENELEC. At present two standards, the EN143

(2004) and EN149 (2001) are valid in the European Union. These standards are concerned with filters and filtrating half-masks for protecting against aerosol particles. These standards are voluminous documents, which describe a full set of measuring methods, among which the following methods estimating the effectiveness of protection against aerosol particles are the most important:

- methods for assessment of the test-aerosol penetration
- method for assessment of the total internal leakage
- method for assessment of the air flow resistance
- measuring method for dust absorbability.

Filters are classified according to their filtration efficiency, whereas filtration half masks are classified additionally according to their total internal leakage. The remaining parameters are defined by additional requirements.

17.5.1   Penetration of the test aerosol

The penetration of aerosol particles through the filtration material is measured by the ratio of the aerosol particle concentration after and before passing the filter. Sodium chloride is used which represents an aerosol with dispersed solid phase. The test with oil mist is a tool for estimation of filtration efficiency against aerosols with dispersed fluid phase. A form of the dispersed medium with known particle sizes has been chosen to estimate the test-aerosol penetration. An aerosol of sodium chloride particles is generated by atomising a 1% aqueous solution of the NaCl salt and evaporating the water. The aerosol produced by this method is polydisperse with a mass mean particle diameter of approximately $0.6\,\mu$m. An aerosol of paraffin oil droplets is generated by atomising heated paraffin oil which results in a median Stoke's particle diameter value equal to $0.4\,\mu$m.

The principle of the sodium chloride aerosol penetration test consists of passing the sodium chloride aerosol through a hydrogen flame before and after its flow through the filter. Sodium chloride particles in air passing through the flame tube are vaporised giving the characteristic sodium emission at 589 nm. The intensity of this emission is proportional to the concentration of sodium in the airflow. The aerosol is drawn through the filter under test with the rate of 95 l/min. A scheme of the measuring equipment is shown in Fig. 17.19.

The measurement of the penetration of an aerosol with dispersed fluid phase is based on atomisation of paraffin oil heated to 100 °C. The oil mist obtained passes the filter and is partly deposited on the fibres. The principle of estimation of the concentration of oil drops in air before and after the filter is based on the assessment of laser light radiation by a photosensitive element. The higher the concentration of oil drops, the higher the dispersion observed. The aerosol with the dispersed fluid phase is led through the filter with a velocity of 95 l/min similar to the foregoing described test. The percentage of aerosol particles that passed the filter, in relation to the total amount of particles fed to the tested

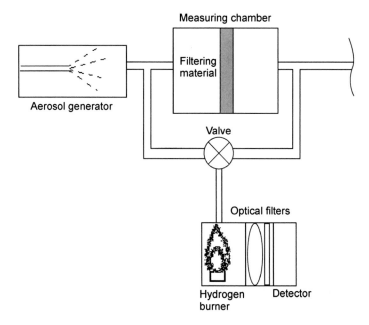

17.19 Scheme of the measuring stand for assessment of penetration with the use of sodium chloride.

object, is the criterion of assigning the filtration materials and filtration half-masks to the particular classes. The requirements, expressed in terms of filter efficiency, for filters assigned to the particular classes in accordance with EN 143 and for half-masks described by EN 149 are presented in Table 17.5.

Table 17.5 The filter efficiency for the particular filter and half-masks protection classes at constant flow of the test-aerosol with a velocity of 95 l/min

Filter class	Filter efficiency (%)	
	Sodium chloride	Paraffin oil mist
P1	80	80
P2	94	94
P3	99.95	99.95

Half-mask class	Filter efficiency (%)	
	Sodium chloride	Paraffin oil mist
FFP1	80	80
FFP2	94	94
FFP3	99	99

17.5.2   Breathing resistance

Breathing resistance means the resistance which the respiratory protective equipment, or its elements, cause to the airflow through them. Breathing resistance is a parameter that determines the usability features of the respiratory tract protecting equipment. It represents the users ability to perform the correct physiological breathing functions while using the filter or the filtration half-mask. The measuring principle is based on passing through the test object air at room temperature, at atmospheric pressure, and at a humidity which does not cause condensation. The air is passed through the filter at a rate of 30 and 95 l/min, and the pressure drop after the filter is measured in relation to the atmospheric pressure. The acceptable air flows to meet the tests correspond with minute lung ventilation during light and hard work, and are related to the inspiration phase. A breathing resistance measurement over the expiration phase is also carried out additionally for filtration half-masks at an air flow of 160 l/min. A special device called the Sheffield head is used for this test (EN 149, 2001); a model of the head is shown in Fig. 17.20. The breathing resistance is a further criterion for dividing filtration materials and filtration half-masks into classes. The requirements for filters of different classes in accordance with the EN 143 standard are presented in Table 17.6, whereas for half-masks in accordance with the EN 149 standard are shown in Table 17.7.

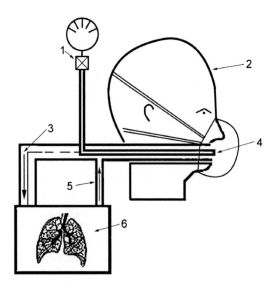

17.20 Scheme of a Sheffield head model: 1 – manometer, 2 – model of the head, 3 – channel to artificial lung, 4 – connector for pressure measurement, 5 – channel from artificial lung, 6 – artificial lung.

Table 17.6 The maximum permissible breathing resistance values for protective filters of different classes

Filter class	Breathing resistance (Pa)	
	30 l/min	95 l/min
P1	60	210
P2	70	240
P3	120	420

Table 17.7 The maximum permissible breathing resistance values for filtration half-masks of different classes

Half mask class	Breathing resistance (Pa)		
	Inspiration 30 l/min	Inspiration 95 l/min	Expiration 160 l/min
FFP1	60	210	300
FFP2	70	240	300
FFP3	100	300	300

17.5.3 Clogging

Clogging, also called dust absorptivity, determines the airflow resistance through the filtration material after treating the filter with 1.5 g of dust. The determined breathing resistance, which is the criterion for dividing the filtration materials into classes, should not be exceeded after the dust has been added. The maximum permissible flow resistance values after clogging with dust for filters and filtration half-masks are listed correspondingly in Table 17.8. A device for carrying out the tests mentioned consists of the following four basic modules. A

Table 17.8 The maximum permissible flow resistance values for the particular classes of protective filtration half-masks and filters after cladding them with 1.5 g of dolomite dust

Class	Breathing resistance (Pa)				
	Filtration half-masks without valves		Filtration half-masks with valves		Filters
	Inspiration 95 l/min	Expiration 95 l/min	Inspiration 95 l/min	Expiration 160 l/min	95 l/min
FFP1 or P1	300	300	400	300	400
FFP2 or P2	400	400	500	300	500
FFP3 or P3	500	500	700	300	700

dust feeder and injector is the first module. Its aim is to feed an appropriate amount of dust over a given time into a chamber in order to obtain a concentration of $400 \pm 100 \, \text{mg/m}^3$. The second module is a system that forces a determined continuous airflow through the dust chamber and filter, at a rate of $60 \, \text{m}^3/\text{h}$ and $95 \, \text{l/min}$ correspondingly. In a case of filtration half masks the artificial lung is used, simulating the inspiration and expiration phase with different airflows. The third module is a holder for fastening the test material and connecting it with a device for measuring the pressure drop as the result of air flow through the porous barrier consisting of the filter. The Sheffield head connected with an artificial lung is the holder for filtration half-masks. The fourth module is the measuring system for airflow parameters.

In the case of filters, the tests are conducted with air at a temperature of $23 \pm 2 \, ^\circ\text{C}$ and relative humidity of $45 \pm 15\%$, whereas in the case of filtration half-masks the following air parameters should be obtained:

- Over the inspiration phase: air temperature of $23 \pm 2 \, ^\circ\text{C}$, and relative humidity of air $45 \pm 15\%$.
- Over the expiration phase: air temperature of $37 \pm 2 \, ^\circ\text{C}$, and relative humidity of air minimum 95%.

The DRB 4/15 dolomite standard dust of exact determined particle size distribution must be used for the tests.

17.5.4   Total internal leakage

An important measure of the performance of the filtration half-masks is their total internal leakage. It examines the object tested under real-use conditions. During the test, the human test participant with filtration half-mask set on the face marches on a treadmill under different exercise conditions. The treadmill is situated in a measuring chamber with sodium chloride aerosol flowing through it. Air is sampled from under the mask's facepart during the inspiration phase, with the aim of assessing the sodium chloride content directly after transmission through the filtration material. The sodium chloride measurement is based on flame photometry, similar to that used in assessing the filtration efficiency of the solid dispersed phase. The average sodium chloride concentration delivered to chamber should be into the range of $8 \pm 4 \, \text{mg/m}^3$. The criterion for dividing the filtration half- masks into classes is the value of total internal leakage determined as:

$$\text{TIL} = \frac{C_2}{C_1} \left(\frac{t_{in} + t_{ex}}{t_{in}} \right) 100\% \qquad\qquad 17.36$$

where C_2 is the concentration of sodium chloride in the air under the mask's facepart, C_1 is the reference concentration of sodium chloride, t_{in} and t_{ex} are respectively time of inspiration phase and time of expiration phase.

The total internal leakage TIL of 46 test results per 50 results of individual exercises should not be greater than 25% for the FFP1 class, 11% for the FFP2 class, and 5% for the FFP3 class. Additionally, for at least eight out of ten average test results for each participant should not be greater than 22% for the FFP1 class, 8% for the FFP2 class, and 2% for the FFP3 class.

17.6   References

Albrecht F (1931), Theoretische Untersuchungen über die Ablegerung von Staub und Luft und ihre Anwendung auf die Theorie der Staubfilter, *Physik, Zeits*, **32**, 48.

Baumgartner H, Loeffler F, Umhauer M (1985), Deep-bed electret filters: the determination of single fibre charge and collection efficiency, *IEEE Transactions E1-21*, No.3, 477.

Brown R C (1984), Air filters made from mixtures of electrically charged fibres, in: *Symposium on Electrical and Magnetic Forces in Filtration*, KVIV, Antwerp, Belgium.

Brown R C (1992), Blended-fibre filter material, *European Patent* EP-B1 0246811.

Brown R C (1993), *Air filtration. An Integrated Approach to the Theory and Applications of Fibrous Filters*, Pergamon Press.

Buntin R R (1973), Melt-blowing a one step web process for new nonwoven products, *Tappi*, **56**, 74.

Davies C N (1973), *Air Filtration*, Academic Press, London.

Einstein A (1905), On the movement of small particles suspended in a stationary liquid demanded by the molecular-kinetic theory of heat, *Ann. Phys.*, **17**, 549.

European Standard EN 143 (2004), Respiratory protective devices – Particle filters – Requirements, testing, marking.

European Standard EN 149 (2001), Respiratory protective devices – Filtering half masks to protect against particles – Requirements, testing, marking.

Exxon Chem. (1989), *European Patent*, EP 0 363 033 A2.

Feldhaus G M (1929), Schutzmasken in vergangenen Jahrhunderten, *Die Gasmaske*, **1**, 104.

Feltham J (1979), The Hansen Filter, *Filtration and Separation*, **16**, 370.

Formhals A (1934), Process and apparatus for preparing artificial threads. *US Patent*, No.1 975 504.

Freudenberg Company (1988), *European Patent*, EP 0 312 687 A2.

Gradoń L, Majchrzycka K (2001), Efektywna ochrona układu oddechowego przed zagrożeniami pyłowymi, CIOP, Warsaw, Poland.

Hansen N L (1931), Method for the manufacture of smoke filters or collective filters, *British Patent*, BP 384 052.

Hilczer B, Małecki J (1992), Elektrety i piezopolimery, PWN, Warsaw, Poland.

Hruza J (2004), Nanospider fibres filtration, *Proc. IVth Autex World Textile Conference*, Roubaix, France.

Jirsak O, Wadsworth L C (1999), *Nonwoven Technology*, Carolina Academic Press, Durham, USA.

Kabsch P (1992), Odpylanie i odpylacze. Mechanika aerozoli i odpylanie suche, T.1, WNT, Warsaw, Poland.

Kaufman A (1936), Die Faserstoffe für Atemschutzfilter, *Z. Verein Deutsches Ing.*, **80**, 593.

Klata E, Babe K, Kucińska I (2004), Preliminary investigation on carbon nanofibres for electrochemical capacitors, *Proc. IVth Autex World Textile Conference*, Roubaix, France, 2004.

Krucińska, I (2002), The influence of technological parameters on the filtration efficiency of electret needled nonwoven fabrics, *J. Electrostatics*, **56**, 143–153.

Krucińska I, Zakrzewski S, Kot J, Brochocka A (1995), Badania nad otrzymywaniem wysoko skutecznych materiałów filtracyjnych, *Przegląd Włókienniczy,* **XLIX,** 25–27.

Krucińska I, Zakrzewski S, Kowalczyk I (1997), Winiewska-Konecka, J., Investigation on Blended Fibre Filtering Material used Against Respirable Dust, *JOSE*, **3**, 141–149.

Krzyżanowski J (1986), Struktura i właściwości włóknin formowanych z polipropylenu techniką pneumotermiczną przy zmiennych parametrach energetycznych, Doctoral thesis, Instytut Włókiennictwa, Lódz, Poland.

Kuwabara S (1959), The forces experienced by randomly distributed parallel circular cylinders or spheres in viscous flow by small Reynolds numbers, *J. Phys. Soc. Japan*, **14**, 522-532.

Langmuir I (1942), Report on Smokes and Filters, Section I, U.S. Office of Scientific Research and Development, no 865, Part IV.

Lunenschloss J, Albrecht W (1985), *Non-woven Bonded Fabrics*, Ellis Horwood Limited, Chichester.

Pitch J (1966), Theory of aerosol filtration by fibrous and membrane filters, *Aerosol Science*, ed. Davies, CN, Academic Press, London.

Schmidt K (1980), Manufacture and use polycarbonate felt pads made from extremely fine fibres, *Melliand Textil.*, **61**, 495–497.

Smith P A, East G C, Brown R C (1988), Generation of triboelectric charge in textile fibre mixture and their use as air filters, *J. Electrostatics*, **21**, 81–98.

Stechekina I B, Fuchs N A (1968), Studies on fibrous aerosol filters, III. Diffusional deposition of aerosols in fibrous filters, *Ann. Occup. Hyg*, **11**, 299–304.

Stechkina I B, Kirsch A A, Fuchs N A (1969), Studies on fibrous aerosol filters, IV. Calculation of aerosol disposition in model filters in the range of maximum penetration, *Ann. Occup. Hyg.*, **12**, 1–8.

Stenhouse J I T (1974), The influence of electrostatic forces in fibrous filtration, *Filtration and Separation*, **11**, 25–26.

Taylor G I (1964), Disintegration of water drops in an electric field, *Proc R Soc Lond A*, **280**, 383–397.

Taylor G I (1969), Electrically driven jet, *Proc R Soc Lond A*, **31**, 453–475.

Trouilhet Y. (1981), *Advances in web formation*, EDANA, Brussels.

Turnhout J van, Bochove C van, Veldhuizen G J van (1976), Electret filters for high efficiency filtration of polluted gases, *Staub*, **36**, 36–39.

Weghmann A (1982), Production of electrostatic spun synthetic microfibre nonwovens and applications in filtration, *Proceedings of the 3rd World Filtration Congress*.

Wehman M (2004), Innovative nonwovens in filtration, *CD Proceedings of 7. Symposium, Textile Filter*, Chemnitz, Germany.

Wente A (1956), Superfine thermoplastic fibres, *Industrial Engineering Chemistry*, **48**, 13.

Zeleny J (1914), The electrical discharge from liquid points, and a hydrostatic method of measuring the electric intensity at their surface, *Phys.Rev*, **3**, 69–91.

17.7 Appendix: notation

α	coefficient of filtering efficiency
δx	increase of thickness of filter
ϵ_o	permittivity of free space
η	dynamic viscosity of fluid
κ	Kozeny constant
λ	free path of molecules
μ	particle mobility
ρ	density of fluid (air $1.20\,\mathrm{kg/m^3}$)
ρ_p	density of particle material
ψ	stream function
ΔP	pressure drop due to flow of fluid through porous medium
a	radius of particle
A, B, C, J, K, k, Q_1	constant
A_f	area of filter
c	volume fraction of fibre, porosity
C_n	Cunningham correction factor
C_1	reference concentration of sodium chloride
C_2	concentration of sodium chloride in the air under the mask's facepart
D	coefficient of diffusion
D_1	dielectric constant of particle
d_f	diameter of fibre
dN	change of particle concentration in air
d_p	diameter of particle
E	filter efficiency
E_D	filter efficiency due to diffusion
E_{DR}	filter efficiency due to diffusion and interception
E_G	filter efficiency due to gravitation
E_{IR}	efficiency by interception with particle inertia
E_R	filter efficiency due to interception
E_S	single fibre efficiency
$f(c)$	function of packing factor
h	thickness of filter
K_1	permeability
k_B	Boltzmann's constant
Kn	Knudsen number
Ku	Kuwabara coefficient
L	total length of fibres per unit volume of filter
m	mass of particle
n	number of cylinders in filter material model
N	particle concentration in air passing the filter
N_0	particle concentration in air entering the filter

N_{qQ} dimensionless parameter describing capture of charged particles by charged fibres

N_{Qo} dimensionless parameter describing capture of neutral particles by charged fibres

N_R ratio of particle diameter to fibre diameter

P particle penetration

Pe Peclet number

Q air flow

Q_f charge per unit length of fibre

q charge of particle

R radius of fibre

r, Θ polar co-ordinates

Re   Reynold's number

St   Stokes' number

t time

T absolute temperature

u velocity of filter particle

v velocity of fluid inside filter

v_g rate of fall of a particle due to gravity

v_x, v_y velocity constituents at cartesian co-ordinates

v_r, v_o velocity constituents at polar co-ordinates

V rate of amount of air per unit time

W flow per unit area of filter

$\bar{x}$ average displacement of a particle due to diffusion

y co-ordinate on streamline upstream of fibres

18
Electrostatic protection

J A GONZALEZ, University of Alberta, Canada

18.1 Introduction

The term 'electrostatic' or 'static electricity' refers to the phenomenon associated with the build up of electrical charges generated, for example, by contact and/or rubbing of two objects. Static electricity is generated by unbalancing the molecular configuration of relatively non-conductive materials.

The word 'electricity' comes from 'electron' (amber in Greek) and it was Thales of Miletus (640–548 BC) who first observed this specific property. It was termed 'electrical' in 1600 by William Gilbert who is said to have begun the scientific study of electricity and magnetism. Our current ideas on the nature of electricity stem from the knowledge of atomic structure and the existence of tiny indivisible particles of both kinds of electricity.

Many years ago the problems arising from static charges were relatively small with natural fibres in high humidity environments, but these problems became recognized as serious when synthetic fibres of a hydrophobic nature were introduced. The need for in-depth understanding of the fundamentals of electrostatics in several industries has been growing fuelled by the proliferation of synthetic fibres, the use of atmospherically controlled environments, high-speed manufacturing, and static-sensitive devices.

Even natural fibres like wool and cotton, when completely dry, are very poor conductors, but their conductivity increases in high-humidity atmospheres because they absorb substantial amounts of moisture (of the order of 10%, calculated on the weight of the fibre). On the other hand, many manufactured fibres absorb little or no moisture and remain poor conductors in atmospheres of more than 60% relative humidity. Some people believe that garments made of cellulose-based fibres are less prone to static electricity than those made of synthetic fibres. This belief is based on the high moisture regain of the former at high relative humidity. For example, at 50% relative humidity, cotton products have lower surface resistivity than aramid fabrics; but at 20% relative humidity, aramid fabrics have just a slightly lower apparent surface resistivity than regular cotton or FR cotton fabrics (Gonzalez et al., 2001).

The major concern with static electricity is that electrostatic discharges (ESD) have proven to be a significant ignition source for flammable gases, vapours, or powders that may be present in certain industrial environments, resulting in fires and explosions and the possible loss of human life. Sources of ESD are as numerous and varied as the processes and material combinations in industry. For example, in cold regions such as Canada, the absolute humidity level declines extremely with very cold temperatures, so the electrostatic threat can be more significant than in warmer regions. It has been shown that clothing made of thermal protective fabrics such as aramids and flame retardant cotton may generate enough energy to ignite a fuel vapour-air mixture (Rizvi *et al.*, 1995).

Considerable research has been carried out on the charge generation and dissipation characteristics of textiles used in clothing. However, several issues remain unanswered, and the industry is still in need of test methods that can strongly correlate with real-life conditions.

18.2   Principles of electrostatics

Static electricity is generated when almost any pair of surfaces is separated. The amount of charge transferred from one surface to another depends on the relative affinities of the materials for a charge of given polarity (Sello and Stevens, 1983). It has been stated that the static electrification between two insulators of the same material is increased by asymmetric friction or temperature difference (Shaw and Jex, 1951; Hersh and Montgomery, 1955). Although Shirai (1984) found that the electric charge of two sheets of polyester depended mainly on their thickness rather than on their asymmetric friction.

The static charge which is involved in a spark phenomenon is often generated on the clothing or footwear of the individual and transferred onto the skin. Hence, the charging characteristics of clothing and shoes play a critical role in determining the possibility to produce a spark which could ignite flammable gases (Berkey *et al.*, 1988). An electrostatic system can be represented by a simple circuit (Fig. 18.1). It contains four elements: a charge generator, I, to represent the charge generating mechanism, a capacitor, C, on which charge is stored, a resistance, R, which represents the charge relaxation mechanism in the electrically stressed insulator, and a spark gap which limits the maximum charge that can be held in the system. Charge storage occurs on the system capacitance while the resistance of the system allows the charge to dissipate. The capacitor may be an insulating material or a person not properly grounded. The magnitudes of the capacitance and resistance determine the decay time (or relaxation time) of the charge in the system:

$$\tau = RC \qquad\qquad\qquad 18.1$$

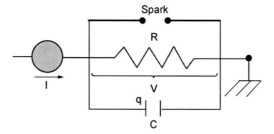

18.1 Representation of an electrostatic system.

18.2.1   Electrostatic properties of materials

Materials differ in their propensity to lose some of their electrons when in contact with another material (Crow, 1991). The energy required to cause the removal of an electron from a metal is called the work function (ϕ_W). When two bodies make contact, that which has the lower work function loses electrons to that with the higher work function. Work function is the minimum energy required to extract the weakest bound electron from its maximum excursion distance from the surface to infinity (Gallo and Lama, 1976). Also, work function is defined as the difference in energy between the Fermi level and the zero of potential energy, and represents the minimum energy required at absolute zero to enable an electron to escape from the potential box (Arthur, 1955).

A solid may be represented energetically as a potential well, where zero potential energy represents the space outside the solid (vacuum level). α is zero of kinetic energy, so that an electron of kinetic energy $\alpha\beta$ has total energy $o\beta$ and is confined to the potential well. An electron in level χ has total energy $o\chi$ and is a free electron. A metal may be represented in this manner, the continuous spectrum of levels being filled up to energy α and empty above that (Harper, 1967; Taylor and Secker, 1994). When the solid is an insulator, the energy spectrum is discontinuous, and energy levels are divided into two bands separated by a forbidden band of width δ. The lower or valence band is full and the upper or conduction band is empty under normal conditions. Since the valence band is full and an electron at the top of the valence has no higher energy level available, no acceleration in an electric field is possible (Arthur, 1955). These models are used to explain contact charging in terms of electron transfer, resulting in three main types of contact: metal-metal, metal-insulator, and insulator-insulator. The last two are important to the phenomenon of charge generation on textiles.

The transfer of electrons to an insulator is somewhat complex. Electrons can only move and carry current if they have sufficient energy to jump out of the filled band, through the forbidden band, into the conduction band. The chance of their acquiring this excess of energy is small, and any electron transferred to an insulator must be transferred into the conduction band (Morton and Hearle, 1975). This would lead to conductivity in charged insulators, which is not true. Gonsalves (1953) has suggested that there may be additional energy levels on the surface of an insulator called surface states (Tamm levels), which enable an electron to remain localized on the insulator surface in an energy state within the range of energies of the forbidden band.

18.2.2   Charge generation

For materials which are poor conductors of electricity (insulators), as are most textiles and polymers, the causes of charging are very complex. In good conductors, the charging is largely electronic in nature, but the surface of textiles is usually contaminated with additives, finishes, dirt and moisture, in all of which resides an abundance of ions (Wilson, 1987). In this case, charging may comprise electrons, ions and charged particles of the bulk materials, or any combination of these (Taylor and Secker, 1994). They also stated that because there might be little information on the ionic population of surfaces before contact is made, it may not be possible to predict the magnitude of the transferred charge.

Charges may be generated between a non-conductor and a conductor by induction. Consider a negatively charged rubber rod (non-conductor) brought near a neutral (uncharged) conducting sphere insulated from ground. The region of the sphere nearest the negatively charged rod will obtain an excess of positive charge, while the region of the sphere farthest from the rod will obtain an equal excess of negative charge. If the sphere is grounded, some of the electrons will be conducted to earth. When the grounding connection is removed, the sphere will contain an excess of induced positive charge (Haase, 1977).

Charging that occurs when two solids come into contact has been referred to as contact charging, frictional charging, tribo-electric charging and tribo-electrification. Usually, but by no means exclusively, contact charging is used to describe simple contacts between surfaces (i.e., contacts in which no sliding or rubbing occurs between the contacting surfaces). Thus, static electricity is generated when almost any pair of surfaces is separated, unbalancing the molecular configuration in the case of relatively non-conductive materials (Sello and Stevens, 1983). Henry (1953) reported that when the two surfaces are separated, either with or without obvious rubbing, charged particles are found to have crossed the boundary, with the usual result that the two surfaces have gained equal and opposite charges.

Although rubbing is not necessary for charge generation, it usually increases the amount of charge produced. 'Triboelectrification' is the term that applies when an electrical charge is generated on a body by frictional forces and is probably the major mechanism for the generation of electrostatic charge in textile materials (Wilson and Cavanaugh, 1972). Experiments have shown that, when an insulating surface is rubbed either by a conductor or another insulator, charge transfer may be several orders of magnitude greater than in a simple touching contact. This may be rationalized by noting that rubbing increases the intimacy of the contacting surfaces. Unless the electrical states of the two materials are extremely well balanced, there will be a large transfer of charge when their surfaces are brought into contact (Morton and Hearle, 1975).

Hersh and Montgomery (1955) found that the manner of rubbing, the length of material rubbed, the normal force, and rubbing velocity (in some cases) affected the amount of charge generated. Haenan (1976) showed that charge transfer increased with rubbing pressure, and Coste and Pechery (1981) reported that charge transfer was greatest when surface roughness was small. Some researchers (Montgomery et al., 1961; Zimmer, 1970; Ohara, 1979) have found some instances where the charging goes through a maximum value. Such effect has been related to local temperature gradients appearing across the contact, resulting in the enhanced diffusion of electrons from the hotter to the cooler surface (Taylor and Secker, 1994).

18.2.3 Charge dissipation

There are substantial differences in the ways a charge is dissipated, depending on whether it is located on an insulated conductor or on an intrinsically insulative material. In both cases, there are also differences that depend on whether the dissipation is carried out by charge carriers already present or if these carriers are created by the process itself. All neutralization processes deal with the movement of charges under the action of electric fields (established by the charges to be dissipated) spreading the charges or re-combining opposite charges, and making the fields decay or at least decrease (Chubb, 1988).

The transport of charge in a decay process is described by the basic relation (Ohm's law):

$$j = \lambda E = (1/\rho)E \qquad\qquad 18.2$$

where E is the field strength in a given point from the charge to be dissipated, j is the current density in the direction of E around the point, and λ and ρ are the conductivity and resistivity, respectively. Although the decay is always described by this basic equation, it is convenient to distinguish between three different types of decay processes: (a) charge decay of a capacitive system, (b) charge decay of a non-conducting system, and (c) charge decay through the air.

An insulated conductor may be characterized electrically by its capacitance C and leakage resistance R, both with respect to ground, and this arrangement is called a capacitive system. It is a special characteristic for the decay of charge on a conducting system that the contact between the conductor and the resistive path only needs to be established at a single point. The capacitance C is an integral measure of the distribution of the electric field from a given charge on the conductor, between the conductor and ground. The capacitance will thus depend upon the location of the conductor and may change somewhat if the conductor itself or other neighbouring conductors are moved (Jonassen, 1991).

In the case of a non-conducting system (insulator or semi-insulator), the charge decay may depend, in a rather complex way, on the geometrical, and dielectric and resistive conditions of the environment (Baumgartner, 1987). If a charge is located on an insulator there is in principle no way by which the charge may ever dissipate. If, however, the charged insulator is surrounded by a conducting fluid in contact with the surface, the charge or the field from the charge may dissipate by oppositely charged ions being attracted to the insulator. This is not so simple when air is the current-carrying medium because atmospheric air is normally neutral, but it may be ionized, that is, made to contain mobile charge carriers. Thus, the dissipation effect depends upon the air containing ions of opposite polarity to the charge on the insulator (Jonassen, 1991).

18.2.4   Effect of environmental conditions

Fibres vary significantly in quantity of water vapour they absorb. Most textile fibres absorb moisture (water vapour) from the air. As the relative humidity (RH) of the air increases, the amount of moisture absorbed generally increases. The amount of moisture a fibre contains has a profound effect on the electrical properties of the fibre. The rate of absorption depends on a variety of factors: temperature, air humidity, wind velocity, surrounding space, thickness and density of material, nature of the fibre, etc. (Morton and Hearle, 1975).

Within a fibre, the material accessible to moisture will be either the surfaces of crystalline regions or the non-crystalline regions. In crystalline regions, the fibre molecules are closely packed together in a regular pattern. Thus, it will not be easy for water molecules to penetrate into a crystalline region, and, for absorption to take place, the active groups would have to be freed by the breaking of some bonds or cross-links. The moisture at any particular relative humidity would be proportional to the amount of effectively non-crystalline material (Hearle and Peters, 1963).

Several researchers have reported the great effect environmental conditions have on electrical characteristics of textile materials. Hearle (1953) found that the moisture content of a textile fibre is the most important factor in determining its electrical resistance. Hearle and others (King and Medley, 1949; Cusick and Hearle, 1955; Sereda and Feldman, 1964) have shown that the resistivity of yarns and other textile materials increases exponentially with decrease in the relative humidity of the environment with which they are in equilibrium.

Not only the electrical resistivity of a fibre is affected by changes in its moisture content. Rizvi et al. (1998) reported that mean discharge potentials and energies from clothed humans decreased to about 50% with some garment systems when the relative humidity increased from 0 to 20%. Crugnola and

Robinson (1959) evaluated several garment systems at different relative humidity levels. They found that below 20% RH all garment systems generated high voltages considered dangerous by the US National Bureau of Standards (NBS), but none of the garment combinations produced potentials required for the detonation of the most active materials studied by the NBS.

Sereda and Feldman (1964) explained that the rapid decrease in charge generation at high humidity is due to an increase in conductivity and/or charge dissipation along a surface-water film. A change in relative humidity produces a change in the equilibrium moisture content of most textile materials, and hence alters their chemical nature (Hersh and Montgomery, 1956). Also, the moisture content of the ambient air plays a role in the generation and dissipation of charges. Onogi et al. (1996) found atmospheric charge dissipation into the air by water molecules through evaporation of water droplets from textile surfaces, resulting in static charge reduction. Furthermore, they reported that the rate constant for the atmospheric charge dissipation depends on the water content of the textile material.

The effect of temperature has not been addressed as thoroughly as the influence of moisture. Hearle (1953) reported that the resistance of a limited number of textile fibres decreased as the ambient temperature increased, and that relationship fitted, to some extent, a logarithmic model. Also, significant increase in body potentials was found when the temperature was decreased from room temperature to −40 °C (Crugnola and Robinson, 1959). Sharman et al. (1953) reported a decrease in conductivity in both drawn and undrawn nylon filaments as the temperature decreased from 45 to 15 °C, but they found that at none of the temperatures studied could the temperature dependence be accounted for as a simple rate process.

Onogi et al. (1997) reported interaction effects of temperature and moisture content in either fibre or ambient air, and their combined influence on static dissipation from textile surfaces. They reported variation in the critical water content of different fibres at various temperatures. Critical water content is a characteristic of the structure of the polymer molecule and also the bulk structure of the fibre. Absorbed water in a sample with less than this critical water content could not carry charge into the air. They also evaluated the slopes of the line relating the rate constant of charge dissipation for various fabrics and the free water, and found that at all temperatures the slopes for each fabric were quite different. Thus, they concluded that the rate constant for charge dissipation into the air does not generally depend on only the amount of free water in the textile fabric, but also on the vapour pressure of water (absolute humidity).

18.3 Electrostatic hazards

Static electricity has long been cited by investigators as a possible cause of accidental ignition of flammable or explosive liquids, gases, dust, and solids.

Many cases have been documented (Scott, 1981; Crow, 1991) where charges generated on an object reach the level at which the resistance of the air gap between the object and a conductor at a lower potential breaks down, producing a spark.

18.3.1   Requirements for an electrostatic discharge (ESD)

Static electricity manifests its destructive nature through ESD. The electrostatic build-up on people or materials, particularly non-conductive materials such as textiles, can be significant in dry conditions. Workers in the oil and gas industries have expressed opinions that some thermal protective clothing they are required to wear may not be safe due to its static propensity. Many still hold a traditional belief that 100% cotton garments are less prone to static electricity than are garments of more thermally stable fibres. This belief is based on measurements of certain electrical properties taken under conditions of relatively high humidity and may be misleading for low-humidity environments.

The average individual walking across a non-conductive floor or sliding off a car seat can generate discharge potentials up to 15 kV, depending on the environment, for example, in low relative humidity (Matisoff, 1986; Rizvi *et al.*, 1995). They usually pose no electrical danger to human beings because the charges generated are too small; depending on the individual, the human body has a threshold for shock of over 3 kV (Sclater, 1990). However, a spark of just less than 50 V can damage ESD-sensitive electronic devices (McAteer, 1987). But the main hazard of ESD, or sparks, is their incendiary properties.

Tolson (1980) reported that the incendivity of a discharge can be estimated once the circumstances of charge accumulation are known. Charge accumulation on an ungrounded conductor (human body or discrete conductive fabric) and charge accumulation on an insulator (synthetic fabrics and plastics) are two very different situations. The former represents by far the greatest risk because it can discharge all the electrostatic energy instantaneously in the form of a spark. In the case of electrically insulating materials (e.g. fabric), however, their high surface and volume resistivity impede the flow of charge to the point of discharge and only a fraction of the total charge on the surface is released in the discharge. The equation to calculate potential energy can not therefore be used to calculate the energy of the discharge because the charged insulator is not intrinsically an equipotential surface (Löbel, 1987). The character of a discharge from an insulator may be described in terms of the total charge transferred in the discharge and its distribution with space and time. Thus, the incendivity of a discharge depends not only upon the total amount of energy or charge released, but also upon the time distribution of the energy (Glor, 1988). A corona discharge extended in time is less incendive than a short-lived spark of the same total energy (Gibson and Lloyd, 1965).

18.3.2 Maximum charge density

The factors which determine the maximum density of charge that can remain on a surface without discharging into the surrounding medium are complicated. They are nevertheless important because discharge into the surrounding medium can set a limit to the charge obtained by friction charging. In the case of ambient air, when the electric breakdown field (EBD) of 2.7 MV/m is exceeded, a corona, brush or even a spark discharge will occur which will dissipate the excess surface charge (Gibson and Lloyd, 1965).

In insulating sheets, an extended charge of uniform density σ on the surface gives a field just outside the charged surface of magnitude $\sigma/2\epsilon_0$. The maximum charge density, σ_{max}, that can be tolerated before this field exceeds the breakdown strength, EBD, of the ambient medium is given by

$$\sigma_{max} = 2\epsilon_0 EBD \qquad\qquad 18.3$$

When one surface of an initially uncharged sheet is tribo-charged, the limiting charge density will be half this value because, as the contacting surfaces separate, all the flux lines, i.e., the electric field, will be bounded by the two separating surfaces. Thus

$$\sigma_{max} = \epsilon_0 EBD \qquad\qquad 18.4$$

which for air yields, $\sigma_{max} = \pm 24\ \mu C/m^2$ (Ji et al., 1989).

18.3.3 Minimum ignition energy (MIE)

The ignition energy of a gas, vapour or dust depends strongly on the percentage of flammable material present. At low concentrations the ignition energy is high but decreases to a minimum at some critical concentration before rising again on further increasing the concentration. The lowest energy required to cause ignition of the material, or a mixture of it in critical concentration with air, is known as the minimum ignition energy (MIE). There is no standard method to measure the energy required to ignite flammable gases. Typically, a known mixture of gas and air is placed in a grounded Plexiglas box which contains an electrode. A charge is discharged through the electrode, either from a charged person or from a capacitor. Wilson (1982) reported that the critical voltage at which ignition occurs decreases with the size of the electrode down to 1 mm and then increases again with the smallest electrode because of corona discharging, and that the lowest voltage for an ignition is independent of body capacitance.

Assessment of the ignition risk from an electrostatically charged body essentially requires comparison of the igniting power of any discharge from the body with the minimum ignition energy of the flammable atmosphere (Glor, 1988; Owens, 1984). Wilson (1977/1978) showed that the minimum ignition energy of coal gas and air is 0.03 mJ, of natural gas and air is 0.3 mJ. The MIE required to ignite methane and air in a closed chamber by a spark between a finger and an earthed electrode has been evaluated as 18.6 mJ (Wilson, 1977/78), 5.9 mJ (Movilliat and Monomakhoff, 1977), 1.1 mJ (Tolson, 1980), and as low as 0.5 mJ (Crugnola and Robinson, 1959). These experiments were performed under different conditions – gas mixtures, electrode sizes, and body capacitance. Rizvi and Smy (1992) found that the minimum energy density thresholds for incendive and non-incendive sparks were 10 J/m^2 and 0.25 J/m^2, respectively.

18.4   Measurement techniques

There have been two main approaches to assessing the electrostatic propensity of textile materials. One is to measure the charge built up on a clothed person or the electrical capacitance of a body (human-body model); the second is to measure some electrostatic characteristics of textiles (e.g., surface resistivity, charge decay rate, peak potential, etc.) in small-scale tests.

18.4.1   Small-scale tests

Several standard methods to measure different static characteristics are utilized, but there is generally a lack of correlation between small-scale and human-body data. Small-scale tests normally measure the electrostatic characteristics of an insulator and do not represent the real phenomenon of a charged clothed human body being discharged through a grounded object.

Measurements in small-scale tests are different from measurements in human-body experiments as their physical quantities are different in value and order of magnitude. Several conditions are involved in real-life sparks: electrostatic discharges are the result of the typical charge generation processes (tribo-electrification, induction and conduction charging), charge accumulation, type of materials involved, capacitance of the system, atmospheric conditions, etc. In order to understand better and control the various parameters which are related to ESD, the use of appropriate instrumentation, measurements, and standardized test methods is necessary.

18.4.2   Electric field

An electrostatic field exists in the region surrounding an electrically charged object. This charged object, when brought in close proximity to an uncharged object, can induce a charge on the formerly neutral object. This is known as an induced charge. Quantitatively, this induced charge is the voltage gradient between two points at different potentials (Matisoff, 1986). In most situations, it is the electric field from the charge which causes electrostatic effects.

One technique for evaluating the possible sparking hazard is therefore to measure the electric field intensity (V/m) at the surface of the charged fabric

(Owens, 1984). It has been demonstrated that field intensities less than 5 kV/cm cannot ignite any fuel that has a minimum ignition energy (MIE) greater than 0.15 mJ (Rizvi and Smy, 1992).

18.4.3   Charge

There has been an ongoing discussion whether measuring voltage from an insulator is correct or not. Some authors have stated that it is not possible to characterize a charged insulator with a single voltage value (Jonassen, 2001a,b; Baumgartner, 1984). Others have stated that measuring surface voltage from a charged insulator is meaningful (Chubb, 2001). The surface voltage of a charged insulator can be measured only in the case of a thin uniformly charge insulator (such as a fabric) backed by a grounded conductor. In this case, the approximate value of the surface voltage can be determined by

$$V_s = E \cdot d = \sigma \cdot d / \epsilon \qquad\qquad 18.5$$

where V_s is the surface voltage, E is the electric field, d is the thickness of the material, σ is the charge density (C/m^2), and ϵ is the insulator permittivity. In all other cases, the surface voltage of an insulator varies from point to point as its electric field inside the material. Therefore, it is recommended to determine the charge density by using a field meter to measure the field strength E (V/m) scanning the surface of the insulator at different points. E is then multiplied by ϵ_0 (8.85×10^{-12} F/m) to determine the charge density and its distribution on the insulator. Another mode of measuring charge is measuring the total charge on any item by a Faraday pail and an electrometer.

In its simple form, a Faraday cage or pail consists of an insulated, initially uncharged metal container, into which a charged insulator is placed. Then, its charge can be measured by an electrometer in nanocoulombs units (1 nC = 10^{-9} C). It is important to note that this measured charge is the net charge on the insulator.

18.4.4   Resistance and charge decay

Measurement of electrical resistivity is a frequently used method for the evaluation of static propensity of textiles (Coelho, 1985, Löbel, 1987). One of the most accepted small-scale methods used is surface resistivity (e.g., AATCC Test Method for Electrical Resistivity of Fabrics Method 76 (AATCC, 2000), and ASTM D257 Standard Test Methods for d-c Resistance or Conductance of Insulating Materials (ASTM, 1999)). The advantages of this kind of measurement over the determination of surface potentials are many. Measurement of electrical resistivity is described as simple and reproducible, and commercial equipment is widely available (Ramer and Richards, 1968). Teixeira and Edelstein (1954) discussed the fundamental principles of resistivity, giving definitions and explaining how charges are developed and how they are dissipated. Wilson (1963) showed experimentally that resistance along and through a fabric was a main factor in determining the rate of leakage of charge from the charged surface and through the structure, respectively, provided that the capacitance was maintained constant.

Measurement of charge decay rate on fabrics is another well known and industry-wide method (e.g. Federal Test Standard 101C Method 4046 Electrostatic Properties of Materials (Superintendent of Documents, 1969), and Federal Test Standard 191A Method 5931 Determination of Electrostatic Decay of Fabrics (Superintendent of Documents, 1990)). In using a charge decay meter to measure the dissipation rate, decay time indicates the ability of the surface to transfer the electrons from a charged body through the work surface to ground. Thus, the greater the resistance, the slower the charge decay rate (Matisoff, 1986).

Taylor and Elias (1987) discussed the problem of measuring the static dissipative properties of materials and proposed a new charge decay meter capable of measuring decay times from 100 ms to 10,000 s. They showed that for many materials surface resistivity did not correctly specify the ability of the material to dissipate static charge. Chubb (1990) described a new approach for the measurement of charge decay on insulating and semi-insulating materials. Charge was deposited on the material and a fast response electrostatic field meter was used to measure the rate of charge decay. Chubb and Malinverni (1993) reported the results of comparative studies on the charge decay characteristics of a variety of materials as observed by Federal Test Standard 101C Method 4046 and by the method developed by Chubb. They concluded that measurements by FTS 101C relate to charge decay by the fastest charge migration component of the materials whereas the method developed by Chubb gives times determined by charge decay on the outermost surface of the material. Chubb and Malinverni stated that since the ability of materials to dissipate static charge depends upon charges generated by tribo-charging the outermost surface it is such decay times which are of practical relevance.

18.4.5   Voltage

To overcome some shortcomings of the standard test methods mentioned above, the NFPA 1992, paragraph 6-22: Material Static Charge Accumulation Resistance Test (National Fire Protection Association, 2000) is used to evaluate the static electrical charge generated by tribo-electrification and the rate of discharge on protective clothing. This tribo-charging device is based on the system developed by the National Aeronautics and Space Administration (NASA) in the late 1960s (Gompf, 1984). Although the method incorporates controlled frictional charging, and reliable measurements of potentials have been obtained in experiments carried out at the University of Alberta, the charge decay results obtained in those experiments have given very low correlation with either triboelectric charge obtained by the method or human-body discharge potentials (Gonzalez *et al.*, 2001). Stull (1994) reported that '. . . although this method may not appraise the hazard under actual use conditions, it does permit the ranking of material performance and identification of potential material problems for a given set of conditions. It seems particularly suited for chemical protective applications' (p. 26).

18.4.6   Electrostatic discharges

Antistatic performance is an essential requirement for clothing worn by workers in the military, oil and gas industries, micro-electronics, etc. The antistatic performance requirement of a textile material may differ according to its end use and can be determined by the risks of explosion, shock, electronic damage, and dust protection.

Due to the diversity of textile materials used and the complexity of end-use applications, the antistatic performance cannot be defined by a single test that evaluates only electrostatic properties such as resistivity, charge, and field strength. To address this limitation, some measuring techniques to evaluate electrostatic discharges (ESD) from either textile surfaces (Wilson, 1985; Kathirgamanathan *et al.*, 2000) or a clothed human body (Wilson, 1979) have been proposed.

A simple modification of the NASA's tribo-charging device has been developed to measure potentials and energies from the discharge of a capacitor, which is previously charged from the tribo-electrification of a fabric system (Fig. 18.2). The system was designed to simulate the phenomenon experienced by a clothed person rubbing an insulated surface, touching a grounded object, and subsequently generating a spark (Gonzalez *et al.*, 1997).

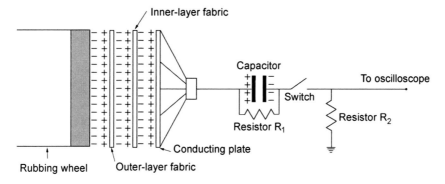

18.2 Rubbing, resistance/capacitor and discharging sections in a modified NFPA 1991, 9-29 tribo-charging device.

18.4.7   Human-body model

There is no standard method for measuring the static charge built up on a person. The generally accepted method is for a person to walk in a controlled fashion into a Faraday cage. This is a wire cage onto which is induced an equal but opposite charge to that on the person entering it. This induced charge is recorded to give a measure of the static electricity on the person.

Most human-body experiments have been conducted in the United States, Canada and the United Kingdom. Measurements of the charges generated between different materials agree somewhat with rankings in existing tribo-electric series, where materials placed close to each other develop less static charge than those ranked further apart. Some conclusions are that the static propensity of clothing depends on such factors as temperature, humidity, the type of textile, the type of charge mechanism, and the nature of the footwear worn (Crow, 1991).

Experiments at The Arctic Aeromedical Laboratory (Veghte and Millard, 1963) were conducted specifically on the accumulation of static electricity on Arctic clothing. In the experiment, three different Arctic clothing outfits made mainly from nylon were worn by fifteen different subjects. The electrostatic charges on the clothing systems and the capacitance of the subjects were measured. The experiments were conducted at ambient temperature ranging from 5 to −43 °C and relative humidity at between 50% and 74%. The research pointed out the dangers of personnel working outside, coming indoors and removing exterior clothing in a dry environment, a situation which tends to produce very high electrostatic charges.

Wilson's study (1977/1978) was intended to investigate the charge generation characteristics of clothing in normal use by workers. The objective of this project was to assist in developing a specification which could be used to identify safe fabrics for use when handling flammable materials. The garments were the type worn by military personnel and were made of fabrics such as polyester and linen/polyester coveralls, aramid and cotton flying suits, and polyurethane coated nylon weather suits. The subject wearing a garment and a pair of rubber-soled shoes, sat down on a covered chair and slid off it into a standing position. The chair cover materials were lambswool, PVC-coated cotton, leather, and cotton canvas. In all cases, the body voltages were discharged to ground via the fingers to produce sparks which were measured. This work was done at relative humidities in the range of 15 to 80%, at 21 °C. The result showed that cotton as well as synthetic fabrics are static prone at low humidity.

Scott (1981) noted that the majority of work on the incendivity of spark discharges from the human body had concentrated on surface resistivity of fabrics and the voltages generated on the body. He found that the capacitance of the human body varies according to size, the footwear being worn and stance, that larger bodies have greater capacitance, and that insulated footwear raises the electrical insulation of the body and lowers the body capacitance.

De Santis and Hickey (1984) measured the static build-up on an extended cold wet clothing system (ECWCS) before and after laundering. They measured the potential accumulated with a Faraday cage enclosing a test participant wearing various ECWCS items. The dressing and removing of garments took place outside the cage at a test temperature of $-40\,°C$. The authors concluded that the donning and doffing of the outer Gore-Tex$^®$ jacket created very little static electricity. The polyester pile shirt showed evidence of static charge which resulted from the shirt brushing against a static-generating material.

Researchers at the University of Alberta (Rizvi *et al.*, 1995, 1998; Crown *et al.*, 1995) developed a method in which a clothed person performs a physical activity, and then touches a grounded electrode. The discharge potential of the resulting spark is measured and monitored by a digital oscilloscope. From the discharge-voltage waveform, other parameters of interest, for example: transferred charge, discharge energy, peak current, duration of event, etc., were calculated. They reported the characteristics of ESD from humans wearing protective garment systems and performing two activities: sliding off a car seat, and walking and removing a garment. The experiments were conducted under very low humidity, at room temperature. It was found that garments made from antistatic fibres (aramid/carbon and aramid/stainless steel) generated static charges of less energy than those made of non-antistatic fibres (aramid and FR cotton), but those charges were still bigger than the MIE of different gases, vapours and mixtures.

18.5   Abatement of static electricity

There are two distinct differences in the electrostatic performance of conductors and insulators. The first difference is that a charged conductor can dissipate all the energy stored in its field in a single discharge, neutralizing its entire charge. The second difference is that a charged conductor needs only to be connected to ground from a single point of its surface through a suitably conductive path to have its charge eventually migrated (ESD Association, 1998a).

A discharge from a charged insulator, on the other hand, neutralizes only part of the charge and hence dissipates only part of the energy stored in the field. Furthermore, charges on an insulator cannot be removed by connecting the surface of an insulator to ground (Jonassen, 1991).

Most static charge removal processes do not involve actual removal of an electric charge from the charged object. The exception is charged conductors. In all other situations, the neutralization consists of oppositely charged carriers, either ions or electrons, being drawn to the excess charge. The field from the neutralizing charge superimposes the original field, and the resulting reduced field is then interpreted as a reduction or removal of the charge (ESD Association, 1998b).

In principle, there are three methods for neutralizing charges on insulators: conductance through the bulk of the material, conductance along the surface of the material, and the attraction of oppositely charged ions from the air. The next three sections are based on one of these methods of neutralization (Onogi *et al.*, 1996).

18.5.1 Materials and their design

There has been an increasing research to provide insulative materials an appropriate conductivity without affecting their other advantageous properties, usually mechanical ones. This is usually achieved by mixing the material with inherently conductive additives. The best-known example of such an intrinsic antistatic agent is carbon black. Carbon black can be added to a variety of polymeric materials and is used when the resulting blackening of the base material is acceptable.

For many years, the most important area of use for carbon black was conductive rubber. Ordinary vulcanized rubber can have a bulk resistivity of $10^{13}\,\Omega \cdot m$, but adding carbon black can lower the resistivity by a factor of up to 10^{15}. Normally, however, a resistivity of about 10^5–$10^6\,\Omega \cdot m$ is low enough to prevent dangerous or annoying charge accumulations (Jonassen, 2001c).

Another use of carbon black is in the manufacturing of textile products and antistatic floor coverings. The textile fibres can be made with either a central core of carbon black and a sheath of polyamide or, conversely, with a central core of aramid fibre and a sheath of carbon black (e.g., DuPont's Nomex IIIA® fibre).

A different approach is to use special high-performance antistatic fibres: metal, metallized, and bi-component fibres containing conductive features are among those used. A small quantity of these fibres is blended with conventional fibres to achieve dissipative (between 10^5 and $10^{12}\,\Omega/sq.$ surface resistivity) or conductive (less than $10^5\,\Omega/sq.$) levels as per standard EIA-541 (Electronic Industries Alliance, 1988).

18.5.2 Manufacturing

The basic rule for fighting the unwanted effects of static electric charges is to ground all conductors that might become charged or exposed to induction from other charged objects. Ungrounded charged conductors can produce discharges ranging from weak current pulses that may harm only the most sensitive electronic components to energetic sparks that may cause explosions and fires.

In the electronics and other industries, grounding through footwear and a floor covering is a widely accepted procedure in many areas of the electronics industry. The device employed for this purpose is a wrist strap, which consists of a band or chain, similar to an expandable watchband and made of metal and conductive plastic or conductive fibres, and a strap that connects the band to ground. The strap is made of either solid conductive plastic or multistrand wire. Normally, the strap includes a series safety resistor of 1 MΩ for minimizing the shock from accidentally touching a live wire while being tied to ground via the strap.

Neutralizing charges on insulators cannot use the grounding method, and charges can only be neutralized with oppositely charged air ions. Air ions are not naturally present, except in environments with high radon and radon-daughter concentrations. They must be produced by high electric fields or radioactive decay and brought to the charge by a field, sometimes aided by airflow. Any ionization process in air starts with an electron being removed from an oxygen or nitrogen molecule. Ionizers currently in the market ionize the surrounding air in different ways. These devices are classified according to their ionization method, as follows: radioactive, field, passive, ac and dc ionizers.

In a radioactive ionizer, alpha particles are emitted with sufficient energy to cause ionization of a large number of air molecules. A passive ionizer is essentially a single grounded emitter or (more often) a row of grounded emitters placed parallel with and close to the charged material. Ac ionizers are connected to an ac voltage supply, forming positive and negative ions alternatively, and the polarity of the charged material determines the polarity of attracted ions. The most effective neutralization is obtained by the use of a dc ionizer, which usually consists of two emitters held at a positive and a negative potential, respectively. When the ionizer is properly balanced, positive and negative ions are provided in the same concentrations in front of the charged material, and the polarity of the charge determines the type of ions used for neutralization.

18.5.3   Finishing

It is often possible to render highly insulative textile materials sufficiently surface-conductive, even at relatively low humidities, by treating the surface with antistatic agents (topical antistatic finishes). Some finishes have been developed that attempt to decrease static build-up by one or more of three basic methods: (1) by increasing the material's conductivity, whereby the charged electrons move to the air or are grounded, (2) by increasing absorption of water by the finish, providing a conductive surface on the fabric that carries away the static charge, and (3) by neutralizing negative and positive charges (Sello and Stevens, 1983).

Chemically speaking, antistatic agents are amphipathic compounds, their molecules containing a hydrophobic group to which is attached a hydrophilic end group. According to the nature of the end group, the agents are divided into cationic, anionic, and nonionogenic agents. Cationic materials are usually high-molecular quaternary ammonium halogenides or ethoxylated fatty amines or amides. Anionic materials can be sulfonated hydrocarbons, and nonionogenic materials can be polyalkylene oxide esters (Joshi, 1996).

Antistatic agents may be added to a polymer either before polymerization or before extrusion; antistatic polyethylene is an example of this manufacturing technique. Ethoxylated fatty amines or amides are mixed with a resin, such as low-density polyethylene, and an antiblock, such as calcium carbonate, to prevent stickiness (Joshi, 1996). After extrusion, the additive has to diffuse to the surface to attract moisture from the air and thus render the material antistatic. Materials with built-in additives maintain their antistatic properties as long as the additive is present on the surface.

Although the vapour pressure of most additives is fairly low, a certain level of evaporation always takes place from the surface. For fresh materials, this evaporation is counterbalanced by diffusion from the interior of the material. As the supply of additive in the solid is depleted, the surface concentration cannot be maintained. The surface is said to 'dry out', resulting in an increasing surface resistivity and the eventual loss of antistatic properties (Sello and Stevens, 1983).

The effective lifetime of a permanent antistatic material depends on many factors, the most important of which are the temperature of the environment and the thickness of the material, which for a given volume concentration determines the amount of additive available for diffusion to the surface. In addition, these finishes may affect other performance properties of a textile such as durability and hand (Holme *et al.*, 1998).

18.6   Future trends

There has been an increasing demand for personal protective equipment (PPE) in the last few decades due to greater awareness about worker's health and safety in hazardous work environments. Significant research has also been carried out in developing protective textile materials and clothing against different hazards such as chemical, thermal, impact, biological and nuclear or electromagnetic radiation. With today's emphasis on safety and security, higher standards of worker protection, the potential of nuclear, biological and/or chemical (NBC) emergencies in highly populated environments, etc., the need to enhance the functionality and performance of protective materials used in personal protective equipment (PPE) is essential and critical. The current challenge is to develop the critical materials and components to enhance the protective performance of PPE without affecting other performance properties such as electrostatic propensity.

Protective textiles of the future will include smart fibres and wearable electronics that interact, inform and assist the wearer. Potential applications include clothes that monitor the medical condition of the wearer, warn of the presence of toxic chemicals, provide global location of a wearer through the global positioning system (GPS), or adjust to suit the wearer's environment. Smart and high-performance fibres are relentlessly replacing traditional materials in many more applications. The integration of smart features into protective textiles and clothing offer great advantage.

However, this optimized performance comes with several drawbacks such as power supply, care and maintenance, dimensional changes, lack of signal integrity, computing reliability and compatibility, effect of sweat, moisture, temperature, mechanical impacts, repeated bending and compression, light, etc.

One area of research that has attracted a substantial amount of attention in the manufacturing of electronic textiles (e-textiles) and other high-tech applications is the development of electrically conductive polymers since they were discovered two decades ago, and new developments are reported regularly in the literature, as follows:

Y. Sano et al. (1997) modified the antistatic behaviour of a poly(ethylene therepthalate) fiber. In this technique, blend polyester fibres containing poly(ethylene terephthalate/5-sulfoisophthalate) (SIP-PET) were prepared by blend spinning and then treated with various cationic surfactants in the process of high-temperature dyeing. The best antistatic properties were attained with the surfactants having a long alkyl group and poly-(oxyethylene) (PEO) chains with a degree of polymerization of ca. 10.

F. Seto et al. (1999) investigated the surface modification of synthetic fiber fabrics via corona discharge treatment and subsequent graft polymerization. Polyethylene (PE) nonwoven fabric and polyamide-6 (PA-6) nonwoven fabric were used as base fabrics. Acrylic acid (AAc) was graft polymerized onto the fabrics via corona discharge pre-treatment. They found that charge dissipation decreased drastically by grafting with PAAc.

Cheng et al. (2001a) studied the development of conductive knitted fabric reinforced thermoplastic composites, with the intention to use them in electrostatic discharge applications. Conductive knitted fabric composites were made using polypropylene as the matrix material, glass fibers as the reinforcement, and copper wires as the conductive fillers. They found that the electrostatic discharge attenuation of knitted fabric reinforced thermoplastic composites can be tailored by changing the fabric knit structure, stitch density, and composition of the knitting yarn.

Cheng et al. (2001b) developed a new stainless steel/polyester woven fabrics for avoiding ESD, which can be used not only in home textiles, electrical and electronic devices, and subsystems, but also in clean-room applications such as the pharmaceutical and optical industries.

Zhou and Liu (2003) studied the surface resistivity of a treated polyurethane (PU) elastomer membrane. Vinyl acetate (VAc) was grafted onto PU using benzoyl peroxide as photoinitiator. Experimental results showed that after modification the surface resistivity of the PU membrane reached the $10^8 \, \Omega$ level.

Jia et al. (2004) developed and characterized conductive composites consisting of an epoxyanhydride matrix with polyaniline-coated glass fiber (PANI-GF) combined with bulk PANI as conductive fillers. They reported that the best conductivity result was obtained with the PANI-GF filler containing 80% PANI, and that PANI-coated glass fibers acted as conductive bridges, giving a higher number of contact points than in the case of PANI particles alone. They also reported that the presence of PANI-coated glass fibers significantly improved the modulus and strength of the epoxy/PANI composite.

With the emerging of new protective textiles where sensors, wireless communication and computing technologies are integrated into conventional protective textile assemblies, the evaluation of their performance may require the capability of measuring other paramaters than typical electrostatic properties such as conductivity/resistivity. These may include radio frequency levels and data transfer rates. In addition, more sensitive testing equipment and stringent product specifications will be required as most of these e-textiles function with tiny voltages. The literature has not reported yet any current research to determine the effect of these integrated technologies on the electrostatic propensity of protective textile materials and products.

18.7   Sources of further information and advice

18.7.1 ESD Information

- Applied Electrostatic Research Centre at the University of Western Ontario, Canada (http://www.engga.uwo.ca/research/aerc/)
- Compliance Engineering (http://www.ce-mag.com/)
- Electrostatics and Surface Physics Laboratory at NASA Kennedy Space Center (http://empl.ksc.nasa.gov/Publications/publications.htm)
- ESD Journal (http://www.esdjournal.com/)
- ESTAT Garment Project (http://estat.vtt.fi/index.html)
- IEEE (http://www.ieee.org/portal/site)
- Institute of Electrostatics Japan (http://streamer.t.u-tokyo.ac.jp/~iesj/homepage-e.html)
- Institute of Physics (http://www.iop.org/)
- International Electrotechnical Commission (http://www.iec.ch/)
- Journal of Electrostatics (http://www.sciencedirect.com/science/journal/03043886)
- Petroleum Equipment Institute (http://www.pei.org/static/)
- Protective Clothing & Equipment Research Facility (PCERF) at the University of Alberta, Canada (http://www.hecol.ualberta.ca/PCERF/)

18.7.2 ESD-related professional organizations

- Electrostatic Discharge Association (http://www.esda.org/)
- Electrostatic Society of America (http://www.electrostatics.org/)
- Electronic Industries Alliance (http://www.eia.org/)

18.7.3   ESD equipment and testing services

- British Textile Technology Group (http://www.bttg.co.uk/)
- Electrostatic Solutions (http://www.static-sol.com/welcome.htm)
- Electro-tech System, Inc. (http://www.electrotechsystems.com/)
- ESD Systems (http://www.esdsystems.com/)
- John Chubb Instrumentation (http://www.jci.co.uk/List.html)
- Keithley (http://www.keithley.com/)
- Monroe Electronics (http://www.monroe-electronics.com/esd_default.html)
- PASCO (http://www.pasco.com/)
- Protective Clothing & Equipment Research Facility (PCERF) at the University of Alberta, Canada (http://www.hecol.ualberta.ca/PCERF/)
- Tektronix (http://www.tek.com/)
- Textile Technology Centre (http://www.ctt.ca/)

18.7.4   ESD standards and specifications

All Military Specifications
Defense Printing Services
700 Robins Avenue
Building 4 Section D
Philadelphia PA 19111-5094
Tel: 215-697-2000
Fax: 215-697-1462 (fax requests accepted)
Web Site: www.dodssp.daps.mil

Domestic and International Documents
Global Engineering Services
15 Inverness Way East
Englewood CO 80112-5776
Tel: 800-854-7179
Fax: 303-397-2740
Web Site: www.global.ihs.com
E-mail: global@ihs.com

ESD – Electrostatic Discharge
Association
7900 Turin Road, Building 3
Rome NY 13440-2069
Tel: 315-339-6937
Fax: 315-339-6793
Web Site: www.esda.org
E-mail: eosesd@aol.com

EIA – Electronic Industries Alliance
2500 Wilson Blvd.
Arlington VA 22201-3834
Tel: 703-907-7500
Fax: 703-907-7501
Web Site: www.eia.org
See Global Engineering Services
(Above)

NFPA – National Fire Protection
Association
11 Tracy Dr.
Avon MA 02322
Quincy MA 02269
Tel: 800-344-3555
Fax: 800-593-6372
Web Site: www.nfpa.org
E-mail: custserv@nfpa.org

ASTM American Society for Test &
Measurement
100 Barr Harbor Drive
West Conshohocken PA 19428
Tel: 610-832-9500
Fax: 610-832-9555
Web Site: www.astm.org
E-mail: customerservice@astm.org

ANSI – American National Standards
Institute
25 West 43rd St.
4th Floor
New York NY 10036
Tel: 212-642-4900
Fax: 212-398-0023
Web Site: www.ansi.org
E-mail: info@ansi.org

ISA – Instrument Society of America
67 Alexander Dr.
RTP NC 27709
Tel: 919-549-8411
Fax: 919-549-8288
Web Site: www.isa.org
E-mail: info@isa.org

FDA – Food & Drug Administration
158-15 Liberty Ave.
Jamaica NY 11433
Fax: 718-662-5431
Web Site: www.fda.gov
E-mail: www.webmail@oc.fda.gov

Telecordia (BellCore)
8 Corporate Place
Room 3A 184
Piscataway NJ 08854
Tel: 1-800-521-2673
Fax: 732-336-2559
Web Site: www.telcordia.com
E-mail: tele-com-info@telcordia.com

18.8   References

AATCC (2000). AATCC Committee RA32. *Electrical Resistivity of Fabrics* (76-2000). Research Triangle Park, NC: American Association of Textile Chemists and Colourists.

Arthur, D. F. (1955). A review of static electrification. *Journal of The Textile Institute, 46,* 721–734.

ASTM. (1999). ASTM Committee D-9 on Electrical and Electronic Insulating Materials. *Standard Test Methods for DC Resistance or Conductance of Insulating Materials* (D 257-99). Philadelphia, PA: American Society for Testing and Materials.

Baumgartner, G. (1984). Electrostatic measurement for process control. In *EOS/ESD Symposium Proceedings* Vol. EOS-6 (pp. 25–33). Rome, NY: The EOS/ESD Association and IIT Research Institute.

Baumgartner, G. (1987). A method to improve measurements of ESD dissipative materials. In *EOS/ESD Symposium Proceedings* Vol. EOS-9 (pp. 18–27). Rome, NY: The EOS/ESD Association and IIT Research Institute.

Berkey, B. D., Pratt, T., & Williams, G. (1988). Review of literature related to human spark scenarios. Plant/Operations Progress, 7(1), 32–36.

Cheng, K. B., Ramakrishna, S., & Lee, K. C. (2001a). Electrostatic discharge properties of knitted copper wire/glass fiber fabric reinforced polypropylene composites. *Polymer Composites, 22*(2), 185–196.

Cheng, K. B., Ueng, T. H., & Dixon, G. (2001b). Electrostatic discharge properties of stainless steel/polyester woven fabrics. *Textile Research Journal, 71,* 732–738.

Chubb, J. N. (1988). Measurement of static charge dissipation. In J. L. Sproston (Ed.), *Electrostatic Charge Migration* (pp. 73–81). Bristol, UK: IOP Publishing Ltd.

Chubb, J. N. (1990). Instrumentation and standards for testing static control materials. *IEEE Transactions on Industry Applications, 26,* 1182–1187.

Chubb, J. N., & Malinverni, P. (1993). Comparative studies on methods of charge decay measurement. *Journal of Electrostatics, 30,* 273–284.

Chubb, J. N. (2001). Voltage of insulators? (Letters to the editor). *Threshold (EOS/ESD Association Newsletter), 17*(4), p. 14.

Coelho, R. (1985). The electrostatic characterization of insulating materials. *Journal of Electrostatics, 17,* 1327.

Coste, J., & Pechery, P. (1981). Influence of surface profile in polymer-metal contact charging. *Journal of Electrostatics, 10,* 129–136.

Crow, R. M. (1991). *Static electricity: A literature review (U)* (Technical Note 91-28). Ottawa, ON: Defence Research Establishment Ottawa.

Crown, E. M., Smy, P. R., Rizvi, S. A., & Gonzalez, J. A. (1995, June 30). Ignition hazards due to electrostatic discharges from protective fabrics under dry conditions. In *Final Report Presented to Alberta Occupational Health and Safety, Heritage Grant Program.* Edmonton, AB: University of Alberta.

Crugnola, A. M., & Robinson, H. M. (1959). *Measuring and predicting the generation of static electricity in military clothing* (Textile series, Report No. 110). Natick, MA: Quartermaster Research & Engineering Command, US Army.

Cusick, G. E., & Hearle, J. W. S. (1955). The electrical resistance of two protein fibres. *Journal of The Textile Institute, 46,* 369–370.

De Santis, J. A., & Hickey, S. L. (1984). *Investigation of electrostatic potential of extended cold wet clothing system (ECWS)* (APG-MT-6018). Aberdeen, MD: Aberdeen Proving Ground.

Electronic Industries Alliance (1988). EIA-541, Packing materials standards for ESD-sensitive items (par. 2.22, 4.2.3). Arlington, VA: EIA.

ESD Association (1998a). Basics of ESD. Part One: An introduction to ESD. *Compliance Engineering, 15*(1), 47–54.

ESD Association (1998b). Basics of ESD. Part Three: An overview of ESD control procedures and materials. *Compliance Engineering, 15*(3), 121–130.

Gallo, C. F., & Lama, W. L. (1976). Some charge exchange phenomena explained by a classical model of the work function. *Journal of Electrostatics, 2,* 145–150.

Gibson, N., & Lloyd, F. C. (1965). Incendivity of discharges from electrostatically charged plastics. *British Journal of Applied Physics, 16,* 1619–1631.

Glor, M. (1988). *Electrostatic Hazards in Powder Handling.* Letchworth, England: Research Studies Press.

Gompf, R. H. (1984). Triboelectric testing for electrostatic charges on materials at Kennedy Space Center. In *EOS/ESD Symposium Proceedings* Vol. EOS-6 (pp. 58–63). Rome, NY: The EOS/ESD Association and IIT Research Institute.

Gonsalves, V. E. (1953). Some fundamental questions concerning the static electrification of textile yarns: Part I. *Textile Research Journal, 23,* 711–718.

Gonzalez, J.A., Rizvi, S.A., Crown, E.M., & Smy, P. (2001). A laboratory protocol to assess the electrostatic propensity of protective clothing systems. *Journal of The Textile Institute, 92 Part 1,* 315–327.

Gonzalez, J. A., Rizvi, S. A., Crown, E. M., & Smy, P. R. (1997). A modified version of proposed ASTM F23.20.05: Correlation with human body experiments on static propensity. In J. O. Stull & A. D. Schwope (Ed), *Performance of Protective Clothing* (pp. 47–61) Vol. 6, ASTM STP 1273. Philadelphia, PA: American Society for Testing and Materials.

Haenen, H. T. M. (1976). Experimental investigation of the relationship between generation and decay of charges on dielectrics. *Journal of Electrostatics, 2*, 151–173.

Harper, W. R. (1967). *Contact and frictional electrification*. London, UK: Oxford University Press.

Haase, H. (1977). Electrostatic hazards: Their evaluation and control. New York: Verlag Chenie-Weinheim.

Hearle, J. W. S. (1953). The electrical resistance of textile materials: Parts I to IV. *Journal of The Textile Institute, 44*, 117–198.

Hearle, J. W. S., & Peters, R. H. (1963). *Fibre structure*. Manchester, England: The Textile Institute.

Henry, P. S. H. (1953). Survey of generation and dissipation of static electricity. *The British Journal of Applied Physics, 4*(2), 6–11.

Hersh, S. P., & Montgomery, D. J. (1955). Static electricity of filaments. Experimental techniques and results. *Textile Research Journal, 25*, 279–295.

Hersh, S. P., & Montgomery, D. J. (1956). Static electrification of filaments. Theoretical Aspects. *Textile Research Journal, 26*, 903–913.

Holme, I., McIntyre, J. E., & Shen, Z. J. (1998). Electrostatic charging of textiles. *Textile Progress, 28*(1), p. 29.

Ji, X., Takahashi, Y., Komai, Y., & Kobayashi, S. (1989). Separating discharges on electrified insulating sheet. *Journal of Electrostatics, 23*, 381–390.

Jia, W., Tchoudakov, R., Segal, E., Narkis, M., & Siegmann, A. (2004). Electrically conductive composites cased on epoxy resin containing colyaniline–DBSA- and polyaniline–dbsa-coated glass fibers. *Journal of Applied Polymer Science, 91*, 1329–1334.

Jonassen, N. (1991). Electrostatic decay and discharge. In *EOS/ESD Symposium Proceedings* Vol. EOS-13 (pp. 31–37). Rome, NY: The EOS/ESD Association and IIT Research Institute.

Jonassen, N. (2001a). Voltage of insulators?. *Threshold (EOS/ESD Association Newsletter), 17*(3), 8–10.

Jonassen, N. (2001b). Surface voltage and field strength: Part I, insulators. *Compliance Engineering, 18*(9), 26–33.

Jonassen, N. (2001c). Abatement of static electricity: Part II, insulators. *Compliance Engineering, 18*(7), 22–25.

Joshi, V K. (1996). Antistatic fibres and finishes. *Man-made Textiles in India, 39*(7), 245–251.

Kathirgamanathan, P., Toohey, M. J., Haase, J., Holdstock, P., Laperre, J., & Schmeer-Lioe, G. (2000). Measurements of incendivity of electrostatic discharges from textiles used in personal protective clothing. *Journal of Electrostatics, 49*, 51–70.

King, G., & Medley, J.A. (1949). DC conduction in swollen polar polymers. 1. Electrolysis of the keratin-water system. *Journal of Colloid Science 4* (1), 1-7.

Löbel, W. (1987). Antistatic mechanism of internally modified synthetics and quality requirements for clothing textiles. In B. C. O'Neill (Ed), *Electrostatics '87* Vol. 85 (pp. 183–186). Bristol, UK: IOP Publishing Ltd.

Matisoff, B. S. (1986). *Handbook of electrostatic discharge controls*. New York: Van Norstrand Reinhold Co.

McAteer, O. J. (1987). An overview of the ESD problem. In B. C. O'Neill (Ed), *Electrostatics '87* Vol. 85 (pp. 155–164). Bristol, UK: IOP Publishing Ltd.

Montgomery, D. J., Smith, A. E., & Wintermute, E. H. (1961). Static electrification of filaments: Effect of filament diameter. *Textile Research Journal, 31*, 25–31.

Morton, W. E., & Hearle, J. W. S. (1975). *Physical properties of textile fibres*. London, UK: The Textile Institute & William Heinemann Ltd.

Movilliat, P., & Monomakhoff, H. (1977). Ignition of gas mixtures by discharge of a person with static electricity. In *Proceedings of International Conference on Safety in Mines* (pp. 230–242). Varna, Bulgaria.

Ohara, K. (1979). Contribution of molecular motion of polymers to frictional electrification. In J. Powell (Ed), *Electrostatics '79* Vol. 48 (pp. 257–264). Bristol, UK: IOP Publishing Ltd.

Onogi, Y., Sugiura, N., & Nakaoka, Y. (1996). Dissipation of triboelectric charge into air from textile surfaces. *Textile Research Journal, 66*, 337–342.

Onogi, Y., Sugiura, N., & Matsuda, C. (1997). Temperature effect on dissipation of triboelectric charge into air from textile surfaces. *Textile Research Journal, 67*(1), 45–49.

Owens, J. E. (1984). *Hazards of personnel electrification: Nomex vs. NoMoStat*. Wilmington, DE: E. I. DuPont de Nemours & Co.

National Fire Protection Association (2000). *Liquid Splash-Protective Ensembles and Clothing for Hazardous Materials Emergencies* (pp. 33–34). Quincy, MA: NFPA.

Ramer, E. M., & Richards, H. R. (1968). Correlation of the electrical resistivities of fabrics with their ability to develop and to hold electrostatic charges. *Textile Research Journal, 38*, 28–35.

Rizvi, S. A. H., Crown, E. M., Gonzalez, J. A., & Smy, P. R. (1998). Electrostatic characteristics of thermal-protective garment systems at various low humidities. *Journal of The Textile Institute, 89*, 703–710.

Rizvi, S. A. H., Crown, E. M., Osei-Ntiri, K., Smy, P. R., & Gonzalez, J. A. (1995). Electrostatic characteristics of thermal-protective garments at low humidity. *Journal of The Textile Institute, 86*, 549–558

Rizvi, S. A. H., & Smy, P. R. (1992). Characteristics of incendive and non-incendive spark discharges from the surface of a charged insulator. *Journal of Electrostatics, 27*, 267–282.

Sano, Y., Konda, M., Lee, C. W., Kimural, Y., & Saegusa, T. (1997). A facile antistatic modification of polyester fiber based on ion-exchange reaction of sulfonate-modified polyester and various cationic surfactants. *Die Angewundte Mukromolekulure Chemie, 251*, 181–191.

Sclater, N. (1990). *Electrostatic discharge protection for electronics*. Blue Ridge: PA: Tab Books.

Scott, R. A. (1981). Static electricity in clothing and textiles. *Thirteenth Commonwealth Defence Conference on Operational Clothing and Combat Equipment* (Malaysia). Colchester, UK: Stores and Clothing Research and Development Establishment.

Sello, S. B., & Stevens, C. V. (1983). Antistatic treatments. In M. Lewin & E. M. Pearce (ed.), *Handbook of Fiber Science and Technology: Volume II, Part B* (pp. 291–315). New York, NY: Marcel Dekker, Inc.

Sereda, P. J., & Feldman, R. F. (1964). Electrostatic charging on fabrics at various humidities. *Journal of The Textile Institute, 55*, 288–298.

Seto, F., Muraoka, Y., Sakamoto, N., Kishida, A., & Akashi, M. (1999). Surface modification of synthetic fiber nonwoven fabrics with poly(acrylic acid) chains prepared by corona discharge induced grafting. *Die Angewandte Makromolekulare Chemie, 266*, 56–62.

Sharman, E. P., Hersh, S. P., & Montgomery, D. J. (1953). The effect of draw ratio and temperature on electrical conduction in nylon filaments. *Textile Research Journal, 23,* 793–798.

Shaw, P. E., & Jex, C. S. (1951). Static electricity on fabrics. *The British Journal of Applied Physics, 3,* 201–205.

Shirai, M. (1984). Electric charges produced by separating two compressed sheets of polyester. *Journal of Electrostatics, 15,* 265–268.

Stull, J. (1994). Measuring static charge generation on PC materials. *Safety & Protective Fabrics, 2* (8), 24–26.

Superintendent of Documents. (1969). *Electrostatic properties of materials* (Federal Test Method Standard 101B, Method 4046). Washington, DC: Government Printing Office.

Superintendent of Documents. (1990). *Determination of electrostatic decay of fabrics* (Federal Test Method Standard 191A, Method 5931). Washington, DC: Government Printing Office.

Taylor, D. M., & Elias, J. (1987). A versatile charge decay meter for assessing antistatic materials. In B. C. O'Neill (Ed), *Electrostatics '87* Vol. 85 (pp. 177–181). Bristol, UK: IOP Publishing Ltd.

Taylor, D. M., & Secker, P. E. (1994). *Industrial electrostatics: Fundamentals and measurements.* Somerset, England: Research Studies Press Ltd.

Teixeira, N. A., & Edelstein, S. M. (1954). Resistivity: One clue to the electrostatic behaviour of fabric. *American Dyestuff Reporter, 43,* 195–208.

Tolson, P. (1980). The stored energy needed to ignite methane by discharges from a charged person. *Journal of Electrostatics, 8,* 289–293.

Veghte, J. H., & Millard, W. W. (1963). *Accumulation of static electricity on arctic clothing* (Technical Report AAL-TDR-63-12). Fort Wainwright, AK: Arctic Aeromedical Laboratory, Aerospace Medical Division.

Wilson, D. (1963). The electrical resistance of textile materials as a measure of their anti-static properties. *Journal of The Textile Institute, 54,* 97–105.

Wilson, L. G., & Cavanaugh, P. (1972). *Electrostatic hazards due to clothing* (Report No. 665). Ottawa, ON: Defence Research Establishment Ottawa.

Wilson, N. (1977/78). The risk of fire or explosion due to static charges on textile clothing. *Journal of Electrostatics, 4,* 67–84.

Wilson, N. (1979). The nature and incendiary behaviour of spark discharges from the body. In *Electrostatics '79* (pp. 73–83). Bristol, UK: IOP Publishing.

Wilson, N. (1982). *Incendiary spark discharge due to static on clothing. Protective Clothing.* Manchester, England: Shirley Institute, 97–113.

Wilson, N. (1985). The nature and incendiary behaviour of spark discharges from textile surfaces. *Journal of Electrostatics, 16,* 231–245.

Wilson, N. (1987). Effects of static electricity on clothing and furnishing. *Textiles, 16*(1), 18–23.

Zhou, X., & Liu, P. (2003). Study on Antistatic Modification of Polyurethane Elastomer Surfaces by Grafting with Vinyl Acetate and Antistatic Agent. *Journal of Applied Polymer Science, 90,* 3617–3624.

Zimmer, E. (1970). Electrostatic charging of high-polymer insulating materials. *Kunststoffe, 60,* 456–468.

Ballistic protection

X CHEN and I CHAUDHRY, The University of Manchester, UK

19.1 Introduction

Ever since the beginning of humanity, warfare has existed. Man has searched for new and improved ways to harm his fellow man in order to impose his will upon his enemies. As long as man has developed weapons, he has simultaneously produced armour; protective clothing that would deflect or cushion the blow of a weapon. The first weapons were the fist and the club. The first type of armour was made from animal skins piled on top of each other to absorb the smack of a club. Warriors of ancient Rome and medieval Europe covered their torsos in metal plates before going into battle.[1] From the use of leather on Grecian shields, layered silk in ancient Japan, to chain mail and suits of armour in the Middle Ages, personnel protection has sought to protect the wearer from the corresponding advances in armaments.

The crude and unsophisticated armour of the Romans to the medieval knights of the Middle Ages established a trend towards armour modernisation. Gunpowder ended the development of armour for centuries until the famous Australian 'bushranger' Ned Kelly introduced effective armour in the 1850s. World War II and the Korean War were a renaissance for body armour.[1] Technological innovation and combat experimentation firmly re-established the requirement for effective body armour.

For many years the textile industry has focused attention and committed resources to technology for developing new materials to improve performance, comfort, efficiency, durability and reliability of body armour. The development of high-strength, high-modulus fibres in the 1960s ushered a new era of body armour that offered protection against small arms munitions and is capable of absorbing enormous amounts of energy quickly enough to change what would normally be a lethal wound into a bruise. Textile armours are by far the newest and most revolutionary materials in the body armour field today. Research and development through the last decades have resulted in 'state of the art' body armour in the hands of the military, police and security personnel, and gives an extra measure of confidence and protection with minimal impairment of their mobility on the battlefield.

19.2   History of body armours

19.2.1   Ancient body armours

Presumably the use of body armour extends back beyond historical records, when primitive warriors protected themselves with various types of materials as body armour to protect themselves from injury in combat and other dangerous situations. As civilisations became more advanced there came the developments of body armours for better protection, which may be classified into three main categories:[1]

1. Armour made of leather, fabric, or mixed layers of both, sometimes reinforced by quilting or felt
2. Chain mail, made of interwoven rings of iron or steel
3. Rigid armour made of metal, horn, wood, plastic, or some other similar tough and resistant material.

In the 11th century BC, Chinese warriors wore armour made of five to seven layers of rhinoceros skin, and the Mongols similarly used ox hides in the 13th century AD. Fabric armour with thick, multi-layered linen cuirasses was worn by the Greek heavy infantry of the 5th century BC. Medieval Japanese warriors used thick body armour made from densely woven silk. Quilted linen coats were worn in northern India until the 19th century.[2]

Ancient Greek infantry soldiers used shields of wood, leather and plate armour consisting of a cuirass. The Roman legionary used leather vestments, chain mail, bronze breastplates and a cylindrical cuirass.[3] Cuirass is body armour that protects the torso of the wearer above the waist or hips. Originally it was a thick leather garment covering the body from neck to waist, consisting of a breastplate and a back-piece fastened together with straps and buckles and a gorget, a collar protecting the throat.

Armour made of large plates was probably unknown in Western Europe during the Middle Ages, and chain mail was the main defence of the body and limbs during the 12th and 13th centuries. However, plate armour of steel superseded chain mail during the 14th century. In about 1510 the Germans used a metal suit with flexible joints covering the wearer literally from head to toe, with only a slit for the eyes and small holes for breathing in a helmet of forged metal.

The plate armour used during the European Middle Ages was composed of large steel or iron plates that were linked by loosely closed rivets and by internal leathers to allow the wearer maximum freedom of movement. Plate armour dominated European armour design until the 17th century, by which time the use of firearms had made body armour in general obsolete. The musket and cannon ball and later the bullet and artillery shell brought a new level of lethality to battlefield.[3]

With the advent of firearms (c.1500), most of the traditional body armour was no longer effective. In fact, the only real protection available against firearms was man-made barriers, such as stone or masonry walls, or natural barriers, such as rocks, trees, and ditches. Gunpowder ended the development of armour for centuries until the famous Australian 'bushranger' Ned Kelly, a formidable foe, developed an effective suit of boiler-plate iron body armour and used it for personal profit in the 1850s. In 1880, the Australian police captured him after he received a few rounds in his lower extremities, as his armour did not cover his legs.[1]

19.2.2   Modern body armours

Projectile shielding for the individual soldier, which was pioneered by ancient civilisations has seen technological rebirth in the 20th century. World War I produced a large number of experiments in the body armour field. France developed a number of chest, thigh and leg protectors, but for some reason never issued them to the field units. The English produced 18 different body shield designs for commercial use, which included some 'soft armours' with padded neck defences and vests with linen, tissue, cotton and silk. From 1917–1918 the British produced a corset known as the E.O.B. Evidently very efficient, it consisted of a metallic breast and back plate with abdomen protection. Germany made by far the most extensive use of body armour. They produced shields weighing 9–11 kg composed of a large metallic breastplate with flat hook-like shoulder harnesses. Secured to the bottom by two long straps were three plates for groin protection. Designed for protection, not mobility, they were primarily used by machine gun crews.[3] During the 1920s and 1930s 'bullet-proof vests', composed of overlapping steel plates sewn to strong fabric garments appeared in America. They were heavy and provided good protection from pistol projectiles but reduced individual mobility.

In World War II, body armour of modern times 'flak jackets' were used by the British Royal Air Force bomber crews and in 1942 these were essentially copied and issued by the US Army Air Corps to its B-17 and B-24 bomber crews flying over Europe. Flak jackets had steel plates sewn into them and were incredibly cumbersome and some aircrew preferred to just sit on them. They could not stop a modern high-velocity bullet but, as their name implied, they could stop shards of shrapnel from exploding German anti-aircraft shells. The 'flak jacket' provided protection primarily from munition fragments, was ineffective against most pistol and rifle threats, and was restricted to military use only.[3] Since then the development of lightweight fibrous material systems to resist penetration by high-velocity bullets has been an important research topic.

19.2.3   Modern lightweight body armours

Modern armours born during World War II and the Korean War became the testing grounds for more capable and lightweight body armour. Research and development through the last decades have resulted in the state of the art body armour of today. Prior to the Korean War, the materials used as armour protection were relatively simple in form and basic in composition. Technical advances in our ability to produce composite materials for ballistic protection have changed rapidly in the past years. Reinforced plastic, ceramics and textile materials have been developed for lightweight body armour. In the 1950s ballistic nylon was originally developed in Korea and over 1,400 armoured nylon vests (T-52-1) and (T-52-2) were tested by aircrews and ground organisations from six different countries. In the early 1960s, ceramic armour provided the first technical breakthrough by reducing weight while improving ballistic protection. It was not until the late 1960s that new textile fibres were discovered which revolutionised the modern generation of concealable lightweight body armours.[3]

In early 1965 Dupont polymer research group developed a new aramid fibre called Kevlar[®]16 which proved stronger than steel and had better ballistic resistance. Kevlar[®] brought revolutionary improvements in body armour performance.[4,23] Since the advent of Kevlar[®] there has been intensive research and development and adaptation of textile fibres and fabrics being used in modern body armour.

19.3   Ballistic protective materials

Modern body armour is divided into two main categories; hard body armour and soft body armour.

19.3.1   Hard body armour

Hard body armour is made from hard ballistic materials like ceramic, plastic or metal plates. Such materials are hard enough to prevent stabbing and slashing injuries from edge weapons. In the early 1960s, a composite material of ceramic (aluminium oxide) with a fibreglass laminate was developed to stop high-energy projectiles. Nowadays armour grade ceramics include aluminium oxide, silicon carbide, and boron carbide. Reinforced plastic armour, also called glass-reinforced-plastic (GRP) was a combination of a glass weave fibre and a chemical resin. It required additional backing for support when used on garments in small overlapping plates. It was good for low-velocity bullets, blast and grenade/mortar fragments but not for rifle calibres. It is lightweight, but expensive. The sleeveless zipper front vest used in the Vietnam era was made of this material.

19.3.2   Soft body armour

Soft body armour is made from manmade polymeric lightweight fibrous materials that exhibit greatly improved ballistic resistance performance. Textile armour materials include aramids (e.g., Kevlar[®], Twaron[®], Technora[®]), highly oriented ultra high molecular weight polyethylene (e.g., Spectra®, Dyneema®), PBO (e.g. Zylon®) is p-phenylene-2-6-benzobisoxazole, polyamide (e.g. Nylon®) and more, such as a new polymeric fibre PIPD (Polypyridobis-imidazole) with high compressive strength (referred to as M5®) are under development. Some of these materials, e.g., aramids are the most widely used ballistic protective materials in today's state of the art body armour, while others have shown great potential for ballistic protection, but are yet to be fielded.

Kevlar®

Developed by Dupont, this is widely used in the modern generation of light-weight concealable body armours with drastically improved ballistic protection over their predecessors. Kevlar® fibres consist of long molecular chains produced from poly-paraphenylene terephthalamide. The chains are highly oriented with strong interchain bonding that results in a unique combination of properties, which include high tensile strength at low weight, low elongation to break, high modulus (structural rigidity), low electrical conductivity, high chemical resistance, low thermal shrinkage, high toughness (work-to-break), excellent dimensional stability, high cut resistance and flame resistance. Kevlar® fibre does not melt or soften and is unaffected by immersion in water, although its ballistic properties are affected by moisture. It is five times stronger than steel on an equal weight basis, and it is lightweight, flexible and comfortable.[5,23] Kevlar® fibres can be processed by textile manufacturers with little difficulty.

Kevlar® 29, introduced in the early 1970s, was the first generation of bullet resistant fibres to make the production of flexible, concealable body armour practical for the first time. In 1988, DuPont introduced Kevlar® 129 that offered increased ballistic protection capabilities against high-energy impact. In 1995, Kevlar® Correctional was introduced, which provides puncture resistant technology to both law enforcement and correctional officers against puncture type threats from knives and other sharp objects, in particular hypodermic needles.[5]

Table 19.1 Different types of Kevlar® fibres

Kevlar type	Characteristics
Kevlar 29	720D, 1000D, 1500D
Kevlar 49	1420D, 2840D
Kevlar 68	Medium type between 29 and 49
Kevlar 100	Coloured Kevlar
Kevlar 119	High durable
Kevlar 129	High tenacity
Kevlar Protera	High tenacity, High absorption
Kevlar 149	Ultra high modulus

In 1996 Dupont made another addition to the Kevlar® line by introducing Kevlar® Protera. Kevlar® Protera is a high-performance fabric that allows lighter weight, more flexibility, and greater ballistic protection in a vest design due to the molecular structure of the fibre. It is reported that its tensile strength and energy-absorbing capabilities have been increased by the development of a new spinning process.[5] Table 19.1 lists different types of Kevlar® fibres with their characteristics.

Twaron®

This is another para-aramid fibre developed by Akzo Nobel (now Teijin Twaron), popularly used in modern ballistic body armours. According to Akzo Nobel, the yarn uses 1,000 or more finely spun single filaments that act as an energy sponge, absorbing a bullet's impact and quickly dissipating its energy through engaged and adjacent fibres. Because more filaments are used, the impact is dispersed more quickly. It is reported that Twaron® CT microfilament is 23% lighter than a vest made of standard aramid yarn. The material has a much higher number of fine microfilaments, enhancing its capacity to absorb impact and bringing distinct benefits in personal protection. Twaron® CT microfilament is believed to be lighter, softer and more flexible, and offers more freedom of movement and comfort. Twaron CT Microfilament 930 dtex has as many as 1,000 filaments, up to 50% more filaments than para-aramid yarns of the same weight used so far.[6]

Technora®

Another para-aramid fibre, this has been developed by Teijin, although there is not much evidence of it being used in modern body armour at the time of writing. Technora offers good tensile strength and modulus, chemical and hydrolytic resistance, excellent fatigue resistance, good dimensional stability and good heat resistance. The stiff and highly oriented molecular structure of Technora fibre enables low creep and stress relaxation.[7]

Spectra®

This fibre, manufactured by AlliedSignal Inc. (now Honeywell), is an ultra-high-strength polyethylene fibre. Ultra high molecular weight polyethylene is dissolved in a solvent and fibres are produced through the gel-spinning process. In general, Spectra® fibres are weight for weight, ten times stronger than steel, more durable than polyester and has a specific strength that is 40% greater than that of aramid fibres.[25] The Spectra® series includes fibre 900, 1000, and 2000, with fibres 1000 and 2000 especially suitable for police and military ballistic protection. Table 19.2 lists some of the physical properties of Spectra fibres.

Table 19.2 Some physical properties of Spectra® fibres

Properties	Spectra® fibre		
	900	1000	2000
Ultimate tensile strength (Gpa)	2.18–2.61	2.95–3.25	3.21–3.51
Modulus (Gpa)	62–79	98–113	113–124
Elongation (%)	3.6–4.4	2.9–3.4	2.8–3.1
Density (g/cm³)		0.97	

DSM Dyneema®

Another HPPE fibre like Spectra, this is made of ultra-high-strength gel-spun polyethylene used in body armour. Dyneema® has an extremely high strength-to-weight ratio and is light enough to float on water. It has high-energy absorption characteristics and dissipates shock waves faster than earlier ballistic materials. Figure 19.1 compares the specific energy absorption of Dyneema® SK60 to other high-energy absorbent fibres. Figure 19.2 illustrates the tensile modulus and strength of Dyneema® fibres. Different types of Dyneema® fibres include SK25, SK6O, SK65, SK66, SK71, SK75 and SK76. DSM has also developed soft ballistic panels, e.g., SB21 and SB31, and hard ballistic panels, e.g., HB2 and HB25 for ballistic vests and inserts.[8]

Zylon® (PBO)

This is a new high-performance fibre developed by Toyobo being used in modern body armour. It consists of rigid-rod chain molecules of poly (p-

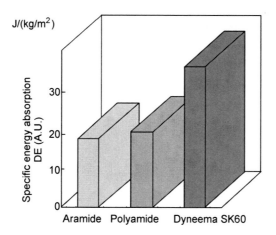

19.1 Energy absorption of Dyneema® SK60.

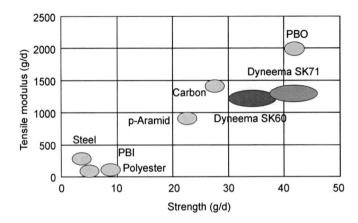

19.2 Tensile modulus and strength of Dyneema® fibres in comparison with other fibres.

phenylene-2,6-benzobisoxazole). The chemical structure is depicted in Fig. 19.3. Zylon® fibre has strength and modulus almost double that of p-aramid fibre and shows 100C higher decomposition temperature than p-aramid fibre. The limiting oxygen index is 68, which is the highest amongst organic super fibres. These properties are displayed in comparison with other high-performance fibres in Fig. 19.4. There are two types of Zylon® fibres, AS (as spun) and HM (high modulus). Zylon® body armour is believed to be lighter, more comfortable and stronger than aramid body armour. However, the tensile strength of Zylon® fibre might be susceptible to degradation under certain extreme temperature and humidity conditions.[8]

Nylon®

This is a polyamide fibre. Nylon fibres were dominant prior to 1972, and these showed considerable non-linearity in stressstrain response, with relatively high strains to failure. Nylon 6-6 (ballistic nylon) has a high degree of crystallinity and low elongation and is still used in body armour.[9–11]

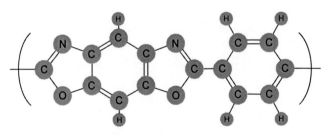

19.3 Chemical structure of Zylon®.

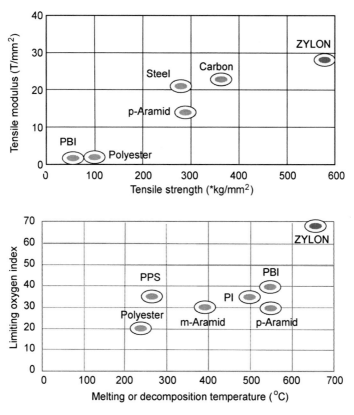

19.4 Comparison between Zylon® and other HP fibres: (a) mechanical properties (b) flame retardancy.

19.4 Fabric structures used for body armour

Bullet-resistant fabric is made from high-strength and high-modulus fibres, such as Kevlar®, Twaron® and Dyneema®. The ballistic performance of a woven fabric is dependent on the dynamic mechanical properties of constituent fibres as well as the fabric geometry (i.e. type of weave, fibres per yarn, weave density, etc.). Usually, the ballistic fabric is densely woven square plain weave (shown in Fig. 19.5) and basket weave. It has been observed that loosely woven fabrics and fabrics with unbalanced weaves result in inferior ballistic performance. The packing density of the weave, indexed by the 'cover factor', is determined from the width and pitch of the warp and weft yarns and gives an indication of the percentage of gross area covered by the fabric.[28] It is recognised that fabrics should possess cover factors from 0.6 to 0.95 to be effective when utilised in ballistic applications. When cover factors are greater than 0.95, the yarns are typically degraded by the weaving process and when cover factors fall below

19.5 Plain weave fabric.

0.6, the fabric may be too 'loose' to be protective.[12–14] Figure 19.6 illustrates the capture of a bullet by a plain structured woven fabric.

Loosely woven fabrics are more susceptible to having a projectile 'wedge through' the yarn mesh. When a projectile strikes a layer of fabric, the fabric deflects transversely and the mesh of yarns is distended, resulting in the enlargement of the spaces between the yarns. If the projectile is relatively small and/or it impacts the fabric at an angle, the projectile can slip through the opening in the fabric by pushing yarns aside instead of breaking them.[15] To prevent yarn sliding when impacted by a bullet, leno structures have also been used for constructing ballistic fabrics. Because of the self-locking effect of such structures, the yarns are less likely to have relative movement to produce openings in the fabric. However, the leno fabrics can only be made to have low yarn packing density in the fabric, which may adversely affect ballistic performance.

In order for the yarns in ballistic fabrics to dissipate impact energy more quickly and to be more protective, unidirectional yarns are used to produce ballistic panels, both soft and hard, most notably by DSM (Dyneema® UD) and Honeywell (Spectra® Shields[25]). In Dyneema® UD all the fibres are laid in

19.6 Illustration of a ballistic fabric capturing a bullet.

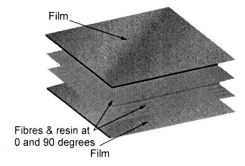

Film

Fibres & resin at
0 and 90 degrees

Film

19.7 Construction of Spectra® Shield.

parallel, in the same plane, rather than being woven together. Dyneema® UD is made of several layers of Dyneema® fibres, with the direction of fibres in each layer at 90 to the direction of the fibres in the adjacent layers. The unidirectional configuration of the fibres in Dyneema® UD allows the energy transferred from the impact of a bullet or other threat to be distributed along the fibres much faster and more efficiently than in conventional woven fabrics. This is because the absorption power of the yarn in woven fabrics is lost at the crossover points, as these points reflect rather than absorb the shockwaves of the impact. In Dyneema® UD, much more of the material is engaged in stopping the bullet, making it more effective against ballistics. Figure 19.7 demonstrates the composition of SpectraShield, where unidirectional Spectra® fibres in one layer are laid at 90° to fibres in its adjacent layer.[8,25]

In recent years research has been carried out to produce ballistic fabric by the nonwoven method and it is reported that nonwoven ballistic fabric have strong weight advantages over woven fabrics. However, this is offset by a significant increase in the bulk of the system, and unidirectional materials are less flexible than their woven counterparts. It is claimed that nonwoven fabrics allow a great deal of moisture and heat transfer compared to light weaves, which would make the ballistic vest more comfortable to wear.[5] However, moisture degrades ballistic performance, so the packs have to be kept dry by encapsulation.

19.5   Working mechanism of body armour

The main principle behind body armour is the rapid conversion and dispersal of kinetic energy from a striking bullet to strain energy of the ballistic body armour system. Modern body armours provide protection by three different methods: (i) the armour totally rejects the projectile by bouncing it off, although this is not common; (ii) the armour retards and stops the projectiles by dissipating kinetic energy along the plane of the impacted material; and (iii) combination of (i) and (ii).[19–20]

In order to describe the ballistic impact on fabric it is necessary to understand the transverse impact on a fibre. When a projectile strikes on a fibre, two waves,

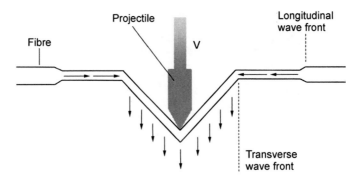

19.8 Projectile impact into fibre.

longitudinal and transverse, propagate from the point of impact. The longitudinal tensile wave travels down the fibre axis at the sound speed of the material. As the tensile wave propagates away from the impact point, the material behind the wave front flows toward the impact point, which is deflected in the direction of motion of the impacting projectile. This transverse movement of the fibre is the transverse wave, which is propagated at a velocity lower than that of the material.[12] Figure 19.8 illustrates the propagation of the two waves.

When a bullet strikes on body armour, it is caught in a 'web' of very strong fibres. These fibres absorb and disperse the impact energy that is transmitted to the vest from the bullet, causing the bullet to deform or to 'mushroom'. Additional energy is absorbed by each successive layer of material in the vest, until such time as the bullet has been stopped. The energy of the impact spreads across the surface of the vest at a tremendous speed (up to 900 m/s). Because the fibres work together both in the individual layer and with other layers of material in the vest, a large area of the garment becomes involved in preventing the bullet from penetrating. This also helps in dissipating the forces that can cause nonpenetrating injuries (what is commonly referred to as 'blunt trauma')

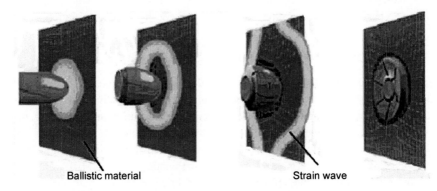

Ballistic material   Strain wave

19.9 Bullet penetration process.

to internal organs.[15–18] This process is illustrated in Fig. 19.9. Metal and reinforced plastic armour protects by partially rejecting the projectile and partially absorbing the impact energy.

19.6 United States NIJ test methods for bullet resistant armours

19.6.1 Brief history of NIJ body armour test standard

Faced with the sharp increase in the number of police fatalities in the 1960s in the USA, the National Institute of Law Enforcement and Criminal Justice (NILECJ), predecessor of the National Institute of Justice (NIJ), initiated a research programme to investigate development of lightweight body armour that on-duty police could wear full time. The investigation readily identified new materials that could be woven into a lightweight fabric with excellent ballistic-resistant properties. In a parallel effort, the National Bureau of Standards (now known as the National Institute of Standards and Technology) Law Enforcement Standards Laboratory (now known as the Office of Law Enforcement Standards) developed a performance standard that defined ballistic-resistant requirements for police body armour. The US Army, the Lawrence Livermore Laboratory, the Federal Bureau of Investigation (FBI) and the Secret Service were also involved in the research program.[21] From 1971 to 1976, more than $3 million of NIJ funds were devoted to the development of body armour. The development of body armour by NIJ was a four-phase effort that took place over several years.

The first phase involved testing Kevlar fabric to determine whether it could stop a lead bullet. The second phase involved determining the number of layers of material necessary to prevent penetration by bullets of varying speeds and calibres and developing a prototype vest that would protect officers against the most common threats. The third phase of the initiative involved extensive medical testing to determine the performance level of body armour that would be necessary to save wearers' lives. The final phase involved monitoring the armour's wearability and effectiveness. The demonstration project armour issued by NIJ was designed to ensure a 95% probability of survival after being hit with a .38 calibre bullet at a velocity of 800 ft/s. Furthermore, the probability of requiring surgery if hit by a projectile was to be 10% or less. Finally NIJ released a report in 1976 concluded that the new ballistic material was effective in providing a bullet-resistant garment that was light and wearable for full-time use.[21]

The NIJ standard for police body armour has gained worldwide acceptance as a benchmark to judge the effectiveness of a given body armour model. In response, NIJ is reaching out to the international community in a co-operative effort for the development of future revisions of the standard.[21]

19.6.2   NIJ classification for body armour

The performance requirements of the latest NIJ Standard-0101.04, which were developed with the active participation of body armour manufacturers and users, ensure that each armour type will provide a well-defined minimum level of ballistic protection. Following are the NIJ's Standard-0101.04 classification types.

Type I (22 LR; 38 Special)

This armour protects against .22 Long Rifle High-Velocity lead bullets, with nominal masses of 2.6 g (40 grains), impacting at a velocity of 320 m/s (1,050 ft/s) or less, and against .38 Special round nose lead bullets, with nominal masses of 10.2 g (158 gr), impacting at a velocity of 259 m/s (850 ft/s) or less. It also provides protection against most other .25 and .32 calibre handgun rounds. Type I body armour is light. This is the minimum level of protection every wearer should have. Type I body armour was the armour issued during the NIJ demonstration project in the mid-1970s.

Type II-A (Lower Velocity 357 Magnum; 9 mm)

This armour protects against 357 magnum jacketed soft-point bullets, with nominal masses of 10.2 g (158 gr), impacting at a velocity of 381 m/s (1,250 ft/s) or less, and against 9 mm full-metal-jacketed bullets, with nominal masses of 8.0 g (124 gr), impacting at a velocity of 332 m/s (1,090 ft/s) or less. It also provides protection against such threats as 45 Auto., 38 Special +P, and some other factory loads in calibre 357 magnum and 9 mm, as well as the Type I threats. Type II-A body armour is well suited for full-time use by police departments, particularly those seeking protection for their officers from lower velocity 357 magnum and 9 mm ammunition.

Type II (Higher Velocity 357 Magnum; 9 mm)

This armour protects against 357 magnum jacketed soft-point bullets, with nominal masses of 10.2 g (158 gr), impacting at a velocity of 425 m/s (1,395 ft/s) or less, and against 9 mm full-jacketed bullets, with nominal masses of 8.0 g (124 gr), impacting at a velocity of 358 m/s (1,175 ft/s). It also protects against most other factory loads in calibre 357 magnum and 9 mm, as well as the Type I and II-A threats. Type II body armour is heavier and more bulky than Types I and II-A. It is worn full time by officers seeking protection against higher velocity 357 magnum and 9 mm ammunition.

Type III-A (44 Magnum; Submachine Gun 9 mm)

This armour protects against 44 magnum, lead semi-wad cutter bullets with gas checks, nominal masses of 15.55 g (240 gr), impacting at a velocity of 426 m/s

(1,400 ft/s) or less, and against 9 mm full-metal-jacketed bullets, with nominal masses of 8.0 g (124 gr), impacting at a velocity of 426 m/s (1,400 ft/s) or less. It also provides protection against most handgun threats, as well as the Type I, II-A, and II threats. Type III-A body armour provides the highest level of protection currently available from concealable body armour and is generally suitable for routine wear in many situations. However, departments located in hot, humid climates may need to evaluate the use of Type III-A armour carefully.

Type III (High-powered Rifle)

This armour, normally of hard or semi rigid construction, protects against 7.62 mm full-metal jacketed bullets (US military designation M80), with nominal masses of 9.7 g (150 gr), impacting at a velocity of 838 m/s (2,750 ft/s) or less. It also provides protection against threats such as 223 Remington (5.56 mm FMJ), 30 Carbine FMJ, and 12-gauge rifled slug, as well as the Type I through III-A threats. Type III body armour is clearly intended only for tactical situations when the threat warrants such protection, such as barricade confrontations involving sporting rifles.

Type IV (Armour-Piercing Rifle)

This armour protects against 30-06 calibre armour-piercing bullets (U.S. military designation APM2), with nominal masses of 10.8 g (166 gr) impacting

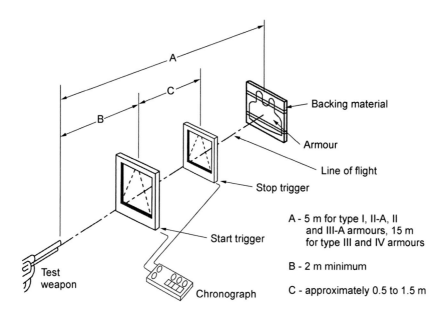

19.10 Ballistic test set-up.

Table 19.3 Summary of the ballistic test variables and the performance requirements

Armour type	Test round	Test ammunition	Nominal bullet mass (grams)	Min. required bullet velocity (m/s)	Required fair hits at 0° angle of incidence	Max. depth of deformation (mm)	Required fair hits at 30° angle of incidence
						Performance requirements	
			Test variables				
I	1	38 Special RN Lead	10.2	259	4	44	2
	2	22 LRHV Lead	2.6	320	4	44	2
II-A	1	37 Magnum JSP	10.2	381	4	44	2
	2	9 mm FMJ	8.0	332	4	44	2
II	1	37 Magnum JSP	10.2	425	4	44	2
	2	9 mm FMJ	8.0	358	4	44	2
III-A	1	44 Magnum Lead SWC Gas Checked	15.55	426	4	44	2
	2	9 mm FMJ	8.0	426	4	44	2
III		7.62 mm FMJ	9.7	838	6	44	0
IV		30-06 AP	10.8	868	1	44	0

Abbreviations: AP – Armour piercing, FMJ – Full metal jacket, JSP – Jacketed soft point, LRHV – Long rifle high velocity, RN – Round nose, SWC – Semi-wadcutter.

at a velocity of 868 m/s (2,850 ft/s) or less. It also provides at least single-hit protection against the Type I through III threats. Type IV body armour provides the highest level of protection currently available. Because this armour is intended to resist armour-piercing bullets, it often uses ceramic materials. Such materials are brittle in nature and may provide only single-shot protection, since the ceramic tends to break up when struck.[21] Figure 19.10 depicts the ballistic test layout where a chronograph is used to determine bullet velocity. Table 19.3 summarises the ballistic test variables and the performance requirements.

19.6.3 Recommended target strikes

In order to evaluate the performance of body armour or ballistic body armour, The NIJ has recommended a scheme of strikes. Figure 19.11(a) suggests the shot positions and the shot sequence on the front panel of ballistic body armour, and Fig. 19.11(b) exhibits the shot positions and sequence for optional second ammunition shot. Based on the NIJ 0101.04, all shots must be at least 7.6 cm from any edge and must be at least 5 cm from another shot.

19.7 Design and manufacture of ballistic body armour

Ballistic body armours are expected to provide protection against projectile penetration and the blunt trauma it causes. The fatal consequences of a bullet hitting the human body are well recognised. Blunt trauma can cause severe injuries to vital organs of the body such as heart, liver, lungs and kidneys, etc., which can lead to death. Neurological consultants believe that blunt trauma impact over the spine of a human might cause immediate damage and contusion to the spinal cord. It can also cause bleeding in the stomach cavity, abdominal pain and fractured ribs.[22]

The design of ballistic-resistant armour requires identification of the threat, selecting a material or combination of materials that will resist that threat, and determining the number of layers of material necessary to prevent both

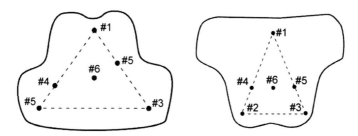

19.11 Body armour test pattern: (a) test ammunition shot series (b) optional second ammunition shot series.

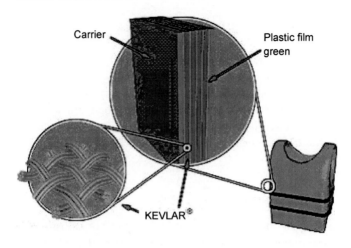

Carrier

Plastic film
green

KEVLAR®

19.12 Typical layered structure of a body armour.

penetration and blunt trauma injury. In designing armour, the armour's final weight is an important factor in the selection of the ballistic-resistant material or materials to be used. The goal is to design the lightest possible unit that achieves the desired protection while still providing comfort and not restricting movement. In a bulletproof vest, several layers of bullet-resistant webbing (such as Kevlar) are sandwiched between layers of plastic film to keep the textiles dry. These layers are then woven to the carrier, an outer layer of traditional clothing material. Figure 19.12 shows a typical layered structure of body armour. There are two principal markets for ballistic body armour, i.e. the military and the police. Other end users of body armour include the security agencies. Body armour is designed according to its applications.

A ballistic panel of body armour may be constructed from a single fabric style or from two or more styles in combination. The location and number of layers of each style within the multiple-layer ballistic panel influence the overall ballistic performance of the panel. Some manufacturers also coat the ballistic fabric with various materials. The manufacturer may add a layer of non-ballistic material for the sole purpose of increasing blunt trauma protection. The manner in which ballistic panels are assembled into a single unit also differs from one manufacturer to another. In some cases, the multiple layers are bias stitched around the entire edge of the panel, whereas in others the layers are tack-stitched together at several locations. Some manufacturers assemble the fabrics with a number of rows of vertical or horizontal stitching; some may even quilt the entire ballistic panel. It is suggested that stitching tends to improve the deformation performance, especially in cases of blunt trauma, depending upon the type of fabric used. It also improves multi-hit capability.

19.7.1   Designs of vests

Basically, ballistic vests are made in general forms, namely, covert vests and overt vests for different purposes. A covert vest is a concealable vest, which is used under a shirt with Velcro/elastic fastening. These are lighter in weight and wash easily and dry quickly. Typically, concealable body armour is constructed of multiple layers of ballistic fabric or other ballistic resistant materials, assembled into the 'ballistic panel'. The ballistic panel is then inserted into the 'carrier', which is constructed of conventional garment fabrics. The ballistic panel may be permanently sewn into the carrier or may be removable. Usual protection levels provided by the covert vests are NIJ II-A, II & III-A. Overt vests are worn over the uniform, heavier in weight, do not wash easily and dry quickly. One common design of overt vest is to fashion pockets on the inside or outside of the vest. When extra protection is needed, ceramic plates are inserted into the pockets. For example, hard armour plates are used for groin protection. These vests are sometimes for protection levels III-A, III, and IV.

19.7.2   Designs for female ballistic vests

For a long time in the past, body armours were produced only for male users, probably because personnel who worked in the forefront were mostly men. Nowadays, female soldiers and female police officers are working side by side with their male counterparts, and modern body armour is designed and manufactured for both male and female users. Male body armour is relatively easy to manufacture, as the ballistic panels are flat pieces. Figure 19.13 shows how the measurements are taken to make comfortable ballistic male and female vests.[24]

In the design and making of female body armour, the contours of the female torso must be taken into consideration in order to make the body armour fit the body. There are two different ways of forming the double curvature to suit the bust. One is by folding the flat ballistic fabric. Although this method is simple and obviates cutting the fibres, it raises a ridge because of the folding. When tens of layers of the ballistic fabrics are layered together, the ridges, which have to be gathered at the armpit area, become substantial and therefore make the body armour less comfortable to wear. The second method is by cutting a dart in the fabric and then sewing the fabric together. This method leads to a neater shape but will inevitably reduce protection especially in the bust area as the fibres become discontinuous in the ballistic fabric layers. It is still a challenge to provide the same level of protection and comfort in female body armour.

19.7.3   Layers and protection level

Depending on protection level and fabric style, the number of layers in soft ballistic vests ranges from 7–50 layers. Table 19.4 relates the number of layers (or thickness) of ballistic materials to the protection level body armour is like to

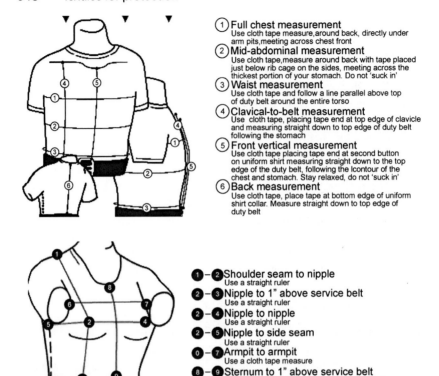

① **Full chest measurement**
Use cloth tape measure,around back, directly under arm pits,meeting across chest front

② **Mid-abdominal measurement**
Use cloth tape,measure around back with tape placed just below rib cage on the sides, meeting across the thickest portion of your stomach. Do not 'suck in'

③ **Waist measurement**
Use cloth tape and follow a line parallel above top of duty belt around the entire torso

④ **Clavical-to-belt measurement**
Use cloth tape, placing tape end at top edge of clavicle and measuring straight down to top edge of duty belt following the stomach

⑤ **Front vertical measurement**
Use cloth tape placing tape end at second button on uniform shirt measuring straight down to the top edge of the duty belt, following the Icontour of the chest and stomach. Stay relaxed, do not 'suck in'

⑥ **Back measurement**
Use cloth tape, place tape at bottom edge of uniform shirt collar. Measure straight down to top edge of duty belt

❶–❷ **Shoulder seam to nipple**
Use a straight ruler

❷–❸ **Nipple to 1" above service belt**
Use a straight ruler

❷–❹ **Nipple to nipple**
Use a straight ruler

❷–❺ **Nipple to side seam**
Use a straight ruler

❻–❼ **Armpit to armpit**
Use a cloth tape measure

❽–❾ **Sternum to 1" above service belt**
Start this measurement from top of Sternum notch Use a cloth tape measure

❿–⓫ **Rear measurement.: side seam to side seam across back 2" above the service belt**
Use a cloth tape measure

⓬–⓭ **Back measurement: from bottom of uniform shirt collar to top of service belt**
Use cloth measure for this measurement

19.13 Measurements for body armours: (a) measurements for male body armour (b) measurements for female body armour.

Table 19.4 Number of layers vs. body armour protection level

Number of layers (thickness)	Protection level
16–20 layers of Kevlar	Level II-A
20–24 layers of Kevlar	Level II
24–28 layers of Kevlar	Level III-A
1/4″ Steel	Level III
1/2″ Steel	Level IV

provide.[21] However, ballistic performance is more accurately related to areal density of fabrics which currently range from 90–250 gm^{-2}.

19.8   Ballistic helmets

The combat helmet is one of the most universal forms of armour and is also among the most ancient forms of military equipment.[2] Steel helmets appeared as a standard item for infantry in the opening years of World War I because it protected the head against the high-velocity metal fragments of exploding artillery shells. The French first adopted the helmet as standard equipment in late 1914 and were quickly followed by the British, the Germans, and then the rest of Europe.[31] The typical helmet was a hardened-steel shell with an inner liner and weighed about 0.5 to 1.8 kg. In 1984 Britain replaced steel helmets by ballistic Nylon helmets for general service. Some specialist helmets for tank crews, bomb disposal teams and Special Forces are made from para-aramid composites (see also Chapter 21).

19.8.1   Modern ballistic helmets

In the early 1970s, the US Army's Natick Research Lab developed a lighter and more protective PASGT (Personal Armor System Ground Troops) helmet using aramid fibre Kevlar, which replaced the old steel pot helmet.[27] PASGT is a one-piece structure composed of multiple layers of Kevlar 29 ballistic fibre and phenol-formal-dehyde (PF) and polyvinyl butyral (PVB) resin.[27,29] Since then, various types of aramid fibres have become the raw materials for making ballistic helmets.

The inner harness of modern helmets is made from a combination of high-quality materials such as leather, polyester fabric, Nylon webbing, PP webbing, Nylon fittings, and brass press studs. The helmet shell can also be fitted with a deformable plastic foam liner to cope with blunt impacts. Each material is selected to provide maximum performance and comfort for each aspect of the harness design. The harness is mounted to the helmet shell using stainless steel screws and inserts that ensure ballistic integrity. The helmet shell consist of a shell body with PVC edge trimming, 5-point suspension system assembly, a head band with brow and back cushioning pads, a perforated crown, a chin strap fitted with quick release buckles, and a chin cup.[35] Ballistic helmets are produced in different sizes, ranging from extra small, small, medium, large, and extra large. Table 19.5 shows the relationship among the helmet sizes, weight, and protection levels.

The head represents approximately 9% of the body area exposed in combat yet receives approximately 20% of all 'hits'.[30] The desirability of protecting this vital structure would appear self-evident. Helmet design is a complex issue. Factors that designers consider include weight, ballistic qualities of the

Table 19.5 Helmet sizes, weight, and protection level

| Size | Level | | |
| | II | II+ | III-A |
		Weight (Kg)	
Small	1.15	1.20	1.25
Medium	1.20	1.25	1.30
Large	1.25	1.30	1.35

construction material, balance, helmet-to-person interface (comfort), maintenance of vision and hearing, equipment and weapon compatibility, ease of modification, available materials and manufacturing techniques, durability, ease of decontamination, disposability, and cost.[36] Figure 19.14 displays the use of inner pads in a helmet as the helmet-to-person interface.

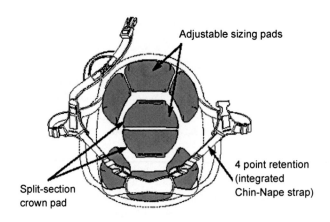

Adjustable sizing pads

4 point retention (integrated Chin-Nape strap)

Split-section crown pad

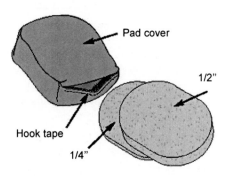

Pad cover

1/2"

Hook tape

1/4"

19.14 Inner pads of helmet.

19.8.2   Test standards for ballistic helmets

NIJ standard 0106.01 has been used to evaluate the performance of ballistic helmets. This standard classifies the ballistic helmet into Type I, Type II-A, Type II and Special Type according to the level of performance. Table 19.6 summarises the helmet types and the test variables and performance requirements.

19.8.3   Test procedure for ballistic helmets

According to NIJ standards, testing is to be conducted with test samples that have been thoroughly temperature conditioned at +70.0 ($\pm$5.0) degrees Fahrenheit for a minimum of four hours. Subsequent to temperature conditioning, the test samples shall be wet-conditioned, so that water spray is applied to the outer helmet shell only in accordance with the wet armour conditioning procedures of NIJ-STD-0101. Helmet samples will be tested in an environment with an ambient air temperature of +70.0 ($\pm$10.0) degrees Fahrenheit.[32-34] Helmet penetration and helmet deformation tests can then be conducted to evaluate ballistic impact on the helmet.

In order to evaluate ballistic performance, five impact areas are to be marked on the helmet during preparation and before firing. The first is the crown area, which is defined as the 6-inch-diameter circle centred about a point on the top of the helmet. The second is the front-45° on either side of the sagittal centreline extending from the crown marking to the lower edge. The third area is the rear-45° area on either side of the sagittal centreline. The fourth and fifth areas are on the right and left sides between the front and rear areas extending from the crown to the lower edge.[32-34] The helmet test samples are positioned on a rigidly mounted head form at a distance of 5 m from the muzzle of a test barrel to produce zero degree obliquity impacts. Figure 19.15 illustrates the set-up for ballistic helmets.

19.9   Future trends

At the time of writing, soldiers, police officers and security personnel are the principal users of ballistic body armour. The coalition forces in Iraq and Afghanistan were wearing state-of-the-art ballistic body armour despite some adverse reports in the media. The police and security personnel are using the lightest, most comfortable and flexible ballistic body armour ever used in body armour history. Since the advent of modern soft body armour in the 1970s the principal market for body armour has been military and police but the cycle of threats necessitating armour protection is progressing and expanding at an accelerated pace against other segments of the population as well. The threat to professionals' lives is increasing with the increase in crime rates. In addition to this, the terrorist events of September 11, 2001, the wars in Afghanistan and Iraq have spurred growth and expansion to the principal markets of body armour. These include close-protection

Table 19.6 Summary of ballistic test according to NIJ Standard 0106.01

| Helmet type | Test variables | | | | Performance requirements | | |
	Test ammunition	Nominal bullet mass	Suggested barrel length	Required bullet velocity	Required fair hits per helmet part	Permitted penetrations
Type-I	22 LRHV Lead	2.6 g	15 to 16.5 cm	320 ± 12 m/s	4	0
	38 Special RN Lead	10.2 g	15 to 16.5 cm	259 ± 15 m/s	4	0
Type II-A	357 Magnum JSP	10.2 g	10 to 12 cm	381 ± 15 m/s	4	0
	9 mm FMJ	8.0 g	10 to 12 cm	332 ± 15 m/s	4	0
Type II	357 Magnum JSP	10.2 g	15 to 16.5 cm	425 ± 15 m/s	4	0
	9 mm FMJ	8.0 g	10 to 12 cm	358 ± 15 m/s	4	0

Abbreviations:
LRHV – long-rifle, high-velocity
RN – round nose
JSP – jacketed soft point
FMJ – full metal jacketed

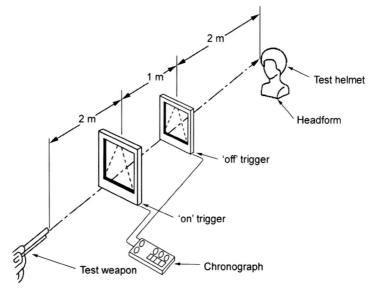

19.15 Test set-up for ballistic helmets.

guards, security guards, prisoner escorts and door supervisors. It also is of interest to vulnerable public figures, war reporters, journalists, aid workers, public transport workers, cab drivers and ambulance paramedic crews.

Given the advanced technological environment of tomorrow's battlefield, what type of threat does the common soldier expect to face? What material will defeat the predicted threat? What is being done now? Answers to these questions require premonition, clairvoyance or prophetic capabilities. With an increase in armour protection comes an increase in weapon penetration. It is reported that in Afghanistan the Mujahideen used a Teflon spray on their bullets and turned regular bullets into armour piercing types that would penetrate Soviet bulletproof vests. Tomorrow's threat seems to fall in the category of 'improved' fragmentation munitions, specifically 'flechettes'. Flechettes are said to be aerodynamic but unstable at velocity. They may be effective at ranges up to several thousand metres. It is thought that Soviet munitions using flechettes may already be in use around the world. The flechettes can penetrate today's aramid fibre body armours. The US Army, with support from the other services, tests and evaluates the state-of-the-art materials to develop effective ballistic protection at the Natick Research and Development Command, Natick, Massachusetts. Working in co-operation with other NATO Nations such as the UK, France, Germany, Netherlands, and Canada regular Personal Armour Systems Symposia (PASS) are held to discuss research and development results.

It can be anticipated that newer materials will be developed providing better protection at lower areal densities against existing and new types of threats. Such materials will be explored for various types of protection, not only body armours and helmets, but also some new forms of protection which will become imperative to fight new threats. As more protective materials are developed, it is reasonable to believe that these materials will be used in conjunction with one another to reinforce the protection level and to optimise other properties that are needed for ballistic protection equipment. Such properties include lighter weight, wearing comfort, and multi-functionality. In the case of soft armours, the ability to minimise trauma will receive attention and there may be new designs coming out. Work will definitely continue to introduce smartness in the design and engineering of ballistic body armours and helmets.

19.10   References and further reading

1. Carothers J. P. (Major) *Body Armour*, A Historical Perspective; USMC CSC 1988.
2. Saxtorph M. (Neils) *Warriors and Weapons of Early Times*, (Macmillan Co., NY, 1972) pp. 11–24.
3. Wilkinson (Frederick) *Battle Dress*, (Doubleday & Co., Inc., Garden city, NY, 1969) pp. 64–71.
4. DuPont, *www.dupont.com/kevlar*
5. DuPont, Kevlar Aramid Fibre, 1994; *www.kevlar.com*
6. Twaron by Azko Nobel, *www.twaron.com*
7. Teijin; *www.teijin-aramid.com*
8. Toyobo Co. Limited; *www.toyobo.com*
9. Lewin M. and Preston J., *Handbook of Fibre Science and Technology*, 1989, Vol. 3 Part B, pp. 2–77.
10. Lewin M. and Preston J., *Handbook of Fibre Science and Technology*, 1993, Vol. 3 Part C, pp. 78–203.
11. Lewin M. and Preston J., *Handbook of Fibre Science and Technology*, 1995, Vol. 4, pp. 74–160.
12. Cunniff P.M., 'An analysis of the system effects of woven fabrics under ballistic impact', *Text. Res. J.* (1992), pp. 495–509.
13. Shim V.P.W., Tan V.B.C. and Tay T.E., 'Modelling deformation and damage characteristics of woven fabric under small projectile impact'. *Int. J. Impact Eng.* (1995), p. 585605
14. Prosser R.A., Cohen S.H. and Segars R.A., 'Heat as a factor in the penetration of cloth ballistic panels by 0.22 caliber projectiles'. *Text. Res. J.* (2000), pp. 709–722.
15. Roylance D., Wilde A. and Tocci G., 'Ballistic impact of textile structures'. *Text. Res. J.* (1973), pp. 34–41.
16. Montgomery T.G., Grady P.L. and Tomasino C., 'The effects of projectile geometry on the performance of ballistics fabrics'. *Text. Res. J.* (1982), pp. 442–450.
17. Kirkland K.M., Tam T.Y. and Weedon G.C., 'New third-generation protective clothing from high-performance polyethylene fibre', in Vigo T.L. and Turbak A.F. (eds), *High-tech Fibrous Materials*, American Chemical Society, Washington DC (1991).
18. Cheeseman and Bogetti T.A., 'Ballistic impact into fabric and compliant composite laminates'. *Compos. Struct.*1 (2003), pp. 161–173.
19. Chocron-Benloulo I.S., Rodriguez J. and Sanchez-Galvez V., 'A simple analytical model to simulate textile fabric ballistic impact behaviour'. *Text. Res. J.* (1997), pp. 34–41.

20. Cunniff P.M., 'A semi-empirical model for the ballistic impact performance of textile-based personnel armour'. *Text. Res. J.* (1996), pp. 45–60.
21. National Institute of Justice, *Resistance of Body Armour*, NIJ Standard-0101.04, Revision A, June 2001.
22. *Journal of trauma-injury infection and critical care*, Some observations relating to behind-body armour blunt trauma effects caused by ballistic impact, Jan 1988, pp. 145–148.
23. DuPont; *Personal Body Armour Facts Book*, June 1994.
24. Second Chance Body Armour, *www.secondchance.com*
25. First Defence, *http://www.firstdefense.com/html/vest_kevlar_vs_spectra.htm*
26. Horroks A.R. and Anand C., *Handbook of Technical Textiles*, The Textile Institute, Woodhead Publishing Limited, pp. 452–457.
27. Adanur S., *Wellington Sears Handbook of Industrial Textiles*, published by Wellington Sears Company 1995, pp. 362–369.
28. Greenwood K. and Cork C.R., *Ballistic Penetration of Textile Fabrics – Phase IV*, UMIST, October 1987, pp. 32–44.
29. Simpson A. (Charles), *The History of PASGT Helmet*.
30. Gurdjian E.S., *Head Injury from Antiquity to the Present with Special Reference to Penetrating Head Wounds*, Springfield, IL: Charles C. Thomas, 1973.
31. *Illustrated War News*, November 17, 1915.
32. NILECJ-STD-0101.01, *The Ballistic Resistance of Body Armor*, National Institute of Justice, U.S. Department of Justice. Washington, DC 20531 (Dec. 1978).
33. NIJ Standard-0104.01, *Riot Helmets*, National Institute of Justice, U.S. Department of Justice, Washington, DC 20531 (Aug. 1980).
34. NILECJ-STD-0105.00, *Crash Helmets*, National Institute of Justice, U.S. Department of Justice, Washington, DC 20531 (June 1975).
35. NIJ Standard-0108.00, *Ballistic Resistant Protective Materials*.
36. Carey M.E., Herz M., Corner B., McEntire J., Malabarba D., Paquette S., and Sampson J.B., *Ballistic helmets and aspects of their design*, Department of Neurosurgery, Louisiana State University Health Sciences Center, New Orleans, Louisiana 70112-2822, USA. Neurosurgery. 2000 Sep; 47(3): 678–88; discussion; 688–9.
37. Sarron J.C., Dannawi M., Faure A., Caillou JP., Cunha J. and Robert R., *Dynamic effects of a 9 mm missile on cadaveric skull protected by aramid, polyethylene or aluminium plate*; an experimental study. Direction Centrale du Service de Sante des Armees, Action Scientifique et Technique – Bureau recherche (DCSSA/AST/REC), France.
38. Cunniff, P.M., 1999. 'Decoupled response of textile body armour'. *Proceedings of the 18th International Symposium of Ballistics*, San Antonio, Texas, pp. 814–821.
39. Cunniff, P.M., 1999. 'Dimensional parameters for optimisations of textile-based body armor systems'. *Proceedings of the 18th International Symposium of Ballistics*, San Antonio, Texas, pp. 1303–1310.
40. Maheux, C.R., 1957. *Dynamics of body armour materials under high speed impact. Part 1: Transient deformation, rate of deformation and energy absorption in single and multilayer armour panels. US Army Chemical Centre Report*, CWLR 2141, Edgewood, MD.
41. Morrison C., 1984. *The mechanical response of an aramid textile yarn to ballistic impact*. Ph.D. Thesis, University of Surrey.

42. Starratt D., Pageau G., Vaziri R., Poursartip A., 1999. 'An instrumented experimental study of the ballistic impact response of Kevlar fabric'. *Proceedings of the 18th International Symposium of Ballistics*, San Antonio, Texas, pp. 1208–1215.

43. Vinson J.R. and Zukas J.A., 'On the ballistic impact of textile armour'. *ASME J. Appl. Mech.* 42 (1975), pp. 263–268.

44. Roylance D. and Wang S.S., 'Penetration mechanics of textile structures'. In: Laible R.C. (ed.) *Ballistic materials and penetration mechanics.* Amsterdam: Elsevier, 1980, pp. 273–93.

45. Freeston Jr. and Claus, Jr. 'Strain-wave reflections during ballistic impact of fabric panels'. *Text. Res. J.* 43 6 (1973), pp. 348–351.

46. Ting C., Ting J., Cunniff P. and Roylance D., 'Numerical characterization of the effects of transverse yarn interaction on textile ballistic response'. *Proceedings of the 1998 30th International SAMPE Technical Conference*, San Antonio, Texas, 20–24 October 1998.

47. Wang Y. and Xia Y. 'The effects of strain rate on the mechanical behavior of Kevlar fibre bundles: an experimental and theoretical study'. *Composites, Part A* 29A (1998), pp. 1411–1415.

48. Wang Y. and Xia Y., 'Dynamic tensile properties of E-glass Kevlar 49 and polyvinyl alcohol fibre bundles'. *J. Mater. Sci. Lett.* 19 (2000), pp. 583–586.

49. Gu B.H., 'Strain rate effects on the tensile behaviour of fibbers and its application to ballistic perforation of multi-layered fabrics'. *J. Dong Hua Univ. (English edition)* **19** 1 (2002), pp. 5–9.

50. Shim V.P.W., Lim C.T. and Foo K.J., 'Dynamic mechanical properties of fabric armour'. *Int. J. Impact Eng.* **25** 1 (2001), pp. 1–15.

51. Laible, R.C., 'Fibrous armor'. In: Laible R.C. (ed.), *Ballistic Materials and Penetration Mechanics*, Elsevier Scientific Publishing Co, New York (1980).

52. Bazhenov S., 'Dissipation of energy by bulletproof aramid fabric'. *J. Mater. Sci.* **32** (1997), pp. 4167–4173.

53. Tan V.B.C., Lim C.T. and Cheong C.H., 'Perforation of high-strength fabric by projectiles of different geometry'. *Int. J. Impact Eng.* **28** 2 (2003), pp. 207–222.

54. Desper C.R., Cohen S.H. and King A.O., 'Morphological effects of ballistic impact on fabrics of highly drawn polyethylene fibres'. *J. Appl. Polym. Sci.* **47** 7 (1993), pp. 1129–1142.

55. Cunniff P.M., 'The V50 performance of body armor under oblique impact'. *Proceedings of the 18th International Symposium on Ballistics*, San Antonio, Texas, 15–19 November 1999, pp. 814–21.

56. Cunniff P.M. 'The performance of poly(para-phenylene benzobizoxazole) (PBO) fabric for fragmentation protective body armor'. *Proceedings of the 18th International Symposium on Ballistics*, San Antonio, Texas, 15–19 November 1999, pp. 814–21.

57. Iremonger M.J. and Went A.C., 'Ballistic impact of fibre composite armours by fragment-simulating projectiles'. *Composites, Part A* **27A** (1996), pp. 575–581.

58. Larsson F. and Svensson L., 'Carbon, Polyethylene and PBO hybrid fibre composites for structural lightweight armour'. *Composites, Part A* **33** (2002), pp. 221–231.

59. Briscoe B.J. and Motamedi F., 'The ballistic impact characteristics of aramid fabrics: the influence of interface friction'. *Wear* **158** 1–2 (1992), pp. 229–247.

20

Chemical and biological protection

Q TRUONG and E WILUSZ, Natick Soldier Center, USA

20.1 Introduction

In many aspects of everyday living it is necessary to provide protection against hazardous chemical/biological (CB) materials. Proper protective clothing is needed during everyday household chores; in industrial, agricultural, and medical work; during military operations, and in response to terrorism incidents. This clothing generally involves a respirator or dust mask, hooded jacket and trousers or one-piece coverall, gloves, and overboots, individually or together in an ensemble. Choices must be made as to which items of protective clothing to select for a given situation or environment. A number of variables to be considered include weight, comfort, level of protection, and the duration of protection required. In addition, the types of challenge to be encountered is of significant consideration. Due to the large number of variables involved, a spectrum of CB protective materials and clothing systems have been developed. Fully encapsulating ensembles made from air impermeable materials with proper closures provide the highest levels of protection. These ensembles are recommended for protection in situations where exposure to hazardous chemicals or biologicals would pose an immediate danger to life and health (IDLH).

20.1.1 History of CB warfare and current threats

Chemical warfare agents (CWAs) such as chlorine, phosgene, and mustard gas (also known as blistering agents) were used in World War I (WWI). There were over a million casualties with approximately 90,000 deaths.[1] CWAs were not used during World War II (WWII) presumably for fear of retaliation, and because CB agents are not easy to control on the battlefield. A sudden shift of wind direction could potentially harm or slow down the advancing army that deployed the agents.[2] Many countries still resorted to the use of CWAs over the years, some examples include the following historical records: Italy sprayed mustard gas from aircraft against Ethiopia in 1935; Japan used CWAs when they invaded China in 1936; Egypt used phosgene and mustard gas bombs in the

1960s in the Yemeni Civil War; during the Iran-Iraq war between 1980 and 1988, Iraq used sulfur mustard and nerve agents on their own Kurdish civilians in northern Iraq and in the city of Sardasht.[3]

The Aum Shinrikyo cult used Sarin nerve agent to terrorize Matsumoto city in 1994, and attacked the Tokyo subway system in 1995. The Tokyo subway incident injured about 5,000 and killed 12 people. On 23 October 2002, 115 people died as a result of the Russian government's use of the BZ chemical, a 'knockout gas,' to subdue about 50 Chechen armed guerillas holding about 800 Russians in a Moscow theater. The use of biological warfare agents (BWAs) has been recorded as early as the 6th century BC when the Assyrians poisoned enemy wells with rye ergot. More recently, BWAs were used by Germany in WWI. In 1937, Japan used aerosolized anthrax in experiments on its prisoners. There were several cases suspected 'yellow rain' incidents in Southeast Asia, and the suspected use of trichothecene toxins (T2 mycotoxin) in Laos and Cambodia. In 1979, there was the accidental release of anthrax at Sverdlovsk. In 1978, ricin was

Table 20.1 CWAs used since WWI

(a)

Year	User	Location
1936	Italy	Ethiopia
1939–40	Japan	Yemen
1960	Egypt	Eritrean Rebels
1970	Ethiopia	Laos
1975–81	Soviets	Kampuchea
1979	Soviets	China
1979	Vietnam	Afghanistan
1979–80	Soviets	Iran
1983–87	Iraq	Iran
1988	Iraq	Iraq (Halabja)
1994	Japan	Japan (Matsumoto)
1995	Japan*	Tokyo
2002	Russia**	Moscow

Source: U.S. Army Chemical School.
* Use by Aum Shirinkyo Terrorists, http://www.emergency.com/wter0896.htm
** Use by the Russian government to solve hostage crisis, http://english.pravda.ru/main/2002/10/28/38794.html

(b) Yellow rain attacks between 1975 and 1981

Location	No. of attacks	Deaths
Laos	261	6,504
Kampuchea	124	981
Afghanistan	47	3,042

used as an assassination weapon in London. In 1991, Iraq admitted its research, development, and BWA weapon productions of anthrax, botulinum toxin, Clostridium perfringens, aflatoxins, wheat cover smut, and ricin to the UN. In 1993, a Russian BW program manager, who had defected, revealed that Russia had a robust biological warfare program including active research into genetic engineering and binary biologicals. Table 20.1 provides a short summary of CWA users and the locations where they were used since WWI.

20.1.2   CB warfare agents and their effects

To design and to fabricate effective CB protective clothing, it is necessary to have an understanding of the hazardous threats that must be prevented from reaching the wearer. The threats comprise the entire spectrum of hazardous CBs. Since the specific characteristics of the threats vary so greatly, effort to develop materials to defeat these threats is an ongoing technological challenge. The military has been concerned since WWI about traditional CWAs, including mustard and organophosphorous nerve agents such as those listed in Table 20.2 and their derivatives, and other toxic chemical groups such as cyanide (causing loss of consciousness, convulsions, and temporary cessation of respiration), pulmonary agents (causing shortness of breath and coughing), and riot control agents (causing burning, stinging of eyes, nose, airways, and skin).[4]

CWAs are defined as natural or synthesized chemical substances, whether gaseous, liquid or solid, which might be employed because of their direct toxic effects on man, animals and plants. They are used to produce death, or incapacitation in humans, animals, or plants. Typical effects of selected CWAs[5] are listed in Table 20.3, and Appendix 20.1 shows the effects of other CWAs and their characteristics.[6] Examples of selected volatile CWAs are: hydrogen cyanide, sarin, soman, tabun, and tear gases; and persistent CWAs include VX, thickened soman, mustard agent, and thickened mustard agent. CWAs are usually classified according to their effects on the organisms (Fig. 20.1).

Table 20.2 Common war chemicals

Nerve chemical warfare agents (CWAs):

VX: O-Ethyl S-2-diisopropylaminoethyl methyl phosphonothiolate
GB: (Sarin) O-Isopropyl methylphosphonofluoridate
GD: (Soman) O-Pinacolyl methylphosphonofluoridate
GA: (Tabun) O-Ethyl N,N-dimethyl phosphoramidocyanidate

Blister CWAs:

L: (Lewisite) 1:2-Chlorovinyldichloroarsine
HD: (Mustard) Bis(2-chloroethyl)sulfide

Table 20.3 Typical effects of toxic chemicals, microorganisms, and toxins

Toxic chemicals

Nerve	affect nervous system, skin, eyes
Blood	prevent oxygen reaching body tissues
Blister	affect eyes, lungs, and skin
Choking	affect nose, throat, and especially lungs
Psycho-chemical	cause sleepiness
Irritant	cause eye, lung, and skin irritations
Vomiting	cause severe headache, nausea, vomiting
Tear	affect eyes and irriate skin

Microorganisms

Anthrax	cause pulmonary complications
Plague	cause pneumonic problems (inflammation of the lung)
Tularemia	cause irregular fever lasting several weeks
Viral encephalitis	affect nervous system (inflammation of the brain)

Toxins

Saxiloxin (STX)	cause shellfish poisoning – highly lethal
Botulinum A (BTA)	cause food poisoning – extremely lethal
Staphenterotoxin B	cause incapacitating effects

Source: *Jane's NBC Protection Equipment, 1990–91*

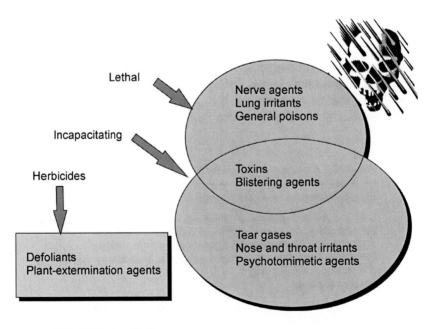

20.1 CWA classification.

BWAs are microorganisms (viruses and bacteria) or toxins derived from living organisms. They are used to produce death, diseases, or incapacitation in humans, animals, or plants. Typical effects of selected BWAs[7] are listed in Table 20.3, and Appendix 20.2 shows the effects of other BWAs and their characteristics in more detail.[8] Toxins are chemical substances extracted from plants, animals or microorganisms, which have a poisonous effect on other living organisms (some can be synthesized in a laboratory.) The sizes of bacterial and viral agents, and toxins which range from submicron to the micron-sized carrier aerosol particulates, make the development of protective materials challenging.

20.1.3   Emerging threats

Recent military concerns include toxic industrial chemicals (TICs) as well as any novel CWAs which may be developed. TICs include chemicals such as common acids, alkalis, and organic solvents. It should be noted that TICs have long been a concern for civilian emergency responders and industrial chemical handlers. The military have become concerned more recently as they have encountered industrial chemicals during various deployments. Protection from TICs needs to be assured.

20.2   Current CB protective clothing and individual equipment standards

There are many different types and designs of CB protective clothing that are available both for military and civilian use in different environments and threat scenarios. Some of these clothing systems are mentioned below.

20.2.1   Military

Currently, the U.S. army uses chemical protective (CP) combat clothing such as the Joint Service Lightweight Integrated Suit Technology (JSLIST) overgarment[9] over the Battle Dress Uniform (BDU) during missions where there is a chemical threat. The JSLIST, when worn with gloves and boots as shown in Appendix 20.3, provides protection against CWAs for 24 hours. Pictures of the JSLIST overgarment and other chemical protective clothing systems are shown in Appendix 20.4. It is designed for extended use, can be laundered every seven days, and is disposable after 45 days wear even if not contaminated. It contains a cleaner and more breathable sorptive liner material than its predecessor, the Battledress Overgarment (BDO).[10,11] The JSLIST overgarment has an integrated hood and raglan sleeve design which allows more freedom of movement. Its integrated suspenders (braces) allows individualized fitting for soldiers of different sizes. Its wraparound hook and loop leg closures allow easy donning and doffing. Other similar carbon-based air-permeable CP overgarments include the French Paul Boyé's NRBC Protective Suit,[12] and the United Kingdom Mark IV NBC Suit. Another U.S. CP clothing system is the chemical protective undergarment (CPU).[13]

The CPU is a two-piece, snug fitting undergarment worn under any standard-duty uniform. It is a stretchable fabric that is designed to provide up to twelve hours of vapor protection, and it has a 15-day service life. Recent U.S. Army R&D efforts in individual CB protection have been on the development of advanced CB protective clothing systems that are launderable, lighter, more comfortable, waterproof, and offer equal or better protection to warfare agents, TICs, and emerging CB agent threats as compared to current clothing systems. The aim is to enable the future soldier to operate longer in a CB contaminated environment comfortably, safely, and effectively. One of the current emphases is the use of selectively permeable membranes (SPM) as a component in future military systems. SPM-based CB protective clothing systems are about a third to half the weight of the standard CP clothing systems, depending on the clothing systems designed for different environments.[14]

The CB protective field duty uniform (CBDU) is a concept of an advanced clothing system that is based on SPM technology. Its development has demonstrated that it is possible to limit or eliminate the need for activated carbon, the use of chemical protective overgarments, the use of chemical protective undergarments, the use of butyl gloves, and the use of overboots. The elimination of any or all of these clothing items would represent a significant weight reduction, reduce logistics concerns, and costs, as well as provide an increase in protection and comfort. Other technologies that are in their early R&D stages are being investigated by the U.S. Army and include electro-spun membranes and reactive membranes.

20.2.2   Civilian

Soldiers as well as civilians use special-purpose clothing such as the Improved Toxicological Agent Protective Ensemble (ITAP) and the Self Contained, Toxic Environment Protective Outfit (STEPO) during domestic emergency operations for chemical spills and toxic chemical maintenance and cleanups in environments with higher threat concentrations. Pictures of the ITAP, STEPO, and other selected civilian emergency response clothing systems are included in Appendix 20.5. The ITAP is used in IDLH toxic chemical environment for up to one hour. It is used in emergency and incident response, routine chemical activity, and initial entry monitoring. ITAP is a CP suit that offers splash and vapor protection, and dissipates static electricity. It can be decontaminated for reuse after five vapor exposures, and it has a five-year minimum shelf life. The U.S. Air Force firefighters use the ITAP with a self-contained breathing apparatus (SCBA), a personal ice cooling system (PICS), and standard TAP gloves and boots. The STEPO is a totally encapsulating protective ensemble that provides four hours of protection against all known CB agents, missile/rocket fuels, petroleum oil and lubricants (POL)s and industrial chemicals. The Explosive Ordinance Disposal (EOD) and Chemical Facility (Depot) munitions personnel engaged in special operations in IDLH environments use the STEPO. It can be decontaminated for reuse after five vapor exposures. Since complete encapsulation is very cumbersome, the work duration in the suit is strictly limited because of the limited air supply, and microclimate cooling is necessary for comfort. The STEPO has four hours of self-contained breathing and cooling capabilities. It has a tether/emergency breathing apparatus option. It also has a built-in hands-free communication system.

If the major concern is only liquid splash protection, then full encapsulation may not be necessary. The use of a coverall or apron may be more appropriate. Such items are typically fabricated from the same type of materials as are used in fully encapsulating suits. Vapor protection is then sacrificed for increased comfort and mobility. For lower threat environments, the suit, contamination avoidance liquid protection (SCALP), and Toxicological Agent Protective (TAP) suits are used. The SCALP is made of polyethylene coated Tyvek and it is worn over the BDO. It is designed to protect the users from gross liquid contamination during short-term operations for up to one hour. Decontamination personnel also use it. The TAP suit is issued to personnel (civilian and military) engaged in monitoring and routine clean up at U.S. CB agent stockpile sites, i.e., chemical activities. The TAP suit offers liquid splash protection. It has an adjustable collar, double sleeves, trouser cuffs and adjustable belt. Its hood has a semi-permanent mounting for an M40 mask. The lower portion of hood is a two-layer shawl. The TAP footwear covers protect its butyl TAP boots from gross contamination. It is used with filtered air or SCBA (for up to one hour).

Similar special-purpose clothing to that being discussed is available commercially. As with combat clothing, special-purpose clothing also has limited wear time. They can be constantly cleaned and reused, and repaired if not contaminated. There are various commercially available suits that are actively being marketed for use in events or incidents involving the use of CB agents. Examples of these suits include the air-permeable Rampart suit[15] and the Saratoga Hammer suit,[16] various air-impermeable Tychem[17] suits, and Kappler's Commander Brigade suit.[18] The weights of these uniforms typically range from 4.10 lbs (Level B Dupont Tychem BR) to 7.05 lbs (Level A Kappler Responder System CPF) or more. Other examples of protective ensembles include suit technologies from Sweden,[19] Germany,[20] and Russia.[21] These suits are different in design and protective capabilities; therefore the potential users must understand their capabilities in order to use them efficiently in different operational environments for specific durations of use. Ongoing efforts by the emergency responders are aimed at developing and field-testing better, lighter, and less costly suits, and SPM-based clothing is being considered and tested for domestic use.

20.3   Different types of protective materials

There are basically four different types of CB Protective Materials.[22] Figure 20.2 illustrates the differences in their protective capabilities.

20.3.1   Air-permeable materials

Permeable fabrics usually consist of a woven shell fabric, a layer of sorptive material such as activated carbon impregnated foam or a carbon loaded nonwoven felt, and a liner fabric. Since the woven shell fabric is not only permeable to air, liquids, and aerosols, but also vapors, a sorptive material is required to adsorb toxic chemical vapors. Liquids can easily penetrate permeable materials at low hydrostatic pressures; therefore, functional finishes such as Quarpel and other fluoro-polymer coatings are usually applied to the outer-shell fabric to provide liquid repellency. Additionally, a liquid and/or an aerosol-proof overgarment such as non-perforated Tyvek protective clothing must be used in addition to permeable clothing in a contaminated environment to provide liquid and aerosol protection. Many users like to use permeable clothing because convective flow of air is possible through the clothing and open closures. This evaporative action cools the body. Examples of air-permeable protective clothing that contain activated carbon include the U.S., British, and Canadian current CP clothing.

20.3.2   Semipermeable materials

There are two different types of semipermeable membranes: porous and solution-diffusion membranes.[23,24] Porous membranes include macroporous, microporous, and ultraporous membrane structures. A macroporous membrane allows a convective flow of air, aerosols, vapors, etc., through their large pores. No separation occurs. A microporous membrane follows Knudsen diffusion through pores with diameters less than the mean free path of the gas molecules allowing lighter molecules to preferentially diffuse through its pores. An

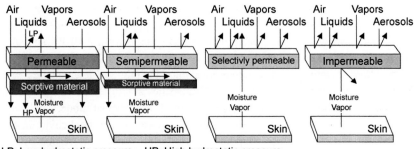

LP: Low hydrostatic pressure   HP: High hydrostatic pressure

20.2 Different types of protective materials.

ultraporous membrane has also been referred to as a molecular sieving membrane where large molecules are excluded from the pores by virtue of their size. A solution-diffusion membrane has also been called a nonporous or a monolithic membrane. This membrane follows Fickian permeation through the nonporous membrane where gas dissolves into the membrane, diffuses across it, and desorbs on the other side based on concentration gradient, time, and membrane thickness. Examples of some semipermeable materials include W.L. Gore & Associates, Inc.'s Gore-Tex®,[25] polytetrafluoroethylene (PTFE) micro-porous membrane, Mitsubishi's Diaplex[TM26] polyurethane nonporous membrane, and Akzo's Sympatex[TM27] copolyester ether nonporous membrane.

20.3.3 Impermeable materials

Impermeable materials such as butyl, halogenated butyl rubber, neoprene, and other elastomers have been commonly used over the years to provide CB agent protection.[28] These types of materials, while providing excellent barriers to penetration of CB agents in liquid, vapor, and aerosol forms, impede the transmission of moisture vapor (sweat) from the body to the environment. Prolonged use of impermeable materials in protective clothing in the warm/hot climates of tropical areas, significantly increases the danger of heat stress. Likewise, hypothermia will likely occur if impermeable materials are used in the colder climates. Based on these limitations, a microclimate cooling/heating system is an integral part of the impermeable protective clothing system to compensate for its inability to allow moisture permeation. ITAP, STEPO, and other OSHA approved Level A suits are examples of impermeable clothing systems. They have been effectively used for protection from CB warfare agents and TICs, but they are costly, heavy, bulky, and incur heat stress very quickly without an expensive and/or heavy microclimate cooling system after donning.

20.3.4 Selectively permeable materials (SPMs)

An SPM is an extremely thin, lightweight, and flexible protective barrier material to CB agents and selected TICs listed in Appendix 20.3, but without the requirement for a thick, heavy, and bulky sorptive material such as the activated carbon material layer being used in current CP protective systems that are discussed above. They allow selective permeation of moisture vapor from the body to escape through the protective clothing layers so that the body of a soldier is continuously evaporatively cooled during missions while being protected from the passage of common vesicant chemical agents in liquid, vapor, and aerosol forms.[29] SPMs have the combined properties of impermeable and semipermeable materials. The protection mechanism of selectively permeable fabrics relies on a selective solution/diffusion process, whereas carbon-based fabrics rely on the adsorption process of activated carbon materials, which has

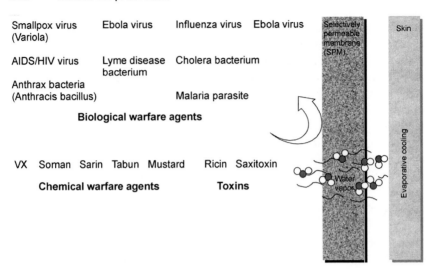

Smallpox virus (Variola)   Ebola virus   Influenza virus   Ebola virus

AIDS/HIV virus   Lyme disease bacterium   Cholera bacterium

Anthrax bacteria (Anthracis bacillus)   Malaria parasite

Biological warfare agents

VX   Soman   Sarin   Tabun   Mustard   Ricin   Saxitoxin

Chemical warfare agents   **Toxins**

Selectively permeable membrane (SPM)

Skin

Water vapor

Evaporative cooling

20.3 SPM material concept.

limited aerosol protection and activated carbon based clothing provides insufficient cooling due to its inherent bulk/insulative properties.

SPMs represent the U.S. Army's pioneering advanced technology.[30] Figure 20.3 shows its material concept. SPMs have been widely used throughout the chemical industry in gas separation, water purification, and in medical/ metabolic waste filtration.[31] There have been many different material technologies co-developed by industry and the U.S. Army Natick Soldier Center (NSC). SPMs consist of multi-layer composite polymer systems produced using various different base polymers such as cellulose, polytetrafluoroethylene (PTFE), polyallylamine, polyvinyl alcohol, among other gas or liquid molecular separation membranes. W.L. Gore & Associates, Inc., Texplorer GmbH, Dupont, and Innovative Chemical and Environmental Technologies (ICET), Inc. are among the leading companies that have been pursuing SPM developments with NSC.

Self-detoxification

Catalysts are under development, which are intended to cause the chemical transformation of warfare agents into less hazardous chemicals. These agent-reactive catalysts, when developed and incorporated into fabric systems, will serve to reduce the hazard from chemical contamination, particularly while doffing the contaminated clothing. The addition of catalysts to CB protective clothing systems is not a trivial matter. The U.S. Army has several R&D efforts to incorporate catalysts (reactive materials) such as OPAA-C18 Organo-

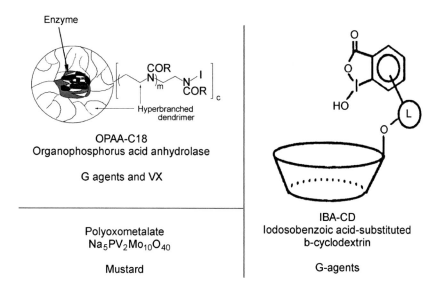

20.4 Agent reactive catalysts.

phosphorus Acid Anhydrolase to neutralize G and VX and Polyoxometalate to neutralize mustard agents (Fig. 20.4).

N-Halamine and Quarternary ammonium salts with alkyl chains are also being investigated for use as biocides, for use in protective clothing via electro-spun fiber processes,[32] in addition, the effectiveness of proprietary copper and silver based nanoparticulates is also being investigated as biocide additives for development of self-decontaminable/biocidal SPMs. Catalytic reactions, by their very nature, are specific to particular types of chemical. Since there are several different types of chemical warfare agents, one catalyst will likely not be sufficient to do the job against the spectrum of possible challenges.

20.4   Proper protective material designs

Material design is critical in the development of a desirable chemical/biological (CB) protective garment. Users often seek material/clothing that is lightweight, comfortable, durable, low cost, easy care, requires little maintenance, and is compatible with existing individual equipment. In order to develop such material and clothing systems, optimal design work involves contributions from multi-disciplinary engineers and scientists from government, academia, and private industry in addition to clothing designers, coaters/laminators, fabricators, and the ultimate end-users for final wear assessments. Several different aspects of the importance of proper material designs must be considered. These aspect include: (i) the different types of protective fabrics which are discussed in section 20.3 and trade-off between protection from toxic chemicals and hazardous microorganisms, and user comfort; (ii) material performance, garment durability, designs of garments and their closure interfaces; (iii) the intended use, environment, productivity, and cost. These aspects are discussed in this section.

CB protective clothing systems have been continually developed and improved over the years. These clothing systems differ in their protective materials, shell and liner fabrics, garment designs, and interfaces/closures (e.g., between gloves and jacket) based on the different levels of protection required and the operating environment. Their common purpose, however, is to provide the user with appropriate protection from hazardous chemicals, toxins, and deadly microscopic organisms. Therefore, the material designers as well as garment designers must understand the protective mechanism(s) in current garments, their components, and functions in order to develop effective CB protective clothing. The users' needs and the intended operational environment are also important. Information on these needs is frequently obtained through interviews, surveys, and/or questionnaires. However, since the protective material is the main component in the design of a CB protective clothing uniform against harmful chemicals and microorganisms, a basic understanding of the following areas is necessary: (i) protection capabilities of current materials; (ii) different concepts of protective materials; (iii) compatibility and integration of protective materials; (iv) wear comfort and material durability; (v) affordability.

20.4.1   Protection capabilities of current materials

The protection capabilities of existing fabrics can be represented in the four different types of materials that were discussed in section 20.3: permeable, semipermeable, impermeable, and SPMs. These material groups are represented in Fig. 20.2, and their typical performances are summarized in Table 20.4.

20.4.2   Different concepts of protective materials

Differences in protective materials are discussed in sections 20.3.1–20.3.4.

20.4.3   Compatibility and integration of protective materials

A CB protective ensemble includes three main components: a textile outer layer material (nonwoven or woven fabric), an inner layer of CB protective material, and a textile liner fabric. These components must be designed to work synergistically with each other. Individual equipment is also an important part of the ensemble which could include gas masks, breathing filters/devices, micro-climate cooling/heating system, and CB agent detection devices. These indivi-

Table 20.4 Typical fabric structures and their performance

Structure	Fabric systems	Aerosol penetration (%/10 min.)	Hydrostatic resistance (psi)	Moisture vapor transmission rate $(g\,m/24\,h)^{-2}$
Permeable	Carbon loaded foam	—	0	1087
	7 oz/yd ^{2}Nylon/Cotton	36	0	915
Semipermeable	Plastolon membrane/ 5 oz/yd ^{2}Nylon/Cotton	0	200	1035
	Gore-Tex II membrane/ 5 oz/yd ^{2}Nylon/Cotton	0	239	713
Sorptive semipermeable	3M Empore membrane	0	52	815
	Soreq NRC membrane	—	290	674
Selectively permeable	'ChemPk Lt-Green'	—	240	764
	'Dehydration' fabric	—	78	824
Impermeable	Parka & Trouser, Wet weather	—	250	<100

dual pieces of equipment and others that are not listed must also be considered for their compatibility and ease of integration into a total CB protective ensemble/garment. Depending on the specific mission or uses, the textile layer could be designed with different functional finishes (e.g., flame protection, water repellency, waterproofing, anti-static, etc.), garment designs, boots, gloves, hood, and closures/interfaces (between clothing and boots, gloves, and hood).

Protective materials can be different in their protective capabilities. Adsorption, reaction, and barrier are three different mechanisms that have been identified. Examples of adsorptive materials are activated carbon, zeolites, aerogels, etc. These materials work by adsorbing chemical vapors in nano-pore structures. The larger the pore surface area, the more desirable the adsorptive material will be. Reactive materials such as Chitosan, chitin, Amberlyst, etc. have been used at Natick to utilize their chemical reactivity in attempts to neutralize chemical agents. The use of reactive materials has been limited because they are reaction-specific to certain chemicals; however, they have great promise for future material development, especially in the area of self-detoxifying and decontaminating clothing.

Barrier materials such as perm-selective membranes and coatings have been gaining acceptance in the user communities as more research and development are carried out by government, industry, and academic institutions. The obvious benefit of a selectively permeable material is that it is extremely light compared to conventional activated carbon based protective materials. Films, skin creams, aerosol sprays, vaccines, etc., are other materials that have been investigated by other U.S. government agencies for different user scenarios. Combinations of adsorptive, reactive, and/or barrier materials have also been used.

20.4.4   Wear comfort and material durability

Comfort is perhaps the next most important concern after CB. Therefore, the development and application of moisture vapor permeable membranes, water and oil resistant coatings, waterproof materials, flexible, elastic, thin, and lightweight materials, as well as those that have high tensile strengths and resistance to tear and puncture damage are being conducted to provide comfortable clothing for the user. Semipermeable (commonly referred to as 'breathable' or moisture vapor permeable) materials are preferred over impermeable materials because they reduce heat stress in warm climates and minimize hypothermia in cold climates. Water resistant coatings/functional finishes are used to minimize weight gain by water or other liquid adsorption. Waterproofing is to keep the individual dry when navigating in wet environments. If waterproof fabrics are used together with waterproof closures, the users will be kept dry when crossing streams and rivers. Materials that are thin, flexible, and lightweight offer textile comforts and ease of garment fabrication. Materials with high tensile strength, resistance to tears and puncture or that possess elasticity offer fabric durability.

20.4.5   Affordability

Life cycle cost, or the total government cost to acquire and own an item or a system over its useful life, is very important in the material design stage. Designers must consider the projected cost of development, acquisition, support, and disposal of the item or system for which the material will be used. Affordability involves life cycle costs in the concept exploration/definition phase, concept demonstration/validation phase, during full-scale development, and finally in operation and support of an end-item. It is a concern in all phases. This report will only emphasize the fact that an increase in competition would result in lower cost materials. Therefore, Natick has been encouraging new material development that would afford comparable CB agent protection, but with lower costs. Affordability is an important consideration in material design because even if a very expensive protective garment provides the best CB protection, it would probably be prohibitively so for common users such as the infantry. In special military or laboratory operations, an expensive, protect-all garment may be purchased and used; however, this would make the CB protective clothing market an unattractive R&D investment by industry's material developers.

20.5   Clothing system designs

The use of excellent protective materials, effective closures, and ergonomic survival equipment for an individual soldier will be meaningless and unproductive without proper garment designs. Therefore, in designing materials, a designer should be familiar with garment design and fabrication. Material designers should understand that garments are designed differently based on the characteristics of the protective materials, different applications, and environment to protect and to maximize the time that a user can operate while wearing the protective garment. There are different garment designs with one-piece garments, two-piece garments, over-garments, under-garments, multi-layer garments, and last but not least important, closures and interfaces.

20.5.1   Coverall or one-piece garments

A one-piece garment eliminates agent penetration through the opening between jacket and trouser/pants. It allows for quick donning and doffing. Another advantage is that it has simplified seaming and sewing in joining fabric pieces during garment fabrication. However, there is no option to open jacket and/or pants for quick release of heat stress/body chill, or exchange of torn/defective jacket or pants. The whole garment must be replaced when it becomes defective and loses its protection.

20.5.2   Two-piece garments

A two-piece garment needs a closure system to seal the opening between jacket and pants. It also requires more seam sealing, sewing, and stitching in joining fabric pieces during garment fabrication. However, it allows donning and doffing for quick release of heat stress/body chill, and exchange of torn/defective jacket or pants. This also allows greater flexibility in sizing users with different dimensions.

20.5.3   Undergarments

Undergarments include underwear and other liner fabrics. They provide protection from the inside and must be worn before the mission. Protection materials/fabric is concealed. They are best used in situations where concealment of protective clothing are required such as in special security operations.

20.5.4   Multilayered garments

Multilayered garments seem to be most popular since they can be oriented toward specific mission(s) in different environments. Clothing layers with specific functions can be donned and doffed for various protection levels (e.g., environmental, chemical, thermal insulation, and/or ballistic protection, etc.). They also provide the option for quick release of heat stress or to alleviate hypothermia, or to exchange torn/defective jacket or pants. However, the users must be conscious of heat stress as more layers are added in order to prevent heat stress injuries. When using multilayer garments, a microclimate cooling system may be necessary. Users must be educated in the protective capabilities of all available layers for maximum protection and environmental adaptability. It will also be time consuming for donning and doffing of clothing and compatibility between different layers may be an issue since they may not have been developed synergistically.

20.5.5   Closure system, components, and systems

Closure interfaces between hood and gas mask, jacket and gloves, jacket and trousers, and trousers and boots are very important in a CB protective garment. Closure systems are very important because protection is a function of fabric, closure/interfaces, activity level, and the motions of the user. To assess these systems, the U.S. Army has developed a test called Man-In-Simulant-Test, commonly referred to as the MIST test. Figure 20.5 confirmed that closure systems are necessary for all CB protective fabric systems to improve protection. Natick has begun to develop a closure system for use with a selectively permeable fabric system. However, there remain technical barriers to overcome.

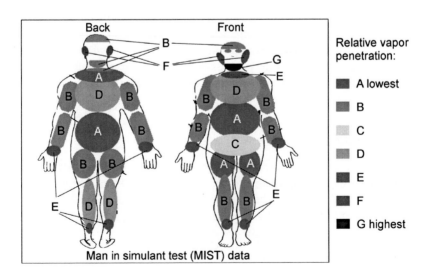

20.5 Vapor penetration resistance of a candidate ensemble.

Current work at NSC addresses concerns for closure system weight, add-on cost to current CP uniforms, and time factor for donning and doffing optimized closures vs. soldiers' comfort and performance. Soldiers' comfort perception of being encapsulated is being studied. Redesign of current gas mask(s) and uniform(s) may be needed and therefore are being addressed by the U.S. Army Edgewood Chemical and Biological Command (ECBC) and the NSC respectively.

20.6   Testing and evaluation of CB protective materials and clothing systems

20.6.1   Material level testing

Table 20.5 shows the performance goals used for research and development of CB protective materials.

20.6.2   Chemical barrier properties

The chemical surety test (known as the live chemical agent test) has been evolved from older methods such as the U.S. Army ECBC's EATM 311-3[33] and the CRDC-SP-84010[34] to the current TOP 8-2-501[35] test method which is used by the U.S. Army to qualify clothing prior to its formal acceptance and classification. These barrier tests include a flooded surface test and laid drop test where the surface of the test sample mounted in the test cells is either saturated with liquid CWAs over the entire surface of the test sample, or the surface is gently laid with droplets of live agent simulants, and the agent permeation is measured over time. The agent contamination density of 10 g/m^2 is often selected in a 24 h test. MINICAMS is used to monitor agent vapor permeation. Vapor permeation (cumulative) will be reported as nanograms/cm^2 versus time. Agent simulant tests[36] with simulants such as trichloroethylene (TCE), methyl salicylate (MeS), dimethyl methyl phosphonate (DMMP), dichloropentane (DCP), dichlorohexane (DCH) and triethyl phosphate (TEP) are often used as 'quick checks' or guides during the material development phase.

NFPA 1994: this is a new performance standard released in August 2001 for testing protective ensembles for CB terrorism incidents.[37] This standard defines three classes of ensembles based on the perceived threat at the emergency scene. Differences between the three classes are based on: (i) the ability of the ensemble design to resist the inward leakage of chemical and biological contaminants; (ii) the resistance of the materials used in the construction of the ensembles to chemical warfare agents and toxic industrial chemicals; (iii) the strength and durability of these materials. All NFPA 1994 ensembles are

Table 20.5 Performance goals for CB protective materials

Chemical protection
Blister (HD, L) agent $\leq 4\,\mu g/cm^2$
Nerve (GB, GD, VX) agents $\leq 10\,\mu g/cm^2$
TOP 8-2-501 (AVLAG test). Preferred.
CRDC-SP-84010 (Mary Jo Waters test used in the past)

Chemical agent deactivation
Diisopropylfluorophosphate (DFP):
Must exhibit significant chemical reactivity
($> 50\,wt\%$ of CWA neutralized)
ECBC CWA TM or NSC TM (will find TM#)

Moisture vapor transmission rate
$\geq 700\,g.m^{-2}/24\,h$
ASTM E96-95, Procedure B.

Water vapor flux @ $32\,°C \geq$
$1800\,g.m^{-2}/24\,h$
ASTM F2298, Procedure B. (DMPC)

Hydrostatic resistance $\geq 35\,lb/in^2$
(water $\rightarrow$ liner)
ASTM D3393-75 or FTMS191A TM 5512

Bonding strength $\geq 10\,lb/in^2$
FTMS191A TM 5512 (water $\rightarrow$ shell fabric)

Weight $\leq 7\,oz/yd^2$ (3-layer fabric laminate)
FTMS191A TM 5041

Torsional flexibility: Pass
FTMS101A TM 2017

Water permeability after flexing
@ $70\,°F$ and $-25\,°F$: Pass
FTMS191A TM 5516

Biological protection: Zero penetration of all microorganisms (10 to 0.001 μm)
USARDEC/NSC's Aerosol Penetration Test Method.

Biocidal activities: Must exhibit biostatic (retard attraction of bacteria and viruses on surface) and sporicidal activities (bacteria and virus kill ability)
ASTM TM or AATCC TM (will find TM#)

Tensile strength @ break: Warp: $> 200\,lb$; Fill: $> 125\,lb$.

Elongation @ break: $>35\%$
FTMS191A TM 5034

Abrasion resistance $> 5,000$ cycles
FTMS191A TM 3884

Delamination: Pass
FTMS191A TM2724

Stiffness $\leq 0.01\,lb$
FTMS191A TM5202

Thickness $\leq 18\,mils$ (3-layer fabric laminate)
FTMS191A TM5030

Dimensional stability (Unidirectional shrinkage $<3\%$)
FTMS191A TM2646

Laundering: Pass > 5 times without delamination
FTMS191A TM2724

Chemical warfare agent simulation permeation $\leq 25\,g/m^2/24\,h$
USARDEC Inhouse Test Method

USARDEC/NSC: U.S. Army Research, Development, and Engineering Command/Natick Soldier Center; ECBC: Edgewood Chemical/Biological Center/SBCCOM; FTMS: Federal Test of Material Standard; TM: Test Method; ASTM: American Standard of Testing Materials

designed for a single exposure (use). Ensembles must consist of garments, gloves, and footwear. Table 20.6 shows the differences between the three classes of NFPA 1994 approved materials.

Toxic industrial chemical testing includes testing by the American Society for Testing and Materials (ASTM) F739/1000, NFPA 1994, and ITF 25 test

Table 20.6 National Fire Protection Agency (NFPA) 1994 Standard

Class	Challenge	Skin contact	Vapor threat	Liquid threat	Condition of victims
1	Vapors Aerosols Pathogens	Not permitted	Unknown or not verified	High	Unconscious, not symptomatic and not ambulatory
2	Limited vapors Liquid splash Aerosols Pathogens	Not probable	IDLH	Moderate	Mostly alive, but not ambulatory
3	Liquid drops Pathogens	Not likely	STEL	Low to none	Self-ambulatory

procedures. These tests measure the permeation of toxic chemicals that are being used by the industry. Although TIC testing is as stringent as the safety protocols of warfare chemicals (nerve and blistering agents) testing, but similar test precautions are taken because in sufficient dosage, TICs could be as deadly as that of CWAs. Appendix 20.6 lists these chemicals. NFPA 1994-test procedure is briefly described in section 6.1.1, and its full text could be requested from the National Fire and Protection Agency or reviewed online.[37] The ASTM Test Method F739 measures the permeation of chemicals through protective materials, and the ASTM 1001-89 lists these chemicals. This method evaluates the materials' chemical resistance to liquids or gases where their breakthrough time and permeation rate are measured. The test results are reported as belonging to indices 0 to 3. Index 0 is the best and most resistant material and is recommended. Index 1 indicates a highly resistant material and may often be accepted by an industrial hygienist for harmful chemicals. Index 2 requires a greater degree of judgement by an industrial hygienist before it will be accepted. Index 3 materials are not usually sufficiently protective to be recommended by industrial hygienists unless there is no other choice or unless the work involves protection only against occasional splashes or compounds that are not very harmful.

20.6.3 Chemical reactive properties

Chemical reactivity testing catalytically and non-catalytically measures the performance of reactive materials to the challenging CWAs or simulants. Although these chemically reactive materials have been in existence for a long time, recent efforts focus on incorporating them into clothing for potential development of self-decontaminable CB clothing systems.

20.6.4   Biological barrier properties

The barrier properties of CB protective fabrics are tested using NSC's in-house test method to measure the aerosolized penetration of MS2 viral and *Bacillus globigii* bacterial spores.

20.6.5   Biocidal activity/properties

The U.S. Army Edgewood Chemical & Biological Command (ECBC) test protocol is used to test the biocide-containing materials and a fabric system's ability to kill BWAs such as Anthrax. The kill rate of BWAs and simulants such as spores of non-virulent *Bacillus anthracis* are measured. The two-test protocols to assess the test material and fabric's sporicidal/bactericidal effects include Protocol A, which involves spore plating on nutrient (DIFCO) plates, while the test material is in close contact with the spores. Testing was done to determine the biocidal activity of the test membrane/fabric to anthrax. The Anthrax spore dilutions were prepared ranging from 10^{-1} to 10^{-5}. The stock spore titer was ~1.5 $\times 10^{8}$/ml. Spore counts (survival or colonies formed) in the presence of the test material are documented. Protocol B involved growth of spores in nutrient broth media (DIFCO) in the presence of test materials. The procedure in this protocol used 2-ml nutrient broth in 12×75 mm tubes. The $1''$ diameter test sample was put at the bottom of the tube. The test samples were completely submerged in the broth. An aliquot of 50 μl from 10^{-2} dilution (~30,000 spores) was added to each tube, and the tubes were shaken in a 'New Brunswick' shaker at 180 rpm at 30 °C for 36 hours. The absorbance was read at 600 nm.

20.6.6   Physical properties

Thickness[38]

The thickness of the membranes, fabrics and fabric systems were measured at 4.1 KPa pressure head using FTMS 191A TM 5030.

Weight[39]

The test samples' weights were measured using FTMS 191A TM 5041.

Aerosol penetration resistance[40]

Figure 20.6 displays the diagram of NSC's in-house aerosol penetration testing apparatus. The apparatus contains two important parts, namely, an aerosol generator and a detector. A potassium iodide salt-water solution is used to generate salt aerosols. The solid particle sizes are in the range of 2 to 10 μm with a 4.5 μm mean size. With 0.5 weight percent, the particle sizes shift to a range of

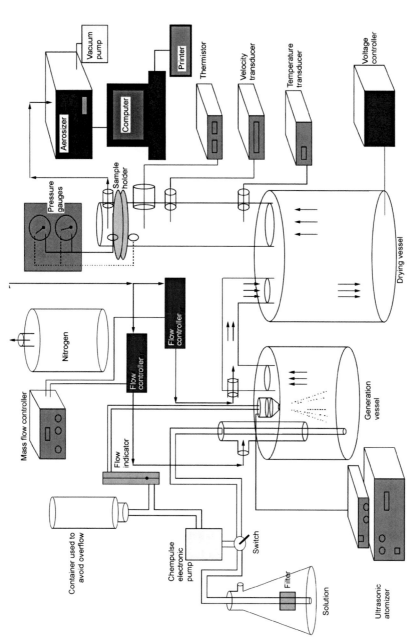

20.6 Aerosol penetration testing apparatus.

1 to 10 μm with a 3.5 μm mean size. An AEROSIZER®, Amherst Instruments Inc. (software version 6.10.09), is used to analyze counts and the size of particles that can range from 0.5 μm to 200 μm.

Hydrostatic resistance

The water penetration resistance of the membrane-fabric was measured by Federal Test Method Standard (FTMS) 191-A, Test Method (TM) 5512.[41] FTMS 191A TM 5514[42] was sometimes used for systems with low-pressure hydrostatic resistance or to test the membranes alone. The membrane faces the water with the fabric reinforcement behind the membrane during testing.

Stiffness[43]

FTMS 191A TM 5202 is designed to determine the directional flex-stiffness of cloth by employing the principle of cantilever bending of the cloth. A Tinius Olsen Stiffness Tester using a 0.46 kg moment, fixed weight is used. The load needed to cause a 60° deflection is measured to calculate the sample's stiffness (flexibility).

Bonding strength[41]

The degree of the cohesion between the fabric and the membrane was measured by the same high-pressure hydrostatic resistance (HPHR) method described above, except that during the tests, water is applied to the shell fabric until the membrane breaks or balloons away from the fabric.

Torsional flexibility[44]

This test is designed to determine the torsional flex-fatigue of cloth by employing twisting and pulling actions to the fabric sample tested. A total of 2,000 cycles is used as passing this test. The test is usually conducted at room temperature and at −25 °C for 2,000 cycles to measure the effectiveness of the test materials. FTMS 191A TM 5514 is used to measure the integrity of the tested fabric materials for water leakage. If the fabric leaks, it is considered to have failed the torsional flex test.

Scanning electron microscopy[45]

Surface and cross-sections of the membranes were viewed and photographed using an AMRAY Scanning Electron Microscope (SEM) model 1000A. The samples were mounted on aluminum tin mounts and sputter coated for three five-minute intervals using gold-palladium. The samples were then viewed in the SEM at 10 or 20 kilovolts. Selected SEMs were also taken using an environmental SEM.

Guarded hot plate[46]

The thermal insulative value and the moisture vapor permeability index were measured as outlined in the American Society for Testing Materials (ASTM) Method D1518-77, 'Thermal Transmittance of Textile Materials Between Guarded Hot Plate and Cool Atmosphere.'

20.6.7   Moisture vapor transport properties

Evaporative cooling potential is measured using the ASTM standard F2298-03.[47,48] This is a new ASTM test method, which determines the amount of water loss over time through a wide range of relative humidity.

20.6.8   Durability testing

System level testing

System level testing of garments as a system is important because these system tests allow the users to see how durably the garments were fabricated, and most importantly how well they protect the wearer. It is also to see how the user is affected by wearing the suit. The experimental suits are usually tested along with commercial off-the-shelf clothing items for test result comparison. Man-in-simulant testing, aerosol testing, physiological testing, rain-court testing, and field exercises are essential and must be performed to find out how well these garments protect the user.

Rain court testing (NSC test facility)

This test is to see how well a garment resists penetration and determines if there are any leakage points. For statistically valid sampling, eight different suits are required in the rain court, and the testing is performed at the rate of one inch per minute. These tests are performed using manikins that will be wearing cotton long underwear, appropriate respirators, and butyl gloves. The manikins will be checked from the start of the test at 5, 10, 15 and 30 minutes. Soldier volunteers can also be used, but with a test protocol that has been approved by the Army Research Institute for Environmental Medicine (ARIEM) Human Use Committee. The test will last one hour. This will give an indication of any leakage, especially at the sewn seams of the suits and at the interface areas (sleeve-to-glove, trouser-to-jacket, and boot-to-leg). Any sign of leakage at each time period will be recorded and reported.

Aerosol system testing (RTI test facility)[49]

This test is to determine how well the chemical protective ensembles protect against penetration by aerosol particulates.

Vapor system testing (man-in-simulant test)

Vapor system testing is a system-level test that measures the amount of vapor that penetrates each suit over a certain period of time. Human test subjects wear each chemical protective garment, along with the appropriate breathing apparatus, and passive adsorption dosimeters (PADs) and enter a man-in-simulant-test (MIST) simulant chamber, and perform a series of physical activities that provide a full range of motion and uniform exposure to a wind stream for two hours. The chamber uses methyl salicylate (MS) as the operative chemical agent simulant. This is used due to its low toxicity and close physical characteristics to those of sulphur mustard (H) vapor. MS is commonly known as oil of wintergreen. The chamber is kept at 27 °C, has a relative humidity of 55%, wind speed of 3–4 mph, and a MS concentration of 85 mg/m³ throughout the test. The PADs are affixed directly to the skin on the areas of the body shown in Fig. 20.7 to determine how much vapor comes in contact with the body. PADs have the same adsorption rate as human skin to give an accurate measure of the amount of simulant that penetrates the suit. They are removed after the tests and analyzed to determine the protection factor of each suit.[50,51] A manikin has also been used at Natick Soldier Center (NSC) to test garments and closure designs to cut down actual human based testing cost and time. Figure 20.7 shows a schematic of NSC MIST chamber.

Physiological testing

The Army Research Institute for Environmental Medicine (ARIEM) conduct physiological testing for NSC using live subjects (soldier volunteers) on each

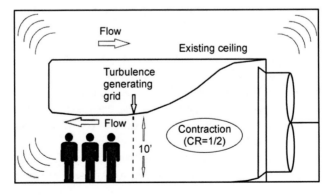

20.7 Schematic of a MIST chamber.

chemical protective suit to determine the effects that wearing the suit has on the user.[52] Initially each suit will be measured on a thermal manikin to get a base line clo (insulative) value. This base line measurement will also give us an idea of the degree of heat stress that the live participants will encounter when they don the suits. Heat stress, core temperature, and other physiological signs will be measured on each participant wearing the various protective suits. It is hoped that the results of these tests will show significant positive differences in the heat stress levels of the SPM technology over the carbon based adsorptive technology for CB protection.[53] The clothing system components such as suit (coverall, or jacket and trousers), mask, underwear, socks, gloves, and boots are procured and sized for the subject volunteers prior to the testing to ensure proper fit.

Limited field experiments

Limited field experiments typically range from one week to as long as four weeks with a maximum of two weeks of test time conducted to assess clothing system designs, durability and user comfort while wearing the experimental clothing systems. Limited field experiments are based on ARIEM HUC's approved test protocol, and with structured questionnaires that are used to interview soldier volunteers at the conclusion of the testing. Control garment(s) are used for comparative purposes. Locations are selected by the program managers based on the intended environments and climates. Examples of a few field test locations include: Aberdeen Proving Ground, Aberdeen, MD and Fort Benning, Georgia, Ft. Lewis, Washington, and the Marine Corps Base, Hawaii.

20.7   Future trends

Current and future efforts are concentrated on: (i) novel closure systems for use with carbon based clothing and that of SPM based clothing; (ii) super activated carbon and reactive materials for potential replacement of the current sorptive material that is being used in carbon based fabric systems such as the JSLIST overgarments; (iii) moisture-permeable butyl rubbers for replacement of the current butyl glove to improve comfort through evaporative cooling; (iv) electro-spun nanofiber-based membranes for lighter weight clothing system; (v) nanoscale materials for improved strength and CB protection; (vi) elastomeric SPMs (eSPMs) for minimizing the number of garment sizes and improved CB protection and comfort; (vii) smart materials such as shape memory polymers for allowing greater comfort when used in high-temperature environments; (viii) self-decontaminable materials such as catalytically reactive SPMs for increased safety and protection of wearers as well as support personnel not in CB protective outfits; (ix) biocidal materials for instant-viral/bacterial kill SPMs; (x) TIC resistant SPMs for use in urban warfare environments; and (11) induction based fluidic moisture vapor transport facilitated CB protective systems for better and more comfortable protective clothing than that of current fabric systems.

20.8   Acknowledgments

The authors would like to acknowledge the contributions from leaders, scientists, and engineers from the United States and foreign governments, and U.S. Army Natick Soldier Center's industry partners who have been working to provide the individual soldier with comfortable clothing and better protection from toxic war and industrial chemicals, deadly microorganisms, and chemically and biologically derived toxins. Special thanks are due to the Woodhead Publishing Company for inviting the authors to contribute this chapter in the book *Textiles for Protection*. The authors would like to personally thank Mr Richard Scott, Editor-In-Chief, Emma, Melanie, and other Woodhead Publishing personnel for their publication guidance, editing, contributions, patience, and understanding which was necessary for the authors' contributions.

20.9   References

1. http://encyclopedia.fablis.com/index.php/Use_of_poison_gas_in_World_War_I
2. http://www.ndu.edu/WMDCenter/docUploaded/2003%20Report.pdf, At the Crossroads Counterproliferation and National Security Strategy, A Report for the Center of Counterproliferation Research, p. 8. Apr 2004.
3. http://encyclopedia.fablis.com/index.php/Iran-Iraq_War
4. Jane's Information Group, *Jane's Chem-Bio Handbook*, 1340 Braddock Place, Suite 300, Alexandria, VA 22314-1651, Quick Reference.
5. *Jane's NBC Protection Equipment*, 1340 Braddock Place, Suite 300, Alexandria, VA 22314-1651, 1990-1991.
6. http://www.nbcindustrygroup.com/handbook/pdf/AGENT_CHARACTERISTICS.pdf, p. VI.
7. http://encyclopedia.fablis.com/index.php/Biological_warfare
8. http://www.nbcindustrygroup.com/handbook/pdf/AGENT_CHARACTERISTICS.pdf, p. V.
9. Military Specification MIL-DTL-32102, JSLIST Coat and Trouser, Chemical Protective, 3 April 2002.
10. Military Specification MIL-C-43858A, Cloth, Laminated, Nylon Tricot Knit, Polyurethane Foam Laminate, Chemical Protective and Flame Resistant, 17 September 1981.
11. Military Specification MIL-S-43926, Suit, Chemical Protective.
12. http://www.paulboye.com/products_nbcf_1.html
13. Military Specification MIL-U-44435, Undershirt and Drawers, Chemical Protective and Flame Resistant.
14. Military Medical/NBC Technology, *NBC Threat Specialist Q&A – Detection, Protection Rank High on the Army's Medical Technology Agenda*, Vol. 5, Issue 3, 2001, pp. 20–24.
15. http://www.approvedgasmasks.com/suit-rampart.htm

16. http://www.nbcteam.com/products_saratoga.shtml
17. http://www.approvedgasmasks.com/protective-suits.htm
18. http://www.labsafety.com/store/product_group.asp?dept_id=18195&cat_prefix=5WA
19. http://www.frenatus.com/
20. http://www.trelleborg.com/protective/template/T036.asp?id=523&lang=
21. http://www.wolfhazmat.de/hazmat_russia.htm
22. Truong Q., M.S. Thesis, Test and Evaluation of Selectively Permeable Materials for Chemical/Biological Protective Clothing, May 1999, University of Massachusetts Lowell, Lowell, Massachusetts.
23. Truong, Q., Rivin, D., *Testing and Evaluation of Waterproof/Breathable Materials for Military Clothing Applications*, NATICK/TR-96/023L, US Army Natick RD&E Center, Natick (1996).
24. W.S. Winston Ho and Kamalesh K. Sircar, eds., *Membrane Handbook*, Chapter 1, Van Nostrand Reinhold, New York, 1992.
25. http://www.gore-tex.com/
26. http://www.diaplex.com/
27. http://www.sympatex.com/
28. Wilusz, E., in *Polymeric Materials Encyclopedia*, J.C. Salamone, ed., 899, CRC Press, Boca Raton (1996).
29. http://www.bccresearch.com/membrane2003/session4.html
30. Truong, Q., U.S. Army Natick RD&E Center, Contract DAAK60-90-C-0105, *Selectively Permeable Materials for Protective Clothing*.
31. Koros, W. and Fleming, G., *Journal of Membrane Science*, 83 (1993).
32. Heidi L. Schreuder-Gibson, Quoc Truong, John E. Walker, Jeffery R. Owens, Joseph D. Wander, and Wayne E. Jones Jr., 'Chemical and Biological Protection and Detection in Fabrics for Protective Clothing', *Material Research Society (MRS) Bulletin*, Volume 28, No. 8, Aug 03. *http://www.mrs.org/publications/bulletin/2003/aug/aug03_abstract_schreuder-g.html*
33. Ciborowski, S., ERDEC Data Report No. 196, US Army Chemical RD&E Center, Edgewood (1996).
34. Waters, M. J., *Laboratory Methods for Evaluating Protective Clothing Systems Against Chemical Agents*, US Army CRDC-SP-84010, June 1984.
35. *Chemical Agent Testing*, US Army TOP-8-2-501.
36. Rivin, D. and Kendrick, C., *Carbon*, 35, 1295–1305 (1997).
37. National Fire Protection Agency (NFPA) 1994, *Protective Ensembles for Chemical/Biological Terrorism Incidents* (2001 edition). Online review is available at: http://www.nfpa.org/itemDetail.asp?categoryID=279&itemID=18172&URL=Codes%20and%20Standards/Code%20development%20process/Free%20online%20access&cookie%5Ftest=1
38. Park, H. B., Rivin, D., *An Aerosol Challenge Test for Permeable Fabrics*, U.S. Army Natick Research, Development and Engineering Center Technical Report, NATICK/TR-92/039L, July 1992.
39. Federal Test Method Standard No. 191A, Test Method 5030, Thickness of Textile Materials, Determination of, 20 July 1978.
40. Federal Test Method Standard No. 191A, Test Method 5041, Weight of Textile Materials, Determination of, 20 July 1978.
41. Federal Test Method Standard No. 191A, Test Method 5512, Water Resistance of Cloth; High range, Hydrostatic Pressure Method, 20 July 1978.

42. Federal Test Method Standard No. 191A, Test Method 5514, Water Resistance of Cloth; Low range, Hydrostatic Pressure Method, 20 July 1978.
43. Federal Test Method Standard No. 191A, Test Method 5202, Stiffness of Cloth, Directional; Cantilever Bending Method, 20 July 1978.
44. Federal Test Method Standard No. 101A, Test Method 2017, Flexing Procedures for Barrier Materials, 13 Mar 1980.
45. Stereoscan 100 SEM, Cambridge Instruments Inc., Eggart and Sugar Roads, Buffalo, New York, NY 14240.
46. American Society for Testing Materials D1518-77, Thermal Transmittance of Textile Materials Between Guarded Hot Plate and Cool Atmosphere.
47. P. Gibson, C. Kendrick, D. Rivin, L. Sicuranza, and M. Charmchi, 'An Automated Water Vapor Diffusion Test Method for Fabrics, Laminates, and Films,' *Journal of Coated Fabrics*, 24, 322–345, 1995.
48. ASTM Standard Test Methods for Water Vapor Diffusion Resistance and Air Flow Resistance of Clothing Materials Using the Dynamic Moisture Permeation Cell, ASTM F2298-03.
49. Aerosol Protection System Testing. US Army TOP 10-2-022.
50. Royal Military College of Canada, 2002, Canadian Standard Vapour Protection Systems Test Standard Protocol.
51. US Army Standard Vapor Protection Systems Test Standard Protocol. US Army Dugway Proving Ground.
52. US Army Research Institute of Environmental Medicine Standard Test Protocol.
53. US Army Dugway Proving Ground MIST Test Report for the Author (Quoc Truong).

Appendix 1   Chemical warfare agent characteristics

			PHYSICAL AND CHEMICAL PROPERTIES					
Agent Type	Chemical Agent; Symbol; Chemical Structure	Molecular Weight	State @ 20°C	Odor	Vapor Density (Air = 1)	Liquid Density (g/cc)	Freezing/ Melting Point (°C)	Boiling Point (°C)
N E R V E	Tabun; **GA** $C_2H_5OPO(CN)N(CH_3)_2$	162.3	Colorless to brown liquid	Faintly fruity; none when pure	5.63	1.073 at 25°C	-5	240
	Sarin; **GB** $CH_3PO(F)OCH(CH_3)_2$	140.1	Colorless liquid	Almost none when pure	4.86	1.0887 at 25°C	-56	158
	Soman; **GD** $CH_3PO(F)OCH(CH_3)C(CH_3)_3$	182.178	Colorless liquid	Fruity; camphor when impure	6.33	1.0222 at 25°C	-42	198
	(Cyclo-sarin); **GF** $CH_3PO(F)OC_6H_{11}$	180.2	Liquid	Sweet; musty; peaches; shellac	6.2	1.1327 at 20°C	-30	239
	VX $(C_2H_5O)(CH_3O)P(O)S(C_2H_4)N[C_2H_2(CH_3)_2]_2$	267.38	Colorless to amber liquid	None	9.2	1.0083 at 20°C	below -51	298
	V$_X$ ("V sub x")	211.2	Colorless liquid	None	7.29	1.062 at 20°C	—	256
B L I S T E R	Distilled Mustard; **HD** $(ClCH_2CH_2)_2S$	159.08	Colorless to pale yellow liquid	Garlic or horseradish	5.4	1.268 @ 25°C; 1.27 @ 20°C	14.45	217
	Nitrogen Mustard; **HN-1** $(ClCH_2CH_2)_2NC_2H_5$	170.08	Dark liquid	Fishy or musty	5.9	1.09 @ 20°C	-34	194
	Nitrogen Mustard; **HN-2** $(ClCH_2CH_2)_2NCH_3$	156.07	Dark liquid	Soapy (low concentrations); Fruity (high)	5.4	1.15 @ 20°C	-65 to -60	75 at 15mmHg
	Nitrogen Mustard; **HN-3** $N(CH_2CH_2Cl)_3$	204.54	Dark liquid	None, if pure	7.1	1.24 @ 20°C	-37	256
	Phosgene oximedichloro-foroxime; **CX** CCl_2NOH	113.94	Colorless solid or liquid	Sharp, penetrating	3.9	–	35 to 40	53–54 at 28mmHg
	Lewisite; **L** $ClCHCHAsCl_2$	207.35	Colorless to brownish	Varies; may resemble geraniums	7.1	1.89 @ 20°C	-18	190
	Mustard-Lewisite mixture; **HL**	186.4	Dark, oily liquid	Garlic	6.5	1.66 @ 20°C	-25.4 (pure)	<190
	Phenyldichlorarsine; **PD** $C_6H_5AsCl_2$	222.91	Colorless liquid	None	7.7	1.65 @ 20°C	-20	252 to 255
	Ethyldichlorarsine; **ED** $C_2H_5AsCl_2$	174.88	Colorless liquid	Fruity, but biting; irritating	6.0	1.66 @ 20°C	-65	156
	Methyldichlorarsine; **MD** CH_3AsCl_2	160.86	Colorless liquid	None	5.5	1.836 @ 20°C	-55	133
B L O O D	Hydrogen cyanide; **AC** HCN	27.02	Colorless gas or liquid	Bitter almonds	0.990 @ 20°C	0.687 @ 20°C	-13.3	25.7
	Cyanogen chloride; **CK** $CNCl$	61.48	Colorless gas or liquid	Pungent; biting; Can go unnoticed	2.1	1.18 @ 20°C	-6.9	12.8
	Arsine; **SA** AsH_3	77.93	Colorless gas	Mild garlic	2.69	1.34 @ 20°C	-116	-62.5
CHOK-ING	Phosgene; **CG** $COCl_2$	98.92	Colorless gas	New-mown hay; green corn	3.4	1.37 @ 20°C	-128	7.6
	Diphosgene; **DP** $ClCOOCCl_2$	197.85	Colorless gas	New-mown hay; green corn	6.8	1.65 @ 20°C	-57	127–128
V O M I T I N G	Diphenylchloroarsine; **DA** $(C_6H_5)_2AsCl$	264.5	White to brown solid	None	Forms little vapor	1.387 @ 50°C	41 to 44.5	333
	Adamsite; **DM** $C_6H_4(AsCl)-NH)C_6H_4$	277.57	Yellow to green solid	None	Forms little vapor	1.65 (solid) @ 20°C	195	410
	Deiphenylcyanoarsine; **DC** $(C_6H_5)_2AsCN$	255.0	White to pink solid	Bitter almond-garlic mixture	Forms little vapor	1.3338 @ 35°C	31.5 to 35	350
Incapa-citating	**BZ**	337.4	White crystal	None	11.6	Bulk 0.51 solid; Crystal 1.33	167.5	320
T E A R	Chloroacetophenone; **CN** $C_6H_5COCH_2Cl$	154.59	Solid	Apple blossoms	5.3	1.318 (solid) @ 20°C	54	248
	Chloroacetophenone in Chloroform; **CNC**	128.17	Liquid	Chloroform	4.4	1.40 @ 20°C	0.23	variable, 60 to 247
	Chloroacetophenone and Chloropicrin in Chloroform; **CNS**	141.78	Liquid	Flypaper	~5	1.47 @ 20°C	2	variable, 60 to 247
	Chloroacetophenone in Benzene and Carbon Tetrachloride; **CNB**	119.7	Liquid	Benzene	~4	1.14 @ 20°C	-7 to -30	variable 75 to 247
	Bromobenzylcyanide; **CA** $BrC_6H_4CH_2CN$	196	Yellow or solid liquid	Soured fruit	6.7	1.47 @ 25°C	25.5	Decomp-oses at 242
	O-chlorobenzylmalonitrile; **CS** $ClC_6H_4CHC(CN)_2$	188.5	Colorless solid	Pepper	—	1.04 @ 20°C	93 to 95	310 to 315
	CR $(C_6H_4)_2(O)(N)CH$	195.25	Yellow powder in solution	Burning sensation	6.7	—	72	335
	Chloropicrin; **PS** Cl_3CNO_2	164.38	Liquid	Stinging; pungent	5.6	1.66	-69	112

Appendix 1 continued

			PHYSICAL AND CHEMICAL PROPERTIES			
Agent Type	Vapor Pressure (mm^Hg)	Volatility (mg/m³)	Heat of Vaporization (cal/g)	Decomposition Temperature (°C)	Flash Point	Stability
N E R V E	0.037 @ 20°C	610 @ 25°C	79.56	150	78°C	Stable in steel at normal temperatures
	2.9 @ 25°C; 2.10 @ 20°C	22,000 @ 25°C; 16,090 @ 20°C	80	150	Non-flammable	Stable when pure
	0.4 @ 25°C	3,900 @ 25°C	72.4	130	High enough not to interfere w/ military use	Less stable than GA or GB
	0.044 @ 20°C	438 @ 20°C	90.5	—	94°C	Relatively stable in steel
	0.0007 @ 20°C	10.5 @ 25°C	78.2 @ 25°C	Half-life of 36 hr at 150	159°C	Relatively stable at room temperature
	0.007 @ 25°C; 0.004 @ 20°C	75 @ 25°C; 48 @ 20°C	67.2	—	—	Relatively stable
B L I S T E R	0.072 @ 20°C	610 @ 20°C	94	149 – 177	105°C; ignited by large explosive charges	Stable in steel or aluminum
	0.24 @ 25°C	1,520 @ 20°C	77	Decomposes before boiling is reached	High enough not to interfere w/ military use	Adequate
	0.29 @ 20°C	3,580 @ 25°C	78.8	Below boiling; polymerizes with heat generation	High enough not to interfere w/ military use	Unstable
	0.0109 @ 25°C	121 @ 25°C	74	Below boiling point	High enough not to interfere w/ military use	Stable
	11.2 @ 25°C (solid); 13 @ 40°C (liquid)	1,800 @ 20°C	101 at 40°C	Decomposes slowly at normal temperature	—	Decomposes slowly
	0.394 @ 20°C	4,480 @ 20°C	58 at 0°C to 190°C	>100	None	Stable in steel and glass
	0.248 @ 20°C	2,730 @ 20°C	58 to 94	>100	High enough not to interfere w/ military use	Stable in lacquered steel
	0.033 @ 25°C	390 @ 25°C	69	Stable to boiling point	High enough not to interfere w/ military use	Very stable
	2.09 @ 20°C	20,000 @ 20°C	52.5	Stable to boiling point	High enough not to interfere w/ military use	Stable in steel
	7.76 @ 20°C	74,900 @ 20°C	49	Stable to boiling point	High enough not to interfere w/ military use	Stable in steel
B L O O D	742 @ 25°C; 612 @ 20°C	1,080,000 @ 25°C	233	>65.5	0°C; ignited 50% of time when disseminated by artillery shells	Stable if pure; can burn on explosion
	1,000 @ 25°C	2,600,000 @ 20°C	103	100	None	Tends to polymerize; may explode
	11,100 @ 20°C	30,900,000 @ 20°C	53.7 @ -62.5°C	280	Below detonation temp.; mixtures w/ air may explode spontaneously	Not stable in uncoated metal containers
CHOK-ING	1.173 @ 20°C	4,300,000 @ 7.6°C	59	800	None	Stable in steel if dry
	4.2 @ 20°C	45,000 @ 20°C	57.4	300 to 350	None	Unstable; tends to convert to CG
V O M I T I N G	0.0036 @ 45°C	48 @ 45°C	56.6	300	350	Stable if pure
	Negligible	Negligible	80	>boiling point	None	Stable in glass or steel
	0.0002 @ 20°C	2.8 @ 20°C	71.1	300 (25% decomposed)	Low	Stable at normal temperatures
Incapa-citating	0.03 @ 70°C	0.5 @ 70°C	62.9	begins at 170°C	246°C	Adequate
T E A R	0.0041 @ 20°C	34.3 @ 20°C	98	Stable to boiling point	High enough not to interfere w/ military use	Stable
	127 @ 20°C	Indeterminate	n/a	Stable to boiling point	None	Adequate
	78 @ 20°C	610,000 @ 20°C (includes solvent)	n/a	Stable to boiling point	None	Adequate
	variable; mostly solvent vapor	Indeterminate	n/a	>247	<4.44°C	Adequate
	0.011 @ 20°C	115 @ 20°C	79.5 @ 20°C	60 to 242	None	Fairly stable in glass, lead, or enamel
	0.00034 @ 20°C	0.71 @ 25°C	53.6	—	197°C	Stable
	0.00059 @ 20°C	0.63 @ 25°C	—	—	188°C	Stable
	18.3 @ 20°C	165,000 @ 20°C	—	>400	Not flammable	Adequate; unstable in light

Appendix 1 continued

Agent Type	Median Lethal Dose (LD50) (mg-min/m³)	Median Incapacitating Dose (ID50)	Eye & Skin Toxicity	Rate of Action	Physiological Action	Detoxification Rate	CWC Schedule
			PHYSIOLOGICAL ACTION				**CWC**
N E R V E	15,000 by skin (vapor) or 1500 (liquid); 70 inhaled	<50 inhaled	Very high	Very Rapid	Cessation of breath -- death may follow	Slight, but definite	1.A.(2)
	10,000 by skin (vapor) or 1700 (liquid); 35 inhaled	25 inhaled	Very high	Very rapid	Cessation of breath -- death may follow	cumulative	1.A.(1)
	2,500 by skin (vapor) or 350 (liquid); 35 inhaled	25 inhaled	Very high	Very rapid	Cessation of breath -- death may follow	Low, essentially cumulative	1.A.(1)
	2,500 by skin (vapor) or 350 (liquid); 35 inhaled	25 inhaled	Very high	Very rapid	Cessation of breath -- death may follow	Low	1.A.(1)
	150 by skin (vapor) or 5 (liquid); 15 inhaled	25 by skin (vapor) or 2.5 (liquid); 10 inhaled	Very high	Very rapid	Produces casualties when inhaled or absorbed	low, essentially cumulative	1.A.(3)
	—	—	Very high	Rapid	Produces casualties when inhaled or absorbed	low, essentially cumulative	
B L I S T E R	900 (inhaled); 5,000 by skin (vapor) or 1,400 (liquid)	500 (skin); 100 (inhaled); 25 (eyes or nose)	Eyes very susceptible; skin less so	Delayed: hours to days	Blisters; destroys tissue; injures blood cells	Very low -- cumulative	1.A.(4)
	1,500 (inhaled); 20,000 (skin)	200 by eye; 9,000 by skin	Eyes susceptible to low concentration; skin less so	Delayed: 12 hours or longer	Blisters; affects respiratory tract; destroys tissue; injures blood cells	Not detoxified; cumulative	1.A.(6)
	3,000 (inhaled)	<HN-1 & >HN-3; 100 by eye	Toxic to eyes; blisters skin	Skin -- delayed 12 hrs or more; Eyes -- faster than HD	Similar to HD; bronchopneumonia possible after 24 hours	Not detoxified; cumulative	1.A.(6)
	1,500 (inhaled); 10,000 by skin (est.)	200 by eye; 2,500 by skin (est.)	Eyes very susceptible; skin less so	Serious effects same as HD; minor effects sooner	Similar to HN-2	Not detoxified -- cumulative	1.A.(6)
	3,200 (inhaled)	very low	Powerful irritant to eyes and nose; liquid corrosive to skin	Immediate effects on contact	Violently irritates mucous membranes, eyes, and nose; forms wheals rapidly	—	
	1,200–1,500 (inhaled); 100,000 (skin)	<300 by eye; >1,500 to 2,000 by skin	Severe eye damage; skin less so	Rapid	Similar to HD, plus may cause systemic poisoning	Not detoxified	1.A.(5)
	15,000 (inhaled); >10,000 (skin)	200 by eye; 1,500 to 2,000 by skin	Very high	Prompt stinging; blistering agent about 13 hours	Similar to HD, plus may cause systemic poisoning	Not detoxified	1.A.(4); 1.A.(5)
	2,600 (inhaled)	16 as vomiting agent; 1,800 as blister	633 mg-min/m³ produces eye casualty; less toxic to skin	Immediate eye effects; skin effects in 30 to 60 minutes	Irritates; causes nausea, vomiting and blisters	Probably rapid	
	3,000–5,000 (inhaled); 100,000 (skin)	5 to 10 by inhalation	Vapor harmful on long exposure; liquid blisters <L	Immediate irritation; delayed blistering	Damages respiratory tract; effects eyes; blisters; can cause systemic poisoning	Rapid	
	3,000 – 5,000 (est.)	25 by inhalation	Eye damage possible; blisters less than HD	Immediate irritation; delayed blistering	Irritates respiratory tract; Injures lungs and eyes; Causes systemic poisoning	Rapid	
B L O O D	Varies widely with concentration	Varies with concentration	Moderate	Very rapid	Interferes with body tissues' oxygen use; accelerates rate of breathing	Rapid: 0.017 mg/kg/min	3.A.(3)
	11,000	7,000	Low; lacrimatory and irritating	Very rapid	Chokes, irritates, causes slow breathing rate	Rapid: 0.02 to 0.1 mg/kg/min	3.A.(2)
	5,000	2,500	None	Delayed 2 hours to 11 days	Damages blood, liver, and kidneys	Low	
CHOK-ING	3,200	1,600	None	Immediate to 3 hr. depending on conc.	Damages and floods lungs	Not detoxified -- cumulative	3.A.(1)
	3,200	1,600	Slightly lacrimatory	Immediate to 3 hr. depending on conc.	Damages and floods lungs	Not detoxified -- cumulative	3.A.(1)
V O M I T I N G	15,000 (est.)	12 (>10 minutes)	Irritating; not toxic	Very rapid	Like cold symptoms, plus headache, vomiting, nausea	Moderate	
	Variable; avg.: 11,000	22 (1 min.); 8 (60 min. exposure)	Irritating; relatively not toxic	Very rapid	Like cold symptoms, plus headache, vomiting, nausea	Rapid in small amounts	
	10,000 (est.)	30 (30 sec); 20 (5 min. exposure)	Irritating; not toxic	More rapid than DM or DA	Like cold symptoms, plus headache, vomiting, nausea	Rapid	
Incapa-citating	200,000 (est.)	112	—	Delayed; 1 to 4 hours depending on exposure	Fast heart beat, vomiting, dry mouth, blurred vision, stupor, increasing random activity	—	2.A.(3)
T E A R	7,000 to 14,000	80	Temporarily severe eye irritation; mild skin irritation	Instantaneous	Causes tearing; irritates eyes and respiratory tract	Rapid	
	11,000 (est.)	80	Temporarily severe eye irritation; mild skin irritation	Instantaneous	Cause tearing; irritates eyes and respiratory tract	Rapid	
	11,400	60	Irritating; not toxic	Instantaneous	Vomiting and choking agent as well as a tear agent	Slow because of effect of PS	
	11,000 (est.)	80	Temporarily severe eye irritation; mild skin irritation	Instantaneous	Powerfully lacrimatory	Rapid	
	8,000 to 11,000 (est.)	30	Irritating; not toxic	Instantaneous	Irritates eyes and respiratory passages	Rapid in low dosage	
	61,000	10 to 20	Highly irritating; not toxic	Instantaneous	Highly irritating; not toxic	Rapid	
	—	0.15	Highly irritating; not toxic	Instantaneous	Irritates skin, eyes, nose, and throat	Moderate	
	2,000	9	Highly irritating	Instantaneous	Acts as tear, vomiting, and choking agent	Slow	3.A.(4)

Appendix 2   Selected biological agent characteristics

Agent Type	Disease/Condition Causative Agent/ Pathogen	Description of Agent	Transmissible Person to Person	Infectivity/ Lethality	Incubation Period	Duration of Illness	Persistence/ Stability
BACTERIA	Anthrax (inhalation) *Bacillus anthracis*	Rod-shaped, gram-positive, aerobic sporulating micro-organism, individual spores ~(1–1.2)x(3–5)⌐m	No	Moderate/ High	1–7 days	3–5 days	Spores are highly stable
	Brucellosis *Brucella suis, melitensis & abortus*	All non-motile, non-sporulating, gram negative, aerobic bacterium; ~(0.5-1)x(1-2)⌐ m	No	High/Low	Days to months	Weeks to months	Organisms are stable for several weeks in wet soil and food.
	Cholera *Vibrio cholerae*	Short, curved, motile, gram-negative, non-sporulating rod. Strongly anaerobic, these organisms prefer alkaline and high salt environments.	Negl.	Low/Mode rate-High	1-5 days	1 or more weeks	Unstable in aerosols and pure water, more so in polluted water.
	Glanders *Burkholderia mallei*	Gram-negative bacillus primarily noted for producing disease in horses, mules, and donkeys	Negl.	/Moderate-High	10-14 days	N/A	N/A
	Plague (pneumonic, bubonic) *Yersinia pestis*	Rod-shaped, non-motile, non-sporulating, gram-negative, aerobic bacterium; ~(0.5-1)x(1-2)⌐m	High	High/Very High in untreated personnel, the mortality is 100%	2 to 6 days for bubonic and 3 to 4 days for pneumonic	1-2 days	Less important because of high transmissibility.
	Shigellosis *Shigella Dysenteriae*	Rod-shaped, gram-negative, non-motile, non-sporulating bacterium	Negl.	High/Low	1-7 days (usually 2-3)	N/A	Unstable in aerosols and pure water, more so in polluted water.
	Tularemia *Francisella tularensis*	Small, aerobic, non-sporulating, non-motile, gram-negative cocco-bacillus ~0.2x(0.2-0.7)μm	No	High/ Moderate if untreated	1-10 days	2 or more weeks	Not very stable
	Typhoid *Salmonella typhi*	Rod-shaped, motile, non-sporulating gram-negative bacterium	Negl.	Moderate/ Moderate if untreated	6-21 days	Several weeks	Stable
RICKETTSIAE	Q-Fever *Coxiella burneti*	Bacterium-like, gram-negative organism, pleomorphic 300-700 nm	No	High/Very low	10-20	2 days to 2 weeks	Stable
	Typhus (classic) *Rickettsia prowazeki*	Non-motile, minute, coccoid or rod shaped rickettsiae, in pairs or chains, 300 nm	No	High/High	6-15 days	Weeks to months	Not very stable
VIRUSES	Encephalitis	Lipid-enveloped virions of 50-60 nm dia., icosohedral nucleocapsid w. 2 glycoproteins					
	-Eastern/Western Equine Encephalitis (EEE, WEE)		Negl.	High/High	5-15 days	1-3 weeks	Relatively unstable
	-Venezuelan Equine Encephalitis		Low	High/Low	1-5 days	Days to weeks	Relatively unstable
	Hemorrhagic Fever						
	-Ebola Fever	Filovirus	Moderate	High/High	7-9 days	5-16 days	Relatively unstable
	-Marburg	Filovirus	Moderate				
	-Yellow Fever	Flavivirus. Isosahedral nucleocapsid 37-50 nm diam., lipoprotein env. w/ short surface spikes	Negl.	High/High	3-6 days	1-2 weeks	Relatively unstable
	Variola Virus (Smallpox)	Asymmetric, brick-shaped, rounded corners; DNA virus	High	High/High	7-17 days	1-2 weeks	Stable
TOXIN	Botulinum Toxin	any of the seven distinct neurotoxins produced by the bacillus, *Clostridium botulinum*	No	NA/High	Variable (hours to days)	24-72 hours/ Months if lethal	Stable
	Ricin	Glycoprotein toxin (66,000 daltons) from the seed of the castor plant	No	NA/High	Hours	Days	Stable
	Staphylococcal enterotoxin B	One of several exotoxins produced by *Staphylococcus aureus*	No	NA/Low	Days to weeks	Days to weeks	Stable
	Trichothecene (T-2) Mycotoxins	A diverse group of more than 40 compounds produced by fungi.	No	NA/High	Hours	Hours	Stable

Appendix 2 continued

Agent Type	Vaccination/ Toxoids	Rate of Action	Symptoms
B A C T E R I A	Yes	Symptoms in 2–3 days; Shock and death occurs with 24–36 hrs after symptoms	Fever, malaise, fatigue, cough and mild chest discomfort, followed by severe respiratory distress with dyspnea, diaphoresis, stridor, and cyanosis
	Yes	Highly variable, usually 6-60 days.	Chills, sweats, headache, fatigue, myalgias, arthralgias, and anorexia. Cough may occur. Complications include sacroiliitis, arthritis, vertebral osteomyelitis, epididymoorchitis, and rarely endocarditis.
	Yes	Sudden onset after 1-5 day incubation period.	Initial vomiting and abdominal distension with little or no fever or abdominal pain. Followed rapidly by diarrhea, which may be either mild or profuse and watery, with fluid losses exceeding 5 to 10 liters or more per day. Without treatment, death may result from severe dehydration, hypovolemia, and shock.
	No	N/A	Inhalational exposure produces fever, rigors, sweats, myalgia, headache, pleuritic chest pain, cervical adenopathy, splenomegaly, and generalized papular/pustular eruptions. Almost always fatal without treatment.
	Yes	Two to three days	High fever, chills, headache, hemoptysis, and toxemia, progressing rapidly to dyspnea, sturdier, and cyanosis. Death results from respiratory failure, circulatory collapse, and a bleeding diathesis.
	No	Symptoms usually within 2-3 days, however, known to demonstrate in as little as 12 hours or as long as 7 days.	Fever, nausea, vomiting, abdominal cramps, watery diarrhea, and occasionally, traces of blood in the feces. Symptoms range from mild to severe with some infected individuals not experiencing any symptoms.
	Yes	Three to five days	Ulceroglandular tularemia with local ulcer and regional lymphadenopathy, fever, chills, headache, and malaise. Typhoidal or septicemic tularemia presents with fever, headache, malaise, substernal discomfort, prostration, weight loss, and non-productive cough.
	Yes	One to three days	Sustained fever, severe headache, malaise, anorexia, a relative bradycardia, splenomegaly, nonproductive cough in the early stage of the illness, and constipation more commonly than diarrhea.
R I C K E T T S I A E	Yes	Onset may be sudden	Chills, retrobulbar headache, weakness, malaise and severe sweats.
	No	Variable onset, often sudden. Terminates by rapid lysis after about 2 weeks of fever	Headache, chills, prostration, fever, and general pain. A macular eruption appears on the fifth to sixth day, initially on the upper trunk, followed by spread to the entire body, but usually not the face, palms, or soles.
V I R U S E S	Yes		Inflammation of the mengies of the brain, headache, fever, dizziness, drowsiness or stupor, tremors or convulsions, muscular incoordination.
	Yes	Sudden	Inflammation of the mengies of the brain, headache, fever, dizziness, drowsiness or stupor, tremors or convulsions, muscular incoordination.
	No		Malaise, myalgias, headache, vomiting, and diarrhea may occur with any of the hemorrhagic fevers May also include a macular dermatologic eruption.
	No / Yes	Sudden	May also include a macular dermatologic eruption.
	Yes	2-4 days	Malaise, fever, rigors, vomiting, headache, and backache. 2-3 days later lesions appear which quickly progress from macules to papules, and eventually to pustular vesicles. They are more abundant on the extremities and face, and develop synchronously.
T O X I N	Yes	12-72 hours	Initial signs and symptoms include ptosis, generalized weakness, lassitude, and dizziness. Diminished salivation with extreme dryness of the mouth and throat may cause complaints of a sore throat. Urinary retention or ileus may also occur. Motor symptoms usually are present early in the disease; cranial nerves are affected first with blurred vision, diplopia, ptosis, and photophobia. Bulbar nerve dysfunction causes dysarthria, dysphonia, and dysphagia. This is followed by a symmetrical, descending, progressive weakness of the extremities along with weakness of the respiratory muscles. Development of respiratory failure may be abrupt.
	Not effective	6-72 hours	Rapid onset of nausea, vomiting, abdominal cramps and severe diarrhea with vascular collapse; death has occurred on the third day or later. Following inhalation, one might expect nonspecific symptoms of weakness, fever, cough, and hypothermia followed by hypotension and cardiovascular collapse.
	Not effective	30 min-6 hours	Fever, chills, headache, myalgia, and nonproductive cough. In more severe cases, dyspnea and retrosternal chest pain may also be present. In many patients nausea, vomiting, and diarrhea will also occur.
	Not effective	Sudden	Victims are reported to have suffered painful skin lesions, lightheadedness, dyspnea, and a rapid onset of hemorrhage, incapacitation and death. Survivors developed a radiation-like sickness including fever, nausea, vomiting, diarrhea, leukopenia, bleeding, and sepsis.

Appendix 2 continued

Agent Type	Treatment	Possible Means of Delivery
B A C T E R I A	Usually not effective after symptoms are present, high dose antibiotic treatment with penicillin, ciprofloxacin, or doxycycline should be undertaken. Supportive therapy may be necessary.	Aerosol.
	Recommended treatment is doxycycline (200 mg/day) plus rifampin (900 mg/day) for 6 weeks.	Aerosol. Expected to mimic a natural disease.
	Therapy consists of fluid and electrolyte replacement. Antibiotics will shorten the duration of diarrhea and thereby reduce fluid losses. Tetracycline, ampicillin, or trimethoprim-sulfamethoxazole are most commonly used.	1. Sabotage (food/water supply) 2. Aerosol
	Few antibiotics have been evaluated *in vivo*. Sulfadiazine may be effective in some cases. Ciprofloxacin, doxycycline, and rifampin have *in vitro* efficacy. Extrapolating from melioidosis guidelines, a combination of TMP-SMX + ceftazidime ± gentamicin might be considered.	Aerosol.
	Early administration of antibiotics is very effective. Supportive therapy for pneumonic and septicemic forms is required.	May be delivered via contaminated vectors (fleas) causing bubonic type, or, more likely, via aerosol causing pneumonic type.
	The antibiotics commonly used for treatment are ampicillin, trimethoprim/sulfamethoxazole (also known as Bactrim* or Septra*), nalidixic acid, or ciprofloxacin. Persons with mild infections will usually recover quickly without antibiotic treatment. Antidiarrheal agents such as loperamide (Imodium*) or diphenoxylate with atropine (Lomotil*) are likely to make the illness worse and should be avoided.	Contaminated food or water
	Administration of antibiotics with early treatment is very effective. Streptomycin – 1 gm I. M. q. 12 hrs x 10 10-14 d. Gentamicin – 3-5 mg/kg/day x 10-14 d.	Aerosol.
	Chloramphenicol amoxicillin or TMP-SMX. Quilone derivatives and third generation cephalusporins and supportive therapy.	Sabotage of food and water supplies.
R I C K E T T S I A E	Tetracycline or doxycycline are the treatment of choice and are given orally for 5 to 7 days.	May be a dust cloud either from a line source or a point source (downwind one-half mile or more).
	Tetracyclines or chlormphenical orally in a loading dose of 2-3 g, followed by daily doses of 1-2 g/day in 4 divided doses until ind. becomes afelorite (usually 2 days) plus 1 day.	May be delivered via contaminated vectors (lice or fleas).
V I R U S E S	No specific treatment; supportive treatment is essential	Airborne spread possible.
	No specific treatment; supportive treatment is essential	Airborne spread possible.
	No specific treatment; intensive supportive treatment is essential	Airborne spread possible.
	No specific treatment; supportive treatment is essential	Airborne spread possible.
T O X I N	(1) Respiratory failure—tracheostomy and ventilatory assistance, fatalities should be <5%. Intensive and prolonged nursing care may be required for recovery (which may take several weeks or even months). (2) Food-borne botulism and aerosol exposure—equine antitoxin is probably helpful, sometimes even after onset of signs of intoxication. Administration of antitoxin is reasonable if disease has not progressed to a stable state. Use requires pretesting for sensitivity to horse serum (and desensitization for those allergic). Disadvantages include rapid clearance by immune elimination, as well as a theoretical risk of serum sickness.	1. Sabotage (food/water supply). 2. Aerosol
	Management is supportive and should include maintenance of intravascular volume. Standard management for poison ingestion should be employed if intoxication is by the oral route.	Aerosol
	Treatment is limited to supportive care. No specific antitoxin for human use is available.	1. Sabotage (food/water supply) 2. Aerosol
	General supportive measures are used to alleviate acute T-2 toxicoses. Prompt (within 5-60 min of exposure) soap and water wash significantly reduces the development of the localized destructive, cutaneous effects of the toxin. After oral exposure management should include standard therapy for poison ingestion.	1. Sabotage 2. Aerosol

Appendix 3   Protective gloves and shoes

Glove set, chemical protective

Worn under standard issue handwear

Attributes

- 25 mil. (0.025 inch thick) butyl rubber glove and cotton inner glove
- 24 hour protection
- 14 day field wear

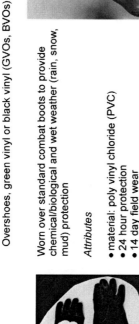

Tactile CB glove

Used in place of the standard CB glove for operations that require greater tactility

Attributes

- lighter and thinner (7 mil and 14 mil)
- 14 mil: 24 hour protection
  14 days wear
- 7 mil: 6 hour protection
  14 days weat

Overshoes, green vinyl or black vinyl (GVOs, BVOs)

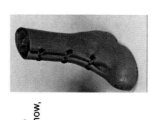

Worn over standard combat boots to provide chemical/biological and wet weather (rain, snow, mud) protection

Attributes

- material: poly vinyl chloride (PVC)
- 24 hour protection
- 14 day field wear

Multipurpose overboots (MULO)

Worn over standard combat boots, jungle boot and intermediate cold/wet boot (ICWB) to provide protection from chemical, biological and environmental hazards

Attributes

- POL, flame and decon solution resistance
- enhanced sole for better traction
- simplified closure system
- operational functional to 0 degrees F
- compatable with standard, jungle and desert combat boots and intermediate cold/wet boot (ICWB)
- 60 days of durability

Appendix 4　Overgarment and other chemical protective clothing systems

Appendix 5   ITAP, STEPO and other selected civilan emergency response clothing systems

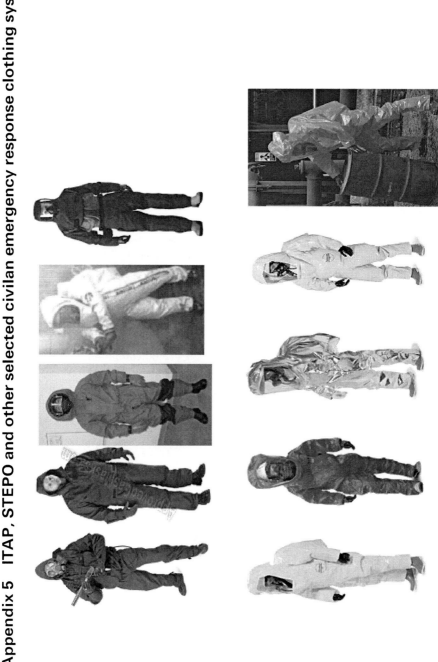

Appendix 6   Selected toxic industrial chemicals (TICs)

NFPA 1994 (C/B warfare agents)
 1. Ammonia (liquid)
 2. Carbonyl chloride (gas)
 3. Chlorine (gas)
 4. Cyanogen chloride (gas)
 5. Dimethyl sulfate (liquid)
 6. Lewisite (liquid)
 7. Hydrogen cyanide (gas)
 8. Sarin (liquid)
 9. Distilled sulfur mustard (liquid)
10. V-agent (liquid)

ASTM F1001 (used in ASTM F739-96 (Toxic Industrial Chemical Permeation Test))

 1. Acetone
 2. Acetonitrile
 3. Ammonia (vapor)
 4. 1,3 Butadiene (vapor)
 5. Carbon disulfide
 6. Chlorine (vapor)
 7. Dichloromethane
 8. Diethylamine
 9. Dimethylformamide
10. Ethyl acetate
11. Ethylene oxide (vapor)
12. Hexane
13. Hydrogen chloride (vapor)
14. Methanol
15. Methyl chloride (vapor)
16. Nitrobenzene
17. Sodium hydroxide
18. Sulfuric acid
19. Tetrachloroethylene
20. Tetrahydrofuran
21. Toluene

Part III

Case studies

21
Military protection*

R A S C O T T , RASCOTEX, UK

21.1   Introduction

Protection of military personnel on land, sea, and in the air differs significantly
in many respects from protection of civilians. Civilians usually face involuntary
accidental situations in the workplace which require protective clothing,
whereas military personnel in war face many complex hazards which are
deliberately aimed at maiming or killing them. Modern weapons are lethal and
sophisticated, whether aimed at individuals, or as weapons of mass destruction
(WMD).

It is still the case in the 21st century that no matter how sophisticated the
weapon systems or how remotely they can be used, we still need humans in
close proximity to detect targets, to fire or launch weapons, and to take, patrol
and hold territory. Among the weapons which military forces face are ballistic
projectiles such as bullets, flying debris, bomb and grenade fragments, flames,
heat and weapon flash, toxic chemical and biological agents, nuclear blast,
radiant heat, and radiation. In addition lives are threatened by detection using
human senses of sight, smell and hearing, or by sophisticated wide spectrum
electronic sensors. In recent years another category of weapons has been named
'non-lethal weapons' which are designed to interfere, disrupt and injure
personnel. These include rubber bullets, water cannon, high-intensity sound,
blinding lasers, and mind-altering drugs. Tactically, an injured soldier on the
battlefield takes up many valuable human resources which are involved in
rescue, evacuation, medical treatment, and recuperation, so non-lethal weapons
will be very effective.

Perhaps the most important and prevalent hazards to troops on the ground are
extreme weather and environmental conditions; cold, heat, rain, snow, dust,
wind and adverse terrain pose a significant risk to the efficient functioning of
troops. Historically, both Napoleon's and Hitler's armies failed to recognise the
folly of attacking Russia in winter. The Germans suffered approximately

350,000 casualties in the winter of 1941–42 due to the cold (Wilmott *et al.*, 2004). It is also thought that more casualties occurred through adverse environmental conditions than through enemy fire in the Korean War (Chappel, 1987). Keeping combatants alive, comfortable and uninjured is therefore of paramount importance.

Military personnel in peacetime are subject to the same Health & Safety legislation under which civilians operate, so that suitably marked and standardised personal protective equipment has to be supplied for most tasks around bases and installations. However, this chapter will concentrate on the unique problems associated with protection of Army, Navy and Air forces during wartime operations.

21.2   General requirements for military protective textiles

These include the basic physical and economic requirements as detailed in Table 21.1.

Table 21.1 Physical and economic requirements for military protective materials

Property	Comments
Physical properties	
Low weight	Items have to be carried by
Low bulk	infantry soldiers
High durability	Must operate reliably in adverse
Low maintenance	conditions for long periods
Good handle and drape	
Low maintenance	Maintenance difficult in the field
Long storage life	Up to 10–20 years
Easy care	Smart, easily laundered in the field without damaging performance
Anti-static	To avoid incendive or explosive sparks
Economic considerations	
Minimal cost	Paid for with public funding from taxes
Readily available	From competitive tendering in industry
Repairable	In the field or in HQ workshops
Decontaminable or disposable	After contamination by nuclear, biological or chemical agents
Easy care	Smart, non-iron, easily cleanable, low shrink

21.3   Textiles for environmental protection

Military ground forces face the most difficult operational conditions of all. Modern UK forces such as infantry, marines and parachutists operate as lightly equipped, highly mobile brigades who are expected to move at short notice to any part of the world. Once in place they have to wear or carry all their personal equipment, and can be exposed to the widest range of environmental conditions. These are detailed in standard publications (Defence Standard 00-35, 1996). Unlike civilians, ground forces cannot choose to operate in good weather, nor can they take control over their work rates. Soldiers typically operate in short bursts of high activity, running carrying equipment and weapons. In between they may have to lie immobile under cover or in trenches for long periods of time. Keeping dry and comfortable is essential. Excessive activity and sweating in a cold climate followed by inactivity can lead to hypothermia (cold stress), whereas high work rates whilst wearing layers of protective clothing in hot climates leads to hyperthermia (heat stress). These conditions can lead to illness or death (see also Chapter 10 by Rossi and Chapter 14 by Holmér). The required functional properties to protect against the environment are detailed in Table 21.2.

21.3.1   Next-to-skin materials

Underwear is primarily worn as a hygiene and tactile comfort layer. The tactile properties are associated with fit, flexibility, surface roughness (frictional properties) and dermatitic skin reactions (Goldman, 1988). The moisture transport properties of underwear are more important than the thermal insulation properties. Perspiration only has a cooling effect if it leaves the skin as a vapour. If this is not possible, it is imperative to remove liquid sweat from the skin, as discomfort is related directly to the area of skin wetted. Modern wicking underwear materials remove sweat to a layer remote from the skin. The 'buffering capacity' of knitted underwear materials can be measured using a skin model or sweating guarded hotplate apparatus (ISO 11092, 1993). This measures the

Table 21.2 Requirements for protection against the environment

Property	Comments
Air permeable	In hot climates
Water vapour permeable	To allow perspiration to escape
Windproof	In cold, wet climates
Waterproof	
Snow shedding	In very cold climates
Thermally insulating	
UV light resistant	In hot, sunny climates
Rot-resistant	If stored in moist conditions
Biodegradable	If discarded or buried

amount of water passed from the plate to the environment, and the amount left in the material. The resulting calculation gives the buffering index (Kf) which has values between 0 (no water transported) and 1.0 (all water transported). Values above 0.7 are indicative of good wicking performance. The UK MOD measured a range of cotton, polyester and polyester/cotton underwear materials (Hobart and Harrow, 1994). Of the eight types tested four had Kf values above 0.7, although the range of results was only between 0.635 and 0.765. Even the UK in-service 100% cotton underwear material (specification UK/SC/4919, 2002) performed as well as specialist wicking fibres made from 100% polyester (Coolmax®). The best knitted materials were composed of blends of hollow polyester with cotton in a double jersey construction. In reality the advantages of wicking materials only manifest themselves over a narrow range of partial sweating rates. Once the material is fully saturated with liquid the wearer feels discomfort. The main advantage of low moisture regain fabrics like polyester is that water is merely mechanically entrained in the fibre interstices, so that these materials tend to dry faster than absorbent cotton.

21.3.2   Thermal insulation materials

Thermal insulation depends primarily on the entrapment of still air in a structure. Fibres, yarns and fabrics offer a very large surface area to trap the maximum amount of still air. Finer fibres tend to trap more still air than coarse fibres for the same bulk, although they tend to produce dense felt-like structures. Nature uses fibrous media in the form of fur, fleeces, down and feathers to keep animals warm. An efficient insulator will comprise about 10–20% of suitably dispersed fibre and 80–90% air. Textiles have the advantages of low density, good resilience, good drape and handle, easy care, and durability. When UK forces were involved in the Cold War a significant amount of research was undertaken to improve the efficiency of cold weather clothing, as ground troops had to carry large amounts of clothing in the arctic. Work by UK MOD studied a wide range of woven, knitted, pile fabrics, fleeces and battings used as insulation media (Scott, 2000a) using a Togmeter, a standard guarded hotplate apparatus (ISO 5085 – Parts 1 & 2, 1996). (See also Chapter 14 by Holmér.) The results were expressed in terms of warmth/thickness ($Tog\, cm^{-1}$) and warmth/weight ($Tog\, m^2\, kg^{-1}$).

The results showed that the warmth to thickness ratio for the majority of woven, knitted and non-woven materials only vary over a small range from about 2.0 to 2.6 tog per cm. Microfibre batting such as Thinsulate® exhibited higher values of 3.0 tog per cm. due to their larger fibre surface area. Comparing the same materials on a warmth to weight basis showed a much larger range of values, from 1 to 2 $tog\, m^2\, kg^{-1}$ for woven and knitted fabrics, up to values between 9 to 18 $tog\, m^2\, kg^{-1}$ for non-woven quilts and battings. Pile fabrics possessed intermediate efficiency values between 6 to 7 $tog\, m^2\, kg^{-1}$. The exception was Thinsulate[®] B which is a dense microfibre felt. It is clear from this work that the critical parameter is the density of the insulating medium rather than the thickness, a hollow polyester (Quallofil[®]) quilted batting was 18 times as efficient as a polyester/cotton woven fabric on a warmth/mass basis.

21.3.3 Waterproof/water vapour permeable materials

Known as WWWW textiles, as they are waterproof, water vapour permeable, windproof, and water-repellent, these modern materials have attempted to solve the perennial problems of keeping active humans dry whilst allowing perspiration vapour to escape freely. Traditional waterproof coated fabrics based upon PVC, rubber, or polyurethane were impermeable to sweat, and caused significant physiological problems to active wearers in cold climates. There are three main types of WWWW materials:

1. High density woven fabrics, the first of which was developed during World War II for pilots who ditched in the sea, and was called Ventile[®]. This is a densely woven Sea Island cotton fabric treated with a stearamido derivative repellent finish Velan PF[®]. It is still used today by RAF pilots, as it has good comfort properties, although the waterproofness is low. Modern analogues are based upon tightly woven microfibre polyester fabrics carrying silicone or fluorocarbon repellent finishes. Many are of Japanese origin and have trade names such as Teijin Ellettes[®], and Unitika Gymstar[®] (Scott, 1995).

2. Microporous coatings and membranes – one of the most well known of these is Gore-Tex[®] developed since the 1970s and based upon biaxially shock expanded microporous PTFE membranes. These are point bonded with adhesive dots to one or more textiles. The membrane has a very high void volume with pore sizes between 0.1 and 5 microns, which allows monomolecular water vapour to pass through, but PTFE is highly hydrophobic, and resists water penetration. It is used by many military forces due to its good performance, although its high cost tends to limit its use to specialist forces. Other microporous coatings and membranes are based upon polyurethane chemistry, typical products have the trade names Aquatex[®], Triple Point[®], Drilite[®], and Entrant[®] (Scott, 1995). The microporous polyurethane coated fabrics tend to have medium waterproofness and vapour permeability.

3. Hydrophilic solid coatings and films – These are usually based upon block copolymers containing hydrophilic functional groupings such as –O-, -CO, -OH or –NH_2. These can form reversible hydrogen bonds with water molecules, which diffuse through the film in a stepwise action along the molecular chains (Lomax, 1990). One of the first of these was based upon segmented polyurethanes with polyethylene oxide adducts, and the initial

Table 21.3 Military uses of waterproof/vapour permeable textiles

Type of material	Clothing item	Fabric specification number
PTFE laminates	CS95 Waterproof Jacket & Trousers, Army, Royal Marines, RAF	UK/SC/5444
	MOD Police Anorak	UK/SC/4978
	Arctic Mittens Mk. 2	UK/SC/4778
	Insock, Boot Liners	PS/04/96
	Smock & Salopettes, Petrol Protective	PS 13/95
Microporous and hydrophilic polyurethanes	CS95 Waterproof Jacket & Trousers, Army, Royal Marines, RAF	UK/SC/5444
	Suit, Waterproof, Aerial Erectors	UK/SC/5070
	Suit, Foul Weather, RN	PS 15/95
	Gaiters, Snow, General Service	UK/SC/5535
Ventile cotton	Coverall Immersion, Mk. 20A	
	Jacket, Windproof, Aircraft carrier Deck	
	Coveralls, Swimmer Canoeist	

research was sponsored by the UK MOD Stores & Clothing Research & Development Establishment in the 1970s. Products now exist under a wide variety of trade names, including Witcoflex Staycool®, Keelatex®, and Isotex®. Another popular product, Sympatex® is based upon a modified polyester film into which polyether groups have been introduced (Drinkmann, 1992). This class of materials tends to have good water proofness, but lower vapour permeability (Scott, 1995).

UK and other NATO forces now use these materials for a wide range of combat and protective clothing. The choice of a particular class depends on the required level of performance, durability and cost. Ground troops with the most demanding roles tend to favour the PTFE laminates for the highest all round performance. In contrast the UK Royal Navy on board ship require high water-proofness, but a lower vapour permeability is acceptable for their operations. The performance of a range of these materials has been measured by Weder (1997). Pressures on costs, and the use of competitively tendered performance specifications tends to drive the procurement process towards cheaper products. Table 21.3 shows the range of WWWW materials used by UK forces.

21.4   Military combat clothing systems

Current UK combat clothing is based upon the layering principle where each garment has a specific function. It comprises a basic system of six layers to which can be added combat body armour, nuclear, biological and chemical

21.1 Combat soldiers clothing system.

(NBC) oversuit, a snow camouflage oversuit in white nylon, and an extra cold weather insulated jacket and trousers. The system includes leather gloves, arctic mitten assembly, snow gaiters, wool/nylon socks, waterproof boot liners, leather boots and knee pads. Figure 21.1 shows soldiers wearing the clothing system. It is used by all Army, Royal Marines, and RAF Regiment personnel, and has undergone continuous development since its introduction in 1995. Table 21.4 shows the main components and materials.

21.5   Thermal and water vapour resistance data for combat clothing systems

The thermal resistance (Rct) and water vapour resistance (Ret) values for the UK combat clothing ensemble have been measured using a sweating guarded hotplate apparatus (ISO 11092, 1993). The values are additive, although the true total is somewhat higher due to the air gaps between layers (Congalton, 1997). From these the water vapour permeability index (imt) can be calculated such

Table 21.4 UK combat clothing system materials

Clothing layer	Material description	Material specification
1. Underwear	100% Knitted Cotton 1×1 rib, olive	UK/SC/4919
2. Norwegian shirt	Knitted Cotton, plush, Terry loop pile, olive	UK/SC/5282
3. Combat trousers & shirt	67/33 polyester/cotton 2×1 twill, lightweight, DPM Woodland, Desert, NIRR	UK/SC/5843
4. Field jacket, windproof	100% cotton Gaberdine, 2×2 twill, DPM woodland, NIRR	UK/SC/5878
5. Fleece pile jacket	100% polyester, knitted fleece pile double faced, olive	UK/SC/5412
6. Waterproof rainsuit	3 layer laminate, nylon/membrane/nylon, Waterproof, water vapour permeable, DPM woodland, NIRR	UK/SC/5444
7. Jacket & trousers, thermal	Mixture of polyester hollow and microfibre batting, 200 g/m², 20 mm thick	UK/SC/5919

Table 21.5 Thermal and water vapour resistance of combat clothing materials

Textile layer	Rct (m^2KW^{-1})	Ret (m^2PaW^{-1})	imt
Cotton underwear	0.03	5.1	0.3
Norwegian shirt	0.05	8.6	0.3
Polyester fleece	0.13	13.4	0.6
Lightweight combat suit	0.01	4.3	0.2
Windproof field jacket	0.005	4.8	0.1
Waterproof breathable rainsuit	0.003	11.2	0.01

that $imt = S \times Rct/Ret$, where $S = 60 \, PaW^{-1}$. imt has values between 0 (totally impermeable) up to 1 (totally permeable) (see also Chapter 10 by Rossi) Table 21.5 shows the results for clothing and footwear. The Ret values for footwear show that the leather boot is the determining factor, and adding socks and boot liners brings the permeability to the level at which condensation will occur inside the footwear.

21.6 Ballistic protection

The main ballistic threats to military personnel are fragmenting projectiles rather than bullets. The projectiles originate from grenades, mortars, artillery shells, mines, and improvised explosive devices (IEDs) the latter used by terrorists. The other threats are low velocity bullets from hand guns, and high velocity bullets from rifles and machine guns. In contrast the main causes of injury to civilians (including Police officers) are low velocity bullets fired from handguns at close range. Ballistic casualties in general war, including World War II, Korea, Vietnam, Israel, and the Falklands were recorded as 59% from projectile fragments, only 19% from bullets, and 22% from other causes (Tobin, 1994). In terms of lethality bullets are more likely to kill than fragments, although the latter may inflict several wounds ranging in severity. There may also be casualties from secondary explosions, including flying debris, and destruction of buildings, vehicles, ships and aircraft.

21.6.1   Casualty reduction using personal armours

Providing adequate ballistic protection for an individual is a complex process. The limiting factors are related to the weight, bulk, rigidity and physiological burden imposed by wearing the armour. It is clear that textile structures can offer advantages of low density, flexibility and comfort over rigid metallic armour. However, textiles alone cannot protect against high-velocity bullets of 5.56 mm, 7.62 mm, 12.7 mm, and sharp projectiles that cut through textiles. In these situations additional rigid plates made from textile composites and ceramics need to be added to protect vital organs such as the heart. Head protection is a particular priority that can be met by the use of shaped, moulded composites using specialist textiles (Shephard, 1986). The following are the estimated reductions in casualties to troops standing in the open and threatened by mortars.

- Wearing a ballistic helmet can reduce casualties up to 19%.
- Wearing body armour alone can reduce casualties up to 40%.
- Wearing both helmet and body armour can reduce casualties up to 65%.

Estimates indicate the synergism caused by wearing both items (Tobin, 1994). These figures are approximate, as the protection level also depends upon the compromise between body area coverage and the acceptable burden that does not limit mobility significantly.

21.6.2   Textile fibres for ballistic protection*

Historically, woven silk fabrics were used for ballistic protection. More recently high modulus aliphatic nylon 6.6 with a high degree of crystallinity and low elongation was developed and is widely used in body armour, and as the textile reinforcement in composite helmets (Marsden, 1994). Since the 1970s a range of aromatic polyamide fibres (p-aramids) have been developed. These are based upon poly-parabenzamide or poly-paraphenylene terephthalamide (PPTA) with trade names Kevlar® (Du Pont) and Twaron® (Akzo Nobel, now Teijin) Another

---

* See also Chapters 5 and 19.

fibre which is a copolymer of >85% PPTA is Technora® (Teijin), although it does not appear to be used at the time of writing.

An alternative route to high-modulus fibres is through copolymerised aromatic polyesters, a commercial product being Vectran® (Celanese). It has tensile properties similar to para-aramids, but the fibres do not creep, and have better abrasion and flex resistance. In recent years a range of high-modulus gel-spun polyethylene fibres has been developed, known as HMPE, HPPE or UHMWPE fibres. These fibres have a low density of 0.97 gml^{-1} which theoretically allows for packs of lower weight, although this can be accompanied by increased bulk The main disadvantage is the low melting point of about 150 °C, and the properties deteriorate above room temperature. Trade names include Spectra® (Honeywell) and Dyneema® (DSM and Toyobo). Fraglight® is a non-woven needle felt version of Dyneema. Research work on the formation of composites for helmets using HMPE fibres indicated that good ballistic performance was possible with significant reductions in areal density compared with ballistic nylon (Morye et al., 1996). However, the structural rigidity of the helmet was compromised. A recent addition to the high modulus fibres was PBO (poly-paraphenylene benzobis oxazole) with the trade name Zylon® (Toyobo). PBO fibres have moduli and strength twice that of para-aramids, but unfortunately the fibre degrades by hydrolysis in warm, moist conditions.

In the 1990s a polymer with a ring structure similar to PBO was synthesised, but with hydroxyl groups hydrogen bonded in all transverse directions. The resulting PIPD or M5 fibres have greatly improved transverse compression and shear properties, and a pilot plant for production is expected in 2005 (see also Chapter 5 by Hearle). A comparison of ballistic textile fibre performance against steel is given in Table 21.6. It is clear that these specialist technical textiles offer the great advantages of much lower density and high modulus compared with steel wire.

21.6.3 Textile structures for ballistic protection

Ballistic protection involves arresting the flight of high-speed projectiles in as short a distance as possible. High-modulus textile fibres possess very high strength and very low elongation, which prevents indentation and subsequent

Table 21.6 Ballistic performance of textiles versus steel

Property	Steel wire	Ballistic nylon	Kevlar 29	Dyneema SK60
Tensile strength (MPa)	4000	2100	3400	2700
Modulus (MPa)	18	4.5	93	89
Elongation (%)	1.1	19.0	3.5	3.5
Density (g ml^{-1})	7.86	1.14	1.44	0.97

bruising and trauma after a body impact. Woven textiles are most commonly used, although non-woven felts are also available. The majority of woven fabrics are of coarse, loose, plain weave constructions. Continuous multifilament yarns with very low twist produce light, flexible fabrics which are the optimum for shaping clothing panels. However, the loose sett means that there is a high probability of projectiles sliding between individual filaments and yarns. Also, ballistic resistance increases with overall areal density, which necessitates the use of between about five and 25 layers of fabric to produce adequate performance. To allow each textile layer to move independently the shaped packs are secured by stitching around the edge, or quilt stitched in lines or squares. It is necessary to seal the shaped packs inside a waterproof, light-tight cover (often made from PVC coated fabrics) as the presence of moisture and UV light can degrade the ballistic performance. Figure 21.2 shows a soldier wearing the ultimate protective ensemble for Explosive Ordnance Disposal operations.

21.2 Explosive ordnance disposal clothing system.

21.6.4   Ballistic testing and evaluation

Military ballistic packs are tested in instrumented firing ranges using fragment simulating projectiles (FSPs) of standardised weight and size. The FSPs are fired at a range of test velocities so that pass/fail criteria in the form of V_{50}, V_0, or Vc can be determined. V_{50} is the velocity in ms^{-1} at which there is an expected probability of penetration of 50% of the FSPs. A more understandable approach for the user is to measure Vo or Vc (critical velocity) at which no penetration of the full pack occurs, since the main objective is to shield the wearer from any of the projectiles. However, V_{50} is still the most commonly used measure, as the V_0 does not statistically exist. Other threats are often tested by a complete pass/fail protection test. There are several standard test procedures: the UK uses its own (Specification UK/SC/5449, 2002). NATO countries should use a Standardisation Agreement (STANAG 2920, 1999), whilst several nations use their own standards.

Models are used to judge how effective a protective armour might be in combat situations. Initial data fed into the model includes Vo values for several sizes of FSP, together with the area of body coverage. Data concerning the use of real weapons against unprotected versus protected individuals are then included. This casualty reduction analysis enables developers to predict the real effectiveness of personal armour. The UK MOD developed a model called CASPER (Couldrick *et al.*, 2002).

21.6.5   In service ballistic materials and end items

UK military forces use a wide range of ballistic protective clothing assemblies and composite helmets using ballistic Nylon and para-aramid fabrics. Table 21.7 details both the materials and the clothing items. Ballistic clothing assemblies vary in protection levels and complexity depending on the type of operation. Highly mobile infantry are equipped with the lightest combat body armour (CBA) weighting about 2.5 to 3.5 kg. The protection level against high-velocity bullets can be increased progressively by covering a larger body area and incorporating rigid, shaped composite or ceramic plates. This may conceivably weigh 13 to 15 kg, or about one-fifth of the weight of an active fit soldier. This does not include helmets, visors, and leg protection (Marsden, 1994). The complete ultimate explosive ordnance disposal (EOD) assembly shown in Fig. 21.2 can weigh up to 30 kg.

A range of protective helmets made from textile composites based upon ballistic nylon or para-aramids are provided. The resin used is a 50/50 mixture of phenol formaldehyde and polyvinyl butyral The majority of combatants wear the General Service Helmet Mk6, whilst specialists such as combat vehicle crews, parachutists and EOD teams have items specifically developed for their particular roles.

Table 21.7 Ballistic textiles and clothing assemblies – UK forces

Item	Description	Specification
Cloth	Ballistic Resistant, Nylon, 290 g/m²	UK/SC/3990
Cloth	Ballistic Resistant, plain woven, Para-aramid, 163 Tex	UK/SC/4468
Clothing		
Combat Body Armour, Lightweight	Uses Kevlar – 12 plies Nylon – 4 plies	UK/SC/5108
Explosive Ordnance Disposal Ensemble, Mk 4	Uses p-Aramid for action Uses Nylon for Training	UK/SC/4900
Suit, Lightweight, Combat EOD	Uses Kevlar Comfort 129	UK/SC/5717
Helmets		
Shell, Helmet, General Service, Mk 6	Uses Ballistic Nylon composite	UK/SC/4796
Helmet, Parachutists, Lightweight	Uses ballistic nylon composite	UK/SC/4513
Helmet Combat Vehicle Crewmen	Uses para-aramid composite	UK/SC/4864

21.7   Camouflage, concealment and deception

Camouflage is predominantly a military form of protection against external threats. The basic philosophy is based upon the premise that if the enemy cannot detect a target then that target is effectively protected, and casualties are avoided. The term camouflage was first introduced by the French in World War I, being derived from 'camoufler' (to disguise). It defined the science of concealment of objects and people by the imitation of their physical surroundings. Effective camouflage must break up the human's characteristic outline, and minimise the contrast between them and the environment in terms of colour, texture, shadow, and electronic signature.

Concealment of personnel to avoid visual detection by eye or photography is still the primary means of military surveillance and target acquisition (Vickers, 1996). However, modern battlefield surveillance systems may operate in a wide waveband of the electromagnetic spectrum, including UV, near infra-red (NIR), far infra-red (FIR), radar in the millimetric or centimetric band, and acoustic ranges. Table 21.8 shows details of the threat wavebands and the detection systems used. Textiles are widely used as camouflage media in the form of light, flexible nets, covers, and garnishing, as well as dyed or printed clothing items. Textiles are lightweight, durable, cheap, and can be manufactured in a wide

Table 21.8 Requirements for camouflage, concealment and deception

Property	Wavelength or frequency	Comments
Visible spectrum	400–800 nm	Match colour, texture, and appearance of background. Woodland, Desert, Arctic. Detection by eye
Ultraviolet	200–400 nm	Match optical properties of snow and ice. Using UV detectors or eye
Near infra-red	750–2000 nm	Match reflectance of background when viewed by image intensifiers, low-light television, IR photography
Far infra-red	2600–5000 nm 8000–14,000 nm	Minimise heat signature from humans or hot equipment
Radar	2–18 GHz 90–98 GHz	Avoid movement – detection by doppler radar
Acoustic	20–20,000 Hz	Detection by ear or microphones

variety of colours, textures, and patterns. This chapter concentrates on camouflage systems for clothing and personal equipment, but see also Scott (2000a) for further details.

21.7.1   The visible waveband

Visible range camouflage is a unique application of textile coloration for functional rather than aesthetic purposes. Natural or artificial backgrounds are mimicked by matching their colours, texture, gloss and patterns. Natural backgrounds vary according to physical geography, seasons, and environment, thus camouflaged clothing has to mimic temperate, tropical, desert and arctic backgrounds, and any pattern has to be a compromise. Each Nation has developed its own distinctive patterns and colour ways, arrived at empirically by measurements and extensive trialling (Vickers, 1996). National differences also aid identification friend or foe (IFF) on the battlefield at close quarters. The UK temperate woodland pattern uses four colours, green, khaki, brown and black in a random disruptively patterned material (DPM) (Blechman, 2004). A separate desert pattern comprising two colours – sand beige and brown is also used. These patterns are used on all outer clothing layers, including helmet covers, load carriage and webbing. Figure 21.1 shows soldiers wearing the UK temperate woodland pattern, which works best when viewed against a background of bushes, scrubland or trees. It is not so effective in open grasslands, but soldiers always seek cover, or try to enhance the effectiveness by adding freshly cut vegetation and adding face paints. In addition, arctic troops wear a lightweight white nylon oversuit with a finish to match the UV reflective properties of snow and ice. Other nations may use up to eight colours, including pink, orange, olive, beige, grey, yellow, black, brown and many shades of green.

21.7.2   Near infra-red camouflage (NIR)

This military threat is posed by optical imaging devices which amplify low levels of light, including moon and starlight. Such image intensifiers (IIs) can be in the form of compact, lightweight binoculars, monoculars, or low-light television cameras. The earliest ones were developed during World War II (Newark *et al.*, 1996; Ivanov and Tyapin, 1965). Infra-red 'false colour' film cameras can also be used, but are useful only for detecting fixed installations, encampments and assets, but not rapidly moving ground troops (Richardson *et al.*, 1998, Fabian *et al.*, 1992). These imaging devices operate at wavelengths from 700–1,300 nm, just above the visible range. In this waveband the reflectance spectrum for leaves, branches, and grasses rises dramatically from less than 10% at 600 nm up to between 40 and 60% at 1,000 nm. Dyes, pigments and colourants used in the textiles must match the reflectance of this 'chlorophyll rise'. This poses a problem, as very few dyes and pigments exhibit this behaviour in the NIR. Additionally, the reflectance of vegetation varies widely; deciduous tree leaves have high reflectance compared with coniferous needles. There is also a seasonal change since deciduous trees lose their leaves, bark and branches have much lower reflectances than leaves.

Figure 21.3 shows the reflectance requirement curves for the UK temperate woodland print, including the envelope for NATO IRR monotone green

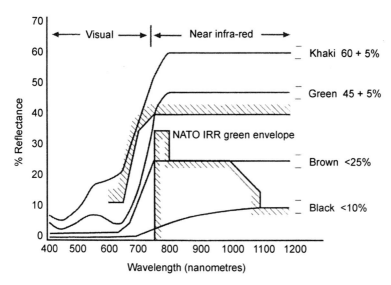

21.3 Camouflage reflectance requirements (four colour disruptive-pattern prints).

(Defence Standard 00-23, 1980). Each colour has to ideally meet the specified reflectance value between 1,000 and 1,200 nm, as shown in Fig. 21.3, or as laid down in MOD specifications. The NATO IRR green value should be 35 ± 5% between 750 and 1,200 nm. Some nations such as Canada, Norway and Sweden have extended the NIR requirements out to 2,000 nm. Moreover, the overall reflectance values integrated with the area of each print colour have to fall within the envelope for NATO IRR green (Fig. 21.3) in accordance with the equation: (0.16 × black) + (0.35 × brown) + (0.34 × green) + (0.15 × khaki) = NATO IRR green.

21.7.3   Dyes and pigments for near infra-red camouflage

There exists a specially selected range of vat dyes for cellulosic fibres and blends thereof, which confer both the correct visual and NIR reflectances (CIBA, 1972; Burkinshaw et al., 1996). Vat dyes are based upon large anthra-quinone–benzanthrone–acridine polycyclic ring systems. They have very good fastness to light, washing, chemicals, and abrasion on cellulosic fibres such as cotton, viscose, and polynosic rayons. Achieving the visual and NIR require-ments on synthetic fibres such as nylon, polyester, aramids and polyolefins is much more difficult, as their compact structures mean that they can only be dyed with small molecules which tend to have very little absorbance in the NIR band.

An alternative technique to increase the absorbance of brown and black colours is to bind carbon black pigments in the printing pastes. However, metering the small amounts of the necessary carbon is difficult to control in production, and the poor rubbing fastness of resin bound pigments means that the reflectance properties can change with wash and wear cycles. Production specifications often allow wider tolerances than those detailed in Fig. 21.3, because of these complications. Another approach is to include carbon pigments in the polymer melt. Thus, Rhone-Poulenc produced a 'grey polyester' for military textiles, containing about 0.01% by weight of carbon.

27.7.4   Far infra-red (thermal) camouflage

The thermal wavebands occur from 3–5 microns and 8–14 microns. Modern lightweight, compact and relatively cheap thermal imagers can now be used on the battlefield to detect hot vehicle tyres, engines and exhausts at a range of several kilometres. Humans can also be readily detected at the longer 8–14 micron waveband, especially if skin is exposed on face and arms. In simple terms the relationship between the amount of emitted radiation and the absolute temperature of the target is governed by Stefan's Law, which states: $E = \eta\sigma T^4$, where E = radiant energy emitted, T = absolute temperature, η = emissivity and σ = constant. There are two parameters that we can change to reduce the thermal signature of the human target as follows:

1. Reduce the temperature of the target by wearing more thermal insulation, putting covers over the face and exposed skin, or increase the external surface area using fur, pile or leafy structures. However, these lead to discomfort in warm climates, and humans are reluctant to wear insulated coverings on the face.

2. Reduce the emissivity of the target human by using shiny, reflective covers. Textiles have high emissivities between 0.92 and 0.98, whereas shiny metallic surfaces have emissivities of between 0.04 and 0.12 (Vickers, 1996).

Practical thermal screening materials tend to be complex laminates which include a textile fabric support carrying a film or foil of aluminium or other shiny metal. The foil must be covered with a dull coloured coating which has the correct visual and NIR characteristics. Thermal screening materials for vehicles and equipment have been in service with UK forces for some years (Specification UK/SC/ 5154, 2000). Providing thermal camouflage in clothing presents problems associated with water vapour impermeability, drape, handle, thermal overload, and laundering. Research continues to provide comfortable, practical thermal camouflage in clothing systems.

21.7.5   Radar camouflage

Defeating doppler radar systems is difficult to achieve, especially since any movement of the human target is detected. It is possible to incorporate stealth measures into textile nets and covers for static vehicles and installations. The most significant threat band is 2–18 GHz followed by 90–98 GHz. Commercial radar nets exist, but have limitations. Howard (2000) calculated that a 22 dB reduction in radar signature was required to make a static target look like background clutter. Radar nets tend to use textile substrates with the inclusion of metallic fibre bundles and granular coatings of stainless steel or silver.

21.8   Flames, heat and flash protection

Personnel operating in confined spaces such as armoured vehicles, ships, submarines and aircraft are at high risk of burns, as are ground troops exposed to nuclear weapons. The main threats to tank crews are detailed in Table 21.9 This equates the threats with their typical heat fluxes and estimated survival times for no injury to occur (NATO Standardisation Group, 1992). In addition, fires in confined spaces produce toxic products which can kill, and smoke which hinders escape. Modern thermoplastic fibres can melt and drip injuring humans, and spreading the fire in furnishings and fittings. See further details in Chapters 11 by Song and Chapter 15 by Horrocks.

Table 21.9 Heat threats to military tank crews

Threat source	Typical heat flux ($kW\,m^{-2}$)	Survival time (sec.)
Burning fuels	~ 150	7–12 for no injury
Exploding munitions	~ 200	<5
Penetrating warheads	~ 500–560	<0.3
Nuclear thermal pulse inside closed vehicle	~ 600–1300	<0.1

21.8.1   Criteria for protection of the individual

Protective clothing must:

- prevent the outer clothing and equipment catching fire by using flame retardant, self-extinguishing textiles. These should retain at least 25% of their original strength, and should not shrink more than 10% after the attack (Elton, 1996).
- prevent conducted or radiated heat from reaching the skin by providing several layers of thermal insulation or air gaps.
- minimise the evolution of toxic fumes by careful selection of materials. Naval submarines pose particular problems, as they have closed cycle air-conditioning and scrubbing systems when submerged. Fibres or coatings such as PVC, PVDC, Neoprene rubber, PTFE laminates, and acrolein from cotton pose common risks.
- for high-risk personnel, prevent clothing in contact with bare skin melting and sticking by avoiding thermoplastic fibres such as nylon, polyester, PVDC, and polyolefins (Staples, 1996).

21.8.2   Flame retardant textiles in military use

The most widely used FR material in the UK forces is Proban® treated cotton, alone or in blends with up to 30% polyester. Its advantages are low cost, wide availability, and low shrinkage in fire. Its disadvantages are that the THPOH treatment can weaken the fibres, it liberates fumes and smoke, and it must not be laundered using soap in hard water, as flammable residues can be left in the fabric. Figure 21.4 shows Royal Navy action clothing made from Proban® cotton, including two-layer anti-flash coverall, hood and gloves. Meta-aramid fabrics such as Nomex®, Conex® and Kermel® are used for specialist, low volume applications listed in Table 21.10, due to their high cost. Their advantages are high durability and low shrinkage when blended with between 5 and 20% of para-aramid fibres. They can be rendered antistatic by incorporating 1–3% of conductive fibres in the blend. They also evolve low toxicity and low smoke combustion products.

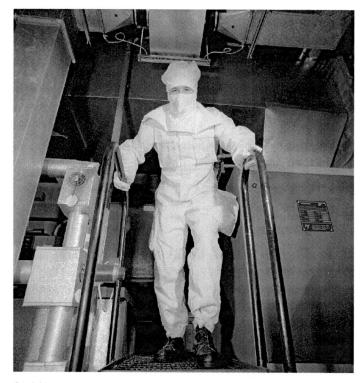

21.4 Naval action stations clothing.

Table 21.10 Military uses of flame retardant textiles

Fibre/fabric type	Cost	Clothing items
Proban® treated cotton	Low	Navy Action Working Dress Shirt and Trousers, Blue Navy Action Coverall, White Navy Anti-flash Hood, White Navy Anti-flash Gloves, White Air Maintenance Ratings Coverall Welders' Coverall
Meta-aramid	High	Tank Crew (AFV) Coverall Submariners' Action Dress Shirt & Trousers Aircrew Coverall, olive Aircrew Immersion Coverall, Outer. Mk. 20A Aircrew Life Preserver Vest, Mk 40 Aircrew Anti-G Trousers, Mk. 10 Explosive Ordnance Disposal Suit, Outer, Mk 4 & 5
Zirpro® treated wool	Medium	Navy Firefighters' coverall RAF Firefighters' suit (old) Foundry Workers' Suits
Modacrylic	Low	Nuclear, Biological and Chemical Oversuit Mk4
PBI gold	High	MOD Firefighters' Suits

Wool can be treated with colourless zircomium and titanium hexa-fluoro-complexes to improve the natural FR properties of wool (see Table 21.10 for applications). Modacrylic fibres in twill fabrics with nylon have been used for the UK's nuclear, biological and chemical (NBC) suits mark 2 to 4 to provide a degree of flash resistance. Poly-benz-imidazole (PBI) fabrics have been used for US Aircrew and UK military firefighters' clothing.

21.9   Nuclear, biological and chemical (NBC) protection

In general NBC weapons are classed as weapons of mass destruction (WMD).

21.9.1   The nuclear threat

This was considered to be high between the 1960s and 1980s, during the Cold War. Since then the threat priority has decreased, despite the fact that up to 20 nations possess nuclear capability. The threats to individuals are blast, radiant heat, flying debris, and radiation in the form of α, β, and χ rays. NBC clothing can protect temporarily against α and β contaminated dusts, but not against hard χ rays which can penetrate several centimetres of metal or concrete.

21.9.2   The chemical threat

This has escalated since the 1990s, particularly since chemical agents are relatively easy to prepare by terrorists (Rudduck et al., 1997, Miltech 8, 2002). Their effects are insidious, lethally horrific, and raise highly emotional fears amongst the general population (see also Chapter 20 by Truong and Wilusz).

The classical chemical agent, mustard, (bis-2-chloroethyl sulphide) was first prepared in 1822, but not used until World War I. It attacks moist skin, tissues and the respiratory tract, causing severe blistering, swelling, and burns. It was recently used in the Iran/Iraq wars.

Nerve agents are so called because they affect the transmission of nerve impulses in the body. They were developed by German chemists in the 1930s (Hay, 1984; Hay et al., 1982). They are all organophosphorus compounds, including phosphonofluoridates and phosphoryl cyanides, which are rapidly absorbed by the skin, although they are primarily a respiratory hazard. They were given the names: Tabun (GA), Sarin (GB) – used by terrorists in the Tokyo underground (Miltech 8, 2002), Soman (GD), while Agent VX, developed by the US in the late 1950s, is one of the most toxic and persistent agents known to man (Iversson et al., 1992). Hydrogen cyanide forms insoluble complexes in the bloodstream, a form of which was used in the Nazi gas chambers in World War II (Zyklon B). Cyanide is also believed to have been used by Sadam Hussein against Kurdish villages in the 1980s and 1990s.

21.9.3   The biological threat

The distinction between biological and chemical agents has become less clear recently, as developments in biotechnology and genetic manipulation have multiplied the possibilities. Classical agents would include bacteria and viruses, including anthrax, botulism, smallpox, cholera, and haemorrhagic fevers. We can now add toxins, peptides and bioregulators to the list. The essential protection devices for individuals are ori-nasal or full face respirators which filter out and deactivate the toxic species (see also Chapter 17 by Krucińska). Full body protection must also be worn, as mustard attacks skin and nerve agents can be absorbed readily at pressure points such as fingers, knees, elbows and seat. One- or two-piece suits with hoods and efficient seals, gloves and overboots complete the ensemble. Clothing, footwear and hand wear can be made from totally impermeable butyl rubber materials, but the physiological load imposed means that they can be worn for only short periods for reconnaissance and decontamination procedures. Most nations now wear air permeable suits utilising activated carbon absorbents on a textile substrate. The carbon can be in the form of powder coatings, small beads (Blucher®), or in carbonised fabric form. In all cases the activated carbon has a highly developed pore structure with a very high surface area, enabling it to absorb a wide range of toxic vapours.

The UK has for many years used a disposable two-layer NBC suit comprising an inner non-woven textile sprayed with powdered charcoal in a carrier/binder, and treated with an oil and water-repellent fluorocarbon finish (Specification UK/SC/3346; Issue 8, 2003). The outer layer is a woven twill fabric comprising a nylon filament warp with a modacrylic weft which carries a water-repellent finish. This layer is designed to wick and spread oleophilic agents to evaporate as much as possible before it transfers to the charcoal layer underneath. The most recent version is designated Suit, Protective, NBC, No. 1 Mark 4, and is used by all three services. The assembly includes butyl rubber overboots and butyl/neoprene dipped gloves. The respirator is designated type S10. At the time of writing a new suit has been developed, Suit Chemical Vapour Protective, No. 1, Mark 5. The suit is described in a technical performance specification that allows manufacturers to use a range of charcoal cloths and outer fabrics to meet the required protection levels.

21.10   Future trends

The main military nations have research programmes geared towards future combat and protective clothing as integrated systems. Thus the US has '21st Century Warrior', the UK has 'Future Integrated Soldier Technology' (FIST), France has 'FELIN', Canada has Integrated Soldier Systems Programme, and Australia has 'Project Wundurra' (Baddely, 2004). The programmes tend to be led by military threats or capability gaps doctrine, rather than the exploitation of new technologies for the sake of it. The systems approach involves all the major stakeholders, including, strategic planners, users, equipment capability managers, operational analysers, R & D scientists, producers, contracts staff, etc. (Sparks, 2004). The general aims of future systems are to

• improve protection against natural and battlefield threats
• maintain thermophysiological comfort, or survival in extreme conditions
• improve compatibility between and within different clothing components
• reduce weight and bulk of materials
• integrate functionality, so that fewer layers provide multi-role protection
• reduce life-cycle costs by making systems more effective, durable, recyclable, and by buying fewer components in the sytem.

The UK recently carried out a research programme which included innovative, smart, reactive textile systems (Scott, 2000b). These were designed to be unobtrusive in normal wear, but would react directly to an imposed threat to provide enhanced protection. Ideas included adjustable pile or inflatable thermal insulation, phase change materials to buffer heat changes, reactive heat and flame protection using shape memory alloys, thermochromic and photochromic dyes, and intumescent treatments for textiles (Scott, 2001). See also Chapter 7 by Van Langenhove *et al.*, and Chapter 15 by Horrocks for additional details. To reduce the number of layers which had to be procured, worn or carried, a project to study multi-role textiles was initiated. The end product was a textile laminate which combined nine levels of functionality in one outer combat layer that was waterproof, water vapour permeable, windproof, water and oil repellent, flame retardant, non-melting, and camouflaged in the visible, near IR and acoustic bands.

Future combat clothing systems will incorporate significant levels of opto-electronic technology. Thus, the UK 'FIST' and the US 'Objective Force Warrior' programmes (Brown, 2004; Newson and Foley, 2004; Tassinari and Leitch, 2004) will include textile wiring, connectors, electronic circuitry, and 'soft switches' in clothing (Tait, 2000; Holme, 2004a) to control such features as GPS positioning, real time terrain maps, remote weapon sighting, and inter/intra-squad video and voice communications through head-up displays or head-down displays (HUD or HDD). Their features will include electro-optical fibres, electrotextiles, and sensors to monitor battlefield hazards, individual injuries and wide spectrum camouflage (Fisher, 2001; Rawcliffe, 2001). Nanotechnology may provide better breathable and self-decontaminating barrier fabrics (Tassinari and Leitch, 2004), improved textiles and camouflage through enhanced micro and nano-particle technology in fibrous polymers (Frankel *et al.*, 2004; Holme, 2004b).

21.11   References

Baddeley A (2004) *21st Century Warrior*. Internet Publications, Sept 04.htm

Blechman H (2004) *DPM – Disruptive Patterned Material*. DPM Ltd. (BVI). ISBN 095434040X.

Brown M (2004) 'Integrated Soldier Technology'. *Proceedings of International Soldier Systems Conference* (ISSC 2004). Boston, Mass. USA, December 13–16, 2004.

Burkinshaw S M, Hallas G, Towns A D (1996) 'Infra-Red Camouflage'. *Review Prog. Colouration*, Vol. 26, pp 47–53.

Chappel M (1987) *The British Soldier in the 20th Century. Part 5, Battledress 1939–1960*. Wessex Military Publishing., Devon, pp 8–16.

CIBA Ltd (1972) *Dyes for Infra Red Camouflage*. CIBA-Geigy, Basle, Switzerland.

Congalton D (1997) 'Thermal and Water Vapour Resistance of Combat Clothing'. *Proc. International Soldier Systems Conference* (ISSC 97), ed. R A Scott, Colchester, Essex, UK, October, pp 389–404.

Couldrick C A, Gotts P L, Iremonger M J (2002) 'Optimisation of Personal Armours for Protection'. *Proc. Personal Armour Systems Symposium* (PASS 2002). The Hague, Netherlands, Nov. 2002.

Defence Standard 00 – 23 (1980) 'NATO Green CamouflageSpectral Curves – Visual & Near Infra Red Requirements'. *Def Stan. Ops.*, Glasgow, UK.

Defence Standard 00 – 35, issue 2 (1996) 'Environmental Handbook for Defence Materials', chapter 1-01, Table 2, p 5. *Def Stan Ops.*, Glasgow, UK.

Drinkmann M (1992) 'Structure & Processing of Sympatex Laminates'. *J. Coated Fabrics*, Vol. 21, pp 199–211.

Elton S F (1996) 'UK Research into Protection from Flames and intense Heat for Military Personnel'. *Fire & Materials*, Vol. 20, pp 293–295.

Fabian J, Nakazumi H, Matsuoka M (1992) 'Infra Red Photography'. *Chem. Rev.* 92, 1197.

Fisher G (2001) 'Intelligent Textiles for Medical and Monitoring Applications'. *Technical Textiles International Newsletter*, June 2001, pp 11–14.

Frankel K, Cowan R, King M (2004) 'Concealment of Warfighters through enhanced Polymer Technology'. *Proc. ISSC 2004*, Boston, Mass. USA, December 13–16, 2004.

Goldman R F (1988) 'Biomedical Effects of Underwear'. In *Handbook on Clothing*, ed. L Vangaard. Chapter 10. NATO Research Study Group 7, Natick, Mass. USA.

Hay A (1984) 'At War with Chemistry'. *New Scientist*, 22 March, pp 12–18.

Hay A, Murphy S, Perry-Robinson J, Rose S (1982) 'The poison clouds hang over Europe'. *New Scientist*, 11 March, pp 630–635.

Hobart A-M, Harrow S (1994) *Comparison of Wicking Underwear Materials*. Defence Clothing & Textiles Agency Unpublished Report, Colchester, UK.

Holme I (2004a) 'Go WEST'. *Textile Horizons*, Nov/ Dec. Textile Institute, Manchester, pp 22–23.

Holme I (2004b) 'Nanotechnology in Textiles – for Now and in the Future'. *Technical Textiles International Newsletter*, Sept., pp 11–14.

Howard A D (2000) *Radar capable camouflage nets – performance requirements*. Unpublished work. Project D/DCTA/ R58, MOD, Research & Project Support, Caversfield, Bicester, Oxon. UK.

ISO 11092 (1993) *Measurement of Stationary Thermal & Water Vapour Resistance by Means of a Thermo-regulatory Model of Human Skin*. ISO, Geneva, Switzerland.

ISO 5085 – Parts I & 2 (1996) *Determination of Thermal Resistance using a Guarded Hotplate Apparatus*. ISO, Geneva, Switzerland.

Ivanov Y A, Tyapin B V (1965) *Infra Red Technology in Military Matters*. English Translation AD 610765. Wright–Patterson Air Force Base, Ohio, USA. Published by Brasseys, London.

Iversson U, Nilsson H, Santesson J (1992) *Briefing Book on Chemical Weapons*. No. 16. Forsvarets Forskingsanstalt (FOA), Sundyberg, Sweden.

Lomax G R (1990) 'Hydrophilic Polyurethane Coatings'. *J. Coated Fabrics*, Vol. 20, pp 88–107.

Marsden A L (1994) *Current UK Body Armour & Helmets*. MOD, DCTA lecture, Colchester, Essex, UK.

Miltech 8 (2002) 'CBRN Terrorism: How Real is the Threat'. *Military Technology*, Aug., pp 8–12.

Morye S S, Hine P, Duckett R A, Car D J, Ward I M (1996) 'A preliminary study of Gel Spun HMPE fibre based composites'. *Proc. PASS 96 Symposium*, Colchester, UK, Sept.

NATO Standardisation Group (1992) *Standardisation of Material & Engineering Practices. Allied Combat Clothing Publication No. 2* (ACCP2).

Newark T, Newark Q, Borsarello J F (1996) *Brasseys Book of Camouflage*.

Newson H, Foley J (2004) 'UK Integrated Soldier Systems Technology'. *Proc. ISSC 2004 Conference*, Boston, Mass. USA, December 2004.

Rawcliffe N (2001) 'Conductive textiles already revolutionising our lives'. *Technical Textiles International Newsletter*, Sept., pp 25–28.

Richardson M A, Luckraft I C, Picton R S, Rodgers A L, Powell R F (1998) 'Surveillance and Target Acquisition Systems'. *Brasseys Battlefield Weapons & Systems Technology*. Vol. VIII. Brasseys, London.

Rudduck M, Wakeford D, Plant P (1997) 'Towards a Safer World'. *Chemistry in Britain*, March, pp 25–29.

Scott R A (1995) 'Coated and Laminated Fabrics'. In *Chemistry of the Textile Industry*, ed. Carr C M. Blackie Academic & Professional, pp 234–243.

Scott R A (2000a) 'Textiles in Defence'. Chapter 16 in *Handbook of Technical Textiles*, eds. Horrocks R and Anand S A, Woodhead Publishing, Cambridge, UK, pp 432–433.

Scott R A (2000b) 'Fibres Yarns and Fabrics for Future Military Clothing'. *Proc. NOKOBETEF 6 conference*, eds. Holmér I and Kuklane K, Stockholm, Sweden, May 7–10 2000, pp 108–113.

Scott R A (2001) 'Textiles for Future Military Clothing'. *Proc. Tomorrows Textiles conference*. UMIST, Manchester, 25th May 2001.

Shephard R G (1986) *The Use of Polymers in Personal Ballistic Protection*. MOD, DCTA lecture, 28th November 1986.

Sparks E (2004) 'A systems approach to the delivery of future UK Operational Clothing'. *Proc. Int. Soldier Systems Conference* (ISSC 2004), Boston, Mass, USA, Dec. 13–16.

Specification UK/SC/3346 issue 8 (2003) *Cloth, bonded, multifibre Antigas*. UK DLO Defence Clothing IPT, Caversfield, Bicester, Oxon. UK.

Specification UK/SC/4919 issue 3 (2002) *Vest and Drawers, Winter Underwear, Olive*. UK DLO, DC IPT, Caversfield, Bicester, Oxon. UK.

Specification UK/SC/5154 (2000) *Cloth, camouflage, low emissivity*. Engineer Support

Systems IPT, DLO, Caversfield, Bicester, Oxon. UK.

Specification UK/SC/5449 (2002) *Ballistic Test Methods for Personal Armour and lightweight materials*. UK DLO, DC IPT, Caversfield, Bicester, Oxon. UK.

STANAG 2920 (1999) *Test Methods for Personal Protective Armours*. NATO Publications, Def. Stan. Glasgow, UK.

Staples R A J (1996) *The melt/burn hazard from cotton/polyester materials*. Unpublished work. MOD, DCTA Tech. Memo. 96/10. MOD, DCTA, Colchester, UK.

Tait N (2000) 'Wearable Electronics hit the market place'. *Australasian Textiles & Fashion*, Sept/Oct, p. 18.

Tassinari T, Leitch P (2004) 'Interactive Textiles for Soldier Systems'. *Proc. ISSC 2004*, Boston, Mass. USA, Dec.

Tobin L (1994) *Military and Civilian Protective Clothing*. MOD, DCTA lecture given at Royal Military College of Science, Shrivenham, Wilts, UK. Jan.

Vickers A F (1996) *Camouflage – what the eye can't see*. MOD DCTA lecture to Physics Dept. Durham University, November 1996.

Weder M (1997) 'Performance of Breathable rainwear materials with respect to protection, Physiology, Durability and Ecology'. *J. Coated Fabrics*, Vol. 27, Oct., Technomic Publications, pp 146–168.

Wilmott H P, Cross R, Messenger C. (2004) *World War Two*. Pub. Dorling Kindersley, London, p134.

22
Firefighters' protective clothing

H MÄKINEN, Finnish Institute of Occupational Health, Finland

22.1 Introduction

The goal of firefighting and rescue work is to get the emergency situation under control as soon as possible and to minimise the eventual human and material losses (Mäkinen, 1991). Wordwide there are many thousands of organisations involved in firefighting activities with millions of firefighters trained and equipped for this goal. This chapter deals with the demands for firefighters' protective clothing, arising from the hazards they face and from their work demands, but focuses on clothing against heat and fire. The performance levels set in standards are summarised. The construction and design of garments, typically used materials, heat stress problems, as well as the effects of moisture on thermal protection are considered. Also selection and care problems and attempts to define the useful life of garments are covered. Aspects of future trends affecting the development of firefighters' protective clothing are reviewed.

22.2 Different tasks and environments

The tasks performed by firefighters have become diverse. Their work on fires includes structural firefighting, aircraft fires and other vehicle fires, wildland firefighting, and a number of other types of fires. But the work is now shifting from firefighting, which takes up 5–10% or even less of the duty time of firefighters, exposing them to extremes of heat and flame (Strickland, 1987; Anon., 2003), more and more towards rescue work. For example, in 1977–1999 inspectorates in the UK have reported an increase of special services (non-fire incidents) in England and Wales, 171% in Scotland and 373% in Northern Ireland (McAllister, 2003). 'Non-fire' activities are road traffic accidents, hazardous materials incidents, and other incidents requiring a broad range of rescue competencies. Firefighters are the first responders in urban search and rescue operations like building/structural collapses that are often associated with earthquakes or bombings, extraction of victims at the scene of vehicle accidents, terrorist attacks, confined space entry, and trench/cave rescue. Also search in air, water, high angle rescue, swift or still water rescue, and contaminated water diving can be part of their work (Stull *et al.*, 1996a). Often they also provide the first emergency medical care in these situations. Today the concept of 'rescue worker' would describe the occupation better.

22.2.1   Hazards in different firefighting tasks

The following list of various tasks gives a picture of the wide variety of environments where firefighters must face different kind of hazards.

Firefighting

Heat and fire in different forms are the most important hazards in firefighting. The thermal hazards in ground fire conditions are usually radiant or convective energy from open flames, explosions, flashing or back-draughts or radiant heat from hot surfaces, objects, etc., falling objects, debris and hot materials or objects exposed to contact heat (Coletta *et al.*, 1976). Hoschke (1981) divides the thermal conditions into three categories, i.e., routine, hazardous and emergency, each defined by a range of air temperature and a range of radiant flux (Fig. 22.1). Figure 22.1 also shows why high thermal protection is needed, e.g., exposure to a radiant heat level of $4\,kW/m^2$, which is quite normal in structural firefighting (Rossi, 2003), causes second-degree burns to bare skin in 30 seconds.

The minimum required function of protective equipment in hazardous conditions, e.g., during entry into a burning room, is to protect the user sufficiently throughout the operation time and to minimize heat stress. In emergency conditions the protective equipment should not hinder escape (Mäkinen, 1991). It has been estimated that it takes 3–10 s to escape from aircraft or vehicle crash fuel spills with heat flux intensities peaking between 167 and $226\,kW/m^2$ (Holmes, 2000a).

The heat flow can today be higher than shown in the Hoschke-Table (Hoschke, 1981) because of increased fuel loads from the use of synthetic building materials, interior finishes and furnishings. Lawson (1997) showed that very high heat fluxes, as high as $30\,kW/m^2$, with a temperature of 175 °C on the ground are possible. Rossi (2003) measured radiant heat fluxes typically between 5 and $10\,kW/m^2$, and the temperature reached 100–190 °C at 1 m above ground level.

The introduction of properly maintained and functioning smoke detectors, residential alarm systems, and modern fire service communications systems have made it possible for firefighters to arrive on the fire scene before a structure is fully involved, and they can start extinguishing at the source of the fire. The increased performance level of personal protective equipment provides a

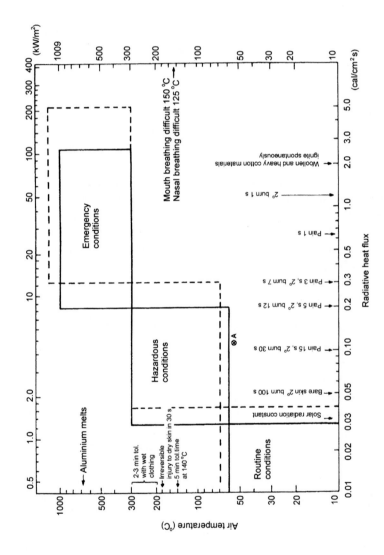

22.1 The thermal environment of firefighters in firefighting (Reprinted from *Fire Safety Journal*, Vol. 4, Hoschke 'Standards and specifications for firefighter's clothing', pp. 125–137 1981, with permission from Elsevier).

sufficient delay in heat transfer for the firefighters to operate longer, and to enter a very hazardous zone where the temperature is high enough to increase the possibility of the structure collapsing. If there is no early warning device in the protective equipment or clothing, the operating time is limited by the air supply of the breathing apparatus, which normally contains an air supply for 20–30 minutes. Thus escape may not be possible in time. In addition to thermal hazards, firefighters face electrical hazards, climatic conditions, water, mechanical hazards, poor visibility, biological and chemical hazards, and physiological/heat stress.

Wildland firefighting

Wildland firefighters generally operate for long periods (8–16 h/day) and are generally exposed to a radiant heat flux of $1 \, kW/m^2$ to $8 \, kW/m^2$. But they can be exposed during increased activities in extreme fire conditions for shorter periods to higher radiant heat fluxes up to $100 \, kW/m^2$. The hazards in wildland fires vary, depending mainly on the following factors (Donarski and Poulin, 2004): fine fuel load, fire danger index, slope, drought factor, air temperature, relative humidity, wind velocity, fuel volume. Heavy working at high temperatures for long periods exposes the firefighters to heat stress. These problems are dealt with in a special issue of *International Journal of Wildland Fire* (Weber, 1997).

Chemical attacks and hazardous environments

The increased use and transportation of toxic chemicals raises the risk of accidents. In such situations the first responders often do not know what chemical(s) are involved, or which hazard level they will encounter. In certain countries, including the UK, 'Hazchem' coded plates are fitted to road tankers. The number and letter codes give rescue teams information on the nature of the contents, the level of hazard, and advice on the safest ways of treating any spills.

Motor vehicle accidents

When removing victims from the accident vehicles many tools with a high level of working power are needed. This creates mechanical hazards as well as flying pieces of sharp broken glass, sharp metal, etc. There is also the risk of the fuel tank exploding. The normal traffic presents an additional risk on the roads and highways.

Other types of rescue work

In urban search and rescue tasks, firefighters face a wide range of hazards, such as: (Stull *et al.*, 1996a)

- physical hazards (metal and masonry debris, working around heavy equipment, floating and submerged debris, slips and falls)
- thermal hazards (exposure to cold water, air and radiant temperatures causing heat stress, subzero temperatures)
- flame and heat related hazards
- chemical hazards (broken gas lines, open solvent containers)
- biological hazards (direct exposure to bacteria-contaminated water, involvement in emergency medical services, contact with victims having infectious diseases).

22.2.2   Accident statistics

Statistics on accidents to firefighters are lacking in many countries. Firemen are often classified as belonging to wider group of workers such as civil servants or municipal public workers. They may also be classified as a part of the armed forces or the police force. Also different statistical systems make it difficult to compare accident data from different countries. In the USA 105 firefighters died while on duty in 2003. Most of the deaths occurred while they were responding to or returning from alarms; 29 (28%) of deaths occurred in the fire grounds. The largest proportion of deaths (45%) was due to heart attacks, and were attributed to stress or over-exertion. There was a sharp increase in the number of heart attacks from 37 in 2002 to 47 in 2003. In 2003 the other major categories were internal trauma (40 deaths), burns (seven deaths), asphyxiation (five deaths), and crushing injuries (four deaths) (LeBlanc and Fahy, 2004).

In an EU-project (Development of the current European standards for firefighters' footwear, 2002) on firefighters' footwear, accident data were collected in 1992 and 1997 from the participating counties, i.e., Austria, France, Finland, Spain, Sweden and the UK. Sprains, strains and dislocations played a major role in all the countries. Burns made up only 1–7% of all accidents. Falling, slipping and tripping accidents constituted the greatest part of the accidents in all the countries. Unfortunately, these general statistics do not give a realistic picture of situations in which protective clothing is used, because the data includes all tasks during working hours, including fitness training.

Lawson (1997) analysed the role of clothing from the burn injury statistics of the United States. Burns on the shoulder and trunk area accounted for about 56% of all burn injuries. Burn injuries to the arm and hand accounted for 29% of all firefighters' burn injuries. Burn injuries to the legs and feet accounted for 9% of all firefighters' burn injuries, being the fourth most common type of burn injury. The thermal protection was often degraded by moisture and compression, because sweat from the head and neck tends to collect in fabrics at the shoulders, and in addition, the SCBA straps extend across the shoulders and back, compressing the clothing layers. Lawson furthermore assumed that non-flame contact injuries with no degradation in the clothing are relatively frequent. Pre-heating before the task contributes to the burns.

22.3 Types of clothing needed for protection

22.3.1 Clothing for firefighting

Firefighters' garments should be designed to perform several functions, the most important of which is protection against heat and flames. Protection against moisture is also important, depending on the type of extinguishing method. The clothing should, additionally, give some degree of protection against mechanical hazards such as cuts and abrasions. In addition, some protection should be given against chemical and biological contact. The tendency to design the level of multi-purpose protection of personal protective equipment for actual fires and against the worst possible scenarios has led to the situation that firefighters are overprotected during the main part of their working time, except for incidents with hazardous materials. This is the normal procedure today in most fire brigades.

ISO TC 94/SC 14/WG 3 specifies the following criteria for firefighters' optimal protective clothing for wildland firefighting. The clothing shall (Donarski and Poulin, 2004):

- permit free evaporation of sweat and be loose-fitting, light, well ventilated and permeable to water vapour
- shield firefighters from radiant heat
- completely dissipate metabolic heat
- allow free evaporation of 1–2 l of perspiration per hour
- sustain a thermal equilibrium and comfort in a wide spectrum of fire intensity, weather, work intensity and duration
- minimise the risk of burn injuries
- minimise episodes of heat exhaustion.

22.3.2 Clothing for chemical attacks and working in hazardous environments

A safe response to leakages of containers or failures in chemical processes, requires a chemical-protective suit. For such purposes, the personal protective equipment must insulate the wearer totally from the hazardous environment, and a breathable air supply independent of the ambient atmosphere is needed. These types of suits are used with a self-contained open-circuit compressed air breathing apparatus (SCBA) worn inside or outside the chemical protective suit. The material and design of the suit must resist the most common liquid and gaseous chemicals. In addition, protection against flames and cold may be needed, depending on the type of chemical and other hazards.

22.3.3   Clothing for different types of rescue work

Because of the diversity of the requirements, no overall specifications are available at the moment, but ISO TC 94/SC14 is working on them. There are, however national specifications for these types of activities, e.g., the Department of Homeland Security has adopted a list of standards for personal protective equipment to protect first responders against chemical biological, radiological, nuclear and explosive threats (http://www.dhs.gov/dhspublic/).

22.3.4   Specifications given in standards

Before the 1980s there were only national requirements for firefighters' clothing. In the 1980s ISO/TC94/SC13/WG2 started the standardisation work for firefighters' protective clothing. Standards for test methods and performance requirements concerning firefighters' protective clothing for structural firefighting (ISO 11613:1999), wildland (ISO15384:2003) and specialised (ISO 15538:2001) firefighting are now available.

In Europe, CEN TC 162/WG 2 prepares specification standards and test methods for activities in which heat and fire are the hazard. European standards are prepared under a mandate given to CEN by the European Commission and the European Free Trade Association, and support the essential requirements of EU Directive 89/686/EEC (Council Directive 1989). European standards for structural firefighting (EN 469:1995), specialised (EN 1496: 1997) firefighting and for fire hoods (EN 13911:2004) are available for firefighters' protective clothing.

The subcommittee ISO/TC 94/SC 14 was established in 2001 for standard-ising the quality and performance of protective clothing and personal equipment intended to safeguard firefighters against the hazards they encounter while performing their duties. The technical work program is divided between working groups representing functional areas of firefighters' activities, including firefighting, rescue operations, chemical and biological hazards. The per-formance standards will cover all parts of the PPE.

In addition to the above international standards, there are still many national standards for firefighters' protective clothing. The USA and Canada follow NFPA (National Fire Protection Association) standards for firefighters' pro-tective clothing. NFPA 1971 (2000) covers minimum design and performance criteria and test methods for protective clothing designed to protect firefighters against adverse environmental effects during structural firefighting. The 2000 edition of NFPA 1971 replaces the 1997 edition. The most important additions are the total heat loss test to measure the protective clothing system's ability to release body heat, changed 'preconditioning' before testing to predict more accurately the durability of some components, and the inclusion of a test to ensure a reasonable degree of thermal protection against compression injuries.

The NFPA 1977 (1998) standard on protective clothing and equipment for wildland firefighting specifies the minimum design and performance criteria and test methods for protective clothing, helmets, gloves, footwear, and fire shelters to protect firefighters against adverse effects during wildland firefighting.

The basic aim of all of the above standards is to provide the minimum requirements for protection performance and durability of protective clothing. Most of the requirements are based on material tests. Table 22.1 lists the requirements of EN 469:1995, of the revision of EN 469 3rd enquiry draft 2002, and of NFPA 1971:2000. They also define the voluntary full-scale protective clothing test. There are facilities to conduct this test in some laboratories in Europe (Bajaj, 2000), North America (Behnke et al., 1992) and Japan. Testing with a manikin provides information about the entire garment design. The testing conditions in the laboratory are restricted, compared to the conditions during actual firefighting. Also the wide variation in the current test equipment limits the use of the method as a standard test method.

The current ISO 11613 defines two performance levels; level 1 is based on the EN 469 from 1995 standard and level 2 on NFPA 1971 from the 1997 standard. The main difference between EN and NFPA standards is that EN defines protection performance separately for radiant and convective heat. NFPA, on the other hand, defines thermal protective performance (TPP) testing to establish the thermal insulation qualities of the material combination. A TPP rating of at least 35 cal/cm^2 is required when the outer surface of the combination is exposed to a heat flux of 84 kW/m^2 (50% radiative and 50% convective heat) for a minimum of 17.5 s. At this level of protection a firefighter would have about 10 s to escape from flashover conditions before getting second-degree burns (Krasney et al., 1988). EN 943-2 (2002) and NFPA 1991 (2000) define the requirements for emergency chemical protective clothing. They specify mechanical requirements for the materials, resistance to permeation against 15 liquid and gaseous test chemicals, requirements for seams, joins and assemblages, and e.g., requirements for tightness, air supply and practical performance of the whole suit.

22.4   Materials used in firefighters' protective clothing

Protective clothing ensembles for structural firefighting typically consist of a flame-resistant outer shell material and of an inner liner which is generally composed of a moisture barrier and a thermal barrier and lining material. The clothing for wildland firefighting is a one- or two-layer garment. The outer layer of the protective garment for specialised firefighting reflects radiant heat, utilising aluminised outer surfaces.

Table 22.1 Material requirements of protective clothing of firefighters according to EN and NFPA standards

Performance property	Test methods	EN 469:1995	revEN 469:2004	NFPA 1971:2000
Flame and thermal resistance				
Flame resistance (applied to both sides of component assembly; also wristlet)	EN 532 ISO 15025 NFPA 1971	Procedure A (face exposure) afterflame ≤ 2 s, afterglow ≤ 2 s no flaming to top or side edge or molten debris, no hole formation	Level 1 and 2 Procedure A (face exposure afterflame ≤ 2 s, afterglow ≤ 2 s no flaming to top or side edge or molten debris, no hole formation	Bottom edge exposure. No melting or dripping, afterflame ≤ 2 s, char length ≤ 100 mm, all layers are individually tested
Heat transfer, flame Flame heat transmission index HTI_{24} and HTI_{12}	EN 367 ISO 9151	$HTI_{24} \geq 13$ s $HTI_{24} - HTI_{12} \geq 4$ s	Level 1 $HTI_{24} \geq 9$ s $HTI_{24} - HTI_{12} \geq 3$ s Level 2 $HTI_{24} \geq 13$ s $HTI_{24} - HTI_{12} \geq 4$ s	—
Heat transfer, radiation	EN 366, Method B in EN 469 ISO 6942, Method B at 40 kW/m	$t_2 \geq 22$ s $t_2 - t_1 \geq 6$ s Heat transmission factor 60%	Level 1 $t_2 \geq 10$ s $t_2 - t_1 \geq 3$ s Level 2 $t_2 \geq 18$ s $t_2 - t_1 \geq 4$ s	—
Thermal protective performance test	Section 6-10 of NFPA 1971–2000	—	—	TPP ≥ 35
Residual strength of material when exposed to radiant heat, 10 kW/m	EN 366 Method A ISO 5081, Method A at 10 kW/m	≥ 450 N	Levels 1 and 2 ≥ 450 N	—

Property	Standard			
Heat resistance and thermal shrinkage (each material used in garment, including wristlet)	EN 469, Annex A ISO 17493 NFPA 1971	5 min 180–190 °C, not melting, dripping or ignition and not shrink more than 5%	Levels 1 and 2 Test at 180 °C – no melting, dripping, separation, or ignition – shrinkage ≤ 5%	Test at 260 °C – no melting, dripping, separation, or ignition – shrinkage ≤ 10%
Conductive compressive heat resistance (reinforced knee and shoulder regions)	NFPA 1971 280 °C and contact pressure of 55 kPa for knees and 14 kPa for shoulders	—	—	$t_{24} \geq 13.5$ s
Thread heat resistance (garment and wristlet)	NFPA 1971	—	—	No melting
Strength and physical hazard performance				
Tensile strength (outer material)	ISO 5081 ISO 13934-1	≥ 450 N	Levels 1 and 2 ≥ 450 N	≥ 623 N
Tear strength (outer material)	ISO 4674 Method A2 NFPA 1971	≥ 25 N	Levels 1 and 2 ≥ 25 N	Outer shell and collar lining ≥ 100 N
Tear strength (moisture barrier, thermal barrier)	NFPA 1971	—	—	Moisture barrier, thermal barrier, winter liner ≥ 23 N
Seam strength (Major A seams)	ISO 13934-2 ASTM D 1683	—	Levels 1 and 2 ≥ 225 N	≥ 675 N
Seam strength (Major B seams)	ASTM D 1683	—	—	≥ 337.5 N

Table 22.1 Continued

Performance property	Test methods	EN 469:1995	revEN 469:2004	NFPA 1971:2000
Seam strength (Minor seams)	ASTM D 1683	—	—	≥ 180 N
Dimensional change (all layers)	ISO 5077, ISO 6330	≤ 3%	≤ 3%	≤ 5%
Water and liquid resistance performance				
Surface wetting (outer shell)	ISO 4920/ EN 24920	≥ class 4	≥ class 4	—
Water absorption resistance (outer shell and collar lining)	FTMS (Federal test method standard) 191A, 5504	—	—	≤ 30%
Penetration by liquid chemicals, (layer combination)	EN 368 –40% NaOH, 20°C –36% HCl, 20°C –30% H_2SO_4, 20°C –white sprit ISO	> 80% runoff and no penetration to innermost surface	Levels 1 and 2 Runoff ≥ 80% no penetration to innermost surface, o-xylene instead of white spirit	—
Water penetration resistance (layer combination)	ISO EN 20811 FTMS 191 A, 5512	Voluntary, no requirement	Level 1, not required Level 2 10 mbar/min rate no water droplets at 20 kPa	60 mbar/min no water droplets at 175 kPa

Property	Method		
Liquid penetration (hydrostatic method)	ASTM F 903, Procedure B, AFFF 3% conc. Battery acid Hydraulic fluid Pool chlorine add. Synthetic gasoline	—	No penetration; moisture barrier continued through 2 cycles of 5X launderings and 10 minute 140°C heat exposures
Viral penetration resistance	ASTM F1671	—	No penetration; moisture barrier continued through 2 cycles of 5X launderings and 10 minute 140°C heat exposures
Thermal comfort performance			
Water vapour resistance (layer combination, testing from inside out)	ISO 11092 (EN 31092)	Voluntary, no requirement	Level 1 $\geq$ 30m^2 Pa/W Level 2 $\geq$ 30m^2 Pa/W
Total heat loss	ASTM F 1868	—	$\geq$ 130 W/m^2

22.4.1   Outer shell

The outer shell is the first line of defence for the firefighter. It provides flame resistance, thermal resistance, and mechanical resistance to cuts, snags, tears and abrasion (Holmes, 2000a). Today normally only inherently flame-retardant fibres, such as aromatic polyamides (aramids) and polybenzimidazole (PBI) are used for the outer layers of firefighters' turnout suits. On the market there are meta-aramids from different manufacturers, e.g., Nomex (DuPont), Conex (Teijin), Fenilon (Russian) and Apyeil (Unitika). Para-aramid fibres like Kevlar (DuPont), Twaron (Akzo Nobel) and Technora (Teijin) are used in blends with meta-aramids to increase durability, e.g., Nomex III (blend of Nomex and Kevlar (95/5%) and X-fire (blend of Teijin Conex and Technora). Especially in France, a polyamide-imide fibre called Kermel from Rhone-Poulenc is used for firefighters' protective clothing. Polybenzimidazole (PBI) fibre was developed by Celanese. Its advantage is that it absorbs more moisture than does cotton, and has a comfort rating from the wearers equivalent to that of 100% cotton (Bajaj, 2000).

The fibres on the market have, e.g., the following trade names: Nomex® III, Nomex® Antistatic (IIIA), Nomex® Outershell Tough (Delta T), Kermel® HTA, PBI®Gold (Ibena). Typical blends are PBI/aramid, Nomex with flame-retardant viscose and flame-retardant wool, Kermel with viscose. Typical constructions of the outer fabrics are twill or ripstop woven fabrics with a mass of 195–270 g/m². In addition to the above fibres in the garments for wildland firefighting, some materials with flame-retardant finishes (FR) (e.g. Proban® and Pyrovatex® for cotton) are used. They must retain the FR properties after 50 launderings (ISO 15384:2003).

22.4.2   Thermal liner

The thermal liner prevents the transfer of heat from the environment to the body. It can consist of a spunlaced, nonwoven felt or batting quilted or laminated to a woven lining fabric. It can also be a knitted fabric between the outer shell and the lining to give the highest insulation against heat, but at the same time allow the escape of moisture due to perspiration. The thermal liner is normally made of inherently flame-retardant fabrics or their blends. A similar fibre content of the thermal liner and outer shell fabric make the laundering of the garment easier. Fibres and yarns are not the real thermal insulators of a garment because fibres conduct heat 10–20 times better than still air. The W L Gore company therefore developed a non-textile insulation material, i.e., an air cushion to replace the traditional textile insulation. Airlock® is a combination of a moisture barrier and thermal protection. 'Spacers' made of foamed silicone on the GORE-TEX® moisture barrier create the insulating air buffer in the material (Hocke et al., 2000).

22.4.3   Moisture barrier

A moisture barrier is obligatory in some countries, whereas in some countries firefighters prefer suits without a moisture barrier because of their thermal comfort. The moisture barrier in firefighters' clothing is (i) laminated or coated to the inside of the outer shell fabric, (ii) is a lightweight knitted material or web, and the structure is inserted loosely between the outer fabric and the liner, or (iii) is on the outside of the thermal liner. The moisture barrier provides protection against water as well as against many common liquids such as common chemicals and bloodborne pathogens. The moisture barrier can be a microporous or hydrophilic membrane or coating (Holmes, 2000b; Anon., 1999). GORE-TEX®, CROSSTECH® and TETRATEX® are textile laminates incorporating microporous polytetrafluoroethylene. PORELLE®, PROLINE® and VAPRO® are microporous polyurethane laminates with textiles. BREATHE-TEX PLUS®, STEDAIR 2000® are hydrophilic polyurethane laminates or coated fabrics, SYMPATEX® is a hydrophilic polyester laminate. The microporous and hydrophilic coatings are normally polyurethane products. ACTION® is example of a polyurethane coating. Neoprene (NEOGUARD®) and polyvinyl chloride (PVC) are non-breathable moisture barrier products.

22.4.4   Accessory materials

Materials to improve visibility are important accessory materials. For instance, fluorescent materials are used to increase day-time visibility and retroreflective materials night-time visibility. Even though there are hundreds of types of fluorescent materials on the market, only a few materials meet the requirements of heat resistance needed for firefighters' protective clothing. The colours of these materials are very sensitive when exposed to the smoky environments in firefighting. The retroreflective material and fluorescent colours are often combined in the trim near the hands, head and feet of garments. Motion increases the visibility. The manufacturers of reflective materials (e.g 3M Scotchlite™, Reflexite®, Unitika®) have heat-resistant products for firefighting applications. The basic requirement of other accessory materials like zippers and braces (suspenders), is resistance to heat.

22.5   Design, sizing and ergonomics

Inadequately designed garments, especially when used together with SCBA, cause extra effort for the firefighters, lowering the mechanical efficiency of moving and breathing, thus demanding a higher energy expenditure, causing discomfort and eventually heat stress (Lusa *et al.*, 1993; Huck, 1991). Restrictive uncomfortable clothing can tempt firefighters to remove their protective garments or to wear them incorrectly, and this increases the potential for injury (Veghte, 1986, Huck and McCullough, 1988). Because of the high physical fitness requirements, the anthropometric dimensions of a population of firefighters may differ significantly from the general population. The same general sizing system is nevertheless applied to firefighters' clothing. The number of female firefighters is on the rise in many countries. If the clothing is too tight and fits poorly, thermal protection will be compromised. Female firefighters therefore need garments specially designed for women (Stirling, 2000).

The design of the garments must take into consideration easy donning and doffing, collar design and closing system, the shoulder, elbow and underarm, the back for increased mobility, pockets, closure system of the whole garment, construction of the knee area and the crotch. Bellows and gussets can be placed in the shoulder and elbow area to allow more movement and flexibility. The design should always take into account the use of breathing apparatus, helmet, fire hood, gloves and footwear. Compatibility with other personal protective devices is one basic requirement of the PPE directive 89/645/EEC on personal protective equipment (Council Directive, 1989). For example, the straps of the SCBA may hinder reaching into the pockets.

The study by Graveling and Hanson (2000) in the UK found problems in the sleeves of garments. They tended to ride up with reaching or stretching movements, and potentially created an unprotected gap at the wrist. Other problems included leggings with insufficient expansion in thigh diameter for squatting or kneeling, insufficient leg length, knee padding incorrectly located in the leggings, and insufficient or excessive body length in a one-piece garment. The standards define some basic aspects of design, and the revisions of the standards (e.g. rev. EN 469 2004) will contain, as annexes, guidelines on how to check the ergonomic features of protective clothing.

22.5.1   Heat stress

Many layers of fabrics limit the evaporation of sweat, even when apparently vapour-permeable materials are used, and reduce heat dissipation from the body. The changeover from an impermeable barrier to a microporous barrier in garments was considered to improve comfort due to lower skin temperature and skin wettedness (White and Hodous, 1988). The ventilation design in the study by Reischl and Stransky (1980) influenced the sleeves and legs most. Also ventilation spaces in the garments have been studied (Reischl et al., 1982). In the study of Dukes-Dubos et al. (1992) the greatest increase was measured at the chest when the turnout gear was opened at the collar, and opening the trousers at the ankles increased ventilation at the legs and the crotch. Replacing the belt by braces (suspenders) increased ventilation at the back, legs and crotch. Mäkinen et al. (1996) compared the thermal strain of firefighters dressed in suits with and without a microporous moisture barrier under simulated firefighting conditions.

The results revealed considerable thermal strain from both types of turnout suits worn in the heat. The body temperature was higher when the suit with a membrane was worn, but the difference was not significant. The difference in mean skin temperature was statistically significant at the end of the hard work, and more sweat condensed in the underclothing when the suit with the membrane was used.

Graveling and Hanson (2000) found that no garment style or fabric combination in the same basic style offered any meaningful advantage in physiological load but they found a significant accumulation of body heat. The SCBA has been found to be more important in determining the physiological load of the clothing system of firefighters than variations in garment design or moisture barrier fabrication (Huck and McCullough, 1988). The multilayer turnout suit based on the European standard EN 469 used with Nordic type underclothing and a SCBA (total weight 25.9 kg) reduced the power output, on average by 25% in maximal dynamic muscle work in a thermoneutral environment (Louhevaara et al., 1995). While wearing firefighting protective clothing and a SCBA at higher ambient temperatures, firefighters' passive recovery may not be sufficient to reduce the rectal temperature to below pre-recovery levels (McLellan and Selkirk, 2004). MacLellan and Selkirk (2004) concluded on the basis of their study that replacing the duty uniform long underpants by shorts reduces the cardiovascular and thermal strain during exercise that lasts over 60 minutes.

One possibility to decrease the physiological heat load is to use an additional cooling system. In Sweden (Smolander et al., 2004) an ice-vest has been developed to decrease physiological load. The material of the vest is cotton, and the inside consists of two removable flat plastic containers with several small pockets for water/ice. The ice-vest weighs about 1 kg; it covers most of the trunk area, and is worn over the underwear. Unfortunately, the physiological test included only four test persons. The authors concluded that the ice-vest reduced circulatory, thermal and subjective strain during heavy work in the heat. The added benefit was about 10%.

Havenith and Heus (2004) propose a test battery related to the ergonomics of protective clothing. The proposed tests with human subjects would evaluate physiological load, protection against heat, ergonomic design, loss of performance, protection against rain/moisture, and visibility of the clothing. In Finland job-related drills have been developed with firefighters and the Emergency Services College to measure physiological parameters of firefighters and to assess the effects of protective equipment (Lusa et al., 1993, Louhevaara et al., 1994, Ilmarinen et al., 2000, 2004). CEN TC 162 WG 2 is working to define a physiological test for firefighters' protective clothing as a part of the EN 469 revision.

22.6   Effect of moisture on thermal protection

The performance levels set in the standards are given for new or prewashed materials or material combinations. In normal use the garments are often wet and dirty. The garments can be wetted from the outside by extinguishing water or from the inside by sweat. The sweating rates for persons carrying out heavy firefighting tasks can rise up to 2,000 g/h (Mäkinen *et al.*, 1996). The extinguishing method determines how much water is used.

There are numerous studies on the effect of moisture on thermal protection, but the results are contradictory (Rossi *et al.*, 2004). The moisture in the fabric layers alters several parameters of the fabrics, such as thermal conductivity and heat retention capacity. The moisture has been found to have both positive and negative effects, depending on the intensity of the heat flux. Moisture in the cloth evaporates when heated, and part of it diffuses and recondenses on different parts of the fabric depending on the local temperature (Prasad *et al.*, 2002). Lawson *et al.* (2004) found that under high-heat-flux exposure, external moisture tended to decrease heat transfer through the fabric system, while internal moisture tended to increase heat transfer. On low-heat-flux radiant exposure, in the other hand, internal moisture decreased heat transfer through the fabric system. Also the type of outer and underwear materials in a clothing system affected the rate of convective and radiative heat transfer.

Veghte (1984) added 1 and 2 g of water to the glove liner to simulate the amount of sweat measured in thermal liners during a field study. The TPP values decreased dramatically. Mäkinen *et al.* (1988) obtained similar results when they measured the effect of wetting on protection against radiant and convective heat. Even though the protection was decreased in wetted fabrics, the changes in the materials were slighter after wetting.

Stull (1997) studied conductive heat resistance of garments and found either a slight or a significant effect, depending on the amount and location of the water in the material system. In some cases, when water was added to the outer shell, slight improvements were noted in threshold times. In many cases, however, relatively no change was noted. Only when large quantities of water were added to the liner alone or to the entire clothing system were reductions in threshold times observed.

The influence of steam has not been discussed widely. Steam flowing from the outside towards the body, or evaporating from the textile layers can lead to steam burns. These may be more severe than dry burns, as the hot moisture can be partly absorbed by the skin and transferred to the deeper skin layers. In general, impermeable materials offer better protection against hot steam than semi-permeable ones. The transfer of steam depends on the water vapour permeability of the samples and their thickness. Increasing the thickness of the samples with a spacer increases the protection capacity of the impermeable samples more than that of semi-permeable materials. Materials with good water vapour absorbency tend to offer better protection against hot steam (Rossi *et al.*, 2004).

22.7 Selection, use and care

The proper selection of protective clothing requires the involvement of the user groups to formulate their own specifications. CEN/TC 162 WG 2 has prepared a technical report (CR/TR 14560:2003) to assist employers (or persons who advise the employer) in making decisions regarding the selection, use, care and maintenance of heat-protective clothing. The document highlights the main points that the employer needs to consider. It is in the form of a checklist and serves as a mnemonic, thus helping one to take into account all aspects when organising the management system for protective clothing in the fire brigade.

22.7.1 Care and maintenance

The most fragile materials are the moisture barrier and the fluorescent and retroreflective materials as they deteriorate easily during use and laundering. The manufacturer is responsible for providing instructions concerning wash and care of the garment (Council Directive, 1989). For example, DuPont stresses that Nomex® garments are to be washed at 60 °C, the washing time should not exceed one hour, overdrying should be avoided and liquid detergent with a near-neutral pH should be used (http://www.dupont.com/nomex/europe/protectiveapparel/nomex/technical/set_wash.html). Lion Apparel gives comprehensive guidance on usage, including cleaning and storage (User Instruction, 2003). Primary routine visual inspection is recommended to check for soiling, contamination, rips, tears, cuts, and thermal damage such as charring, burn holes and melting of materials. More knowledge is needed on the ideal method for removing typical firefighting contaminants to ensure a longer life for the garment (Vogelpohl, 1996). The effect of use and launderings has been studied by many researchers, but there are no objective and non-destructive methods for checking the performance level of used garments..

Vogelpohl (1996) evaluated different properties of garments which had been used in fire departments for firefighting or training purposes. Generally the suits had maintained their thermal properties. Only some moisture barrier fabrics failed to meet the specifications for char length and after-flame length. The water resistance of the used turnout coats had decreased; the outer shell fabrics absorbed 10.3–55.9% water. A similar decrease was also seen in the moisture barrier fabrics. Loss of strength was apparent in the seam construction, and also the fabric strength of outer shell and moisture barrier fabrics decreased. Vogelpohl (1996) pointed out that the used coats were moderately to extremely soiled, which may have contributed to the loss of fabric strength and flame resistance. Long after-flame and after-glow times were measured for used aramide garments. They had been laundered together with garments made of other types of fibres, e.g., cotton, with may have left lint on the fabric surface (Mäkinen, 1992).

Rossi and Zimmerli (1996) measured the effect of heat on water vapour permeability and the water-tightness of moisture barriers. The waterproof properties remained unchanged as long as no visible damage had occurred, but the breathability decreased by more than 30%. Routine inspection after each wash is thus needed to ensure the effectiveness and water repellency of the moisture barrier.

Stull *et al.* (1996b) have studied clothing contamination levels in old discarded firefighter turnout gear and evaluated the effectiveness of current laundering practices in decontaminating them. The laundering affected performance of the clothing materials. The most significant change was the reduction of water penetration resistance of non-breathable moisture barrier material. They also found that conventional cleaning methods followed by aeration might be the most effective form of decontamination for all kinds of chemical contamination.

Fire brigades should be able to record categorically the number of times a garment has been worn and cleaned. This is important in accident situations and for evaluating whether the garment is fit for the purpose (Stirling, 2000). Torvi (1999) sees possibilities for developing a computer program which uses the gathered information on use and the data from laboratory tests to indicate when the garments should be retired. At the moment there is guideline NFPA 1851 (2001); if the cost of decontamination is more than 50% of the replacement cost of the PPE, then replacement is generally recommended.

Thorpe and Torvi (2004) have studied more objective non-destructive test methods to assess when to retire protective clothing. Three different non-destructive test methods (Raman luminescence, digital image analysis and colorimetry) were developed and correlated with established destructive tests to indicate a useful service time. The Raman technique of luminescence was considered promising, but the data were too inconsistent. The digital image analysis method is a promising low-cost technique. The colorimeter method produced data that appeared useful, but the scatter indicated that more work is needed to refine the method. When using these techniques they found difficulties in assessing brown turnout gear.

22.8   Future trends

Veghte (1986) summarised the historical development of firefighters' protective materials and standards in 1986. Since then more functional materials have been developed to give high thermal protection and to increase wear and comfort properties. The biggest change has taken place in standards development; in 1986 only national standards were available. Now the goal is to have common international standards indicating levels of performance for firefighters' protective clothing and other protective equipment. The correct performance level is then defined by risk analysis in the fire brigade. More flexible clothing systems need to be developed to better meet the diverse requirements of the changing role of firefighters.

22.8.1   Test method development

Research is going on to model the heat transfer in firefighters' protective clothing, and to develop and design a new test apparatus for evaluating the thermal performance of firefighters' protective clothing over a wide range of thermal exposures. Such studies would help to define the degree of protection, as regards burn injury and heat stress, and thus help to analyse the anticipated performance of new materials and clothing systems economically and quickly (Prasad et al., 2002; Mell and Lawson, 1999).

In the current test methods for heat protection the energy transferred to a copper disk sensor placed behind a fabric specimen is measured. Other types of sensors are skin simulants whose combination of thermal properties is such that they absorb heat at about the same rate as human skin. Tests are being developed to evaluate the stored energy in a garment and the effects on human skin when this energy is released (Torvi et al., 1997; Torvi, 1999). A test fixture for thermal properties developed by the Ktech Corporation, provides a small-scale test apparatus for calculating the thermal properties of firefighting clothing materials at low heat flux exposure levels allowing also the evaluation of wet materials. Further development of the apparatus permits testing under various compression loads, and also non-destructive testing can be done to track the thermal properties of in-service firefighting coats and jackets (Gagnon, 2000). Furthermore, test method development in the area of thermal comfort, e.g., sweating thermal manikins and a sweating torso (Zimmerli, 1998) will contribute to the design of more functional protective clothing for firefighters. These methods are discussed in other chapters of this book.

22.8.2   Possibilities of intelligent fabrics

In the area of nanotechnology, interactive textiles (smart textiles) could offer a lot of possibilities for firefighters' protective clothing when trying to compromise between protection and comfort. These materials interact with human/environmental conditions to change the properties of the material. The most important intelligent material groups today are phase change materials, shape memory materials, thermochromic materials, and conductive materials. Hydroweave® has fire-resistant applications for firefighting. This material can dissipate large amounts of radiant heat energy, and in cases of high level heat fluxes can provide a barrier to shield the wearer from injury (http://www.tut.fi/units/ms/teva/projects/intelligenttextiles/index3.htm).

22.8.3   Garments with wearable electronic technology

Garments with wearable technology open interesting possibilities for enhancing firefighters' safety, and help the management of an operation during an alarm situation. In these e-textiles electronics (wires and sensors) and fabrics are woven together. Firefighters can get information about the temperature and other hazards in the environment which they are entering, and a fire officer can monitor each firefighter's situation in real-time, e.g., elapsed time, air pressure, remaining air time, ambient temperature, heat stress level, etc. He could order his team out when the sensors they are wearing transmit data to his command centre telling him that the firefighters are, for instance, inhaling hazardous fumes, or too much smoke or that the fire is too hot to handle. This new sensor technology may have a positive impact on warning firefighters of impending untenable conditions (Lawson, 1997).

22.9   Conclusion

Protective clothing alone cannot guarantee the safety of firefighters. They must be able to deal safely with a wide range of fires, must have an understanding of how a fire progresses, must use safe and effective firefighting procedures, and must have a general understanding of the limitations of the protective clothing and equipment (Lawson 1997). Firefighters must be trained to avoid preheating, and how to deal with wet protective clothing. After exposure to heat, the first procedure is always to start cooling off by taking off at least the head protectors, the coat of the fire suit, and the gloves. This helps one to recover and may prevent some burns.

22.10   References

Anon. (1999), 'Firefighters protective clothing: Moisture barrier alert and recall', *Stay Safe* 12/15/99, (*http://www.iaff.org/safe/health/alerts/alert06.html*)

Anon. (2003), 'Conference report, Survival 2003 – Functional fabrics were the focus of a recent UK conference', *Textile Month*, Aug., 44–47.

Bajaj P (2000), 'Heat and flame protection', in Horracks A R and Anand S C *Handbook of technical textiles*, The Textile Institute, Woodhead Publishing Limited, 223–263.

Behnke W P, Geshury A J and Barker R L (1992), 'Thermo-man and Thermo-leg: Large scale test methods for evaluating thermal protective performance', in McBriarty J P and Henry N W, *Performance of protective clothing: Fourth volume, ASTM STP 1133*. Philadelphia: American Society for testing and materials, 266–280.

CEN/TR 14560 (2004), *Guidance for selection, use, care and maintenance of protective clothing against heat and flame*, European Committee for Standardization, Rue de Stassart 36, B-1050 Bruxelles.

Coletta G C, Arons I J, Ashley L E and Drennan A P (1976), *The development of criteria for firefighters' gloves, Volume I, Glove requirements*, NIOSH, Cincinnati, Ohio 45202, 86 p.

Council directive of 21 December 1989 on the approximation of the laws of the Member States relating to personal protective equipment (89/686/EEC), *Official Journal of European Communities* 30.12.89.

Development of the current European standards for firefighter's footwear: Focus on the compatibility of functional and protective properties (2002). CONTRACT N : SMT4-CT98-2275. Final Technical Report Project funded by the European Community either under the 'IMT/SMT' Programmes (1994–1998), 105 p.

Donarski R and Poulin S (2004), Wildland fire environment paper prepared by ISO/TC 94/SC14/WG 3/PG 3 for support in the development of ISO 16073 – wildland firefighting personal protective equipment standard, ISO/TC 94/SC14/WG 3/PG 3 N022. Secretariat: Standards Australia, GPO Box 5420 Sydney, NSW 2001, 286 Sussex Street, Sydney, NSW 2000.

Dukes-Dubos F N, Reischl U, Buller K, Thomas N T and Bernhard T E (1992), 'Assessment of ventilation of firefighter protective clothing', in McBriarty J P and Henry N W, *Performance of protective clothing: Fourth volume, ASTM STP 1133.* Philadelphia: American Society for testing and materials.

EN 469 (1995) *Protective clothing for firefighters – Requirements and test methods for protective clothing for firefighting*, European Committee for Standardization, Rue de Stassart 36, B-1050 Bruxelles.

EN 469 (2004-07) *Protective clothing for firefighters – Performance requirements for protective clothing for firefighting*, Document for formal vote, Secretariat: DIN Deutches Institut für Normung e.V., Burggrafenstraße 6, 10787 Berlin.

EN 943-2 (2002), *Protective clothing against liquid and gaseous chemicals, including liquid aerosols and solid particles. Part 2: performance requirements for 'gas-tight' (Type 1) chemical protective suits for emergency teams (ET)*, European Committee for Standardization, Rue de Stassart 36, B-1050 Bruxelles.

EN 1486 (1997), *Protective clothing for firefighters. Test methods and requirements for reflective clothing for specialized firefighting.* European Committee for Standardization, Rue de Stassart 36, B-1050 Bruxelles.

EN 13911 (2004), *Protective clothing for firefighters – Requirements and test methods for fire hoods for firefighters.* European Committee for Standardization, Rue de Stassart 36, B-1050 Bruxelles.

Gagnon B R (2000), 'Evaluation of new test methods for firefighting clothing', a thesis submitted to the Faculty of the Worcester Polytechnic Institute, 142 p.

Graveling R and Hanson M (2000), 'Design of UK firefighter clothing', in Kuklane K and Holmer I, *Proceedings of NOKOBETEF 6 and 1st European Conference on Protective Clothing* held in Stockholm, Sweden, May 7–10, 2000, *Arbete och Hälsa* (8), 227–280.

Havenith G and Heus H (2004), 'A test battery related to ergonomics of protective clothing', *Applied Ergonomics*, 25 3–20.

Hocke M, Strauss L and Nocker W (2000), 'Firefighter garment with non textile insulation', in Kuklane K and Holmer I, *Proceedings of NOKOBETEF 6 and 1st European Conference on Protective Clothing* held in Stockholm, Sweden, May 7–10, 2000, *Arbete och Hälsa* (8) 293–295.

Holmes D A (2000a), 'Textiles for survival', in Horrocks A R and Anand S C *Handbook of technical textiles*, The Textile Institute, Woodhead Publishing Limited, 461–489.

Holmes D A (2000b), 'Waterproof breathable fabrics' in Horracks A R and Anand S C

Handbook of technical textiles, The Textile Institute, Woodhead Publishing Limited, 281–315.

Hoschke B N (1981), 'Standards and specifications for firefighter's clothing', *Fire Safety Journal*, 4, 125–137.

Huck J (1991), 'Restriction of movement in firefighter protective clothing; evaluation of alternative sleeves and liners', *Applied Ergonomics*, 22 (April), 91–100.

Huck J and McCullough E A (1988), 'Firefighter turnout clothing; Physiological and subjective evaluation', in Mansdorf S Z, Sager R, Nielsen A P, *Performance of Protective Clothing: Second Symposium, ASTM STP 989*, American Society for Testing and Materials, Philadelphia, 439–451.

Ilmarinen R, Lindholm H, Koivistoinen K and Helistén P (2000), 'Physiological strain and wear comfort while wearing a chemical protective suit with breathing apparatus inside and outside the suit in summer and in winter', in Kuklane K and Holmer I, *Proceedings of NOKOBETEF 6 and 1st European Conference on Protective Clothing* held in Stockholm, Sweden, May 7–10, 2000, *Arbete och Hälsa*, 8, 235–238.

Ilmarinen R, Lindholm H, Koivistoinen K and Helistén P (2004), 'Physiological Evaluation of Chemical protective suit systems (CPSS) in hot conditions', *International Journal of Occupational Safety and Ergonomics (JOSE)*, 10 (3), 215–226.

ISO 11613 (1999), *Protective clothing for firefighters – Laboratory test methods and performance requirements*, International Organisation for Standardization.

ISO 15384 (2003), *Protective clothing for firefighters – Laboratory test methods and performance requirements for wildland firefighting clothing*, International Organisation for Standardization.

ISO 15538 (2001), *Protective clothing for firefighters – Laboratory test methods and performance requirements for protective clothing with a reflective outer surface*, International Organisation for Standardization.

Krasney J F, Rockett J A and Huang D (1988), 'Protecting fire-fighters exposed in room fires: comparison of results of bench scale test for thermal protection and conditions during room flashover', *Fire Technology*, 24 (Feb), 5–19.

Lawson J R (1997), 'Firefighters' Protective Clothing and Thermal Environments of Structural Firefighting', in Stull J O and Schope A D, *Performance of protective clothing: Sixth volume, ASTM STP 1273*, American Society for Testing and Materials, 335–352.

Lawson L K., Crown E M, Ackerman M Y and Dale J D (2004), 'Moisture Effects in Heat Transfer through Clothing Systems for Wildlands Firefighters', *International Journal of Occupational Safety and Ergonomics (JOSE)*, 10 (3).

LeBlanc P L and Fahy R F (2004), *Full report Firefighter fatalities in the United States – 2003*, National Fire Protection Association 1 Batterymarch Park Quincy, MA 02169-7471, (June), 34 p.

Louhevaara V, Soukainen J, Lusa S, Tulppo M, Tuomi P and Kajaste T (1994), 'Development and evaluation of a test drill for assessing physical work capacity of fire-fighters', *Int J Ind Erg*, 13,139–46.

Louhevaara V, Ilmarinen R, Griefahn B, Künemund C and Mäkinen H (1995). 'Maximal physical work performance with European standard based fire-protective clothing system and equipment in relation to individual characteristics', *Eur J Appl Physiol*, 71, 223–229.

Lusa S, Louhevaara V, Smolander J, Pohjonen T, Uusimäki H and Korhonen O (1993), 'Thermal effects of fire-protective equipment during job-related exercise protocol', *Safe Journal*, 23(1),36–39.

Mäkinen H (1991), 'Analysis of problems in the protection of firefighters by personal protective equipment and clothing – development of a turnout suit', Ph.D.Thesis, Tampere University of Technology, 201 p.

Mäkinen H (1992), 'The effect of wear and laundering on flame-retardant fabrics', in: McBriarty J P and Henry N W, *Performance of protective clothing: Fourth volume, ASTM STP 1133*. Philadelphia: American Society for Testing and Materials, 745–65.

Mäkinen H, Smolander J and Vuorinen H (1988), 'Simulation of the Effect of moisture Content in Underwear and on the Skin Surface on Steam burns of Fire-Fighters', in Mansdorf S Z, Sager R, Nielsen A P, *Performance of Protective Clothing: Second Symposium, ASTM STP 989*, American Society for Testing and Materials, Philadelphia, 415–421.

Mäkinen R, Ilmarinen R, Griefahn B and Künemund C (1996), 'Physiological comparison of firefighter turnout suit with and without a microporous membrane in the heat', in Johnson J S and Mansdorf S Z, *Performance of protective clothing: Fifth volume, ASTM STP 1237*, American Society for Testing and Materials, 396–407.

McAllister A (2003), 'Following suit', *FEJ & FP*, July, 56–57.

McLellan T M and Selkirk G A (2004), 'Heat stress while wearing long pants or shorts under firefighting protective clothing', *Ergonomics* 47 (1) 75–90.

Mell W E and Lawson J R (1999), 'A heat transfer model for firefighter's protective clothing, *NISTIR* 6299, 27 p.

NFPA 1977 (1998), *Protective clothing and equipment for wildland firefighting*, 1998 edition. National Fire Protection Association, 1 Batterymarch Park, PO Box 9101, Quincy, MA 02269-9101.

NFPA 1971 (2000), *Standard on Protective Ensemble for Structural Firefighting*, 2000 edition, National Fire Protection Association, 1 Batterymarch Park, PO Box 9101, Quincy, MA 02269-9101.

NFPA 1851 (2001), *Standard for selection, care, and maintenance of Structural firefighting protective ensembles*, National Fire Protection Association, 1 Batterymarch Park, PO Box 9101, Quincy, MA 02269-9101.

NFPA 1991 (2000), *Standard on vapor-protective ensemble for hazardous materials emergencies*, National Fire Protection Association, 1 Batterymarch Park, PO Box 9101, Quincy, MA 02269-9101.

Prasad K, Twilley W and Lawson J R (2002), 'Thermal Performance of Firefighters' Protective Clothing. 1. Numerical Study of Transit Heat and Water Vapour transfer', *NISTIR* 688, 32 p. *http://www.fire.nist.gov/bfrlpubs/fire02/PDF/f02077.pdf*

Reischl U and Stransky A (1980), 'Assessment of ventilation characteristics of standard and prototype firefighter protective clothing', *Textile Research Journal*, 50, 193–201.

Reischl U, Stransky A, DeLorme H R and Travis R (1982), 'Advanced prototype firefighter protective clothing: heat dissipation characteristics', *Textile Research Journal,* 52 (1), 66–73.

Rossi R (2003), 'Firefighting and its influence on the body', *Ergonomics,* 46 (10), 1017–1033.

Rossi R M and Zimmerli, T (1996). 'Influence of Humidity on the Radiant, Convective and Contact Heat Transmission through Protective Clothing Materials', in Johnson J S and Mansdorf Z S, *Performance of Protective Clothing: Fifth Volume, ASTM STP 1237*, American Society for Testing and Materials, 269–280.

Rossi R, Indelicato E and Bolli W (2004), 'Hot steam transfer through heat protective clothing layers', *International Journal of Occupational Safety and Ergonomics (JOSE)*, 10 (3), 239–245.

Smolander J, Kuklane K, Gavhed D, Nilsson H and Holmér I (2004), 'Effectiveness of a light-Weight Ice-Vest for Body Cooling while wearing firefighter's protective clothing in heat', *International Journal of Occupational Safety and Ergonomics (JOSE)*, 10 (2), 111–117.

Stirling M (2000), 'Aspects of firefighter protective clothing selection', in Kuklane K and Holmer I, *Proceedings of NOKOBETEF 6 and 1st European Conference on Protective Clothing* held in Stockholm, Sweden, May 7-10, 2000, *Arbete och Hälsa* (8), 269–272.

Strickland B (1987), 'Fire risk assessment', Part I, *Fire Command*, 54, 34–37.

Stull O J (1997), 'Comparison of conductive heat resistance and radiant heat resistance with thermal protective performance of firefighter protective clothing', in Stull J O and Schope A D, *Performance of protective clothing: Sixth volume, ASTM STP 1273*, American Society for Testing and Materials, 248–268.

Stull O J, Connor M B and McCarthy R T (1996a), 'Protective clothing and equipment performance requirements for fire service urban search and rescue teams', in Johnson J S and Mansdorf S Z, *Performance of protective clothing: Fifth volume, ASTM STP 1237*, American Society for Testing and Materials, 558–572.

Stull O J, Dodgen C R, Connor M B and McCarthy R T (1996b), 'Evaluation of the effectiveness of different laundering approaches for decontaminating structural firefighting protective clothing', in Johnson J S and Mansdorf S Z, *Performance of protective clothing: Fifth volume, ASTM STP 1237*, American Society for Testing and Materials, 447–468.

Thorpe P A and Torvi D A (2004), 'Development of non-destructive test methods for assessing effects of thermal exposures on firefighters' turnout gear', *Journal of ASTM International*, 1 (6), www.astm.org

Torvi D A (1999), 'Research in Protective Clothing for firefighters: State of the Art and Future Directions', *Fire Technology*, 35 (2), 111–113.

Torvi D A, Dale D, Mark Y A and Crown E (1997), 'A study of new and existing bench top tests for evaluating fabrics for flash fire protective clothing', in Stull J O and Schope A D, *Performance of protective clothing: Sixth volume, ASTM STP 1273*, American Society for Testing and Materials, 108–123.

User Instruction, Safety and Training Guide (2003), Lion Apparel NFPA 1971 Compliant Structural Firefighter Garment, Lion Apparel P.O. BOX 13576, Dayton, Ohio, 45413-0576, *http://www.lionapparel.com/psg/pdfs/StructuralTurnoutUserGuide.pdf*

Veghte J H (1986), 'Functional integration of firefighters protective clothing', in Barker R L, Coletta G C, *Performance of protective clothing: Fifth volume, ASTM STP 900*, American Society for Testing and Materials, Philadelphia, 487–496.

Veghte, J H (1984), 'Effect of Moisture on the Burn Potential in Firefighters' Gloves', *Fire Technology*, 23 (4), 313–322.

Vogelpohl T L (1996), 'Post-use evaluation of firefighters' turnout coats', M.Sc. Thesis, University of Kentucky, Lexington, KY, 124 p.

Weber M G (1997), 'Special issue: Project Aquaris, Stress, strain, and productivity in wildland firefighters', *International Journal of Wildland Fire*, 7 (2).

White M and Hodous T K (1988), 'Physiological responses to the wearing of firefighter's turn out gear with neoprene and Gore-tex barrier liners', *American Industrial Hygiene Association Journal*, 49 (Oct), 523–530.

Zimmerli T, (1998), Schutz und Komfort von Feuerwehrbekleidung', *Textilveredlung*, **33** (3/4), 51–56.

23

Protection against knives and other weapons

P F E N N E , Metropolitan Police, UK

23.1 Introduction

This chapter introduces the subject of protection against stabbing weapons and in particular the ways textiles have contributed to protection schemes in the past, how they are used now, their problems and possible developments in the future. Currently, stab resistant armour is required by a number of agencies. One such group is Police Officers and the threat to them is briefly assessed. Much of the development of modern stab body armour solutions, testing and standards is linked to the requirements of Police Officers.

Human history has been characterised by conflict. Amongst the earliest recorded history of particular relationships is the murder of Abel by Cain. It is likely that Cain used a hand-wielded weapon and inflicted a stab wound for we read 'Abel's blood ran into the ground.'[1] Since that time, weapons have developed to the complexity of the nuclear arsenal, military aircraft, battlefield rocketry and tanks. Personal weapons have developed into complex firearms and grenades. However stabbing weapons have been in use down the centuries of human history.

The development of weapons and the development of protection schemes have proceeded in parallel. The first weapons were either types of club to inflict blunt impact trauma injuries or stabbing implements to impale internal organs. However, the problem of design of personal protection schemes has remained difficult since that time. The human body is not designed to protect itself against physical conflict with others and as a consequence, many vulnerable areas remain. The human body also requires a great deal of flexibility of movement and so the task of designing armour that will 'drape' over a body has been at the forefront of body armour designs. Being hot-blooded we also need the ability to maintain body temperature and this means that a body armour design must not cause the core body temperature to rise significantly. The management of heat dissipation has been a requirement of body armour design over the years.

23.2   History

Early stabbing weapons were made of wood, stone and then metal. The historical record of armour developments extends back to Mesopotamia in approximately 2500 BC. 'The Standard of Ur', an exhibit in the British Museum, shows troops wearing studded cloaks and copper or bronze helmets. They also appear to have armoured skirts, which may be composed of metallic or leather lamellae. These were similar to armour used in Assyria during the reign of Tiglath Pileser III during 745–727 BC. Lamellar armour was made of a large number of individual armour plates. In early designs the plates were joined by cords which, when tightened, closed the gaps between the plates and formed a series of overlaps. Later, the plates were riveted to a leather backing garment. This design of a flexible backing and small plates continued into the Middle ages.

The first time that 'textiles' were used was by the Assyrians who quilted multiple layers of linen into an armour. This is interesting because it appears that the designers realised that the armour was providing some problem of 'wearability' and so a compromise between protection and comfort was made. It is likely that armours made of quilted linen were used for protection against archers and the lammellar armour retained for the greater protection against spears.

The next advance came with the development of chain-mail. It is thought that the first mail was probably produced by the Celts during the 4th century BC. Mail was produced with plain butted joins or with riveted ends. The material used was wrought iron wire. In the 2nd century BC the Roman legions started to invade Gaul. The Romans found that the Gauls were wearing chain-mail armour, so the Romans developed this design and deployed the Lorica Hamata armour. The interesting feature of this armour was that half of the links were solid rings, punched from metal sheets. This technique continued in some later European examples. Another example of chain-mail with punched links came from Persia and India and is called 'Theta' or 'Bar link'. These punched links had a bar across their centre, which makes them resemble the Greek letter 'Theta'. Later on in Europe chain-mail armour with different size links in one product was developed. Thick heavy links over the vital organs in the chest, lighter, thinner links for the arms and areas that needed less protection.[2]

By this time the three main elements of protection materials were being used, textiles (woven materials and including leather), plates (generally metallic) and chain-mail (Fig. 23.1). Designs incorporating these have been developed to the present. In the 1990s the threat from knife stabbing increased in the law enforcement community. As police forces responded to this requirement the body armour industry developed a number of novel solutions to the requirement for stab protection.

23.1 Archers at Agincourt with textile armour.

23.3   Police requirements

In 1994, the requirement for stab protection for routine patrolling officers in the Metropolitan Police Service (MPS) was established and approval for a project team to realise this requirement was set up. Several off-the-shelf examples were trialled and all were found to have significant problems of wearability and were therefore not suitable for issue. The balance of requirements of appropriate protection and wearability was re-examined and a new specification compiled in an attempt to stimulate innovation into the stab-resistant body armour industry. The specification was designed to define a balance between protection level and wearability that would meet the protection requirements and be operationally acceptable. Many novel solutions were offered for evaluation. The evaluation involved assessing protection performance and also wearability in an extensive wearer trial. The armour first issued to routine patrolling officers in 1995 was called Metvest 1 and was made up of a multi-layer aramid textile pack and small plates (four layers of thin stainless steel). The currently issued design, Metvest 2 uses a multi-layer aramid textile pack and stainless steel chain-mail.

In recent years, attacks have been directed at an increasing variety of individuals. No longer are the law enforcement agencies the only targets. Stab-resistant body armour is issued to a number of official agencies. These include the Prison Service, the Home Office Immigration and Nationality Directorate, HM Customs and Excise, the Department for Work and Pensions, the Department of Constitutional Affairs, various trading standards offices, Parks Security, British Transport Police, the London Ambulance Service and various security companies.

Crime that has a direct connection with physical impact to a police officer ranges from public order disturbances, firearms-related robbery, gang culture and domestic violence to terrorist attacks. The types of attack range from impacts from thrown missiles like bricks and improvised spears to fire threats from Molotov cocktail petrol bombs in public-order disturbances. In other circumstances attacks range from handguns, shotguns and rifles to knives, including combat knives to the kitchen knife and terrorist weapons, including improvised explosive devices.

Various protection schemes are required. Public-order protective clothing includes a helmet/visor, fire protection overall, blunt trauma protection and ballistic protection. Firearms officers are issued with ballistic body armour and ballistic helmets. Various shields are deployed to provide specific protection according to the threat. Explosive ordnance disposal (EOD) officers require protection from the blast of improvised explosive devices that they are called out to make 'safe'. The large proportion of routine patrolling officers are vulnerable to the threat from concealable handguns and knives.

To appreciate the threat to police officers relating to knife attack, an analysis of relevant statistics is undertaken. Violent knife crime is reported and categorised under violence against the person, sexual offences, robbery and burglary. The total of violent knife crimes across London is approximately 13,000 per annum or about one every 45 minutes.[3]

Our attention will be focused on knife stabbing attack. In the first instance it is necessary to understand the nature of the problem. The MPS undertakes a regular knife analysis to review the distribution of the types of knife amongst those that are associated with crime. At regular intervals, all such knives and other weapons are collected so as to be securely disposed of. These include those used in attacks, seized during raids, confiscated and handed in or collected during an amnesty. Several thousand knives will be examined at such an analysis. This analysis enables us to understand if particular trends are developing.

Knives fall into a number of general categories. The categories used for a knife analysis are as follows (Fig. 23.2):

- Domestic/kitchen. Your own kitchen will contain a variety of knives. The blade is made from a good quality stainless steel and, when new, is generally very sharp. Most kitchen knives are not re-sharpened properly during their

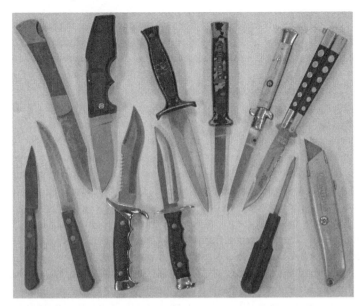

23.2 Knife types: left to right, domestic, lock knives, sheath knives, combat blades and miscellaneous.

life and so end up relatively blunt. However, as a stabbing weapon kitchen knives are very effective.

- Lock. Folding lock-knives differ from penknives in that the blade locks in the extended position to prevent it closing during use.
- Sheath. Outdoor-activity knives with good hilts and strong blades. However, the blade is rarely sharp.
- Combat. Combat knives are produced in a variety of styles from flick-knives, switch blades, butterfly knives and daggers.
- Miscellaneous. Utility knives generally having disposable blades, various sharp tools, penknives and ceremonial blades of all sorts.

Several knife analysis assessments have been undertaken over the years and the results indicate a similar characteristic. Typical distribution percentages are listed in Table 23.1.[4] From these figures it can be seen that domestic blades found in the kitchen make up the greatest proportion of blades, lock knives and sheath knives together make a significant proportion and combat blades all together are maybe less than would be anticipated.

23.4   Knife performance

Knives are generally manufactured to meet requirements for domestic or industrial use or as weapons. Each design consists of a blade and a handle. The design of both the handle and the blade are important to the overall performance

Table 23.1 Percentage distribution of knife types

Type of knife	Percentage
Domestic/kitchen	40
Lock knives	15
Sheath knives	10
Combat (Switch, flick, butterfly and dagger)	12
Miscellaneous (Craft, sharp tools, penknives and ceremonial)	23
Total	100

of the implement. The important features of a blade are the material it is made of, its thickness and the angles that make up the point and sharp edge. A very acute angle of point makes the knife potentially most penetrative. However, a very acute angle point is weaker and more vulnerable to damage whilst the cut is being made. A typical point angle may be about 10° for a surgical scalpel to 60° for a working lock-knife.

Although the variety of knife styles is very wide, they all have blades made of hardened steel to suit their design. Blade metallurgy and heat treatment is a complex technology. In short, the requirement for a blade is sufficient hardness to maintain a sharp edge during use, but not too high as to make it vulnerable to fracture because of being too brittle. Also, a blade that is too hard will be more difficult to grind and whet into an effective point and edge. The steel used for most blades is hot rolled stainless steel. Alloying metals may be added, such as chromium with various combinations of molybdenum, vanadium, manganese or nickel. The process for manufacturing a blade is firstly a forging operation to produce a rough shape. After this is a hardening operation involving quenching from a very high temperature of about 2,000 °C to produce a very hard surface. Then comes a tempering operation to relieve the hardness to a level that will give sufficient point and edge strength but enable the material to be prepared to its final shape/size by the grinding process. The grinding process is an essential part of the overall effectiveness of the knife. A knife must not only be sharp but also strong. Blades come in different thicknesses. For example the craft knife disposable blade is about 0.5 mm thick whereas a survival knife could be 4 mm thick. To achieve the suitable edge angle for the particular usage it may be appropriate a use a plain flat grind or a hollow grind.

The effectiveness of the blade in terms of cutting is that it is both a wedge and a lever. As a simple wedge the sharpness of the edge affects the ability to make the incision and the angle between the blade edges governs the degree to which the material being cut will be forced apart. These two effects generally work together. The physical characteristics of the material being cut will influence how these effects work together. Thus the fibres of a textile, a metal or a brittle material like glass will cut in a different way. However, the initial incision is purely a function of sharpness. Sharpness can be defined as 'the attribute, which allows the instrument to perform the cutting operation with the minimum effort'. The fundamental mechanism whereby a knife cuts is compressive fracture caused by high pressure from the very small area of the edge. Most knives rely on a point and an edge to be able to cut. Some only rely on one, for example, the razor blade requires an edge only, whereas an ice pick only has a point. To achieve maximum sharpness of a blade the edge angle must be low and the tip radius small. Sharpness can be assessed in two ways, firstly a method to physically measure the edge angles, thickness and tip radius or secondly to cut something in a controlled and repeatable manner.

A sharpness-testing machine has been developed by the Cutlery and Allied Trades Research Agency (CATRA) located in Sheffield. This machine works by measurement of angles, tip radii and blade thickness. The blade is clamped vertically and a horizontally mounted laser beam is directed onto the edge. Reflections from the tip and edges appear on a circular graduated scale. The thickness is measured with a specially adapted micrometer. An alternative method to evaluate edge sharpness is the cutting test machine. Again CATRA have developed such a machine. The blade is mounted in the machine, a load is applied and a test sample of silica paper is cut. The sharpness is measured as the depth of cut. Repeated strokes create wear on the edge and give a sharpness/life characteristic. This is suitable for blade types from Stanley knives, kitchen/ butchers knives to large industrial blades. A more delicate instrument for very sharp/fine edges cuts into silicon rubber. Sharpness is measured as the cutting force required, the lower the force the sharper the blade. This method is suitable for all small delicate blades from eye surgery scalpels, razor blades to Stanley knives.

The finished blade is to be fitted into a handle and the design of a knife handle is essential to the function of the knife. The energy required for the cutting or stabbing action must come through the handle and the handle design is a key factor in the effectiveness of the knife. The design of the handle will range from the beautifully carved and decorated handle of some ornamental knives to simple plastic handles for many kitchen knives. However, many combat knives have carefully designed and comfortable hand-sized handles with a large hilt to enable a high stabbing force to be applied. In order to transfer the maximum amount of energy from the hand into the blade, a good coupling between the hand and handle is essential. In fact, in many knife handle designs the coupling is not very good. Take a kitchen knife for example, the typical materials to be cut are meat and vegetables. These materials yield easily to being cut and the load is generally almost perpendicular to the axis of the handle. However, as a stabbing weapon and having a high load applied along the handle axis, the attacker's hand is likely to slide on to the blade if significant resistance is met. So, in knife handle designs where it is important for the axial force to be transferred with the minimum of loss, various designs of hilt are incorporated. A knife with an ergonomic handle and a significant hilt therefore makes an effective stabbing weapon.

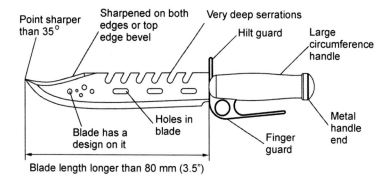

23.3 CATRA combat knife definition.

As a result of the proliferation of potential stabbing weapons in circulation there have been calls for greater control of availability of what is termed the 'combat knife'. One agency involved in this is CATRA. The CATRA definition of a combat knife (Fig. 23.3) is based on a knife possessing more than eight of the following attributes:[5]

- length to weight ratio greater than 8 gm/cm
- blade not made from stainless steel
- length to width ratio is greater than 12:1
- blade length greater than 8 cm (3½″)
- blade thickness is in excess of 3 mm
- tip point sharper than 35°
- sharpened on both edges or has top edge bevel
- holes in the blade
- engraved, etched or marked pattern or design on the blade other than the manufacturer's mark
- back edge serrations
- hilt guard
- finger guard
- large circumference handle
- metal or plastic pommel (handle end)
- supplied with a durable carrying sheath
- multi-functional facilities
- marketed as a weapon.

23.5   Fundamental principles of knife impact

From the knife itself, we turn now to what happens when a knife impacts a material. There are generally two phases. Firstly, the point contacts the material and three actions start. The 'target' material begins to move away under the force of the knife, the target material begins to fail (to open up) and the knifepoint begins to fail also (to blunt). The second phase begins when the knife has pierced through the target material. After this stage the knife opens up the hole already made and penetrates; this is called the 'run through'. It is useful to divide up these actions in this way when designing stab-proof body armour. The aim of a good design is to absorb all of the stab energy before run-through occurs. Each of the actions described dissipates energy and an armour panel design using various materials can be optimised to maximise this energy dissipation.

To begin to design a body armour stab-resistant concept it is useful to grasp the nature and magnitude of the problem to be solved. We are familiar with the concept of 'pressure' or force over a particular area. The technology related to ballistic protection is well established and we will use it to compare and contrast with the requirements for stab attack. Handgun bullets travel at, typically 300–400 m/s and are generally blunt or round-nose shape. Take a 9 mm calibre bullet fired from a self-loading pistol weighing 8 g, with a strike velocity of 360 m/s. The energy of such a bullet is

$$\frac{mv^2}{2} = \frac{0.008 \times 360^2}{2} = 518 \text{ Joules}$$

If we assume that the effective area of the bullet impact is the cross-sectional area of the bullet, i.e., $\pi R^2 = \pi 4.5^2 = 64 \text{ mm}^2$, then the energy/mm^2 = 518/64 = 8 Joules/mm^2. Although the energy of a stab attack may not be excessive as a blow to the body it is the fact that it is applied over such a small area of a knife point that it becomes such a problem to solve. Take a typical stab of impact velocity 5 m/s and energy of 25 Joules. The tip of a sharp blade may have an area of about ¼ mm^2. This represents an energy density of 100 Joules/mm^2, which is higher than most common bullet energy density calculations.

Flexible ballistic-resistant panels are made up of a series of layers of a woven fabric of a very high-strength fibre. Typical fibres are aramids and ultra-high molecular weight polyethylene. The name aramid refers to fibres of aromatic polyamides, which have similar molecular groups as nylon, but with a benzene ring. Aramid fibres are used in the majority of ballistic body armours produced at present. High modulus polyethylene fibres were developed by a gel spinning process in 1979 and provide an alternative. Chapters 3 and 19 give more details on the textile materials used. The process by which the woven fibre pack works is that the bullet deforms the pack and stretches the individual fibres, which absorb energy in tension. A 9 mm pistol bullet will require about 20–25 layers of woven aramid to absorb the energy without penetration. However, this type of aramid pack will not necessarily provide protection against a stab with a sharp knife. Most of the recent developments in stab armour have consisted of combinations of metals as plates or chain-mail in a flexible carrier with a resilient backing.

However, fibre and fabric suppliers have been developing woven materials in the hope of achieving sufficient stab resistance without the use of metals. One such solution is the Dupont product Kevlar®Correctional TM. A yarn of 220 dtex is woven with a sett of 28×28 cm to produce a very fine tight weave of 120 g/m². This is very resistant to perforation by sharp rounded spikes.[6]

We will look at the fundamental principles of a sharp point impacting a number of typical target types, namely a thin metal plate, a metal chain-mail and a woven fabric. The mechanics of perforation of sheet metal by a knife have been investigated by Ian Horsfall in his PhD thesis 'Stab Protection Body Armour',[7] and Blyth and Atkins in their paper 'Stabbing of metal sheets by a triangular knife. An archaeological investigation,'[8] both of which adapt an analysis by Wierzbicki and co-workers.[9] Blyth and Atkins applied the analysis to discuss the effectiveness of ancient Greek armour. The energy required to perforate sheet metal involves contributions from fracture, sheet curling and friction. In the cases studied, the make-up was roughly in the following proportions: fracture 36%; curling 24%; and friction 40%. They performed perforation experiments on many sheet metals, determining the forces, energy and deformation patterns. They demonstrated that the material properties which control resistance to perforation are not only hardness (yield strength) but also ductile fracture toughness. This latter property is rarely quoted for armour but has to be known for a full analysis of any problem involving fracture of material. Since the metallurgy of ancient Greek bronzes is not too dissimilar from modern bronzes, they were able to take independent measurements of the hardness and toughness of bronzes and employ those properties predictively in their analysis. The same is not true of other ancient metals and although microstructure and micro-hardness may be determined from very small samples of ancient artefacts, the determination of fracture toughness from small samples remains a challenge.

Chain-mail is defeated when one or more of the interwoven rings are fractured by the point of a knife passing through. The forces and energy required for this process are currently being investigated by Atkins.[10] The amount of deformation of a chain-mail ring before possible fracture depends upon its dimensions and the mechanical properties of the metal from which it is made (yield strength, ductility, toughness). During elongation of a ring into an oval shape, the sharp edge of a knife blade will additionally cut into the ring and reduce the load-bearing area at that point. While fracture of a ring may occur at this indented region, the actual mechanism by which a ring breaks, and where it fractures, depends on circumstances, such as whether the ring is made by riveting or welding, or is stamped out of the solid. The problem is not yet fully understood, particularly the effect of ring dimensions and the mechanical properties of the ring metal. A full analysis of the problem should lead to optimisation of chain-mail design.

In the case of woven fabrics, a knife may penetrate a small distance merely by pushing fibres aside, but deep penetration and perforation of layers of fabric can only occur by cutting through fibres, particularly in the case of tightly-woven fabrics. The forces and energy to cut through individual fibres of all sorts of materials are of interest to biomimeticists. Lucas and co-workers have used instrumented scissors to cut through materials.[11] It is possible to determine the fracture toughness of the material (the work/area required to separate the material) from the measured forces. The toughness value can then be employed predictively to estimate the forces required to cut fibres by other forms of blade. The analysis accounts for friction, and assumes that the cut parts are 'floppy' so do not store elastically, nor dissipate plastically, any strain energy. That would be a reasonable assumption for cutting through woven fabrics, but is not correct for cutting into ductile sheet metal or into chain-mail where the material is permanently deformed before, and possibly during, the initiation and propagation of the crack leading to failure. The forces and energy for the point of a knife to perforate woven materials can be estimated once the cross-sectional fracture toughness of the aramid filaments (for example) have been determined by means of the scissors, or other appropriate, test.

When, as with woven fabrics, simple cutting is the principal mode of deformation during perforation, the sharpness of the cutting blade becomes important. Sharpness is an ill-understood concept; it is sometimes described in terms of blade flank angle and tip radius of the cutting edge, or sometimes in terms of the cutting force required to slice a standard sheet of material (often paper). However, 'sharpness' really should be considered in conjunction with the thickness of the material to be cut. That is, when the tip radius is much larger than the material thickness, it is easier for the material to deform in bending under the edge of the blade than for an indentation and cut to be achieved. Eventual breakage of a fibre, for example, in these conditions will be by tensile fracture as the filament is stretched around the blade rather than by cutting through the fibre. On the other hand, when the tip radius of the blade is much smaller than the thickness of the material, indentation occurs followed by progressive cutting beneath the blade. It is found that the cracks, which give eventual fracture, propagate at their own 'natural sharpness' characteristic of the material (this property is the 'critical crack opening displacement' (COD) of fracture mechanics, which is an alternative measure of the fracture toughness of the material). The more ductile the material, the greater the COD. There is probably no benefit to be gained by having the blade sharpness (in terms of the tip radius of the cutting edge) smaller than the material COD, since once the crack begins to propagate, it will do so at its COD. 'Sharpness' is relatively less important the more ductile the target.

Whatever the form of the protection, the angle of incidence of the attacking knife will be important; the surface of the armour may determine whether the tip of the knife skids or is otherwise deflected. Trials using an instrumented knife, show that the dynamics of the thrust by the combined arm and weapon is important in determining the effectiveness of the protection against attack.

Furthermore, there is the possibility that on contact with the protection, the tip of the blade may 'get caught' and the blade then bend sideways (a complicated problem linked to the compliances of the attacker's hand at the knife handle, the wrist, elbow and shoulder). It should be remarked, perhaps, that conventional proof testing arrangements for candidate body armour constrain blade motion, after first contact with the body protection, to the original direction of impact and do not permit the possibility of blade bending/tumbling that sometimes occurs in practice.[12]

23.6   Knife protection design principles

The main design characteristics for a stab resistant armour panel are summarised below. Most, if not all users requiring stab protection also require ballistic protection and so almost all stab resistant products also include ballistic protection. This is useful because the metallic stab resistant solutions will perform better with a resilient backing to spread the energy absorbed over a large area and the ballistic element of the armour panel can be utilised for this purpose. To maximise the energy absorbed in deflecting the panel, a flexible, springy material may be placed nearest to the body. Typical materials may include high-density plastic foams and various forms of rubber. The deflection or compression of this material absorbs some energy. This effect is limited because it results in an increase in bulk (thickness) and weight and this affects wearability adversely.

To minimise the target material beginning to fail by opening up we will look first at the plate or chain-mail solution and then at treated aramids. The mechanical properties of the material chosen for the plates or chain-mail should be strength, hardness and toughness. Strength is the ability to resist a tensile, compressive or shear force. Hardness is the resistance to abrasion, deformation or indentation. Toughness is the ability to resist cracking, often used in the context of shock loading. Stainless steel and titanium are key candidates for this application. One of the solutions from the fibre and fabric suppliers is to apply a resin to the woven fabric. When a sharp point contacts a plain aramid multi-layer pack the weave tends to open up and if the implement has a sharp edge, some of the fibres will begin to cut. If the fabric is resinated, the individual fibres cannot separate so easily.

To dissipate energy by causing the knife tip itself to fail, either by chipping or blunting, the fundamental requirement for an armour material is hardness. The design aim is to have an outer surface that is hard, preferably harder than the knife material. An ideal material is a ceramic of typically aluminium oxide or boron carbide. One solution using fabrics is the Teijin Twaron patent involving coating the surface of the aramid fabric with a novel coating design. Their product is called Twaron SRM. The product consists of a fabric made from Twaron aramid yarn and to this is bonded fine silicon carbide particles. The material works by dulling the point of the blade and rendering it far less penetrative.[13]

Having suffered the first phase, that of perforation, we need to limit the second phase of run through. The interaction of the knife with the armour panel has now changed. In the previous phase the knife-point and adjacent part of the edge(s) are most significant. Now, the full edge length and overall profile of the blade are the main controlling factors. The more parallel the overall shape of the blade is, the easier the blade will run through. This is because the resistance of a knife to penetrating body tissue is far less than the resistance of a good design of armour panel. The best material to limit run through is a material that applies a very high friction load as it fails. A typical material is a polymer called polycarbonate. This plastic material requires a high energy to pierce and a high energy to deform. This means that, even if the knife penetrates, the material resists any further opening of the whole by locking onto the blade. In these circumstances, even if a blade pierces the material it can be almost impossible to withdraw the knife. This is quite different to metals, which generally do not offer such high resistance to being 'opened up' after being pierced. Similarly with fabrics, even with those treated, the fabric will offer very little resistance to a knife blade following through once the perforation has been made. The fundamental design parameters relevant to body armour are as follows:

1. protection level required and acceptance test methodology
2. coverage
3. wearability.

23.7   Protection levels

We have established that the range of types of knife is very broad and it is virtually impossible to predict what type of knife we may encounter in a particular situation. In addition the technique of stabbing is variable and each method affects how the knife interacts with the target. The main variables in respect of method are, overhand, underhand, thrust and slash.

Overarm stabs tend to have the highest energy and generally contact the victim in a downwards direction in the upper chest, back or shoulder region. Underarm stabs impact at a slight rising angle to the horizontal in the lower chest or back. If the knife handle can be grasped strongly against the palm of the hand a thrust stab can be used. The knife is in line with the arm and a very high force can be applied. Knives with a short blade and large handle will tend to be used in an attack with a slashing motion. Slash attacks produce wounds where the length is greater than the depth. If the wound involves major blood vessels, it can be life threatening but, in general, they are not as serious as stab wounds. Stab attacks produce wounds where the depth of injury is greater than the length. They penetrate more deeply and tend to come into contact with vital organs in the chest and abdomen. If the blade is rocked in the wound, because the assailant moves the knife around or the victim moves in relation to the knife far more internal injury is likely. The depth of a stab wound is often longer than the length of the knife because of the compressibility of skin and underlying structures, particularly in the abdominal cavity.

We come now to consider what level of resistance to knife stab is required in a successful body armour design. The stab action consists of an arm swing with the knife, the contact with the victim and a 'follow through'. Several studies have been undertaken to measure the physical characteristics of force and velocity during the stab motion.

The MPS undertook a programme of research with the Defence Evaluation Research Agency (DERA) Farnborough, but now called Qinetiq, Materials Division. The first phase was to measure physical parameters of stab events under conditions which were reproducible, and which simulated the stab environment. The test arrangement comprised three main components; an instrumented knife, a fully jointed manikin and support system design, which allowed movement. A 50 percentile size male Ogle Opat fully jointed manikin was used featuring a simulated rib cage using metal strips. The thorax stiffness was measured on the prostrate manikin in a screw-driven tensile/compression test machine under compression. Comparison of the stiffness with published data from *in-vitro* cadavers indicated that the rib cage stiffness was rather high. In order to reduce the stiffness, 'ribs' were removed to give typical force/deflection characteristics. Suitably positioned rubber blocks for the lungs and abdominal organs simulated the weight distribution of internal organs in the abdomen. The thorax was covered with a soft foam cover to simulate a layer of skin and fat. Commercial stab-resistant body armour was used to protect the manikin and to prevent penetration of the knife blade. A steel sub-frame supported the manikin across the shoulderblade region. This sub-frame was attached to the base of a wooden support frame using a flexible rubber mount, which simulated the floor loading typical of an upright subject. To maintain the erect posture, the manikin was supported by elastic bungee cords attached near each shoulder to the steel support frame. The bungee cords gave a degree of recoil after the stab event.

A double-edged symmetric 'commando dagger' blade 175 mm in length was selected. The shank was threaded to enable accurate positioning and locking into a tubular handle which facilitated fitting and changing of blades. The blade shank was locked into position such that it butted against a piezoelectric load cell. The blade edges and point were blunted to minimise danger to the subjects. A tri-axial accelerometer with maximum acceleration limits of 1,000 g was attached. A high data capture rate was achieved using a four-channel PC-based data acquisition system running at 150 MHz capture rate. The outputs from the three axes of the accelerometer were conditioned using a charge amplifier.

Stab trials were performed with 50 fit volunteers, who were instructed to perform three overarm, three underarm and three thrust action stabs. The position of each stab site was annotated on a diagram of the manikin torso on the log sheet, and recorded using a video camera. Subject data such as gender, height, weight and age, were logged. The data was processed using a commercial software package. The blade velocity was calculated by integrating the acceleration, and displacement calculated by double integration. Integration of the force-time measurements gave impact energy.

The overarm stab velocity showed approximately twice the velocity for the maximum compared to the average velocity for all stab types. The velocity levels were on average between 4–5 m/s, and peaked at 10.6 m/s. The data showed a significant difference between the maximum value of force and the average value. This reinforces the difficulty in setting a 'standard' value to a complex event such as a stab. The maximum overarm peak force, for example, was twice as high as the average, and three times as high for the thrust (or lunge) and underarm types. The type of stab action is seen to influence the energy of the stab. The energy to peak force showed the widest range for the overarm action. The distribution of the stab energies showed a nearly normal distribution for the overarm stab, compared to a skew distribution for underarm and thrust. The highest average stab energy was recorded for the overarm stab action, as might be expected. This value of 14.3 J was approximately 20% higher than for thrust and underarm stabs, which both showed an average of 11.4 J. Consistent with this trend was the observation of the highest stab energy being recorded for the overarm stab (i.e. 42.4 J), although the highest thrust stab showed 41.9 J. The highest underarm stab showed an energy of 31.9 J, and this was consistent with the observation that overarm and thrust actions were more easy for an untrained assailant to perform.

The overall stabbing potential is controlled by the wide range of variables summarised below:

- Knife
    - blade geometry
    - tip sharpness
    - steel hardness
    - edge sharpness
    - knife grip/shank.
- Assailant
    - gender
    - fitness
    - body mass and physical size
    - mental state
    - training/experience.
- Stab event
    - over/under/thrust/slash
    - arm and blade velocity

- impact force
- impact energy
- blade impact angle
- post-impact blade twist/turn
- single or multiple stabs.
- Victim
  - body mass
  - reaction
  - site of stab.

The force-time plots derived show the complexity of the stab event with time. The plots demonstrated a double peak in force; a higher frequency-higher force initial peak followed by a lower frequency-lower force peak. It is a point of interest to consider what effect these two peaks might have on the penetration of a knife blade through body armour. The stab event interacting with the body armour is controlled by two parameters, damage initiation and damage propagation.

It is reasonable to assume on the basis of fracture mechanics considerations that the threshold for damage initiation is higher than that for damage propagation. Considering the observed characteristic double peak, the initial higher force peak would be likely to cause damage initiation, and the secondary lower force but more sustained input energy peak could facilitate damage propagation. If, as is likely, the critical parameter for damage initiation is the initial maximum force attained, the resistance of the body armour to this parameter will determine whether or not damage is initiated. If penetration is resisted during this initial transient, then it is likely that the sustained higher energy event will have less effect, in which case the body armour will show a higher resistance to attack.[14]

Other studies have involved forensic examination of body armour actually involved in an attack to predict the approximate level of effort applied during the attack to produce the particular damage pattern sustained. The conclusions of all of this work are that the maximum stab ability of a range of individuals of different gender, size, fitness and strength is extremely variable; the stabbing ability of certain individuals as a result of their own high levels of strength, skill, practice and determination is very high. To put some typical figures to this, stab impact velocity as measured in the laboratory environment with an instrumented knife ranges from of 5–10 m/s and the maximum energy applied by any particular individual may range from 10 J to over 100 J.

23.8   Test methodology

The design of test methodologies including test rigs has developed amongst test agencies since the early 1990s. The Scientific Service of the Zürich Police

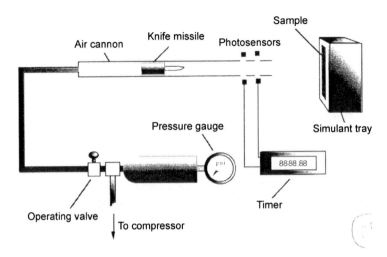

23.4 PSDB Air cannon test rig.

developed a test rig using a circular section sabot that mounted a test blade. This was dropped down a square-section steel tube inclined at a very small angle to the vertical (1°) to prevent axial oscillations onto a sample laid horizontally and backed by a box of modelling clay. The knife used was a stiletto blade from a German manufacturer. The pass/fail criterion was a maximum penetration of 20 mm at 35 J. The use of a square-section guide tube and circular section sabot at 1° to vertical was analysed by Kneubuehl of the Science and Technology Branch of the Swiss MoD and he concluded that this design negated the need for velocity measurement of the sabot as the velocity could be established accurately by calculation.[15] In 1993 the Police Scientific Development Branch (PSDB) developed a methodology using an air cannon (Fig. 23.4). The test knife was mounted in a plastic sabot that was fired horizontally onto an armour sample mounted vertically on a modelling clay backing.[16]

The pass/fail criterion was set at 42 J. This was derived thus. A number of hand stabs using specific knives impacting into a polymeric foam were made and the maximum depth of penetration and velocity measured. The mass of the sabot was then adjusted and fired at a similar sample of polymeric foam at the measured velocity until the same depth of penetration was obtained. This occurred with a mass of 400 g and velocity of just over 14 m/s giving an overall energy of 42 J.[17] Two tests from each of two different blade sizes were required. The blades used for the test were standard production sheath knife blades. In 1995, the MPS had not been able to find a suitable body armour that met this level of protection and that could be worn covertly. The Metvest project team was set up and their remit included investigating the threat more fully and producing an appropriate armour solution. To evaluate the penetration characteristics of various knife blades a swinging arm rig was developed (Fig. 23.5).

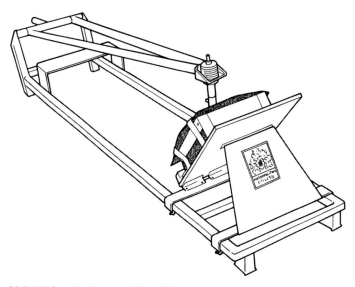

23.5 MPS swinging arm stab test rig.

Different knives could be mounted easily in the swinging head and various masses could be added to control the impact energy. An engineered blade designed to exceed the cutting performance of all other blades was developed. It consisted of finely ground edges and a very strong point with triangular section made of very hard steel (62 Rockwell C). Several body armours available at the time had been tested with real knives in a laboratory test with overarm stabs onto armours supported on modelling clay. A test was undertaken to compare the perforation characteristics of real blades with the MPS triangular blade. The triangular section knife was more penetrative than the real blades over a range of energy levels.

Stab energy tests were undertaken to ascertain the amount of energy that it is possible to inject into a blade. Several fit young males used the triangular section test blade in a custom-made high coupling handle (i.e. good grip, knurled handle, big hilt) to allow maximum energy injection into the blade, and their energies recorded. The maximum energy that was seen to be imparted was 53 J and the average was approximately 32 J. The same subjects then repeated the tests with various real knives. The amount of energy that could be imparted to the real knives was found to be, on average, approximately 50% of that attainable with the MPS test blade and handle. The maximum amount of energy that could be transferred into the real blades, on average, was more in the region of 25 J. This was due in part to the extra mass of the MPS handle and also to the reduced hand to handle coupling with the real knives. Having identified a realistic energy level (25 J), further tests were conducted to assess what penetration was allowable with the triangular section blade when impacted at

25 J on a sample, to ensure no perforation at a similar energy level from realistic knives.

A particular armour that had a protection level that allowed a perforation of 20 mm with the triangular section blade was tested with real blades and no perforation was demonstrated. This level was adopted as being suitable. Subsequent to this the programme of work was begun with DERA Materials Division with a view to investigate novel materials to improve stab-resistant solutions. The first part of this was to investigate the physical characteristics of a stab motion (as described in the previous section). At the same time the Royal Military College of Science (RMCS), Shrivenham[18] was conducting similar research. Their tests were demonstrating generally a higher level of stab energy but the force profile was confirmed between the two series of tests. Meanwhile a committee of representatives across Europe met during 1994–2000 to develop a CEN Body Armour Standard. The stab protection part of this standard included a test methodology based on a drop tower rig. The significant novelty was the test blade that was a specially designed engineering blade. The intention was to reduce the cost and improve the consistency of performance over 'real' blades.

The tip was formed into a small chisel shape to provide greater strength at the point of impact. A series of three protection levels were defined; see Fig. 23.6 for illustrations of test blades. In 2003, PSDB published a new Standard replacing the original one.[19] This standard used a drop tube method with a plastic sabot unit. However, as a result of the instrumented knife work at RMCS, a new design of sabot was introduced. The instrumented knife work had shown a double-peak of applied force in some stab motions. The design of the sabot incorporated a double mass with a spring/damper connection in an attempt to replicate the characteristic double-peak force to the test sample. The spring/damper component comprised a closed cell plastic foam cylinder between the two masses. The knife is mounted in a central mass that slides within the outer sabot cylinder. The foam cylinder is placed between the two. The test knife is an engineered blade designed to replicate one of the blades used in the original PSDB Standard. A sharpened circular section spike impactor is used to test for spike protection. The backing material is a multi-layer pack consisting of neoprene foam, a polymeric closed cell foam and natural rubber. The pass/fail criterion is based on a maximum allowed perforation depth at two specific energy levels. There are three levels of knife resistance, KR 1, 2 and 3, and three additional levels of spike resistance SP1, 2 and 3 (see Table 23.2).

To be certified, an armour under test must meet one of the levels of stab resistance. If a user requires a spike resistance test it may be added as an option. Many body armours have been in use in the police forces of the UK for several years. Most are tested to either KR1 or 2. These armours have saved lives. Considering now the design parameters of a simple design of test rig consisting of a test blade supported in a solid mass dropped vertically onto a horizontally laid test sample mounted on a backing material. In order to identify the physical

23.6 Test blades: top to bottom right, German Standard blade, PSDB P5, PSDB P1 original, MPS Triangular section test blade, CEN Standard engineered blade and PSDB P1 current test blade.

parameters of the test rig we will examine the key factors determining the outcome of a stab action.

23.8.1   The effectiveness of the knife

The geometry, grinding and material from which a knife is made determine its effectiveness. Most 'real' blades are not pristine as far as the point and edge condition are concerned and 'real' blades are very variable in type, shape and general condition. These variations are so wide that it is impossible to produce a representative test blade as far as effectiveness is concerned. The requirement for a test blade should concentrate on consistency of manufacture and therefore consistency of effectiveness. In practice, a simple shape blade of closely

Table 23.2 PSDB test levels

Protection levels	Energy level E1 (Joules)	Maximum penetration at E1 (mm)	Energy level E2 (Joules)	Maximum penetration at E2 (mm)
KR1	24	7	7	7
KR1+SP1	24	KR1=7,SP1=0	KR1=36,SP1=N/A	KR1=20,SP1=N/A
KR2	33	7	7	7
KR2+SP2	33	KR2=7,SP2=0	KR2=50,SP2=N/A	KR2=20,SP2=N/A
KR3	43	7	7	7
KR3+SP3	43	KR3=7,SP3=0	KR3=65,SP3=N/A	KR3=20,SP3=N/A

specified alloy heat-treated steel to close dimensional and grinding surface finish tolerances is required. Such a test blade will tend to give a higher penetration characteristic than real blades.

23.8.2   The knife to hand coupling

Most knife handles are either too small or of the wrong shape to transmit all of the force that could be applied by the hand whereas a test blade rigidly supported in a sliding mass will transmit all of the available force. This in effect will give a higher penetration depth than a real hand stab.

23.8.3   The velocity and force of the stab

The velocity and force will determine the momentum and kinetic energy that will be applied to the target. Analysis of instrumented knife tests has led to a consensus among test agencies to use a test mass of about 2 kg. The combination of mass and height of drop provide the test energy. The 2 kg mass provides a static force to simulate a follow through force. The energy may be varied by using a range of drop heights. In practice, the mass of about 2 kg and a range of heights from 0.5 to 3 m give a useful range of energy to investigate the protection performance of a candidate armour panel.

23.8.4   How the victim (target) moves on impact

Energy will be absorbed by the deflection of the outer armour surface. This is a result of the deflection characteristic of the armour panel itself and that of the underlying material, the body. However, the body viscosity and elasticity are not constant over an individual or between individuals. Again, it is important to specify a backing material with a consistent force against deflection characteristic material to be able to undertake a comparative test. Certain plastic foams and rubbers are suitable for this. The aim of the test is to produce a graph of energy against perforation depth. This will provide the energy at which perforation starts and the characteristic associated with the run through. Each of these is a distinct quality of an armour panel (Table 23.3).

Table 23.3

Level	Maximum allowed perforation of 7 mm for given energy in Joules	Maximum allowed perforation of 20 mm for given energy in Joules
KR1	24	36
KR2	33	50
KR3	43	65

Coverage

The traditional body armour design representing most products in use at present consists of two panels, front and rear. The front panel may in fact be made up of two halves overlapping to form a front-opening join. At the front, the area from the top of the sternum and rising slightly at the shoulders down to the navel and a similar area to the back. The sides often do not quite meet although on some there is a small overlap. This is the result of an attempt to balance protection requirements with wearability. Protection for 'vital' organs of the chest, heart, liver, lungs, stomach, kidneys, spleen (Fig. 23.7) are generally achieved by such a design and practical products can be manufactured successfully. Looking at the stab attack scenario specifically it is useful to review the design parameters in respect to coverage. Firstly, the vulnerability of specific body areas with respect to the likelihood of being hit, secondly the vulnerability in respect of level of injury and thirdly the practicality of manufacturing an armour covering specific body areas.

Several studies have been undertaken to assess the vulnerability of anatomical areas to the probability of a stab strike based on population samples admitted to hospital. The summarised conclusions from these is that the chest and abdomen area suffered the greatest proportion of strikes. The head, including face, and arms accounted for the next two likely strike areas and the legs, neck and shoulder regions the least. In terms of vulnerability according to injury, the most significant threat to survival in the short term is a major neurovascular injury. The heart and major arteries are particularly vulnerable. Severing of a major artery reduces the chance of survival significantly unless immediate treatment is available. Major arteries run relatively near the body surface either side of the neck and down the inner thigh. Cutting damage to other organs, lungs, liver, kidneys, spleen, pancreas, although not in general so life threatening at the scene pose significant medical treatment problems in the hospital.[20] From a design point of view, the chest and abdomen are generally included in the traditional armour. It is not practical to cover the neck and shoulder area due to physiological or operational constraints. As materials improve, extending coverage may become feasible.

Wearability

As well as resisting penetration of blades and with appropriate coverage, a good personal armour will need to be flexible to move with the body, lightweight, of minimum thickness and low thermal insulation so that the wearer does not overheat. Wearability is very dependent on how well the armour fits the individual user. To provide armour for a number of individuals a process of measuring will be required. Manufacturers need to be aware that a large range of sizes will be required. The requirements from the wearer's point of view, or

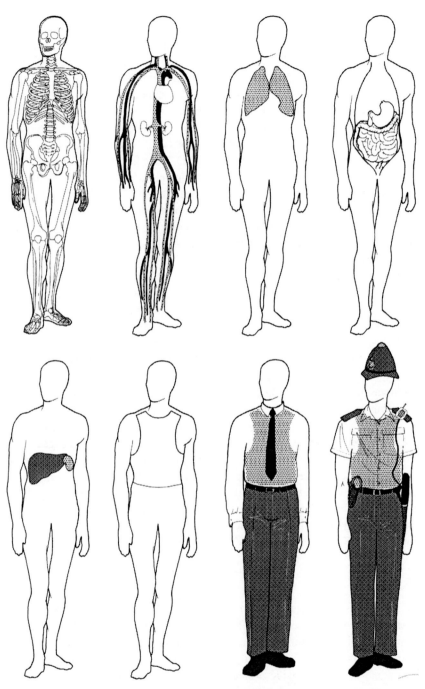

23.7 Vulnerable body areas: skeleton, blood vessels and kidneys, lungs, stomach and intestines, liver and spleen, coverage areas.

'wearability', and the requirements for protection tend to oppose each other, so a very careful balance of the wearer and protection requirements has to be made to produce a good design.

23.9   Stab resistant body armour construction and manufacture

The recent patents relating to stab-resistant solutions illustrate some of the ideas in use in stab-resistant body armour. Some examples are summarised here beginning with Patent no. WO9213250 published in 1992 (Fig. 23.8). This concept involves small rectangular metallic plates with corner holes. Each row of plates comprise alternate overlapping plates fixed at the corners with loose-fitting plastic rivets. Adjacent rows are fixed together with narrow rectangular strips fixed on the same plastic rivets.

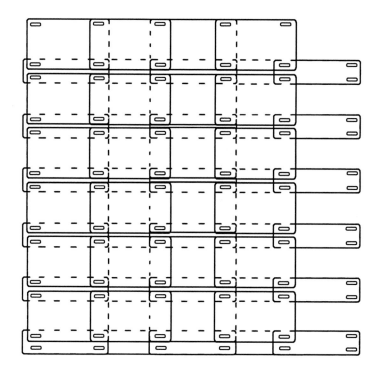

23.8 Drawing of Patent WO9213250.

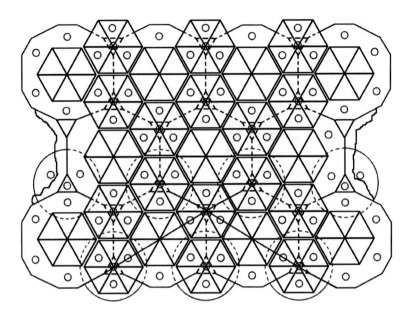

23.9 Drawing of Patent EP0541563.

Patent no. EP0611943 was published in 1994. This solution is made up of several assemblies. Each assembly is an array of thin metal plates enclosed between two layers of fabric and stitched around the periphery of each metal plate. Several of these assemblies were used, each one offset from the previous one so that the weaknesses along each intersection are covered.

Patent no. EP0541563 was published in 1993 (Fig. 23.9). This design consisted of hexagonal plates forged, for example, from aluminium, of which a first kind has a hexagonal central region and a dodecagonal flange which is overlapped by a second form of hexagonal plates which incorporate locating pins, some of which loosely locate in apertures in the flanges, and others carry a disc-shaped retaining member on the obverse side of the flange.

Patent no. EP0640807 was published in 1995 (Fig. 23.10). This invention involves the use of thin flexible metallic strips mounted in a basket-weave configuration.

Patent nos HR20000455, US6586351 and US2003190850 were published in 2001–2003. These patents involved the treating of aramid fibre woven textiles. One side of the aramid material has silicon carbide particles bonded to the fabric. In effect this side resembles emery cloth. The idea is intended to blunt the point of a knife before it reaches the aramid material. This is marketed by Tiejin Twaron as Twaron SRM (Fig. 23.11).

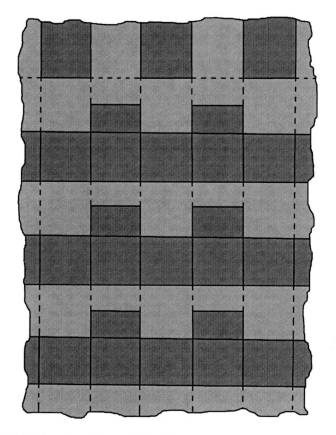

23.10 Drawing of Patent EP0640807.

Patent no. US5515541 was published in 1996 (Fig. 23.12). This patent comprises a flexible woven sheet with formed metal discs riveted by the centre to the flexible sheet. One array of discs is placed on one side and an array is placed on the other side oriented so that it covers the area between the discs on the front side on the rear. The discs had cupped edges facing the threat to prevent a knife tip slipping off.

23.11 Drawing of Twaron SRM.

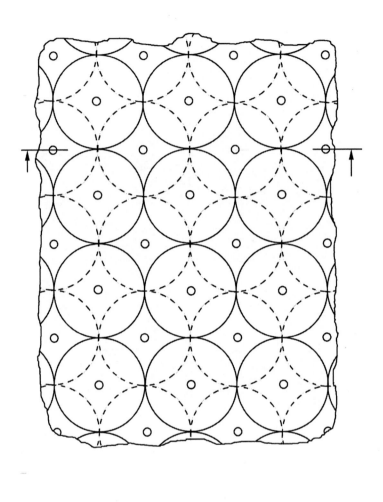

23.12 Drawing of Patent US5515541.

It can be assumed that almost all users will require a multi-layer ballistic pack. Resinated aramid materials like Protexa, or special weave materials like Kevlar®Correctional™, or Twaron SRM may be used as a combined ballistic and stab panel. Another interesting development that provides a degree of slash resistance is a technique for incorporating fine tungsten wire in the knitting process. This is applicable for knitted pullovers and gloves. For higher levels of stab resistance chain-mail, wire-mesh or plate will need to be used in conjunction with the ballistic element. Chain-mail is manufactured from

Table 23.4 Wire diameter/areal density characteristic

Wire diameter mm	Areal density kg/m^2
0.6	1.7
0.7	2.5
0.8	3.5
0.9	4.8
1.0	6.5
1.1	8.5

stainless steel or titanium, with wire diameter of typically 0.6 mm and ring diameter of 7 mm, which is suitable for many knife-resistant armour panel applications. It is important not to over specify the diameter of the chain-mail link wire diameter. The overall wire diameter to areal density of stainless steel chain-mail is shown in Table 23.4 for a 7 mm outside diameter ring.[21]

For a single link, the weight increases as the square of the wire diameter. However, as the wire diameter increases the link centre pitch reduces as a result of the reduction in effective internal ring diameter and because the angle that adjacent rings lie to each other becomes greater. For practical purposes, the maximum ratio of ring/wire diameter is about 6. The relationship between link diameter and wire diameter defines the maximum diameter spike that can penetrate. This critical diameter is approximately $D - 4d$ where D is the ring outer diameter and d is the wire diameter. For 7×0.6 chain-mail this becomes 4.6 mm. To attach the chain-mail to the ballistic element within the armour panel, a fixing arrangement will be required. As chain-mail has no inherent stiffness it will need fixing all around the periphery. Clips or stitching can be used although machine stitching through chain-mail with a sewing machine can be difficult and the risk of needle breakage is increased. An alternative solution is to immerse the chain-mail in a rubber or plastic foam to provide a degree of stiffness to eliminate the need to fix all around the edge.

Wire-mesh of stainless or other heat-treated steel wire is available but is generally less flexible than chain-mail. Platelets may be constructed into an array. Several patents refer to solutions from platelets fixed within a sewn fabric material structure to those fixed with rivets, wire rings or special clips. However, considerable ingenuity is required to eliminate gaps between the plates where a knife may penetrate. Plates may need to overlap and be mounted in several layers to achieve this. Materials suitable for a platelet design range from alloy heat-treated steels, titanium, ceramic, polymer or composites. Thin titanium or polycarbonate sheet materials in one piece may be used but they suffer from reduced flexibility because they bend in only one direction.

A typical knife and ballistic armour design incorporating chain-mail would weigh slightly over 6 kg/m^2, made up of 4.4 kg/m^2 for the ballistic element and 1.7 kg/m^2 for the chain-mail. Including the cover and other fixings in the pack, the complete armour would weigh about 2.5 kg for a medium male size. The armour panel assembly as an operation is generally a labour-intensive production process and appropriate QA procedures are required to ensure production consistency and the reduction of human error effects.

23.10 Future trends

As can be seen, the current optimisation of stab-resistant body armour comprises a textile backing with some form of additional element, for example, chain-mail, to overcome the weakness of most textiles to cutting. Developments to improve textiles and possibly eliminate the need for a metallic content for this application may include:

- improvements in blunting the knife on contact (the Twaron SRM principle)
- increasing the fibre/knife friction coefficient (the PROTEXA principle)
- reducing the stiffness of resinated textiles to improve drape
- increasing the thermal conductivity of protection textiles to reduce heat build up of the user.

23.11 References

1. Genesis 4. 1–11. The Bible
2. Ian Horsfall (2000), Stab Resistant Body Armour, PhD Thesis, Royal Military College of Science, Shrivenham.
3. *www.mpa.gov.uk* Feb 2005.
4. Fenne (2000), 5th International Seminar on Ballistic Protection, Geneva.
5. Roger Hamby, CATRA (2001), Sharpness Measurement and Standards, Sharp Weapons Armour Technology Symposium, RMCS.
6. Fenne-Rob Price, Dupont (2005), Private conversation.
7. Ian Horsfall (2000), Stab Resistant Body Armour, PhD Thesis, Royal Military College of Science, Shrivenham.
8. Blyth and Atkins (2002), Stabbing of metal sheets by a triangular knife. An archaeological investigation, *J Impact Engineering* vol 27 459–473.
9. a. Wierzbicki and Thomas (1993), Closed-form solution for wedge cutting force through thin sheets, *Int J Mech Sci* vol 35, 209–229, b. Cerup-Simonsen and Wierzbicki (1997), Plasticity, fracture and friction in steady-state plate cutting. *Int J Impact Eng* vol 19, 667–691.
10. Fenne-Prof. Tony Atkins, Reading University (2005), Private conversation.
11. B P Pereira, P W Lucas and T Swee-Hin (1997), Ranking the fracture toughness of mammalian soft tissues using the scissors cutting test, *J. Biomechanics* vol 30, 91-4.
12. Fenne-Prof. Tony Atkins, Reading University (2005) Private conversation.
13. Christian Bottger, Teijin Twaron GMBH (2001), Twaron SRM, a novel Type of Stab Resistant Material, Sharp Weapons Armour Technology Symposium, RMCS.
14. Fenne and Dr Martin Kemp (1998), MPS Covert Body Armour Development, Personal Armour Systems Symposium 1998.
15. Dr Beat Kneubuehl (1996), Determination of Tolerances, paper written for CEN,

TC162, WG5.

16. G. Parker (1993), *PSDB Stab Resistant Body Armour Test Procedure*, Police Scientific Development Branch, Publication No. 10/93.

17. Dr J Tan and G Parker (1992), *Stab Resistant Vests* (Part 2), PSDB Publication No 20/92, 1992.

18. Horsfall, Prosser, Watson and Champion (1999), An assessment of human performance in stabbing, *RMCS, Forensic Science International* 102, 1999.

19. John Croft (2003) *PSDB Body Armour Standards for UK Police, Part 3 Knife and Spike Resistance*, Publication No. 7/03/C.

20. Dr L D Payne, Ballistic Injury Archives (2002), The Biological Response Revisited, paper given at Inst of Physics Body Armour Symposium.

21. Fenne-Derek Smith, Mehler Vario Systems (2005), Chain-mail areal density data, private conversation.

24
Flight suits for military aviators

E M C R O W N and L C A P J A C K, The University of Alberta,
Canada

24.1   Introduction

Personal protective equipment for military personnel is needed during non-combat, combat and emergency survival operations. Military aviators' uniforms must meet specific protective performance requirements related to their use of aircraft, including hazards such as gravitational forces during high acceleration, extreme-temperature ambient conditions while in flight or on the ground, immersion hypothermia, and high-heat-flux environments resulting from airplane crashes. Studies of thermal protective garments for operational military personnel have been conducted since the Second World War, when the wide range of temperatures encountered prompted extensive military research programs on clothing for protection against such extremes. Since then, related technologies have developed considerably.

In this chapter we focus on military pilots and flight personnel, including those operating helicopters and transport planes, both combat operational and personnel transport, with less attention given to bombers or fighter aircraft. After summarizing the hazards associated with military aviation, we outline garment system requirements to deal with such hazards, review performance of materials, and emphasize the performance benefits of garment design parameters. Although there are many related considerations in designing protective clothing for military aviators, we focus primarily on thermal protective aspects.

24.2   Hazards of military aircraft operation

24.2.1 Fire hazards associated with plane crashes

Fire hazard has been one of the most severe dangers faced by military aviators. Barring death upon impact or by asphyxiation, the most common cause of death in a crash situation has been immobilization due to injury from burns (Schulman and Stanton, 1970; Albright *et al.*, 1971). Resulting from the unavoidable need for large quantities of highly flammable fuel on board and limited egress facilities (National Materials Advisory Board, 1977; McLaren, 1985), fire associated with military aircraft accidents has been a major cause of mortality and morbidity. Despite technological advances, improved military specifications for aircraft, and increased attention by aircraft designers since publication of these earlier reports, fire associated with aircraft accidents continues to be of concern. Therefore, an aviator's flight suit or garment system must protect against the hazardous environment of a post-crash fire. In addition, as other requirements such as chemical and biological (CB), cold immersion, and anti-gravitational protection are incorporated into aviators' clothing systems, care must be taken that materials for such do not add to the fire load or toxicity of combustion products should the aviator's garments catch fire.

National Materials Advisory Board (1977) report outlined several actual and simulated crash scenarios, including ramp fires, in-flight fires and post-crash fires, but did not include analyses of aviators' clothing. McLaren (1985) described a Canadian Forces helicopter accident in which three survivors suffered varying degrees of burn injury. The jacket of one had been unfastened and the loose lower edge caught fire, causing his coverall to burn as well. The authors are aware of at least two other accidents involving fires on board Canadian Forces aircraft. In all three cases the aviators (pilot, flight engineer, loadmaster, and/or navigator) were wearing wool/polyester flight suits that, when directly exposed to flame had melted and torn away (Crow, 1995). Personnel wearing long underwear were better protected than those without, receiving serious burns primarily where there was no underwear beneath the flight suit. In one case, an aramid mesh life preserver survival vest was worn over the flight suit and protected it from scorching beneath the vest.

Several reports provide estimates of the actual thermal hazards associated with aircraft fire. Difficulty of egress from the aircraft was outlined by the National Materials Advisory Board (1977). Following egress, the aviator often needs to escape a large fuel fire surrounding the plane. Knox et al. (1979) suggested that an aviator, especially if unprotected, must escape such a conflagration within ten seconds of ignition because after 20 seconds the steady-state thermal environment could include temperatures up to 1,260 °C. Although temperatures can be higher during the first transient period of a fire (20 seconds) if the fuel is sprayed as droplets or mist, the steady state temperatures are likely of greatest concern because crash victims may be protected in the cockpit and are unlikely to start running during the transient period (Albright et al., 1971). Stoll's (1962) estimate, that approximately three seconds may be sufficient time for an uninjured person wearing protective clothing to break out of a 50- to 60-foot circle of fuel flames at 1,200 °C, may be overly optimistic given the fluxes involved. Conn and Grant (1991) noted that several estimates of thermal hazards associated with burning fuel have been made, but that few are definitive. Temperature and heat-flux estimates have ranged as high as 2,200 °C and from approximately 20 to approximately 225 kW/m² (Albright et al.; Knox et al.; NATO, 1992; Schulman and Stanton, 1970; Stoll, 1962).

In many crash scenarios, simply the movement of flight personnel out of the aircraft into a fuel-rich external environment may create the hazard of fuel ignition due to friction or electrical spark, if the aviator's garment system contributes to charge generation and/or the initiation of an electrical discharge from either the wearer or the clothing system.

24.2.2   Other hazards

Since the Second World War it has been known that extreme forces of gravity (G) can have ill effects on the human body. Pilots of high-performance aircraft, especially when performing sharp turns, can experience multiple G-forces, causing the blood to move to the lower extremities of the body and depriving the brain of essential oxygen. As a result vision is impaired and the pilot may lose consciousness, which may lead to fatality. Wood (1987) reviewed several aspects of this phenomenon and attempts to counteract it through provision of anti-G suits.

Aircraft crashes on both land and sea can expose aviators to extremely cold environments that in turn can lead to hypothermia and significant impairment of cognitive functions if appropriate immersion and/or cold-protective garments are not worn. Such hazards were documented by several authors at a 1993 conference on the support of air operations under extreme hot and cold weather conditions. For example, deGroot (1993) described land rescue attempts complicated by temperatures ranging from -20 to $-60\,°C$, and Kramer's (1993) report of post-mortem results indicated that death was more frequently caused by hypothermia than by drowning or blunt trauma in personnel who must ditch in water. Immersion in water is particularly hazardous, as water is approximately 25 times more conductive than air, so that the body may reach a critical limit of heat loss even if the water temperature is $20\,°C$ (Mayr and Kramer, 1993). Both movement in the water and fast-moving water increase heat loss due to convection effects.

In other scenarios, exposure to ambient and cockpit temperatures as high as $59\,°C$ have been reported, leading to several medical problems (Cornum, K., 1993; Cornum, R.L.S., 1993). Such conditions are aggravated by wearing either nuclear/biological/chemical (NBC) protective clothing (Macmillan, 1993; Vallerand et al., 1991) or anti-G protection systems (Sowood and O'Connor, 1993), or both.

24.3   Performance requirements for military flight suits

Requirements for military clothing in general are covered elsewhere in this volume (Ch. 21). Due to the focus of this book, the primary performance requirement discussed here is that of safety, and more specifically, protection from high-heat-flux thermal exposures. To determine performance requirements and needs of the user group, Tan *et al.* (1998) followed a systematic functional apparel design process developed by DeJonge (1984) building on work by Jones (1981). Designing a protective garment that incorporates relevant factors is a complex task that involves objectifying the design process so that the resulting garment meets specifications. DeJonge's process is a comprehensive, externalized, strategy-controlled approach that attempts to overcome complex design problems and allows a thorough exploration of the design situation including problem structure, development of design criteria and specifications, exploration of the interactions of the specifications, and development and critical evaluation of a prototype. As part of this design process, Tan *et al.* conducted a market analysis and material and garment analyses, observed video tapes and photographs of previous manikin testing of thermal protective clothing, and analyzed a television program about pilots and planes. They conducted focused group interviews with eight helicopter pilots and 12 transport pilots, and observed, videotaped and analyzed the movements of some of these aviators carrying out the various activities in their jobs.

Flight personnel have specific performance requirements regarding thermal protection and functionality of their flight suits (Tan *et al.*, 1998; Lewyckyj and Reeps, 1978). An aviator's flight suit must be safe and protect against hazardous environments including the high heat-flux environments that result from post-crash fires. The garment must also be comfortable, fit correctly and allow mobility for all necessary activities. In addition, the military flight suit must function for the needs of the person and job, meet psychological requirements such as professional appearance and identity, and be easy and cost efficient to produce and maintain. A garment can be designed to protect very well, but if the intended user is uncomfortable, cannot move around easily, or cannot perform necessary tasks, it will be cast aside or not worn properly. In their focused group interviews, Tan *et al.* found that fit and comfort were considered more salient attributes than safety. Pilots who personally had experienced fire accidents, however, considered safety most important following the accidents; one even purchased thermal protective underwear for himself. In the remainder of this section, we detail several of these salient requirements.

24.3.1   Safety/protection

Requirements for thermal protection, covered more generally in (Chs 11 and 15) are relevant here. In the event of a fire, the garment system should provide sufficient protection to allow the aviator enough time to escape serious injury. Thus, garments must provide insulation values high enough that, for the time of exposure, the heat transferred through the clothing does not raise skin temperature to levels that will cause serious burn damage. Skin burn injuries that result from such accidents are covered in more detail elsewhere in this volume (Chs 11 and 15). Authors generally agree that to contribute to thermal protection, materials/garments must be flame retardant (i.e., do not ignite readily, burn slowly if at all, and self-extinguish), must not melt or shrink, and must be insulative and keep their integrity under high-heat-flux exposures. Materials for thermal protective clothing systems for aircraft personnel are therefore expected to meet specifications based on one of many possible standard bench-scale tests for each of flame resistance, thermal insulation/ protection, and heat resistance (see Ch. 15).

Albright *et al.* (1971) pointed out that tests using only radiant heat sources are useful models for aircraft fuel fires only in that fuel fires are largely radiation sources. They stressed that the energy levels in these fires are such that the convective component is also a severe threat for living tissue. They suggested that if purely radiant heat sources are used for testing because they are controllable, extreme care must be taken to match the spectrum of radiation from the fire to minimize errors due to a fabric's reflectance.

Dating back to the 1960s, full-scale garment tests have been conducted on military flight suits in the U.S.A., initially using dressed manikins exposed to open-pit fuel fires. Manikin tests show the effects of garment design factors as well as material properties. Albright *et al.* (1971) reported that since open-pit fires are characterized by extreme turbulence, the temperatures within the fire and between two different fires cannot be expected to remain constant. More recently, a standard manikin test using a simulated flash fire has been published (ASTM, 2000) and a revised version of it is being considered as an ISO standard. Only a few test systems meeting this standard exist world-wide. In this test, the dressed manikin is exposed to flames of a standard, reasonably consistent heat flux, and the energy transferred through the garment system to skin-simulant sensors is translated into percent second- and third-degree burns on the manikin surface. The estimate of skin injury in this test is based on work by Henriques and Moritz (1947). The model and computer code for the test specified in the standards are detailed in a report by Crown and Dale (1992b). Dale *et al.* (1992) provided details about the construction of one test system that meets the standard. Although the ASTM standard test calls for an exposure of up to $84\,kW/m^2$ for up to five seconds, more severe exposures may be relevant to military aviation crash scenarios, as is the case for other military requirements outlined in a NATO (1992) report.

Another requirement contributing to safety is reduced static propensity of the garment system. As noted above, the possibility of fuel ignition due to electrical discharges from personnel or their clothing is very real. In this case clothing not only fails to protect from a hazardous environment but also may actually contribute to the hazard. Crow (1991) and Scott (1981) both noted that there have been few official reports of static electricity in clothing causing fires or explosions in military operations, yet anecdotal evidence and a worse-case philosophy has meant that the ability of a material to dissipate electrical charge as it is generated is considered an important requirement for military aviators. Such requirements are outlined in detail in Ch. 18.

For some military operations the concept of protection may include chemical and biological protection. Because this requirement is not specific to flight personnel, and is covered elsewhere in this volume (Ch. 20) further details will not be discussed here. It should be noted, however, that any requirement for such protection may increase the difficulty of achieving satisfactory flame-retardant properties (Crown, 1991, 2001). It may also contribute significantly to decreased comfort of garment systems (Macmillan, 1993; Vallerand *et al.,* 1991) as well as poorer fit and reduced mobility.

Garment systems for military aviators must also include some component to protect against hypothermia if flying over large bodies of water or extremely cold land areas. Finally, devices to counteract G-forces are important safety elements for pilots of high-performance aircraft.

24.3.2   Fit and mobility

Tan *et al.* (1998) reported that fit, mobility, and ease of donning and doffing were extremely important to the pilots they interviewed. The pilots expressed many complaints about these attributes relative to their current flight suits. Movement analyses were conducted to understand these requirements. Two helicopter pilots and two transport pilots were videotaped wearing both one-piece and two-piece flight suits while simulating their typical daily activities in and around their aircraft. When pilots are in the cockpit operating aircraft, they must stretch their arms ahead of them and above their heads to reach the aircraft's instruments. The arm movement observed can be described as flexion around the shoulder (Watkins, 1995), in which the arms make a half circle from the vertical rest position through front horizontal to vertical above the head. The garments therefore require enough ease for each arm movement. Video recordings showed the flight suits strained, both across the back and at the underarm, preventing the pilots from moving their arms comfortably, especially when raised. As well as flying the aircraft, helicopter pilots perform many tasks such as squatting to check the helicopter's mechanical operation or climbing to the top of the helicopter. Accommodation of mobility in such activities is therefore required in flight suit design, and any loose-fitting clothing must be carefully controlled so that garments do not catch on aircraft controls or impede escape in any way.

24.3.3   Thermal comfort and heat strain

In focused group interviews flight personnel indicated that their garment systems should be appropriate for both cold and hot weather, should allow evaporation of perspiration, and should have closures and roll-up sleeves that allow for cooling off between active work periods (Tan *et al.*, 1998). Although there is a general understanding among both producers and users of protective clothing, and among researchers in the field, that tradeoffs between protection and comfort are inevitable, the aviators' comments suggest that at least some accommodation of comfort requirements can be met through improved garment design.

Vallerand *et al.* (1991) reported that cockpit environments can be very hot (e.g. dry bulb temperatures as high as 40–50 °C and globe temperatures as high as 50–60 °C). They stated that environmental control systems do not always have the capability to handle the heat stress associated with solar radiation, high ambient temperature and reduced heat dissipation with aviators' ensembles. They reported a growing concern that the interaction of such heat stress and protective clothing could produce an unacceptable level of thermal strain, reduced comfort and deterioration of aircrew performance. Even without the added problems associated with requirements for nuclear, chemical and/or biological protection, aviators' garment systems must accommodate potential conflict between thermal comfort and protection. Achieving thermal comfort and reduced heat strain requires an understanding of physiological heat balance between a person and environment (see Ch. 10). Garment systems that provide cooling or allow heat and moisture to be transferred away from the body are required. Current research at several institutions is directed toward addressing simultaneously the heat and moisture transfer issues associated with the related phenomenon of comfort and protection.

24.3.4   Other functional requirements

Aviators' concerns about the functionality of several features of their flight suits were reported by both Lewyckyj and Reeps (1978) and Tan *et al.* (1998). A major issue in both studies was the requirement for numerous pockets to hold many tools and accessories that must be carried during flight. In essence, the pockets may act as the aviator's briefcase and/or tool box. Some indicated that the presence of numerous pockets with bulky equipment detracted from a professional appearance; nevertheless, having functional pockets seemed more important than neat appearance (Tan *et al.*). Of greater concern was the fact that garments with many accessories in the pockets could interfere with essential movement in the cockpit, so that strategically locating pockets was of utmost importance. Participants in these studies also expressed the need that pockets have closures appropriate to their purpose. Participants in both studies also stressed the importance of having close-fitting closures at ankles and cuffs, not only for thermal protection, but also to ensure that sleeves and legs do not get caught on equipment.

24.4   Contribution of materials to meeting performance requirements

Much research on thermal protective clothing for military personnel has emphasized the properties of fabrics or fabric systems, especially comparisons of different fibres and/or flame retardant finishes. The protection offered by a garment or clothing system, however, depends not only on fibre properties or the presence of finishes, but also on a wider range of interrelated factors including yarn and fabric structure, garment design and assembly of garment layers. We discuss the contributions of garment design in the next section of this chapter. A fabric's retention of structural integrity following heat exposure, its resistance to thermal shrinkage, its mass and thickness, as well as its combination with other materials in layered garment systems, are all important factors in determining a garment's thermal protective properties. A detailed discussion of such factors and their contributions to thermal protective performance of materials in general is found in Ch. 15. Here we summarize some key findings regarding material performance that are most relevant to the requirements outlined above for military aviators.

Early research on aviators' protective clothing focused primarily on comparing conventional non-FR flight suits to those of aramid or aramid-blend materials, usually in combination with cotton underwear. Crown and Dale (1992a) compared the thermal protective performance (TPP) of 35 two-layer fabric systems comprising combinations of seven candidate flight suit materials and five underwear materials. Outer layers included aramid, aramid/PBI, aramid/FR rayon, FR wool/aramid, and cotton fabrics, all with varying masses and woven fabric structures. Underwear layers included cotton, aramid, aramid/ FR viscose, and FR wool/aramid fabrics of either a lighter-weight interlock knit construction or a thicker and somewhat heavier tuckstitch knit construction. The systems were tested with and without a 6.4 mm air gap. Based on the TPP results and the desire of the military to have lightweight outer garments, six combinations were selected for fabrication into garments for manikin testing. The lightweight aramid (203 g/m^2) and aramid/PBI (153 g/m^2) outer fabrics that had performed well on TPP tests in combination with either of the tuckstitch underwear materials, and, despite its tendency to break open during TPP testing, the FR wool/aramid blend fabric (166 g/m^2), also combined with the tuckstitch underwear materials.

The six garment systems were tested on an instrumented manikin where they were exposed to a heat flux of 80 kW/m^2 for 4.5 seconds. The percent manikin surface reaching second-degree burns or greater was lowest for the lightest aramid/PBI flight suit in combination with the aramid underwear. Regardless of which type of underwear was worn, the aramid/PBI flight suits afforded the greatest protection, followed by the aramid flight suits. The FR wool/aramid flight suits had the largest percent manikin surface burn (11–20%), the longest after-flame, and the greatest fabric shrinkage, and had large portions of the outer fabric disintegrate completely. On the basis of these and the TPP results, such a lightweight wool blend material was not recommended for flight suits. In all cases, the aramid tuckstitch underwear performed better than the cotton underwear, which when worn under the FR wool/aramid exhibited localized smouldering. If the flash fire had lasted longer or the heat flux been higher, this underwear might have ignited in such a combination. Thus, the aramid underwear was recommended over the non-FR cotton underwear.

The results summarized above were reported by Crown et al. (1993), along with a series of experiments on non-military single, two- or three-layer FR and non-FR fabric and garment systems. Based on the whole series of experiments, Crown et al. concluded that (i) the protection provided by multiple-layered garment systems is significantly greater than would be expected from the additive effects of the layers used singly, and (ii) the outer layer of a garment system must be flame retardant, as wearing a flammable garment of any type over a flame-retardant one clearly negated the benefits of the FR layer. While it could not be concluded for certain from these experiments that undergarments must also be flame retardant, the results demonstrated that wearing FR underwear offered more protection than did non-FR underwear, especially when worn under the relatively light-weight fabrics that are often used for military flight suits.

The fact that the outer layer in any system must be flame retardant has implications for the design of over-garments, whether for CB protection, buoyancy or warmth. Uglene and Farnworth (1993) reported on the construction and evaluation of a fire-resistant, water-vapour permeable, buoyant and thermally insulating material. Holes were punched in a layer of closed-cell PVC foam. Hydrophilic fabrics were placed on each side of the foam and sewn together to form fabric and thread pathways for transporting liquid sweat through the foam to the outer shell for evaporation. When fabricated into constant-wear aviation coveralls meant to provide buoyancy and hypothermia protection in the case of accidental cold water immersion, and tested on an instrumented manikin in a simulated flash fire, the suit offered excellent protection against a heat flux of $80\,kW/m^2$ for four seconds.

The importance of the integrity of the fabric after exposure to high heat fluxes must be emphasized, as any movement such as running away from a post-crash fire environment will cause a weakened fabric to break open. The charring and break-open property of wool fabrics noted above has been noted elsewhere (Braun et al., 1980). Likewise, light-weight meta-aramid fabrics are known to shrink upon exposure and may break open under pressure of movement. The development of a cylindrical thermal protective performance test that accounts for this phenomenon more accurately than the usual flat, horizontal configuration has been reported (Dale et al., 2000; Crown et al., 2002). Torvi and Dale (1998) discussed energies associated with thermal decomposition reactions. While suggesting the general desirability of having decomposition reactions occur at a lower temperature, they caution that if these reactions severely damage or destroy the mechanical integrity of the fabric, cause it to shrink toward the skin, or release large quantities of decomposition reaction products, then reactions that occur at lower temperatures will decrease the protection a fabric offers.

Almost all aviator flight suits considered in reported research have comprised woven outer garments. Lewyckyj and Reeps (1978) described the development and evaluation of double-knit aramid flyers' coveralls. In addition to improved comfort and mobility, such coveralls were expected to provide better thermal protection than those fabricated from conventional woven aramids. In fire pit testing of the knit coveralls, an average of 20.4% of manikin surface area was severely burned during three-second exposure compared to 70.5% for woven coveralls. The knit fabric was more than twice the mass and more than three times as thick as the woven fabric, accounting for its increased thermal protection. However, it also had more than double the air permeability and almost double the elongation. Therefore, when evaluated by aviators during operational flights in a variety of aircraft, the knit coveralls were deemed to have better comfort, unrestricted entry/egress and arm mobility once secured in the cockpit. It is unlikely that such knit coveralls were adopted due to possible concerns about durability and maintenance. Nevertheless, further work with knits should be encouraged. Today, aramid fleece fabrics are available with excellent thermal protection and comfort characteristics; some have been evaluated for use by the Canadian military (PCERF, 2000–2004).

24.5   Contribution of garment design parameters

As reported above, Tan *et al.* (1998) followed a systematic design process to develop eight different prototype flight suits that varied on style, fit, and closure system (Table 24.1). The suits were evaluated without underwear on a thermally instrumented manikin exposed to a simulated flash fire for 3.5 seconds (Crown *et al.*, 1998). It was expected that garments with controlled fullness would provide better protection than closer fitting styles through the incorporation of air gaps between the garment and the body. Controlled fullness was achieved by adding a flange over the shoulder area, a close fitting waist, longer sleeves with cuffed closures to the wrist and ankle, and a close-fitting stand-up collar. In the initial phase of testing, all garments were fabricated in a light-weight (193 g/m^2) meta-aramid/carbon twill weave fabric. All exhibited shrinkage and brittleness, typical of the fabric used. Although there were significant differences (p<0.05) between one- and two-piece styles and between closure systems A and B (Table 24.1), no significant differences were found between looser and closer-fitting garments, due to the shrinkage and the consequent reduction of air spaces (Fig. 24.1). In a second phase of testing, garments fabricated in a low-thermal-

Table 24.1 Instrumented manikin evaluation of eight flight suits of different style and fit (from Crown *et al.*, 1998, and Tan *et al.*, 1998)

Garment	Style	Fit	Collar and closure system	Mean % burn on manikin surface (std deviation)
1	one-piece	close	A (stand-up collar, exposed front zipper, zipper on sleeves and legs)	35.8 (1.5)
2	one-piece	close	B (convertible collar, hidden front zipper, cuffs with FR Velcro®)	30.6 (2.1)
3	one-piece	loose	A	36.3 (0.9)
4	one-piece	loose	B	33.0 (1.2)
5	two-piece	close	A	31.3 (2.3)
6	two-piece	close	B	30.2 (1.6)
7	two-piece	loose	A	32.2 (1.1)
8	two-piece	loose	B	29.5 (0.8)

shrinkage FR rayon/aramid fabric (293 g/m^2) were evaluated without underwear. In this case, loose-fitting two-piece garments with controlled fullness (controlled at waist, collar and cuff closures) provided significantly ($p<0.025$) more protection to the upper torso, arms and legs than did closer fitting garments. The mean percent manikin surface reaching the burn criteria was 22.05 for the close-fitting garment and 16.55 for the loose-fitting garment. The controlled looseness allowed billowing of the garment during exposure, trapping air and providing better protection (Fig. 24.2).

24.5.1   Incorporation of air gaps

It is well known that air is one of the best insulators and, as illustrated by the experiments just discussed, incorporating air gaps in clothing affects heat transfer. However, air gap size is critical to effectiveness (Kim *et al.*, 2002; Torvi *et al.*, 1999). Prediction of thermal protection while taking into account protection from air gaps and construction details is difficult. Lee *et al.* (2002)

24.1 Garment no. 4 in 193 g/m² fabric, showing reduction of back fullness due to thermal shrinkage after 3.5 seconds exposure.

compared the results of manikin tests and different bench-scale tests using radiant heat sources. They found that areas with zero and small air gaps between the garment and the manikin show good correlation to bench-scale tests conducted with no air gap at similar incident heat flux levels whereas, for the large air gap areas on the manikin, temperatures are lower than those of the bench-scale tests. Song *et al.* (2004) also found that numerical modelling of a garment's thermal protective performance, including the effects of thermal shrinkage, is improved with incorporation of air gaps into the model.

Air gap size is difficult to control in clothing due to the dynamic nature of clothing on the moving body. In concave areas of the body, such as the side torso, there is more air trapped than in convex areas such as the shoulders or upper back (Kim *et al.*, 2002), making the convex areas, especially those that bear the weight of the garment, more vulnerable to burn injury. Crown *et al.*

24.2 Two-piece garment in 293 g/m² FR rayon/aramid fabric, showing no reduction of fullness after 3.5 seconds exposure.

(1998) recommended that double layering of garment material be considered for such vulnerable areas.

24.5.2   Effects of layering

McCullough and Noel (1979) investigated flammability and heat transfer characteristics of layered fabric assemblies (men's dress shirt fabrics and knit undershirt fabrics) in different spatial separations and compared results to the behaviour of single layers of the fabrics. The materials tested were not those used for military garments, nevertheless, their findings indicated that the number of fabric layers, the fibre content of the layers, and the space between them influence garment flammability. The space between the two fabric layers also significantly affected all heat transfer variables (maximum heat transfer rate, time to reach maximum heat transfer rate, and total heat transfer). McCullough and Noel's results pointed to the importance of testing garment assemblies, rather than single layers of fabric, in predicting potential burn hazards.

As noted above, Crown *et al.* (1993) found that protection provided by multiple layered garment systems, where the outer layer is flame retardant, was significantly greater than would be expected from the additive effects of the layers used singly. Layering can also be achieved by wearing long thermal protective underwear under the flight suit. For example, Crown *et al.* (1998) reported that testing garments over long aramid tuckstitch underwear masked effects of garment style. Wearing such underwear may be impractical in hot climates, however, in this case the effects of garment design factors are even more important.

Oakes and Maj (1967) recommended the use of two-piece suits over one-piece coveralls. Crown *et al.* (1998) also reported that two-piece flight suits provided somewhat better protection than one-piece, but concluded this effect was mainly due to tucking in the shirt to create a double layer below the waist. They recommended that double layered yokes or pockets be incorporated in strategic locations for extra protection. Oakes and Maj reported positive evaluations of coveralls in which the entire back was double-layered.

24.5.3 Closures

The role of closure systems on thermal protective garments requires special attention. Lewyckyj and Reeps (1978) found that a tapered, snug fit at wrists and ankles provided better protection than looser styles without cuffs. Crown *et al.* (1998) compared closure systems for thermal protection on one- and two-piece flight suits. It was concluded that cuffed closures offered better protection than metal-zippered closures. Thermal shrinkage of the material in the arms and legs was more evident with zippered closures, while the sleeves and legs with close-fitting cuffs were held in place better during exposure. In addition, the cuffed closures likely controlled air circulation inside the garment, reducing convective heat transfer. Thus, while limited fullness in a garment may add protection through incorporation of air spaces, the role of closures is crucial in controlling such fullness and the air circulation normally associated with looser clothing.

Pilots complained that a convertible collar (a basic shirt collar with no collar band) leaves the neck exposed in a fire accident (Tan *et al.*, 1998). They also believed the convertible collar interfered with fastening the shoulder harness worn during flight. Comparisons of the burn injury patterns of convertible and stand up collars suggest that the stand up collar design offered more thermal protection, especially if the collar ends overlap (Crown *et al.*, 1998). Lewyckyj and Reeps (1978) also found that a neck closure strap on a collar provided improved protection.

24.5.4   Anti-immersion suits and cooling mechanisms

Research on factors that increase survival time in water includes design improvements to immersion suits in attempts to increase thermal insulation and thus reduce the core cooling rate. In general, researchers have found that dry suits are more effective than wet suits and that extra insulation is essential. Light *et al.* (1980) tested three types of immersion suits for protection from hypothermia. They found that a dry suit made of 100 g polyurethane siliconed nylon with an inner surface of metallized material, combined with integral foot pieces, integral wrist and face seals, and a foam neoprene hood provided the best thermal protection from icy water. Such protection came with related problems of water vapour impermeability, inflation at altitude and buoyancy due to the trapped air and may pose exit problems for a pilot trying to escape from a submerged helicopter. Light *et al.* emphasized the importance of a valve or venting system for such events.

The UK adopted an immersion suit made from Ventile cotton with water-repellent finish to obviate many of the problems associated with impermeable coated fabric suits (see Ch. 21). Bramham and Tipton (1993) reported the development of an integrated survival system including an emergency underwater breathing device, a buoyant self-righting lifejacket, and a survival suit that provides thermal comfort and low buoyancy when flying, but a high insulation value and horizontal flotation in water. Uglene and Farnworth (1993) described the development of a fire-resistant, water-vapour permeable, buoyant and thermally insulating material for constant wear aviation coveralls to provide buoyancy and hypothermia protection in the case of accidental cold water immersion. Reinertsen (2000) demonstrated the positive effects of regional insulation on the body including extremities and the posterior region.

While clothing systems can provide protection from fire and other hazards, they also increase the likelihood that the wearer will experience heat strain. Vallerand *et al.* (1991) explored the effectiveness of liquid-cooled and air-cooled vests in reducing heat strain while wearing chemical protective clothing. These vests were worn over underwear but under the chemical protective clothing, and thus were close to the skin. Their results showed that wearing the liquid-cooled vests was an effective means to significantly reduce heat strain for aircrew. The air-cooled vests were even more effective and provided drier skin conditions than the liquid-cooled vests, thereby increasing the comfort level of the wearer. On the other hand, Thornton *et al.* (1993) found that a liquid system was somewhat better than the air system in prevention of heat strain, but the air system was better than the liquid system in reducing flight performance error. Frim *et al.* (1993) and Browne (1993) reported on the success of liquid-cooled vests worn by Canadian Forces pilots in the earliest known deployment of aircrew personal cooling devices, and later by UK aircrews, during combat in the Persian Gulf.

24.5.5   Anti-gravity suits

The first workable anti-G suit, developed by a Canadian team led by Wilbur Franks, was successfully tested in 1941, and many current suits are based on Frank's original approach (Leary, 2000; Wood, 1987). Wood summarized the early development of anti-G suits up to the mid-1980s and reported that in addition to wearing the anti-G suit, researchers found that pilots could withstand much higher gravitational pressures when assuming the prone position. Very little has been done to improve the pilot's position in the aircraft. Instead, scientists today are working on improved versions of the early gravity-fighting gear, as well as radically different flight suits that simulate immersing the body in water to counteract acceleration forces (Leary, 2000; Hess, 1999). Anti-G suits that incorporate positive pressure breathing systems and cooling systems as described above are also under development. The UK RAF have brought into service Trousers Anti-G Mk10 (Air Publication, 2004). This is an overgarment which contains an inflatable bladder made from butyl rubber coated nylon. This extends across the abdomen and down the front of the legs. The outer cover of this suit is made from flame-retardant plain woven Aramid fabric dyed sage green. To overcome the sweat trapping problems caused by rubber bladders developments using Gore-Tex® and other water vapour permeable barriers have been examined, but not yet introduced into service.

24.6   Future trends

As aircraft performance continues to evolve, more and more demands will be placed on the aviators' clothing systems, and such systems will evolve in response to this challenge. As requirements for protection from NBC threats, fire, static electricity, immersion, and gravitational forces increase, and as aviators are expected to operate in extreme hot and cold environments, it is logical that multifunctional textile composites will be increasingly used and that these materials will incorporate developments in high-performance fibres, technical textiles and intelligent textiles (see Chs 18, 5, 6 and 7). One goal will be to incorporate these multiple requirements in as few layers of clothing as possible, reducing weight and bulk and increasing operator efficiency. The incorporation of intelligent textiles could drastically change design requirements as communication devices and warning systems regarding exposures to various hazards will be built into the garments. One consideration that has received little attention in the literature to date is the potential maintenance problems with such clothing systems.

An area that merits extensive research is clothing for female aviators. Very little work has been done specifically on this topic. Since body size, shape and fatness affect such parameters as electrical capacitance, thermoregulatory responses and even garment flammability, research in this area is long overdue.

Improving the fit of protective clothing, whether for males or females, will be aided by body-scanning and other digital technologies.

24.7   Conclusions

In this chapter we have reviewed research on military aviator's clothing systems. We relied heavily on our own research as a framework and incorporated the work of others into this framework. It is our contention that, although understanding materials is essential to the provision of appropriate protection, so too is understanding all garment and system design parameters. Meeting the clothing system needs of aviators in the future will be no less challenging than it has been in the past. Incorporation of new technologies such as intelligent textiles will require that special attention be paid to user needs and to potential interactions with all system components.

24.8   Bibliography

Air Publication AP108F-0404-123 (2004). 'Trousers Anti-G Mk 10'. UK MOD, DLO Strike, Wyton, Cambs. UK.

Albright, J.D., Knox, F.S. III, DuBois, D.R. and Keiser, G.M. (1971), *The testing of thermal protective clothing in a reproducible fuel fire environment: A feasibility study.* Report for U.S. Army Aeromedical Research Laboratory, Fort Rucker, Alabama.

ASTM Committee F23 (2000), *Standard test method for protection against flash fire simulations using an instrumented manikin.* Standard Test F 1930-00, West Conshohocken, PA: American Society for Testing and Materials.

Bramham, E. and Tipton, M.J. (1993), 'Advanced integrated cold protection for aircrew', in *AGARD Conference Proceedings 540, Aerospace Medical Panel Symposium: The Support of Air Operations under Extreme Hot and Cold Conditions* (pp. 17-1 to 17-8). Neuilly Sur Seine, France: NATO Advisory Group for Aerospace Research and Development.

Braun, E., Cobb, D., Cobble, V.B., Krasny, J.F. and Peacock, R.D. (1980), 'Measurement of the protective value of apparel fabrics in a fire environment', *Journal of Consumer Product Flammability*, 7, 15–25.

Browne, P.A. (1993), 'Recent advances in active thermal protection', in *AGARD Conference Proceedings 540, Aerospace Medical Panel Symposium: The Support of Air Operations under Extreme Hot and Cold Conditions* (pp. 35-1–35-9). Neuilly Sur Seine, France: NATO Advisory Group for Aerospace Research and Development.

Conn, J.J. and Grant, G.A. (1991), *Review of test methods for material flammability,* (Contract No. W7714-9-5932/01-ST). Ottawa: National Defence.

Cornum, K. (1993), 'Deployed operations in the heat: A desert shield experience', in *AGARD Conference Proceedings 540, Aerospace Medical Panel Symposium: The Support of Air Operations under Extreme Hot and Cold Conditions* (pp. 27-1–27-5). Neuilly Sur Seine, France: NATO Advisory Group for Aerospace Research and Development.

Cornum, R.L.S. (1993), 'Medical support of attack helicopter battalions during the Gulf War', in *AGARD Conference Proceedings 540, Aerospace Medical Panel Symposium: The Support of Air Operations under Extreme Hot and Cold Conditions* (pp. K2-1–K2-7). Neuilly Sur Seine, France: NATO Advisory Group for Aerospace Research and Development.

Crow, R.M. (1991), *Static electricity: A literature review.* Defence Research Establishment Ottawa Technical Note 91-28. Ottawa, ON: National Defence.

Crow, R.M. (1995), Personal communication with author, Human Protective Systems Division, Defence and Civil Institute of Environmental Medicine, North York, ON.

Crown, E.M. (1996), *Study to determine ways of incorporating flame retardance in nuclear biological chemical (NBC) clothing systems,* SSC File No. XSG94-00165-(610), prepared for Defence Research Establishment Suffield, Medicine Hat, Alberta (72 pp.).

Crown, E.M. (2001), *Investigation of flame retardance in nuclear biological chemical (NBC) protective clothing,* PWGSC File No. EDM-7-00489, report prepared for Defence Research Establishment Suffield, Medicine Hat, Alberta (55pp.).

Crown, E.M. and Dale, J.D. (1992a), *Flammability of clothing,* SSC File No. 202PW.W7714-1-9606, report prepared for Environmental Protection Section, Defence Research Establishment Ottawa (91 pp.).

Crown, E.M. and Dale, J.D. (1992b), *Evaluation of flash fire protective clothing using an instrumented mannequin,* Report prepared for Alberta Occupational Health and Safety Heritage Grant Program. Edmonton, AB: Protective Clothing and Equipment Research Facility, University of Alberta.

Crown, E.M., Ackerman, M.Y., Dale, J.D. and Rigakis, K.B. (1993), 'Thermal protective performance and instrumented mannequin evaluation of multi-layer garment systems', in *AGARD Conference Proceedings 540, Aerospace Medical Panel Symposium: The Support of Air Operations under Extreme Hot and Cold Conditions* (pp. 14-1–14-8). Neuilly Sur Seine, France: NATO Advisory Group for Aerospace Research and Development.

Crown, E.M., Ackerman, M.Y., Dale, J.D. and Tan, Y. (1998), 'Design and evaluation of thermal protective flightsuits. Part II: Instrumented mannequin evaluation', *Clothing and Textiles Research Journal, 16,* 79–87.

Crown, E.M., Dale, JD. and Bitner, E. (2002), 'A comparative analysis of protocols for measuring heat transmission through flame resistant materials: Capturing the effects of thermal shrinkage', *Fire and Materials, 26,* 207–213.

Dale, J.D., Crown, E.M., Ackerman, M.Y, Leung, E. and Rigakis, K.B. (1992), 'Instrumented mannequin evaluation of thermal protective clothing', in McBriarty, J.P. and Henry, N.W., *Performance of Protective Clothing, 4th Volume, ASTM STP 1133,* Philadelphia, PA: American Society for Testing and Materials, pp. 717–733.

Dale, J.D., Ackerman, A.Y., Crown, E.M., Hess, D., Tucker R. and Bitner, E. (2000). 'A study of the geometric effects of testing single layer fabrics for thermal protection', in Nelson, C.N. and Henry, N.W. III, *Performance of Protective Clothing: 7th Volume, ASTM STP 1386* (pp. 383–406). West Conshohocken, PA: American Society for Testing and Materials.

deGroot, W.H. (1993), 'Survival from a C130 accident in the Canadian high Arctic', in *AGARD Conference Proceedings 540, Aerospace Medical Panel Symposium: The Support of Air Operations under Extreme Hot and Cold Conditions* (pp. K1-1–K1-5). Neuilly Sur Seine, France: NATO Advisory Group for Aerospace Research and

Development.

DeJonge, J.O. (1984), 'Forward: the design process', in Watkins, S.M., *Clothing: The portable environment*. Ames: Iowa State University Press.

Frim, J., Bossi, L.L., Glass, K.C. and Ballantyne, M.J. (1993), 'Alleviation of thermal strain in the CF: "Keeping our cool" during the Gulf conflict', in *AGARD Conference Proceedings 540, Aerospace Medical Panel Symposium: The Support of Air Operations under Extreme Hot and Cold Conditions* (pp. 33-1–33-10). Neuilly Sur Seine, France: NATO Advisory Group for Aerospace Research and Development.

Henriques, F.C. and Moritz, A.R. (1947), 'Studies of thermal injury: I. The Conduction of heat to and through skin and the temperatures attained therein. A theoretical and experimental investigation', *American Journal of Pathology*, 23, pp.531–549.

Hess, C. (1999, August), 'High-tech anti-G suits', FLUG Revue, p.84, retrieved 12/3/04 from *http://www.flug-revue.rotor.com/FRheft/FRH9908/FR9908d.htm*

Jones, J.D. (1981), *Design methods: Seeds of human futures*, New York: J. Wiley.

Kim, I.Y., Lee, C., Li, P., Corner, B.D. and Paquette, S. (2002), 'Investigation of air gaps entrapped in protective clothing systems', *Fire and Materials*, 26, 121–126.

Knox, F.S. III, Wachtel, T.L. and McCahan, G.R. Jr. (1979), 'Bioassay of thermal protection afforded by candidate flight suit fabrics', *Aviation, Space and Environmental Medicine*, 50, 1023–1030.

Kramer, M. (1993), 'Aeropathological diagnosis of lethal hypothermia', in *AGARD Conference Proceedings 540, Aerospace Medical Panel Symposium: The Support of Air Operations under Extreme Hot and Cold Conditions* (pp. 21-1–21-6). Neuilly Sur Seine, France: NATO Advisory Group for Aerospace Research and Development.

Leary, W.E. (2000, August 22), 'High-tech suits help pilots avoid gravity's perils', *New York Times*. Retrieved 12/3/2004 from *http://dustbunny.physics.indiana.edu/~dzierba/hp221_2000/NTY/NYT6.html*

Lee, C., Kim, I.Y. and Wood, A. (2002), 'Investigation and correlation of manikin and bench-scale fire testing of clothing systems', *Fire and Materials*, 26, 269–278.

Lewyckyj, J.Z. and Reeps, S.M. (1978), *Development of the CWU-48/P high temperature resistant aramid knit flyer's coverall*, Report No. NADC-77290-60. Washington, DC: Naval Air Development Center.

Light, I.M., McKerrow, W. and Norman, J.N. (1980), 'Immersion coveralls for use by helicopter passengers', *Journal of the Society of Occupational Medicine*, 30, 141–148.

Macmillan, A.J.F. (1993), 'Implications of climatic extremes in aircrew NBC operations', in *AGARD Conference Proceedings 540, Aerospace Medical Panel Symposium: The Support of Air Operations under Extreme Hot and Cold Conditions* (pp. 30-1–30-6). Neuilly Sur Seine, France: NATO Advisory Group for Aerospace Research and Development.

Mayr, B. and Kramer, M. (1993), 'Effectiveness of protection clothing for cold weather conditions after ejection over sea – a case report', in *AGARD Conference Proceedings 540, Aerospace Medical Panel Symposium: The Support of Air Operations under Extreme Hot and Cold Conditions* (pp. 16-1–16-8). Neuilly Sur Seine, France: NATO Advisory Group for Aerospace Research and Development.

McCullough, E.A. and Noel, C.J. (1979, June). 'Flammability characteristics of layered fabric assemblies', *J. of Consumer Product Flammability*, 6, 119–135.

McLaren, M.C. (1985), 'Flame retardant (FR) materials for operational military clothing', Paper presented at the Fourteenth Commonwealth Defence Conference on Operational Clothing and Combat Equipment, Australia.

National Materials Advisory Board, (1977), *Fire Safety Aspects of Polymeric Materials, Vol. 6, Aircraft: Civil and Military.* Washington, DC: National Academy of Sciences.

NATO (1992, June), *Protection from flame and heat resistance provided by AFV crewman's clothing.* NATO Allied Combat Clothing Publication ACCP-2 (Edition 1), prepared by the Group on Materiel (ACSM) Standardization.

Oakes, K.W. and Maj, T.C. (1967), *Evaluation of crew member's improved fire resistant flight coveralls,* ACTIV Project No. ACA-45/67I. San Francisco, California: Army Concept Team in Vietnam.

PCERF (2000–2004), Unpublished Reports 20-019-30, 20-004-32, 20-066-33, & P23-009-02. Edmonton, AB: Protective Clothing and Equipment Research Facility, The University of Alberta.

Reinertsen, R.E. (2000), 'The effect of the distribution of insulation in immersion suits on thermal response', in Kuklane, K. and Holmér, I., *Ergonomics of protective clothing,* Proceedings of Nokobetef 6 and 1st European Conference on Protective Clothing. Stockholm: National Institute for Working Life, pp. 259–261.

Schulman, S. and Stanton, R.M. (1970), '*Nonflammable PBI fabrics for prototype air force flight suits',* Technical Report AFML-JR-70-178, prepared for Fibrous Materials Branch, Nonmetallic Materials Division, Air Force Materials Laboratory, Wright-Patterson Air Force Base, Ohio.

Scott, R. (1981), *Static electricity in clothing and textiles,* Paper presented to the Thirteenth Commonwealth Defence Conference on Operational Clothing and Combat Equipment, Malaysia.

Song, G., Barker, R.L., Hamouda, H., Kuznetsov, A.V., Chitrphiromsri, P. and Grimes, R. (2004), 'Modeling the thermal protective performance of heat resistant garments in flash fire exposures', *Textile Research Journal,* 74, 1033–1040.

Sowood, P.J. and O'Connor, E.M. (1993). 'Thermal strain generated by an enhanced anti-G protection system in a hot climate', in *AGARD Conference Proceedings 540, Aerospace Medical Panel Symposium: The Support of Air Operations under Extreme Hot and Cold Conditions* (pp. 29-1–29-13). Neuilly Sur Seine, France: NATO Advisory Group for Aerospace Research and Development.

Stoll, A.M. (1962), 'Thermal protection capacity of aviator's textiles', *Aerospace Medicine,* 33, 846–850.

Tan, Y., Crown, E.M. and Capjack, L. (1998), 'Design and evaluation of thermal protective flightsuits. Part I: The design process and prototype development', *Clothing and Textiles Research Journal,* 16, 47–55.

Thornton, R., Caldwell, J.L. and Guardiani, F. (1993), 'The use of liquid and air microclimate conditioning systems to alleviate heat stress in helicopter NBC operations', in *AGARD Conference Proceedings 540, Aerospace Medical Panel Symposium: The Support of Air Operations under Extreme Hot and Cold Conditions* (pp. 37-1 to 37-23). Neuilly Sur Seine, France: NATO Advisory Group for Aerospace Research and Development.

Torvi, D.A. and Dale, J.D. (1998), 'Effects of variations in thermal properties on the performance of flame resistant fabrics for flash fires', *Textile Research Journal,* 68, 787–796.

Torvi, D.A., Dale, J.D. and Faulkner, B. 1999, 'Influence of air gaps on bench-top test results of flame resistant fabrics', *J. of Fire Protection Engineering,* 10, 1–12.

Uglene, W. and Farnworth, B. (1993), 'Fire-resistant water vapour permeable buoyant insulation', in *AGARD Conference Proceedings 540, Aerospace Medical Panel Symposium: The Support of Air Operations under Extreme Hot and Cold Conditions* (pp. 12-1–12-8). Neuilly Sur Seine, France: NATO Advisory Group for Aerospace Research and Development.

Vallerand, A.L., Michas, R.D., Frim, J. and Ackles, K.N. (1991), 'Heat balance of subjects wearing protective clothing with a liquid- or air-cooled vest', *Aviation, Space and Environmental Medicine,* 62, 383–391.

Watkins, S. M. (1995), *Clothing: The portable environment,* 2nd edn. Ames: Iowa State University Press.

Wood, E.H. (1987), 'Development of anti-G suits and their limitations', *Aviation, Space, and Environmental Medicine,* 58, 699–706.

25

Protection for workers in the oil and gas industries

E M CROWN and J D DALE, The University of Alberta, Canada

25.1 Introduction

Workers in the oil and gas sectors are exposed to a number of hazardous environments against which they require protection. The specific hazards, the level of risk and the severity of any hazard depend on a worker's specific job classification, the tasks performed and the environment in which the job is carried out. Nevertheless, despite an overall concern about safety in these sectors, the majority of workers carry out their jobs in an explosive or potentially explosive environment, and despite well documented safety procedures, flash fires and other explosions occur, so that personal protective equipment (PPE) is required as a last line of defence. The explosive environment also demands that attention be paid to the static propensity of materials used for PPE in these sectors. Since use of steam is expanding for both extraction and production processes, protection against steam and condensate is another important factor.

The case study discussed in this chapter primarily follows research conducted at the University of Alberta over the last two decades, with references to related research elsewhere. Researchers here have worked collaboratively with oil and gas firms in Alberta, where the Canadian industry is centred. The research combines an engineering design approach with a human ecological approach wherein the needs of workers are emphasized, and clothing is considered both as protection from hazardous environments and as interacting with both the person and his or her various environments. The chapter discusses primary considerations in selection, use and maintenance of PPE in the oil and gas sectors. It focuses on protective clothing rather than other forms of PPE, and on the needs of workers in the land-based sector rather than the off-shore sector. Emphasis is given to field workers on the rigs and at wellheads, pipelines and gas plants rather than those in refineries or petrochemical plants, although many of the protective clothing properties discussed apply to workers in all sub-sectors.

25.2   Hazards in the work environment

Workers in the oil and gas sectors are exposed to hazards that are common to many industries – impact from machinery parts or falling objects, exposure to toxic or injurious gases or chemicals, exposure to conductive and radiant heat transfer from machinery or pipes in processing, and exposure to thermal stress (heat and/or cold) from climatic conditions or their specific work environment. However, the primary hazard that mandates protective clothing specific to the oil and gas sectors is the potential for flash fires resulting from gas leaks at wellhead sites, collection points, compressor stations, refineries and petrochemical plants. Ignition of gas leaks can produce accidents resulting in injury and loss of life through both physical impact due to overpressure and skin burns from engulfment in flash fires. Protective clothing for this sector is designed primarily to prevent skin burns. Investigation of numerous accidents suggests that workers are typically exposed to intense heat flux of approximately $80 \, kW/m^2$ for short durations, normally five seconds or less, with occasional exposures of higher heat flux or longer duration. In such scenarios, clothing must protect the worker for the duration of the initial flash fire and allow the worker to escape any ensuing fire, likely of lower heat flux.

In such potentially explosive environments, the danger of ignition and flash fire is increased when static electricity builds up on any materials in the environment, including the PPE worn by workers. The hazard posed by static electricity is heightened considerably in cold, dry environments such as winters in Canada and several other regions where oil and gas industries are located. In Alberta, for example, outdoor temperatures easily drop to $-40\,°C$ or colder. In such conditions, the absolute humidity level is close to zero and problems with static electricity are very common.

The use of steam in processing is long-standing in the oil sector and related heavy industries. In recent years, use of steam in both extraction and processing has increased, generating concern about the safety of workers. Incidents have been documented by firms in the industry in which workers were seriously injured when exposed to steam or condensate. Some workers handle steam on a routine basis (see Fig. 25.1), while others risk accidental exposure. When leaks occur, workers are not only exposed to steam under very high pressure (up to 800 kPa), but may also be exposed to condensate, which due to impurities, may be found at temperatures over $100\,°C$. Such exposures impose requirements for protective clothing that are quite different from those imposed by flash fires.

25.3   Requirements and performance of protective clothing

Our work with oil and gas firms and industry associations to determine the requirements for protective clothing in their sectors began in the 1980s at a time

25.1 Worker handling steam wearing impermeable over-garment.

when little was documented about specific requirements. In Canada, as in most jurisdictions at the time, no specific regulations existed to mandate thermal/flash fire protection for these sectors. Safety officers in many firms were concerned about conflicting information received from marketers of protective clothing, and were interested in the development of a performance standard on flash fire protective clothing, addressing the specific needs of their industry. An initial study was conducted that included management and worker interviews, a worker survey focusing on attitudes and practices regarding protective clothing, wear trials of selected garments including winter and summer trials as well as evaluation of comfort, fit and mobility in an environmental chamber, and laboratory evaluations of potential protective clothing materials (Crown *et al.*, 1989). Work was also begun on the development of a full-scale simulated flash fire/instrumented thermal manikin test system for thermal protective garments.

The study referred to above resulted in a recommended evaluation protocol and decision framework (Crown *et al.*, 1989, Appendix 19), which formed the basis for the first specific industry standards and guidelines in North America (Canadian Petroleum Association, 1991a,b). Because the petroleum industry was North American if not global in scope, the CPA standards, in turn, formed the basis for both Canadian National Standards (CGSB, 2000a,b) and a National Fire Protection Association (2001) performance standard. To our knowledge, no similar industry-specific standard for flash-fire protective clothing exists elsewhere.

Overall evaluative criteria determined by the initial study included many aspects of safety/protection, comfort/fit/mobility, ease of maintenance, durability/appearance retention, functionality/compatibility with other equipment, aesthetics and cost. It was recommended that, when deciding on criteria for protective clothing specifications, individual workers' job descriptions and work environments be taken into account (Crown et al., 1989). This chapter focuses on the requirements for safety and protection. Maintenance issues, especially as they affect protection, are also addressed. Comfort issues are equally important but are covered in Chapter 10. Discussion on meeting these requirements includes both material properties and aspects of garment system design.

25.3.1 Flash fire protection

As indicated above, the most significant hazard requiring protective clothing in the oil and gas sectors has been the potential for flash fires. Thus, in these sectors, protection from flame engulfment is more important than protection from purely radiant exposure. To provide protection from flash fires, a garment system, and thus any material comprising the system, must have the ability to:

- resist ignition and self-extinguish when ignition source is removed
- limit heat transmission during a short-term exposure to high heat flux
- not melt or shrink upon exposure
- maintain its structural integrity and flexibility upon exposure.

As well, the material ideally should not generate much smoke or toxic combustion product when exposed to flame.

There are many bench-scale tests that address flame resistance and heat transmission. Such tests are covered in detail by Horrocks in Chapter 15. One factor that is important in considering flash fire protection, but upon which little agreement exists, is whether materials should be tested in contact with the sensor or with air gaps, especially when testing single-layer materials. Torvi et al. (1999) used numerical modelling and flow visualization to study the effects of such air gaps, and unlike others who had speculated various optimum gaps, concluded that there is no optimum value.

Dale et al. (2003) discussed several limitations of bench scale tests and issues regarding their use. While flame resistance tests may take into account shrinkage and structural integrity after exposure, most heat transmission tests do not do so adequately. Dale et al. (2000) described the effect of changing the test geometry from planar to cylindrical on the thermal protective properties of single-layer fabrics. If shrinkage occurred during exposure it shortened the time to the end point, indicating poorer fabric performance. Some materials showed over 35% shorter times than when tested in the planar configuration. In addition, Crown et al. (2002) demonstrated a better correlation with full-scale manikin test results and tests of single-layer fabrics when using the cylindrical, rather than a planar, test.

Researchers generally agree that only a complete garment test can come close to reality when testing thermal protection (Dale *et al.*, 1992; Prezant *et al.*, 2000; Zimmerli, 2000). Such full-scale tests of garments or, more realistically, layered garment systems on an instrumented manikin exposed to a simulated flash fire not only take into account shrinkage and structural integrity better than do bench scale tests, but also account for the effects of air gaps or, more generally, of garment style and fit. Facilities for such tests exist in several countries (Zimmerli, 2000). The design and construction of one such facility built primarily for testing protective clothing for the oil and gas sectors are described in detail in Crown *et al.* (1989) and Dale *et al.* (1992). A recent upgrade of this facility providing capability of extending the test duration up to 20 seconds for multi-layered garment systems is described by Ackerman *et al.* (2004).

Several test variables affect the results of instrumented manikin testing of garments and garment systems typically used in the oil and gas sectors. These include, among others, the preconditioning of the specimens, the shape, size and stance of the manikin, the duration and the heat flux distribution of the simulated flash fire exposure, the type, distribution and calibration of sensors used to measure heat transmission, the skin burn injury model and the data analysis routines used. In an attempt to standardize these and other exposure variables a standard test method was developed by ASTM (2000) and a revision of this method is under consideration as an ISO test method. Although procedures are standardized, exposure conditions may vary but must be reported. The ASTM standard includes specification of a standard garment for evaluating material properties and allows for comparisons of different garment designs with the same material, as well as testing actual end-use garment specimens. End-use garment testing is generally of most interest both to protective apparel producers and their clients in the oil and gas sectors.

Material evaluation

When evaluating garment systems for the oil and gas industries on an instrumented manikin, a 3 or 4 s exposure at 80–84 kW/m^2 is normally considered appropriate because such conditions simulate real-life flash fires. Garments typically used by the oil and gas sectors include those fabricated from meta-aramid fabrics, flame-retardant (FR) treated cottons and viscose, FR viscose/aramid blends and para-aramid/pbi blends. Although several fabric and garment system design variables affect results, it has generally been found that when tested with or without underwear at 3 s exposures, typical FR cotton and viscose fabrics and viscose/aramid blends of approximately 300 g/m^2 provide somewhat better protection than do the lighter (approximately 200 g/m^2) aramid and aramid/pbi fabrics, due primarily to their greater thickness and mass. At 4 s exposures, however, the FR treated fabrics may disintegrate and the lighter aramid and aramid/pbi tend to offer as good or better protection. The FR treated materials also generate more combustion products. (It should be noted here that some laboratories will not test modacrylic garments on their manikin systems due to the possible generation of acidic, toxic combustion products.) At both 3 and 4 s exposures, many meta-aramid garments shrink and, depending on the fit of the garment, may lose protective quality when air gaps are decreased due to shrinkage.

Garment system design

The design and fit of the garment system are as important as the material used to fabricate the garments. Oil and gas sector employees work both in extremely hot environments (outdoor climate or indoor plant conditions) and extremely cold outdoor environments. For the former, many workers may choose to wear nothing but light underwear under the lightest possible single-layer protective garment. For these workers, choice of underwear is important for both comfort and protection. If wearing light-weight outer protective garments, especially if the material may shrink and come into closer contact with the skin, undergarments can provide a needed extra layer of protection for vulnerable areas of the body. It is important that such undergarments do not melt when exposed to the heat that may be transmitted through the outer garment during a flash fire exposure. For that reason, synthetic thermoplastic materials such as polyester, nylon, polyethylene and PVDC are not recommended for undergarments. Although it has not been demonstrated that close-fitting underwear needs to be flame resistant, Crown *et al.* (1993) demonstrated that aramid underwear offers more protection than does similar cotton underwear.

Those who work in cold environments need to wear multi-layer protective ensembles such as parkas or insulated coveralls. Likewise, many workers need to wear water- or chemical-resistant garments over their thermal protective garments, to protect against drilling mud or processing chemicals. In the early days of wearing protective clothing, it was the practice of many oil sector workers to wear flammable winter garments, rain gear, or disposable coveralls over an FR protective coverall. It has been demonstrated very clearly, however, that the outermost layer of protective clothing systems for the oil and gas sectors must be flame resistant (Crown *et al.*, 1993). This research also demonstrated the important benefits of garment layering in general. Protection provided by multi-layered garment systems was significantly greater than would be expected from the additive effects of the layers used singly. This result is due mainly to the added insulating effect of air between the fabric layers. Thus, for vulnerable areas of the body such as shoulders where there is no room for air between the garment and the body, at least two layers of fabric are recommended, even in what are otherwise considered single-layer garments. The potential of multi-layered garments to store and later release energy is recognized and has been studied recently by Song (see Chapter 11).

Research on military flightsuits has demonstrated the benefits of incorporating air gaps through controlled fullness in garment design. The protective value of air gaps is lessened, however, in garments fabricated of materials that shrink upon exposure (Crown *et al.*, 1998; see also Chapter 24). More recently, the effects of different widths of air gaps within garment systems have been studied through the use of body scanning technology to quantify the gaps (Lee *et al.*, 2002; Kim *et al.*, 2002; Song *et al.*, 2004). These studies confirm the benefit of incorporating air gaps through garment design.

25.3.2   Static propensity of thermal protective garment systems

Because of the potentially explosive atmosphere in the workplace, the potential for ignition from high-energy spark discharges from clothing or the clothed body is a constant hazard in the oil and gas sectors. Two main types of activities can lead to incendiary discharges. In the first, the worker's clothing is charged triboelectrically through friction by rubbing over or against some object. The charge is induced from the clothing onto the worker's body which acts as a capacitor, and is later discharged through a spark from the body when the worker comes into close proximity to a conductor. The second type of activity is simply removing one garment from over another in a flammable environment. Separation of the two materials can generate sufficient charge to cause a discharge that can ignite flammable gases. The basic phenomena involved are discussed in detail by Gonzalez in Chapter 18.

Workers have expressed fears that some of the thermal protective clothing they are required to wear might not be safe due to its static propensity, and many hold the traditional belief that 100% cotton garments are less prone to static electricity than are garments of inherently flame-resistant materials. Such beliefs, based on the properties of cotton and other natural fibres at higher humidities, are misleading for low-humidity environments, however. In cold dry environments such as experienced in the Alberta oilfields, absolute humidity is close to zero so that problems with static electricity are common even when high-regain materials are used in protective garments. Thus, anti-static materials are required.

Much research on static electricity in textiles has been conducted at room temperature and a relative humidity of 40% or greater, where it may not be a serious problem. Measuring static propensity at zero to 15% RH is more realistic in determining the static propensity of materials in dry environments, whether by measuring resistivity or charge decay or using a human body model. A series of human body experiments at zero to 20% RH were conducted, measuring the electrostatic discharges from humans wearing multi-layer protective clothing systems and either sliding over a vinyl truck seat or walking and removing a garment (Crown *et al.*, 1995; Rizvi *et al.*, 1995, 1998). Although garments of aramid fabrics that incorporated anti-static fibres generated static charges of less energy than those made of either aramid or FR cotton without antistatic fibres, the energy of some discharges was greater than the minimum ignition energy of various gases and mixtures.

The incendivity of an electrostatic discharge, however, depends not only on its total energy but also on its time distribution. In the experiments where the subject slid over a truck seat, results were affected more by the fabrics in the outer layer, although the fabrics in the inner layer(s) also had an effect. As might be expected, in the experiments where garment layers were separated, the inner layer had as much effect as the outer layer. One outcome of these experiments, therefore, is the realization that testing for the static propensity of protective clothing systems needs to include all layers. More recent work aimed at simulating these experiments in small-scale tests (Gonzalez *et al.*, 2001; Grant and Crown, 2001a,b) as well as other research on static protection is discussed in more detail by Gonzalez in Chapter 18.

25.3.3   Protection against steam and condensate

Concern about use of steam and hot water condensate for different processes in the oil sector and related heavy industries has brought into consideration questions regarding the level of protection flame-resistant (FR) clothing can provide against these two elements, the need for materials specifically designed to protect against them, and appropriate methods to evaluate materials intended for this application. Currently, workers who handle steam on a routine basis normally wear over-garments that completely cover the body and are made from liquid- and vapour-impermeable materials (Fig. 25.1). These tend to be uncomfortable, may be stiff especially in cold weather, and often interfere with carrying out tasks. Textile materials used in manufacturing FR/steam-protective clothing systems must be evaluated in terms of steam/hot water permeation or penetration as well as heat transfer. Specifications for these clothing systems should be developed to prevent partial or full-thickness burns from heat transfer onto the skin during or after an exposure incident. The development of appropriate test methods and protocols, evaluation of existing protective materials, and establishment of specifications for the design of protective clothing against steam and hot water are urgently needed.

Few studies have been found in the literature that focused on the steam/hot water permeability of thermal protective clothing, its effects on FR properties of fabrics, or consequences for the wearer. Most research on steam effects has focused on firefighters, where the steam is generated through exposure of water to a high-energy heat source, and does not deal with industrial steam under very high pressures. Watkins *et al.* (1978) reported on their design for firefighting apparel that included an outermost layer that was FR and waterproof, and a liner comprising three fabric layers to protect the firefighter from heat transferred through the outer shell. These researchers hypothesized that, although steam would be able to penetrate the system, it would gradually move through the layers so that the skin would be acclimated rather than causing the pain threshold to be reached. This is unlikely to be the case, however, for steam at pressures found in the oil industry. Watkins (1995, p.16) stated that when condensation occurs, a clothing system for environments where high heat and moisture are present must insulate the wearer from the point at which condensation takes place. She also pointed out that to avoid injury a clothing system is required that can control heat flux or the rate at which vapour reaches the body surface.

Rossi *et al.* (2004) studied the transfer of steam through various layered textile systems as a function of sample parameters such as thickness and permeability. The influence of different sweating rates on the heat and mass transfer during steam exposure was assessed. To simulate perspiration from the human body, a cylinder releasing defined amounts of moisture was used. They determined that impermeable (waterproof) materials normally offer better protection against hot steam than do semi-permeable ones. The transfer of steam depended on the water vapour permeability of the samples, their thermal insulation and their thickness. Increasing the thickness of the samples with a spacer gave a larger increase in protection for the impermeable samples compared to the semi-permeable materials. Measurements with pre-wetted samples showed a reduction in steam protection while measurements with a sweating cylinder showed a beneficial effect of sweating.

The need for more research regarding steam permeability is evident as several factors relevant to the steam/hot water permeability of FR protective clothing have not been thoroughly addressed. Furthermore, the effects of hot condensate, especially with contaminants, do not appear to have been studied. It seems likely that vapour impermeable clothing will be required under many conditions where steam is involved. Research focusing on garment design parameters is therefore needed to deal with potential conflicts between protection and concerns about comfort and heat stress.

25.4 Maintenance of thermal protective performance properties

Flame-resistant (FR) protective clothing can protect a worker in flash fire situations if the clothing is kept clean and is free of combustible materials, but keeping it clean under field conditions is often difficult. Most field workers' garments are contaminated with oily dirt during each wearing, and contamination is often not removed thoroughly in maintenance. Helpful advice on appropriate maintenance is provided by most FR clothing producers, but such advice is not always practical in the field. Also, the degree of cleanliness required to maintain the protective qualities of FR clothing is not well known.

Most laundry research on thermal protective clothing has addressed the issue of durability of FR finishes, or has focused on firefighters' protective clothing. Stull *et al.* (1996) studied the effectiveness of several cleaning methods for structural firefighting protective clothing and found that dry cleaning and even aeration were most effective in removing common contaminants in such garments. Dry cleaning facilities are often not available close to field sites in the oil and gas industries, however. In studying the effect of wear and laundering on structural firefighters' worn garments, Mäkinen (1992) included measures of flammability and heat transmission. Worn, dirty garments had greater flammability than new garments, but heat transmission was not significantly affected.

Crown and Chandler (2003) conducted a multi-phase research project in cooperation with major firms having field operations in Alberta. The purpose was to develop appropriate, practical care procedures to help maintain the protective properties of FR protective clothing over its useful lifetime. To gain an understanding of the working conditions and laundering procedures typical for many oil and gas sector workers who wear thermal protective clothing, and of the barriers to keeping FR clothing clean, field sites were visited and employee interviews and a survey were conducted. The results of this early phase indicated that, given typical working conditions, it may not be possible to keep some FR garments completely clean but that many improvements in consistency and quality can be made to the current cleaning procedures, including locating laundry facilities at all plant/field sites with clearly posted instructions for cleaning.

The second phase of the study (Crown and Chandler, 2003) comprised a five-month wear trial designed to determine if garments worn in the field in normal work environments could be cleaned satisfactorily under controlled laboratory conditions using currently recommended procedures. New garments fabricated from both aramid and FR viscose/aramid blend fabrics (as normally worn by workers in the firms involved) were worn by five employees according to a planned schedule, and after each day of wearing were sent to our laboratory for inspection, laundering and mending. Garments were removed from the trial for flame resistance testing following CAN/CGSB 4-2 No. 27.10 (CGSB, 2000c) after one, five, ten and fifteen wear/wash cycles. The aramid garments that were rated the dirtiest before laundering were the most likely to fail the flame resistance test after laundering, while none of the FR viscose/aramid blend garments failed the same test after laundering. However, it was not clear that the blend garments were cleaner.

Based on observations from the first two phases, a series of small-scale and larger-scale contamination/decontamination laboratory experiments simulating domestic laundry procedures were conducted to compare the effectiveness of various laundry procedures (Crown and Chandler, 2003; Crown *et al.*, 2004). Fabric, contamination (oil or none), detergent and laundry pre-treatment were varied systematically. As in the wear study, laundered specimens were tested for flame resistance to determine effectiveness. In the experiments using small specimens in simulated laundry equipment (Launder-Ometer), results reflected those from the wear trial, in that contaminated and laundered aramid specimens were more likely to fail the flame resistance test. The effect, however, was dependent on the detergent/pre-treatment combination, suggesting that careful selection of laundry supplies and procedures is an important consideration that is sometimes overlooked in the field situation. On the other hand, results for the FR viscose/aramid blend specimens again reflected those from the wear trial, in that oily specimens after laundering were consistently as flame resistant, or more so, than the uncontaminated specimens.

It was hypothesized that oily dirt on the specimens burns with high enough energy to activate the FR mechanism of the FR treatment on these specimens and extinguish the flame, even from the burning oil. This hypothesis was tested further. First, specimens contaminated with oil were tested without laundering and also met flame resistance criteria. Second, an experiment was conducted using additional FR treated fabrics with similar results (Crown *et al.*, 2004). One tentative conclusion from these experiments is that fabrics with an FR-treated cotton or viscose component are more likely to maintain flame resistance when contaminated with oily dirt than are fabrics made from inherently flame-resistant fibres. The proposed mechanism, however, requires further study. An alternative hypothesis based on known oily soiling mechanisms is that the oil is bound more internally in the cotton and viscose but remains on the surface of the aramid fibres.

In one further experiment, larger fabric samples were laundered in a domestic washing machine and then cut into smaller specimens for testing (Crown *et al.*, 2004). Possibly due to the larger water:fabric ratio, this procedure was somewhat more effective in removing oily dirt and preventing redeposition. Although the contaminated and laundered aramid specimens were less flame resistant than the uncontaminated ones, and some individual specimens failed, none of the contaminated samples failed as a group. This result suggested that with careful attention to procedures, improvements in laundry effectiveness could be achieved. Again, as in earlier phases, the FR viscose/aramid blend specimens maintained their flame resistance.

The studies outlined briefly above led to recommendations that more attention be paid to standardizing maintenance procedures at field-site laundry facilities. It was also recommended that dirty, oily garments receive careful pre-treatment prior to laundry, and that wherever possible, heavily contaminated aramid garments should be dry-cleaned or commercially laundered following recommended procedures from suppliers. Such practices, although logical, are not always the norm in the field, where individual workers may be left to launder their own garments without standard procedures. This research has also highlighted the need to develop FR thermal protective materials that are oil repellent and soil releasing.

25.5   Future trends

There are many alternative solutions available for the provision of basic flash fire protection. There is no one best solution for all situations; rather selection should be based both on risk assessment and an evaluation of the workers' specific tasks and work environments. There are many outstanding issues that require resolution before workers will have protective garments that meet all of their needs. As is the case with protective clothing for many sectors, trade-offs between protection and comfort still require attention. To date, thermal protection and thermal comfort usually have been addressed as separate phenomena, but both are related to heat and mass (moisture) transfer and need to be addressed as related phenomena.

Integrating anti-static protection and steam protection into comfortable FR thermal protective garments will continue to challenge fibre, fabric and garment producers. A requirement for ease of maintenance increases the complexity of this challenge. The use of multi-functional composite materials with different components providing different types of protection or comfort may offer solutions to several concurrent demands. Incorporation of nano-particles into some of the layers or surfaces may be possible in the future. For example, it is quite possible that nanotechnology may be used to provide surfaces that will readily repel oily dirt, if compatibility with FR requirements can be achieved.

Future research should also pay more attention to functional garment design. For example, it is possible that steam protective garments could incorporate both vapour-impermeable and vapour-permeable components to protect the most vulnerable locations on the body while allowing for thermal comfort. This requires careful attention to, and understanding of, the worker and how specific tasks are performed. All future research and development efforts in the protective clothing area must pay such close attention to the needs of the worker. Only then will appropriate solutions to very complex problems be found.

25.6   Bibliography

Ackerman, M.Y., Bailey, R., Crown, E.M., Dale, J.D. and Fleck, B. (2004), 'Design of a flash fire simulator', in *Proceedings of the 2004 Spring Technical Meeting of The Combustion Institute Canadian Section* (pp. C-3, 1–6), Kingston, ON.

ASTM Committee F23 (2000), *Standard test method for protection against flash fire simulations using an instrumented manikin*, Standard Test F 1930-00, West Conshohocken, PA: American Society for Testing and Materials.

Canadian Petroleum Association (1991a), *Performance standard for thermal protective clothing*. Calgary, AB: CPA (currently Canadian Association of Petroleum Producers).

Canadian Petroleum Association (1991b), *Recommended practices for provision and use of thermal protective clothing*. Calgary, AB: CPA (currently Canadian Association of Petroleum Producers).

CGSB (2000a), *CAN-CGSB-155.20-2000, Workwear for protection against hydrocarbon flash fire.* National Standard of Canada. Ottawa, ON: Canadian General Standards Board.

CGSB (2000b), *CAN-CGSB-155.21-2000, Recommended practices for the provision and use of workwear for protection against hydrocarbon flash fire.* National Standard of Canada. Ottawa, ON: Canadian General Standards Board.

CGSB (2000c), *CAN-CGSB-4.2, No. 27.10, Flame resistance – vertically oriented textile fabric or fabric assembly test.* National Standard of Canada. Ottawa, ON: Canadian General Standards Board.

Crown, E.M. and Chandler, K. (2003), *Maintaining the protective qualities of thermal protective workwear.* Report prepared for BP Canada Energy Company. Edmonton, AB: Protective Clothing and Equipment Research Facility, The University of Alberta.

Crown, E.M., Rigakis, K.B. and Dale, J.D. (1989), *Systematic assessment of protective clothing for Alberta workers.* Report prepared for Alberta Occupational Health and Safety Heritage Grant Program. Edmonton, AB: Protective Clothing and Equipment Research Facility, The University of Alberta.

Crown, E.M., Ackerman, M.Y., Dale, J.D. and Rigakis, K.B. (1993), 'Thermal protective performance and instrumented mannequin evaluation of multi-layer garment systems', in *AGARD Conference Proceedings 540, Aerospace Medical Panel Symposium: The Support of Air Operations under Extreme Hot and Cold Conditions* (pp. 14-1–14-8). Neuilly Sur Seine, France: NATO Advisory Group for Aerospace Research and Development.

Crown, E.M., Smy, P.R., Rizvi, S.A. and Gonzalez, J.A. (1995), *Ignition Hazards due to Electrostatic Discharges from Protective Fabrics under Dry Conditions.* Report prepared for Alberta Occupational Health and Safety Heritage Grant Program. Edmonton, AB: Protective Clothing and Equipment Research Facility, The University of Alberta.

Crown, E.M., Ackerman, M.Y., Dale, J.D. and Tan, Y. (1998), 'Design and evaluation of thermal protective flightsuits. Part II: Instrumented mannequin evaluation', *Clothing and Textiles Research Journal,* 16, 79–87.

Crown, E.M., Dale, JD. and Bitner, E. (2002), 'A comparative analysis of protocols for measuring heat transmission through flame resistant materials: Capturing the effects of thermal shrinkage', *Fire and Materials,* 26, 207–213.

Crown, E.M., Feng, A. and Xu, X. (2004), 'How clean is clean? Maintaining thermal protective clothing under field conditions in the oil and gas sectors'. *International Journal of Occupational Safety and Ergonomics,* 10, 247–254.

Dale, J.D., Crown, E.M., Ackerman, M.Y., Leung, E. and Rigakis, K.B. (1992), 'Instrumented mannequin evaluation of thermal protective clothing', in McBriarty, J.P. and Henry, N.W., *Performance of Protective Clothing, 4th Volume, ASTM STP 1133* (pp. 717–733). Philadelphia, PA: American Society for Testing and Materials.

Dale, J.D., Ackerman, A.Y., Crown, E.M., Hess, D., Tucker R. and Bitner, E. (2000), 'A study of the geometric effects of testing single layer fabrics for thermal protection', in Nelson, C.N. and Henry, N.W. III, *Performance of Protective Clothing: 7th Volume, ASTM STP 1386* (pp. 383–406). West Conshohocken, PA: American Society for Testing and Materials.

Dale, J.D., Crown, E.M., Ackerman, M.Y., Lawson, L.K. and Perkins, H. (2003), 'Fundamental issues in bench scale testing of fabrics for thermal protection', in *Challenges for Protective Clothing,* Proceedings of the 2nd European Conference on

Protective Clothing (ECPC) and NOKOBETEF 7 [CD]. Montreux, Switzerland: EMPA and European Society of Protective Clothing.

Gonzalez, J.A., Rizvi, S.A., Crown, E.M. and Smy, P. (2001), 'A laboratory protocol to assess the electrostatic propensity of protective clothing systems', *Journal of the Textile Institute*, 92 Part 1, 315–327.

Grant, T.L. and Crown, E.M. (2001a), 'Electrostatic properties of thermal-protective fabric systems: Part 1, Simulation of garment layer separation'. *Journal of the Textile Institute*, 92 Part 1, 395–402.

Grant, T.L. and Crown, E.M. (2001b), 'Electrostatic properties of thermal-protective fabric systems: Part 2, Effect of triboelectric parameters', *Journal of the Textile Institute*, 92 Part 1, 403–407.

Kim, I.Y., Lee, C., Li, P., Corner, B.D. and Paquette, S. (2002), 'Investigation of air gaps entrapped in protective clothing systems', *Fire and Materials*, 26, 121–126.

Lee, C., Kim, I.Y. and Wood, A. (2002), 'Investigation and correlation of manikin and bench-scale fire testing of clothing systems', *Fire and Materials*, 26, 269–278.

Mäkinen, H. (1992), 'The effect of wear and laundering on flame-retardant fabrics'. In McBriarty, J.P. and Henry, N.W. (eds), *Performance of Protective Clothing: Fourth Volume, ASTM STP 1133* (pp. 754–765). Philadelphia, PA: American Society for Testing and Materials.

National Fire Protection Association (2001), *NFPA 2112, Standard on flame resistant garments for protection of industrial personnel against flash fire*. Quincy, MA: NFPA.

Prezant, D.J., Barker, R.L., Bender, M. and Kelly, K.J. (2000), 'Predicting the impact of a design change from modern to modified modern firefighting uniforms on burn injuries using manikin fire tests', in Nelson, C.N. and Henry, N.W. (eds), *Performance of Protective Clothing: Seventh Volume, ASTM STP 1386* (pp. 224–232). West Conshohocken, PA: American Society for Testing and Materials.

Rizvi, S.A.H., Crown, E.M., Osei-Ntiri, K., Smy, P.R. and Gonzalez, J.A. (1995), 'Electrostatic characteristics of thermal protective garments at low humidity', *Journal of the Textile Institute*, 86, 549–558.

Rizvi, S.A., Crown, E.M., Gonzalez, J.A. and Smy, P.R. (1998), 'Electrostatic characteristics of thermal-protective garment systems at various low humidities', *Journal of the Textile Institute*, 89, Part 1, 703–710.

Rossi, R., Indelicato, E. and Bolli, W. (2004), 'Hot steam transfer through heat protective clothing layers', *International Journal of Occupational Safety and Ergonomics, 10,* 239–245.

Song, G., Barker, R.L., Hamouda, H., Kuznetsov, A.V., Chitrphiromsri, P. and Grimes, R. (2004), 'Modeling the thermal protective performance of heat resistant garments in flash fire exposures', *Textile Research Journal*, 74, 1033–1040.

Stull, J., Dodgen, C.R., Connor, M.B. and McCarthy, R.T. (1996), 'Evaluating the effectiveness of different laundering approaches for decontaminating structural fire fighting protective clothing', in Johnson, J.S. and Mansdorf, S.Z. (eds), *Performance of Protective Clothing, Fifth Volume, ASTM STP 1237* (pp. 447–468). West Conshohocken, PA: American Society for Testing and Materials.

Torvi, D.A., Dale, J.D. and Faulkner, B. 1999, 'Influence of air gaps on bench-top test results of flame resistant fabrics', *J. of Fire Protection Engineering*, 10, 1–12.

Watkins, S. M. (1995), *Clothing: The portable environment*, 2nd edn. Ames: Iowa State University Press.

Watkins, S.M., Valla, M. and Rosen, L. (1978), *The development of two protective apparel systems for firefighting*. A report submitted to the National Fire Prevention and Control Administration. Washington, DC: U.S. Department of Commerce.

Zimmerli, T. (2000), 'Manikin testing of protective clothing – a survey', in Nelson, C.N. and Henry, N.W. (eds), *Performance of Protective Clothing: Seventh Volume, ASTM STP 1386* (pp. 203–211). West Conshohocken, PA: American Society for Testing and Materials.

Motorcyclists

P V A R N S V E R R Y, PVA Technical File Services Limited, UK

26.1 Introduction

Textiles have been widely used in motorcyclists' clothing providing protection from the elements – wind, rain, heat and cold – since the origins of the transport form. Latterly, addition of impact protection components to localised areas of garments provided the impression that these were safety products capable of preventing injuries should a rider or pillion passenger fall from the motorcycle. The introduction of European Standards for motorcyclists' protective clothing has clearly demonstrated that many such products are neither suitable nor safe as protective clothing, but the availability of the standards has also motivated and enabled development of alternative textile technologies which meet with or exceed the requirements of the documents.

The crumple zones, airbags, side impact bars and seatbelts routinely fitted to motor cars in order to safeguard the vehicle occupants from injury in a road traffic accident are entirely absent from motorcycles. In stark contrast, the only protection a rider or pillion passenger has is the safety helmet and clothing they are wearing. The Transport Statistics Bulletin *Road Casualties in Great Britain: Main results: 2003* reveals that in the period studied, 693 motorcyclists were killed, 6,959 seriously injured and 20,759 slightly injured.

Despite the best of efforts in terms of rider training and engineering safer roads, it is not possible to prevent all accidents. Adequate, protective motorcycle clothing, worn by a greater number of motorcyclists, might provide a significant reduction in the level of slight injuries to this vulnerable road-user group, and might furthermore commute a serious level of injury to that of a slight injury.

26.2 Motorcycle clothing in the past

Whilst leather is the material of choice generally associated with motorcyclists' clothing, courtesy of films such as *The Wild One* and *Girl on a Motorcycle*; in fact textiles provided the original outer wear and were only later usurped by leather. Since the very earliest, pioneer days of motorcycling, riders have been seeking the 'Holy Grail' of clothing that represents the optimum balance of comfort, all-season climate control, protection from the elements and protection in a fall from the motorcycle. Sturdy outdoor clothing was pressed into service as motorcycle apparel which, over a period of time, evolved into purpose-designed garments.

A key example of this is the heavy-duty waxed cotton jackets and trousers manufactured and marketed by the Barbour and Belstaff companies. Originally conceived for mariners' foul weather clothing, and later adopted by outdoorsmen, waxed cotton was also ideal for motorcyclists' rainwear. Additionally, the paraffin-wax impregnation resulted in a fabric surface with a reduced coefficient of friction, imbuing the treated garment with enhanced resistance to abrasion in a slide along a tarmac surface. Additionally, the 'thorn-proof' properties of the fabric made it a favourite with competitors in off-road motorcycle competition such as trials, which are often held in densely wooded areas. Furthermore, the fabric could be stripped of its wax coating and retreated, enhancing its durability. The author knows one owner of a waxed cotton jacket which is at least fifty years old and has seen service through three generations of motorcyclists within the same family.

The downside of the use of waxed cotton was its unpleasant feel and the manner in which road grime became embedded in the impregnation, leaving the wearer with dirty hands when handling the garment. Additionally, whatever they were wearing underneath suffered from the soiled collar and cuffs which became associated with users of garments manufactured from this material.

With the advent of man-made materials, garments such as the Belstaff 'Black Prince' suit represented use of this technology in its infancy. This garment was manufactured from a laminate of PVC over heavyweight cotton, was rather heavy and very stiff in cold ambient conditions. As time progressed, other innovative new fibres, such as air-textured nylon, found their way into motorcycling garments, and the specialist demands of motorcycling led to the creation of still more specialist materials. Many of these materials claimed to combine the lightness and weather protection of textile garments with the injury protection afforded by leather. Early efforts included the 'Fackelmann Safedress' suit of the late 1970s. Despite use of dramatic images of a stuntman wearing the suit being dragged along the ground behind a speeding vehicle, this suit failed to establish itself in the market and, apart from briefly revisited efforts to return to the market in the mid-1980s, it faded into obscurity.

One material which was specifically conceived for use in motorcyclists' clothing, and is now well established as a component of premium products, is 'Keprotec®'; a combination of Cordura®, Kevlar® and Lycra® manufactured by Schoeller Textiles AG, of Switzerland (Fig. 26.1(a),(b)). At the time of writing, this fabric is promoted on the Schoeller website as 'Extremly [sic] tear and abrasion resistant' and 'crash proof' in the context of its use in motorcyclists' protective clothing. A number of variants of this fabric have been developed

(a)   (b)

26.1 Two versions of the same design of motorcycle suit. (a) The British Superbike Championship colours of the Rizla Suzuki team as manufactured in Schoeller Keprotek by GTS Racing, of Telford, United Kingdom, and (b) the leather variation produced by Scott Leathers, of Barnard Castle, United Kingdom.

over the intervening years, and it is in widespread use in premium brand motorcycling garments.

It is believed the motivation behind development of 'Keprotec®' was a requirement expressed by multiple motorcycle sidecar World Champion, the Swiss Rolf Biland, for a motorcycling suit which would offer the same skin-tight fit and streamlining as the suits worn by downhill racing skiers, but with the protection of leather. Additionally, sponsors' logos would be screen-printed onto the new garment, avoiding disruption of the otherwise aerodynamic profile of the rider generally created by raised leather lettering on conventional motorcycling suits.

In order for Keprotec® to be used in suits for motorcycle competition, in accordance with the rules of the sport's governing body, the FIM (Fédération Internationale de Motocyclisme), required a certificate to be provided by a technical institute confirming that the material met the criteria which were in force at the time and which are essentially unchanged at the time of publication. Non-leather material may be used if it meets with the following requirements and must be at least equivalent to 1.5 mm of cowhide (not split leather):

- fire retardant
- resistant to abrasion
- low coefficient of friction against all types of asphalt
- perspiration-absorbing qualities
- medical test – non-toxic and non-allergenic
- of a quality that does not melt
- non-flammable.

Amongst the methods by which Keprotec® was evaluated against the FIM's Technical Regulations[1] the abrasion machine developed by Darmstadt University was used. The more recent introduction of European Standards[2–5] for protective motorcycling clothing has shown that a single layer of Keprotec® does not provide adequate performance to defeat the specified tests for impact abrasion and impact cut resistance, and in the case of the former is little better than Cordura alone. Interestingly, the reverse side of Keprotec generally exhibits superior impact abrasion resistance to the aesthetically pleasing 'face' of the material. Whilst it remains a novel and innovative effort at producing the optimum fabric for motorcyclists' clothing, unfortunately Keprotec® does not currently live up to the claims made about it and these are key to the safety of motorcyclists.

Foul-weather garments for motorcyclists have moved away from the early PVC and nylon base constructions to air-textured nylons which provide enhanced durability in extended wear. Early, direct lamination of waterproof treatments, such as PU coatings to the reverse face of the fabric, have given way to waterproof and breathable interlinings, and more recently have come full circle to advanced direct laminates, such as the latest Gore-Tex® XCR® fabrics which are claimed to provide a more stretchable garment with enhanced vapour permeability.

26.3   Development of European Standards

Whilst developmental work in the field of improved weather protection has continued, the single most motivating factor for the integration of protective capabilities in advanced textile motorcycle clothing solutions has been the development and publication of the European Standard EN 13595 Parts 1–4 *Protective clothing for professional motorcyclists – Jackets, trousers, one-piece and divided suits*. BMW describes the construction of its Rallye 2 textile suit of jacket and trousers as 'Tear resistant Cordura reinforced at exposed areas for added protection', which requires the whole garment to comply with the requirements of the PPE Directive, but only the fitted impact protectors conform to a standard and are CE marked (Fig. 26.2).

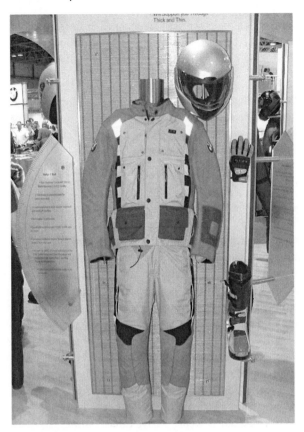

26.2 BMW's Rallye 2 textile suit.

These European Standards, plus others for gloves and footwear, resulted from motorcyclists' protective clothing falling within the scope of the Personal Protective Equipment Directive.[6] For a period, industry and riders' groups questioned the validity of this decision, and lobbied the European Commission and European Standards agency CEN for the work programme to be halted, claiming that motorcyclists' clothing is merely outerwear intended to shield the wearer from non-extreme ambient conditions of heat, cold, wind and rain. It was, however, subsequently agreed that inclusion of protective components – impact protectors for the major joints and long bones of shoulders, elbows and forearms, hips, knees and shins, plus back protectors – and claims in advertising that the garments featured certain characteristics which would prove of benefit in a fall from a motorcycle, such as abrasion-resistance and reinforced areas, meant that motorcyclists' clothing with a protective function did indeed exist and standards for such products were therefore entirely necessary. Specialists in motorcyclists' impact protection Planet Knox have ventured into the manufacture of garments featuring built-in limb, back and chest protection.

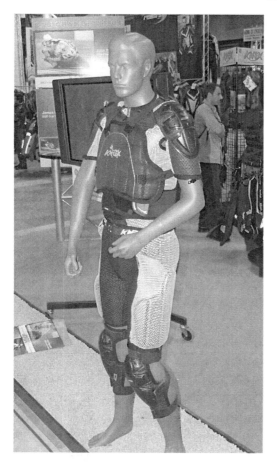

26.3 Protective undergarments by Planet Knox.

Pictured in Fig. 26.3 is a shirt and shorts manufactured from a combination of perspiration-wicking materials and mesh for ventilation. The knee and shin protectors strap on and are held in place by Velcro fastenings. The whole ensemble is intended to be worn under outer garments providing protection from abrasions and cuts, but has the potential for applications in action sports such as downhill mountain biking and skiing.

The European riders' groups initial display of a healthy ambivalence towards the standards was also due to concerns that the documents might pave the way for compulsory protective clothing for motorcyclists; however, this was tempered by their broad support for some form of independent and recognisable mark of fitness for purpose which would enable consumers to differentiate between competing products in the marketplace. For a brief period in the late 1990s, the European industry and riders' groups collaborated on an alternative product accreditation programme – the Industry Quality Label, or IQL – which placed its focus on 'quality'-driven assessment regimes and the IQL project was quietly shelved as work on the CEN standards accelerated. European Standard EN 13595 incorporates the following tests for the protective performance characteristics of sheet materials and assemblies:

- impact abrasion resistance
- impact cut resistance
- determination of bursting strength
- tensile strength

26.4   Impact abrasion resistance tests (Fig. 26.4)

A variety of test methodologies – including the Darmstadt device briefly touched upon above – were evaluated by European Standards Committee CEN/ TC 162/WG9 'Protective clothing for motorcycle riders' in order to identify an acceptable and appropriate procedure. Established test machines such as the Martindale, revolving drum and Taber abrader were considered, but rejected on the basis that they were not capable of testing the multitudinous single and combination materials and constructions present in motorcyclists' protective

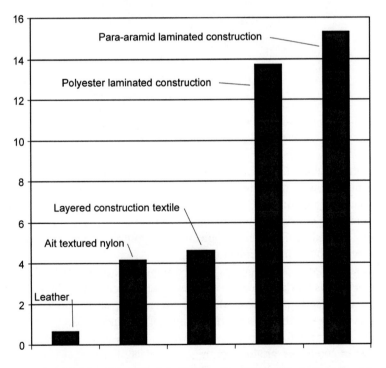

26.4 Comparison of typical relative impact abrasion resistance (in seconds) of various materials used in the construction of motorcyclists' clothing.

clothing. The Darmstadt machine appeared ready to be adopted as the EN13595 impact abrasion test method by default, and is described in documentation submitted to WG9.[7]

26.4.1   The Darmstadt tester

The Darmstadt machine consists of a 'doughnut' of concrete, in the centre of which is situated an electric motor. Attached atop the drive shaft of the motor, on an apparatus which can be locked onto or released from the shaft, are three (necessary for stability when the sample holder is spinning) specimen holders. In testing, the motor is run up to a specified speed at which point the specimen holders are released and freefall the short distance down the axis of the shaft into contact with the surface of the concrete. The specimens continue to spin freely around the drive shaft and in contact with the concrete until coming to a halt. The mass of test specimens both before and after testing is recorded, and the difference established.

26.4.2   The Cambridge tester

In the meantime, however, Dr Roderick Woods at Cambridge University UK had conceived, built and developed an alternative test device which combined the dynamic characteristics of both impact and abrasion. This became known as the Cambridge machine and formed an integral part of 'The Cambridge Standard';[8] an alternative technical specification produced by Dr Woods for testing and EC Type-Examination of motorcyclists' protective clothing prior to the publication of the European Standards.

The Cambridge machine consists of an OP60 aluminium oxide abrasive grit belt which passes around cambered rollers, one of which is driven by a 750 W motor, to provide a belt speed of 8 m/s. The test specimen is mounted on a holder which is attached to the free end of a horizontal, rigid pendulum which pivots at the opposite end to the specimen holder. In testing, the pendulum is released, usually electromagnetically, and falls from a height of 50 ± 5 mm onto the moving belt. A fine copper wire of 0.14 mm diameter, located across the outer face of the specimen, is cut upon contact with the moving belt and this starts an electronic timer. A second wire is exposed and cut when the specimen is abraded through, which stops the timer and records the time taken to perforate the specimen. The more prolonged the period between contact and perforation, the better-performing the material. The belt is continually cleared of debris from the test specimen throughout the test process by a combination of electrically driven rotating brushes and vacuum cleaning. A reference fabric of canvas is used in order to establish and check the calibration of the abrasive belt before testing and after each group of test specimens.

26.4.3   Comparison of two test methods

The design of the Cambridge machine addressed reservations about the mode of operation of the Darmstadt device, and the veracity of the data generated by it, which were held by Dr Woods and other experts. Firstly, there were concerns over difficulties in repeatability of the concrete surface's specification, since the chemical constituency of concrete is entirely dependent upon the geological make-up of the materials from which it is formed. Secondly, there was the problem that the concrete was not cleared of debris from the test specimens between each revolution which, it was considered, continually altered the coefficient of friction properties of the surface and so entirely compromised the credibility of any data generated.

Thirdly, the free play between the specimen holders and the drive shaft of the central shaft necessary to permit the specimen holders to rotate freely around the shaft during the test process allowed a rocking effect to manifest itself at the point of the specimen holders, such that there was an inconsistent loading on the test specimens throughout testing. Furthermore, there were concerns from some experts that the design of the Darmstadt machine enabled the Bernoulli effect to occur – particularly at higher revolutions – with the face of the test specimen only in light contact, or out of contact altogether, with the surface of the concrete.

Fourthly, the method of recording test results – by establishing the amount of the original mass lost between the point of contact and the end of the test – was unable to establish whether test specimens which had failed catastrophically (perforated to a hole) had done so immediately or within a short period of contacting the concrete, or immediately before specimens had come to rest. Finally, data from the Darmstadt device produced a hierarchy of relative abrasion resistance of materials which contradicted experience, rating some weak materials known to be entirely inappropriate for use in motorcyclists' protective clothing above those with an established and proven track record.

By contrast, the design and function of the Cambridge machine ensures the face of the test specimens is in contact with the abrasive belt under an invariable loading and at a constant – not a fluctuating, uncontrolled and decreasing – speed, and the end point of the test is fixed. It is a test to destruction with all specimens abraded to a hole.

26.4.4   Crash simulations using manikins

Data generated by the Cambridge machine also correlated with that produced in real road simulation tests conducted during the early 1980s[9] using the 'Metal Mickey' manikin. Garments and specimens were attached to Metal Mickey's anatomy at the three key impact points – shoulder, knee and hip – identified from clothing damaged in real-life crashes. The manikin was then placed on a support manufactured so as to rotate and release the dummy in a position resembling a skid or 'low side accident' common in single vehicle accidents, where the rider falls off the side of the motorcycle and drops not more than one metre to the ground. The opposite scenario is a 'high side', common where one or both tyres of the motorcycle lose adhesion with the road surface, slide and then grip again, causing a violent pendulum effect which launches the rider into the air, often to a height of several metres in higher-speed accidents such as those common in motorcycle competition, resulting in a fall to earth from a significantly greater height and a dramatically increased risk of impact injuries. In the Metal Mickey tests, the support was released at a predetermined road speed and the manikin would pivot and fall onto the tarmac, before sliding to a halt.

Within WG9, there was a pro-Darmstadt machine lobby, largely of manufacturers who had invested in the device and had developed materials and based their marketing on data generated by the apparatus. For financial and commercial reasons, they had an understandable reluctance to see their past efforts and investments overtaken by the adoption of the Cambridge machine as the specified test method.

Darmstadt University was provided with an opportunity to address the technical criticisms levelled at their device. Similarly, validation of the Cambridge machine was sought; however, whereas Dr Woods attempted to answer each of the questions tabled concerning his machine, the Darmstadt technicians failed to provide any form of response. The decision was therefore finally taken to adopt the Cambridge machine as the impact abrasion test method for EN13595.

26.5   Other test methods

By comparison, selection of the test methods for determination of burst strength, impact cut resistance and tear strength was achieved much more easily. Established test equipment (or, in one instance, a minor variation thereof, in the case of the burst test method) were readily pressed into service.

26.5.1   Bursting strength test

The burst test uses a modified Mullen-type apparatus, in which the test specimen is clamped into position above a heavyweight rubber diaphragm. Water is pumped behind the diaphragm, which distends against the test specimen, stretching it until it fails. Failure can be a tear in the material, breaking of the thread in seams or parting of the two sides of zip fasteners. In this test, the higher the water pressure required to cause failure, the better the performance of the material or construction assembly.

26.5.2   Impact cut resistance test (Fig. 26.5)

The impact cut test involves dropping a weighted blade from a specified height onto the test specimen. In this test, the lower the extent of blade penetration, the better (reduced blade penetration equating to higher impact cut resistance). In all the above tests, the standard's requirements for protective performance vary depending on the area of the garment in which the materials or constructions are found. The risk of damage is different in the various areas of motorcyclists' clothing, with the limb joints and immediately surrounding areas at higher risk than, for example, the chest or armpits.

26.5.3   Tear strength tests

Finally, tear strength of non-leather materials – excluding elasticated and knitted fabrics – is evaluated according to ISO 4674:1977 *Fabrics coated with rubber or plastics – Determination of tear resistance.*

26.6   Manufacture of textile garments

With the availability of test methods for evaluation of materials and constructions of motorcyclists' protective clothing presented by the evolving draft standards, a number of manufacturers attempted to develop textile

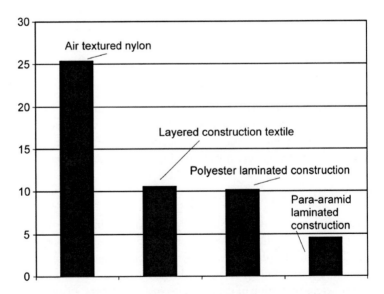

26.5 Comparison of typical impact cut resistance of various textiles used in the construction of motorcyclists' clothing (expressed in millimetres of penetration). The lower the result, the more cut resistant and, thus, better performing the material.

protective clothing. Unfortunately, it quickly became apparent that none of the high-tech fabrics which were available from the major specialist suppliers could withstand the rigours of the EN 13595 test apparatus. Whilst the burst strength requirements for materials and assemblies, evaluated by the method described in EN 13595 Part 3, were relatively easy to meet, the continuing inability of the so-called 'market leading fabrics' to address the requirements for impact abrasion resistance and the critical impact cut resistance test, described in Parts 2 and 4 respectively, were leaving clothing suppliers despondent.

Some manufacturers have reportedly made official representations to Notified Bodies, stating that they have found the EN 13595 tests too severe for their textile garments to meet, and have sought changes in the standard's requirements or the test methodologies. However, what they need to do is break away from developing and trying to use materials and fabric weights which can never provide a suitable and safe level of protection. The requirements of the standard should not be reduced to a level which unsafe products can meet, but should continue to provide a benchmark to which protective clothing for motorcyclists should aspire.

The challenge of defeating the standardised tests with textiles has, however, motivated other manufacturers, who have seized the initiative to investigate different and innovative approaches to garment construction, and to experiment with materials supplied by weavers and textile mills not ordinarily active in this specialist market. To date, three technological approaches have been used to deliver textile motorcyclists' garments which are capable of satisfying the requirements of EN 135985. These are:

• a heavyweight, single layer construction
• a layered construction
• a laminated construction.

26.6.1   The heavyweight single layer construction

The heavyweight, single layer construction employs the identical manufacturing methodology utilised in non-protective motorcycling garments, but uses a fabric of significantly increased thickness, stiffness and mass in order to increase its impact abrasion and impact cut resistance to a level which will meet the requirements of the standard. As a consequence of their increased mass and stiffness, the resultant garments are not as comfortable to wear as non-protective garments or those protective items manufactured using the alternative approaches described below, and to date it has been possible only to meet the lowest, Level 1 requirements of EN 13595, using the heavyweight, single layer system. Typical impact abrasion resistance results are 16.5% above the standard's requirements for Level 1, but 33% lower than the threshold for Level 2 categorisation. The two further approaches listed above were conceived by Dr Roderick Woods, formerly of Cambridge University's Department of Physiology, during the early 1990s.

26.6.2 The layered construction

The layered concept consists of an outer, cosmetic layer, generally of polyamide construction, behind which is inserted a layer of loom state, plain weave geotextile of Nylon 6,6, with a fabric weight over 400 grams per square metre, which serves as a 'mechanical interlining'. The plain weave ensures that single fibres of yarn are abraded or cut by the road surface, leaving the underlying fibres in place to support the mechanical structure of the fabric. If a similar composition of fabric featuring a twisted yarn were to be used, then every fibre in the yarn would be exposed to being cut at several points along its length, and these cut sections might then be torn away from the structure of the fabric, resulting in appreciably reduced abrasion resistance and the prospect of rapid onset of catastrophic failure. This mechanical interlining is described as the 'Structurally Strong Layer' (SSL) in EN 13595-1. All seams are sewn with a compatible thread of a breaking strain commensurate with the properties of the SSL fibre. Specialist handling of cut pieces is required, plus specific seam constructions, due to the propensity of the cut edges of the fabric to fray substantially.

The outer layer fabric may feature a coating designed to render it waterproof and breathable, or a proprietary drop liner, such as Gore-Tex® or Sympatex®, may be present between the outer layer and the SSL. An inner lining of polyester or polyamide is present, which should be soft and comfortable in order to be worn next to the skin. The choice and use of a lining fabric is important if skin injuries caused by shear force interception are to be prevented, since in an accident there will be movement between the various layers comprising the garment, and a lining which does not move across the skin will enable these forces to be distributed amongst the outermost layers. The layering concept also delivers products which wearers perceive as exhibiting significantly improved flexibility compared to a heavyweight, single layer garment. Although the overall mass of the garments may be similar, or the layered garment may even be slightly heavier, this is masked by the improved comfort of the layered approach.

In cosmetic terms, the outer layer of the garment may be visually identical to non-protective garments, and of a similar fabric type and weight (typically, above $215\,g/m^2$). Whereas these fabric types and weights are not adequate to pass the standardised test – typical impact abrasion test results falling in the range of 0.48–0.65 s, for example – their performance is enhanced by the 'supportive' effect provided by the SSL, and it has been observed in examples of such garments worn in motorcycle accidents, that the outer layer may be abraded or torn, but the SSL has sustained little or no visible damage. The mode of failure of garments featuring this layering system is a progressive attenuation and dissipation of the abrasion and cut-related forces in an accident through successive layers, not unlike the function of ballistic fibres in bullet-resistant clothing.

By simply adding an appropriate SSL, garments manufactured from fabrics which would otherwise significantly fail in the EN 13595 tests are in fact elevated to achieve a pass at the highest, Level 2 performance class. Typical results for relative impact abrasion resistance are in the range of 52–85% above the standard's requirements, with impact cut resistance and burst strength similarly impressive at 41% and 25% above Level 2 minimum requirement respectively.

To date, the layered approach has been adopted by six companies with whom the author has worked and developed products. Recently, the Metropolitan Police in London have purchased two styles of jackets incorporating the layered system, one of which features a waterproof and breathable drop liner for use in cold and wet weather while the other incorporates ventilation features to keep the wearer cooler and more comfortable in warm, dry weather.

Another product featuring this layered construction, which the author also assisted in the development and testing of, is a motorcyclists' jacket with an inbuilt airbag system intended to provide impact protection to the wearer's torso in a fall from a motorcycle. The airbag is inflated by a cylinder of carbon dioxide, which is operated by a lanyard connecting a triggering device to the motorcycle. When the rider separates from the motorcycle in an accident, the lanyard is pulled and the triggering device releases the carbon dioxide, which inflates the airbag system in less than one second. In this jacket, the SSL protects both the wearer and the airbag from abrasions and cuts which occur when impacting and sliding along the road surface. A number of other manufacturer's 'airbag' jackets for motorcyclists fail to incorporate a structurally strong layer, and the risk is that the airbag will be punctured upon contact with the road, suffer a catastrophic failure and therefore prove quite unable to provide the protection which is claimed by the manufacturer.

26.6.3   The laminated construction

The third type of technology utilises a laminated construction comprising a terry-knit para-aramid with a fabric weight above $700\,g/m^2$. This material exhibits excellent abrasion and cut resistance but relatively poor tensile strength. To address this deficiency, a high-tenacity polyamide or polyester mesh is used to back the terry-knit layer. This material has poor impact abrasion and impact cut resistance, but is protected from these threats by the terry-knit outer layer. This combination of fabrics forms the structurally strong layer described in EN 13595 and is subject of UK patent number GB2306390, where it is described as follows:

A protective material from which a protective garment such as is used by motorcyclists is made, and has an outer layer of fibres having a high softening temperature of 400 °C. and a mass (580–620 g/m^2) to provide high-quality abrasion resistance. The outer layer is preferably two separate sheets one preferably being of Terry Loop knitted fabric, and the other of high-tenacity Polyester having a 300 g/m^2 upwards warp knit permitting high air flow therethrough. An inner layer conveniently of a Polyamide Rachel knit supports an impact attenuating layer preferably made from one of Polyurethane, Polynorbornene, Nitrile/PVC or synthetic rubber. The impact attenuating layer is preferably solid and flexible, is attached to the inner layer between the outer and inner layers. The material so formed provides sufficient abrasion resistance and high tensile strength together with permeability through the material which is ensured by providing through holes directly through the impact attenuating material.

A variation of this concept replaces the para-aramid with polyester and is described in Swedish patent SE0401891-7. This combination of materials exhibits a level of impact abrasion and burst resistance which rivals that of the para-aramid version. However, although the impact cut resistance of the polyester terry-knit is sufficient to meet the demands of EN 13595, the para-aramid version maintains a distinct advantage. There is a major benefit in the use of polyester in terms of cost (a saving of approximately 80% compared to para-aramid) and also in that it is not affected by UV light. Polyester has a significantly lower melting point, however, (258–263 °C), so in a fuel fire the wearer is at risk of sustaining burn injuries. There is still the risk of melt/burn/sticking injuries from man-made fibre underwear, but in practical terms such events are so rare as to represent a minuscule risk.

Conceived, developed and patented by Dr Roderick Woods, the laminate of terry-knit fabric and high-tenacity mesh permits air to flow through the structure with minimal impedance. By adding an external, cosmetic layer which is air-permeable (for example, by using mesh fabric), and which shields the para-aramid from the detrimental effects of UV light, the airflow past the motorcycle can be allowed to permeate through to the wearer, providing a beneficial, cooling effect to reduce the deleterious effects of heat stress in hot and humid conditions. A wind and waterproof outer layer ensures the wearer is kept both warm and dry in inclement conditions.

In an impact with a hard, abrasive road surface, the loops of the terry knit are cut, effectively instantaneously, and the resultant 'string' of fibres, which are attached at the base of the knit, can move more readily between the raised profiles of road aggregate. This is analogous to running a comb through one's hair. Were human hair formed from loops in a similar manner to the terry-knit fabric, it would be impossible to comb, but in the absence of loops, a comb can easily pass between the strands of hair. In the terry-knit system, this mechanism reduces the extent of contact between the fibres and the road surface, significantly prolonging the period before the fabric is abraded through to a hole.

26.7   Police motorcyclists clothing trials (Fig. 26.6)

A motorcyclist's clothing system employing this terry-knit fabric/high-tenacity technology was first developed and produced during the mid-1990s. Known as 'Tritector', and utilising the original, para-aramid terry-knit, it was tried out by a number of police forces across the United Kingdom before the licensed manufacturer ceased trading after being taken over by a larger competitor. For several years between the late 1990s and 2003, the concept lay dormant. In 2003, however, the author was working within the UK West Midlands Police motorcycle clothing project and had taken up his role as advisor to the ACPO Motorcycle Project Group. The United Kingdom police service was reviewing its operational requirements for motorcyclists' clothing and particularly the need for motorcycle section officers to be provided with protection from ballistic and stab threats. The combination of the leather jackets and trousers (ordinarily in the form of a two-piece suit, zipped together around the circumference of the waist), topped with fabric high-visibility jacket and, in wet weather, over-

26.6 An example of a textile suit for police motorcyclists: the Halvarssons Police Safety suit, produced by Jofama AB, of Sweden, and modelled by PC Andrew Burden of Thames Valley Police. Thames Valley Police are just one of eleven forces who have been assisting in the development of advanced textile protective clothing for police motorcyclists, benefiting from an initiative of West Midlands Police which has been further progressed by the ACPO Motorcycle Project Group.

trousers, was well-established, with the majority of criticisms from officers relating to the heat stress suffered by them in warm, hot and humid ambient weather conditions.

Adding a ballistic and stab-resistant vest increased heat stress still further and reduced mobility; the leather jacket and the vest imposing severe ergonomic burdens on the wearer. Furthermore, police motorcyclists' suits were in the main made to measure and had been manufactured before the need to consider use of a ballistic/stab vest had been raised. The fit of the jacket did not provide sufficient room for the vest to be worn underneath, close to the wearer's body as is required in order for the vest to perform to the fullest extent of its design capability. Leather trousers did not seem to affect the outcome either way, sitting as they ordinarily do below the lowest level of the vest. It was immediately apparent that a textile jacket of sufficiently generous cut to permit use of the vest would overcome the mobility restriction difficulties and if ventilation points were built into the garment these would enhance the movement of air around the inside of the garment, cooling the wearer.

As indicated above, the Metropolitan Police Service was already evaluating the layered system and consequently was unable to consider alternative products due to the restrictions imposed by the European tender process. From May 2004, however, a number of other UK police forces were presented with the opportunity to evaluate competing para-aramid and polyester-based garment systems, based on the original Tritector principles, which had been developed by two experienced manufacturers of motorcyclists' clothing. Although the individual manufacturer's choices of material for the terry-knit layer differed, each had followed the same modular garment concept which had its roots in the original, 1990s product, as follows:

- Jacket
  - terry-knit/HT mesh SSL jacket
  - wind-permeable outer jacket
  - windproof/waterproof outer jacket
- Trousers
  - terry-knit/HT mesh SSL trousers
  - wind-permeable outer trousers
  - windproof/waterproof outer trousers

Feedback from all forces involved in the trials was encouraging, with the jacket receiving entirely positive comments. Views on the trousers were mixed, however, with a number of officers of the opinion that they did not match the promise of the jacket and provided no benefit to the existing issue leather trousers. End users reported that the textile jacket appeared to be significantly lighter than leather jackets offering comparable protection, whereas the comparative weights of textile and leather trousers was believed to be little different. Part of the reason for this lies in the risk category zoning principle which is explained in EN 13595-1 Annex C. Examination of motorcycle clothing damaged in accidents has shown that the risk of impact and abrasion varies across the surface area of the garments. This principle is also explained in *Performance of Protective Clothing – Fifth Volume*,[10] which also constitutes the peer review for many of the requirements and test methodologies described within the Cambridge Standard and, by extension, EN 13595.

The major joints – shoulders, elbows, hips and knees – are at high risk of impacts, and these are protected by impact protectors specified in EN 1621-1. EN 13595 categorises this area as Zone 1. Zone 2 encompasses the Zone 1 areas and also extends across the width of the buttocks and along the lateral side of the legs to the knee joint. Zones 1 and 2 are considered to be at high risk of abrasion damage. Zone 4 encompasses limited areas at low risk of abrasion damage, and Zone 3 covers the remainder of the garment, which is considered at moderate risk.

It is clear that whereas the Zone 2 areas of the jacket are relatively small, covering as they do the regions of the shoulder joints, elbows and forearms, the Zone 2 areas of the trousers encompass a significantly greater surface area of the garment. Since the performance requirements specified in EN 13595 are higher for Zones 1 and 2, and lower for Zone 4, the materials and constructions must reflect the demands placed on them in testing and in real life. Consequently, a significantly greater proportion of the surface area of trousers will feature these more robust materials and constructions, with the attendant weight and movement restriction penalties (Fig. 26.7).

When a typical leather police jacket and trousers and the polyester version of the laminated system were weighed, there was a rather surprising result. The leather jacket weighed 2.26 kg, the textile jacket 3.25 kg, the leather trousers 4.07 kg and the textile trousers 3.15 kg – entirely the opposite of what end users' impressions had indicated would be the case! It should be noted, however, that an additional, high-visibility textile jacket is generally worn on top of the leather jacket, which might erode or eradicate the leather garment's apparent weight advantage, whereas the outer layer of the laminated textile jacket is already manufactured to a high-visibility specification.

It is perhaps in mass reduction where the next stage of development of the laminated system might take place, with research into whether lighter-weight terry-knit fabric can be introduced into the Zone 3 and Zone 4 areas. It has been established that fabric weight and performance in the EN 13595 tests do not follow a linear path, and there is a critical point at which reductions in fabric weight will cause a disproportionately adverse effect on test performance. Whether significant reductions in mass can be made before that critical point is reached is unknown at the time of writing, but it is suspected that developments will hit a brick wall before the overall weight of the trousers is at a noticeably lower level.

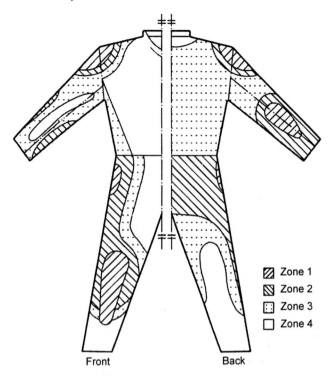

26.7 Risk category zoning for motorcyclists' protective clothing as defined by EN 13595 Part 1.

All three areas of textile protective clothing for motorcyclists are recent and ripe for further development. As more manufacturers recognise the need for lateral thinking in terms of fabric design and specification, other innovations will undoubtedly follow.

26.8   Acknowledgement

The author expresses his thanks to Dr Roderick Ian Woods, for his assistance over many years, and his contributions to the content of this chapter.

26.9   References

1. Fédération Internationale de Motocyclisme – 'Road Racing International Meetings Appendices & Technical Appendices for International Road Racing Meetings', a version of which can be downloaded from: http://www.fim.ch/en/rules/Sportifs/ccr/ Inter_Events/2004/04-APP%20MEETING%20ROAD_2.pdf
2. EN 13595-1 *Protective clothing for professional motorcycle riders – Jackets, trousers and one-piece or divided suits – Part 1 – General requirements.*

3. EN 13595-2 *Protective clothing for professional motorcycle riders – Jackets, trousers and one-piece or divided suits – Part 2 – Test method for determination of impact abrasion resistance.*

4. EN 13595-3 *Protective clothing for professional motorcycle riders – Jackets, trousers and one-piece or divided suits – Part 3 – Test method for determination of burst strength.*

5. EN 13595-4 *Protective clothing for professional motorcycle riders – Jackets, trousers and one-piece or divided suits – Part 4 – Test method for determination of impact cut resistance.*

6. *Council Directive of 21 December 1989 on the approximation of the laws of the Members States relating to personal protective equipment (89/686/EEC)* (as amended).

7. *Das FZD – Prüfverfahren für Motorradfahrerschutzkleidung* – Dipl.-Ing. Alois Weide und Dipl.-Wirtsch-Ing. Martin Schmeider – eine Broschüre des Fachgebiets Fahrzeugtechnik (Leitung: Prof. Dr.-Ing. B. Breur), Technische Hochschule Darmstadt, Ausgabe 1991.

8. Woods R I (1999), *The Cambridge Standard for Motorcyclists Clothing – Part 1: Jackets, Trousers, one-piece suits and two-piece suits intended to provide mechanical protection against some injuries on metalled road surfaces* – Issue 2, 12.8.99.

9. Prime D M and Woods R I, 'Tests on the Protection Afforded by Various Fabrics and Leathers in a Simulated Impact of a Motorcyclist on a Road Surface'. *Proceedings of the 1984 International IRCOBI Conference on the Biomechanics of Impacts.* IRCOBI, Bron, France 1984.

10. *Performance of Protective Clothing: Fifth Volume.* James S. Johnson and S.Z. Mansdorf, editors. ASTM Publication Code Number (PCN): 04-012370-55.

26.10 Appendix

List of manufacturers of type-approved and CE-marked textile protective clothing for motorcyclists (correct and verified as at date of publication).

Bickers plc (www.bikestyle.co.uk)

BKS Leather (www.bksleather.co.uk)

D.P.I. Safety s.r.l. (www.motoairbag.com)

Fowlers of Bristol Limited (www.fowlers.co.uk)

Jofama AB (www.jofama.se)

Rukka L-Fashion Group (www.rukka.com)

Scott Leathers International Limited (www.scottleathers.com)

Sumitomo Corporation Europe Limited (www.sumitomocorpeurope.com)

Lightning Source UK Ltd.
Milton Keynes UK
UKOW012002260911

179323UK00001B/5/P